Studies in Logic

Volume 63

Argumentation and Reasoned Action

Proceedings of the 1st European Conference on Argumentation, Lisbon 2015

Volume II

Volume 52
Inconsistency Robustness
Carl Hewitt and John Woods, eds.

Volume 53
Aristotle's Earlier Logic
John Woods

Volume 54
Proof Theory of N4-related Paraconsistent Logics
Norihiro Kamide and Heinrich Wansing

Volume 55
All about Proofs, Proofs for All
Bruno Woltzenlogel Paleo and David Delahaye, eds

Volume 56
Dualities for Structures of Applied Logics
Ewa Orłowska, Anna Maria Radzikowska and Ingrid Rewitzky

Volume 57
Proof-theoretic Semantics
Nissim Francez

Volume 58
Handbook of Mathematical Fuzzy Logic, Volume 3
Petr Cintula, Petr Hajek and Carles Noguera, eds.

Volume 59
The Psychology of Argument. Cognitive Approaches to Argumentation and Persuasion
Fabio Paglieri, Laura Bonelli and Silvia Felletti, eds

Volume 60
Absract Algebraic Logic. An Introductory Textbook
Josep Maria Font

Volume 61
Philosophical Applications of Modal Logic
Lloyd Humberstone

Volume 62
Argumentation and Reasoned Action. Proceedings of the 1st European Conference on Argumentation, Lisbon 2015. Volume I
Dima Mohammed and Marcin Lewiński, eds

Volume 63
Argumentation and Reasoned Action. Proceedings of the 1st European Conference on Argumentation, Lisbon 2015. Volume II
Dima Mohammed and Marcin Lewiński, eds

Studies in Logic Series Editor
Dov Gabbay

dov.gabbay@kcl.ac.uk

Argumentation and Reasoned Action

Proceedings of the 1st European Conference on Argumentation, Lisbon 2015

Volume II

Edited by

Dima Mohammed

and

Marcin Lewiński

ISBN 978-1-84890-212-1

College Publications
Scientific Director: Dov Gabbay
Managing Director: Jane Spurr

http://www.collegepublications.co.uk

Original cover design by Orchid Creative www.orchidcreative.co.uk
Printed by Lightning Source, Milton Keynes, UK

Table of contents

Thematic Panels:

Regular Papers:

PANEL I

Argumentation in Institutionalized Contexts

A Pragma-Dialectical Approach of Argumentative Patterns in Academic, Legal, Medical, and Political Contexts

I.1

Prototypical Argumentative Patterns in the Justification of Judicial Decisions: A Pragma-Dialectical Perspective

EVELINE T. FETERIS
University of Amsterdam, The Netherlands
e.t.feteris@uva.nl

ABSTRACT: In my paper I give a pragma-dialectical analysis of prototypical patterns in the justification of judicial decisions. I clarify the nature and rationale for the argumentative patterns from the perspective of the institutional function of legal justification and I distinguish different argumentative patterns in clear cases and hard cases in the justification of judicial decisions.

KEYWORDS: Argumentative pattern, clear cases, hard cases, legal decision, legal interpretation, legal justification, legal rule, method, prototypical argumentative pattern

1. INTRODUCTION

In different institutional contexts, different argumentative patterns can be distinguished that are related to the function of argumentation in the specific context. In pragma-dialectical research of argumentation the prototypical argumentative patterns are considered to be related to the institutional point of the argumentative activity in a particular context. As van Eemeren and Garssen (2014, p. 6) point out, such argumentative patterns consist of specific standpoints, specific argumentation structures and specific argument schemes. Recent studies of argumentative patterns in institutional contexts investigate which patterns are prototypical of the argumentation in a particular type of argumentative activity, in the sense that they are to be expected, given the institutional point of the activity and the institutional constraints.[1] In this paper I will describe the prototypical argumentative patterns in

[1] See, for example van Eemeren and Garssen (2014), Feteris (2015), Garssen (2013).

D. Mohammed & M. Lewiński (eds.) (2016). *Argumentation and Reasoned Action: Proceedings of the 1st European Conference on Argumentation, Lisbon, 2015. Vol. II, 3-14.* London: College Publications.

the justification of judicial decisions that are required in light of the institutional goal of legal justification.

In legal justification different argumentative patterns can be distinguished. The kind of pattern that is used is related to the different decisions courts have to give in their official capacity, that is to terminate a dispute about the application of a legal rule in accordance with the law. Courts must take a decision in a dispute about the question whether a particular legal claim which is based on the application of a legal rule to certain facts, is justified or not. The different decisions court must give in their official capacity can be illustrated on the basis of the elements of a legal decision that are schematically represented in *Figure 1a* (a positive decision) and Figure 1b (a negative decision). The different elements of the decision are (1.1') the formulation of the applicable legal rule, (1.1) the legal qualification of the facts, and (1) the legal decision.

1.1'	Formulation of the applicable legal rule R:	If legal facts X1, X2, etcetera, then legal consequence Y must follow (for example: 'If someone acts unlawfully and the act causes damages to someone else, then he is obliged to pay the damages that are caused by the act')
1.1	Legal qualification of the facts:	Legal facts X1, X2, etcetera (for example: 'Mr. X has caused damages of 1000 euros to the car of Mr. Z by an accident that was the fault of Mr. X')
1	Legal decision:	Legal consequence Y must follow (for example: 'Mr. X is obliged to pay 1000 euros to Mr. Z')

Figure 1a. Elements of a legal decision (positive form)

(1) Is the decision in the strict sense, and the other elements are the reasons for the decision, the grounds on which the decision is based. Because these grounds (1.1') and (1.1) are, in their turn, the result of a decision process, I call them also decisions, but in a broader sense. These decisions in a broader sense depend on the nature of the dispute that is presented to the court and the decisions the court must take in its official capacity.

1.1'	Formulation of the applicable legal rule R:	If legal facts X1, X2, etcetera, then legal consequence Y must follow (for example: 'If someone acts unlawfully and the act causes damages to someone else, then he is obliged to pay the damages that are caused by the act')
1.1	Legal qualification of the facts:	Legal fact X1 and not X2 (for example: 'Mr. X has caused damages of 1000 euros to the car of Mr. Z by an accident but the accident was not the fault of Mr. X')
1	Legal decision:	Legal consequence Y must not follow (for example: 'Mr. X is not obliged to pay 1000 euros to Mr. Z')

Figure 1b. Elements of a legal decision (negative form)

In the context of a legal process a dispute may concern the legal qualification of the facts of the case or the applicability of the legal rule to these facts. In the case the legal qualification of the facts is at issue, the court will have to make further decisions about the proof that can be put forward in support for the facts. In the case the applicability of the legal rule is at issue, because there are different views with respect to the meaning of the rule (for example the meaning of the term 'unlawfully'), the court will have to make further decisions about the interpretation of the meaning of the rule. The different decisions courts have to make in these cases require different types of justification and result in different prototypical argumentative patterns.

To clarify the nature and rationale for these prototypical argumentative patterns, in section 2 start with explaining the institutional function of legal justification as argumentative activity and I describe the different decisions courts have to take and account for in differences of opinion that are prototypical in the application of legal rules. I explain the distinction between so-called clear cases and hard cases and I explain the different types of hard cases. In section 3 I specify the prototypical argumentative patterns in such clear cases and hard cases that are required in light of the institutional function of legal justification. Then, in section 4, I concentrate on a specific form of hard cases in which courts must establish the meaning of the legal rule and I describe the prototypical argumentative patterns in the various differences of opinion.

2. LEGAL JUSTIFICATION AND THE APPLICATION OF LEGAL RULES: CLEAR CASES AND HARD CASES

As has been indicated, the institutional task of a court is to give a final decision in a dispute about the application of the law. It is the task of the court to apply the law to specific cases. The institutional function of legal justification as argumentative activity is to account for the way in which the court has used its discretionary power in applying the law. This justification enables the parties, higher courts (in the case of appeal) and the legal community to check the correctness of the decision so that they can criticize them on the basis of the grounds that are put forward in support of them.

In the justification courts must specify the different decisions they have given and the grounds on which these decisions are based. As indicated, courts must decide whether a particular legal rule (that can be formulated as 'If legal facts X1, X2, etcetera, then legal consequence Y', for example 'If someone acts unlawfully and the act causes damages to someone else, then he is obliged to pay the damages that are caused by the act'), is applicable to the facts of the case. In its decision the court must establish first what the facts of the case are and how they must be qualified in terms of legal facts. Then it must establish which legal rule is applicable to the legal qualification of these facts and whether the conditions for applying the legal rule are satisfied.

When the situation with respect to the facts and the law is clear in the sense that neither the facts nor the application of the law are disputed by the parties and also for the court in its official capacity there is no reason to investigate the 'truth' of the facts and the applicability of the law, the case can be considered as a *clear case* and no further discussion is necessary. It can also be the case that the facts or the legal rule are disputed by one of the parties, or that the court has reasons to question one of these points. When the facts are disputed by one of the parties, or might be questioned by the court in its official capacity (for example in a criminal case) a decision about the support for the statement about the facts is necessary. And when the applicability of the legal rule is disputed by one of the parties (or by the court in its official capacity because different interpretations of the rule are possible), the court must establish what the meaning of the rule is in light of the facts of the case and must decide whether the rule is applicable in that meaning. The last two types of cases in which further decisions are necessary are considered as a *hard case*.

In a clear case in which the facts or the rule are not disputed, it may be sufficient for the judge to put forward argumentation that refers to the facts of the case and the applicable legal rule (in the legal literature also called *first-order argumentation*). However, in a hard case when there is a difference of opinion about the facts or the rule, a further justification, also called *second-order argumentation*, consisting of subordinate arguments will be required. When the facts or the legal rule are disputed or may be disputed from a legal perspective, different forms of second-order argumentation can be distinguished that result in different argumentative patterns. In the following section I shall discuss the various argumentative patterns in clear cases and hard cases.

3. PROTOTYPICAL ARGUMENTATIVE PATTERNS IN CLEAR CASES AND HARD CASES

In this section I specify the prototypical argumentative patterns that can be expected on the basis of the institutional point of legal justification: the obligation to account for the discretionary power to apply the law, implying that a court must specify the grounds that constitute the different elements of the decision. I start with the prototypical pattern of the justification in clear cases. Then I give a general characterization of the justification in hard cases and specify the argumentative pattern that is required in a hard case in which there is a difference of opinion about the facts and the argumentative pattern that is required in a hard case in which there is a difference of opinion about the law.

The justification of the decision in a clear case implies that the court must specify the factual and legal grounds of the decision. The argumentative pattern must consist of the following elements of the decision mentioned earlier:

1: The standpoint specifying the decision that legal consequence Y must or must not follow
1.1: An argument specifying the legal qualification of the facts of the case in terms of the conditions for applying the legal rule R (as X1, X2 or not X1, X2)
1.1' An argument specifying the applicable legal rule R

The prototypical argumentative pattern in a clear case can be schematically represented as in *Figure 2*, below.

1
Legal decision: legal consequence Y must (not) follow
↑

1.1	&	1.1'
Legal qualification of the facts facts X1, X2 (or not X1, X2)		Formulation of the applicable legal rule R: If facts X1, X2, then legal consequence Y must follow

Figure 2. Prototypical argumentative pattern in a clear case

In the decision-making process, different problems may arise that create different types of hard cases, and the different types of hard cases require different argumentative patterns in the justification. The problems that may occur can concern the facts of the case or the applicability of the law. In both situations the court will have to give a further justification of the decisions it has made. In the first type of hard case the court must give a further justification for the decision about the facts, in the second type of hard case the court must give a further justification of the decision about the law that is to be applied.

In the first type of hard case in which the difference of opinion concerns the facts of the case, on the level of the main argumentation the same justification as given in a clear case must be given. In addition, as a support for the argument in which a statement is made about the facts (1.1), a subordinate second-order argumentation (1.1.1) is required that consists of proof for the facts. The prototypical argumentative pattern in such a hard case can be schematically represented as in figure 3.1 (where the second-order argumentation 1.1.1 is printed in bold).

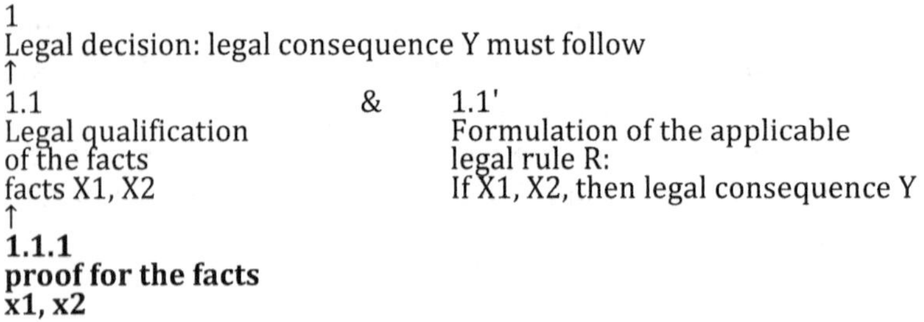

Figure 3.1. Argumentative pattern in a hard case in which the facts are disputed

The subordinate argumentation advanced in (1.1.1) can be considered as a reaction to 'institutional' criticism with respect to the legal basis of the facts.[2] The parties in dispute can appeal the decision, and in appeal the decision about the facts can be submitted to further scrutiny by a higher courts. For this reason, the court must give a further support for its decision about the facts. Depending on the legal system and the field of law, different criteria for the 'truth' of the facts apply. A form of argumentation that can be used to support the facts is

[2] For the discussion of the critical reactions of the court in its official capacity see Feteris (1993).

for example proof by means of written documents, testimonies, expert reports, etcetera.

In the second type of hard case in which the difference of opinion concerns the applicable law (in civil law systems the applicability of a particular legal rule) the court must decide whether the facts (1.1) can be considered as an implementation of the conditions for applying the legal rule (1.1'). The difference of opinion about the legal rule implies that there are two (or more) views with respect to the meaning of the rule in light of the facts of the case. The task of the court is to establish the meaning of the rule (choosing between the alternative formulations, for example R' and R") for the specific case in order to be able to decide whether the rule is applicable to the facts of the case at hand.

In its justification the court must give the argumentation as given in a clear case (see figure 1) and must specify also on which grounds the choice for one of the two versions of the rule is based. The argumentative pattern in such a hard case can be represented as in figure (3.2) in which 1.1'.1 forms the second-order argumentation (that is printed in bold). In this second-order justification the court justifies the (re)formulation of the rule (a justification that can, in its turn, be the result of a chain of further arguments):

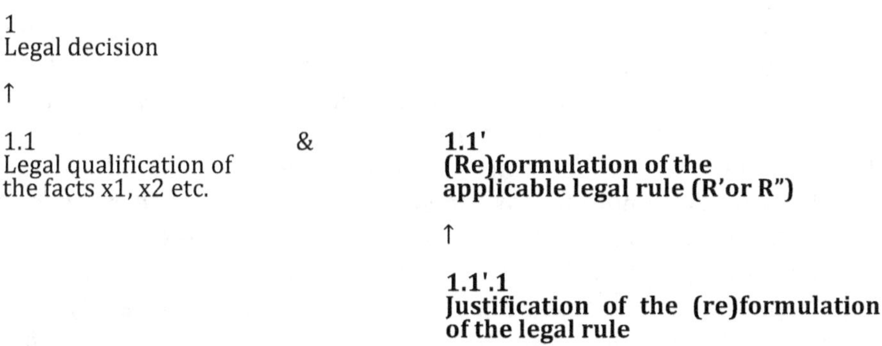

Figure 3.2. Prototypical pattern in a hard case in which the meaning of the rule for the specific case is disputed

The argumentation 1.1'.1 can be considered as a reaction to institutional criticism (for example by higher courts) with respect to the legal basis for the formulation of the rule. The court must give a further support for its decision about the applicable legal rule. Depending on the legal system and the field of law, different criteria for the 'soundness' or 'acceptability' of the grounds for establishing the meaning of a legal rule apply. In continental law systems the support will have to be given by referring to legal sources such as legal rules, legal principles etcetera. In common law systems courts may refer to precedents, legal principles, etcetera. In European law and international law, courts may refer to legal rules, legal principles, certain goals etcetera.

4. PROTOTYPICAL ARGUMENTATIVE PATTERNS IN HARD CASES ABOUT THE MEANING OF THE LEGAL RULE

In a hard case in which the meaning of a legal rule needs to be established for the specific situation, as I have indicated, depending on the nature of the difference of opinion, different decisions have to be made. In this section, I will give an overview of the kinds of difference of opinion and the decisions that have to be made that are relevant for distinguishing different prototypical argumentative patterns. In 4.1 I shall discuss the argumentation in a case in which the court makes an exception to a legal rule. In 4.2 I shall discuss the argumentation in a case in which the court gives an interpretation of a legal rule.

4.1 The argumentation in a hard case in which the court makes an exception to a legal rule

In principle, for judges it suffices to explain that a particular legal rule should be applied to the facts of the case. If the conditions for application of the legal rule are satisfied the legal consequence follows *prima facie*. In a clear case a judge does not have to explain that there are no reasons that could provide a reason not to apply the rule. However, there can be situations in which a judge is confronted with the problem that the conditions for applying the rule are fulfilled, but one of the parties claims that there are reasons not to apply the legal rule in the specific case. The party can, for example, state that there are reasons not to apply the rule that weigh heavier than the reasons for applying it. In such cases the reasons for not applying the rule require a further justification.

When a court decides that the rule is not applicable although the conditions have been fulfilled, this decision takes place in the context of a difference of opinion about the question whether a legal rule R should be applied in the strict standard meaning R' (as 'If facts X1, X2, then legal consequence Y') or an exception should be made for the specific case and the rule should be formulated with an exception, in the meaning R" (as 'If facts X1, X2, and not-Z, then legal consequence Y'). In such a case, the court decides to formulate the rule in the adapted meaning with an exception R" (see figure 3.2.1).

1
Legal decision: legal consequence Y should not follow
↑
1.1 Legal qualification of the facts X1, X2 and not-Z
&
1.1'
Reformulation of the applicable legal rule as R" ('if legal facts X1, X2 and not-Z, then legal consequence Y')
↑
1.1'.1
Justification of the reformulation of the rule as R"

Figure 3.2.1. Prototypical pattern in a hard case in which the court makes an exception to the rule

In legal theory and legal philosophy, authors distinguish various considerations why a legal rule should not be applied although the conditions for applying the rule are fulfilled. For example because an exception must be made to the rule or because the rule should not be applied on the basis of considerations of reasonableness in the specific case. The various considerations for not applying the rule require a different type of justification. In the justification of the (re)formulation of the rule as R" with an exception the court can, for example, refer to the consequences of application of the rule in light of a particular legal goal or value. The court must specify also why the reformulated/ adapted version of the rule R" would have consequences that are more desirable from the perspective of that goal or value than the standard strict formulation R'.

The argument 1.1'.1 can, for example, be based on the desirability of the consequences of applying the rule in an adapted version in which an exception is included or on the undesirability of the consequences of applying the rule in the strict version without an exception. The (un)desirability of these consequences must be justified further by referring to certain legal principles, goals and values underlying the legal system. This implies that for specific implementations the prototypical argumentative pattern must be extended further by specifying the possible argumentative extensions of the general pattern.[3]

[3] See Feteris (2015) for an analysis is given of the complex argumentation put forward as a support for the decisions that constitute further extensions of the general pattern in the case of pragmatic argumentation.

4.2. The argumentation in a hard case in which the court gives an interpretation of a legal rule

When a court gives an interpretation of a legal rule, this decision takes place in the context of a difference of opinion about the question whether a legal rule R should be applied in the meaning R' (often the standard meaning) or formulated in a different meaning. In such a case, the court makes a choice for a formulation of the rule in the meaning R' or R".

The prototypical argumentative pattern in a hard case in which there is a difference of opinion about the formulation of the legal rule in light of the facts of the specific case can be represented as in figure 3.2.2.

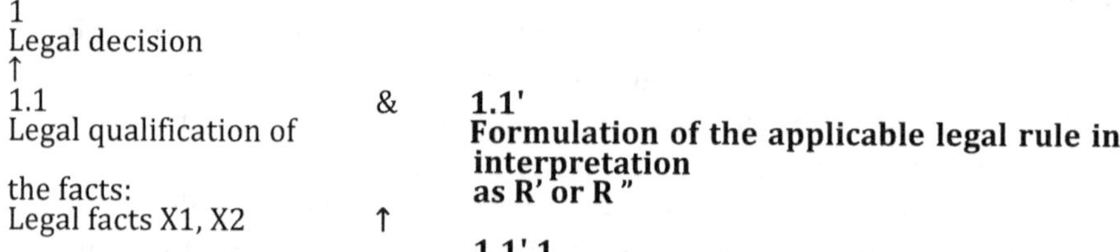

Figure 3.2.2. Prototypical argumentative pattern in a hard case in which the court gives an interpretation of the legal rule

To justify argument 1.1', in which a court gives an interpretation of the meaning of (a term in) a legal rule, the court can, in argument 1.1'.1, refer to various kinds of considerations to justify the interpretation. In the literature various interpretation methods are distinguished that refer to different kinds of considerations such as the grammatical or linguistic method (when the judge refers to the meaning of the term in ordinary language), the historical method (when the judge refers to the meaning intended by the legislator as expressed in parliamentary documents), the systematic method (when the judge refers to the place of the rule in the legal system and the relation with other rules), and the teleological method (when the judge refers to the goal or purpose of the rule).

From the perspective of legal certainty and predictability, certain methods of interpretation and arguments that are based on these methods are to be preferred or may carry more weight than other methods. The idea is that the grammatical/linguistic interpretation method that would reflect the intention of the historical legislator when formulating the rule has the highest position in the hierarchy. If this method does not lead to an acceptable solution, other methods such as the system of the law, or the goal of the rule may be suitable for establishing the meaning of the rule. In hard cases often a combination of different kinds of interpretation methods and arguments are used. Since the different interpretation methods have a different status in the hierarchy, the arguments that are based on the different methods also carry a different weight.[4]

For the translation of the hierarchy of interpretation methods into an argumentative perspective, the question is how the argumentation that is based on the interpretation methods (the argumentation 1.1'.1 and further extensions of this argumentation) can be reconstructed as prototypical argumentative patterns. The different arguments that are based on an interpretation method may occur in different positions in a justification and for this reason different prototypical patterns with different argumentative structures, consisting of different argumentative moves, can be distinguished.

5. CONCLUSION

In this paper I have described how the various forms of legal justification can be characterized as prototypical argumentative patterns from the perspective of the institutional function of legal justification. I have explained how these prototypical patterns reflect the argumentative obligations courts have when they must account for the different decisions they have made in the application of legal rules. I have pointed out that the complexity and the structure of the argumentation also depend on the kind of difference of opinion. If the difference of opinion concerns a clear case the argumentation is less complex than in a hard case. And in different kinds of hard cases, various differences of opinion require different argumentative patterns. If the difference of opinion concerns the facts the argumentative pattern differs from that in the case of a difference of opinion about the applicable law. For the hard cases about the applicable law I have described the various prototypical patterns as a characterization of the various argumentative obligations that are influenced by the decisions courts make in the interpretation and formulation of legal rules. For hard cases about the interpretation of a legal rule I have specified how the argumentative patterns are influenced by the institutional constraints that concern the interpretation methods and the status of

---

[4] See for a discussion of the hierarchy of interpretation methods for example MacCormick and Summers (1991, p. 525).

the various legal sources in various legal systems that constitute the institutional conventions of the legal domain in a particular legal culture.

REFERENCES

Eemeren, F.H. van (to be published). Pragmatic argumentation in stereotypical argumentative patterns.

Eemeren, F.H. van & Garssen, B. (2014). Argumentative patterns in discourse. In D. Mohammed and M. Lewiński (Eds.), *Virtues of argumentation. Proceedings of the 10th OSSA Conference at the University of Windsor, May 2013.* Windsor, ON: OSSA. (CD-rom).

Feteris, E.T. (1993). The judge as a critical antagonist in a legal process: A pragma-dialectical perspective. In R.E. McKerrow (Ed.), *Argument and the Postmodern Challenge. Proceedings of the eighth SCA/AFA Conference on argumentation* (pp. 476-480). Annandale: Speech Communication Association.

Feteris, E.T. (2015). The role of pragmatic argumentation referring to consequences, goals and values in the justification of judicial decisions. In B. Garssen, D. Godden, G. Mitchell & A.F. Snoeck Henkemans (Eds.), *Proceedings of the 8th ISSA conference.* Amsterdam: Sic Sat. (CD-rom).

Garssen, B. (2013). Strategic maneuvering in European Parliamentary debate. *Journal of Argumentation in Context, 2*(1), 33-46.

MacCormick, N., & Summers, R. (Eds.) (1991). *Interpreting statutes: A comparative study.* Aldershot: Dartmouth.

I.2

Strategic Maneuvering in Administrative Judicial Decisions: Groundwork for Argumentative Patterns

H. JOSÉ PLUG
University of Amsterdam, The Netherlands
h.j.plug@uva.nl

This paper focuses on strategic maneuvering that takes place in Dutch administrative judicial decisions. These decisions may be seen as a distinct argumentative activity type. Starting from the characteristics that are pertinent to this activity type, I will explore how implications of current discussions on the changing task of the administrative judge may become manifest in the judge's strategic maneuvering by means of the presentation of complex argumentation and the introduction of additional standpoints.

KEYWORDS: institutional point, judicial advice, legal argumentation, mediator, multiple roles, obiter dictum

1. INTRODUCTION

In order to analyse argumentation as it takes part in a specific communicative practice, the analyst should take into account the institutional conventions of that particular communicative practice. The way in which institutional conventions of the legal domain should be taken into account for the analysis of argumentation in judicial decisions from a pragma-dialectical perspective, has been the central focus of studies by, amongst others, Feteris (1989), Kloosterhuis (2002) and Plug (2000). These studies, however, came into being before van Eemeren and Houtlosser (2002, 2010) developed the extended pragma-dialectical argumentation theory by including a rhetorical dimension that provided for the analysis of strategic maneuvering. To this rhetorical extension, van Eemeren (2010, pp. 138-146) added a more defined and detailed approach to account for the communicative practices in which strategic maneuvering takes place. In this approach, different communicative activity types as they manifest themselves in reality are defined by the goals that are pursued in realizing their

D. Mohammed & M. Lewiński (eds.) (2016). *Argumentation and Reasoned Action: Proceedings of the 1st European Conference on Argumentation, Lisbon, 2015. Vol. II, 15-27.* London: College Publications.

institutional point, their conventionalization, and the main properties of their format. With regard to these three defining components, considerable differences may be observed, depending on the communicative domain in which the discourse takes place. Communicative activity types in the legal domain, for instance, can be regarded as strictly conventionalized, whereas activity types in the political domain or the medical domain may be considered less conventionalized. This does not mean, however, that the defining components as they manifest themselves in a communicative domain should be considered as characteristics of a communicative activity type that are absolute or static. As a result of, for example, discussions in society, new legislation or scientific insights, the institutional point of an activity type may need to be revised. Such a revision may consist of a modification of the original institutional point or of an extension of the institutional point to a multiple institutional point. Due to changes with regard to the institutional point, the goal, the conventionalization and the main properties of a format may be altered as well. The importance of including these changes in the characterization of the activity type does not only pertain to the analysis and the evaluation in general of the argumentation, it may also be relevant for researching the argumentative patterns in these texts. According to van Eemeren and Garssen (2014, p. 7) such an argumentative pattern will consist of a particular constellation of argumentative moves in which in a particular kind of argumentation structure a particular combination of argument schemes is exploited in defense of a particular type of standpoint.

In this contribution I will demonstrate how conventions in the legal communicative domain may change and what consequences these changes may have for the characterisation of the communicative activity type. The activity type I will discuss is that of judicial decisions that are produced by the Dutch administrative court. The changes that are discussed concern the role of the administrative judge and how these changes may be of influence on the institutional point of (administrative) judicial decisions.

In section 2, I will first sketch the communicative activity type of administrative judicial decisions as it is formally institutionalized by means of legal procedural rules. In section 3, I will discuss the way in which altered perspectives on the demands that are made on administrative judges may affect the conception of the judge's role. In section 4 I will discuss how changes in the conception of the judge's role may have their influence on the formulation of the institutional point and hence on (the analysis of) the argumentation.

2. THE ACTIVITY TYPE OF ADMINISTRATIVE JUDICIAL DECISIONS

The communicative activity type of administrative judicial decisions refers to motivated, binding decisions by judges in cases pertaining to differences of opinion between citizens and civil authorities. Administrative law not only entitles civil authorities to govern, but it also limits the scope of its administrative activities. Administrative law safeguards citizens against improper actions on the part of civil authorities. As part of administrative law, the law of administrative procedure prescribes the way(s) in which citizens may lodge complaints against government decisions. Once an administrative authority has taken a decision, a disagreeing citizen, in most cases, may lodge a complaint against this ruling at the administrative body in question. If a citizen disagrees with the so-called 'decision on complaint' by the said administrative body, he may contest this decision before the judge in administrative law (the administrative court). The judge, then, has to decide whether or not the (final) ruling by the administrative body is legitimate. Subsequently there are the possibilities of appeal at the Council of State or at the Central Appeals Court. The general provisions of administrative law apply to both the realization of government decisions and to the judge's evaluation on decisions on complaint and decisions on appeal. In the context of administrative decisions seen as a communicative activity type, the general provisions of administrative law, therefore, are of considerable importance to the institutional goal(s), the conventions and the specifications of both the argumentation put forward in the decision and in the legal discussion preceding the final ruling.

Taking the outline of argumentative activity types as presented by van Eemeren (2010, p. 143) as a starting point, I characterised (Plug, 2015) the activity type of administrative judicial decisions as in Figure 1 below.

Figure 1 demonstrates that a contextualised argumentative analysis or evaluation of a concrete speech event such as the judicial decision by the administrative (district) court on December 21, 2012, presupposes a text to be considered as a representation of a particular subtype of (argumentative) communicative activity: the activity type of administrative judicial decisions by the (district) court. This communicative subtype in its turn belongs to the domain of legal communication and makes use of the prototypical genre of adjudication.

The institutional point of this communicative subtype can be formulated as providing a binding decision in a difference of opinion about whether a citizen is being treated unlawfully by the actions of an administrative authority.

domains of communicative activity	*general genres of communicative activity*	*specific communicative activity types*	*concrete speech events*
[= more or less institutionalized macro-contexts]	[= families of conventionalized communicative practices]	[= **sub**types of conventionalized communicative practices]	[= instantiations of communicative activity types]
legal communication	adjudication	*Judicial decisions by the District Court (Administrative Law*)	*Dutch judicial decision [by the District Court Utrecht 21-12-2012]*

Figure 1

3. CHANGES IN ADMINISTRATIVE LAW AND THE ROLE OF THE JUDGE

The role of the judge in administrative law, traditionally, focuses on evaluating or testing government decisions. The evaluation, in reaction to the appeal lodged by an interested party, means that a judge examines a decision for its legitimacy. In case the judge is of the opinion that a decision is indeed legitimate, the appeal is denied and the original ruling is maintained. If, however, the judge deems a decision not legitimate, he will declare the appeal well founded and consequently quash the original ruling. The Dutch General Administrative Law Act (chapter 8) states that the administrative body in question is to provide the judge with all written material pertaining to the decision making process, in order to enable the judge to reach a conclusion. Moreover, the parties to the process should be allowed to put forward their standpoints and arguments on the (un)lawfulness of the decision, both verbally and in writing. It is on the basis of these standpoints and arguments that the judge will form his opinion on the administrative ruling.

The difference of opinion between citizens and civil authorities for the judge to rule on, is, then, limited by the scope of administrative law. The propositional content of the judge's ultimate standpoint, on the grounds of which he will decide whether or not to allow the appeal, pertains to the lawfulness of the authorities' decision. The judge's speech act with which he will or will not allow the plaintiff's appeal, may be analysed as an assertive-declarative speech act with which the former adopts a positive or a negative standpoint. The speech acts with which the judge justifies his ruling are analysed as assertive speech acts with which he completes the complex speech act of argumentation.

On the occasion of the 2007 evaluation of the General Administrative Law Act, however, the present committee established to evaluate the general provisions of administrative law (*the evaluation committee Awb III*) recommends to no longer limit the role of the judge in administrative law to providing a ruling as to the lawfulness of a decision. The final evaluative report on project differentiation of tasks in administrative law *[Eindrapport Project Differentiatie van Werkstromen Bestuursrecht]* of 2010 and many other publications indicate that judges may be expected to contribute to social preconditions relevant to administrative law, such as speed, cogency, finality and a focus on solutions. The concept of finality refers to the idea that the judge in administrative law reaches a conclusion or understanding that actually ends the underlying conflict or judicial dispute. With reference to this discussion, Marseille (2009, p. 67; 2010) indicates that the scope of a difference of opinion between parties may reach beyond the decision against which the original appeal was lodged. The parties' decision to go to court need not be based on the contents of the decision alone, but on the course of the decision-making process as well. As Marseille points out (2010, p. 222), it is important for judges to distinguish between the decision against which an appeal was lodged on the one hand and the dispute between the parties on the other. The introduction of *a new approach to case proceedings* [the *Nieuwe zaaksbehandeling]* is aimed at shedding light, already during the court hearing, on the fact that the difference of opinion is possibly not just about contesting the decision, or even about something other than the decision at hand. The judge could, then, explicitly bring forward the ways in which the law could, or could not solve the conflict, investigate whether there are way(s) in which a solution of the conflict between parties could be solved, irrespective of the outcome of the appeal procedure, and in how far he could contribute to such a solution.

These changes, aimed at achieving a more active role on the part of the judge and at increasing the chances of swift and definitive solutions of cases, have an impact on the complexity of the role judges play.[1] Polak (2010, p. 97), for instance, states that these changes in administrative law have resulted in the judge being allotted the role of mediator in disputes on top of his role as controller. However, if a judge is expected to not only contribute to swift and definitive solutions of

---

1 Studies on the judicial roles by Hanson (2002) and Swanson (2011) focus on how judges themselves differ in their views as to the proper functions of courts and the norms of judicial decision-making.

legal disputes but also of non-legal disputes between parties, we should, in my opinion, clearly distinguish between his role as the judge who settles a dispute and his role as alternative mediator.[2] This means that it would be more adequate to distinguish between the following three roles:

1. Controller (who checks for the legitimacy of decisions)
2. Settler of disputes (who aims at contributing to a swift and definitive solution of a legal dispute)
3. Alternative mediator (who aims at contributing to the solution of a non-legal difference of opinion)

Literature on administrative law (such as Marseille, 2010, Verburg, 2013) focuses on the changes pertaining to the judge's role in court hearings. In my contribution, however, I will, henceforth, not focus on the judge's performance in these court hearings but rather on whether the administrative judge's complex role may be reflected in the judicial decisions he brings forward.

4. MANIFESTATIONS AND IMPLICATIONS OF THE COMPLEX ROLE OF THE JUDGE

On the basis of analyses of some examples of administrative judicial decisions published after the introduction of *a new approach to case proceedings* [the *Nieuwe zaaksbehandeling]*, I will demonstrate the possibilities judges avail themselves of in order to convey their role of alternative mediator in non-legal differences of opinion, apart from their role as authorities who check decisions for their legitimacy.[3]

I will first discuss argumentative choices which could be considered as strategic manoeuvres by the judge in his role as alternative mediator. Then I will discuss the implications these observations may have for the characterisation of the communicative activity type of administrative judicial decisions by the district court.

---

[2] The term 'mediation', as used in law, is a form of alternative dispute resolution (ADR), where a third party, the mediator, assists the parties to negotiate a settlement. Since during the proceedings, the contribution of the judge to a settlement is more limited and differs in some aspects fundamentally from that of a mediator, I use the term 'alternative' mediator.

[3] In Plug (forthcoming), I demonstrate how the administrative judge may make use of concessive argumentation and of anticipating argumentation in his role as settler of legal disputes.

4.1 Strategic maneuvering by the judge as alternative mediator

Analyses of administrative judicial decisions show that decisions which aim at solving an underlying, non-legal difference of opinion, take the shape of an advice or recommendation. These recommendations are put forward by the judge in his role as alternative mediator. Since judges do not usually assume that parties accept their recommendations as a matter of course, they will put forward arguments in support of these recommendations. From a pragma-dialectical point of view, the recommendation should, therefore, not be considered as a directive speech act but as an assertive speech act by means of which the judge expresses a positive or a negative standpoint. When clinching matters in cases like these, judges cannot derive authority from the law. These standpoints, unlike standpoints put forward by a judge in his role as controller of the law or as settler of disputes should, therefore, not be analysed as an assertive-declarative speech act. The propositional contents of the standpoint underlying the judge's advice could, first of all, refer to the acceptability of an intrinsic advice (x). In a case like this the standard form is: the advice to (not) do (x) is acceptable. Secondly, the proposition could refer to a procedural advice (y): the recommendation for parties to find an alternative solution for their dispute. In this case the standard form would be: the advice to do (y) is acceptable.

Apart from the standpoint in relation to a non-judicial disagreement, a judicial decision always puts forward a standpoint referring to the judicial dispute as well. In cases like these, therefor, there is always a multiple difference of opinion. Within the context of the non-judicial disagreement, the judge, in his role as alternative mediator, is the protagonist of the (main) standpoint. Within the context of the judicial dispute, the judge, in his role as settler of disputes or the one to control legitimacy, is the protagonist of the (main) standpoint.

In British and English judicial decisions, for instance, the different roles of judges may be expressed by alternating direct references in the form of the singular or plural personal pronouns "I" or "we" with using references in the third person singular, "the court". Kaehler (2013, p. 17) demonstrates that various references to judges offer the opportunity to distinguish, in the decision, between the judges' standpoints and arguments in 'their private role as citizens' and in 'their official role as judges'.

In Dutch, as well as in German and French law, for instance, the (administrative) judge will always refer to himself in the third person: both the single-judge section and the three-judge section refers to itself as 'the court' (Verburg, 2008, p. 208). The administrative judge, therefore, will have to apply different verbal means to express the fact that, in addition to implementing the law, he also puts forward arguments and a standpoint in his role alternative mediator.

Judges often employ the *obiter dictum* ('additional consideration') to indicate that the argumentation which he brings forward does not relate to a judicial standpoint. The phrase 'additional consideration', in pragma-dialectics, is regarded as an indicator of multiple argumentation (van Eemeren et al., 2005). In judicial contexts, however, the use of the 'additional consideration' is not quite so unequivocal (Plug, 2000a, p. 190). If argumentation is put forward, the arguments do not always support the judicial dispute at hand. The judge can make use of this ambiguity to bring forward argumentation for his standpoint as alternative mediator. The following may serve as an example of such a manoeuvre.

In a dispute over a building permit for temporary accommodations for a school in the Municipality of Velsen, the administrative judge decides to accept the objections of the plaintiffs. The decision of the municipality to grant a provisional exemption from a zoning plan for the purpose of temporary accommodation is not to be maintained. The fact of the matter is that the municipality, the defendant, has failed to demonstrate the temporariness of the schoolrooms. The judge, in his decision, puts forward the following considerations.

> The judge deciding on the statutory provision sets store to, quite superfluously, considering the following. (...) The location at [address A] proposed by plaintiffs by means of temporary accommodation would, in case that particular location were to be given a permanent character, mean teaching children next to a building site and can hardly be taken seriously. Neither do alternative locations passed in review during court hearings stand out for their attractiveness compared to the one chosen today. If plaintiffs are also concerned about the interests of the pupils and their parents - and they are, as appears from their attitude during court sessions - it would show flexibility if they, the outcome of these proceedings notwithstanding, would demonstrate the willingness to refrain from permanently dropping the location at [address B] from their list. (...)
> (Court Haarlem 28-2-2011, ECLI:NL:RBHAA:2011:BP6223)

The phrase 'quite superfluously, considering' used by the judge in this fragment indicates that what follows does not refer to the preceding judicial standpoint. The provision is even intensified by the word 'quite'.

The communicative status of each of the following speech acts is, however, not entirely clear. Nevertheless, it becomes apparent that the judge creates an opportunity to continue his argument in his role as alternative mediator. Although the label *obiter dictum* ('additional consideration') is not an exclusive indicator for argumentation, consecutive speech acts should be regarded as argumentative. The speech acts consist of arguments brought forward by the judge in support of his standpoint that 'plaintiffs [should] demonstrate the willingness to refrain from permanently dropping the location at [address B] from their list.' His attempts to convince parties of the acceptability of his intrinsic advice (to not do x), 'the outcome of these proceedings notwithstanding', could be interpreted as an attempt on the judge's part to contribute, in the context of his judicial decision, to the settlement of the underlying conflict between parties.

The following example demonstrates that the judge's may be a twofold advice: he provides an intrinsic advice (to do x) as well as a procedural advice (to do y). In this case the dispute between a citizen of the town Boxtel, the plaintiff, and the city council of Boxtel, the defendant, concerns the rejection of the plaintiff's request to take action against noise caused by an air conditioner placed on their neighbor's roof. A license to place the air conditioner has been suspended by the defendant. The court determines on the grounds of a research by experts, that the noise level of the air conditioner is kept within the noise limit. Accordingly, the court decides that the defendant does not need to take action against the neighbor's air conditioner. In the decision the court brings forward the following.

> Apart from this case, the court finds reason, in the dispute between the plaintiff and [name A], to remark the following. (...) It seems to the court, that the air conditioner can easily be replaced. This replacement will of course involve expenses. Considering that both parties, in view of their disturbed relationship, benefit from a solution in this conflict, the court deems it entirely reasonable for the plaintiff to contribute to the costs of the replacement. The court urges the plaintiff and [name A] and their proxies to once again try to come to an amicable solution, in the knowledge that the administrative law cannot offer a solution to problems between neighbours, such are at issue here.
> (Court 's-Hertogenbosch, 14-12-2012, ECLI:NL:RBSHE:2012: BZ0544)

In this fragment, the judge uses the formulation 'apart from this case' and the expression 'to remark' to indicate that what follows should not be interpreted as part of the justification of the legal standpoint that has been provided above. Despite the indication that what follows should be interpreted as remarks, the judge presents in fact argumentation for the acceptability of both an intrinsic advice as well as for a procedural advice. The argumentation that can be reconstructed as being in support of the acceptability of the judge's (intrinsic) advice for the plaintiff to contribute to the costs of the replacement consists of that it would be reasonable to do so because both parties benefit from the replacement. The argument that the administrative law cannot offer a solution to the problem may be reconstructed as support for the acceptability of the (procedural) advice to try to come to an amicable solution.

4.2 Implications of the judge's changing role for the activity type

The above examples demonstrate that a judge who has perceives a non-legal difference of opinion that cannot be solved during the court hearing, can make a final attempt to successfully contribute to the settlement of this conflict in the administrative judicial decision. If these attempts have to be regarded as part of the altered institutional demands, this would have the following consequences for the characterization of the activity type of judicial decisions by the administrative (district) court.

In contrast with the traditional activity type, the institutional point should be considered to be multiple. Next to the traditional institutional point of providing a binding decision in a difference of opinion about whether a citizen is being treated unlawfully by the actions of an administrative authority, the second institutional point may be formulated as providing a non-binding advice about the resolution of a non-legal difference of opinion that underlies a legal difference of opinion. As a consequence, these decisions are to be characterized as a hybrid activity type in which problem-solving communication is incorporated in the domain of legal communication.

domains of communicative activity	*general genres of communicative activity*	*specific communicative activity types*	*concrete speech events*
[= more or less institutionalized macro-contexts]	[= families of conventionalized communicative practices]	[= **sub**types of conventionalized communicative practices]	[= instantiations of communicative activity types]
legal communication		Judicial decisions by the District Court (Administrative Law)	Dutch judicial decision [by the District Court Utrecht 21-12-2012]
		(Multiple institutional points)	
	adjudication	*(1). Providing a binding decision in a difference of opinion about whether a citizen is being treated unlawfully by the actions of an administrative authority*	
Problem-solving	*mediation*	*(2). Providing a non-binding advice communication about the resolution of a non-legal difference of opinion that underlies a legal difference of opinion about whether a citizen is being treated unlawfully by the actions of an administrative authority*	

Figure 2

5. CONCLUSION

Developments in administrative law have resulted in a more complex role to be played by administrative judges. The judge no longer only assesses the legitimacy of decisions made by administrative authorities, but is expected to also actively apply every possibility to achieve a final settlement of the judicial dispute as well as the underlying conflict. Beside his role as the judge who applies the law or assesses the legitimacy of administrative decisions, a distinction should be made between two additional roles: the one of settler of disputes and the one of alternative mediator.

Ever since the introduction of *a new approach to case proceedings [*the *Nieuwe zaaksbehandeling]* attention for the changing role of the judge has been mainly focussed on his performance during court hearings. On the basis of examples of administrative justice I demonstrated that the increasingly complex roles judges play may also be expressed in judicial decisions. By means of strategic choices as to the presentation of arguments put forward in judicial decisions, the judge can contribute to the solution of the judicial dispute as well as to the underlying, non-legal conflict. By making use of the possibility, within a judicial decision, to put forward a non-legal standpoint, beside a judicial standpoint, by way of advice or recommendation, and in doing so interpreting the dispute as a multiple difference of opinion, the judge, in his role as alternative mediator may contribute to the settlement of the non-legal conflict between parties. Subsequently, I have shown that interpreting the altered institutional demands not only applicable to the court hearing but also to the judicial decision, implies that decisions by the administrative (district) court should be characterised as a hybrid activity type that aims to realise multiple institutional points. This characterization may function as groundwork for future research on argumentative patterns in hybrid legal activity types such as administrative rulings.

REFERENCES

Eemeren, F.H. van & Garssen, B. (2014). Argumentative patterns in discourse. In D. Mohammed & M. Lewiński (Eds.), *Virtues of argumentation. Proceedings of the 10th OSSA Conference at the University of Windsor*, May 2013. Windsor, ON: OSSA. (CD-rom).

Eemeren, F.H. van, Houtlosser, P., & Snoeck Henkemans, A.F. (2007). *Argumentative indicators in discourse. A pragma-dialectical study.* Dordrecht: Springer.

Eemeren, F.H. van & Houtlosser, P. (2006). Strategisch manoeuvreren, het model van een kritische discussie en conventionele actietypen. *Tijdschrift voor Taalbeheersing, 28*(1), 1-14.

Eemeren, F. H. van (2010). *Strategic maneuvering in argumentative discourse. Extending the pragma-dialectical theory of argumentation.* Amsterdam: John Benjamins.

Eindrapport Project Differentiatie van werkstromen Bestuursrecht (2010). www.rechtspraak.nl/Organisatie/Publicaties-En-Brochures/rapporten-en-artikelen.

Feteris, E.T. (1989). *Discussieregels in het recht. Een pragma-dialectische analyse van het burgerlijk proces en het strafproces.* Dordrecht: Foris.

Hanson, R. (2002). The changing role of a judge and its implications. *Court Review, 38*(4), 10-16.

Kaehler, L. (2013). First-person perspectives in legal decisions. *Law and Language, 15*, 1-32.

Kloosterhuis, H.T.M. (2002). *Van overeenkomstige toepassing: De pragma-dialectische reconstructie van analogie-argumentatie in rechterlijke uitspraken*. Amsterdam: Thela Thesis.

Marseille, A.T. (2007). De bestuursrechter en diens vrijheid. Van actief naar lijdelijk (en weer terug). *Trema, 10*, 423-431.

Marseille, A.T. (2009). *De zitting bij de bestuursrechter. Een onderzoek naar het belang van de zitting voor een adequate afdoening van bestuursrechtelijke beroepsprocedures*. Den Haag: Boom.

Marseille, A.T. (2010). Effecten van informalisering van bestuursrechtspraak. *NTB, 8*, 221-229.

Plug, H.J. (2000). *In onderlinge samenhang bezien. De pragma-dialectische reconstructie van complexe argumentatie in rechterlijke uitspraken. Amsterdam*. Amsterdam: Thela Thesis.

Plug, H.J. (2000a). Indicators of *obiter dicta*. A pragma-dialectical analysis of textual clues for the reconstruction of legal argumentation. *Artificial Intelligence and Law, 8*(2-3), 189-203.

Plug, H.J. (2015). Transparency in legal argumentation: Adapting to a composite audience in administrative judicial decisions. In F.H. van Eemeren & B. Garssen (Eds.), *Scrutinizing Argumentation in Practice* (pp. 121-133). Amsterdam: John Benjamins.

Plug, H.J. (forthcoming). Argumentative consequences of the changing role of the judge.

Polak, J.E.M. (2010). Veranderende perspectieven van de bestuursrechter. In T. Barkhuysen, W. Den Ouden & J.E.M. Polak (Eds.), *Bestuursrecht harmoniseren: 15 jaar Awb*. (pp. 97-110). Den Haag: Boom Juridische uitgevers.

Poorter J.C.A. de, & de Graaf, K.J. (2011). *Doel en functie van de bestuursrechtspraak: Een blik op de toekomst*. Den Haag: Raad van Staten, 385-405.

Schueler, B.J., Drewes, J.K., et al. (2007). *Definitieve geschilbeslechting door de bestuursrechter*. Amsterdam: WODC.

Swanson, R.A. (2011). Judicial roles in state high courts. *Judicature, 94*(4), 169-177.

Verburg, D.A. (2008). *De bestuursrechtelijke uitspraak - en het denkmodel dat daaraan ten grondslag ligt*. Zutphen: Uitgeverij Kerckebosch.

Verburg, D.A. (2013). De nieuwe zaaksbehandeling van de bestuursrechter. *Tijdschrift Conflicthantering, 3*, 19-23.

I.3

Anticipating Critical Questions to Pragmatic Argumentation in Over-The-Counter Medicine Advertisements

FRANCISCA SNOECK HENKEMANS
University of Amsterdam, The Netherlands
a.f.snoeckhenkemans@uva.nl

In this paper, a number of prototypical argumentative patterns in over-the-counter (OTC) medicine advertisements will be investigated, consisting of a pragmatic main argument and supporting arguments by means of which the arguer attempts to deal with critical questions concerning his pragmatic argument. First, the institutional point and regulations for OTC medicine advertisements will be discussed. Next, a basic argumentative pattern for this type of advertisement as well as possible extensions of this pattern will be presented.

KEYWORDS: argumentative pattern, critical question, over-the-counter medicine advertisement, pragmatic argumentation

1. INTRODUCTION

In particular communicative practices specific patterns of argumentative moves can be observed. According to van Eemeren (2016a), such argumentative patterns consist of a combination of a particular type of standpoints, a particular argument scheme or combination of argument schemes and a particular argumentation structure. In recent studies within pragma-dialectics, study is made of such argumentative patterns in order to find out which patterns are prototypical of particular activity types, in the sense that they are to be expected, given the institutional constraints of the communicative

D. Mohammed & M. Lewiński (eds.) (2016). *Argumentation and Reasoned Action: Proceedings of the 1st European Conference on Argumentation, Lisbon, 2015. Vol. II, 29-42.* London: College Publications.

practice (Andone, 2016, van Eemeren & Garssen, 2014; Garssen, 2013; Feteris, 2015; Wierda, 2015).[1]

According to the extended pragma-dialectical theory, arguers attempt to design their argumentative discourse strategically, aiming to be not just reasonable but also effective (van Eemeren, 2010). This balancing between reasonableness and effectiveness, or, "strategic maneuvering", has to take place within the extrinsic institutional constraints imposed by the communicative activity type in which the argumentation occurs.

Achieving the institutional point of a particular communicative activity type may require the use of a particular type of argumentation, which means that the critical reactions pertaining to that type of argumentation will have to be anticipated by the arguers if they want to convince their audience of the acceptability of their standpoint. The institutional point and the rules and conventions may further influence which critical reactions are relevant in the context concerned and how they might be dealt with. As a result of arguers' attempt to maneuver strategically within the prevailing institutional constraints, specific argumentative patterns will be formed.

In this paper, I will discuss an argumentative pattern that can be shown to be prototypical for the communicative practice of over-the-counter medicines advertisements, which involves the use of a pragmatic argument as a main argument for the standpoint. Until now, in the research undertaken by argumentation theorists on the argumentative characteristics of drug advertisements, the focus has generally been on prescription drug advertisements (Goodnight, 2008; Rubinelli, Nakamoto, & Schulz, 2008; van Poppel & Rubinelli, 2011; Wierda & Visser, 2014). Apart from medicines that are only available on prescription because a medical professional needs to supervise their use, there are also medicines that are available to consumers without prescription, the so called 'over-the-counter medicines'. For the advertising of this type of medicines, different regulations apply than for prescription medicines. In the UK (and in some other countries) traditional herbal medicines are included in the regulations for over-the-counter (OTC) medicines. Traditional herbal medicines require a traditional medicines registration instead of a so-called 'marketing authorisation,' and there are a number of other specific requirements

[1] This research can eventually be followed up by investigating empirically to what extent such patterns are also *stereotypical* of those activity types, that is, whether they occur with a certain frequency in practice (cf. van Eemeren, 2016b, for a further explanation of the distinction between 'prototypical' and 'stereotypical' argumentative patterns).

they need to fulfil, in addition to the general requirements for over-the-counter medicines.

Just as in prescription drug advertisements, in over-the-counter medicines advertisements pragmatic argumentation plays a central role. In this type of argumentation the standpoint that a certain action should be carried out is defended by pointing out that the result of that action is desirable. By claiming that their medicinal product will have a beneficial effect on a particular health condition, advertisers try to convince consumers that they should buy the product. Since claims concerning the efficacy of the medicinal product play such a key role in the attempt to convince the consumer that the product is worth buying, there are a number of advertising guidelines that aim to regulate the presentation of such claims as well as the support that should be given for them.

In this paper I shall first discuss the institutional point of the communicative activity type of over-the-counter medicines advertisements and give some examples of how specific advertising regulations pose constraints on the argumentation that can be advanced. Next, I shall present a basic argumentative pattern for over-the-counter medicines advertisements and provide a non-exhaustive overview of types of extensions that may be expected to occur in practice.

2. INSTITUTIONAL POINT AND INSTITUTIONAL REQUIREMENTS OF OVER-THE-COUNTER MEDICIEN ADVERTISEMENTS

Over-the-counter medicine advertisements seem to have the same institutional point as prescription drug advertisements. Wierda and Visser (2014) analyse direct-to-consumer prescription drug ads a hybrid genre of communicative activity in which promotion and consultation are combined. The institutional point of this type of ad is twofold: on the one hand the advertiser attempts to persuade the patient-consumer to use the medical product, and on the other hand the advertiser is legally required to inform patient-consumers in such a way that they can make an informed choice on whether to use the product or not. Although the regulations for over-the-counter medicines consumer advertising are less stringent than those for prescription medicines advertising, the advertising guidelines for this type of medication still aim to allow consumers to make an appropriate and informed choice, and to guarantee their health and safety as much as possible.[2]

[2] These are the advertising aims mentioned in the regulations for over-the-counter drugs and natural remedies issued by Health Canada (2006). The UK

Moreover, since in the case of over-the-counter medicines there is no supervision by a physician, it is all the more important that consumers are indeed able to make an informed choice.

In what follows I shall discuss some examples of regulations and guidelines that apply to over-the-counter medicines advertisements, to show in what kind of ways such regulations may have an influence on the emergence of prototypical argumentative patterns in this type of ad. The regulations for over the counter medicines differ to some extent from country to country, although there is also quite a lot of overlap. Within Europe, there are a number of general guidelines which apply to medicines advertisements in all European countries, but there is also room for specific additional guidelines for each individual country. In this paper, as a frame of reference, I shall use the Medicines advertising codes for over-the-counter medicines developed by the Proprietary Association of Great Britain (2013) as part of a system of self-regulation.

As I hope to make clear, the over-the-counter medicines advertising regulations pose certain limitations on the type of main argument and the types of sub arguments that may be used in such ads. They also have consequences for the content and presentation of those arguments. And finally, some of the regulations require advertisers to answer particular critical questions and to do so in accordance with specific requirements.

In view of the institutional aim: getting consumers for whom this is suitable (i.e. who suffer from a particular condition and for whom this is not contra-indicated) to use a certain medicinal product, the main standpoint in OTC-medicines advertisements always concerns a recommended action: "Consumers for whom it is suitable should use product X."[3]

According to the advertising code, an over-the-counter medicines ad must always include the product's therapeutic indication (rule 20 and 57). This means that some beneficial effect on the consumer's health (for instance that the product relieves dry coughs) must always be presented in the ad.[4] Since we may take it that the main

---

regulations for OTC medicines have been developed by a group of pharmaceutical manufacturers "to protect the public from misleading medicines advertising" and to "ensure that their advertising was balanced and responsible" (PAGB Medicines Advertising Codes 2013).

[3] As Wierda (2015) explains, it is not in the interest of advertisers to get consumers for whom this is not suitable to use their products.

[4] Mentioning the therapeutic indication is important in advertisements for over-the-counter medicines, since consumers have to diagnose themselves and make an unsupervised decision on whether a certain product might relieve

reason for consumers to use a certain medicinal product is that the product has some beneficial effect on their health, the main argument in this type of advertisement is necessarily a pragmatic argument by means of which it is argued that the action recommended in the standpoint will lead to a desirable result.

Underlying this type of argument is the so-called 'positive version' of the pragmatic argumentation scheme. The pragma-dialectical characterization of this scheme is:

> 1 Action X should be carried out, because
> 1.1 Action X will lead to positive result Y, and
> 1.1' If an action of type X [such as X] leads to a positive result of type Y [such as Y], then that action should be carried out (van Eemeren, 2016b)

The following three critical questions are associated with pragmatic argumentation (van Eemeren, 2016b):

> 1. Do actions of type X lead to results of type Y?
> 2. Is result Y really positive (i.e. desirable)?
> 3. Does action X not have any major negative (i.e.) undesirable side-effects?

According to van Eemeren (2016b), this list of critical questions may have to be expanded if the circumstances in which the argumentation occurs call for it. In the context of medicines advertisements, two further questions are relevant:

> 4. Are there are any other actions that need to be taken together with action X to achieve result Y?
> 5. Could result Y not be achieved more easily by other actions?

Question 4 is relevant, since the effectiveness of a medicine is often dependent on other factors, such as a healthy life style. Question 5 needs to be added because consumers generally have a choice between several similar products, so that comparing other ways of achieving the

---

their complaints or not. Advertisements for conditions that are difficult to self-diagnose are usually also required to state that people should first obtain a doctor's diagnosis before they use the product for the first time (rule 57). Apart from allowing individual consumers to make an informed choice of medication, the guidelines also aim to prevent the overuse and unwarranted use of drugs in society at large.

beneficial result is almost always a relevant issue in making the decision on whether to use a certain product or not.[5]

Questions 1, 2 and 4 can be used to criticize premise 1.1, since they either question the causal relation between X and Y, or the desirability of result Y. Questions 3 and 5 can be seen as ways of criticizing the bridging premise 1.1', since they refer to possible "rebutting" circumstances which would make this premise less acceptable.

In addition to the main pragmatic argument, the guidelines also allow advertisers to mention other, non-therapeutic effects (for instance cosmetic effects) or other benefits (for instance ease of use) as direct support of the main standpoint, but the medicinal benefits should always be presented as the primary reason for purchasing the medicine (rule 34).

On the basis of the regulations for OTC-medicines, we can therefore explain that a basic argumentative pattern in advertisements for this type of medicines consists of pragmatic argumentation for a prescriptive standpoint. The regulations do not only require advertisers to use pragmatic argumentation as the main argumentation in the ad, they also influence the content of the premise in which the beneficial effect is mentioned: the effect claimed may not go beyond what is allowed according to the product's market authorisation or, in the case of traditional herbal medicines, the traditional herbal medicines registration.[6] For instance, if the official indication is that a medicine relieves allergy symptoms, a claim that the product relieves allergies would go too far.[7] Furthermore, the guidelines forbid the advertiser to claim or imply that a product's effects are guaranteed (rule 17). Instead of "gets rid of pain", for instance, only a weaker claim such as "relieves pain" is allowed.

It can further be predicted that extensions of the basic argumentative pattern in OTC-medicines advertisements will mainly

---

[5] In the Dutch Code for the consumer advertising of medicines, this characteristic of advertisements is explicitly acknowledged by claiming that almost every advertisement contains a comparative element (cf. rule 33, *Code voor de publieksreclame voor geneesmiddelen)*.

[6] A product's Marketing Authorisation (or Registration) also includes the socalled "Summary of Product Characteristics". All claims made (concerning the indications, speed of action, duration of action, absorption etc.) must be consistent with this summary (rule 3).

[7] The Canadian guidelines for OTC-drugs require the beneficial effect of the product to be formulated in concrete terms (so not for instance: "product X lets you get on with your day," without a specification of the therapeutic indication)

consist of anticipations of the critical reactions to the main pragmatic argumentation, resulting in further support for the premises of the main pragmatic argument. In the next section, I will discuss what constraints the UK regulations for OTC-medicines advertising pose on the further support of the main argument.

3. EXTENSIONS OF THE BASIC PATTERN

In principle, the advertiser is not obliged to support premise 1.1 in the advertisement itself: it is sufficient if the advertiser is able to substantiate all claims made in the advertisement if required to do so by the authorities concerned. There has to be sufficient evidence that using the product will indeed have the positive effect claimed in the argumentation. Whether this is indeed the case is something that is, in principle, checked beforehand as part of the authorisation procedure. If an over-the-counter medicine has received a Market authorisation, this means that evidence has been provided of the medicine's effectiveness. With traditional herbal medicines this is different: these products are not able to obtain a Marketing Authorisation due to there being insufficient scientific evidence of their effectiveness. Their efficacy is only based on traditional use and they are registered as a traditional medicine.

Even if an over-the-counter medicine has obtained a Market Authorisation, advertisers will often still anticipate that potential consumers may have doubts about whether using the product will indeed produce the beneficial results on their health. Advertisers may then provide further support for the efficacy of their product. This means that they anticipate the *first critical question* for pragmatic argumentation: "Do actions of type X lead to results of type Y?"

To make it clear that this question can be answered positively, advertisers could, of course, refer to the scientific evidence for the efficacy that they have had to provide in order to get their medication authorised. This may, however, often make the argumentation too technical for a lay audience. An alternative may be to use an argument from scientific authority, for instance by pointing out that the effect has been scientifically or clinically proven, without going into the details of the scientific evidence. In the case of traditional herbal medicine products, the European Commission has forbidden this type of authority argument. Instead, the statement that the efficacy of the product is exclusively based upon long-standing use as a traditional remedy should always be included in the advertisement (rule 20).

Instead of a scientific authority argument, advertisers may also choose to make use of other types of authority argumentation. A reason for doing so might be that they anticipate that a scientific authority argument may not be convincing to everyone. Consumers might be somewhat distrustful of advertisers claiming that the effectiveness of their product is scientifically or clinically proven. If such doubt is anticipated by the advertiser, a better option might be to make use of other types of authority argumentation, such as the argument from expert opinion. By referring to the opinion of others instead of making claims themselves, advertisers may seem more convincing.

In some countries, using an argument from expert opinion is allowed in OTC-medicines advertising. The Canadian guidelines for health products, for instance, allow advertisers to use endorsements by recognized groups, such as the Canadian Dental Association. The UK's advertising code, however, forbids advertisers to refer directly or indirectly to recommendations by scientists or health professionals, or even by actors playing a doctor in a television show (rule 45). It is only acceptable to state that a particular health professional recommends the use of a specific ingredient, formulation or preparation but not a particular brand (rule 46). It is also allowed to state that a medicine is 'licensed' or 'authorised' as long as it is not suggested that it "has been specifically endorsed or approved" by authorities such as the Department of Health or the Medicines and Healthcare products Regulatory Agency (rule 19). The UK guidelines also forbid recommendations by celebrities (rule 47). The only endorsements that are allowed are testimonials, or, experience based authority arguments, by users of the product who are not health professionals or celebrities (rule 50, 52).

Apart from authority arguments, causal arguments may also be used to support the efficacy of the medicinal product. In the context of over-the-counter medicines advertisements, this will often be a type of causal argument in which it is claimed that specific ingredients produce the product's beneficial effect. An example is the following argumentation for the effectiveness of a gel against muscle pain: "Product X contains three active ingredients, camphor, menthol, and methyl salicylate, which provide relief against muscle pain." Another type of causal argument that is used in over-the-counter medicines advertisements refers to the process that is involved in producing the health benefits. An example is the following argument given for the standpoint that a certain kind of gel helps to get rid of warts: "helps to kill the virus within the wart."

The premise that Action X will lead to positive result Y can also be criticized in a different way, by raising the question of whether there are any other actions that need to be taken together with action X to achieve the desirable effect Y (*critical question 4*). In the guidelines reference is made to this critical question by the regulation that advertising should not undermine current healthy life-style advice (rule 7 and rule 16). This would, for instance, be the case if it were suggested that the use of the advertised product is a substitute for a healthy lifestyle. In an advertisement for weight loss pills this critical question is anticipated by stating that the product helps obtaining an optimal metabolic rate "as part of your healthy diet and exercise program." In this way, the advertiser concedes that just using the medicine is not sufficient to produce the desired result.

Apart from the critical questions concerning the efficacy of the medicinal product, other critical questions may be anticipated in advertisements as well. A first example is *critical question 2*: is the result that the action is claimed to produce indeed positive? Although in general, one may take it that people will not need much convincing that doing something that is beneficial to their health is positive, there are exceptions. Sometimes the advertiser first needs to convince the consumer that there is a need to do something about a particular condition or ailment. For instance, in an advert for a wart treatment it is explained that warts can grow and spread, particularly if left untreated. In arguing for the desirability of treating a particular condition, advertisers have to take care, however, not to cause consumers unwarranted anxiety by falsely suggesting that consumers' health could be adversely affected if the consumer chooses not to use the medicine advertised (rule 10 and 11).

Further critical questions may be raised with respect to the bridging premise: "If an action of type X leads to a positive result of type Y, then that action should be carried out." In the first place, doubt may be cast on this premise by raising *critical question 3*: does action X not also have any major negative side-effects? If advertisers want to anticipate this kind of doubt, they have to do so within the constraints posed by the regulations. According to the guidelines, it is unacceptable to claim that a product is "side-effect-free" (rule 26). It is also forbidden to state without qualification that a product is "safe". It is allowed however, to state that there are "no known side effects" or that a product has a "good safety profile", or is "suitable for use" in specific categories of people, such as children (rule 26).

Another question that might be used to criticize the bridging premise 1.1' is *critical question 5*: could result Y could not be achieved more easily by other actions, such as for instance using a different type of medication. The UK guidelines forbid advertisers to suggest that another medicine is unsafe, of poor quality or ineffective (rule 38).[8] It is also not allowed to suggest that a product's effects are better than or equal to those of another identifiable product (rule 39). Comparisons with other products are acceptable, however, if they refer to qualities such as product palatability, speed of action or duration of action (rule 37). It is also allowed to highlight that a specific side effect that is common among similar products is absent in the advertised product. An acceptable example could be "Product X relieves your allergy symptoms and it won't make you drowsy". It is not permitted to state that a product does not contain an active ingredient present in competitor products thereby implying that this ingredient should be avoided. It is, however, allowed to highlight the absence of other types of ingredients where this may be of benefit to a particular group of consumers, for instance, by stating that a product is 'caffeine-free', 'perfume-free' or 'sugar-free' (rule 40).

4. ARGUMENTATIVE PATTERNS IN OVER-THE-COUNTER MEDICINE ADVERTISEMENTS

From the discussion of the regulations for over-the-counter medicine advertisements it has become clear that the goal of persuading the consumer to use the product in a responsible way minimally requires the use of pragmatic argumentation as a main argument for the standpoint that consumers for whom the medicinal product is suitable should use it. The specific standpoint in combination with the pragmatic argument in which beneficial health effects are claimed can therefore be seen as a basic argumentative pattern in over-the-counter medicines advertisements. This basic pattern can be further extended by the arguer in order to deal with anticipated critical reactions to his main argumentation.

The arguer can advance sub argumentation for all the premises of his pragmatic argument, thereby attempting to deal with the relevant critical questions for pragmatic argumentation. As a result, the basic pattern may be extended in a number of ways, and such extensions, in combination with the main argumentation, can be seen as prototypical patterns in the argumentation of over-the-counter medicine ads. In *Figure 1* an overview is provided of the basic argumentative pattern of

---

[8] The reason for forbidding this is that medicines are licensed on the basis of safety, quality and efficacy, so that one may take it that all authorised medicines are safe, of good quality and effective.

over-the-counter medicines advertisements and possible extensions of this pattern.

Basic pattern: main argument is a pragmatic argument

1 Consumers for whom it is suitable [i.e. who suffer from a relevant condition and for whom it is not contra-indicated] should use medicinal product X.

1.1 Using medicinal product X will have beneficial effect Y on the consumer's health.

1.1' If using a medicinal product of type X will have a beneficial effect of type Y on the consumer's health, then the consumer should use product X.

Possible extensions: supporting arguments as attempts to deal with the critical questions for pragmatic argumentation:

A Support for premise 1.1:

1.1.1a Achieving effect Y is a means of realising the desirable effect of preventing or diminishing a health risk. [*pragmatic argument*]

And/Or: 1.1.1b That using X will produce effect Y has been clinically proven. [*science based authority argument*; not allowed in the case of traditional herbal medicine]

Or: 1.1.1b That using X will produce effect Y is based upon a long-standing use as a traditional remedy.
[*argument from the authority of tradition*; must be used in the case of traditional herbal medicine]

Or: 1.1.1b That using X will produce effect Y is a claim by (group of) Expert(s) G.
[*argument from expert opinion*, not allowed in the UK, but allowed in, for instance, Canada]

Or: 1.1.1b That using X will produce effect Y is a claim by endorser E based on his experience.
[*experience-based authority argument*]

Or: 1.1.1b Product X is a licensed/authorised medicine.
[*legal/competent authority argument*]

Or: 1.1.1b Active ingredient I of product X produces the beneficial effect Y.
[*causal ingredient argument*]

Or: 1.1.1b Product X influences physiological process P in such a way that beneficial effect Y is produced.
[*causal physiological process argument*]

B Support for bridging premise 1.1':

1.1'.1 X has no known side effects/ X has a good safety profile.

Or: 1.1'.1 There is no better way than X to achieve the beneficial effect.

1.1'.1.1 X is more effective than competing types of medication.
[not allowed in the UK, but allowed in, for instance, Canada]

Or: 1.1'.1.1 Unlike competing products, X has (non-therapeutic) advantage A/X does not have negative side-effect S/ X does not contain (non-active) ingredient I.

Figure 1. Basic argumentative pattern of OTC-medicine advertisement and extensions

Which critical questions should be dealt with and how this should be done, is partly up to the advertiser, and partly prescribed by the regulations. In some cases, advertisers are obliged to deal with a particular critical question, and even to do so in a particular way. The requirement regarding traditional herbal medicines to include the information that the product's indication is exclusively based upon long-standing use as a traditional remedy is a good example. Although this requirement aims to prevent advertisers from unwarrantedly claiming that the effectiveness of the medicine has been demonstrated, it is likely that for many consumers the long-standing use will function as an argument in support of the efficacy of the medication. Viewed from this perspective, the requirement obliges advertisers to use a particular type of argument with a particular content in support of the efficacy of the traditional herbal product.

As we have seen, there are also regulations which forbid the use of particular types of arguments, thereby restricting the possible ways in which advertisers may deal with anticipated criticism. An example is the prohibition in the UK guidelines of the use of arguments from expert opinion in support of the effectiveness of the medicinal product.

Finally, there are regulations which only allow dealing with particular critical questions if the advertiser meets specific requirements when formulating the argument by means of which it is made clear that the criticism does not hold. A case in point is the critical question "Does X not also have negative side-effects?" According to the guidelines, in anticipation of this question it is not allowed to make any absolute claims (such as that product X is side-effect free or safe), but only qualified claims (such as that product X has no known side-effects).

5. CONCLUSION

In this paper, I hope to have made clear that the institutional point of the argumentative activity type of medicinal product advertisements in combination with the regulations for this type of advertisements can be used to provide an explanation for a basic argumentative pattern in such advertisements consisting of a particular type of standpoint supported by pragmatic argumentation as a main argument. Typically, further supporting arguments will refer to the critical questions pertaining to pragmatic argumentation. For those criticisms that are the most relevant in the context at hand, guidelines are provided by the authorities that pose specific restrictions on how they may be dealt with. On the basis of these restrictions it can partly be explained why certain types of support do not occur, why certain types of support will always be present with particular types of medicinal products and why specific formulations of answers to critical questions have been chosen. The basic argumentative pattern for over-the-counter medicines advertisements, in combination with the prototypical patterns that are formed by specific extensions as ways of dealing with critical questions, can be used as a starting-point to investigate how advertisers in practice attempt to maneuver strategically within the institutional constraints of the activity type of OTC-medicines advertisements.

REFERENCES

Andone, C. (2016). Argumentative patterns in the political domain: The case of European parliamentary committees of inquiry. *Argumentation, 30*(1), 45-60.

Eemeren, F.H. van (2016a). Bingo! Promising developments in argumentation theory. In F.H. van Eemeren & B. Garssen (Eds.), *Reflections on theoretical issues in argumentation theory* (pp. 3-25). Dordrecht: Springer.

Eemeren, F.H. van (2016b). Identifying argumentative patterns. A vital step in the methodical development of pragma-dialectics. *Argumentation, 30*(1), 1-23.

Eemeren, F.H. van & Garssen, B. (2014). Argumentative patterns in discourse. In D. Mohammed & M. Lewiński (Eds.), *Virtues of argumentation. Proceedings of the 10th OSSA Conference at the University of Windsor,* May 2013. Windsor, ON: OSSA. (CD-rom).

Feteris, E.T. (2015). The role of pragmatic argumentation referring to consequences, goals and values in the justification of judicial decisions. In B. Garssen, D. Godden, G. Mitchell & A.F. Snoeck Henkemans (Eds.), *Proceedings of the 8th ISSA conference.* Amsterdam: Rozenboom. (CD-rom).

Garssen, B. (2013). Strategic maneuvering in European Parliamentary debate. *Journal of Argumentation in Context, 2*(1), 33-46.

Goodnight, G.T. (2008). Strategic maneuvering in direct to consumer drug advertising: A study in argumentation theory and new institutional theory. *Argumentation, 22,* 359-371.

Health Canada (2006). *Consumer Advertising Guidelines for Marketed Health Products* (for nonprescription Drugs including Natural Health Products) http://www.hc-sc.gc.ca/dhp-mps/advert-publicit/pol/guide-ldir_consom_consum-eng.php.

PAGB Medicines Advertising codes (2009 edition, updated July 2013). http://www.pagbadvertisingcode.com/_common/pdf/PAGBAdvertisingCodes.pdf.

Poppel, L. van & Rubinelli, S. (2011) 'Try the smarter way'. On the claimed efficacy of advertised medicines. In E.T. Feteris, B. Garssen & A.F. Snoeck Henkemans (Eds.), *Keeping in touch with Pragma-Dialectics* (pp. 153-164). Amsterdam: John Benjamins.

Rubinelli, S., Nakamoto, K., & Schulz, P. J. (2008). The rabbit in the hat: Dubious argumentation and the persuasive effects of Direct-To-Consumer Advertising of prescription medicines. *Communication & Medicine, 5*(1), 49-58.

Wierda, R. (2015). *Experience-based authority argumentation in direct-to-consumer medical advertisements. An analytical and experimental study concerning the strategic anticipation of critical questions.* Dissertation University of Amsterdam.

Wierda, R., & Visser, J. (2014). Direct-to-consumer advertisements for prescription drugs as an argumentative activity type. In S. Rubinelli & A.F. Snoeck Henkemans (Eds.), *Argumentation and Health* (pp. 81-96). Amsterdam: John Benjamins.

I.4

Criteria for Deciding what is the 'Best' Scientific Explanation

JEAN H.M. WAGEMANS
University of Amsterdam
j.h.m.wagemans@uva.nl

In justifying their choice of the 'best' scientific explanation from a number of candidate explanations, scientists may employ specific theoretical virtues and other criteria for good scientific theories. This paper is aimed at providing an inventory of such criteria and at analyzing how they function argumentatively by indicating their systematic place within the pattern of argumentation based on abduction.

KEYWORDS: abduction, academic communication, argumentative pattern, inference to the best explanation, oxygen theory, phlogiston theory, pragma-dialectics, scientific explanation

1. INTRODUCTION

The nature of scientific explanations has been the subject of extensive research. Scholars from fields as diverse as philosophy, psychology, and sociology have described the form of reasoning that is involved in finding a scientific explanation, the cognitive circumstances under which such explanations are formed, and the social aspects that play a role in scientific practices respectively.[1]

Within the field of argumentation theory, the interest in academic discourse in general and scientific explanations in particular is relatively new. In comparison to the research that is carried out in the fields mentioned above, argumentation theoretical research concerning

[1] For an overview of the research on this subject see for example Hacket et al. (2008).

D. Mohammed & M. Lewiński (eds.) (2016). *Argumentation and Reasoned Action: Proceedings of the 1st European Conference on Argumentation, Lisbon, 2015. Vol. II, 43-53.* London: College Publications.

scientific explanations does not focus on the *genesis* of such explanations but on their *justification*, i.e. the way in which scientists support their claims with arguments. In this paper, I aim to describe the content as well as the argumentative function of the criteria scientists employ when they justify their selection of the 'best' scientific explanation from a number of candidate explanations of an observed phenomenon.

Philosophers of science have labeled the form of reasoning underlying scientific explanations as 'abduction' or 'inference to the best explanation'. In section 2, I summarize the layout of the argumentative pattern that is generated when scientists justify their explanation of an observed phenomenon by making use of this type of reasoning.

Philosophers of science have also contemplated the theoretical virtues or criteria for determining whether an explanation can genuinely be called a 'good' scientific explanation. In section 3, I present an inventory of these criteria and determine their argumentative function in justifying the choice of the 'best' scientific explanation.

The history of science provides a plethora of examples of scientific theories and explanations being replaced with better ones. In order to illustrate the use of the analytical tool developed, I give in section 4 an analysis of the arguments involved in the justification of the replacement of the so-called 'phlogiston theory' by the 'oxygen theory' within the field of 18th century chemistry. The analysis makes clear which criteria are involved in preferring the one theory to the other as well as how these criteria function argumentatively.

Finally, in section 5, I will discuss my findings and indicate the need for making an additional inventory of decision rules in order to enable a more complete analysis of the way in which scientists justify their explanations of observed phenomena.

2. THE PATTERN OF ARGUMENTATION BASED ON ABDUCTION

In order to determine the argumentative function of criteria regarding 'good' scientific theories, it is necessary to dispose of a general account of the way in which scientists justify their explanations. In this section, I provide such an account by summarizing the layout of the pattern of argumentation based on abduction.[2] The description of the pattern in this section is an amended version of the description I presented in an earlier paper (Wagemans, 2014).

---

[2] See van Eemeren (2016) for a general explanation of the pragma-dialectical starting points regarding the research into argumentative patterns in various communicative domains.

According to the received opinion within the field of the philosophy of science, abduction plays a key role in scientific practice since it is the form of reasoning that scientists employ in generating an explanation for an observed phenomenon. The philosopher Peirce is acclaimed for having provided a seminal description of this type of reasoning, which runs as follows:

> Long before I first classed abduction as an inference it was recognized by logicians that the operation of adopting an explanatory hypothesis - which is just what abduction is - was subject to certain conditions. Namely, the hypothesis cannot be admitted, even as a hypothesis, unless it be supposed that it would account for the facts or some of them. The form of inference, therefore, is this:
>
> The surprising fact, *C*, is observed;
> But if *A* were true, *C* would be a matter of course,
> Hence, there is reason to suspect that *A* is true.
>
> Thus, *A* cannot be abductively inferred, or if you prefer the expression, cannot be abductively conjectured until its entire content is already present in the premiss, "If *A* were true, *C* would be a matter of course." (Peirce, 1974, 5.189)

In argumentation theoretical terms, this 'generative' definition of abduction can be translated into a pattern consisting of the following elements:

1 We may assume that *A* is true
1.1 It is observed that *C* is the case
1.1' If it is observed that *C* is the case, we may assume that *A* is true
  1.1'.1 If *A* were true, *C* would be a matter of course

Apart from the description given by Peirce, one may also encounter a somewhat different description of abductive reasoning. Starting from the idea that abduction plays a key role in the process of finding explanations of observed facts, some philosophers have argued that abduction does not only involve the process of generating hypotheses, but also the consecutive process of selecting the 'best' candidate from the hypotheses that have been generated. This has led to a tradition in which abduction is described as a form of reasoning that involves the process of selection. Within this tradition, abduction is defined as 'inference to the best explanation'.[3]

---

[3] See Aliseda (2007, p. 267) for an explanation of the distinction between the process of 'generation' and the process of 'selection' of explanations. On the

The main difference between the two definitions of abduction is that while the generative definition only involves reasoning from observed facts to a possible explanation of those facts, the 'selective' definition also involves making a choice between a number of candidate explanations:

> In textbooks on epistemology or the philosophy of science, one often encounters something like the following as a formulation of abduction:
>
> ABD1 Given evidence E and candidate explanations $H_1, \ldots, H_n$ of E, infer the truth of that H_i which best explains E.
>
> An observation that is frequently made about this rule, and that points to a potential problem for it, is that it presupposes the notions of candidate explanation and best explanation, neither of which has a straightforward interpretation. (Douven, 2011, pp. 10-11)

Like the generative definition of abduction, the selective definition involves a standpoint expressing a specific explanation of an observed fact and a main argument expressing that fact. Using the same words, except for the variables expressing the explanation and the observed fact, these two elements of the pattern can be expressed as "We may assume that H_i is true" and "It is observed that E is the case" respectively.

Different from the generative definition of abduction, the selective definition mentions the notions of 'candidate explanation' and 'best explanation'. By incorporating these notions into the formulation of the argument supporting the justificatory force of the main argument, the following argumentative pattern can be identified:

1 We may assume that H_i is true
1.1 It is observed that E is the case
1.1' If it is observed that E is the case, we may assume that H_i is true
1.1'.1 Of candidate explanations H_1 – H_n, H_i is the best explanation of E

Within this second pattern, the argument supporting the justificatory force of the main argument expresses an evaluative sub-standpoint concerning the choice of the best explanation from a number of candidate explanations (1.1'.1). In the next section, I present an

differences between the two definitions of abduction see for instance Campos (2011), Minnameier (2004), and Paavola (2006).

inventory of theoretical virtues functioning as criteria that scientists use in order to justify this evaluative sub-standpoint.

3. JUSTIFYING THE CHOICE OF THE 'BEST' EXPLANATION

In the case that scientists support their explanation of an observed phenomenon by arguing that this is the 'best' explanation selected from a number of candidate explanations, they may further justify their choice of the best explanation in anticipation or reaction to doubt or criticisms raised by their peers. As Douven remarks, in such justifications specific theoretical virtues may play a role:

> ... it is often said that the latter [i.e., the best explanation, JW] must appeal to the so-called theoretical virtues, like simplicity, generality, and coherence with well-established theories; the best explanation would then be the hypothesis which, on balance, does best with respect to these virtues. (Douven, 2011, pp. 10-11)

How can this 'appeal to the theoretical virtues' be described in argumentation theoretical terms? And what does 'on balance' mean in this respect? In order to answer these questions, I now turn to describing an extension of the pattern of argumentation based on selective abduction identified in the previous section.

Within the pattern, the argument expressing the choice of the best explanation is formulated as 'Of candidate explanations H_1 - H_n, H_i is the best explanation of *E*'. Since making a choice always involves criteria on the basis of which it is decided which of the available options is the best, this argument may be supported by arguments expressing the scores of the candidate explanations on the decision criteria employed. Such arguments can be represented in the form of a decision matrix, the dimensions of which depend on the number of candidate explanations (H_1 - H_n) and the number of criteria (C_1 - C_n) involved in making the decision, which contains the scores ($S_{1,1}$ - $S_{n,n}$) of the candidate explanations on these criteria:

	C_1	C_2	C_3	C_4
H_1	$S_{1,1}$	$S_{1,2}$	$S_{1,3}$	$S_{1,4}$
H_2	$S_{2,1}$	$S_{2,2}$	$S_{2,3}$	$S_{2,4}$
H_3	$S_{3,1}$	$S_{3,2}$	$S_{3,3}$	$S_{3,4}$

Figure 1 – Decision matrix concerning the choice of the best explanation

The argument expressing the scores of the candidate explanations on the decision criteria employed may then be formulated as follows: 'H_1 – H_n meet criteria C_1 – C_n with scores $S_{1,1}$ – $S_{n,n}$'.

As to the content of the criteria involved, Douven in the quote above mentioned three specific 'theoretical virtues' that may play a role in deciding what is the best explanation: 'simplicity, generality, and coherence with well-established theories'. A survey of the literature shows that there are many of such virtues and other types of criteria and that their application may vary considerably from one field to the other. The philosopher of science Kuhn, just to mention one other example, distinguishes between five criteria for evaluating the adequacy of a theory:

> Among a number of quite usual answers I select five, not because they are exhaustive, but because they are individually important and collectively sufficiently varied to indicate what is at stake ... These five characteristics – accuracy, consistency, scope, simplicity, and fruitfulness – are all standard criteria for evaluating the adequacy of a theory ... Together with others of much the same sort, they provide *the* shared basis for theory choice. (Kuhn, 1998, p. 103, original italics)

Below I present the result of my survey of literature concerning theoretical virtues and criteria for good scientific theories, in which for clarity's sake I grouped the criteria under three different headings. The headings express the relation of the criteria to three of the key aspects of the situation in which the decision takes place (without claiming exhaustiveness of the list nor mutual exclusiveness of the categories):

related to the observed phenomenon	- accuracy - scope, genericity, fruitfulness, explanatory force, subsumptive power
related to the scientific context	- refutability, empirical content (testability, observability) - coherence - consistency
related to competing explanations	- simplicity, elegance - parsimony (fewest assumptions) - consilience (convergence of evidence)

Figure 2 – Inventory of criteria for 'good' explanations

While philosophers of science more or less agree on the content of the criteria, they do not seem to fully understand how scientists use these criteria in order to arrive at the 'best' explanation. Kuhn observes two problems regarding this issue. One problem relates to the potential vagueness of the individual criteria used to determine the acceptability of scientific theories and the second relates to the conflicts that may arise when there is more than one criterion at stake:

> Individually the criteria are imprecise: individuals may legitimately differ about their application to concrete cases. In addition, when deployed together, they repeatedly prove to conflict with one another; accuracy may, for example, dictate the choice of one theory, scope the choice of its competitor. (Kuhn, 1998, pp. 103-104)

Moreover, even if scientists agree on the criteria to be employed, they may attribute different weighing factors to them and therefore end up with a different decision: 'When scientists must choose between competing theories, two men fully committed to the same list of criteria for choice may nevertheless reach different conclusions' (Kuhn, 1998, p. 105).

From these problems in understanding the use of criteria in reaching a conclusion as to what is the best explanation, we may conclude that apart from the scores of the candidate explanations on the criteria employed, scientists may further support the justificatory force of the argument expressing these scores by an argument expressing the specific decision rule involved. This leads to the following extension of the argumentative pattern:

1.1'.1 Of candidate explanations H_1 – H_n, H_i is the best explanation of E
1.1'.1.1 H_1 – H_n meet criteria C_1 – C_n with scores $S_{1,1}$ – $S_{n,n}$
1.1'.1.1' If H_1 – H_n meet criteria C_1 – C_n with scores $S_{1,1}$ – $S_{n,n}$, then of possible explanations H_1 – H_n, H_i is the best explanation of E
1.1'.1.1'.1 Decision rule R applies

In this extension of the pattern, the argument expressing the choice of the best explanation (1.1'.1) functions as a sub-standpoint that is supported by an argument expressing the scores of the candidate explanations on the criteria involved (1.1'.1.1), the justificatory force of which is supported by an argument expressing the decision rule applied (1.1'.1.1'.1).

4. ANALYSIS OF AN EXAMPLE

In order to illustrate how the extended pattern of argumentation based on selective abduction can be used for reconstructing justifications of scientific explanations, I now provide an analysis of an example taken from the history of chemistry. The example concerns the overthrowing of the 'phlogiston theory' of combustion by the 'oxygen theory'.

The explanation given by late 17th century scientists for the observation that substances are able to combust and rust (i.e., in modern terms, to oxidize) is the existence of a fire-like element that normally inheres the substance but disappears during the oxidation process. The theory was put forward for the first time by Becher in 1667, while a variant in which the element concerned was named 'phlogiston' was circulated by Stahl in 1703. In the course of the 18th century it became clear that, among other problems, the phlogiston theory could not account for the empirical observation that substances gain weight when burning. At the end of the same century, the theory was replaced by the 'oxygen theory' put forward by Lavoisier. This theory explains combustion by assuming the existence of an element called 'oxygen'.

In a paper read in 1783 and known under the title *Reflections on phlogiston*, Lavoisier puts forward a number of arguments against the phlogiston theory and also defends his own oxygen theory. In the introduction of the paper, he employs the theoretical virtues of explanatory power, simplicity, and parsimony:

> In the series of papers that I have submitted to the Academy, I reviewed the principal phenomena of chemistry, I emphasised those that accompany combustion, the calcination of metals and, in general, all the processes where there is absorption and fixation of air. I have deduced all explanations from one simple principle: that is, that pure air (vital air) is composed of a special characteristic principle—which forms its base and which I have named the *oxygen principle*—combined with the matter of fire and heat. Once this principle was accepted, the principal difficulties of chemistry seemed to fade and dissipate and all the phenomena were explained with astonishing simplicity.
>
> But if all of chemistry is explained in a satisfactory manner without the help of phlogiston, it is by this fact alone infinitely probable that this principle does not exist, that it is a hypothetical entity, a gratuitous supposition and it is indeed a principle of a good logic to not multiply entities unless necessary. Perhaps I could have restricted myself to these negative proofs and contented myself with having proven that one can give a better account of the phenomena without phlogiston than with phlogiston. But it is time I explained in a more formal and precise manner an opinion that I regard as a disastrous error in chemistry and that seems to me to have

> retarded [624] its progress considerably by introducing a bad style of philosophising into the science. (Best, 2015, p. 139, original italics)

In a later passage, Lavoisier explains that the 'phlogiston theory' is at odds with the observation that the process of combustion involves an increase of weight:

> According to Stahl, phlogiston (the inflammable principle) is a heavy substance. Indeed, one cannot form any other idea of an earthy principle or even of a composite containing an earthy element. He even tried to determine its weight in his treatise on sulphur.
>
> This theory of Stahl's on the calcination of metals and on combustion in general did not account for a very anciently observed phenomenon, verified by Boyle, that has become today an incontrovertible truth—that all combustible substances increase in weight when they burn or calcine. This is what is observed in an especially striking manner in metals, sulphur, phosphorus etc. Conversely, in Stahl's system, phlogiston (which is a heavy principle) escapes from [626] metals when they are calcined and from combustible substances that burn; they must therefore lose a portion of their weight instead of gaining it. (Best, 2015, p. 141, original italics)

Finally, after having discussed several attempts of other scientists to save the phlogiston theory by making adaptations to Stahls version, Lavoisier summarizes his objections in the conclusion of the paper in the following way:

> All these reflections confirm what I have proposed, what I had aimed to prove, which I am going to repeat again—that the chemists have made phlogiston a vague principle that is not rigorously defined and that, consequently, adapts itself to all the explanations into which one wants it to force it. Sometimes this principle is heavy, sometimes it is not; sometimes it is free fire, sometimes it is fire combined with an earthy element; sometimes it passes through the pores of vessels, sometimes they are impervious to it. It explains at the same time causticity and non-causticity, transparency and opacity, colours and the absence of colour. It is a veritable Proteus that changes form at each instant. (Best, 2015, p. 149)

In this summary, Lavoisier states that the theory should be abandoned because the principle is 'vague' and 'not rigorously defined', thereby meaning that phlogiston is attributed many different opposite properties. This criticism employs the criteria of explanatory force and consistency. Although the assumption of phlogiston enables explanation of many different observations, this comes with a loss of consistency.

Making use of the argumentative patterns identified above, Lavoisier's choice of the 'oxygen theory' above 'phlogiston theory' can be reconstructed in the following way:

1 We may assume that there is an element called oxygen
1.1 It is observed that some substances combust and rust (oxidize)
1.1'.1 Of candidate explanations oxygen and phlogiston, oxygen is the best explanation of these substances oxidizing
1.1'.1.1a Phlogiston theory explains combustion, rusting, and other phenomena, but does not explain why all combustible substances increase in weight when they burn or calcine. Oxygen theory predicts that combustible substances become heavier when they burn, which is confirmed by experiments.
1.1'.1.1b Oxygen theory explains the observed phenomena with astonishing simplicity.
1.1'.1.1c Phlogiston theory unnecessary multiplies entities.
1.1'.1.1d Phlogiston is attributed contradictory properties.

In this reconstruction, argument 1.1'.1.1a contains the scores of the candidate explanations with regard to the criterion 'explanatory force' or 'subsumptive power', argument 1.1'.1.1b the score of one of the candidates with regard to 'simplicity', and arguments 1.1'.1.1c and 1.1'.1.1d the scores of the other candidate with regard to 'parsimony' and 'consistency'. The reconstruction makes clear that not all the candidates have been scored on all the criteria and that an explicit decision rule is lacking (although it may be assumed that the explanatory force regarding combustible substances gaining weight is deemed very important or even decisive).

5. CONCLUSION

In this paper, I have described the pattern of argumentation based on abduction by translating influential accounts of this type of reasoning that have been put forward in the field of philosophy of science into argumentation theoretical terms. The pattern has a particular scientific explanation as the standpoint and the observed fact that is explained as the main argument. In the situation where several candidate explanations have been taken into consideration, the pattern can be extended with an argument that supports the justificatory force of the main argument and that expresses the choice of the best explanation: 'Of candidate explanations H_1 – H_n, H_i is the best explanation of *E*'. After having given an inventory of theoretical virtues and other criteria that scientists may use to justify this choice, I have indicated the argumentative function of these criteria by further extending the pattern. Also, I have pointed at the need for extending the pattern even further with an argument expressing the decision rule used to determine what is the best explanation. A future exploration of theories for selecting and weighing the criteria involved in the decision may help describing the specific decision rules involved. Finally, I analyzed an example taken from the history of science in order to illustrate how the argumentative pattern can be used as an analytical tool for reconstructing justifications of scientific explanations.

REFERENCES

Aliseda, A. (2007). Abductive reasoning: Challenges ahead. *Theoria, 22*(3), 261-270.

Best, N.W. (2015). Lavoisier's "Reflections on phlogiston" I: Against phlogiston theory. *Foundations of Chemistry, 17*(2), 137-151.

Campos, D.G. (2011). On the distinction between Peirce's abduction and Lipton's inference to the best explanation. *Synthese, 180*, 419-442.

Curd, M., & Cover, J.A. (1998). *Philosophy of science: The central issues.* New York / London: Norton.

Douven, I. (2011). Abduction. In E.N. Zalta (Ed.), *The Stanford Encyclopedia of Philosophy (Spring 2011 Edition).* URL = <http://plato.stanford.edu/archives/spr2011/entries/abduction/>.

Eemeren, F.H. van (2016). Identifying argumentative patterns. A vital step in the methodical development of pragma-dialectics. *Argumentation, 30*(1), 1-23.

Hacket, E.J., Amsterdamska, O., Lynch, M., & Wajcman, J. (Eds.) (2008). *The handbook of science and technology studies* (3rd edition). Boston, MA / London: MIT Press.

Kuhn, T.S. (1998). Objectivity, value judgment, and theory choice. Reprinted in M. Curd and J.A. Cover, *Philosophy of science: The central issues* (pp. 102-118). New York / London: Norton.

Minnameier, G. (2004). Peirce-suit of truth: Why inference to the best explanation and abduction ought not be confused. *Erkenntnis, 60*, 75-105.

Paavola, S. (2006). Hansonian and Harmanian abduction as models of discovery. *International Studies in the Philosophy of Science, 20*(1), 93-108.

Peirce, C.S. (1974). *Collected papers of Charles Sanders Peirce.* Edited by Charles Hartshorne and Paul Weiss. Cambridge, MA: Harvard University Press.

Wagemans, J.H.M. (2014). The assessment of argumentation based on abduction. In D. Mohammed & M. Lewiński (Eds.), *Virtues of argumentation: Proceedings of the 10th International Conference of the Ontario Society for the Study of Argumentation (OSSA), 22-26 May 2013* (pp. 1-8). Windsor, ON: OSSA.

PANEL II

Visual Arguments and Beyond

*II.*1

Kinds of Visual Argument

IAN J. DOVE
University of Nevada, Las Vegas, USA
ian.dove@unlv.edu

There are (at least) two kinds of visual arguments. One has a model or base in non-visual argument. For example, visual argument from analogy is modeled upon standard argument from analogy. The other has no non-visual base; I offer two such examples. The analysis and evaluation of such arguments requires novel schemes to explain their (apparent) structure, and critical questions from which to assess argument strength.

KEYWORDS: abduction, argument from analogy, argumentation schemes, inference to the best explanation, visual argument

1. INTRODUCTION

There is some controversy regarding the existence of visual arguments. Indeed, at this very conference, Igor Z. Zagar is giving a paper entitled "Against Visual Argumentation." Such visual argumentation scepticism typically depends upon an unnecessarily restrictive definition of visual argument or visual argumentation. Zagar's view seems to be that so long as we allow for the interplay between the modes—visual and verbal, say—there's no problem with meaning. I prefer a rather open definition for visual argument that is a descendant of one from Leo Groarke (Groarke, 1996, p. 107). A visual argument is an argument at least one component of which is *carried* or *conveyed* by a visual rather than a verbal element. I'm likewise fairly open to what counts as a visual element; but, to corral this notion somewhat, let's take visuals to be pictures, photographs, maps, graphs, and their kin. This definition of argument requires that only one element be visual. Hence, given the kind of scepticism that seems to be engendered in Zagar's talk or that is explicit in, say, David Flemming's early work (Flemming, 1996), this definition skirts the supposed problem by allowing that some of a visual argument's parts are verbal.

D. Mohammed & M. Lewiński (eds.) (2016). *Argumentation and Reasoned Action: Proceedings of the 1st European Conference on Argumentation, Lisbon, 2015. Vol. II, 57-68.* London: College Publications.

Leaving any remaining scepticism regarding the existence of visual arguments to one side, I'm interested in how it is that we seem to evaluate arguments that partake of visual elements. Insofar as Argumentation Schemes have become one (of many potential) sources of evaluation, I endeavour to distinguish two kinds of visual arguments in terms of the kinds of schemes used to evaluate such arguments. On the one hand, some visual arguments will best be assessed using schemes based upon verbal argumentation schemes. For example, as I show below, it makes sense to assess arguments from visual analogy using a scheme developed from the scheme for arguments by verbal analogy. On the other hand, there are some visual arguments that don't have verbal counterparts. Such arguments require wholly new schemes. I present two such schemes here.

The distinction between visual schemes based on verbal schemes and visual schemes that are not so based isn't meant to capture some essential feature of the argument types. I fully expect that if there is more than a handful of the second type, such schemes could be wildly different from each other. The only commonality is a negative—that they aren't based upon or derived from verbal schemes. Rather, the distinction is important mostly for what it suggests about the practice of evaluating visual argumentation generally, viz., that we need to investigate visual reasoning with an eye towards developing novel schematic types.

2. WHY SCHEMES?

Insofar as *logic* has come to mean *formal logic* in much of philosophy, formal languages, with their requirement for well-behaved denizens, restrict the possibility to analyse and evaluate arguments that appeal to visual elements. The problem isn't that such arguments are odd or unfamiliar. Rather, visuals, except in rare instance, are not as well behaved as recursively defined formulae. There are, therefore, few attempts to spell out a formal semantics for images. This means that any theorist wishing to explain and evaluate reasoning that appeals to visuals will need a different account. Fortunately, argumentation theory generally, and informal logic in particular, aren't limited to formal analyses. In particular, argumentation schemes have become one of the standard methods for assessing arguments.

Although there is a rich history for argumentation schemes running perhaps all the way back to Aristotle or before, I take recent work by Douglas Walton, Chris Reed, and Fabrizio Macagno as my model (Walton et al., 2008). A scheme, on my view, is a recipe for making sense of a standard argument pattern or type. Hence, there are well-rehearsed schemes for argument types such as Argument from Analogy, Appeal to Authority, and the like. The scheme, however, isn't just a demarcation of the structure of the reasoning. Along with a presentation of the typical components of such reasoning, a scheme offers evaluative criteria, in the form of *critical questions*, which aid in the appraisal of said arguments. This combination of analytical elements—a listing of the typical components of an argument type—along with evaluative elements—the critical questions—makes schemes so useful.

To see the general value of using schemes, it is best to see them in action. Let's start with an example I analysed many years ago for a different purpose (Dove, 2008).

> During excavations of the Bronze Age levels at El Mirador Cave, a hole containing human remains was found. Tapaphonic analysis revealed the existence of cutmarks, human toothmarks, cooking damage, and deliberate breakage in most of the remains recovered, suggesting a clear case of gastronomic cannibalism. [...]The identification of cannibalism in archaeological contexts is based on taphonomic criteria. Turner (1983) suggests fourteen indicators to establish the existence of cannibalistic practices, subsequently reducing them to a minimal set of five (Turner and Turner, 1992). These are: (a) deliberate bone breakage, (b) cutmarks, (c) evidences of cooking, (d) abrasions caused by anvils, and finally (e) absence or crushing of vertebrae, resulting from the extracion of fat and marrow from the vertebral bodies (Turner, 1983). This revision, however, took no notice of what would be the most direct skeletal indicator of anthropophagy: the presence of human toothmarks (Botella and Alema´n, 1998; Botella et al., 2000). Other indicators (such as cutmarks, for example) may also be caused during body cleaning processes for funerary rituals. (Caceres et al., 2007, p. 899)

One relevant scheme for the appraisal of this argument is Argument from Sign (cf. Walton et al., 2008). An argument from sign has two premises. One of the premises typically asserts the existence of a sign, symptom or indication. The other premise asserts some connection between the sign and what it purportedly signifies, the symptom and some associated disease, or the indication and what it indicates. I call these the *sign* and *connection* premises respectively. The conclusion, then, is that the signified, the disease, or the thing indicated obtains as well. The critical questions for arguments from sign are: (1) how strongly are sign and signified correlated, and (2), are there other, better explanations for sign than what it supposedly signifies?

Comparing the scheme with the passage, we can reconstruct the reasoning thusly. Sign: [The human remains contained] cutmarks, human toothmarks, cooking damage and deliberate breakage. Connection: [These items are indicative of gastronomic cannibalism.] Conclusion: This is a clear case of gastronomic cannibalism. The article answers both of the critical questions explicitly. In the passage quoted above, the correlation between these signs and what they signify is asserted to be strong. And later alternative explanations for the signs are discounted. Thus, and following from the schematic appraisal, one ought to view this argument as *good*. Note well, though, that the power of the scheme is the combination of analytic and evaluative components. Thus, the scheme helped to identify the components of the argument, and then it aided in the appraisal of quality of the reasoning.

3. SCHEMES WITH A VERBAL BASE

For the proponent of visual argumentation, argumentation schemes could be a useful source for argument appraisal. Yet, few currently available schemes seem available to analysing, much less evaluating visual components of argumentation. To remedy this, we should look to repurpose some of the current verbal schemes in ways that will make them amenable to visual elements. That is, for some of the schemes one is likely to encounter, there seem to be visual counterparts to the argument types.

As an example, let's consider the scheme for Argument from Analogy. This scheme has two premises, though one of them could contain multiple elements, and a conclusion. The first kind of premise is comparative or analogical—this is where the argument gets its name. Two items are compared and judged similar across some set of dimensions. I call this the *comparison* premise. Next, one of the items is claimed to have some further characteristic. This premise is sometimes called the *base*, but I prefer to call it the *source*. Finally, this characteristic is *projected* from the one item, the source, onto the other, the target, in the conclusion. The critical questions for this scheme include the following. (1) Are there relevant dissimilarities between the items that would tend to undermine the comparison? (2) Is the comparison relevant to stated conclusion? We can visualize the structure of the argument as two columns containing similar components, along with a projection from one column to the other connecting a further element. In this case, the arrow represents the projection, and the question marks signify that the result of the projection is inferred from the premises rather than asserted in them.

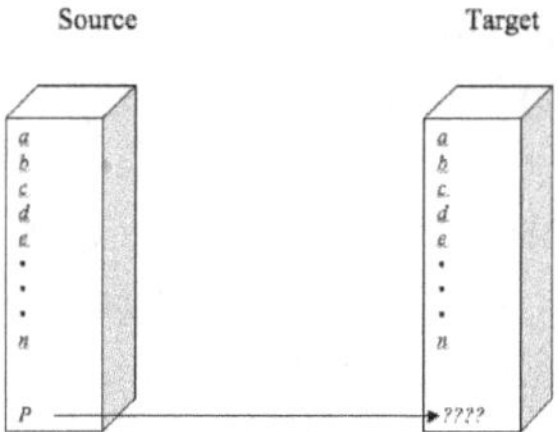

Figure 1 – The Structure of an Argument from Analogy

There is nothing in this presentation of the scheme that explicitly rules out the assessment of visual argumentation. Hence, though it may be theoretically unnecessary to construct an explicit scheme entitled Argument from Visual Analogy, such a scheme could unassumingly reiterate the same elements and critical questions as the verbal except with explicit permission for, say, the comparison to be visual rather than verbal. Consider the following argument paraphrased from an essay regarding primate breeding strategies as a function body types (Martin & May, 1982, p. 238).[1] The information could be presented verbally, but my *visual paraphrase* eschews some of the verbiage for visual/graphical elements.

	Gender Dimorphism		Pair-Bond Type
Humans	♂ ♀	→	?????
Gibbons	♂ ♀		Monogamy
Bonobos	♂ ♀		Polygyny
Chimps	♂ ♀		Polygyny
Gorillas	♂ ♀		Polygyny
Orangutans	♂ ♀		Polygyny

Figure 2 – An Argument from Visual Analogy

The comparison premise likens the category of gibbons to the category of humans. These categories are compared and judged similar as regards gender dimorphism—the average size difference between the genders of the species. Of primates and regarding only gender

[1] I previously analyzed this example for this purpose (Dove, 2011, p. 5).

dimorphism, humans and gibbons are similar. Furthermore, gibbons, unique among non-human primates, practice pair-bonded monogamy. Given the similarity and the source, pair-bonded monogamy is projected to be natural for humans. Moreover, by presenting the average body dimorphism of other primates, the argument at least partially answers the critical questions associated with the scheme. That is, not only are humans and gibbons similar as regards the rate of gender dimorphism, but this similarity is uncommon among primates. Indeed, greater gender dimorphism seems to correlate with polygyny as a mating strategy.

Unfortunately for proponents of monogamy, this argument isn't *good*. The problem, of course, comes from the real answers to the critical questions. There are many ways in which humans and gibbons differ that are relevant to determining mating strategies.

The point of giving the argument wasn't to argue for polygyny (or monogamy or anything); rather, the key was that information was presented visually that was necessary for the argument. Of course, there may be interpretational issues for such information, but that is true of information presented verbally as well. I don't take the fact that there can be competing paraphrases of some visual information to imply that such visual information is uselessly ambiguous. The visual presentation might be both more perspicuous and less open to interpretational variation for some audiences. The male/female pictograms depicting relative size are like a kind of technical vocabulary that once learned, allow straightforward communication of gender dimorphism for given species among *speakers* of this technical (aka *visual*) language.

Such extensions of already existing schemes to cover related visual argumentation should be straightforward and no more controversial than the existence of visual argumentation itself. I would argue that some schemes already seem ripe for exploitation as visual schemes. Take, for instance, Argument from Sign. The examples of signs, symptoms, and indications given by Walton, Reed, and Macagno are verbal descriptions of visual data: bear tracks in the one case. My suggestion for this scheme is that there is no need to see the paraphrase of the visual evidence of bear tracks into the sentence, "these are bear tracks," as essential in understanding the argument. It seems likely that in at least some cases, folks who tokened this type of argument didn't bother to verbalize it for themselves. Rather, upon seeing bear tracks, they simply became more wary.

4. NOVEL VISUAL SCHEMES

Repurposing accepted schemes is one thing; the more interesting question concerns the existence of uniquely visual patterns of reason. If there are such beasts, then the old schemes wouldn't help in their appraisal. Towards this end, I consider two visual argument types as potential novel schemes.

Consider a photo of my dog in Figure 1. Suppose I cut the photo into four irregular pieces. Then I let someone mix the pieces. I attempt to reassemble the photo. How do I know whether I've gotten it right? Obviously, by looking.

Figure 3 – A Correct Reassembly

As I slide these pieces together, they fit both in shape and in content. If I were only considering the shape, say by looking at just the backside of the cut photo, I could still reassemble the photo correctly by appealing to the way they fit together. What's important to note here, is that, were I asked to *justify* my claim that my reassembly put the pieces in the right place, my answer would simply be to point at the result and say, "look." It is from the goodness of the fit that I infer the correctness of the position.

This isn't just a one-off claim. For example, in the practice of fracture matching, the inference from goodness of fit to connection is used to associate items that may place a suspect at the scene of a crime. In the history of physical geography, this inference was used by Alfred Wegener to justify his theory of continental drift (cf. Dove, 2013). And, I regularly use this to justify not using all the parts when assembling various items that are labelled "some assembly required." There is only one premise type for this argument: some claim that some items fit together. The conclusion is justified by a visual inference license claiming that if two items fit together with some level of specificity, then they were from the same object (or they are the same object). I've hypothesized that the critical questions for this scheme are: (1) Is the fit/match *close* enough? (2) Can the appropriate level of fit be determined from the visual? (3) Is there another, better explanation for the apparent fit? (Adapted from Dove, 2013). Since I've discussed this scheme at length elsewhere, I'll move on to another potential scheme.

Why do advertisements, for example, use photographs of the things they are selling? One thing that these photos seem to do is to allow potential customers to *imagine* the products as their own. This *imagining* can act as a kind of *plausibility boost* for the product. I did not know that I wanted a green linen jacket until I saw the photo of one in an advertisement for the American clothier, Brooks Brothers. It wasn't that I didn't know of the *possibility* of such a purchase. Rather, the idea of a green jacket seemed unlikely to fit with, say, my ideal wardrobe. The photo allowed me to re-examine this initial plausibility claim–really an implausibility claim. In re-examining and imagining it as my own, the plausibility that such an item would fit in my wardrobe increased; though in this case, the plausibility of wanting the item did not also coincide with affording the item. So far, though, this doesn't seem much like an argument, it is more like an enticement.

To see that it is more than mere enticement, I want to consider how a similar use of visuals might operate argumentatively in scientific contexts. Start with the manufactured debate regarding the feeding habits of the most famous dinosaur, T-Rex. In a popular book on the subject (Horner & Lessem, 1993), Jack Horner and Don Lessem stirred the pot by suggesting the T-Rex was more likely to be a scavenger than a predator—much to the chagrin of tiny T-Rex fans the world over, and me. The alternative, of course, is that T-Rex was the top predator of the age. The reason that the latter seems so much more plausible to fans of the predator T-Rex hypothesis is that we have been inundated by images of T-Rex *as predator*. These images don't merely entice one to believe that T-Rex was a predator. Instead, the images associate the predatory elements of T-Rex with other characteristics of our beliefs regarding the times of dinosaurs. By filling out an otherwise limited notion of T-Rex, the images make the plausibility of a predatory T-Rex *more real* by being *more complete*.

This isn't so different from the case of the green linen jacket. In both cases, the images facilitate connections. In the case of the photograph of a jacket, the photo facilitates the connection between what was considered an initially unstylish choice with an idealized version of what I considered an acceptable wardrobe; thus allowing me to construe a green linen jacket as acceptable. In the case of various popular images of T-Rex, the images allow us to connect our view of predatory behaviour and the potential for predatory behaviour with T-Rex's environmental conditions, body type, possible prey items, etc. The inference is to the acceptability of an option.

As a third example of this kind of inference, consider the still open question regarding the earliest migrants to what we now call the Americas. Leaving aside interesting, but implausible hypotheses such as sea farers from Polynesia or ice jumpers from Europe, two early theories depend upon lower sea levels due to glaciation as part of the explanation for how people migrated from what is now Asia to what is now North America. The theory I learned from a geography class is what we might now call The Overland Hypothesis. On this view, because of the lower sea levels, a *land bridge* formed across the Bearing Strait. Migrating people *walked* across the land bridge following prey. The continued walking overland until they had reached and populated the lands to the bottom of America. Another theory, The Kelp Highway Hypothesis, has it that early migrants followed the coastline across Beringia—the land bridge from Siberia to Alaska—and down the west coast of America. On this view, the migrants both walked some of the journey as well as used simple boats for other parts. There are now, apparently, good reason for thinking the Kelp Highway Hypothesis is better than the Overland Hypothesis for reasons relating to the possibility of traversing the necessary route by foot given the possible ice coverage for the time frames (cf. Erlandson & Braj, 2011).

Figure 4 – Two Routes to the New World (Made with Google Earth.)

But before the discoveries that jeopardized the plausibility of the Overland Hypothesis, the Kelp Highway Hypothesis was considered more controversial partly because there was no evidence that the ancient people making this journey had the requisite boat-building skills. It wasn't that boat building was impossible for ancient peoples—there was evidence that people in the southern hemisphere had such technologies. Rather, there just wasn't evidence pointing to that technology in the north. And without the evidence, the theory was considered less plausible. For example, though he was a proponent of the theory, E. James Dixon, relegates his acceptance of the hypothesis to a section entitled, "Summary and Speculation" (Dixon, 1999, pp. 243ff.).

To overcome the implausibility, museums, it seems, used images of apparently ancient people in boats close to shore. For example, Simon Fraser University Museum of Archaeology and Ethnology uses a painting of ancient people coming ashore in a stretched skin canoe as part of their exhibit of the earliest migrations.[2] So, though there wasn't sufficient evidence in favour of boat-building technology, the images of ancient people in simple boats undermined the implausibility.

These arguments have a similar structure. First, there is a claim regarding the initial implausibility of an option, theory, or explanation. Next, an image is used to rebut, perhaps not decisively, the initial implausibility. The conclusion is the reassessment of the plausibility or subjective probability of accepting the option, theory or explanation. The critical questions are: (1) Does the image cohere with accepted elements? (2) Are the alternative explanations undercut by the rebutting image?

Figure 5 – P.Z. Makes the Case for What?

---

[2] One can view the image at http://www.sfu.museum/journey/an-en/multimedia/illustrations/bateau-boat.

Even under the best circumstances, arguments with this structure are very weak. These are used only to increase the plausibility of claims minimally. On their own these arguments don't offer much guidance. For example, there is a humorous image of noted atheist blogger and biologist, P.Z. Meyers, riding a dinosaur at Ken Hamm's "museum."[3]

I think the purpose of putting such an exhibit in a "museum" dedicated to the furtherance of belief in Creationism fits very well within this argument type. Indeed, the very notion that humans existed at the same time as dinosaurs is implausible, to say the least. The reason it is effective for, say, the average visitor to the "museum" has more to do with the uncritical acceptance of the rebuttal the exhibit offers than any structural weakness with the argument type. For, if the apparent rebuttal offered by the notion of riding a dinosaur were queried using the critical questions, the argument would be judged bad. The image of a human riding a dinosaur does not cohere with our best scientific evidence. And, since this is precisely what is at issue for the Creationists, using this kind of image is akin to begging the question.

Still, though the critical questions do allow us to judge bad arguments bad and good arguments good, the scheme seems to work, even if the best result one could hope for with such an argument is an incremental increase in plausibility. Again, what is important to recall is that these arguments make essential use of visual images. And, insofar as they are amenable to schematic appraisal, we can at least offer some normative appraisal of visual reasoning that is not merely the weighing of emotional appeal.

5. CONCLUSION

People reason from and with visuals all the time. We need some method to appraise when such reasoning is reasonable versus when it is not. Argumentation schemes offer just such a method for verbal cases. I have, in this paper, made a case for repurposing some of the verbal schemes to appraise some verbal arguments. In this regard, both Argument from Analogy and Argument from Sign seem likely candidates for such repurposing. On the other hand, I've argued that some visual arguments don't have verbal counterparts. In such cases, no repurposing of verbal schemes is likely to aid in the efforts of appraisal. Hence, for such cases, one must invent new schemes. Towards this end I've suggested two such schemes: Argument from Fit and Visual

---

[3] This example was first suggested to me by David Beisecker. A similar example was suggested during the Q&A of this talk at ECA 2015.

Plausibility. If there are visual arguments, then there are likely to be many more uniquely visual argument types. And this seems a ripe area for further research.

ACKNOWLEDGEMENTS: I thank member of my home philosophy department at UNLV for useful discussion during an early presentation of these results.

REFERENCES

Cáceres, I., Lozana, M., & Saladié, P. (2007). Evidence for Bronze Age cannibalism in El Mirador Cave (Sierra de Atapuerca, Burgos, Spain). *American Journal of Physical Anthropology, 133*, 899–917.

Dixon, E. J. (1999). *Bones, boats, bison.* University of New Mexico Press.

Dove, I. (2008). Towards a theory of mathematical argument. *Foundations of Science, 14*, 137–152.

Dove, I. (2011). Visual arguments and analogies. In F. Zenker (Ed.), *Argument Cultures: Proceedings of the 8th International Conference of the Ontario Society for the Study of Argumentation* (OSSA) (pp.1-16). May 18-21, 2011. Windsor, ON (CD ROM)

Dove, I. (2013). Visual arguments and meta-arguments. In D. Mohammed & M. Lewiński (Eds.), *Virtues of Argumentation. Proceedings of the 10th International Conference of the Ontario Society for the Study of Argumentation (OSSA)*, (pp. 1-15). 22-26 May 2013. Windsor, ON: OSSA.

Erlandson, J., & Braje, T. (2011). From Asia to the Americas by boat? Paleogeography, paleoecology, and stemmed points of the northwest Pacific. *Quaternary International, 239*, 28-37.

Fleming, D. (1996). Can pictures be arguments? *Argumentation & Advocacy, 33*(1), 11-23.

Groarke, L. (1996). Logic, art and argument. *Informal Logic, 18*(2-3), 105-129.

Horner, J., & Lessem, D. (1993). *The complete t-rex.* New York: Simon & Schuster.

Macagno, F., & Walton, D. (2009). Argument from analogy in law, the classical tradition, and recent theories. *Philosophy and Rhetoric, 42*(2), 154-182.

Martin, R. D., & May, R. M. (1982). Outward signs of breeding. In J. Maynard Smith (Ed.), *Evolution Now: A Century After Darwin* (pp. 234-239). Oxford: W.H. Freeman and Company.

Walton, D., Reed, C., & Macagno, F. (2008). *Argumentation schemes.* Cambridge: Cambridge University Press.

II.2

Visual Argument: Content, Commensurability, and Cogency

DAVID GODDEN
Michigan State University, Philosophy, USA
dgodden@msu.edu

Visual arguments can seem to require unique, autonomous evaluative norms, since their content seems irreducible to, and incommensurable with, that of verbal arguments. Yet, assertions of visual-verbal incommensurability seem to preclude counting putatively irreducible visual content as functioning argumentatively. By distinguishing two notions of content, informational and argumentative, I contend that arguments differing in informational content can have equivalent argumentative content, allowing the same argumentative norms to be rightly applied in their evaluation.

KEYWORDS: argument appraisal, argumentative content, autonomy thesis, evaluative equivalence, mode of argument, normative revisionism, visual argument, visual-verbal incommensurability

1. INTRODUCTION

The work reported on in this paper develops work that I initially presented at the Ontario Society for the Study of Argumentation 2013 conference (Godden, 2013), and which is further sketched out my commentary appearing in the special 2015 issue of *Argumentation* on visual argument, guest edited by Jens Kjeldsen (2015a) (Godden, 2015).

For the purposes of this paper, I assume that there are visual arguments, and that we have a suitable way to properly "extract" (Dove, 2013) their argumentative content. The question I am interested in is: Do visual arguments require the revision of our normative theories, methods, criteria, or standards of argument appraisal (Godden, 2013)?

D. Mohammed & M. Lewiński (eds.) (2016). *Argumentation and Reasoned Action: Proceedings of the 1st European Conference on Argumentation, Lisbon, 2015. Vol. II, 69-82.* London: College Publications.

2. PRELIMINARY MATTERS: SOME WORKING ASSUMPTIONS

Here I set forth two assumptions about arguments and norms that inform the work to follow. While I don't find these assumptions to be objectionable, like the assumption that there are visual arguments, I recognize that they may be controversial and are not universally shared. As such, they may be taken to demarcate the theoretical space in which I wish to work. The two assumptions are:

> (1) ***The rational nature of argument:*** Arguments (whatever else they are or do, and however they are presented) necessarily involve (contain, express, or convey) reasons.
> (2) ***Rational appraisal of argument:*** The evaluation of argument involves assessing the probative or rational support claims are provided by reasons.

3. TWO DISTINCTIONS

I proceed now to make two distinctions that will further help to conceptually demarcate the normative terrain in which we are working. These distinctions map out a "disagreement space" that different normative theories of visual argument inhabit.

The first distinction, between normative revisionism and normative non-revisionism, concerns the aptness of our existing ways of argument appraisal for evaluating visual arguments (Godden, 2013). By "*ways* of argument appraisal" and "evaluative *ways*" I mean the theories, methods, criteria, and standards of argument appraisal, broadly understood.

> (NR) ***Normative revisionism:*** Visual arguments require their own evaluative ways. (Johnson, 2010; Gilbert, 1994, 1997)[1]

> (NNR) ***Normative non-revisionism:*** Visual arguments do not require their own evaluative ways. They can be properly appraised using existing, non-specialized evaluative ways. (Groarke, 1996, 2015; Blair, 1996, 2004, 2015)

The second distinction, between evaluative equivalence and evaluative non-equivalence, concerns whether content-equivalent arguments,

---

[1] Gilbert's (1994, 1997) work on multi-modal argumentation does not take a position on whether the visual counts as a mode of argument. Recent correspondence confirms that, while multi-modal argumentation is intended to be open to such a possibility, Gilbert has not taken a position on this question.

presented in different modes, should receive the same rational evaluation. By "*mode* of argument" I mean the manner of presentation (Gilbert, 1997; Blair, 2015), rather than the material ingredients (Groarke, 2015) of an argument.[2]

> (EE) ***Trans-modal evaluative equivalence***: The same argument, no matter how it is presented, should receive the same rational or probative evaluation, *ceteris paribus*. (Blair, 2015)

> (ENE) ***Trans-modal evaluative non-equivalence***: An argument, presented in one mode (e.g., visual) can properly receive a different rational or probative evaluation than the same argument presented in some other mode (e.g., verbal), *ceteris paribus*.

Importantly, then, there are at least two ways that visual arguments might be normatively unique, or distinct, from non-visual arguments. The first is methodological, and concerns *how* they should be evaluated, while the second concerns the result of the method—i.e., *what* evaluation they should receive. Although related, the questions are quite different.

3.1 Equivalence and revisionism: Entailments

To appreciate how they are related, consider the relations of entailment, inconsistency, and consistency that obtain between them.

So long as the evaluation an argument receives is the result of some application of the theories, methods, standards, and criteria of appraisal, then that two arguments cannot be appraised using the same evaluative ways entails they cannot receive the same evaluation:

> (1) Normative revisionism (NR) entails trans-modal evaluative non-equivalence (ENE).

---

[2] Considered functionally, arguments are composed of reasons and claims, not of words or images. Without these functional roles, neither words nor images are arguments; with them either can be. Argument appraisal involves evaluating reasons, rather than collections of words or images. Claims and reasons can be identified by their argumentative content, and categorized by their mode, such that the same content can be presented in different modes. Thus the function, content, and mode of an argument's components in can be distinguished a way suggested by Groarke's (2015) *key component tables* (Godden, 2015).

Conversely to (1),

> (2) Trans-modal evaluative equivalence (EE) entails normative non-revisionism (NNR).

Again, so long as the evaluation an argument receives is the result of some application of the theories, methods, criteria, and standards of appraisal, then that two arguments receive the same evaluation entails that they were appraised using the same evaluative ways. For the same reason,

> (3) Trans-modal evaluative equivalence (EE) is inconsistent with normative revisionism (NR).

Lastly,

> (4) Normative non-revisionism (NNR) is consistent with trans-modal evaluative non-equivalence (NNE).

The case of (4) recognizes the possibility that the modality of an argument can affect its proper evaluation without changing how it should be evaluated. This final position allows for a theoretical space in which visual arguments may receive different evaluations than their non-visual counter-parts, while retaining the view that ordinary ways of argument appraisal properly apply to visual arguments.

4. INTUITIONS ABOUT VISUAL-VERBAL CONTENT NON-EQUIVALENCE

What inclines us to normative revisionism? Often, it seems to me, it is the compelling intuition that *what is expressed* in the image is somehow different from *what is expressed* with the words. Intuitions like this seem to inform claims against "verbal repackaging" (Groarke, 2013) or "linguistic imperialism" (Roque, 2009; Groarke, 2009) as a method of interpreting the argumentative content of visuals by "reducing" (Groarke, 2007, p. 139), "translating," or "converting" (Roque, 2009, pp. 5-6) visual content to verbal content.

Intuitions like this can incline us to accept something like the *autonomy thesis*:

> *Autonomy thesis* "Visual argument is a distinct and autonomous type of argument, and is not to be treated as an extension of verbal argument." (Johnson, 2010, p. 2)

And, intuitions like this, it seems to me, inform the further intuition that the image must be evaluated differently, and with different ways, from its verbal repackaging. This route to normative revisionism accepts a visual-verbal incommensurability—where the central idea of incommensurability is that of incomparability, or the absence of a common metric or standard against which the incommensurable things (arguments, their components and their properties, in this case) can be described, related, or evaluated.

4.1 Content non-equivalence: particular and universal

To begin to critique this incommensurability argument for normative revisionism, consider two degrees of content non-equivalence:

> ***Particular content non-equivalence:*** that *this* particular visual argument (or component thereof) is not content-equivalent to *that* specific verbal argument (or component thereof).
>
> ***Universal content non-equivalence***: that there is *no* verbal argument (or component thereof) that is content-equivalent to *any* visual argument (or component thereof).

Visual-verbal incommensurability commits the normative revisionist not to the weaker claim of particular content non-equivalence, but to the stronger one of universal content non-equivalence. It's not as though there were some *other* verbal argument with the same content as some particular visual argument. Rather, the visual and the verbal are categorically different from each other such that there is a general and complete absence of content-equivalence between them.

4.2 From content non-equivalence to incommensurability: A mistaken inference

There are at least two problems with arguments of this sort. First, the premises rejecting particular verbal-repackagings do not support the conclusion of incommensurability. Second, the claim of visual-verbal content incommensurability is not consistent with other commitments of the normative revisionist.

First, recall the intuition motivating the incommensurist: that *what is expressed* in the image is somehow different from *what is expressed* with the words. Intuitions of this sort about visual-verbal content non-equivalence readily support the particular, but not the general, claim. All too often, though, being dissatisfied with a number of attempted verbal reconstructions of the visual, we hastily jump to the strong, and unwarranted, conclusion of a universal content non-equivalence. Yet, this is a non-sequitur. Establishing the stronger claim does not involve several failed attempts to capture visual content verbally—it requires a positive argument demonstrating that no such translation is possible. And, importantly, any such argument would involve specifying the visual content that is putatively verbally inexpressible. At the very least, any such specification would make the visual content commensurable with some verbal content.[3]

A second problem with the incommensurist position is that it is not consistent with a central commitment of the normative revisionist: that the visual elements of argumentation are argumentatively relevant. To be argumentatively relevant, the visual features of visual arguments must function *like* reasons. Reasons, whether visual or verbal, must be capable of supporting claims by (minimally) making their acceptability more apparent than it was previously. So, either: (i) the visual elements of visual arguments really are incommensurable with the verbal such that there is a universal content non-equivalence between them, in which case the visual is entirely unlike a reason and hence is not argumentatively relevant. Or, (ii) the visual elements of visual arguments really can function as reasons, thereby making them argumentatively relevant but also commensurable with verbal reasons. Either way, there are no features of visual arguments that are both incommensurable with the verbal and argumentatively relevant.

5. CONTENT: INFORMATIONAL AND ARGUMENTATIVE

A more general problem for the intuitive visual-verbal content non-equivalentist and the normative revisionist is that they have provided no account of the argumentative content of visual images.

To retain our intuitions about the content of visuals while avoiding the problems and mistakes (just identified) that can arise from these intuitions, I suggest we need a way of talking about the content of

---

[3] As an inoculation against "translation" problems of this sort, I recommend letting the individual audiences supply, i.e. articulate, what they take the argumentative content of the image to be, rather than impose some reconstruction on them. Empirical studies (Morgan, 2005; Kjeldsen, 2015b) suggest that that audiences readily supply some verbal description of the argumentative contents of persuasive visual advertisements, and that, having done so, they tend to be satisfied that their own verbal description adequately captures the argumentative content of the image.

visual images that recognizes their differences from collections of words, and yet accounts for, and explains, their argumentative content.

5.1 Informational content

The notion for the job, I suggest, is *information*. Information is meaningful data (Floridi, 2010), or a "difference which makes a difference" (Bateson, 1973, p. 428), where data is a "lack of uniformity," which can be natural or contrived, intentional or accidental. The content of visual images, whether argumentative or otherwise, is informational.

The idea of information allows us to explain the differences in content between visuals and their verbal reconstructions without resorting to any mysterious, ineffable, visual somethings-we-know-not-what. There is nothing mysterious or ineffable about the content of visuals: they are (entirely!) comprised of, and convey, information. And, typically, that information is visual, not verbal, information.

Thus, visual and verbal arguments tend to have different informational content. Typically, even the simplest visual images contain and convey vastly, indeed infinitely, more information than any finite set of sentences (Kitcher & Varzi, 2000). Thus, we can easily account for the ordinary lack of fidelity between some persuasive visual and its infinitely elliptical, enthymematic, informationally impoverished verbal reconstruction. Indeed, any verbal reconstruction that approached the information content of even a basic visual construction would be so cognitively unwieldy as to be entirely argumentatively ineffectual.

So, an informational account of the content of visuals explains and validates our intuitions of visual-verbal content non-equivalence. Yet, equally importantly, it does not justify any incommensurist conclusions. Both visual and verbal arguments contain and convey information. So, in principle, they are commensurable, even if, in practice, a content-equivalent verbal expression of some visual is entirely impractical.

5.2 Argumentative content

Yet, an informational content non-equivalence between the verbal and the visual needn't be the infinite, unbridgeable chasm that it might at first seem. After all, not all of the informational content of a visual is argumentatively relevant. (Indeed, not all of the information in a verbal argument is argumentatively relevant—e.g., the font in which a textual argument is presented is, typically, argumentatively irrelevant.)

This gives us a notion of argumentative content as distinct from informational content, where:

> ***Argumentative content***: [is] content that constitutes, or ineliminably contributes to, a reason or claim.

To have argumentative content, an (informational, content) expression (whatever its modality) must satisfy, or contribute to, the *functional* role of either a claim or a reason (or some essential part thereof).

Hence, to specify the argumentative content of some (informational, content) expression (whatever its modality) we must articulate its *functionality* in the space of reason—that is, we must specify what it counts as a reason for, and what counts as a reason for it (Brandom, 2000), or how it contributes to the rational workings of a reason or claim. Unless the inferential operation of the image (or visual element thereof) can be articulated, what sense can be made of the claim that the image (element) acts as, or contributes to, a reason or claim—i.e., that it has argumentative content?

6. CONCLUSIONS

If what I have said to this point is on the mark, then there are important consequences for theories of visual argumentation, concerning both the analytical, interpretive aspects of visual arguments as well as their appraisal and evaluation.

6.1 Consequences for the analysis of visual argument analysis

Two important consequences follow concerning the analysis of visual arguments. First, just as it does not follow from the mere fact that *some* specific visual argument is not content equivalent to *some* particular verbal argument that *no* verbal argument is content equivalent to *any* visual argument, similarly, it does not follow from the mere fact that some verbal argument is *informationally* content non-equivalent to some visual, that they are *argumentatively* content non-equivalent. So long as they express the same argumentative content, they may rightly be considered content-equivalent in all aspects bearing upon (i.e., relevant to) the rational appraisal of argument.

Second, there is good reason to think that the argumentative content of any visual can be verbally expressed. This might happen in one of two ways: at the level of the object language, or in the meta-language. In the object language, there may well be sentences, or manageable collections of sentences, that equivalently express all of the argumentatively relevant information in some image (see, fn. 3, above). And, even in cases where we remain resistant to such a "translation" of the visual into the verbal, the meta-linguistic rules governing the argumentative functioning of the visual can be articulated in natural language.

> Even if the object language is an imagistic one of pictures, the meta-linguistic rules governing the inferential use of images as entitlement-establishing, entitlement-preserving, and entitlement-defeating are expressible in natural languages. Indeed these meta-linguistic rules often constitute the warranting principles that arguers must invoke when justifying their argumentative moves if challenged. As such, if the inferential use of images cannot be specified or codified in a more-or-less rigorous manner adequate to the communicative and rational needs of arguers, significant doubt is cast on whether the images really are functioning argumentatively. (Godden, 2015, pp. 236-237)

Thus, not only is visual-verbal incommensurability false, so is a universal visual-verbal content non-equivalence. Any visual whose argumentative content is governed by a set of meta-linguistic rules specifying its argumentative functionality, or proper use in argument, will be argumentatively equivalent to any verbal expression whose proper use is governed by those same rules. (It should be noted that this is a consequence of the view (cf. Groarke, 2014) that the visual and the verbal gain their meaning in the same way—namely, according to their use.) And, even if there is no verbal expression currently in use that operates according to the same meta-linguistic rules, one could readily be coined and specified to have the same argumentative functional role as the image.

Despite the falsity of visual-verbal incommensurability our intuitions that images often convey something different than their verbal repackaging can be validated and explained by differences in their informational content—the very thing that makes the visual and the verbal commensurable. Similarly, by distinguishing informational from argumentative content, the falsity of a universal visual-verbal content non-equivalence has also been demonstrated.

6.2 Consequences for the appraisal of visual arguments

Two important consequences also follow for the appraisal of visual arguments. First, despite any intuitions we might have towards Johnson's *autonomy thesis*, normative revisionism is mistaken. At the very least, no compelling reason—indeed no good reason whatsoever—has been offered that the visual require any special evaluative ways. Quite the contrary. Recent work (Godden, 2013; Blair, 2015; Dove, forthcoming) suggests that our existing theories, methods, criteria, and standards of argument appraisal apply just as well to visual arguments as they do to non-visual, verbal ones. And this should come not as a surprise. These evaluative ways were designed to appraise the probative force of reasons, not words. Thus, just as incommensurability is an unwarranted and untenable position about the content and analysis of visual arguments, normative revisionism is an unwarranted and untenable position about visual argument appraisal.

Second, although normative revisionism is false, our intuitions that the visual may be acting upon us differently than the verbal, and hence should receive a different appraisal may have some merit. And, perhaps the modality in which argumentative content is presented can make a difference to its probative merits or our ability to take cognizance of those probative merits. That is, there may be something to the idea of a visual-verbal evaluative non-equivalence.

Suppose that the same argumentatively relevant informational content is presented visually and verbally. That is, let us stipulate that there is a visual-verbal equivalence of argumentative content in certain cases. Perhaps, in some of those cases, we might more easily apprehend, detect, understand, grasp, parse or appreciate, either the informational content itself or its probative qualities because of its manner of presentation.

A wide variety of examples merit consideration for a full investigation of this hypothesis. Here, I will briefly consider only a small few. Blair (2015) considered four samples, and seemed to argue not only for normative non-revisionism but also for an evaluative equivalence between the visual examples and their verbal surrogates. Godden (2013) considered two examples. In one, he argued that a syllogism stated verbally ought to receive the same rational appraisal as a content-equivalent syllogism expressed visually, say in Venn diagrams. Yet, perhaps he was mistaken. As Dove (forthcoming) observes, students of logic often apprehend the validity or invalidity of a-typical syllogisms expressed visually well before they recognize the same probative qualities in the verbal expressions of those same syllogisms. Similarly, Guarini (2011) (responding to Dove, 2011), considered an example that contrasts visual and verbal analogies comparing skeletal bird feet. (In an attempt to ascertain the species of bird from one of the samples, the question was: which of the remaining samples is the skeletal structure of the target sample more similar to?)

Let's assume that the argumentative contents of those visual and verbal analogical arguments are equivalent. As Guarini notes:

> Extended sentential descriptions of the feet of that many different types of creatures may not render perspicuous the [possibly quite subtle and complex] similarity relation that holds between them. The images, though, do render the similarity in a perspicuous fashion. (Guarini, 2011, p. 4)

Our cognitive resources dedicated to the processing of visual information is vast, allowing us to process much more visual information, and far more quickly, than we can process verbal information. (Hence, it's no wonder that we often feel as though there's a lot more going on in a persuasive visual than in any verbal rendering of it. Often, there is! And, we can detect, process, and act on that information far more efficiently and effectively, often without ever articulating it to ourselves, than we can information delivered verbally.) Thus, comparing the feet visually, we recognize and gesture at, without explicitly articulating, the salient qualities (similarities and differences) constituting the evidential basis of the analogy. Because of this, not only do we tend to experience the visual as more informative than the verbal, but we also often experience the visual as being more persuasive than the verbal. And, perhaps, in some cases, we ought to. Perhaps we ought to find the visual to be more, or at least differently, persuasive than its repackaged verbal translation, even if that translation is informationally equivalent in all the argumentatively relevant aspects.

If this is the case, then there is a space for a certain measure of normative difference of the visual, even if normative revisionism is false. As already noted, normative revisionism is consistent with trans-modal evaluative non-equivalence, and perhaps that is the proper way of accounting for the normative differences that we intuitively apprehend between visual and verbal arguments.

ACKNOWLEDGEMENTS: Research and travel for this paper were supported by Old Dominion University, Norfolk, Virginia, U.S.A., with whom I was affiliated when the paper was presented. In addition to the European Conference on Argumentation (ECA) conference audience for their pressing and constructive discussion and questions, several people deserve my thanks and recognition. I would like to thank my co-panellists, Ian Dove and Leo Groarke, each of whom, together with Tony Blair, have inspired and influenced my thinking on this topic in more ways than I could list. I also extend my thanks to Jens Kjeldsen, whose invitation to contribute a commentary to his special issue of *Argumentation* (Kjeldsen, 2015a) on visual argumentation provided me the opportunity to develop and articulate my position on these matters in a way that I would never have had otherwise. Lastly and most importantly, I offer my sincere thanks to the ECA conference organizers and proceedings editors, Dima Mohammed and Marcin Lewiński.

REFERENCES

Bateson, G. (1973). *Steps to an ecology of mind.* Frogmore, At. Albans: Paladin.

Blair, J.A. (1996). The possibility and actuality of visual arguments. *Argumentation and Advocacy, 33*, 23-39. Reprinted in J.A. Blair. (2012). *Groundwork in the theory of argumentation* (pp. 205-223). Amsterdam: Springer.

Blair, J.A. (2004). The rhetoric of visual arguments. In C.A. Hill & M. Helmers (Eds.), *Defining visual rhetorics* (pp. 41-61). Mahwah, NJ: Lawrence Erlbaum. Reprinted in J.A. Blair. (2012). *Groundwork in the theory of argumentation* (pp. 261-279). Amsterdam: Springer.

Blair, J.A. (2015). Probative norms for multimodal visual arguments. *Argumentation, 29*, 217-233. doi: 10.1007/s10503-014-9333-3

Brandom, R. (2000). *Articulating reasons: An introduction to inferentialism.* Cambridge, MA: Harvard UP.

Dove, I. (2011). Visual analogies and arguments. In F. Zenker et al. (Eds.), *Argumentation: Cognition and community: Proceedings of the 9th international conference of the Ontario Society for the Study of Argumentation (OSSA*), May 18-21, 2011, CD ROM (pp. 1-16). Windsor, ON: OSSA.

Dove, I. (2013). Visual arguments and meta-arguments. In D. Mohammed & M. Lewiński (Eds.), *Virtues of argumentation. Proceedings of the 10th international conference of the Ontario Society for the Study of Argumentation (OSSA)*, 22-26 *May 2013*, CD ROM (pp. 1-15). Windsor, ON: OSSA.

Dove, I. (forthcoming). Visual scheming: Assessing visual arguments. *Argumentation and Advocacy*, special issue: "Twenty years of visual argument."

Floridi, L. (2010). *Information: A very short introduction.* Oxford: Oxford UP.

Gilbert, M. (1994). Multi-modal argumentation. *Philosophy of the Social Sciences, 24*, 159-177.

Gilbert, M. (1997). *Coalescent argumentation.* Mahwah, NJ: Lawrence Erlbaum.

Godden, D. (2013). On the norms of visual argument. In D. Mohammed & M. Lewiński (Eds.), *Virtues of argumentation. Proceedings of the 10th international conference of the Ontario Society for the Study of Argumentation (OSSA)*, 22-26 *May 2013*, CD ROM (pp. 1-13). Windsor, ON: OSSA.

Godden, D. (2015). Images as arguments: Progress and problems, a brief commentary. *Argumentation, 29,* 235-238. doi 10.1007/s10503-015-9345-7

Groarke, L. (1996). Logic, art and argument. *Informal Logic, 18,* 105-129.

Groarke, L. (2007). Beyond words: Two dogmas of informal logic. In H.V. Hansen & R.C. Pinto (Eds.), *Reason reclaimed: Essays in honor of J. Anthony Blair and Ralph Johnson* (pp. 135-152). Newport News, VA: Vale Press.

Groarke, L. (2009). Commentary on G. Roque's "What is visual in visual argumentation?" In J. Ritola (Ed.), *Argument Cultures: Proceedings of the 8th international conference of the Ontario Society for the Study of Argumentation (OSSA),* CD-ROM (pp. 1-3). Windsor, ON: OSSA.

Groarke, L. (2013). On Dove, visual evidence and verbal repackaging. In D. Mohammed & M. Lewiński (Eds.), *Virtues of argumentation. Proceedings of the 10th international conference of the Ontario Society for the Study of Argumentation (OSSA),* 22-26 *May 2013,* CD ROM (pp. 1-8). Windsor, ON: OSSA.

Groarke, L. (2014). Visual argument, Wittgenstein and Patterson: How to do things without words. International Society for the Study of Argumentation (ISSA), 8th International Conference on Argumentation, at the University of Amsterdam, the Netherlands, July 1-4, 2014.

Groarke, L. (2015). Going multimodal: What is a mode of arguing and why does it matter? *Argumentation, 29,* 133-155. doi: 10.1007/s10503-014-9336-0

Guarini, M. (2011). Commentary on Ian Dove's "Visual analogies and arguments." In F. Zenker et al. (Eds.), *Argumentation: Cognition and community: Proceedings of the 9th international conference of the Ontario Society for the Study of Argumentation (OSSA),* May 18-21, 2011, CD ROM (pp. 1-5). Windsor, ON: OSSA.

Johnson, R.H. (2010). On the evaluation of visual arguments: Roque and the autonomy thesis. [Unpublished conference paper, presented to] Persuasion et argumentation: Colloque international organisé par le CRAL à l'Ecole des Hautes Etudes en Sciences Sociales, 105 Bd. Raspail, 75006 Paris, Salle 7, 7-9 Septembre 2010.

Kitcher, P., & A. Varzi. (2000). Some pictures are worth 2ℵ0 sentences. *Philosophy, 75,* 377-381.

Kjeldsen, J. (2015a). The study of visual and multimodal argumentation. *Argumentation, 29,* 115-132. doi: 10.1007/s10503-014-9342-2

Kjeldsen, J. (2015b.) Where is visual argument? In F.H. van Eemeren, and B. Garssen (Eds.), *Reflections on Theoretical Issues in Argumentation Theory* (pp. 107-117). Dordrecht: Springer.

Morgan, S. (2005.) More than pictures? An exploration of visually dominant magazine ads as arguments. *Journal of Visual Literacy, 25,* 145-166.

Roque, G. (2009). What is visual in visual argumentation? In J. Ritola (Ed.), *Argument Cultures: Proceedings of the 8th international conference of the Ontario Society for the Study of Argumentation (OSSA)*, CD-ROM (pp. 1-9). Windsor, ON: OSSA.

II.3

The Semantics of Multimodal Arguing and the Fallacy of Independent Meaning

LEO GROARKE
Department of Philosophy, Trent University, Canada
leogroarke@trentu.ca

Extending Blair, Groarke, van den Hoven, and others I develop an account of multimodal meaning designed for the analysis and assessment of multimodal arguments that employ visual images and other non-verbal phenomena. My account has its roots in Wittgenstein, and is built on the notion that *language games* can include *multimodal games* (picture games, image games, performance games, and so on).

KEYWORDS: Fleming, Johnson, language-games, meaning as use, multimodal arguing, Patterson, visual arguments, Wittgenstein

1. INTRODUCTION

This short essay addresses the question whether there is an account of meaning that can explain non-verbal modes of arguing. Visual arguing – arguing that uses pictures, photographs, diagrams or other non-verbal visual phenomena to supporting a conclusion – is a specific case in point. In proposing an account of meaning that explains such arguing, I appeal to two aspects of Wittgenstein's later philosophy of language: his suggestion that meaning is use and his account of language games. I argue that these two notions accommodate "picture games" and other multimodal forms of meaning, and in this way explain how non-verbal elements are components of arguing, reasoning and reasoned action.

The questions of meaning I address are important because many commentators (among them, Fleming, 1996; Johnson, 2005; Patterson, 2010) have maintained that non-verbal visuals cannot function as arguments. In this article, Patterson's views are especially relevant, for he has argued against visual arguments by maintaining that the meaning of pictures is radically undetermined, and that Wittgenstein holds that this is a crucial difference that distinguishes pictures from

D. Mohammed & M. Lewiński (eds.) (2016). *Argumentation and Reasoned Action: Proceedings of the 1st European Conference on Argumentation, Lisbon, 2015. Vol. II, 83-96.* London: College Publications.

words. In sharp contrast, I argue that Wittgenstein believes that non-verbal elements in communication function like words and sentences, and in doing so provides a ready account of the semantics of visual and other non-verbal modes of arguing.

Patterson's views are a useful criticism of my own, but it bears noting that his remarks overstate my views and the views of other commentators who have defended visual arguing. In his own description of his arguments, he writes that:

> I shall draw on Wittgenstein's remarks ... in order to argue that although visual images may occur as elements of argumentation, broadly conceived, it is a mistake to think that there are purely visual arguments, in the sense of illative moves from premises to conclusions that are conveyed by images alone, without the support or framing of words. (Patterson, 2010, p. 15)

Suffice it to say that the radical conception of "purely visual arguments" Patterson rejects here is not, so far as I am aware, one that characterizes any author who defends the notion of visual arguments. Many commentators have argued that there are visual arguments, but I cannot find any commentator who understands them as arguments conveyed "by images alone, without the support or framing of words."

Most visual arguments mix verbal, visual and other modes of communication. When making a point with a photograph, for example, arguers typically use words to elaborate what it presents (much as they may use photographs to elaborate what they try to say in words). This widespread inclination to mix the visual, verbal and multimodal makes quixotic attempts to find "purely" visual arguments inherently artificial and of marginal interest in any attempt to account for arguing as it is actually practiced in the world of argument today.

In light of this I want to emphasize, at the very beginning of this essay, that it makes no claims about purely visual arguments understood in the radical way that Patterson suggests. The conception of visual argument it assumes encompasses attempts to use non-verbal visual means to convey evidence, make an inference, or draw a conclusion. It leaves open the possibility that most (or even all) of these attempts are supported or framed by verbal discussion of one sort or another.

2. SOME PRELIMINARIES

I understand an "act of arguing" as an attempt to justify a point of view by providing evidence in support of it. The classical account understands such acts as something that we perform with words. This makes an argument a set of sentences (or the propositions they refer to) which includes a conclusion and premises in support of it.

In broadening this traditional account, I follow Pinto, who understands an argument as "an invitation to inference" (Pinto, 2001, pp. 68–69) that may involve premises and conclusions which are not verbal and may not be propositional. Like Groarke (1996) and Blair, (1996) (and like many others following them), I understand visual acts of arguing as attempts to construct arguments using non-verbal, visual components. As Groarke (2014) suggests, other kinds of multimodal arguing may invoke non-verbal sounds, physical demonstrations, music, smells, and other non-verbal elements. As Roque (2012) points out, most instances of visual arguing are multimodal in the sense that they mix visual and verbal (and possibly other) modes of expression.

Examples of visual argument are much easier to find than one would think when one reads those who have rejected the idea of visual arguments. Propaganda films (like Leni Riefenstahl's *Triumph of the Will*) uses images as well as words to try and justify political points of view. Criminal trials increasingly use videos to prove that a crime was heinous, committed by the accused, or carried out in a particular way. In arguing that you should consider buying a particular house, a real estate agent may use photographs and maps. In medical diagnosis, a doctor's viewing of a skin condition or the image on an MRI may lead to the conclusion that a patient needs a particular treatment. In other situations, smells, tactile sensations and other experiences play a role in arguing, but I will focus on instances of visual arguing in this paper.

A good example of visual arguing is tied to the illustration I have reproduced below.

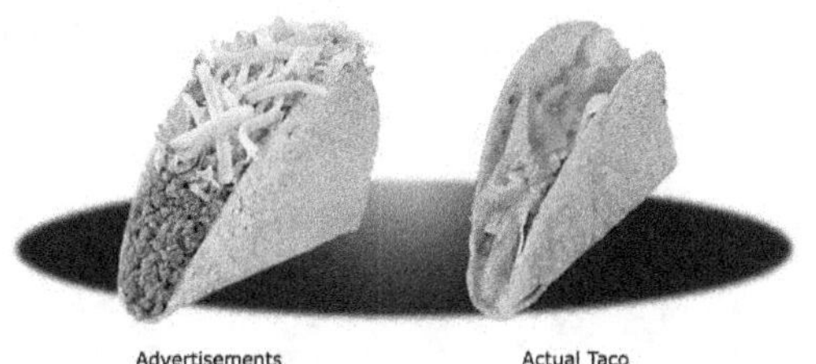

It is one of a series of similar images featured on a web blog entitled *Fast Food FAILS: Ads vs Reality* (BoredPanda, 2011). In it the author argues that fast food advertisements fail to present reality. At one point they wonder how such egregious misrepresentation can be legal. In defence of their views they present a series of photographic comparisons which contrast advertising images of fast food items with what the items look like when purchased in a restaurant. The photograph above illustrates the result in the case of Taco Bell's "Crunchy Taco".

This is a clear case in which someone is offering us evidence as the basis of an inference. In verbal arguments evidence is presented in sentences that function as the premises of an argument. In visual arguments like the present one, visual images play that role. The argument that results can be summarized in the table and diagram in Figure 1, below.

Act of arguing: Key components	*Role in the argument*	*Mode of Expression*
Taco Bell Crunchy Taco Advertisements Actual Taco	Premise (*a*)	Visual (photograph)
Taco Bell Crunchy Taco Advertisements Actual Taco	Premise (*r*)	Visual (photograph)
[The Ad for Taco Bell's 'Crunchy Taco' fails to accurately represent it.]	Conclusion (*f*)	Implicit (implicit in the broader conclusion that fast food ads fail to represent reality)

Figure 1—Table and Diagram

Tables of this sort ("Key Component" or "KC" tables) are a convenient and effective way to summarize the meaning of visual and multimodal arguments. They identify an argument's key elements and the modes of communication they depend on. Visual elements may be replicated visually (as above); or by pointing to them, verbally or otherwise. In an act of arguing like the one summarized here, it is the visual inspection of the visual elements that lends support to the proposed (and invited) inference.

As in this case, visual acts of arguing are typically embedded in chains of verbal and visual reasoning which feature interconnected arguments and conclusions. In this example, the conclusion serves as a subconclusion which is used in a much broader argument for the general conclusion that "Fast food ads fail to represent reality". Other visual subarguments invoke photographic comparisons that compare ads to the reality of products available at Burger King and McDonalds. As in most cases of visual arguing, the appeals to visual evidence are intertwined with verbal comments and discussion.

3. WITTGENSTEIN, LANGUAGE, AND PRAGMA-DIALECTICS

Wittgenstein's later view of language is a response to his views in the *Tractatus Logico-Philosophicus.* It is famous for its "picture" theory of meaning, a theory inspired by Wittgenstein's reading of an account of Paris legal proceedings in which model cars and dolls were used to mimic (to picture) what happened in an accident (Grayling, 1988, p. 40). Wittgenstein similarly suggests that language is made up of statements that convey facts about the world by picturing the states of affairs that they assert. As we read in the *Tractatus* (Wittgenstein, 1961):

> What is the case—a fact—is the existence of states of affairs. (2.01)
> A state of affairs (a state of things) is a combination of objects (things). (2.131)
> In a picture the elements of the picture are the representatives of objects. (2.14)
> The fact that the elements of a picture are related to one another in a determinate way represents that things are related to one another in the same way. (2.15)
> The simple signs employed in propositions are called names. (3.203)
> A name means an object. (3.21)
> The configuration of objects in a situation corresponds to the configuration of simple signs in the propositional sign. (3.22)

In his later work, Wittgenstein moves in a different direction, rejecting the idea that the essence of language is statements which are verbal pictures of the states of affairs that they refer to. His new views are rooted in the notion that "meaning is use". His move in this direction is already evident in *The Blue Book*, where he is moving away from the notion that language and meaning should be understood as word-pictures in a Tractarian way, preferring the notion that "if we had to name anything which is the life of the sign, we should have to say that it was its use" (Wittgenstein, 1958, p. 4). Wittgenstein confirms his move in this direction in his *Investigations*, writing that "For a large class of cases in which we employ the word meaning it can be explained thus: the meaning of a word is its use in the language." (Wittgenstein, 1953, 1.43) As Blitezi and Matar (2014) note, "This basic statement is what underlies the change of perspective most typical of the later phase of Wittgenstein's thought: a change from a conception of meaning as representation to a view which looks to use as the crux of the investigation."

According to this new view, language needs to be understood as a means of doing different things.

> Think of the tools in a tool-box: there is a hammer, pliers, a saw, a screw-driver, a rule, a glue-pot, glue, nails, and screws. The functions of words are as diverse as the functions of these objects. (Wittgenstein, 1953, 1.11-12)

One finds another metaphor which emphasizes the functions of words in *On Certainty*, where Wittgenstein suggests that we "Compare the meaning of a word with the 'function' of an official. And 'different meanings' with 'different functions'." (Wittgenstein, 1963, p. 64).

In an attempt to accommodate the multifarious ways in which words are used (and can have meaning), Wittgenstein suggests that language is a collection of different "language games" which comprise the different ways that words are used. "Here the term language game is meant to bring into prominence the fact that the speaking of a language is part of an activity..." (Wittgenstein, 1953, 1.23). The activities in question are bound by very different rules which tie the meaning of a particular use of language to a specific context and the social practices and conventions (the "form of life") with which it is associated.

> [H]ow many kinds of sentence are there? Say assertion, question, and command? – There are countless kinds: countless different kinds of use of what we call 'symbols', 'words', 'sentences'. And this multiplicity is not something fixed, given once for all; but new types of language, new language games, as we may say, come into existence, and

> others become obsolete and get forgotten. (We can get a rough picture of this from the changes in mathematics.)
> Review the multiplicity of language game in the following examples, and in others:
> * Giving orders, and obeying them –
> * Describing the appearance of an object, or giving its measurements –
> * Constructing an object from a description (a drawing) –
> * Reporting an event –
> * Speculating about an event –
> * Forming and testing a hypothesis –
> * Presenting the results of an experiment in tables and diagrams –
> * Making up a story; and reading it –
> * Play-acting –
> * Singing catches –
> * Guessing riddles –
> * Making a joke; telling it –
> * Solving a problem in practical arithmetic –
> * Translating from one language into another –
> * Asking, thanking, cursing, greeting, praying.
> – It is interesting to compare the multiplicity of the tools in language and of the ways they are used, the multiplicity of kinds of word and sentence, with what logicians have said about the structure of language. (Including the author of the *Tractatus Logico-Philosophicus.*) (Wittgenstein, 1953, 1.23)

We can summarize these aspects of Wittgenstein's later view of language by saying that we must, in most cases, look for the meaning of words in their use, and that an understanding of their use requires an understanding of the language games in which they occur – and the circumstances and social practices this implies. In attempting to understand a particular communicative act, this means that we must pay attention to its verbal content but also, with equal importance, to the context and the language game in which it is embedded.

Wittgenstein's account of language has useful affinities to the principles of communication that are the basis of pragma-dialectics and its account of speech acts. These fundamental principles can be summarized as the stipulation that we must interpret speech acts in a manner which does not make them incomprehensible, insincere, superfluous, futile, or inappropriately connected to other speech acts (van Eemeren & Grootendorst, 1992, pp. 49-55). There may be cases when it is impossible to interpret a speech act in a way that satisfies all of these demands (because someone using language makes a mistake of one sort or another), but the meaning of language would be arbitrary and indeterminate if the principles of communication could not be counted on as a good general guide to meaning.

In the case of both pragma-dialectics and the later Wittgenstein, we must determine meaning by paying attention to the situation in which an act of communication occurs. In both cases, this means that we must interpret in a way that makes sense of its connections to other speech acts, and the circumstances that they respond to. Wittgenstein adds the suggestion that this requires an understanding of the language games – the activities and social practices – in which words and language are embedded. It is this view of meaning I will appeal to in explaining how arguments can meaningfully employ pictures and other kinds of non-verbal elements.

4. ARGUING WITH PICTURES

In applying Wittgenstein's later views to argumentation, we might begin by noting that arguing is an activity – a language game – which usefully illustrates his claim that the meaning of words is a function of their use. A mere description or list of claims is not, therefore, an act of arguing, but becomes one when it is *used* as evidence for some conclusion.

Deciding between those lists of claims which are only lists and those which constitute acts of arguing is a standard topic in introductory logic textbooks. As such texts point out, there is, in natural language, no verbal formula which definitively identifies what is and is not an argument. In part, this is because standard verbal indicators (words like *therefore, so, because*, etc.) are vague and ambiguous and can be used in different ways. More importantly, there are many circumstances in which real life arguers do not rely on indicator words.

In many cases, arguers don't use indicator words because the context makes it clear that they are arguing. Typical examples occur:

- when someone answers an accusation;
- when a lawyer speaks in court;
- when a real estate agent gives reasons why a house is a good buy;
- when arguers are engaged in formal debate;
- when an advertiser highlights the advantages of a particular car;
- when someone explains something to a police officer;
- when a religious person outlines why one should believe in God;
- and so on and so forth.

Ultimately, the way to decide whether a set of sentences is an argument is by asking whether it is an attempt to justify a conclusion -- i.e. by asking how sentences are *used* in the case in question.

In his rejection of visual arguments, Patterson accepts Wittgenstein's point that the meaning of words – in arguing and elsewhere – is established by use, and by the conventions, practices, and activities that are associated with it. He rejects visual arguments because he thinks that this account of meaning cannot be extended to pictures. As he puts it:

> The ways in which we might interpret a sentence are bounded by the sentence's being embedded in the rule-governed, communal activity of language. The conditions under which a sentence, uttered by a speaker, will be intelligible to an audience of the same linguistic community restrict the possible meanings of the sentence. Importantly, these conditions restrict not just the meanings that the audience is likely to 'take away' from the speaker's utterance, but the meanings that the speaker may coherently intend by what he says.... Whereas rules and communal criteria of meaning keep us from falling into humpty-dumptyism with language, there are no such checks on picturing. (Patterson, 2010, pp. 111-112)

In answer to these remarks, many things might be said. As art historians and others continually point out, many kinds of pictures do conform to well-established conventions (see, e.g., Gombrich, 1960). And many kinds of pictures are embedded in language games which seem to make it perfectly clear that they are elements of arguments. The examples I have already given demonstrate this point. In sorting through these examples, it is not difficult to apply the standard pragma-dialectical principles of communication (Groarke, 2002).

Consider once again the case of a real estate agent who I hire to help me purchase a new house. When we first meet we go through a list of houses she would like to show me. In each case she reports some key elements of the property which provide evidence for the conclusion that I should view it ("It is on the water," "It has a very reasonable price," "It has a beautiful, state of the art, kitchen"). In each case I consider what she says and agree or do not agree to view the house in question.

In this case, the language game I play with the real estate agent can be described as a series of arguments. Each argument in the series provides evidence for the conclusion that I should see a particular house. The agent does not need verbal indicators to make this clear. Her role as sales agent and my role as a buyer of a house makes it clear that she is, in each case, providing me with reasons for a conclusion.

Now consider a second version of this language game in which the real estate agent intersperses her verbal reasons for the conclusion that I should view a house with non-verbal visual reasons. She says, for example, "It is on the water," "It has a good price," and hands me photos to show me what the kitchen looks like and the view from living room. Here again, it is clear that she is offering me a series of arguments.

What establishes this in the second case is exactly what establishes it in the first: i.e. the activity in which words (and in this case, pictures) are embedded. What matters is that words and pictures are being *used* as evidence in favour of a conclusion. In the case of each argument, we could construct a KC table that summarizes its verbal and non-verbal components. A diagram of the argument will probably include verbal and visual elements. To think that we don't have arguments when because we have a visual rather than a verbal ways of presenting evidence is to fail to see that it is the use of pictures and sentences (not simply their internal elements) that determines whether they are an instance of the language game that we call "arguing".

5. THE FALLACY OF INDEPENDENT MEANING

In the case of words and language, a fallacy I shall call *the fallacy of independent meaning* occurs when one assumes that the meaning of words is independent of the context, language games, activities, and form of life in which they are embedded.

Wittgenstein's later philosophy is in many ways an attack on the fallacy of independent meaning, which he associates with the way that most philosophers practice philosophy. As he puts it, "Philosophical problems arise when language goes on holiday." (Wittgenstein, 1953, 1.116) Here language goes on a holiday because philosophers mistakenly take words out of the activities and contexts where they do the work they are designed to do. The way to correct the problem is by placing words back into the mundane activities which define their meaning: "What we do is bring words back from their metaphysical to their everyday use." (Wittgenstein, 1953, 1.116)

Patterson is, in a different way, guilty of the fallacy of independent meaning when he claims that pictures used in communication are characterized by a "humpty-dumptyism" which makes their meaning radically indeterminate. His conclusion that pictures cannot be arguments is contradicted by the examples I have given in this paper, which clearly show that the argumentative meaning of pictures is, like words, often established by the contexts and the language games with which they are associated.

One finds an even more egregious instance of the fallacy of independent meaning in Fleming, who suggests that a picture cannot be an argument because:

> ...it lacks the requisite internal differentiation; it is impossible to reliably distinguish in a picture what is position and what is evidence for that position. The distinction at the heart of argument, the difference between that which asserts and that which supports, is thus collapsed. (Fleming, 1996, p. 13)

From a Wittgensteinian point of view, such comments betray a fundamental misunderstanding of how meaning works. In answer to Fleming it can simply be said that he looks for the meaning of pictures in the wrong place – i.e. inside the picture, rather than outside of it, in a picture's use within a particular form of activity. I have argued that it is the latter that explains how pictures can be arguments.

5. CONCLUSION: WITTGENSTEIN, AGAIN

I want to conclude this paper by noting that my account of the meaning of pictures in arguing is very much in keeping with Wittgenstein's later views. When he rejects the picture theory of meaning, he is not rejecting the use of pictures in language games. What he rejects is the idea that sentences should be understood as verbal pictures. As he moves away from this notion, toward meaning as use, he opens the door to a role for pictures and non-verbal visual elements in language. He does so by recognizing that such elements are often used in the language games that are the core of language.

A good example is the language game that initiates in his discussion in the *Philosophical Investigations.* It quickly evolves into a variant which reaches beyond words and employs colour samples (in comment 8). As the discussion continues he asks "What about the colour samples that A shews to B: are they part of the language?" and answers "It is most natural, and causes least confusion, to reckon the samples among the instruments of the language" (Wittgenstein, 1953, 1.16)

As the discussion continues, Wittgenstein explicitly includes "presenting the results of an experiment in tables and diagrams," "play-acting" and "constructing an object from a description (a drawing)" as examples of language games, suggesting that they include non-verbal elements, and can incorporate what we might call "picture games".

In keeping with this, Wittgenstein continually mixes, compares and slides between discussions of the meaning of words and pictures and visual representations (see 1.86, 48, 70, 108, 108, 166, 169, 216, 280, 398, 454, 520, 539, 548, 563; 2.iii, 2.xi 2.xii). As he complains that we fail to appreciate that there are a multitude of different kinds of words and sentences, he complains that we fail to distinguish between different kinds of pictures (1.291, 522, 526; 2.xi). At one point he suggests that we are attracted to the analogy between words and pictures:

> because one can find a word appropriate [as one can find a picture appropriate]; because one often chooses between words as between similar but not identical pictures; because pictures are often used instead of words, or to illustrate words; and so on. (Wittgenstein, 1953, 1.140)

Wittgenstein himself illustrates the point that the meaning of a picture depends on its particular use in a note to the *Investigations*:

> Imagine a picture representing a boxer in a particular stance. Now, this picture can be used to tell someone how he should stand, should hold himself; or how he should not hold himself; or how a particular man did stand in such-and-such a place; and so on. (Wittgenstein, 1953, 1.23)

However intently one looks at the picture in question, this looking will not reveal what the picture means on a particular occasion. Considered on its own, the picture displays a kernel of its meaning (what Wittgenstein calls a "proposition-radical") but we cannot interpret it on a particular occasion until we know how it is being used.

What is true of the picture is, of course, also true of a verbal description of the boxer. Here too a description can be used to explain to someone how they should stand or hold themselves; how they should not stand or hold themselves; or the way that a particular boxer (say Mohammed Ali) stood in a particular boxing match. In a case of arguing we might use descriptions *or* photographs, drawings or videos to convince a boxer we are training that their stance is mistaken – say, that they hold their feet too close together. In providing evidence for this claim we might verbally describe how as series of renowned fighters (Joe Louis, Mohammed Ali, Joe Frazier, etc.) positioned their feet but we might alternatively (and probably more effectively) show our boxer a series of photographs that illustrate the way these boxers held their hands.

In the latter case we have a visual argument, in the former a verbal counterpart. In both cases, and in a situation in which we mix our verbal and visual descriptions, the meaning of our argument can be understood by considering the way we are using our descriptions (in this case in support of an attempt to convince a boxer how to stand).

As soon as we recognize that meaning is determined by use, it is not difficult to understand how there can be visual arguments. A broader version of the same account of meaning can explain, in a more general way, explain how there can be multi-modal arguments.

ACKNOWLEDGEMENTS: Earlier versions of this paper were presented at the Centre for Research on Reasoning and Rhetoric at the University of Windsor (June 3, 2014); and at the tenth Conference of the International Society for the Study of Argumentation at the University of Amsterdam on (July 2, 2014). I am grateful to those audiences for comments on my ideas, in particular to Chris Tindale, Tony Blair and Alan Wright.

REFERENCES

Blair, J. (1996). The possibility and actuality of visual arguments. *Argumentation and Advocacy*, *33*(1), 23-39.

Biletzki, A., & Matar, A. (2014). Ludwig Wittgenstein. In E. N. Zalta (Ed.), *The Stanford Encyclopedia of Philosophy* (Spring 2014 Edition). URL = <http://plato.stanford.edu/archives/spr2014/entries/wittgenstein/>.

BoredPanda. (2011). Fast food ads vs. reality. Blog Post. URL = http://www.boredpanda.com/fast-food-ads-vs-reality/?image_id=fastfoods-ads-vs-reality-taco.jpg

Eemeren, F. H., van, & Grootendorst, R. (1992). *Argumentation, communication, and fallacies: A pragma-dialectical perspective*. New York: Routledge.

Fleming, D. (1996). Can pictures be arguments? *Argumentation and Advocacy* *33*(1), 11-22.

Gombrich, E. (2000, 1960). *Art and illusion. A Study in the psychology of pictorial representation*. Princeton: Princeton University Press.

Grayling, A.C. (1988). *Wittgenstein*. New York: Oxford University Press.

Groarke, L. (1996). Logic, art and argument. *Informal Logic, 18* (2-3), 116-131.

Groarke, L. (2002). Toward a pragma-dialectics of visual argument. In F. H. van Eemeren (Ed.), *Advances in Pragma-dialectics* (pp. 137-151). Amsterdam/Newport News: Sic Sat/Vale Press.

Groarke, L. (2014). Going multimodal: What is a mode of arguing and why does it matter? *Argumentation, 28*(4), 133-155.

Johnson, R. H. (2005). Why 'visual arguments' aren't arguments. In H. V. Hansen, C. Tindale, J. A. Blair & R. H. Johnson (Eds.), *Informal Logic at 25* (CD-ROM). Windsor: University of Windsor.

Patterson, S. W. (2010). 'A picture held us Captive': The later Wittgenstein on visual argumentation. *Cogency*, 2(2), 105-134.

Pinto, R. C. (2001). *Argument, inference and dialectic: Collected papers on informal logic*. Dordrecht: Kluwer.

Roque, G. (2012). Visual argumentation: A further reappraisal. In F. H. van Eemeren & B. Garssen (Eds.), *Topical Themes in Argumentation Theory: Twenty Exploratory Studies* (pp. 273-288). New York: Springer.

Wittgenstein, L. (1961). *Tractatus logico-philosophicus.* Tr. David Pears and Brian McGuinness. New York: Humanities Press.

Wittgenstein, L. (1953). *Philosophical investigations.* G.E.M. Anscombe & R. Rhees (Eds.), G.E.M. Anscombe (trans.). Oxford: Blackwell.

Wittgenstein, L. (1958). *Preliminary Studies for the 'Philosophical Investigations' (The Blue and Brown Books).* London: Blackwell.

Wittgenstein, L. 1969. *On Certainty.* G. E. M. Anscombe & G. H. von Wright (eds.), G.E.M. Anscombe and D. Paul (trans.). Oxford: Blackwell.

Panel III

Argumentation, politics and controversy in Mexico

III.1

What Personality Traits Should a Governor Have?

JOSE MARIA INFANTE BONFIGLIO
Universidad Autónoma de Nuevo León, México
jminfanteb@hotmail.com

MARIA EUGENIA FLORES TREVIÑO
Universidad Autónoma de Nuevo León, México
meugeniaflores@gmail.com

In this paper, we analyze a video from the web, where a federal Congressman for the State of Nuevo León México, Javier Treviño (2014) sets the features that a future Governor for Nuevo León should have, and to each, he opposes a rebuttal. In the analysis we present an interpretation of the argumentative processes used from an expanded rhetorical perspective, which includes critical discourse analysis, political discourse analysis, and the semiotics of culture[1].

KEYWORDS: argumentative process, critical discourse analysis, expanded rhetorical perspective, political discourse, semiotics of culture

1. INTRODUCTION

The analysis of personality traits in politicians generally causes controversy, especially when promoting a candidate to the governance. On July 13th, 2014, Javier Treviño, a federal Congressman of Nuevo Leon, Mexico, uploaded a video on *YouTube* in which he stated the candidate to the government of the 2015 State elections needed to possess three characteristics: administrative and executive capabilities, the ability to simplify the resolution of problems, and political capacity to unify efforts, and finally, State vision. To each of these characteristics he countered an opposite condition, according to his personal focus.

---

[1] The text was translated by Germán Domínguez, and reviewed by Ma. Eugenia Martinez.

D. Mohammed & M. Lewiński (eds.) (2016). *Argumentation and Reasoned Action: Proceedings of the 1st European Conference on Argumentation, Lisbon, 2015. Vol. II, 99-115.* London: College Publications.

This fact caused an immediate reaction and the replies were published in diverse media, using arguments and discussions that weren't always in the same line of argument. This paper belongs to other large-scale projects[2], but it analyzes the events.

The problem that concerns us refers to the discussion surrounding the political imaginary which the aspirant to candidate sustains and projects as desirable for the Governor and the State. Therefore, we guide our objective to the description and interpretation of argumentative strategies used in those imaginaries of the analyzed speech.

The theoretical and methodological strategy includes the semiotics of culture (Lotman, 1996, 1999), and the political discourse (Chilton & Schäffner, 2008). We examine the imaginary construction (Pêcheux, 1970; Goffman, 1967, 1986; Jodelet & Moscovici, 1991) from the simulation (Baudrillard, 1981), of the future governor of Nuevo Leon and other figurations in the transcription of a 14 minute length video (1942 words), as well as journalistic speeches, to get an approach on power and ideology (Foucault, 1970; Reboul, 1986). For the examination of the sample we segmented the video transcription to find discursive actions[3] established as operational units.

Austin's proposal (1962) is used in the study of the actions, likewise Hutcheon's (1992)[4] to define the speech act as something placed (p. 174). This implies a distancing from the Jakobson static model and an approach to a system with coordinates that would be less exclusively linguistical lowing to move towards the socio-pragmatic and

---

[2] It's assigned to the projects: *Discurso politico mexicano*, composed of 600 texts directed by José Infante (2008), where the discourse appearing in the Mexican media is studied and also the project directed by María Eugenia Flores: '*Estudios lingüísticos, discursivos y didácticos en corpus orales y escritos*'. The projects were founded by the UANL Scientific and Technological Research Support Program in Mexico (*Programa de Apoyo a la Investigación Científica y Tecnológica de la Universidad Autónoma de Nuevo León, México*) (PAICYT) from 2009 to date. (As a product, the Flores and Infante publications: 2009, 2012, 2013, 2014).

[3] For this study the coincident in the Austin's speech act is conceived, but expands its interpretation to the semiotic-discursive perspective.

[4] It refers to the idea of Catherine Kerbrat Orecchioni (1980) who suggests an expansion of the Jakobson model and points out that in a situation of communication (as the case being revised) the speaker and receiver are simultaneously executing their respective competencies by producing and interpreting verbal and non-verbal messages in which the roles of the speaker and receiver are transferred (Hutcheon, 1999, p. 174).

semiotic-discursive dimension. The model of the study is presented below:

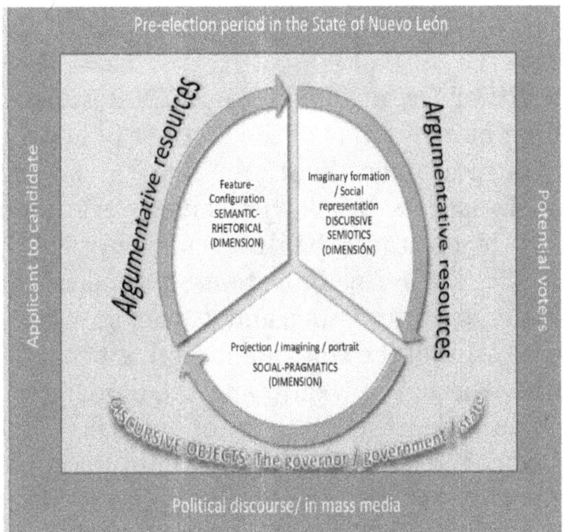

Figure 1. Operative model of the study.

As shown in figure 1, we examined the features proposed in the discursive portrait from the rhetoric-semantic perspective; then articulated the incidences of the projected representation with the argumentative resources in the socio-pragmatic dimension to propose their significance in construction of the imaginary formations (Pêcheux, 1970) which are linked to the social representation (Jodelet, 1986) likewise the constituted framework.

2. BACKGROUND

Margaret Hermann stated not long ago that sooner or later when a dialog centers in politics and politicians, the discussion focuses on the personalities (Hermann, 2002). Although this would seem a common custom of political discourse, it doesn't appear in the models of political theory, especially of political psychology.

For instance, when referring to psychopathological phenomena, where the topic turns into taboo. In the United States of America, if a person expects entering to the political career, the individual must hide records of being involved in any type of psychotherapeutic experience and by not doing so the risk of losing all chances of a candidacy are highly probable.

Recently in Mexico just trying to comment on a specific psychopathological condition of a president was dangerous, putting any individual safety at risk.

In the previous six-year term in Mexico, a rumor about an addiction of the President caused a scandal in the Congress with media coverage that never adequately solved the situation in the context of public opinion. However, the topic has been broached since immemorial times.

In book VIII of *The Republic*, Plato (2008 translation), analyzing the corresponding human types of the distinct political regimes signals that "timocracy is about men thirsty for riches, who escape the law as children from their parents, fierce with slaves, very submissive with the rulers, and this leads to a badly organized State" (Plato, 2008 translation, pp. 386-388). The reflection she makes about the human type of democracies are interesting, and as we know, for Plato this was a corrupt regime, a result of the contradiction that occurs in the oligarchic state between the worship of wealth and the moderation in the pursuit of wealth itself. Our interest here is not to make an analysis of the validity or relevance of these concepts, but to point out how the association between personality traits and political structures appeared in the platonic discourse. Since times of Plato to our period, many authors associated personality traits to forms of government. Aristotle, Polybius, Machiavelli, Hobbes, Montesquieu, Hegel, Marx, and many others established with more or less explanatory ability, relations in governance and the attributes or personal qualities of those who rule or execute power, and also the ruled ones.

Lasswell (1930/1977) thought that common language was full of types (or stereotypes) coming from the politicians behavior and shared various common aspects with scientific typologies. For example, the private motive is fused with public objectives that are one "martyr" connotation: implicit in the 'bureaucrat' is the idea of desk job activities for molding a human being and this activity attracts people who have certain qualities related to the "bureaucrat". A distinction of Lasswell, in which we won't stop to fully unwrap, is the one to separate the "politician" from the "business man" where the latter is related to the satisfaction of his private interests while the politician pursues the integration of different interests in the social life.

Raquel Ferrario (1972) stated that the difference of a truly democratic politician and one who isn't dwells on the great quantity of psychopathic constructions in the discourse of the latter one. The psychopathy has been discussed by some authors as the structure of personality, but George Devereux (1970/1973) characterizes psychopathy[5]as a special and complex vicissitude in the culture of an individual who in some way develops an appealing systematic combat against culture. The psychopath presents a systematic madness in most cases, it makes him incapable of embracing certain meanings and cultural values known by common folk. He is unable to experience the same empathy to them as the rest of the people; therefore, the psychopath manipulates in a cynical and selfish way what others love more. Is the ability of empathy, the capacity of placing yourself in other people's shoes and being fond to others. This distinction helps us to know if a person shows psychopath traits or not.

Many of those who achieved a great success in the capitalist system at the expense of the suffering of others have, without a doubt, a high degree of psychopathy, although the social system praises and qualifies them as great winners.

Referring to Nuevo Leon, there are certain formal requirements which, although they are not directly associated with personality traits, they condition the candidacy for elections. Constitutional article 35 of the Political Constitution of the United Mexican States[6]establishes that voting and be voted is a basic right of Mexican citizens and the right to request the candidates registration corresponds to political parties. In addition, the citizens applying for registration in an independent way must fulfil the requirements, conditions and terms provided by the respective legislation. It's not stated in this article, but the right to be voted is restricted to the condition of being a Mexican citizen by birth. In addition, the extensive constitutional article 41 regulates the functioning conditions of the political parties and the National Institute Electoral (a federal institution able to interfere with the State) grants powers to the organizations, control of parties and electoral processes, as well as the operation rules of the above mentioned Institute.

The conditions to be a Representative are established in the Constitutional article 55: being a Mexican citizen by birth, 21 years of age and native of the State for who tries to represent or to have an immediate home with more than six months of residence. In order to be a Senator, the same requirements are entailed, with the condition of being older than 25. Constitutional article 82 provides the requirements

---

[5] We must clarify that there's a difference between the structure of psychopathy and psychosis, due to the confusion in some media, in Mexico especially.

[6] Both the Political Constitution of the United Mexican States and the electoral laws are public domain documents, published in the respective official newspapers. The versions mentioned here and the corresponding quotes come from the website of the Presidency of the Republic.

to become President. In addition to the formalities regarding age, either one of the candidate parents must be Mexican and very peculiar conditions only required in Mexico, for instance "not belonging to the Ecclesiastic State or being a minister of any cult", and "not to be a Secretary or Undersecretary of the State, General Attorney General of the United Mexican States, Governor of any State or Head of the Governor of the Federal District, unless the individual leaves from the position six months before elections day"(sic).

The Nuevo Leon Electoral State Law, also of extensive articles, establishes in the Constitutional Article 133:

> candidates or pre-candidates to the candidacy, or popular election positions, who participate in the internal selection processes convened by each party will not engage in proselytism activities or diffusion of propaganda, for any reason, prior to the commencement date of the pre-campaigns.

The Baroque writing won't prevent anything, many of those who were finally appointed candidates carried out hundreds of proselytism acts, and propaganda dissemination, but they surely had reasons for it.

Our study examines the proposals made by Javier Treviño, who in July 13th, 2014, uploaded a video on *YouTube* with an analysis of the social and political conditions of the country, also related to the State of Nuevo Leon. After criticizing various aspects of the government management in the State and pointing out his opinion on the following project in the future government, indicated that there were two options for choosing who should be capable of facing those challenges (obviously, the future Governor of the State). Particularly, in this work we limit ourselves to this discourse, without analyzing the replies that emerged and not getting into detail about other aspects of personality structures in the political action.

3. SEMIOTIC-DISCURSIVE PERSPECTIVE

In this section we review the processes concerning the construction of the imaginary formations[7], the social image[8] and social representation[9] (Moscovici, 1986).

---

[7] According to Pêcheux theory, it operates in such way that the participants of a communicative situation (Speaker-Message-Receiver) don't designate the physical presence of individual human organisms, but determined places in the structure of the social formation where they belong. The *Imaginary formations* designate the place of the speaker and receiver, also the one attributed to

3.1 *Discursive Positioning. Functions and Strategies.*

In the public servant discourse we found certain strategies that favor the argument of the desired discursive purposes. The assumption of the imaginary formation, which the speaker embraces, is remarkable in its positioning through the implicit personal index (Benveniste, 1991, p. 85), its political hierarchy comes into play and distances it from the rest and intends to endorse the truth of the asserted. Warns one of the qualities of the Eristic, pointed out by Haidar: "the aim is not to reach any consensus (...) but on the contrary, what is sought is the destruction of the adversary" (2006, p. 293):

> (1) And of course, now people trust less and all due to the poor performance of the politicians.
> (2) [...] we must say things as they are: a great part of the political class in the State is not up to the challenges presented, the potential of Nuevo Leon has not been reached yet, they're only concentrated on the elections and what's coming next for them.

In the example 1, we observe how the speaker initiates his intervention with an alignment expression, as it projects the coincidence with a point of view ("and of course"), it employs a distancing resource in the nominal usage, latent in "the people"[10], where the speaker alludes the

---

another speaker within the social formation. There are mechanisms and projection rules that establish the relationships between the situations -which can be objectively defined -and the positions- representations of those situations (1978, pp. 48-52).

[8] Pechêux (1978) stresses that in all discursive process a series of imaginary formations operate and refer to the location in which the partners placed themselves and the other, image they build upon their own space. Therefore, there are mechanisms in all societies and they settle the relationships among the situations, they could be defined in an objective way, along with the positions and imaginary representations. Further reading, Goffman (1963).

[9] On the other hand, in Denise Jodelet's view, is the designation of multiple observed and studied phenomena at various levels of individual, collective, social and psychological complexity (Jodelet, 1986, p. 469). We selected aspects from the theory due to its relevance for this work, the proposal of social representations include a way to interpret and analyze the everyday reality, a form of social knowledge and at the same time a mental activity developed by individuals and groups to fasten their position in relation to situations, objects and communications that concern them (470).

[10] Corresponds to Reboul sense displacements among which is the enriching hyperbole. Ideologies employ this procedure when they refer about "mass", "all

generality, not any social community in particular. Then he adds a pejorative adjective: "the poor performance of politicians", he excludes himself from the statement, as if he did not belong to the system. In the aforementioned fragment 2, where the "complaint" takes place: expressions as "political class", "is not up to", "is concentrated", "what's coming next for them" through which the implicit (Ducrot, 1982) that the speaker does not belong to the guild is projected. In this way, he blocks himself by making a legitimacy loss (Chilton & Schäffner, 2008) of the group's social image. The speaker probably seeks to show some objectivity and an aseptic social representation; from there he manipulates his position in the front.

It's necessary to note that by assuming the role that he grants to himself in the speech, the materiality of the simulation is verified (Haidar, 2006)[11] since there is an auto positioning as the one who holds the truth. We could link it with the ritual of circumstance of Foucault (1970) it defines the qualifications of individuals that speak through a projected discursive alignment with interests of the people.

3.2 Employed resourcing for the construction of the referent social image

For the examination carried out in this section, we rely on Reboul (1986) who proposes several processes to create a referent and we take the ones present in this speech.

3.2.1. The referent creation by presupposition

With Reboul the discourse is able to create it's referent by presupposition, this* author backs it with Ducrot[12]and notes that the presupposed results more persuasive if it's expressly manifested, since

---

the people", "will of the people". This process is about the expressive function "since it's a way tore in force the 'us', subject of discourse" (Reboul, 1986, pp. 66-68).

[11] The author considers the speech is formed by materials which are 'layers, and such layers constitute what Foucault calls 'the archaeology of knowledge' and they transfer the semiotic-discursive construction that exceeds the linguistic and semiotic dimension of the surface" (Haidar, 2006, p. 82).

[12] *Cf.* Reboul (1986, pp. 62-67) and Ducrot define the presupposition as 'the one which is presented as evidence, as an indisputable framework where the conversation must register [...] is given as a proper contribution of the utterance and engages the responsibility of the one who chose the statement; [...] the speaker tries to share this responsibility with the listener by disguising what is being said, under the appearance of a common belief' (1982, pp. 22-23).

attention is placed on what is expressed and the presupposed is not questioned, which tends to be out of the topic. On the receiver, this strategy creates the impression of having freedom of choice[13]. Reboul proposes that a last advantage of presupposition remains unconscious for those who employ it because it uses the metalinguistic function to create a referent or a reference frame, the free world, these ideas are illustrated in the following segments:

> (3) There's no place for improvised governments or to start thinking what to do at the moment they reach the desired position.
> (4) There is no space for short-term governments with no sense of public responsibility, these kind of governments are concentrated on seeing where the pennies at.
> (5) Because that translates into frivolous, corrupt, incompetent governments.

The speaker performs the role of coercion[14] by placing his peers, other politicians, through a profuse adjectivization within a non-positive imaginary before the eyes of the voters: "improvised governments", "short-term", "frivolous", "corrupt", "incompetent", at the same time avails himself of presupposition ("there is no room") to create in the listener the image of the current Government. Also attends to certain expressions ("what do you have in mind? /no sense of responsibility") to denigrate his colleagues performance.

As the speaker holds the position of Representative in the Government builds an inconsistent imaginary with his own socio-discursive position, which transversely affects the macrostructure (Van Dijk, 1984) of the speech. The social representation of the speaker is exempt from it due that is normally perceived as inherent of the political sector, and intends to form a new imaginary formation, starting from an unusual discursive position that wouldn't regularly assume.

At socio-pragmatic level, the representative uses strategies of autonomy in affirmation trait of himself, since he projects "the desire of

---

[13] The ideology states: "to be able to say yes or no, even if the fact of saying it makes him admit the essential without knowing it", "it has a true pedagogy sense, consists of letting people respond questions to make them admit ipso facto [sic] the presuppose" (Reboul, 1986, p. 62).

[14] It's defined by the authors as a process in which the others [...], institutional opposition [...] must be presented in a negative way. Techniques such as: ideas of difference and borders, acts of speech like blame, accusing, insulting and others' are used (Chilton & Schäffner, 2008, p. 305).

the person to be distinguished from the group as a front man, aware of his positive social attributes, and this will allow him to express his views in a persuasive and powerful way" (Bravo, 1996, p. 93). We found resistance, opposition and coercion (Chilton & Schäffner, 2008) because the speaker reflects opposition to the government and also resists to it, while it places him in a non-favorable position for his social image.

3.2.2. The creation of the referent by amalgam

Regarding the idea for the State of Nuevo Leon, we propose according to Reboul, every word that states, denies, explains, etc., is found on its referent, and it can be real or imaginary. The reference is what is being discussed, the topic, what is being said. His proposal is really valuable for this study, the referent is not the world as it is in itself, but the world as it is perceived by a given culture (a subject from its place of enunciation), likewise his opinion on ideologies: "ideologies play on the relativity of the referent, but without mentioning it" (1986, pp. 57-58). The theory of Reboul related to the presence of minimum effort in the ideological mechanisms seems appropriate: "creates all sorts of concepts that immediately accept as realities" (1986, pp. 57-58). It's about a particular case which he calls *amalgam* "the fact of using a reductive term to assimilate different realities" (1986, pp. 57-58). Following the quote here are some examples of this:

> (6) Nuevo Leon deserves a new destination. We have to imagine a different Nuevo Leon and work to make it a reality.
> (7) We have great opportunities and great potential;
> (8) We are a State that can be one of the most important economies of America.
> (9) We can become the energy capital of the country.
> (10) We have the calling to be the entrepreneurial center of Latin America.
> (11) A State that could become one of the best places to live in the world.
> (12) To take advantage of the monetary waste generated by the foreign investment and the consolidated national capitals, and take the most of the squandering of being the Houston of Mexico, the energy capital of the country.

Through his speech the speaker projects and invites the audience to build an imaginary ideal of the State ("we have to imagine"); it uses isotopies[15] (in bold) to build the semantic associations that form the referent in the way that is suitable for the speech. It also employs verbal forms (we have/imagine/work/we are/can be/be/we can become/could become/take advantage/generated). There are modal, stative, operational and dynamic verbs, they help to the purpose of building the referent. At the same time, the speaker projects his image as one of a visionary, by this way, places himself in an asymmetric imaginary formation, reflecting superiority to his other political colleagues.

3.2.3. Simulation and other resources

In this work we propose that the simulation is a transverse materiality to the speech. This is why they are relevant to the fulfillment of the study objectives, the considerations uttered by Jean Baudrillard in his studies on the simulation refer to the fact:

> *Aujourd'hui l'abstraction n'est plus celle de la carte, du double, du miroir ou du concept. La simulation n'est plus celle d'un territoire, d'un être référentiel, d'une substance. Elle est la génération par les modèles d'un réel sans origine ni réalité: hyperréel'*[16] (Baudrillard, 1981, p. 17).

This proposal benefits the investigation because it establishes the procedure of the simulation in a development of fiction and the creation of a referent that goes beyond reality and does not derive from it either duplicates it, but it's the creation of other reality, which is evidenced in the analyzed speech. In addition, the simulation is ideal for the purposes the ideology seeks, due it allows the speaker to manipulate the referents to his convenience through the speech.

The simulation is based on the construction of fiction, a referent absent from the natural inferences which the speech calls created by the procedures that these phenomena involve. They combine and manifest

---

[15] The Greimas´ theory about *discourse isotopy* is revisited, its function consists in clarifying the message unit seen as significance whole. It's about the study of iterations and recurrence of meaning that's projected through the discourse signs. They translate into temporary ephemeral equivalences based on the existence of one or more common marks in the juxtaposed segments for the analysis (Greimas, 1973, pp. 104-113).

[16] Currently the abstraction is not of a map, a double, a mirror or a concept. The simulation is not of a territory, a referential being, a substance. Is the generation from models of a realm without origin or reality: it's the hyper reality of our translation.

to cover up the desires shown by the speaker in the studied text. We rely on Baudrillard's idea, it's a complex process in which the simulation is opposed to the representation, but is plausible (1981, p. 12). We proceed to review the ideas mentioned.

In the analyzed speech built by the representative, among other processes, a praise of the speaker social image is formed using implications and other relations which constitute an explicit appreciation of the referent. Both are used to achieve persuasion in favor of social representation. The politician starts his speech mentioning the current situation of the State, then performs managerial acts and carries out rhetorical questions that will be answered along the speech:

> 13) We have to think about the transformation Nuevo Leon needs.
> (14) Nuevo Leon, in what do we want to transform it?
> (15) What's the thing that makes Nuevo Leon citizens different and strong? The geographic location, industrial capacity, health infrastructure or Universities?
> (16) We must have clear what we want to be as a State and whoever takes control has to adopt the vision.
> (17) Need to define and set a common purpose for Nuevo Leon based on the vision shared by the State.

The process of fiction, the simulation, is developed when the Representative Treviño states the "common purpose" which he has in mind for Nuevo Leon, a State that would transform, as it is show in the following statements:

> (18) A place in Latin America where is easier to start and make a fruitful company
> (19) The platform of the new Mexican entrepreneur: competitive, globalized, innovative with social awareness
> (20) Where work and property are sacred
> (21) Where conditions are created to influence the initiative of the private sector
> (22) Where corruption is severely punished, along with attacks against property
> (23) Where bureaucrat is a synonym of facilitator
> (24) Where the governing person strives and works equally or even more than the ordinary citizen

In the examples is possible to perceive how the simulation of a referent is constructed in the speech, by stacking reiterations and images representing an ideal for any society. A simulation of a daily reality takes place (the Nuevo Leon status quo is far from resembling the proposal) and - such as Baudrillard states (1981, p. 17)- that reality is altered (projecting an utopia), it conceals the referent (a mask of prosperity, it masks under a full law State) and in this way, the referent dissociates itself from the original ties of sense according to the speaker convenience.

In addition, the materiality of the simulation is judged with the Reboul proposals (1986). It's related to what this author defines as an apparent function of language. The author relies on the classification of the language functions elaborated by Roman Jakobson[17] to review them in the ideological language. It indicates a difference between the implicit purpose of the speaker and the verbal form used may exist.

Using this function, the transmitter transgresses the literalness and diverts the sense of the speech performing a decoded simulation at pragmatic level. According to Reboul, "the apparent function of a message may not correspond to its real function. A message [...] will hide behind other form [...] erasing all signs of enunciation in the ideological discourse" (1986, p. 52)[18]. Therefore, a message is not necessarily recognizable by the structure. Reboul mentions the diverse functions of the language —are not among the six stated by Jakobson— can abide to one or more of the six. For the work taking place here is of interest to mention the *critical function* that 'is either referential or metalinguistic' (1986, p. 44). Both functions are used by the official, from the beginning of his speech we find expressions like:

> (25) In order to see things clearly, if Nuevo Leon was a country, we would be the 48th place in math and reading, 50th in science, from a total of 65 countries participating in the international student evaluation program known as PISA and the way things are going it would take 57 years to be 1st place and 19 years to get the average of countries from the OECD, assuming the average will remain the same and the first country won't improve in that time.

Possesses referential function used by the speaker to base the proposals included in his speech. Also manifests the metalinguistic function

[17] General Linguistic Essays, passim.

[18] Reboul collects the Jakobson proposal on the language functions and confers them distinct values, which acquire when they manifest in the ideological discourse (1986, p. 49). This reformulation results in a contributor when the materiality illustrates the functioning of the simulation in the analyzed speech since it allows examining the double level, in which the figuration is constructed, due to the duplicity of the functions.

generated from the questioning and guidelines that we illustrate in the examples 15 to 17 located in the fragments 18 to 24, the speaker focuses on the definition and establishment of the concept through the same language code. Therefore, the Representative Treviño builds his speech as if he was worried about the future and the solution of problems in the State when in fact he's praising his social image. In this way a *legitimization* takes place (Chilton & Schäffner, 2008) of his social representation by placing himself as the person who knows what must be done, and the one that sets the vision of the shared state (implicitly creating a support for his social image). Simultaneously limits the Government and the political class by means of *delegitimization* (Chilton & Schäffner, 2008) through the contrast of the current situation and the desirable circumstance.

The epilogue also fore shadows of some imaginaries:

> (26) Popularity can be created in relatively short time, but talent and capacity can't.
> (27) To govern properly you need to be capable, not popular.
> (28) We are Nuevo Leon, and we all have the right and duty to imagine the State we desire, to make the decisions needed and work hard for its creation.

In which he attacks the social image of his probable contenders – they are more popular than him – making use of syllogisms (marked in bold type) through them leaves the insinuation (Ducrot, 1982) that he is capable of governing.

At the end of the speech the subject has placed himself in the position of the knowledge holder, "the one with capacity and talent". On the implicit, he has sanctioned and discredited the Government and its governors through coercion of the image in the speech. Functions of *coercion, resistance and delegitimization* appear in it.

4. THE FINDINGS

The political semiosphere (executive, legislative) (Lotman, 1991), the personal (attributes, qualities, desires) and the social (economic and educational conditions, human rights, security, legislation) are intersected. The *explosion* takes place, and the *(un) predictability* (Lotman, 1999) occurs by the fact that a member of the political semiosphere resists - in appearance - abruptly against the system he belongs.

The rhetorical strategy of the *simulation* goes across the speech and thanks to the simulation in the *ductus subtilis* the speaker oscillates between autonomy and affiliation in the social sector to which he belongs. It distances from it to legitimize his image, but it positions itself as a member of that guild to propose paths of government.

The rhetorical resources favor the creation of the imaginaries: the desired State in the explicit level, the subject who is "capable" of conducting it in the implicit level, the politicians, Government and the Governors. He distances himself from all of them to validate his social representation.

Adjectivize, semantic associations, isotopies, verbal uses, reiteration, analogies and comparison are some of the rhetorical means used by the speaker with the purpose of persuading. He also employed ideological-discursive strategies such as the creation of the referent by presupposition, amalgam and displacement of sense, and in this manner the enriching hyperbole was found.

The strategies that transversely cross this discourse are the simulation, the presupposition, the implicit and the unspoken statements. These manipulated the latent level of the speech and favored the projection of the language apparent function in the semantics surface. This function was constituted by the referential and metalinguistic function.

This work perspective is intended to address with great detail the refutative and argumentative strategies in the discourses, to provide an exact approach in the structuring of such unique communicative event, to review the articulation of the socio-discursive interfaces and their connections; as well as, proposing the intersemiotic incidents produced when articulating the aforementioned semiospheres. In theory, we can say the speech doesn't keep a coherent argumentative line mixing personality traits with the presentation of work programs on structural conditions in the decision-making process of politics.

REFERENCES

Ascher, W., & Hirschfelder-Ascher, B. (2005). *Revitalizing political psychology. The legacy of Harold Lasswell.* Mahwah, N.J.: Lawrence Erlbaum Associates.

Ayala, V. (2014). El tema no soy yo. Javier Treviño. *El Norte*, year LXXVI, n° 27607, July 16th.

Baudrillard, J. (1981). La précession des simulacres. *Simulacres et simulation* (pp. 9-68). Paris: Galilée.

Bravo, D. (1996). *La risa en el regateo: estudio sobre el estilo comunicativo de negociadores españoles y suecos.* Stockholm: Academitryck AB.

Chilton, P., & Shäffner, C. (2008). Discurso y política. In T. Van Dijk (Ed.), *El discurso como interacción social* (pp. 297-330). Barcelona: Gedisa.

Devereux, G. (1970/1973). *Ensayos de etnopsiquiatría general* (Monge F., Trans.). Barcelona: Barral (original text published in 1970).

Ducrot, O. (1982). *Decir y no decir. Principios de semántica lingüística.* Barcelona: Anagrama.

Ducrot, O. (1986). *El decir y lo dicho. Polifonía de la enunciación.* Barcelona: Paidós.

Ferrario, R. (1972). *Comunicación personal.* Italy: CNR Press.

Foucault, M. (1980). *El orden del discurso.* Barcelona: Tusquets.

Freud, S., & Bullitt, W. (1973). *El presidente Thomas Woodrow Wilson. Un estudio psicológico* (Najlis E., trans.). Buenos Aires: Letra Viva (original text published in 1966).

Greimas, A. J. (1973). La isotopía del discurso. *Semántica estructural* (pp. 104-155). Madrid: Gredos.

Haidar, J. (2006). El campo del análisis del discurso y de la semiótica de la cultura. *Debate CEU-Rectoría. Torbellino pasional de los argumentos* (pp. 63-117). México: UNAM.

Hermann, M. (2002). *Assesing leadership style: A trait analysis.* S.d.: Social Science Automation.

Jodelet, D. (1986). La representación social: Fenómenos, concepto y teoría. In S. Moscovici,(Ed.), *Psicología Social II. Pensamiento y vida social. Psicología social y problemas sociales* (pp. 469-494). Barcelona: Paidós.

Lasswell, H. (1948). *Power and personality.* New York: W.W. Norton.

Lasswell, H. (1977). *Psychopathology and politics.* Chicago: The University of Chicago Press (original text published in 1930).

Lykken, D. (2000). *Las personalidades antisociales* (Ferrer I. Trans.) Barcelona: Herder (original text published in 1995).

Mondak, J. (2010). *Personality and the foundations of political behavior.* Cambridge: Cambridge University Press.

Owen, D. (2010). *En el poder y en la enfermedad. Enfermedades de los jefes de Estado y de gobierno en los últimos cien años* (Cóndor M. Trans). Madrid: Siruela (original text published in 2009).

Pancer, S. M., Brown, S., & Barr, C. W. (1999). Forming impressions of political leaders: A cross-national comparison. *Political Psychology, 20*(2), 345-368.

Platón (2008). *Diálogos. IV. La república* (del Pozo, A., trans.). Madrid: Gredos.

Reboul, O. (1986). *Lenguaje e ideología* (Milton, trans.). México, D.F.: F.C.E.

Todorov, T. (2014). *Los enemigos íntimos de la democracia* (Sobregués N., trans.). Barcelona: Gutenberg Galaxy (original text published in 2012).

Treviño, J (2014).Conferencia de Javier Treviño - Nuevo León en 2015 - Una visión de futuro. [Retrieved from: https://www.youtube.com/watch?v=A7bmF5JYLyc].

Van Dijk, T.A. (1997). Macroestructuras semánticas. *Estructuras y funciones del discurso,* (Myra Gann/ Mur, M., trans). México: Siglo XXI.

Winter, D. (2003). Personality and political behavior. In D. Sears, L. Huddy & R. Jervis (Eds.), *Oxford handbook of political psychology* (pp. 110-145). Oxford: Oxford University Press.

Winter, D. (2013). Personality profiles of political elites. In L. Huddy, D. Sears & J. Levy (Eds), *The Oxford handbook of political psychology* (pp. 422-458). Oford: Oxford University Press.

Ramos, M. (2014). Cuestionan que Treviño critique hasta ahora. *El Norte*. year LXXVI, n° 27606, July 15th, 2014.

III.2

Refusal Argumentative Strategies in a Telephone Interview: The Mayor in Monterrey, Mexico, Hands Over the Keys of the City to Christ.[1]

MARÍA EUGENIA FLORES TREVIÑO
Universidad Autónoma de Nuevo León, México
meugeniaflores@gmail.com

ARMANDO GONZÁLEZ SALINAS
Universidad Autónoma de Nuevo León, México
armandogsalinas@yahoo.com

A dialogue between Monterrey City Mayor, Margarita Arellanes (2012-2015) and the host of a local television newscaster is looked over about a controversial speech given by Arellanes in Monterrey, where she handed over the city keys to Jesus Christ. The conflict is discussed and viewed from social representations, argumentations carried out in conversational interaction, to power and ideology; also the construction and operation of imaginary formations, discourse control and ideology are taken into account.

KEYWORDS: argumentation, dialogue, imaginary formations, semiotic discourse, speech acts, social representations

1. INTRODUCTION

The Mexican Constitution of 1857 proclaimed the separation of Church and State, and the one of 1917, in its Article 130 lays down the rules and regulations of churches and other religious groups, and state the establishment of Mexico as a secular state is emphasized. The fact that a public governing woman, in this case, establishes commitments with divinity makes it controversial. Such is the case in this paper, we review a telephone conversation between the mayor of Monterrey, Margarita

[1] This article was originally written in Spanish by Flores Treviño, and was fully translated, including quotes, by González Salinas.

D. Mohammed & M. Lewiński (eds.) (2016). *Argumentation and Reasoned Action: Proceedings of the 1st European Conference on Argumentation, Lisbon, 2015. Vol. II, 117-136.* London: College Publications.

Arellanes acting for 2012-2015 and the host of a local television news, Maria Julia Lafuente. The interview arises from the speech delivered by the mayor Alicia Margarita Arellanes Cervantes on June 8, 2013, in a religious event called "Monterrey Prays" (Monterrey Ora) organized by Alliance of Pastors which took place in Zaragoza Square of this Mexican city of Monterrey. The official headed a massive event that weekend in order to hand over the keys of the city to Jesus Christ, which caused controversy in the city. For this study, this dialogue was chosen because it was broadcasted on local television, which implies an interesting complex structure of enunciation (Benveniste, 1977) and provides an opportunity to review the confrontation of imaginary/notional discursive involved in communicative interaction.

The aim of this paper is to describe the strategies used in argumentation and power relations that are established in the dialogue between the interlocutors, whose interaction is determined by coercion exerted from the imaginary formations and affects social representations involved in the singular declarative framework which is reviewed. We study the linguistic acts (Austin, 1969; Searle, 1969) which are manifested through the materialities in the semiotic-discursive practices (Haidar, 2006), and speech acts in argumentation (van Eemeren & Grootendorst, 2013). Other supporting material in this paper is: for gender perspective study (Lamas, 2000); discursive control, power (Foucault, 1970), ideology (Reboul, 1986; van Dijk, 1999); habitus and capital (Bourdieu, 2000, 2002); co-construction (Koike, 2002) and operation of imaginary formations (Pêcheux, 1970); social representations (Moscovici, 1986; Jodelet, 1986), (im)politeness (Kaul de Marlangeon, 2005, 2006) and simulation (Baudrillard, 1981). This is a case study[2] with a qualitative and descriptive approach. As a methodological strategy, we take the discursive act[3] as an operational unit. We study the actions by taking advantage of Austin (1962) proposal, and that of Hutcheon (1992) who considers the speech act as something located (p. 174), because it involves a departure from Jakobsonian static model, and an approach to a system whose

---

[2] In general we can consider it as research methodology based on a specific inter / subject / object that has a singular performance, and despite its particular character, it must also be explained as an integrated system. It is in this sense that we are talking about a unit that has a specific function within a given system; then it is the expression of an entity that is the subject of inquiry and for this reason it is called a case (Diaz et al., 2011, p. 5).

[3] For this study, we agree with Haidar (2006) as it is seen as coinciding with Austinian speech act theory, but she expands her interpretation to the semiotic-discursive perspective.

coordinates would be less exclusively linguistic. We value his proposal that a research on semantic contextualization of actions [(im) polite], are necessarily interested in this notion of situated act. Upon analysis of performative and perlocutionary acts, the sample was segmented into the speech acts that form the interaction, then proceeded to the examination and interpretation according to theory. The process used[4] permitted to describe actions and their effects on the semiotic-discursive level.

2. BACKGROUND

This paper is assigned to these projects: 1. Mexican political discourse, led by Jose Infante (2008), where the speech is studied as it appeared in Mexican media, and 2. "Linguistic studies, discourse and teaching in oral and written corpora" whose responsible is María Eugenia Flores, both projects were financed by the Support Program for Scientific and Technological Research by Autonomous University of Nuevo Leon, Mexico (PAICYT) from 2009 to date. These projects have studied the following concepts: irony, polyphony, social image and verbal and nonverbal resources (*cfr.* Flores & Infante, 2010a, 2010b, 2011, 2012, 2013, 2014). The topic comes from a school term paper by Myriam Ibarra (2013) as part of a Sociolinguistics course, taught by one of the authors.

It is appropriate to say that the problematic issue here studied entails a direct link to a gender perspective because in the eighties, it was not very common in our society to see women in the US Congress, neither this invisibility of women was questioned, because in the common imaginary, there was no such possibility, it was awarded only to men, despite the law of equality and citizenship granted in 1953 to Mexican women. For this reason, it is important to consider it since it involves recognizing that one thing is the sexual difference and the other, powers, ideas, representations and social prescriptions that are constructed with reference to that sexual difference. The Gender Perspective is meant as an explanatory look at the kinds of relationships

---

[4] First, in order to perform segmentation, within each speaker's discourse, speech acts are identified, those constituting the macroact in that fragment of enunciation; then, their structure is reviewed and defined following the model proposed by Searle, who "presents a minimum of three rules to study speech acts: Preparatory, which refers to the presupposed or pre-constructed one on which the given speech act is built and executed; Sincerity, the subsisting degree of accuracy; and Essential, central quality of the communicative act" (Acero et al., 1992, p. 74).

and behaviors that are given between men and women in social interaction, the functions and roles they each take in their respective living environments, work and family, which contributes to an explanation of the social position of women and men. In the XXI century, educational, cultural, political, and economic development sharpens the contradiction between the traditional female role, that of a mother and a housewife - and new roles, as a citizen and a worker. In order to design innovative projects to attract, promote and retain more women in public spaces, whether occupational or political, an implementation of the gender perspective is essential, it helps to understand and decipher the cultural codes to eradicate prejudices and stereotypes that harm women, and yet, as we show in this paper, it serves as a resource for social image manipulation.

3. ENUNCIATION AND SOCIO-DISCURSIVE POSITIONING. INCIDENCE IN ARGUMENTATION

To examine the speaking position of the participants in this discourse, the concept of *imaginary formations* (IF) that "designate the place the transmitter and receiver each have of themselves and the place attributed to another speaker within the social formation" is articulated (Pêcheux, 1970, pp. 48-52), as well as with *social representations* (SR) (Jodelet, 1984) that "include a way to interpret and think about everyday reality, a form of social knowledge and also mental activity that individuals and groups develop in order to secure their position in relation to situations, objects and communications that concern them"(p. 473)[5]. Marta Lamas (2000) comments that culture introduces sexism, that is, that discrimination arises according to sex by means of gender and each culture establishes a set of practices, ideas, discourses and social representations that are attributed to specific characteristics of women and men. This symbolism is what is called gender that regulates and determines the objective and subjective behavior of people. Women and men are "domesticated" by a process of what a man and a woman must be which is based on ideas preconceived by the state and the traditions of what is proper to each sex. The relationship is reviewed in the studied event because the IF coerce the discursive

---

[5] Denise Jodelet defines these as "multiple phenomena that are observed and studied at various levels of complexity, both individual and collective, psychological and social (1984, p. 469); to Van Dijk, it is a matter of sociocultural knowledge of other shared beliefs that offer a "common basis" of every social discourse or interaction. He proposes that all representations are social as well as mental" (1999, p. 23).

behavior and determine the resulting social representations, as we show in the following figure:

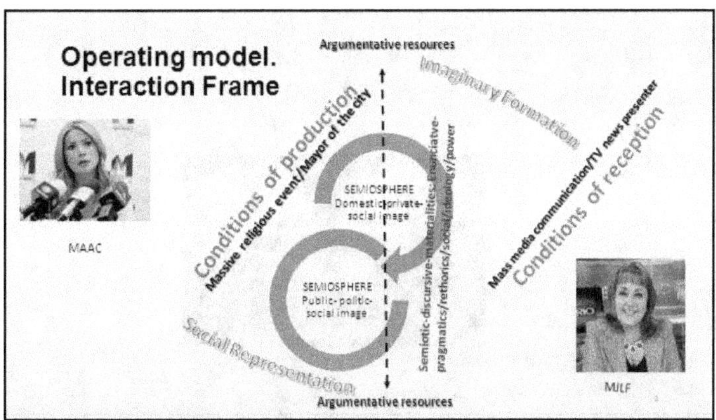

Figure 1. Operating model.

We also consider the variables of *distance*[6], power[7], and range[8] which, in the pragmatic situation of interaction, determine the symmetry or asymmetry context that allows the exercise of power speaker S over listener L, or vice versa. The process is described in Figure 2.

The illustration shows how the different formations that constrain the discourse of dialoguers fall upon the measure of power that the range of each of the speakers - determined by the pragmatic situation - allows them to project the social distance that constrains his/her speech and causes (im)-polite actions. Thus, when different IFs and speakers SR[9] are involved, different levels of power are unleashed on the exchange studied. Below we show the characteristics of the communicative situation (Figure 3).

---

[6] "The concept of distance as an asymmetrical individual measure of difference under which the S aims the purposes of his/her act, we found that it is revealed by (...) lack of respect when impoliteness is directed to cause damage to the negative image of L" (Kaul de Marlangeon, 2005, p. 309).

[7] "The symbolic power can translate into the ability of individuals to act in the world through language" (Vizcarra, 2002, p. 66).

[8] To Kienpointner "the 'imposition range ' of a speech act in a given culture involves emotions such as fear of intrusion into the English culture, or the desire of affiliation in the Spanish culture. All this also justifies the explicit integration of emotions in a theory of (im)politeness " (2006, p. 26)

[9] Social Representations.

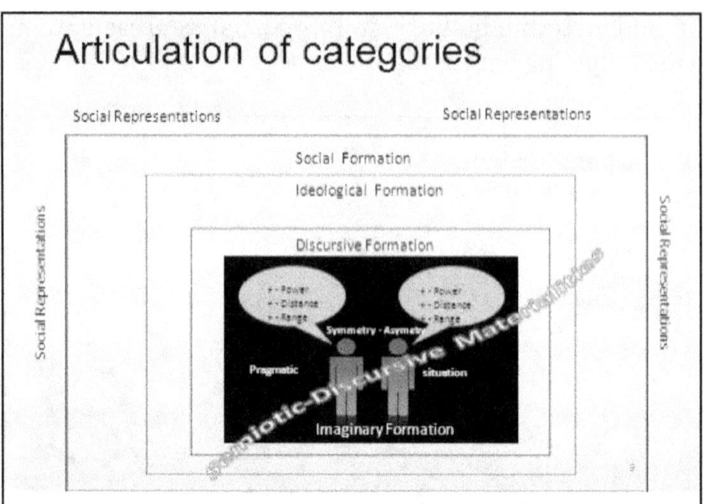

Figure 2. Articulation of the analysis of categories.

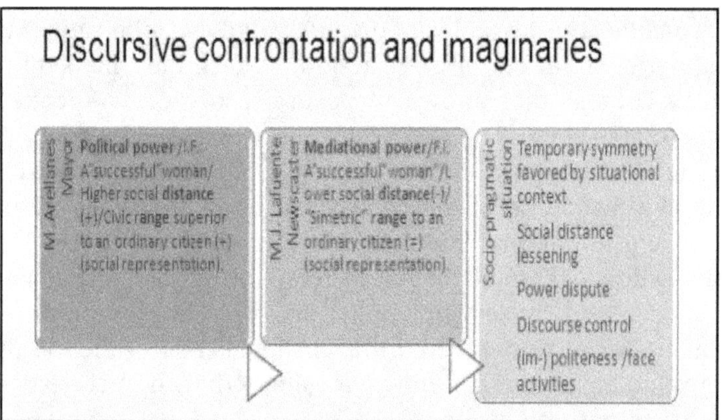

Figure 3. The pragmatic situation of confrontation

Here, within the first two frames, we show the outlining of elements that both dialoguers have in the pragmatic situation: a higher range of ordinary citizens due to the political power of one, MAC[10] mayor of Monterrey[11], and the media power of the other, MJL[12]. The third frame

---

[10] Margarita Arellanes Cervantes.

[11] Feminism and gender studies provided new meanings to the definitions of culture, politics, power and identity. From modernity on, women still have not been able to define within the category of subjects, organizing axis at their current identity, modernity has generated what Serret calls "torn identity," since women debate between what they are and what they should be according to what symbolic of womanhood designates. However, MAC uses these dissociated identities to argue his rebuttal.

in Figure 3 includes features that allow the confrontation we study, namely: MAC is called by MJL to respond, in a telephone interview aired on the news at noon at national level, to certain questions about her behavior in the event "Monterrey prays." This circumstance temporarily gives a wider range of power to the interviewer. However, the mayor will resist the coercion of it.

Regarding IFs held by MAC in this speech, we can see that the newscaster, as an argumentative strategy in search of persuasion, alternatively draws the various IF held at that event in order to get their argumentative purposes.

The development in question is shown in Figure 4:

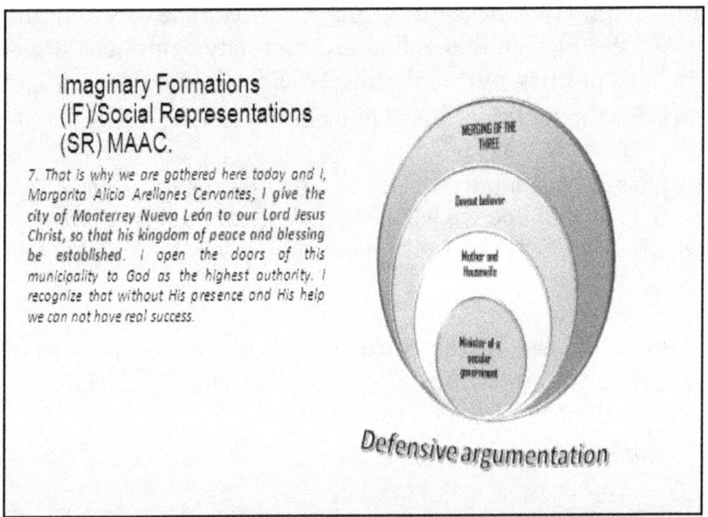

Figure 4. Margarita Arellanes's imaginary formations.

In this figure we present the various IF that the civil employee combines and the final guidance in her speech that validates in her power to "deliver/offer the city to Christ." We found that the Mayor (MAAC) uses an oscillating position and a merging of the imaginary formations (Minister of a secular government, a devout believer/mother, and a housewife) as a refusal argumentative resource, as it is showed.

4. THE DISCOURSE AND ARGUMENTATIVE RESOURCES: "MONTERREY PRAYS" 8 OF JUNE, 2013

---

[12] María Julia Lafuente.

We next examine the speech addressed by MAC in the event of religious associations in the city which is the origin of the telephone interview, the reason for this work.

> (1) Almost 417 years ago, a few steps from here, our city was founded. Taking the first seven words of its founding document the support and destination of Monterrey: in the name of Almighty God[13].

In Example 1, the beginning of her speech, the mayor goes to the strategy of intertextuality in the form of allusion[14], using the original formula of the Charter (Foundation Act), and the religious linking to Monterrey from the beginning of its existence, as argumentative justification; the phrase is used as an authority argument. Possibly she does it to link politics with religion, because it would be a convenient framework for the justification of her person in that event.

> (2) Yes, it was in the name of God that our city was born. And in the name of God we will keep it, defend it and love it for the good, prosperity and development of all the families of Monterrey.

In fragment 2, the use of the third person plural pronoun is observed, whereby it is assumed that the S is expressed in her official SR since she speaks on behalf of the city she represents.

In the following example:

> (3) We have been in the last months, and I say it humbly, witnesses of a positive change, each time more evident in our city and we can say that this has happened because we have opened our doors to God.

From what Reboul (1986) called false causality, the discourse of the magistrate creates the implicit assumption that the development of the city is due to the religious relationship that she has allowed in her

---

[13] All fragments of this speech correspond to the Tab: Radio Capital City DF (2013), which is disclosed in the references.

[14] "A take on undeclared loan, but also literal; under even less explicit and less literal, the allusion, that is, a statement whose full insight involves the perception of a relationship between that statement and another that necessarily leads to either of its inflections" (Gennete, 1997, p. 54).

government. This process contributes to project a hybrid image[15] that links the two powers: political and religious.

4.1 Pronominal strategies and isotopies

In this section we describe other strategies used such as the use of pronominal and semantic and semiotic weave in MAC discourse to manipulate her IF / SR. For example:

> (4) We have open the doors to God / I open the doors of this municipality to God as the highest authority/

It can be observed that in this segment the mayor has changed the pronoun to use the first person singular with which she projects to carrying out the enunciation through her IF as a mayor:

> (5) Humbly I ask God before this community as witnesses, that He may enter this city / that He may enter this city and may make it His lodging/ may the Lord dwell in the hearts of each and every one of Monterrey people /[16]

Besides, she creates a metaphor with which she makes Monterrey the "house of God", and, that she, as the "owner" of that house (hybridity in her representation), asks the divinity to enter and gives possession of the city, she hands over the capital city to the divinity:

> (6) Lord Jesus Christ, welcome to Monterrey, the house that you have built us. This is your house, Lord Jesus / Lord of Monterrey.

The IF is transformed, and the distance is shortened, as now the Lord Jesus Christ is addressed as an equal "(TÚ familiar pronoun) **have built** / **your** "(TU familiar possessive adjective) **home**." Below in:

---

[15] Face is the term originally proposed by Goffman (1955) to the concept of social presentation or image of a person. However, as *Žižek* says, the human face in everyday relationships with others is fetishized as hiding the "true face" which is the ultimate reality of our neighbour (Žižek, 2002, p. 23).

[16] In Spanish the third person singular and the formal second singular forms of the verbs in the subjunctive mood is the same, that is why we use the modal may and comment: "humildemente le pido a Dios ante esta comunidad como testigos, **que entre (may He enter)** en esta ciudad/que entre en esta ciudad y **la haga su (may He enter and make it his lodging)** / que el Señor **habite (may the Lord dwell)** en los corazones de cada uno de los regiomontanos/

(7) In the name of God the almighty. Yes, it was in the name of God that our city was born. And in the name of God we will keep it...

She goes to the source of reiteration to emphasize that the link created since the beginning of speech exists.

4.1 Lexical strategies

In this section, we provide information on uses and lexical recurrences, which provide an idea of the semic substratum of that speech. For example, Graph 1 shows the prominent frequencies in the MAC's discourse:

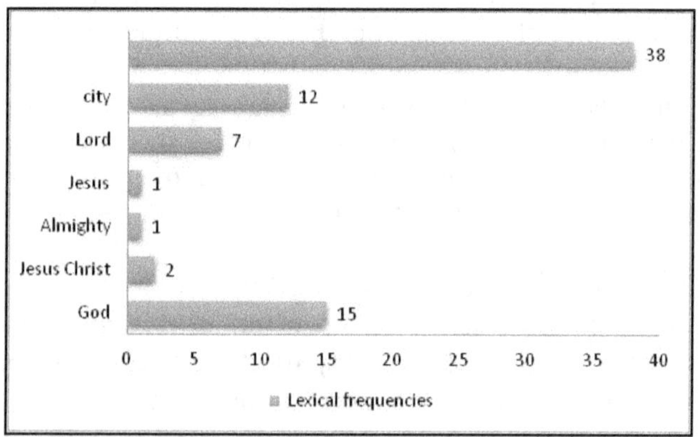

Graph 1. Lexical frequencies in MAC discourse.

Out of a count of 38 words repeated in the speech, we see in Graph 1 that the designations by proper name or by adjectives are minimal, 1 to 2 frequencies; however, the terms referred to "God" are repeated 15 times, 39.4%; while references to the city were counted in 12 repetitions, 31.5%. The data obtained confirm the articulation proposal of the leaders and imaginaries discussed above.

5. ARGUMENTATION STRATEGIES IN THE INTERVIEW: MJL – MAC

In this section, we approach the communicative interaction between the newscaster: María Julia Lafuente (MJL) on June 12, 2013 through Multimedios a TV broadcast nationwide chain, and acting Mayor Margarita Arellanes Cervantes (MAC).

5.1. Discursive confrontation

In the dialogic process (Bakhtin, 1970) reviewed, to the question asked about the reason for their attendance at "Monterrey prays", MAC replied:

> (8) MAC: [...] and I went **in a personal capacity**, and **in a personal capacity** is that within this symbolic and spiritual act I participated. I reiterate my respect for all faiths. And also with the responsibility that I have over the city. But in this act, in which I participated on Saturday at eight in the evening, and I went with my family it was **in a personal capacity**.

In the reply given by MAC, there is the resistance function (Chilton & Schäffner, 2008), expressed in a reiteration of the IF she wants to prevail in the interpretation of the issue, because, she uses it as argumentative resource *ad nauseam*[17]. The presenter attacks the social image of the official[18] that she is determined to merge with her family mother imaginary formation:

> (9) MJL: [...] Margarita, you have an investiture as a mayor, you are a renowned lawyer from the Autonomous State University of Nuevo León. You never thought you- uh – about all the kind of disturbance ... that was going to come over you? Because you are a Mayor for three years and you are a Mayor 24 hours a day all week, every month, so there are no days on and days off. So you still think the same way, that you made no mistake, I am just asking...

As noted in Example 9, and according to Kaul of Marlangeon, impoliteness occurs when there is a disparity of power between the interlocutors (2006, p. 255). In this case, such asymmetry arises from: the political and protagonist position of the one being questioned in the

---

[17] This type of fallacy is based on "the wrong belief that it is possible that a statement is true or accepted as true, the more times is heard (...) she uses constant repetition, by saying the same thing over and over, until one gets sick of listening (Arias, 2004, p. 127).

[18] In the example, one can observe how social practices regarding gender are given worldwide and have generated a division: private sector equals female and public sphere equals male (De Miguel, 2012). Maffia (2007) notes that as a result of these parallel spheres women do not participate fully in the scope of any activity allegedly predestined for men. The inclusion of women in the spaces of institutionalized knowledge is reduced to its nature as object, and in general, to account for their condition of inferiority. Such is also the case of law, politics, and power, at least until about three decades ago, since this area has been equated with male in the collective imaginary (Olsen, 2000).

communicative situation, and the fact that the interviewer has the turn of speaking in the speech by representing the media; the attention of society to the content of the speech; the opportunity to make public statements, and others. MJL emphasizes the social distance through the use pronominal "you"(formal) and constructed, by repetition, the official's profile: "inauguration as a mayor / recognized lawyer", with the intention to intensify (Albelda, 2005)[19] the politician's SR which were breached in the event in question. Also she uses amplification[20] as an intensifying resource by which, in the implicit (Ducrot, 1982), she reminds her about the duties of her ministry: "for three years and you're a mayor 24 hours a day all week, every month, so there are not days on and days off". Finally she closes her inquiry with a mitigating verb form[21], used to conceal the directivity of the act performed, because she presents it in a different, less threatening way: "I'm just asking." The generated macro discursive act[22] becomes covert impoliteness, conducted by superficial, insincere, politeness strategies (Kaul of Marlangeon, 2005). In Figure 5, we present the process.

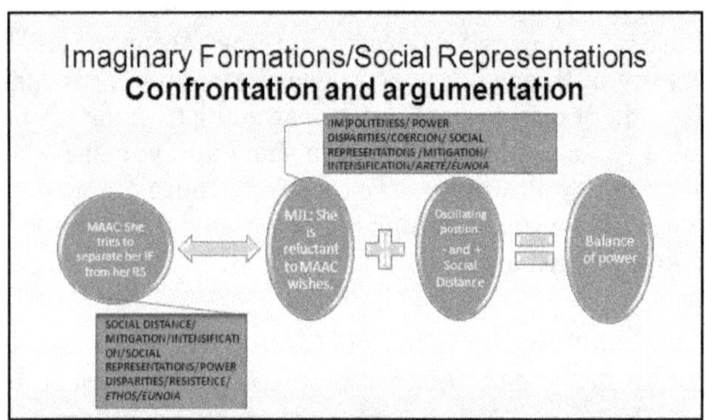

Figure 5. Confrontation of Imaginary formations versus Social Representation

---

[19] According to Albelda "through intensification one or some elements of the statement or the enunciation are enhanced" (2005, p. 95).

[20] Beristáin (1997) defines it as "a *metábola* whose class membership is directly related to the procedures applied to the expansion of the initial idea in order to gradually encompassing it, so it often covers more than one linguistic level" (1997, p.33).

[21] It corresponds to variable 8 of Mitigation Studies Data Sheet by Albelda, *et alia* (2014, p.8)

[22] A discursive macro-act must be understood sd a succession of language acts that are not limited to a simple linear sum of all of them, nor linked to sequences of interrelated acts, since they constitute a linguistic and discursive act of completeness (Adam & Chabrol, 2005).

In the illustration we show how MAC intends to untie her SR from her IF (mother / Christian / mayor) to protect her social image injury (Example 8); however, in Example 9, MJL resists MAC's intention and emphasizes the impossibility of separating the SR supported by the official, she exercises coercion. An oscillating position of both interlocutors arises which goes from lower to higher social distance and vice versa. Therefore, there is at certain times of the exchange, a balance of power that favors covert discourtesy by the TV presenter.

5.2 Gender-image-impoliteness and simulacrum in argumentation

The speech has the ability to articulate the reality on the social value that is given to words with its pragmatic context (Foucault, 1970), and the language is linked to power through which one femininity is established within the sex-gender system and its ability to form identities and beliefs. Look at the example:

> (10)MJL: Margarita, I consider you **part of my gender**, I know that you are **a very intelligent and beautiful woman**, and you are not there just ...because. But because you have worked hard to become a Mayor. I would ask you these questions: do you regret all this? Did you plan well? **Because women are quite smart**, Margarita, or because all this is to your own benefit?

In Example 11, we watched how MJL as a strategy assumes a symmetrical position by referring to existing gender equality. She uses affiliation (Bravo, 1999)[23] as a strategic politeness resource. Then performs an intensification of the feminine attributes of the mayor in which sarcasm is perceived, thus it performs the function of delegitimization (Chilton & Schäffner, 2008, p. 305) of the Mayor social image, and a covert impoliteness (Kaul of Marlangeon, 2005), as her purpose is to damage the Mayor's SR by an *ad hoc* argument[24].

The journalist makes attacks on the image of the official targeted to their political claims, as in:

> (11) MJL.: Margarita, and before finishing this interview, people in the government say that you are also doing it for

---

[23] The author establishes it as "everything that can be identified with the group, that is, to perceive oneself and be perceived by the entity as someone who is part of the group" (1999, p. 40).

[24] It is "when the argument applies solely to the case to be explained, or defend, using the argument" (Ferrater & Terricabras, 1994, p. 60).

> politics, that you're aiming at, you are seeking the governorship in 2015. What do you say to these people?

MJL returns to the asymmetric position that the situation allows, and reduces the social distance between her and the mayor by questioning about her aspirations, however, to mitigate the attack, she uses an impersonalization[25]: "people in the government say / What do you say to these people?"; then reduces the strength of the directive act, which the question encloses, by using a reformulation[26]: that you're aiming at, you are seeking the governorship in 2015.

Strategies show scalarity since delegitimization of the mayor's image is intensified, "that you did it [hand over the keys to Christ] also for political reasons".

To rescue her image of the injuries inflicted, the mayor goes to rebuttal resource, and assumes a position:

> (12) MAC.: Look, they can think whatever they like, the truth is that I am very reassured that I have always participated **in a genuine way** and I will continue to participate **in a genuine way, in an individual way, in a personal capacity**, and, well, to the benefit, first of all, **to each and every person** who wants to participate, to uplift their spirit, the values of **each person**, but also to have a better city.

In 12, the delimitation is made by the mayor with respect to responding to all the questionings and emphasizes their individual position - she goes back to the resource used in her initial defense - through an anaphoric construction, which we show the lexical recurrences in Figure 6.

We have analyzed the speeches based on the rules of critical discussion (Van Eemeren & Grootendorst, 1992, pp. 222-232), and our findings concerning interlocutors are:

a) **The Mayor**

- Violation of Rule 2: she comes back to a point of view immune to all criticism
- Violation of Rule 4: she manipulates the audience emotions.
- Apology: implicit on her own and explicit on the referent.

---

[25] Considered in variable 15 Mitigation Studies Sheet (Albelda et al., 2014): "To appeal to the judgment of the majority by impersonal verb forms and discursive particles that depersonalize the deictic origin of the statement".

[26] Variable 17 Mitigation Studies Sheet (Albelda et al., 2014).

- Violation of Rule 6: she falsely presents a premise as itself evident. She presents the premise as if it were an accepted starting point.
- Violation of Rule 9: she turns the success of the defense as absolute.
- Violation of Rule 10: she makes use of formulations that are confusingly ambiguous.

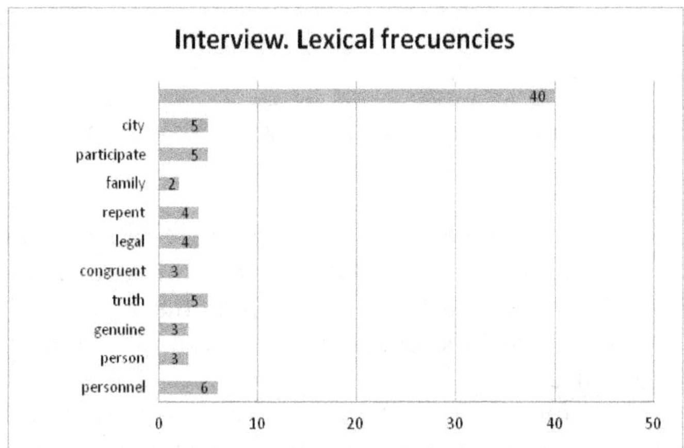

Figure 6. Lexical recurrences in MAC's last refutation

On the one hand, we show Rules violation frequencies in Figure 7, The speech.

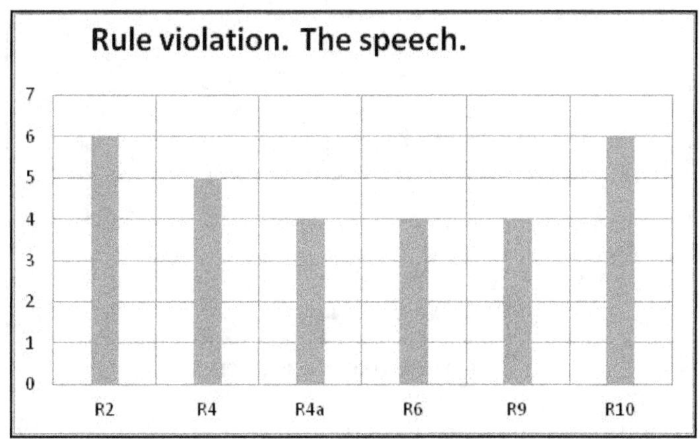

Figure 7. Rule violation Mayor's speech

On the other hand, the following Rule violations were found in the Mayor's discourse in the interview.

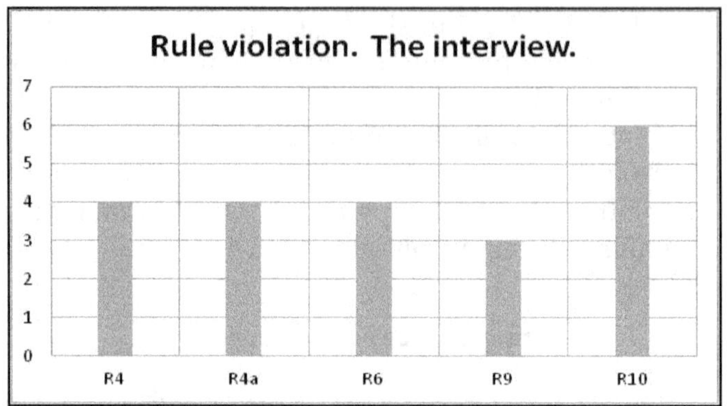

Figure 8. Rule violation: The interview. Mayor's discourse

b) **The TV Host**:

- Violation of Rule 2: she presents her opponent as unreliable.
- She comes back to a point of view immune to all criticism.
- Violation of Rule 4: she manipulates the audience emotions.
- Violation of Rule 6: she hides a premise within an implicit premise.
- Violation of Rule 10: there is semantic-referential ambiguity.

Use frequencies are shown in Figure 9.

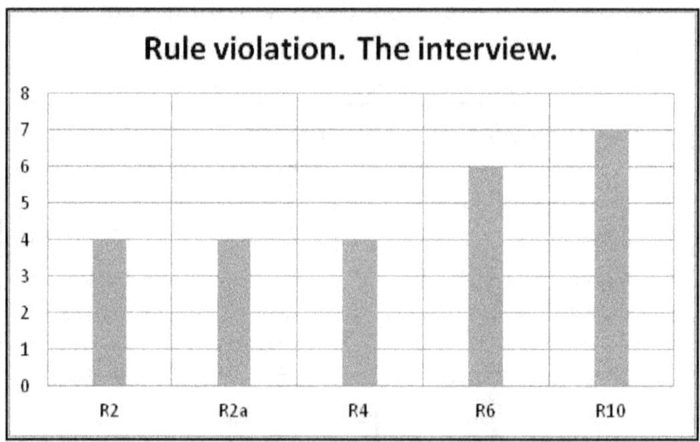

Figure 9. Use frequencies

Therefore, violations of the Rule of this critical discussion are like this:

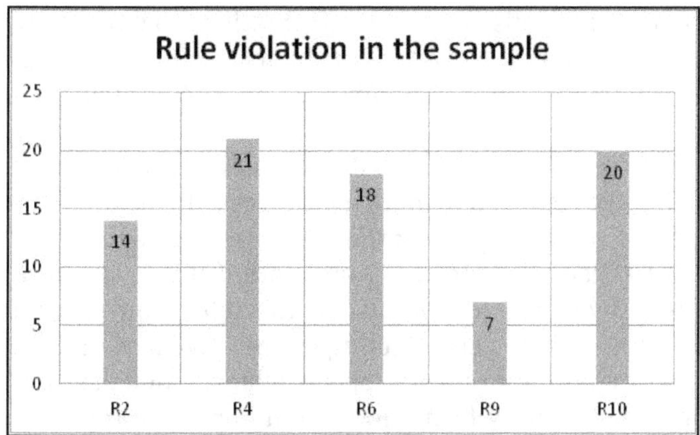

Figure 10. Rule violation in the sample

As we can notice, in Rule 4 (she manipulates the audience emotions), in 10 (there is semantic-referential ambiguity) and in 6 (she falsely presents a premise as itself evident. She presents the premise as if it were an accepted starting point). These were the most frequently transgressed.

6. PRELIMINARY CONCLUSIONS

It is well known that women struggling allowed them to criticize and reevaluate their work, their world and their own being, to define -as a wish - their own humanity, and with this, emerging space and rights were opened in order to venture in politics. The woman before versus the woman now who has to fight to eradicate gender stereotypes and discriminatory traditions (Bourdieu, 1998).

Also, to have women recognize their oppression status of male hegemonic system to perpetuate the social order that keeps them still far from the agora, and their true social representation as active and free subjects that modern society needs. The revised case does not happen like that, because we are spectators of two social representations, one, the television host, who has taken the ideas expressed above, and the other that comes and takes refuge in gender stereotypes to excuse strategies used in political campaigning.

Although the examination of one case does not allow us to offer definite conclusions, but it does help us anticipate the findings made up to this phase of the study: In the discourse of both speakers, there exist, transversely, linked to the communicative-pragmatic materialities of ideology and power, the materiality of simulacrum (Haidar, 2006), that can help to implement the strategic politeness, covert impoliteness and attenuation in discourse.

There is a discursive confrontation between two female speakers with different social rank, which is possible thanks to the temporal symmetry and asymmetry of the pragmatic-discursive situation.

Social representations that speakers claim in the context of everyday life, are reversed from those expressed during the interview; a situational and temporary parenthesis occurs and favors the discursive struggle that affects the type of argumentation strategies there used.

The provisional positioning of the television presenter allows her to perform impolite acts affecting the positive face of the mayor and damage her social representation. Arellanes' defense is based on delimiting her imaginary formations from her social representation. As shown in Lafuente's refutations, it is an unlikely process, socially, as it is the case of a public official.

There are covert impoliteness acts made by the interviewer, in the form of protocol and strategic politeness. At a socio pragmatic level, we can find attenuation of the attacks on the social image as well as polite and impolite intensifications. Thanks to lexical frequency count. we have some basis to track isotopies that constitute the semantic axes of the event.

7. PROSPECTS

Addressing the rhetorical strategies of argumentation in both speeches, in order to offer an approach to the structuring of such an original communicative event. Review from gender studies, socio-discursive interfaces at stake, from the intersection of semiospheres appearing in the discourse. And also propose inter-semiotic incidents that occur in articulating the stated edges.

REFERENCES

Adam, J.M., & Chabrol, C. (2005). Macro acto de lenguaje. In P. Charaudeau & D. Maingueneau (Eds.), *Diccionario de análisis del discurso* (pp. 367-368). Buenos Aires: Amorrortu.

Albelda, M., Briz, A., Cestero, A.M., Kotwica, D. & Villalba C. (2014). Metodología para el análisis sociopragmático de la atenuación en corpus discursivos del español. *Oralia*. Retrieved from: http://preseea.linguas.net/Portals/0/Gu%C3%ADa%20de%20Estudios%20de%20la%20Atenuaci%C3%B3n%20en%20corpus%20PRESEEA%202014.pdf

Albelda, M., Briz, A., Cestero, A.M., Kotwica, D., & Villalba C. (2014). *Metodología para el análisis sociopragmático de la atenuación en corpus discursivos del español. Oralia.* Retrieved from: http://preseea.linguas.net/Portals/0/Gu%C3%ADa%20de%20Estudios%20de%20la%20Atenuaci%C3%B3n%20en%20corpus%20PRESEEA%202014.pdf

Arias, O. (2004). *Silogismos. In Acopios de comunicación.* Raleigh:*https://www.google.com.mx/url?sa=t&rct=j&q=&esrc=s&source=web&cd=1&cad=rja&uact=8&ved=0CB0QFjAA&url=http%3A%2F%2Fen.wikipedia.org%2Fwiki%2FRaleigh%2C_North_Carolina&ei=2rsdVdL0KsfmsAWZpYGgAw&usg=AFQjCNE_c-3RGYlAJ9Nds2W9m8ljL75O9A&sig2=s6hcUTglkWZ_Kmu-uHxjyw&bvm=bv.89947451,d.b2w* Lulu Press Inc.

Austin, J.L. (1962). *How to do things with words.* Oxford: Oxford University Press.

Bajtín, M. (1970). *Estética de la creación verbal.* México: Siglo XXI,

Baudrillard, J. (1981). La précession des simulacres. *Simmulacres et simulation* (pp.9-68). París: Editións Galilée.

Benveniste, E. (1991). El aparato formal de la enunciación. *Problemas de lingüística general* (pp. 82-93). T. II. México: Siglo XXI.

Beristáin, H. (1997). *Diccionario de retórica y poética.* México: Porrúa.

Bourdieu, P. (2000). Espacio social y poder simbólico. *Cosas dichas* (pp.127-142). Barcelona: Gedisa.

Bourdieu, P. (2002). El habitus y los espacios de los estilos de vida. *La distinción. Criterio y bases sociales del gusto* (pp. 169-226). México: Aguilar, Altea, Taurus, Alfaguara.

Bravo, D. (1999). ¿"Imagen positiva" vs. "Imagen negativa"?. *Oralia* (pp. 155-184).

Bravo, D. (2005). Hacia una redefinición de la "cortesía comunicativa". In D. Bravo (Ed.) *Estudios de la (des)cortesía en español. Categorías conceptuales y aplicaciones a corpora orales y escritos* (pp. 21-52). Buenos Aires: EDICE/Dunken

Ferrater, J., & Terricabras, J.M. (1994) Diccionario de filosofía, Vol. 1. Barcelona: Ariel.

Flores, M.E. (2010). Ironía, autoironía y autorrepresentaciones discursivas femeninas en un corpus oral. In Norma Gutiérrez *et alii*, (Eds.) *Voces en ascenso. Estudios sobre mujeres y perspectiva de género* (pp.561-577). Zacatecas, México: UAZ/INMUZA/UG/SPAUAZ/AZECM.

Foucault, M. (1970). *El orden del discurso.* Barcelona: Tusquets.

Gennete, G. (1997). La literatura a la segunda potencia. In *Intertextualité.*

Goffman, E. (1955). On face-work. An analysis of ritual elements in social interactions. *Psychiatry: Journal of Interpersonal Relations, 18*(3), 213-231

Haidar, J. (2006). El campo del análisis del discurso y de la semiótica de la cultura. *Debate CEU-Rectoría. Torbellino pasional de los argumentos* (pp. 63-117). México: UNAM.

Jodelet, D. (1986). La representación social: Fenómenos, concepto y teoría. In S. Moscovici (Eds.), *Psicología Social II. Pensamiento y vida social. Psicología social y problemas sociales* (pp.469-494). Barcelona: Paidós.

Kaul De Marlangeon, S. (2005). Descortesía de fustigación por afiliación exacerbada o refractariedad. In D. Bravo (Ed.), *Estudios de la (des)cortesía en español. Categorías conceptuales y aplicaciones a corpora orales y escritos* (pp. 299-318). Buenos Aires: Dunken.

Kaul De Marlangeon, S. (2006). Tipología del comportamiento verbal descortés en español. In Briz, et al (Eds.), *Cortesía y conversación. De lo escrito a lo oral. Actas del III Coloquio Internacional del Programa EDICE* (pp. 254-266). Valencia: Depto. De Filología Española, Universitat de Valencia.

Kienpointner, M. (2006). Cortesía, emociones y argumentación. In Briz, *et al.* (Eds.), *Cortesía y conversación. De lo escrito a lo oral. Actas del III Coloquio Internacional del Programa EDICE* (pp. 36-52). Valencia: Depto. De Filología Española, Universitat de Valencia.

Lamas, M. (2000). La antropología feminista y la categoría 'género'. *El género. La construcción cultural de la diferencia sexual* (pp. 97-125). México: UNAM/PUEG.

Moscovici, S. (1986). *Psicología Social II. Pensamiento y vida social. Psicología social y problemas sociales.* Barcelona: Paidós.

Pêcheux, M. (1970). *Hacia el análisis automático del discurso.* Madrid: Gredos.

Radio Capital D.F. (2013). Discurso Margarita Arellanes. In *Monterrey Ora* [Retrieved from: https://www.youtube.com/watch?v=H3ux_4oHA40]

Reboul, O. (1986). *Lenguaje e ideología.* (Milton, Trans.), México: F.C.E

Searle, J. (1990). *Actos de habla. Ensayos de filosofía del lenguaje.* Madrid: Cátedra.

Telediario. (2013). Entrevista telefónica de Margarita Arellanes con Ma. Julia Lafuente [Retrieved from: https://www.youtube.com/watch?v=TvmlYFIXTi4]

Van Dijk, Teun A. (1999). *Ideología. Una aproximación multidisciplinaria* (Berrone L. Trans.). Barcelona: Gedisa.

Van Eemeren, F., & Grootendorst, H. (1992). *Argumentación, comunicación y falacias: una perspectiva pragma-dialéctica.* Santiago: Universidad Católica de Chile.

Vizcarra. F. (2002). Premisas y conceptos básicos en la Sociología de Bourdieu, *Estudios sobre las Culturas Contemporáneas* (p. 66). II, Vol. VIII. (16). Colima: Universidad de Colima-CONACULTA, december.

III.3

Eristic Argumentation in CEU-Rectory Debate

JULIETA HAIDAR
Escuela Nacional de Antropologia e Historia, México
jurucuyu@gmail.com

This paper aims to analyze the arguments presented in the eristic argumentation produced in the development of the University Student Council (CEU) movement in 1987 at the Universidad Nacional Autónoma de México. The polemic and the controversy between the rectory authorities and student representatives, produced in the month of January 1987, had a great social, political and historical impact, mainly because the university crisis is articulated with other crisis such as the economic one in the country.

KEYWORDS: argumentation, argumentative models, complexity, discursive strategies, eristic, refutation, transdisciplinarity

1. INTRODUCTION

This paper aims at analyzing the eristic argumentation occurring in the movement of the University Student Council (CEU) in the year 1987, at the National Autonomous University of Mexico (UNAM). The controversy that occurred in January of 1987, between the authorities of the rectory and leaders student, had a great social, historic, political impact, mainly because the university crisis is articulated with other crises in the country, such as the economic and policy.

This critical juncture was conducive for this debate to break academic borders, and expands nationwide, in all the social fields, involving many sectors of political and civil society. In this article, will develop three sections.

In the first one, deals with the field of the argumentation from complexity and transdiscipline, which implies a redefinition of the same in their extensions, scopes and models. In this sense, we are dealing with the argumentation from Aristotelian logic, formal logic, natural logic, informal logic, pragmadialectic, pragmatics and communicative

D. Mohammed & M. Lewiński (eds.) (2016). *Argumentation and Reasoned Action: Proceedings of the 1st European Conference on Argumentation, Lisbon, 2015. Vol. II, 137-150.* London: College Publications.

interactions, rhetoric/logic, integrated rhetoric, among other trends, with the aim of presenting the enlargement with the contributions of various analytical models (Haidar, 2006). In the second section, include proposals to work the controversy in the argumentative field, as examples are of the approaches of Fedoseiv, Popov, Kotarbinski, Kopperschmidt, among others. In addition, expose other proposals that can use to analyze the controversial component in the discourses, strategies of refutation in public debates, like from Toulmin and Grize/Vignaux. In the third section, we select some more concrete proposals to analyze the eristic procedures of debate CEU-Rectory.

2. FIELD OF ARGUMENTATION: PROBLEMS, ENLARGEMENTS, TRENDS

The enlargement of the field of argumentation, from the epistemology of complexity (Morin, 1997, 1999) and transdisciplinarity (Nicolescu, 1999), implies redefining argumentation from various angles, for which we assume the need for the convergence of several trends, to analyze in greater depth argumentative patterns and processes. We can´t fail to mention that in many approaches are present the classical philosophers Greeks, but redefined such as Aristotle and others. The various types of logic, the re-founded rhetoric, the dialectic articulated with rhetoric, the eristic, the several trends of pragmatic, the linguistics, the discourse analysis, the semiotics constitute the main analytical routes of the expanded field of argumentation. The foregoing leads to rethink the relationships between logic and rhetoric, dialectic and eristic, among others, what can be observed in the following figure 1:

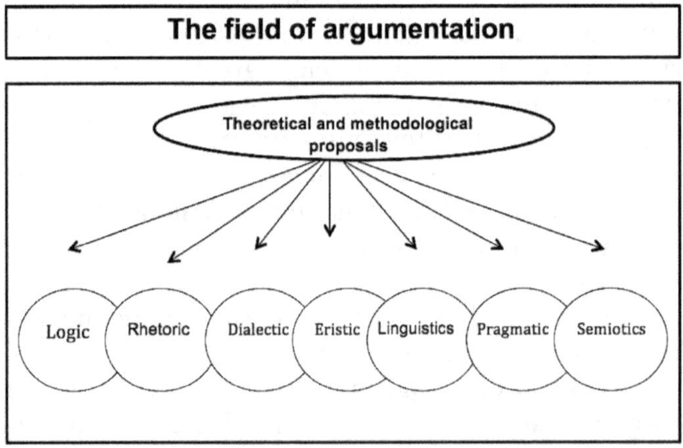

Figure 1 – The field of argumentation (Haidar 2006, p. 283)

A second important contribution of enlargement can be seen in the types of argument, derived from the joint with the discourse analysis and semiotics, as we can see in the figure 2 below:

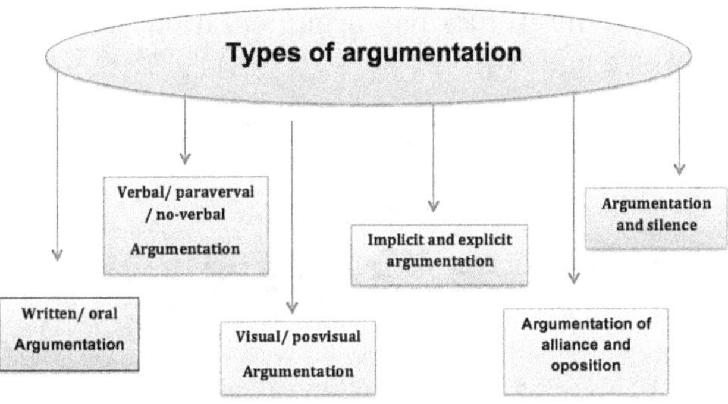

Figure 2 - Types of argumentation (Haidar 2006, p. 305)

A third proposal is related to the enlargements of the argumentative functions that we realize from the proposals of Portine (1973, pp. 149-153), Grize (1982), which can see in the figure 3 below:

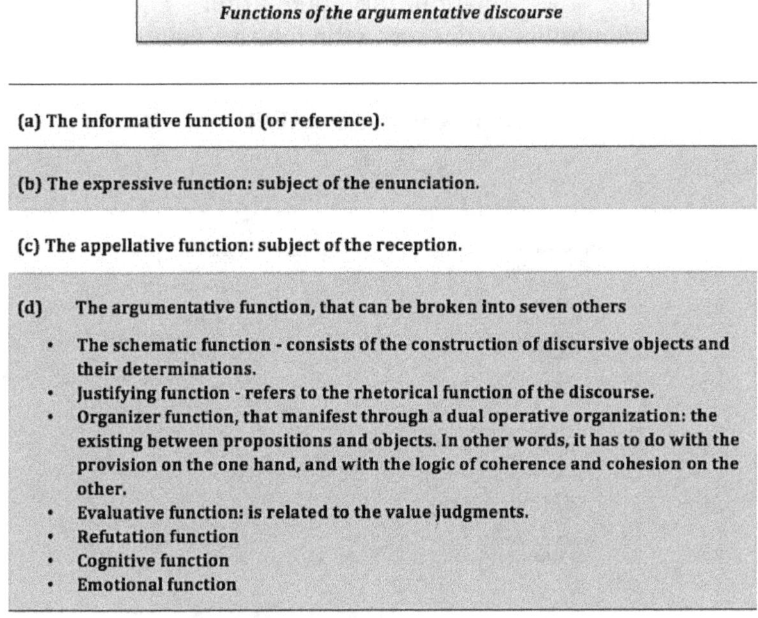

Figure - 3 Function of argumentative discourse (Haidar2006, p 312)

In this figure (3), we highlight that the appellative function is related with persuasion/seduction. In addition, it is important to integrate the emotional and cognitive function from the transdisciplinarity, to analyse further the functions of the argumentative discourse.

We finish this section with a summary of the main models of argumentation, of course not the only ones, because missing to integrate the proposals of the refutative argumentation and the emotional argumentation, see figure 4

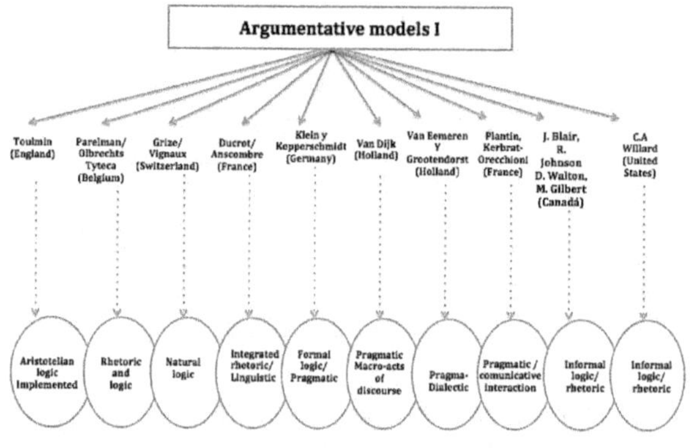

Figure 4 - Argumentative models I (Haidar, 2006, p. 319)

To complement the Figure 4, we place other argumentative models, which integrate the relation argumentation / refutation and argumentation/emotion, as can see in the figure 5 below

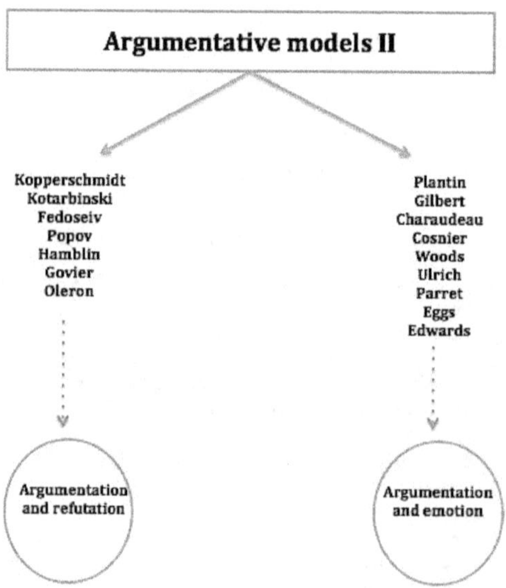

Figure 5 – Argumentative models II (Haidar, 2006)

3. THE POLEMIC, THE CONTROVERSY IN THE ARGUMENTATION

In today's world we live, by continuous and intense conflicts that are not resolved, already cannot be sustained a culture of argumentation in which prevails the classic functions of persuasion, of the conviction, and becomes central the refutation and the counter-refutation, at the national and international levels. Therefore, in the refoundation of the field it is necessary to give a nuclear place to the conflict, to confrontation, to debate, to the controversial in all areas.

In the analyzed corpus, nuclear discursive object for the CEU is "the University democracy", while for Rectory are "University reforms". Changes of discursive nuclear objects, during the debate, and the movements of the explicit and implicit meanings had significant consequences, such as the intensifications in strategies of refutation and the increased presence of the emotional component in the last days of student movement. In these, from the eristic, find two gladiators fighting with words, with discourses, making use of "symbolic violence" in all senses.

Among the diverse models explained, we select some to analyse strategies of the controversy between the students and the rectory. First of all, we synthesize the proposals of Kotarbinski, that we extend with the approaches of Oleron (1983), Ducrot (1983), Puig (1991), with which the following rules can be proposed:

- First rule: in the general theory of the struggle, the starting point is to take an attitude that is disadvantageous and surprise for the adversary, rule that must be accomplished to achieve the victory.
- Second rule: is the threat that is to say the adversary the ability to recall events which today would harm the opposite thesis test.
- Third rule: it is know order the adversary of the proof of his thesis, which will be located in a defensive position and the partner can then take the offensive position.
- Fourth rule: - the third derivative - refers to the use of an offensive or defensive position. In the domain of the controversy, the attack focuses on the cancellation of the thesis of the adversary, while the defense is the form of resistance.
- Fifth rule: is the anticipation. In the dispute must take the initiative, either to establish the order of the discussion or to take the floor in the first place. At the end of this rule, is the possibility that the controversy be developed based on the terms laid down by the party who spoke first, while the arguments of the opposing party must adapt to, considering that it has been argued. However, this rule suffers modifications according to circumstances, because you can use the privilege which gives to be the last speaker.

- Sixth rule: refers to the cancellation of the argumentation of the adversary. In verbal combat care must be taken to attack the fundamental propositions of all the argumentative structure of the adversary; a skillful polemicist not visa attack all statements of the adversary, but the most important, because by destroying the axioms of the adversary, the entire system is destroyed. In addition, the refutation of the nodal points requires argumentation to rotate around them.
Seventh rule: try to take the place of the refuter in the controversy, the dominant position. It is interesting to analyze in a dispute who takes the place of refuter, why it is and how to impose the rules of the game. In the case of the debate, the CEU assumes the role of refuter in the public debate and an offensive, not defensive attitude as it is that of the Rectory. However, say this has no relevance if not explained: a) why the University authorities - state institutional representatives - fail to occupy this place; and (b) why the CEU takes the place of refuter and successfully attacks the power of the authorities, managing to reverse the rules of the game in their favor.

In our research, these issues were considered. To continue exposing models, we stop at the proposal of Kopperschmidt, which complements the other proposals significantly.

3.1. Model of Kopperschmidt

Kopperschmidt (1985, pp. 159-162) proposes a model of global argumentative analysis that integrates the controversy and therefore the refutation operations. The macrostructural analytical framework of any argumentation must contain the following steps:
1. the definition of the problem - the quaestio: problems can be theoretical or practical. Establish what problems, conflicts that motivate argumentation and consistent operations of refutation.
2. formulation of the controversy thesis: establish the basic thesis that defend themselves and which are attacked. If it is a dispute there are competitive thesis that objects with greater or lesser force depending on the extent of the controversy.
3. segmentation of the arguments: the segmentation and identification of statements that academic work in a discourse, can be based on linguistic signals of argumentation, such as connectors.
4. reconstruction of the argumentative threads: an argument - its logic, its grammar - tissue serves to elucidate and evaluate the argumentative potential that can be realized with an argument.
5. reconstruction of the "global argumentative structure"

The author presents his proposal at Figure 6, which can see below:

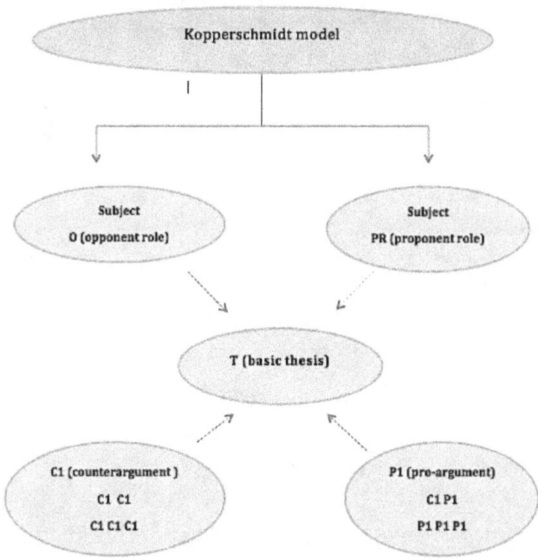

Figure 6 - Kopperschmidt model (Kopperschmidt, 1985 p. 163-164)

I think that it is necessary to return to this model to work on it, because it will recognize that the author had its own objectives. For example, the role of proponent and opponent are subjective places, which have emerged from the objective theory of the subject with other scopes; the basic thesis becomes a set of basic thesis, which arise from the nucleus and are divided into two subsets: a group from the side of the argumentation, and another on the side of the counteragumentation, or refutation argumentative, as I preferred to be called; in addition, the subjective place of the third is introduced: the place of the doxa, of the judge. In the Figure 7, below, the changes can be observed.

As you can see in the Figure 7, the following changes were made:

- Changed the order of the proponent and opponent, because you could not explain the order given by the author, by placing the proposer to the right.
- Introduced three subjective places, instead of the two proposed, highlighting the third, which already appears in Toulmin & Rieke & Janik (1979) and is developed by Kerbrat-Orecchioni & Plantin (1995)
- In the nucleus, there are proposed nuclear thesis, which are divided into two sets: the protagonist and the antagonistic ones.

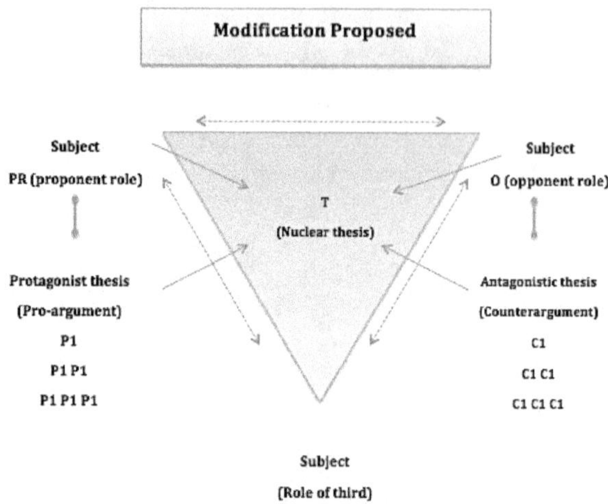

Figure 7 - Modification Proposed (Haidar 2006, p. 411)

In synthesis, from a transdisciplinary perspective, refutation strategies which may be more or less direct, more or less implicit, according to the conditions of production, circulation, reception, linked to all types of power, are as follows (Haidar, 2006, pp. 412-413):

1. initial disqualification of the opponent as a subject. Been disqualified anyway its representativeness, their knowledge, their skills, settle differences, etc.
2. disqualification of the main thesis of the opposition presenting a counterthesis or a set of counterthesis, as usual. The disqualification can have graduations.
3. disqualification of the arguments of the main thesis of the opposition presenting rebuttals, total or partial, of explicitly or implicitly.
4. refutation of the implicit of the thesis of the opponent. This strategy is more complex, more subtle.
5. imputation of thesis and arguments that the subject attributed to the opponent, and vice versa, in which there is no support for veracity.
6. create implied that they were not considered, be distorted inferences. A type of argument distortion.
7. use distortion arguments, i.e., assign a completely contrary argument, distorting what had been raised.
8. rebuttal by the rupture of the discursive silences - Foucault. Case of the CEU: questioning university political power, which was a taboo
9. rebuttal by the silence, in the other direction. It is muted, ignored the speech of the other, it is not. It is not a muted for granted, but ignore, diminish the importance.
10. reformulation of arguments or thesis in favor of his argument.

11. refute evidence of contradictions in the argumentation of the adversary. Quite effective strategy because discursive logic and coherence are fundamental. A discourse that may not prove to be coherent and not contradictory, loses the battle.
12. Refutation arguments in fact. This is one of the strongest strategies, very difficult to refute.
13. Refutation arguments of authority - are also very strong.
14. Rebuttal by the use of fallacies. Strategy very used because these are continuously present in any speech, and only you can give an account of them with a thorough analysis. The fallacies are a mechanism widely used in political discourse, in the public, mass media, and so on.
15. Refutation in the paraverbal and non-verbal (the visual, the posvisual); these are dimensions very little or not considered in analytical models argumentative, in spite of its importance.

Of course, as in the debate CEU- Rectory develops an public, oral, face to face argumentation, refutation strategies must be analyzed not only in the discursive, but in a future study, in the paraverbal and in the non-verbal dimension, which would yield other strategies not covered in this enumeration. It should also be noted that all strategies rebuttal used abundantly rhetorical figures: of the construction, the diction, the thinking, etc. The polemicist subjects make use of different rethoric figures of discourse, and another interesting research will study the types of figures used by both polemicists, which could cover in greater depth the innumerable senses related to discursive objects of controversy.

4. THE FUNCTIONING ERISTIC IN THE DEBATE CEU-RECTORY

To analyze the refutation eristic in this debate, we return to the trilogue argumentatif (Kerbrat-Orecchioni & Plantin, 1985), that integrates three subjective places: the proponent, the opponent and the third: it is important to emphasize that places are dynamic and are mainly determined by the nature of the quaestio, that plays a fundamental role in the polemic situation and the determination of the argumentative guidelines. The discursive subject can occupy different places: can be proponent, opponent, or the third as it is the case of this debate. With this proposal, we can expand the reflections on the controversial argumentative subjects, as well as on the complex site of the third, which is located in the area of reception.

In synthesis, the trilogue is a structure more open and unpredictable than the dialogue; correlatively, the trilogue is potentially more controversial than the dialogue, to which we add that placed a third more complicated analysis of the communicative interactions, but it also allows to explain them more deeply, more so when it comes to procedures argumentative. In Figure 8, below, you can see the application of the trilogue argumentative in the CEU-Rectory debate.

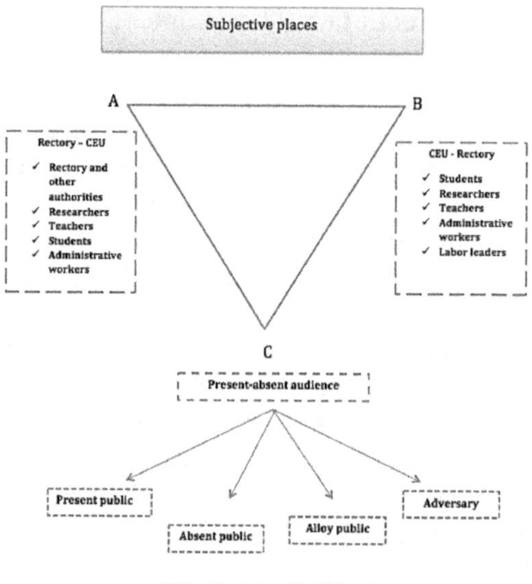

Figure 8 - Subjective places (Haidar 206, p. 311)

Other eristics aspects that arise in the debate, is materialized in two peculiar characteristics discursives, as you can see in the figure 9 below:

Rectory Discourse	CEU Discourse
• Established power • Masking language and discourse • More authoritarian • Nondirectional discursive acts	• Emerging power • Language and discourse authentic • More democratic. • Directional discursive acts

Figure 9 - *Eristics aspects* (Haidar, 2006,pp 267-270)

In this figure, you can see contradictory characteristics involving the interrelation of the implicit/explicit. In the discourse of the Rectory, there are simulations by the use of a language of concession in explicit, but not in the implicit dimension. This produces a reversal, because the concessive rectory language is opposed to the inflexible language of the CEU, but this is more democratic and it of Rectory more authoritarian.

In synthesis, the characteristics of the speech of Rectory, excepting the first one, spend all for the process of the masking and of the simulacrum in the argumentation, functioning that exists in much less grade in the speech of the CEU. In this sense, in the Rectory speech almost explicit managing discursive acts do not appear and in that of the CEU they are abundant.

In the debate CEU Rectory, more that to answer, predominates over the hyperbolic eristic component. Nevertheless, although the predominant act should be that of the refutation, it is not possible to allow emphasizing the act of support, of alliance, which takes place in the intergroup relations, that is to say, between the speeches of the commissions of the CEU and of Rectory, respectively. But, in the debate, like hierarchy, the refutation is predominant.

4.1. Polemic movements of the debate and refutation argumentative

The main thesis of two commissions condense the deep structure of the sense of two discourses, and the refutation processes are:

The (leading) rectory thesis (protagonist): 1. The Rectory reforms are to solve the problems of the UNAM; 2. The regulations of Inscription, Examinations and Payments are necessary to achieve the academic excellence and 3. Therefore, we defend the proposed reform and the regulations in particular. This set of nuclear thesis of the debate contains an argumentative logic that is not syllogistic because it implies the subjectivity, the emotional thing, the fallacy; this set of thesis is the leading ones, because the Rectory commission proposes them.

The thesis of the CEU (antagonistic): 1. The Rectory reforms do not solve the problems of the university system; 2. The regulations of Inscription, Examinations and Payments do not solve the academic excellence and 3. Therefore, we request the abolition of the regulations and of the proposed university reform, for being undemocratic. Therefore, intends the realization of a decisive and democratic University Congress.

In the same way that in the previous set, in this one also a logical argumentative not syllogistic is present, because in her there are involved the subjects, the emotional component, the deceits; these thesis are antagonistic, because they derive from the position of the refuter subjects of the CEU.

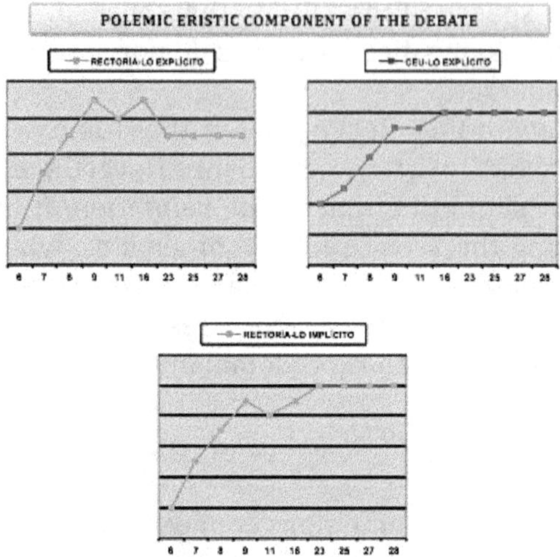

Figure 10 - Polemic eristic Component of the Debate (Haidar, 2006)

The three graphs show argumentative refutation and the emotional component in the CEU - Rectory debate. As we can see in the graphs 1, 2 and 3 the greatest degree of controversy is in the 16th, when the CEU presents its counter-proposal to the Rectory, delivered on day 11, and rises from the table of talks: producing a semiotic-discursive rupture unquestionable. In the mentioned graphs, it is possible to observe the eristic functioning, where the argumentative refutation is demonstrated constantly.

In the oscillatory lines it is observed that the polemic - emotional component ends in the first stage, which is the 9th, with a refutation high degree. Later, in the second stage, shaped for the 11th and 16th, when there appears the proposal of Rectory and the counterproposal of the CEU, opponents condense the positions and there remain very clear the thesis and the strategies of the argumentative refutation of both sides. In the third stage — of the 23rd to 28 – there is a maintenance of the grade of the eristic polemic and its culmination it is in last day, with the declaration of the student strike.

5. CONCLUSION

The discursive skein was unraveled to show the intricate situation of the construction threads of this debate. Threads full of processes of interdiscursivity, which leave open a lot of ways and routes for the investigators of the sense. In the debate there breaks the discursive rainbow that constitutes a complex polyphonic set, of chromatic -

sonorous character — the synesthesia — and in the end only there stay two domineering, polar, irreconcilable positions that break the possible dialectics, for the eristic.

REFERENCES

Blair, J.A., & Johnson, R. (1980). *Informal logic. The first international symposium*. Point Reyes: Edge Press.

Hamblin, C.L. (1970). *Fallacies*. Canada, Vancouver: Vale Press.

Cosnier, J. (1977). Gestes et situation conversacionnelle. In *Stratégies discursives*, Actes du Colloque du Centre de Recherches Lingüistiques et Sémiologiques de Lyon, Lyon: Presses Universitaires de Lyon.

Charaudeau, P. (2000). Une problématisation discursive de l' émotion- A propos des effets de pathémisation à la télévision. In C. Plantin, M. Doury & V. Traverso (Eds.), *Les émotions dans les interactions* (pp. 125-155). Lyon: Presses Universitaires de Lyon.

Ducrot, O., & Anscombre, J. C. (1983). *L'argumentation dans la langue*. Bruxelles: Editeur Pierre Mardaga.

Edwards, D. (1999). Emotion discourse. *Revue Culture & Psychology*, *5*(3), 271-291.

Eggs, E. (2000). Logos, ethos, pathos: l'actualité de la rhétorique des passions chez Aristote. C. Plantin, M. Doury & V. Traverso (Eds.), *Les émotions dans les interactions* (pp. 15- 31). Lyon: Presses Universitaires de Lyon.

Fedoseiv, P.N & Popov, S.I. (1982). *El arte de la polémica*. México: Editorial Cartago.

Gilbert, M. A. (1997). Prolegomenon to a pragmatics of emotion. *Proceedings of the Ontario Society for the Study of Argumentation*. St. Catherine's, Canada: Brock University.

Grize, J.B. (1982). *De la Logique à l'Argumentation*, Genève : Librairie Droz.

Govier, T. (1999). *The philosophy of argument*. NewPort News, Virginia: Vale Press.

Haidar, J. (2006). *Debate CEU-RECTORÍA. Torbellino pasional de los argumentos*. México: Universidad Nacional Autónoma de México.

Kerbrat-Orecchioni, C., & Plantin, C. (Eds.) (1995), *Le trilogue*. Lyon: Presses Universitaires de Lyon.

Klein, W. (1980). Argumentation und Argument *Zeitschrift für Literaturwissenschaft und Linguistik*, *38/39*, 9-57.

Koppershmidt, J. (1985). An analysis of argumentation. In: T. van Dijk (Ed.), *Handbook of Discourse Analysis*, Vol 2. Florida: Academic Press.

Kotarbinski, T. (no date). L'Eristique, cas particulier de la théorie de la lutte. *La théorie de l'Argumentation*. Louvain, Paris: Centre National Belge de Recherches de Logique.

Morin, E. (1997). *Introducción al pensamiento complejo*. Barcelona: Editorial Gedisa.

Morin, E. (1999). *El método. El conocimiento del conocimiento*. Madrid: Ediciones Cátedra

Nicolescu, B. (1999). O Manifesto da transdisciplinaridade, Sao Paulo, TRIOM. Original: Bulletin Interactif du Centre International de Recherches et Études transdisciplinaires (CIRET) n° 15 - Mai.

Oleron, P. (1983). *L'Argumentation*. Paris: Presses Universitaires de France.

Parret H. (No date). *Enunciación, sensación, pasiones*. Madrid: Editorial Edicial.

Perelman, C., & Olbrechts-Tyteca, L. (1989). *La Nueva Retórica: Tratado de la Argumentación*. Madrid: Editorial Gredos.

Plantin, C. (1990). *Essais sur l´Argumentation –Introduction a l´étude linguistique de la parole argumentative*. Paris: Editions Kimé.

Plantin, C. (1997). L´Argumentation dans l´ émotion. *Revue Pratique - Enseigner l´Argumentation, 96*, CRESEF, 81-100.

Portine, H. (1978). *Analyse de discours et didactique de langues*. Paris: Bureau pour l'enseignement de la Langue et de la Civilisation Française.

Puig, L. (1991). *Discurso y argumentación. Un análisis semántico y pragmático*. México: Instituto de Investigaciones Filológicas, UNAM.

Toulmin, S., Rieke, R., & Janik, A. (1979). *An introduction to reasoning*. New York, Macmillan Publishing Co., Inc.

Ulrich, W. (1992). In defense of the fallacy. In W. L. Benoit, D. Hample & P. J. Benoit (Eds.), *Readings in argumentation, Studies in argumentation in pragmatics and discourse analysis* (pp. 337-356). Berlin/New York: Foris Publications.

van Dijk, T. A. (1980). *Texto y contexto*. Madrid: Editorial Cátedra.

van Eemeren, F. H, & Grootendorst, R. (1992). *Argumentation, communication and fallacies: A pragma-dialectical perspective*. New Jersey/London: Lawrence Erlbaum Associates Publishers.

Vignaux, G. (1976). *La argumentación: Ensayo de lógica discursiva*. Buenos Aires: Editorial Hachette.

Walton, D. (1992). *The place of emotion in argument*. University Park, PA: The Pennsylvania State University Press.

Walton, D. (1996). *Argument structure: A pragmatic theory*. Toronto: University of Toronto Press.

Willard, C. A. (1991). L´argumentation et les fondements sociaux de la conaissance. In Different Edit, *L´Argumentation - Colloque de Cerisy*, (pp. 91-106). Liège, Belgique: Pierre Mardaga Editeur.

Woods. J. (1992). Who cares about the fallacies. In F. H. van Eemeren, R. Grootendorst, J. A. Blair & C. A. Willard (Eds.), *Argumentation Illuminated* (pp. 23-48). Amsterdam: SICSAT.

Regular Papers

1

Understanding the Competence Involved in Constructing Argumentative Contexts

MARK AAKHUS
School of Communication and Information, Rutgers University, USA
aakhus@rutgers.edu

Communicative contexts are not simply given but develop and in many cases are actively developed to achieve particular purposes. This practice, which is the focus of much contemporary work in society, entails the active shaping of argumentation. To further understand this argument practice, findings from field interviews with practitioners involved in managing disagreement among stakeholders in policy disputes are reported. The analysis examines the practical reasoning involved in constructing contexts for argumentation and reasoned action.

KEYWORDS: communication design, design thinking, disagreement space, policy argumentation, policy practice

1. ARGUMENTATION, DESIGN, AND PRACTICE

Much innovation in argumentation theory in the past few decades has been around the development of analyses that take communicative context (e.g, dialogue, institutional genre) into account in reconstructing and evaluating arguments. These advances, as realized in van Eemeren's *Pragma-Dialectics* and Walton's *New Dialectics*, aim to understand reasonableness by taking account of the practical activity context in which arguments are made. The rules of a dialogue game enable the analyst to see whether a move was reasonable or not. The new premise for evaluation is that the rationality of what is said cannot be adequately performed without first interpreting what is said in light of the pragmatic conditions of its uttering; thus, enabling the analyst to see how arguments are used in less than ideal conditions and in strategic maneuvering within institutional contexts. Even with their incorporation of dialogue and institutional talk, these advances take communicative context as a given basis from which to interpret and

D. Mohammed & M. Lewiński (eds.) (2016). *Argumentation and Reasoned Action: Proceedings of the 1st European Conference on Argumentation, Lisbon, 2015. Vol. II, 153-161.* London: College Publications.

evaluate the argumentative quality of what takes place within that dialogue. While contexts can be taken as given, they are often under construction. What happens when we open the important issue about the construction of communicative contexts as it relates to argumentation?

One strand of normative-pragmatic research on argumentation has developed a *design stance* toward argument (see Jackson, 2015) that engages this question. Within the design stance significant attention has been given to the "invention of means to control the way people pursue controversies and differences of opinion" (Aakhus, 2003, p. 281). This line of work has focused on the practices and procedures for articulating "disagreement space" (van Eemeren, Grootendorst, Jackson, & Jacobs, 1993), such as found in the work of dispute mediators, meeting facilitators, professional practitioners, information and communication media, and macro-actors (e.g., organizations). The competence in constructing communicative context communication has been shown as consequential for what is argued, how that argumentation is pursued, and what epistemic and actional consequences that argumentation generates (e.g., Jacobs & Aakhus, 2002). Examining the design thinking of practitioners who take responsibility in constructing communicative contexts can shed light on the relationship between communication's design and argumentation (e.g., Aakhus, 2009).

2. INQUIRY INTO COMMUNICATION DESIGN WORK AND DISAGREEMENT MANAGEMENT

In depth interviewees about public policy involvement and controversies were conducted with 18 policy practitioners. The practitioners were located in North America, primarily in the Northeastern United States. They included public agency managers, state/local elected officials, non-governmental organization leaders, and policy consultants. They were selected for their varied and significant experience around a range of matters including environment, land use, water, education, health, and transportation. The interviews were part of a larger sample including third party neutrals specializing in policy conflict.

The policy practitioners were interviewed to collect their accounts and explanations about situations where they were involved in managing disagreement among stakeholders in policy disputes. The interviews were conducted using the "interview by comment" technique that elicited accounts (e.g., stories) and explanations (e.g., reasons). The key prompt in the interviews focused on their personal experience in handling decision-making in complex policy processes. The interviews ranged from 115 to 43 minutes in length and averaged 68 minutes (H = 1'55"; L = 43") and were transcribed for coding and analysis. The word count for each interview transcript ranged from 5,729 to 13,374 words with an average word count of 9,923 words.

An initial coding of the interviews marked off segments that were either accounts or explanations about accounts about policy events and the choices made about how stakeholder differences and disagreement were managed. Accounts refer to the interviewee offering a description of an event they were involved in that elaborates the elements of a scene and the dramatic unfolding of what happened. Explanations refer to the interviewee offering rationales for what they highlighted about a scene and why things unfolded as they did, especially including their own actions. The 42 accounts and the 87 explanations from the 18 interviews were then coded, using an open coding approach, as to types of accounts and explanations regarding the management of differences and disagreement. Axial style coding was used to specify the implicit rules of inclusion and exclusion for each type of account and explanation. In so doing, emergent classification began to reveal patterns in reasoning about the construction of communication contexts for managing disagreement. Two key themes are addressed here.

2.1 The materiality of communication for disagreement management and argumentation

Policy professional's (PP) could not talk about policy without talking about meetings, so the interviews were further analyzed using meetings as a point of entry into the way PPs understand their work. PPs experience an ecology of meetings, as illustrated by an elected town official's account below. The sections of the passage are in bold to highlight the network of meetings and meeting products designated as input for other meetings:

> [09:269.077] Elected Official – Town Council: "The process redevelopment **goes through the planning board**. So the **planning board;** so we, there's **a master plan that designates** some areas in need of redevelopment-or revitalization- and then **the council economic development committee does a lot of the work, makes recommendations to the council**. Then **the council talks things through and then redirects to the planning board**. So the **planning board was tasked with looking at redevelopment and making a recommendation**. So **they did a year and a half worth of meetings**. So **there's a legal**

> **requirement for what the minimum is, and they decided to go beyond that, and they had as many meetings as they thought they needed to, to hear everybody who would ever come to a meeting**. . . ."

A meeting, and the ecology of meetings in which a meeting takes place, is a significant, obdurate reality for the policy professional. It is a reality that they recognize as both given and made, such that they are attentive to how to turn a given encounter into a preferred encounter and the consequentiality of making that happen.

What is noteworthy is that their talk about meetings reveals an awareness of the layers of the built up environment for communication about policy. The PPs are quite sensitive to the immediate demands of a meeting as an encounter where the local, turn-by-turn, emergence of the form and quality of that encounter are at stake. That awareness extends to the implications of that local unfolding for what has happened in other encounters and what could happen in subsequent encounters. While meetings are a given fact of life for PPs, they are also a tool, albeit a recalcitrant tool, for reconfiguring interactivity among participants in an encounter to achieve different states of knowing and acting within the policy process. This is illustrated in the continuation of the interview with an elected town official. Selections of the passage are bolded to highlight how encounters are framed as particular kinds of meetings by crafting particular modes of engagement among stakeholders to generate particular content and outcomes:

> [09:269.077 – continued] Elected Official – Town Council: "...**We would have charrettes**, which are sort of; at the very beginning, we would have pictures of what the original, what the town looked now, and people would comment – 'I'd love to see this, I'd love to see this, I'd love to see that.' **We had three charrettes, where each one built upon the comments of the one before, so people could see, could literally be writing down what they wanted**. Then things would get redrawn with some of their ideas actualized. **And then people kept commenting on the plans, and commenting on the plans. The plan would revise. They'd comment on the plans again to a point where the architects and planning group that we hired said, 'This is what we think this represents, what your town is telling us they want and need.'** So then we would look at that, and look at how much things cost, and what's realistic, and what we can do in the short term and the long term, and then we would work on that."

There is pervasive recognition among PPs that meetings, as Goffman so aptly noted, are fateful events. That fatefulness is experienced in the local, turn-by-turn, emergence of the form and quality of an encounter for those involved but also in the implications of that local unfolding for what has happened in other encounters and what could happen in subsequent encounters. While PPs are attentive to these possibilities, it is also the case that these possibilities are on the horizon of awareness. The attention, however, discloses that they make inferences about how specific events and the ecology of these events work, how they could work, and what is possible in any given moment or sequence.

2.2 The prospects of communication's design for managing disagreement

PPs accounts orient toward how meetings can be used to shape disagreement. PPs recognize that while meeting are stages for the social and political drama of the broader conflict, they see themselves as playing a significant role in directing that drama and finding ways for it to play out. Disagreement features prominently in their ideas about shaping the quality and direction of policy communication. The uses of clash and expertise are two strategies evident in the accounts of PPs for articulating disagreement space among actors in policy circumstances.

The PPs recognize that the content, direction, and outcomes of policy communication can take many different forms, and are concerned with what form it takes. They do not necessarily seek to diffuse the sources of conflict among the many stakeholders to a policy, as illustrated in the following two accounts. Selections are bolded to highlight attention given about generating conditions for differences to be expressed.

> [8:150.37] Elected Mayor: "It also happened to be one of our most affluent neighborhoods, so I had a lot of people complaining about why we weren't doing more to cut down on the speed and all this traffic. 'Why did we allow all this traffic to go through on their street? Why didn't we redirect it to other streets?' And, of course, if you redirected it, then you'd get 'why are you catering to those snobs over there and moving it from their neighborhood,' all that sort of stuff. And sometimes, quite frankly, you let that happen. **In other words, you make sure that if let's say a group of people made the suggestion that we redirected the traffic so that it mainly went on the parallel street, I'd make sure that the parallel street knew about that.** So then that creates; I guess to do conflict resolution, **you've got to get people to appreciate the fact that there really is a conflict.**"

> [5:202.24] Transportation Consultant: "And so they took this easy road out, in which they just simply held public meetings and revved up the neighborhood groups, and what you ended up with was a completely one-sided effort. **And a government agency really has to do more than that....But you have to get them all arrayed, and if the government has business allies that could be very powerful and helpful, and they allow them to disappear, then the project has very little hope, because you're not really getting the kind of parties that ought to be butting heads butting heads, which is creative.**

Both accounts attend to how clash could be realized by the arrangement of communication and thus use clash to bring disagreement to the surface, to create an occasion for particular lines of argumentation to feature in deliberation, and to activate a meeting network.

The PPs also incorporated experts into meetings as a way to discipline opinion formation, constrain the grounds for raising disagreement, and to exploit expert neutrality.

> [09.269.081] Elected Official Town Council: "So we often use the experts to either give us some facts, Identify some options, react or review to something we're thinking about. And then sometimes we'll actually have them be the lightning rod, so they'll introduce the idea and walk people through the public meeting....So the governing officials aren't seen as the bad guy, if you will, but there's someone not associated with the town, who could be seen as a more neutral, objective person providing information... because we want to be seen as the one hearing, and listening and caring about the constituents. But sometimes constituents don't realize we have to think about everybody, and not just the individual. So it helps to have some of those experts deal with the individual and be the ones that say 'no.'"

3. REASONING ABOUT COMMUNICATION IN DISAGREEMENT MANAGEMENT

While a recalcitrant tool, meetings are means for PPs to use formalities of interaction (e.g., agenda topics and sequencing, inviting participation) and their official role to shape features of interactivity to articulate the disagreement space. Their accounts revealed strategies and their explanations of what they do lend insight into how their attention to the communication pragmatics of encounters, especially the epistemic and actional consequences of encounters.

The prospects for design and attention to communication pragmatics reveal concerns about the demands of constructing communicative contexts. Throughout the accounts and explanations of interventions by PPs, there is an underlying instrumental rationality about intervention regarding the designability of communication - that is, by tinkering with the features and principles of language-use and interaction, preferred forms of engagement can be generated to address situational demands and circumstantial goals:

> If an encounter is organized like X (or Y or Z) then it will be possible to achieve Q (or E or D) while overcoming A (or B, C, or D) and avoiding L (or M, N, O, or P).

The equation highlights the concern with making something work, similar to the pragma-dialectical concept of "problem-solving validity," but equally concerned that what-works is a normative judgement, similar to the pragma-dialectical concept of "intersubjective-validity." Beliefs about how communication works and how it ought to work enter into what might otherwise appear as straightforward means-end rationality. Instead, there is a bounded rationality framed by descriptive and normative assumptions used in reasoning about communication's design in the management of disagreement.

The decision-making explanations revealed the PP's attention to the local management of interactivity and vigilance for the validity of the interaction. Their concerns with problem-solving validity were most explicitly addressed through explanations of local management issues such as (1) arranging the form of activity to properly signal the nature of the encounter, (2) shaping the participation status of the stakeholders into a particular participation framework, and (3) ensuring the process-outcome relationship (i.e., a meeting led to the relevant outcome such as a recommendation). Their concerns with intersubjective validity were most explicitly addressed through explanations of validity-vigiliance about the uses of persuasion to remove doubts about the viability of the encounter within the socio-political context and the legitimacy of the encounter's outcomes.

The reasoning about communication's design ranges from awareness of matters about problem-solving and intersubjective validity to more full-fledged practical theories used in interpreting what's problematic and what's solvable in a setting and in warranting the intervention into communication to manage disagreement. Aakhus (2009) spells out two such practical theories by PPs that suggests there are design logics about handling communicative events that can become reflected theories of practice about making a given situation into a preferable situation through the particular orchestration of interaction.

4. CONCLUSION

While the argumentation field has given considerable attention to the evaluation of arguments, and what counts as good arguments within a given context, considerably less attention has been given to the construction of those contexts and how argumentative possibilities get defined. By comparison, consider Searle's (2001) criticism of standard decision theory as having too much to say about assessing choices given certain desires and too little to say about the essential feature of decision making - which is, determining what desire(s) to have.

Argumentation scholars can go further in argumentation analysis and evaluation by seeing that resolutions to contemporary problems do not simply lie in figuring out how to verify a good argument, but also lie in figuring out what argumentation to have. Indeed, time spent trying to figure out what a good argument is without attention to how argumentation contexts are constructed, is time powerful and creative actors have for strategies influencing what is argued about. An important part of the argumentation studies enterprise is to understand how these communicative contexts are constructed and thus shaping argumentative potential, and it is imperative to develop imagination in communication design practices that shape argumentation.

REFERENCES

Aakhus, M. (2003). Neither naïve nor critical reconstruction: Dispute Mediators, impasse, and the design of argumentation. *Argumentation, 17*(3), 265–290.

Aakhus, M. (2009). The experiences of policy professionals in designing deliberation. In S. Jacobs (Ed.), *Concerning argument: Selected papers from the 15th Biennial Conference on Argumentation Biennial Conference on Argumentation* (pp. 33–40). Washington, DC: National Communication Association.

Eemeren, F. H. van, Grootendorst, R., Jackson, S., & Jacobs, S. (1993). *Reconstructing Argumentative Discourse.* Tuscaloosa, AL: University of Alabama Press.

Jackson, S. (2015). Design thinking in argumentation theory and practice. *Argumentation, 29*(3), 243-263. doi: 10.1007/s10503-015-9353-7.

Jacobs, S., & Aakhus, M. (2002). What mediators do with words Implementing three models of rational discussion in dispute mediation. *Conflict Resolution Quarterly, 20*(2), 177–204. doi: 10.1002/crq.3890200205.

Searle, J. R. (2001). *Rationality in Action*. Cambridge, MA: MIT Press.

2

Using Abstract Dialectical Frameworks to Argue about Legal Cases

LATIFA AL-ABDULKARIM
Department of Computer Science, University of Liverpool, UK
latifak@liverpool.ac.uk

KATIE ATKINSON
Department of Computer Science, University of Liverpool, UK
katie@liverpool.ac.uk

TREVOR BENCH-CAPON
Department of Computer Science, University of Liverpool, UK
tbc@liverpool.ac.uk

Recent work has shown how to map factor hierarchies for legal reasoning into Abstract Dialectical Frameworks (ADFs), by defining acceptance conditions for each node. In this paper we model as ADFs bodies of case law from various legal domains, rewrite them as logic programs, compare the results with previous legal reasoning systems and propose improvements by increasing the scope of reasoning downwards to facts.

KEYWORDS: ADF, case based-reasoning, factors, legal reasoning

1. INTRODUCTION

A recent development in computational argumentation has been Abstract Dialectical Frameworks (ADFs) by Brewka and Woltran (2010). ADFs can be seen as a generalisation of standard Argumentation Frameworks (AFs) (Dung, 1995) in which the nodes represent statements rather than abstract arguments, and each node is associated with an acceptance condition that determines when a node is acceptable in terms of whether its children are acceptable. Thus links in AFs express only one relationship, namely defeat, but ADFs can represent a

D. Mohammed & M. Lewiński (eds.) (2016). *Argumentation and Reasoned Action: Proceedings of the 1st European Conference on Argumentation, Lisbon, 2015. Vol. II, 163-180.* London: College Publications.

variety of attack and support relations. In consequence, whereas nodes in an AF have only the single acceptance condition that all their children are defeated, nodes in ADFs can have different acceptance conditions specifically tailored for each node. In Al-Abdulkarim *et al.* (2014) it was argued that ADFs are very suitable for representing factor based reasoning with legal cases as found in the CATO system (Aleven, 1997) and as formalised in Prakken and Sartor (1998).

The key idea in Al-Abdulkarim *et al.* (2014) is that the abstract factor hierarchy of CATO (Aleven, 1997) (an extract given in Figure 1) corresponds directly to the node and link structure of an ADF, or rather (since the links are labelled "+" or "−") a Prioritised ADF (PADF) Brewka *et al.* (2013) which partitions links into supporting and attacking links, and so corresponds to the labels on the links in the factor hierarchy. To express CATO's factor hierarchy as an ADF, acceptance conditions need to be supplied for each of the nodes. Finally the logical model of IBP (Brüninghaus & Ashley, 2003) can be used to tie the various parts of the factor hierarchy together to supply decisions for particular cases. In Al-Abdulkarim *et al.* (2014) it was suggested that the acceptance conditions could be expressed as Prolog procedures. These could then be used directly to form a Prolog program that could be executed to classify cases (as to which side they are decided for) represented as sets of factors.

In this paper we will evaluate this approach. We will use US Trade secrets as the domain, allowing us to use the analysis of CATO which will permit direct comparison with CATO, IBP, the various systems used as comparators in Brüninghaus & Ashley (2003) and the AGATHA system of Chorley and Bench-Capon (2005). We will firstly consider a quantitative analysis, in terms of performance and how easily the program can be refined to improve performance, and then consider the system in terms of the transparency of its outputs, the relation to case decision texts, and the relation to formal frameworks for structured argumentation such as Prakken (2010). Finally we will consider whether the method can be readily applied to other domains, by briefly describing an application of the ADF approach to *Popov v Hayashi* and related cases as modelled in Bench-Capon (2012).

2. BACKGROUND

In this section we will recapitulate the essentials of ADFs (Brewka *et al.*, 2013), CATO (Aleven, 1997) and IBP (Brüninghaus & Ashley, 2003).

2.1 *Abstract Dialectical Frameworks*

An ADF is defined in Brewka *et al.* (2013), as:

> Definition 1: An ADF is a tuple $ADF = < S,L,C >$ where S is the set of statements (positions, nodes), L is a subset of $S \times S$, a set of links, and $C = \{C_{s \in S}\}$ is a set of total functions $C_s : 2^{par(s)} \rightarrow \{t, f\}$, one for each statement s. C_s is called the acceptance condition of s.

In a Prioritised ADF, L is partitioned into L+ and L−, supporting and attacking links, respectively. Although the acceptance conditions are often expressed as propositional functions, this need not be the case: all that is required is the specification of conditions for the acceptance or rejection of a node in terms of the acceptance or rejection of its children.

2.2 *CATO*

CATO (Aleven, 1997), which was developed from Rissland and Ashley's HYPO (1990), takes as its domain US Trade Secret Law. CATO was primarily directed at law students, and was intended to help them form better case based arguments, in particular to improve their skills in distinguishing cases, and emphasising and downplaying distinctions. A core idea was to describe cases in terms of factors, legally significant abstractions of patterns of facts found in the cases, and to build these base-level factors into an hierarchy of increasing abstraction, moving upwards through intermediate concerns (abstract factors) to issues. An extract from the factor hierarchy is shown in Figure 1.

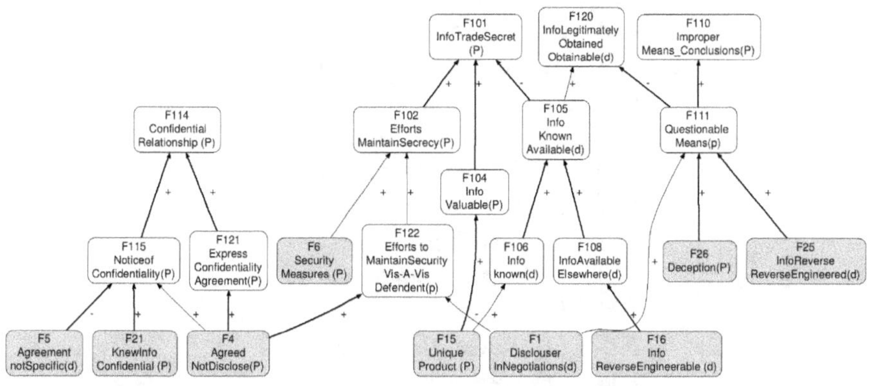

Figure 1 – CATO Abstract Factor Hierarchy from Aleven(1997)

Each factor favours either the plaintiff or the defendant. The program matches precedent cases with a current case to produce arguments in three plies: first a precedent with factors in common with the case under consideration is cited, suggesting a finding for one side. Then the other side cites precedents with factors in common with the current case but a decision for the other side as counter examples, and distinguishes the cited precedent by pointing to factors not shared by the precedent and current case. Finally the original side rebuts by downplaying distinctions, citing cases to prove that weaknesses are not fatal and distinguishing counter examples. CATO used twenty-six base level factors (there is no F9), as shown in Table 1.

ID	Factor
F1	DisclosureInNegotiations (d)
F2	BribeEmployee (p)
F3	EmployeeSoleDeveloper (d)
F4	AgreedNotToDisclose (p)
F5	AgreementNotSpecific (d)
F6	SecurityMeasures (p)
F7	BroughtTools (p)
F8	CompetitiveAdvantage (p)
F10	SecretsDisclosedOutsiders (d)
F11	VerticalKnowledge (d)
F12	OutsiderDisclosuresRestricted (
F13	NoncompetitionAgreement (p)
F14	RestrictedMaterialsUsed (p)
F15	UniqueProduct (p)
F16	InfoReverseEngineerable (d)
F17	InfoIndependentlyGenerated (d
F18	IdenticalProducts (p)
F19	NoSecurityMeasures (d)
F20	InfoKnownToCompetitors (d)
F21	KnewInfoConfidential (p)
F22	InvasiveTechniques (p)
F23	WaiverOfConfidentiality (d)
F24	InfoObtainableElsewhere (d)
F25	InfoReverseEngineered (d)
F26	Deception (p)
F27	DisclosureInPublicForum (d)

Table 1-Base Level Factors in CATO

There is, however, no single root for the factor hierarchy as presented in Aleven (1997): rather we have a collection of hierarchies, each relating to a specific issue. To tie them together we turn to the Issue Based Prediction (IBP) system of Bruninghaus and Ashley (2003).

2.3 *Issue Based Prediction*

In IBP, which is firmly based on CATO, the aim is not simply to present arguments, but to predict the outcomes of cases (find for plaintiff/defendant). To enable this, the issues of CATO's hierarchy are tied together using a logical model derived from the Uniform Trade Secret Act and the Restatement of Torts. The model is shown in Figure 2.

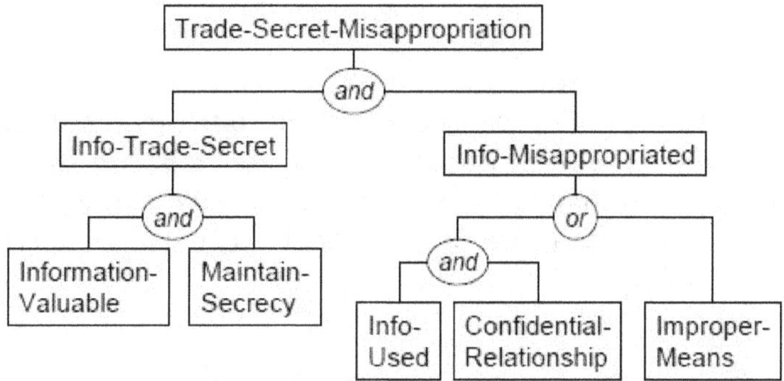

Figure 2 – IBP Logical Model from Brüninghaus & Ashley (2003)

Now consider the factor hierarchy, part of which is shown in Figure 1. We can now regard this as an ADF by forming the set S from the issues, intermediate concerns and base level factors, L+ from the links labelled "+" and L- from the links labelled "-". Using the complete factor hierarchy given in Figures 3.2 and 3.3 of (Aleven, 1997) we will have an ADF which has as its leaf nodes the base level factors of CATO. This is described in tabular form in Table 2.

ID	*S*	*L+*	*L−*
F102	EffortstoMaintainSecrecy	F6, F122, F123	F19, F23, F27
F104	InfoValuable	F8, F15	F105
F105	InfoKnownOrAvailable	F106, F108	
F106	InfoKnown	F20, F27	F15, F123
F108	InfoAvailableElsewhere	F16, F24	
F110	ImproperMeans	F111	F120
F111	QuestionableMeans	F2, F14, F22, F26	F1, F17, F25
F112	InfoUsed	F7, F8, F18	F17
F114	ConfidentialRelationship	F115, F121	
F115	NoticeOfConfidentiality	F4, F13, F14, F21	F5, F23
F120	LegitimatelyObtainable	F105	F111
F121	ConfidentialityAgreement	F4	F23
F122	MaintainSecrecyDefendant	F4	F1
F123	MaintainSecrecyOutsiders	F12	F10
F124	DefendantOwnershipRights	F3	

Table 2 CATO as ADF

The roots of CATO's hierarchies correspond to the leaves of the logical model: we can therefore form them into a single ADF by using this structure. The relevant additions to the ADF needed to integrate the IBP model are shown in Table 3 (note that F124 is not discussed in Brüninghaus & Ashley, 2003).

ID	*S*	*L+*	*L−*
F200	TradeSecretMisappropriation	F201,F203	F124
F201	Info-Miasappropriated	F110,F112,F114	
F203	Info-Trade-Secret	F102,F104	

Table 3-IBP Logical Model as ADF

IBP used 186 cases, 148 cases analysed for CATO and 38 analysed specifically for IBP. Unfortunately, these cases are not all publicly available and so we will use the 32 cases harvested from public sources by Alison Chorley and used to evaluate her AGATHA system (2005). As part of the evaluation in Brüninghaus and Ashley (2003) nine other systems were also considered to provide a comparison. Most of these were different forms of machine learning system, but programs representing CATO and HYPO were also included. IBP was the best performer: results reported in IBP (Brüninghaus & Ashley, 2003), Naive Bayes (the best performer of the ML systems), CATO, HYPO and a version of IBP which uses only the model, with no CBR component, are shown in Table 4[1].

	correct	error	abstain	accuracy
IBP	170	15	1	91.4
Naive Bayes	161	25	0	86.5
CATO	152	19	22	77.8
HYPO	127	9	50	68.3
IBP-model	99	15	38	72.6

Table 4- Results from [12]

Direct comparison with AGATHA is hampered by the fact that evaluation in AGATHA was directed towards evaluating the different heuristics and search algorithms used in that system, and so no version can be considered "definitive", and, of course, many fewer cases were used in the experiments. However, typically 27-30 of the 32 (≈ 84–93%) cases were correctly decided by the theories produced by AGATHA (Chorley & Bench-Capon, 2005).

3. ACCEPTANCE CONDITIONS

We now supply acceptance conditions for each node, to supply the elements of *C* required for the ADF. We will rely only on the definitions of the factors in Aleven (1997). We do not use precedents at this stage; as Aleven remarks:

> for certain conflicts, it is self evident how they should be resolved. For example, the fact that plaintiff's product was unique in the market (F15) arguably supports a conclusion that plaintiff's information (which is used to make the product) is not known outside plaintiff's business (F106), but not if it is also known, for example, that plaintiff disclosed its information in a public forum (F27). Common sense dictates that in those circumstances, the information is known outside plaintiff's business. It is not necessary to look to past cases to support that point. CATO's use of link strength enables a knowledge engineer to encode inferences like this. (p. 47).

From Tables 2 and 3 we can see that we have eighteen nodes to provide with acceptance conditions. One (F124) has only a single supporting

[1] No explanation for using a different number of cases for CATO and IBP model is given in Brüninghaus and Ashley (2003).

child: thus the acceptance condition will be Parent $\longleftrightarrow$ Child. We will write this (and the other acceptance conditions) as a set of tests for acceptance and rejection, to be applied in the order given, which allows us to express priority between them. The last test will always be a default. We choose this form of expression because we find it easier to read in many cases, because it corresponds directly to the defeasible rules with priorities used in formalisms such as ASPIC+ (Prakken,2010), and because it is directly usable as Prolog code. Thus we write Parent $\longleftrightarrow$ Child as

Accept Parent if Child.
Reject Parent.

Where NOT is required we use negation as failure. The tests are individually sufficient and collectively necessary, ensuring equivalence with the logical expression (see Clark, 1978). Six nodes (F201, F203, F105, F108, F114 and F124) have only supporting links: these can be straightforwardly represented using AND and OR. We followed the IBP model for the two nodes taken from that model (F201 and F203), and used OR for the other four. The most complicated was InfoMisappropriated (F201):

Accept InfoMisappropriated if F114 AND F112.
Accept InfoMisappropriated if F110.
Reject InfoMisappropriated.

Five nodes have one supporting and one attacking link. These are best seen as forming an exception structure: accept (reject) the parent if and only if supporting (attacking) child unless attacking (supporting) child. Note that the exception may be the supporting or the attacking child: in the former case the default will be reject, and in the latter the default will be accept. Thus:

Accept Parent if Support AND (NOT Attack).
Reject Parent.
Reject Parent if Attack AND (NOT Support).
Accept Parent.

For F110, F120 and F121 the attacking child is the exception, while for F122 and F123 the supporting links are the exceptions. This leaves seven nodes. For F200 we regard the attacking link as an exception to the case where the conjunction of the supporting links holds.

For F104 and F112 we see the supporting links as offering disjoint ways of accepting the parent, and the attacking child as a way of establishing that the factor is not present. We default to yes because in many cases there are no factors for either side present relating to this point. We take it that this factor was often simply accepted on the facts and uncontested, and so there was no discussion on the point. A full description of all the truth conditions is given in Al-Abdulkarim *et al.* (2015).

4. PROLOG PROGRAM

The Prolog[2] program was formed by ascending the ADF, rewriting the acceptance conditions as groups of Prolog clauses to determine the acceptability of each node in terms of its children. The tests were restated using the appropriate syntax, with some reporting to indicate whether the node is satisfied (defaults are indicated by the use of "accepted that"), and some control to call the procedure to determine the next node, and to maintain a list of accepted factors. Examples can be found in Al-Abdulkarim *et al.* (2015).

Each of the tests in the acceptance condition is applied in a separate clause, using the set of factors currently identified as present in the case, before proceeding to the next factor, with the current factor added to the applicable factors if it is accepted. To allow completion of the database (Clark, 1978), a final clause is added to catch any case not covered by any of the preceding clauses. These defaults may favour either side. In some cases, the default is accept because few case descriptions related to these abstract factors, although they are a *sine qua non* for any claim. Our belief is that these aspects were uncontested and so the factors were not explicitly discussed in the trial, and so do not appear in the CATO analysis. Where we felt it was clear that the factor needed to be explicitly established, the default was reject.

The above demonstrates that it is a straightforward and reasonably objective process to transform a factor based analysis such as is found in (Aleven, 1997) to an executable program via an ADF. Although judgement was sometimes required to form the acceptance conditions, we would suggest that such judgements were not difficult to make. Moreover if there are difficult choices, the effect of the alternatives can be compared on a set of test cases. Overall the relatively small number of factors relevant to particular nodes greatly simplifies the task.

---

[2] Prolog was used because of its closeness to the acceptance conditions, and made the implementation quick, easy and transparent.

5. RESULTS

We can now run the program on the cases. We represent the cases as a list of base-level factors. For example, the Boeing case[3] is represented as case(boeing,[f4,f6,f12,f14,f21,f1,f10]).

giving output:
1 ?- go(boeing).
accepted that defendant is not owner of secret
efforts made vis a vis outsiders
efforts made vis a vis defendant
there was a confidentiality agreement
defendant was on notice of confidentiality
there was a confidential relationship
accepted that the information was used
questionable means were used
accepted that the information was not available elsewhere
accepted that information is not known
accepted that the information was neither known nor available
accepted that the information was valuable
not accepted that the information was legitimately obtained
improper means were used
efforts were taken to maintain secrecy
information was a trade secret
a trade secret was misappropriated
find for plaintiff
boeing[f200, f201, f203, f102,f110,f104,f111, f112,f114,f115,f121,f122,f123, f4,f6,f12,f14,f21,f1,f10]
decision is correct

The initial program correctly classified 25 out of the 32 cases (78.1%). While all ten of the cases won by the defendant were correctly classified, seven of the 22 cases won by the plaintiff were not. The figure for correct answers is remarkably close to the 77.8% reported for the version of CATO used in (Brüninghaus and Ashley, 2003), which, of course, uses exactly the same analysis of the domain and cases that we have adopted here. Thus as a first conclusion we can tentatively suggest that executing the analysis in Aleven (1997) as an ADF produces very similar results to those obtainable using the original CATO program (albeit we are using a smaller set of cases). We can now investigate how the initial program might be improved.

---

[3] The Boeing Company v. Sierracin Corporation, 108 Wash.2d 38, 738 P.2d 665 (1987).

The wrongly predicted cases were:

```
case(spaceAero,[f8,f15,f18,f1,f19]).
case(televation, [f6,f12,f15,f18,f21,f10,f16]).
case(goldberg,[f1,f10,f21, f27]).
case(kg,[f6,f14,f15,f18,f21,f16,f25]).
case(mason,[f6,f15,f21,f1,f16]).
case(mineralDeposits,[f18,f1,f16,f25]).
case(technicon,[f6,f12,f14,f21,f10,f16,f25]).
```

Examination of the cases showed that five of the seven had F16 (ReverseEngineerable) present and that these cases were the only cases found for the plaintiff with F16 present. The problem in these five cases is that the program finds for the defendant because the information is available elsewhere (F105). This is established by the presence of ReverseEngineerable and is unchallengeable. Examination of the ADF shows that F16 is immediately decisive: if that factor is present, there is no way the plaintiff can demonstrate that the information is a trade secret. Goldberg[4] also fails through F105 (information known or available), since disclosure in a public forum (F27) is sufficient to deny the information trade secret status. It would appear that we could significantly improve performance by refining this branch to allow the plaintiff some way to defend against, in particular, F16. See Al-Abdulkarim *et al.* (2015) for discussion of the texts of the decisions and possible refinements. The refined program can equal or better the performance of any of the existing systems.

5.1 Discussion

By using the ADF we can readily explain the points at which the acceptance conditions do not concur with the decisions taken in the actual cases. We can then return to the original decisions and use them to determine possible refinements to the representation. In some cases, the problem seems to lie with the attribution of the factors. Such matters were contested in the actual case, and ascribing the presence or absence of particular factors requires interpretation of the case by the analyst. The interpretation cannot be disputed without descending to the level of facts as advocated by Atkinson *et al.* (2013) and Al-Abdulkarim *et al.* (2014). Addition of the fact layer has been the subject of work subsequent to that reported here. Other decisions suggest that we may wish to modify the description of factors intended to guide the

[4] Goldberg v. Medtronic, 686 F.2d 1219 (7th Cir. 1982).

analyst. Adding or removing a factor to or from a particular case provides a local solution which will solve a problem with a particular case. Our results, however, indicated a general problem which was applicable to several cases: the dominant affect of F16, reverse engineerable. It seemed clear to us that the presence of F16 should not by itself be sufficient for a finding for the defendant. Again the decisions themselves suggested several possible ways of arguing against F16: in particular the use of restricted materials and the uniqueness of the product. Either or both of these exceptions could be incorporated in the ADF without adversely affecting any of the test cases, but we would need to have a reasonably large set of new cases in order to evaluate the different solutions and to guard against over fitting.

Finally it should be conceded that the decisions themselves may be erroneous. Assuming that there are least some poor decisions which we would not wish to serve as precedents, we should be willing to tolerate a certain number of divergences from our results.

To summarise:

- Simply translating the analysis of (Aleven, 1997) into an ADF and executing the resulting program gave results almost identical to those found for CATO in the IBP experiments reported in Brüninghaus and Ashley (2003). Note that this is achieved without need for balancing of pro and con factors central to existing case based reasoning systems.
- The reasons for the "incorrect" decisions can be readily identified from the output and the ADF, as we saw from the discussion of the wrongly decided cases above.
- Examination of the texts of the decisions readily explained why the results diverged, and suggested ways in which the analysis could be improved, either at the case level by changing the factors attributed, or at the domain level by including additional supporting or attacking links.

From this we conclude that use of ADFs provides good performance, and has a number of positive features from a software engineering (and domain analysis) standpoint, which would enable the ADF to be refined and performance improved. We also believe that we do need to include a fact layer to permit increased transparency in the ascription of factors to cases.

6. QUALITY OF EXPLANATIONS

As the Prolog program proceeds it reports on the acceptability or otherwise of the various abstract factors and the resolution of issues. As shown above, this provides an excellent diagnostic for divergent decisions, but how does it measure up the actual decisions found in cases?

Of course, without facts, we will not be able to follow the decision very closely. But consider a reordering of the elements of our decision for, say Boeing. We also omit some elements, and add a little linking text. Recall too that we wrote the program used thus far to "decide" the cases: in a version to supply explanation we would want to customise the text reports to indicate the particular clause being used for a node by giving the base level factors used. Below is what a decision might look like: we show the current program output in boldface, possible clause-specific customisations in italics and linking text in ordinary font.

> We **find for plaintiff**. The **information was a trade secret**: **efforts were taken to maintain secrecy**, *since disclosures to outsiders were restricted and the defendant entered into a non-disclosure agreement and other security measures were applied.* The **information was unique**. It is **accepted that the information was valuable** and it is **accepted that the information was neither known nor available**.
> **A trade secret was misappropriated**: **there was a confidential relationship** *since the defendant entered into a non-disclosure agreement* and it is **accepted that the information was used.**
> Moreover **improper means were used** *since the defendant used restricted materials.*

This seems to have the makings of a reasonable explanation. There are two problems: it does not indicate what the defendant contended, since the clauses of the program which were not reached do not feature in the report, and, of course, the facts on which the finding are based are not present. None the less, we find the output a distinct improvement on previous work such as Chorley and Bench-Capon (2005). We believe that the output from the current program could be readily used to drive a program of the sort envisaged by Branting (1993), and that this will become even more useful when we have added a fact layer to allow the explanation of the attribution of factors.

7. APPLICATION TO A SECOND DOMAIN

In the above we have considered the approach with respect to a single domain. If the approach is to be of general significance, however, it needs to be applicable to other domains. This section describes a further exercise designed to show that the approach is more generally applicable. We will apply the method to a domain which has often been used as an illustration of factor based reasoning: the wild animals cases and *Popov v Hayashi*. The wild animals cases were introduced into AI and Law in Berman and Hafner (1993) and extended to the baseball case of Popov in Wyner *et al.* (2007). We will use the factor-based analysis of Bench-Capon (2012) as our starting point.

Briefly the wild animals cases concern plaintiffs chasing wild animals when their pursuit was interrupted by the defendant. Post was chasing a fox for sport. Keeble was hunting ducks, Young fish and Ghen a whale, all in pursuit of their livelihoods. *Popov v Hayashi* concerned disputed ownership of a baseball (valuable because it had been hit by Barry Bonds to break a home run record). Popov had almost completed his catch when he was assaulted by a mob of fellow spectators and Hayashi (who had not taken part in the assault) ended up with the baseball when it came free. The wild animals cases were cited when considering whether Popov's efforts had given him possession of the ball.

Thirteen, base-level, factors are identified in Bench-Capon (2012). The first task is to form them (together with appropriate abstract factors) into a factor hierarchy, to use as the node and link structure of our ADF. This factor hierarchy is shown in Figure 3: some adaptations have been made; for example, we include a factor Res (Residence Status) to indicate the attachment of the animals to the land, since it appears to make a difference whether they are there permanently, seasonally, habitually, occasionally, or whatever. The nodes and links are given in Table 5.

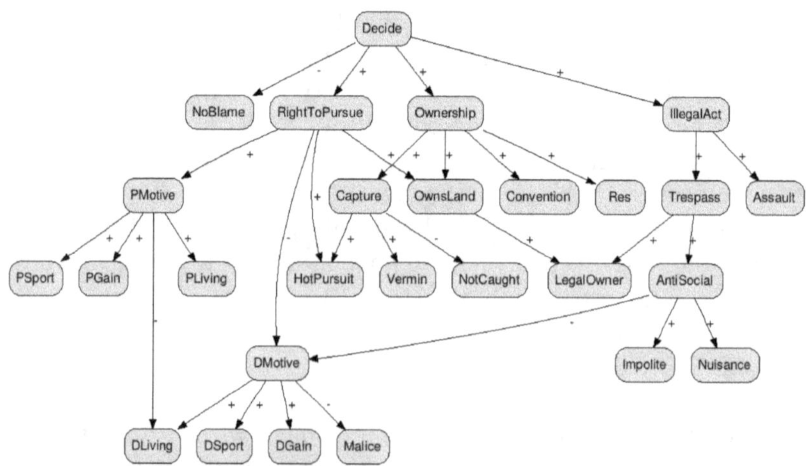

Figure 3-Factor Hierarchy/ADF for Popov

S	L+	L-
Decide	Ownership,RightToPursue,IllegalAct	NoBlame
Capture	HotPursuit, Vermin	NotCaught
Ownership	Convention, Capture, OwnsLand, Res	
PMotive	Pliving, PSport, PGain	DLiving
DMotive	DLiving, DSport, DGain	Malice
OwnsLand	LegalOwner	
RightToPursue	OwnsLand, Pmotive, HotPursuit	DMotive
AntiSocial	Nuisance, Impolite	DMotive
Trespass	LegalOwner, AntiSocial	
IllegalAct	Assault, Trespass	

Table 5-Popov as ADF

We now supply acceptance conditions for the nine non-leaf nodes.

1. Decide for Plaintiff if NOT (NoBlame) AND (Ownership OR (RightToPursue AND IllegalAct))
2. Ownership if (OwnsLand AND Resident) OR Convention OR Capture
3. Capture if NOT (NotCaught) OR (Vermin and HotPursuit)
4. RightToPursue if OwnsLand OR (HotPursuit AND PMotive AND (NOT (better) DMotive))
5. PMotive if PLiving OR (PSport OR PGain) AND (NOT DLiving)
6. DMotive if NOT Malice AND (DLiving OR DSport OR DGain)
7. IllegalAct if Trespass OR Assault
8. Trespass if LegalOwner AND AntiSocial
9. AntiSocial if (Nuisance OR Impolite) AND (NOT DLiving)

The only real controversy here is with the determination of Right to Pursue when both the plaintiff and the defendant have good motives. Essentially we want to say that if the land is not owned by one of them, the right to pursue is given to the party with the better motive. The remainder seem fairly uncontroversial. The acceptance conditions can easily be expressed as Prolog procedures and then embedded in code as was done for CATO. We can now execute the program. Running the case for Young v Hitchens produces the output (note that the program abbreviates factor names):

```
1 ?- go(young).
the plaintiff had not captured the quarry
the plaintiff did not own the quarry
plaitiff has good motive
defendant has good motive
```

plainiff did not own the land
plainiff had a right to pursue the quarry
defendant committed no antisocial acts
defendant committed no trespass
no illegal act was committed
do not find for the plaintiff
find for the defendant
young[rtToPursue,dMotive,pMotive,nc,hp,imp,pliv,dliv]

We produce correct results from all five cases discussed in Bench-Capon (2012), and on this basis we believe that the ADF representation can be used to encapsulate the knowledge of the domain. We cannot evaluate it as a decision making program since there are insufficient cases available, but this suggests that the method can be applied straightforwardly to a second domain to construct an executable program. In general we believe that the method can be applied to any domain for which factor based reasoning in the CATO (or HYPO or IBP) style is appropriate. This has encouraged us sufficiently to attempt to apply the method to a larger scale problem in the domain of the US automobile exception to the fourth amendment rule for which there is no accepted analysis into factors available, so we that need to start from the case decision texts: we will also incorporate a fact layer in this domain. This is the subject of the next stage of our project.

8. CONCLUDING REMARKS

In this paper we have evaluated an approach to reasoning with legal cases described in terms of factors using Abstract Dialectical Frameworks, as described and advocated in Al-Abdulkarim *et al.* (2014). We find that:

- The success of the implementation depends to a large extent on the quality of the analysis. Making a direct translation of the analysis of CATO (Aleven, 1997) yields a success rate almost identical to that found for CATO in (Brüninghaus & Ashley, 2003). This is a creditable 78.1% of the cases decided "correctly".

- The ADF does, however, provide very transparent output that identifies precisely where the outcomes suggested by the implementation diverge from the actual outcomes. Now reading the original decision texts suggests one of four solutions. These are, in ascending order of divergence from the original analysis:

1. Removing a factor wrongly attributed to the case
2. Adding a factor wrongly omitted from the case
3. Modifying an acceptance condition: e.g. changing the priorities

4. Modifying the ADF: e.g. adding a supporting or attacking node for the problem node. Often several of these modifications can potentially solve the problem, and the choice is made according to the context provided by the other divergent cases we are trying to accommodate.

- The ADF approach provides a good way of using a set of test cases to refine an initial analysis.
- The output from the program provided good diagnostics and a reasonable explanation of the outcome. Our output does, however, currently lack the citations and facts which are prominent in actual decisions.
- The method emphasises reasoning with portions of precedents, rather than whole cases. We believe that this does correspond to legal practice as manifest in real decisions.
- The method can be applied to different domains. We believe that any domain for which factor based reasoning would be appropriate would be amenable to this method.

We find all of this encouraging. The next important step will be to extend the method to the fact level, so as to permit argument about the ascription of factors, and to be able to ground our explanations in the particular facts of a case. Once the method has been extended to include the facts of particular cases at the lowest level of the ADF, a program to present the output in a form resembling the texts of actual decisions can also be considered.

REFERENCES

Al-Abdulkarim, L., Atkinson, K., & Bench-Capon, T. (2014). Abstract dialectical frameworks for legal reasoning. *Proceedings of Jurix 2014*, 61–70.

Al-Abdulkarim, L., Atkinson, K., & Bench-Capon, T. (2015). Evaluating the Use of Abstract Dialectical Frameworks to Represent Case Law. *Proceedings of the Fifteenth International Conference on Artificial Intelligence and Law*, 156–160.

Aleven, V. (1997). *Teaching case-based argumentation through a model and examples.* PhD thesis, University of Pittsburgh.

Ashley, K. (1990). *Modelling Legal Argument: Reasoning with Cases and Hypotheticals.* Cambridge, MA. : MIT Press.

Atkinson, K., Bench-Capon, T., Prakken, H., & Wyner, A. (2013). Argumentation schemes for reasoning about factors with dimensions. *Proceedings of JURIX 2013*, 39–48.

Bench-Capon, T. (2012). Representing Popov v Hayashi with dimensions and factors. *Artificial Intelligence and Law, 20*(1), 15–35.

Berman, D., & Hafner, C. (1993). Representing teleological structure in case-based legal reasoning: The missing link. *Proceedings of the Fourth International Conference on Artificial intelligence and Law*, 50–59.

Branting, L. K. (1993). An issue-oriented approach to judicial document assembly. *Proceedings of the 4th ICAIL*, 228–235. ACM.

Brewka, G., Strass, H., Ellmauthaler S., Wallner, J., & Woltran S. (2013). Abstract dialectical frameworks revisited. *Proceedings of the Twenty-Third international joint conference on Artificial Intelligence*. AAAI Press.

Brewka G., & Woltran S. (2010). Abstract dialectical frameworks. In *Principles of Knowledge Representation and Reasoning: Proceedings of the Twelfth International Conference*.

Brüninghaus, S., & Ashley, K. (2003). Predicting outcomes of case-based legal arguments. In *Proceedings of the 9th ICAIL*, 233–242.

Chorley, A., & Bench-Capon, T. (2005). Agatha: Using heuristic search to automate the construction of case law theories. *Artificial Intelligence and Law*, *13*(1), 9–51.

Clark, K. L. (1978). Negation as failure. *Logic and data bases*, 293–322. Springer.

Dung, P. M. (1995). On the acceptability of arguments and its fundamental role in nonmonotonic reasoning, logic programming, and n-person games. *Artificial Intelligence*, *77*, 321–357.

Prakken, H. (2010). An abstract framework for argumentation with structured arguments. *Argument and Computation*, *1*(2), 93–124.

Prakken, H., & Sartor, G. (1998). Modelling reasoning with precedents in a formal dialogue game. *Artificial Intelligence and Law*, *6*(2-4), 231–287.

Wyner, A. Z., Bench-Capon, T. J. M., & Atkinson, K. (2007). Arguments, values and baseballs: Representation of popov v. hayashi. *JURIX2007*, 151–160.

3

Non-verbal Arguments in *War Requiem*

JESÚS ALCOLEA-BANEGAS
Universitat de València, Spain
jesus.alcolea@uv.es

Groarke has responded to the possibility of musical arguments, providing some relatively simple examples of ads, in which non-verbal sounds become "flags" that act as resources in argumentative exchanges. Our goal is to show that in the film *War Requiem* Derek Jarman tried and succeed in getting non-verbal arguments against war, and that the strength of the beautifully articulated images, music and other sounds can justify and persuade us of the futility of war.

KEYWORDS: analogy, Derek Jarman, example, multimodal argumentation, non-verbal argument, pragma-dialectics, visual rhetoric

1. INTRODUCTION

The study of visual imagery from a rhetorical perspective has grown with the emerging recognition that visual images provide access to a range of human experience not always available through the study of discourse. Thereby, in argumentation theory it has been argued that images and other non-verbal elements can play an important role in laying out the arguments. A quarter-century ago, Willard (1989, p. 8) wrote that "arguers, like all communicators, employ the full range of available communication modalities, verbal and nonverbal, explicit and implicit". The theory he defended

> defines argument in terms of encounters based on dissensus and regards any communications occurring therein as objects of epistemic and critical interest. It locates argument's epistemic effects in the total package, not simply in implicatures among statements.

D. Mohammed & M. Lewiński (eds.) (2016). *Argumentation and Reasoned Action: Proceedings of the 1st European Conference on Argumentation, Lisbon, 2015. Vol. II, 181-193.* London: College Publications.

Between the non-verbal elements, it could be controversial to include musical elements. However, Groarke (2003) has responded to the possibility of musical arguments inspired by the pragma-dialectics and some examples of advertisements, where the non-verbal sounds become "flags" that act as resources in argumentative exchanges. Interest in the subject has continued as evidenced by a recent issue of *Argumentation* (Kjeldsen (Ed.), 2015), dedicated to the study of the visual and multimodal argumentation, in which Groarke (2015) tries to demonstrate that the modes of arguing can be useful contributions when analysing acts of arguing, many of which are intrinsically multimodal.

Our goal is to show that, in his film *War Requiem* (1989), Derek Jarman (1942-1994) weaves concrete arguments against the war with non-verbal resources. From it, we aim to contribute to the idea that no-verbal or multimodal argumentation must have a right place in the studies of argumentation. The strength of the images beautifully articulated can persuade us of the futility of war, especially for young people, to the extent that they suffer the conflict. As usual in his films, Jarman fuses elements from different universes: war poetry by Wilfred Owen (1893-1918); the eponymous composition by Benjamin Britten (1913-1976), *War Requiem* (1962) —with the Latin Mass interwoven with some poems of Owen—; paintings by Piero della Francesca and some British artists; the biblical world; silent movies; documentaries, and newsreels, without forgetting certain homo-erotic overtones, given his homosexuality, shared by (the work of) Owen and Britten. This in not casual because visual rhetoric requires human action either in the process of creation or in the process of interpretation.

2. A FILM WITH CRITICAL MEANING

Jarman's moving anti-war appeal and homage to Owen and Britten's works is so relevant today as it was on its release in 1989. Jarman shoots a cinematic portrait of the World War I and its consequences in an urgent and emotional manner, creating a visual accompaniment to Britten's impressive choral masterpiece.

Jarman is not making assumptions about historical reality. He is interested in the consequences of the conflict for human beings. He insists on the primacy of the image to create not historical but critical meaning. Owen is at the forefront of the historical process as a referent and a fictional character as reported in his poems included in the Requiem's lyrics. Through his eyes, we see and feel conflict. Doing so, the film is more a structure to be interpreted than a structure to be faithful to reality, although it invokes this reality using some evidences.

Jarman wants us to think and to understand through *images that argue for us against war*, that challenge us in present issues: war, militarism, ideology, nationalism, pacifism, individual identity, etc.

Reminiscent from the golden age of silent, the film stirs the emotions not only through a classical narration, but through the selection, framing, and juxtaposition of still and moving images from film archives. Britten's *Requiem* is the soundtrack that overflows with language, sound effects and music the recreation of some of Owen's vital moments and the clips that generalize the pain of war. The images are gathered to create the broader and critical meaning of the film. Old photos and clips give us direct access to reality, mainly in the case of wounded soldiers from different wars. The indexical relationship to dead men provides an unmediated experience more direct than the sequences staged for Jarman's camera. The people in those photos and clips show as objective elements how times have changed, but the pain of war is still the same: death and destruction. The filmmaker also introduces and ideological subjective element already present in Owen's and Britten's work and life: homosexuality.

The import of the text integrated by the Latin Mass and poetry is clearly pacifist. Requiem's sonorities reinforce the sorrow of war's human waste until the concluding duet between two enemy soldiers as reported by Owen's "Strange Meeting" poem. As G. Watkins says (2003, p. 427), "Britten's music also clarifies the underlying sentiment of Owen's line from a private and highly personal perspective". In fact, "the war conferred on its fighting men the authority to cross the boundary of gender in the expression of affection" to the recognition of the common humanity in the sacrifice and in the conflict.

3. NON-VERBAL ARGUMENTS

It should be clear that with "non-verbal arguments" we are referring to those argumentative configurations in which images, gestures, music and other non-verbal sounds play an essential role, and that can intervene in the argumentative exchanges between actual or potential agents, to the extent that bring their meaning through their juxtapositions and for the conjoint way in which they operate. Therefore, non-verbal arguments would be those arguments we weave to persuade someone of a conclusion providing reasons for accepting it, and in which those elements play a prominent and eventually *irreducible* role.

It is useful to distinguish between the non-verbal elements that accompany an argument and those who are part of its content. Groarke (2003) notes that musical "flags" exploit the way a piece of music can set a tone, announce an occasion or comment on a situation, to the point of captivating us. That's why they can become important resources in argumentation, because without being themselves arguments contribute indirectly to orient us in the argument. Now, to show that music can play an essential role in an argument, it is necessary to find examples that demonstrate that music can directly provide reasons in favour of a conclusion. As we talk about a film, the soundtrack should play a key role in conveying the meaning of a scene or of the entire narrative. If this is achieved, then we should consider the possibility of analysing and evaluating the supposed argument, as it is the case with verbal arguments. Thus, as Groarke (2003, p. 420) says, "recognizing musical arguments allows us to extend the critical eye of argumentation theory to a new domain that has traditionally ignored". This can be done with instruments that pragma-dialectics makes available, to the extent that it highlights the *principles of communication* that govern the (indirect) speech acts included in argumentation (cf. Alcolea-Banegas, 2011).

According to pragma-dialectics, we weave an argument with the aim of defending a certain position, without that aim to be always explicit in practice, because the explicit realization of the argument as speech act "is the exception rather than the rule" (van Eemeren & Grootendorst, 1992, p. 44). Sometimes indirect speech acts constitute implicit assumptions or conclusions (of possible arguments) that can help us understand the content of non-verbal arguments supported, for example, by images or sounds. However, this is only possible when we use a "maximally argumentative interpretation" (van Eemeren & Grootendorst, 2004, p. 117) which ensures us that its argumentative function may be fully recognized. Since it is not always possible to ascertain the intentions of a person that holds a position or is arguing, we must be careful in interpreting the speech acts. For that matter, we turn to the pragma-dialectical principles of communication that allow to reach the meaning of indirect speech acts, and that stipulate that these acts should not be incomprehensible, insincere, superfluous, futile or inappropriately connected to other speech acts (van Eemeren & Grootendorst, 1992, pp. 50-52). Following Groarke (2002, p. 145; 2003, p. 421), we can accommodate these principles in a positive way to non-verbal arguments:

(1) The interpretation of the non-verbal elements in the argument is a communicative act in principle *understandable*.
(2) Images, music, sounds and all non-verbal elements should facilitate an interpretation that makes sense globally and makes clear its *internal coherence*.

> (3) Images, music, sounds and all non-verbal elements should facilitate an interpretation that makes clear its *external coherence*, i.e., in relation to the argumentative discourse they belong to, and to its context.

Therefore, in determining the specific meaning, we must find an interpretation of the non-verbal elements from an internal and external point of view and in a way that make sense of its formal qualities and the context in which they appear. It should be also noted that the context is important in the attribution of intentions to argumentative agents, intentions that are normally persuasive, as Willard says. In fact, that attribution is far more decisive than the illocutionary force that involves the formulation of an argument, and some attributions are based on non-discursive elements or contextual assumptions that are not present in the discourse. That's why he invites us to consider this argument (1989, pp. 108-109):

> [I]f we restrict argument to the exclusively discursive, we rule out interesting genres of persuasive communication. Television commercials, for example, combine music, dance, animation, special effects, visual images, and words to produce (...) claims on our attention and belief. Persuasive appeals can be designed so that words take their meanings from music or from visual effects. (...) persuasive appeals are claims on attention and belief; they are as logical or illogical as other claims. The intentions of arguers are often (if not paradigmatically) persuasive.

4. ARGUMENT BY EXAMPLE

In his *Rhetoric*, Aristotle holds that two common modes of persuasion are the *enthymeme*, a deductive argument with missing parts, and the *paradigm* or *example* (1393a24), a sort of induction. Good speakers use enthymemes rather than complete arguments because, Aristotle says, not everything needs to be stated, and the speaker must avoid saying what is obvious. However, in our opinion, films are better understood as examples or arguments by example.

How does the example work in persuasion? Aristotle tells us that examples are similar to an induction (*Rhet.*, 1393a26). In *Topics* (105a13-16), he offers this definition:

> Induction is a passage from particulars to universals, e.g. the argument that supposing the skilled pilot is the most effective, and likewise the skilled charioteer, then in general the skilled man is the best at his particular task.

Thus, the example can serve as a beginning of an induction to a general truth. In daily life, particular facts are more important and useful than universals. The particular establishes the universal by expressing *likeness*, for the vast majority of people "secure the universal by means either of induction or of likeness" (160a38-39). When we know the points of likeness, it is very easy to induce the universal.

So, although deduction is more cogent and more effective against argumentative opponents, induction, i.e., the use of examples, "is more convincing and clear", more easily "learnt by the use of the senses" and grasped by sense-perception. As a particular case, one can imagine or picture the example, what is probably meant "by the use of the senses" and that's why is available "generally to the mass of men" (105a17-19). In fact, "though one perceives the particular, perception is of the universal [and] instils the universal" (*An. Post.*, 100a17, 100b5). It could be argued, however, that it is just an isolated fact. We must not forget that an example is always an example *of* something, a sort of instantiation of a universal truth.

Aristotle says that enthymemes are based on different things, in particular on examples. When based on examples they "proceed from one or more similar cases, arrive at a general proposition, and then argue deductively to a particular inference" (*Rhet.*, 1402b13-1402b21). But the example is not a source of propositions. It just establishes a persuasive nexus through a relation of likeness that one can appreciate with the concurrence of sense-perception. If the audience believes the example, then the audience will also believe the universal truth. It is indifferent to use past facts or fictitious examples (*Rhet.*, 1393a28).

It is in this context important to notice that the fictional aspect of an example is irrelevant to its persuasive force. Although Jarman's ideas could be easily inserted in the framework of Rapoport's researches on war and peace, showing that he does not simply offer a fictitious example, he does the inductive generalization for his audience. That is, Jarman uses his example as a means of persuasion, and he draws out the lesson of his film quite explicitly, although it is rather easy if not necessary provided one can observe the likeness, as Aristotle points (*Top.*, 105a25). Could not Jarman have also thought about argumentation without explicitly telling us so? Indeed, that *War Requiem* is a film against war and that its main character's position is in consonance seems incontrovertible. Reading the film with some ideas from multimodal argumentation theory is not a departure from the narration, but a new approach on it. Thereby, we hope to be in harmony with the filmmaker's intentions.

5. "THE TRUTH NEVER SAID, THE PITY OF WAR" IS SHOWN AND ARGUED

Agreeing with Willard and with the idea that the media often combine a variety of images of joyful aesthetic, we must add that the concatenations of music, pictures and texts should be understood as arguments that use the referred principles of communication to uncover its implicit propositional content through an adequate interpretation. In fact, when an audience goes to see a film, in some way they are participating in it, insofar as they become emotionally involved (cf. Alcolea-Banegas, 2009, p. 260). The soundtrack is meant to multiply the effect of the image. Japanese filmmaker Akira Kurosawa (1983, pp. 107-108) had as his pet theory that

> cinematic strength derives from the multiplier effect of sound and visual image being brought together. (...) Depending on how the sound is put in, the visual image may strike the viewer in many different ways. (...) The sound powerfully altered the visual image to create a whole new impression.

In our case, Britten's *Requiem* enhances the emotional content by providing the voices and sound effects of Jarman's film. We are permeable to its strength, delicacy and softness, to its sharp, heavy or light qualities, in a way that we are impressed by the meaning of the scenes and we can understand more easily the filmmaker's argumentative strategies. According to this and the aforementioned principles, in the interpretation of the filmic text, viewers should develop a sense of the film applying their own experience, collaborating in assigning particular meaning, and taking into consideration

> 1) the knowledge of the *context*, but not so much of the context from which the film emerged as of the context from which can be rebuild, exposing an argument by example, by analogy, etc.;
> 2) its *intertext*, i.e. from the relationship, dialogue or dependency that there is among the text to be analysed (film) and other texts; and
> 3) the *enunciative strategies* articulated by the filmmaker, counting among them the ones that allow to arouse emotions, i.e. rhetorical strategies.

So attending to other sensory registers and forms of intelligibility can a richer understanding of the film emerge from an argumentative standpoint. When we do so, we can understand that sounds and images immerse us in a multisensory environment, one pressing with strong affective and persuasive forces that extend beyond our immediate cognitive approach. Thus, one can develop the ability to defend the existence in the film of a non-verbal argument that can move.

Considering the three perspectives on the problem of war collected by Rapoport (1960, p. 14), the overall narrative structure of *War Requiem* contains a *non-verbal argument by example* on how war causes physical suffering and moral anguish to death. Jarman takes the *cataclysmic* perspective on war without verbalizing it. There is no difference between enemy combatants —Owen and the German Soldier—, against a *political* perspective on the war, coercive and enforcing, as it is highlighted in the scene in which the German Soldier is charmed to hear the Unknown Soldier interpret a piano melody, which later give way to play with snowballs. At that moment, the schizophrenia of war motivates the unintended death of the German Soldier at Owen's hands. When Owen dies in his turn, and goes to hell, because Heaven has closed its eyes to the victims, he finds his victim as a friend "with compassionate eyes". In hell, everyone gathers around a pond, and the atmosphere is not very different from the trenches. The resemblances are seeing immediately and the metaphor, rhetorical figure, bursts with its persuasive effects in support of the non-verbal argument by example. The trenches with their shells and gases *are* hell. The hell with its despair and its abolition of time *is* the trenches. The pity of war: death of joy that all might have shared. At this point, it is denounced and masterfully refuted the *eschatological* perspective on war. That is, the old ideology related to the privilege that is to die for the fatherland and to become part of History is reduced to a simple truth: to die for one's country is not sweet nor honourable. Dying for one's country is painful and distressing. Owen's declarations go in line with the sham of war in hands of supposed patriots and leaders who do not value human life and seek only their own interests, but also in the moral problems posed to the soldiers to recognize in the enemies equals who suffer and share the same fate.

So much for Jarman's argument by example, an argument that follows a clearly rhetorical strategy, according to its typology. Therefore, it is hardly difficult to generalize in a straightforward way. However, after the representation of Owen's death, to which we will return later on, Jarman's argumentative craftiness is shown in the successive collation of real sequences from all the wars that came after the Great War. The *Libera Me*, from Britten's *Requiem*, that has not stopped a moment along the narration, sounds along these real sequences, including two inserts with two choirs singing parts of requiem. This complementary strategy between images and music launches the content of what is show to the viewer, emotionally insisting on it, as a work of art and as an argument against any war.

They are real wars with real victims who lie in graves, real soldiers who lead others to the grave, or in which a soldier is mortally wounded, while we can see the suffering and anguish of his colleagues, in a clear and not hasty generalization of Owen's "sacrifice". Thus the mythical narrative, that covers the sacrifice, gives it meaning (ironically, historical meaning), connects it with other examples and places the event in a meaningful chain: all wars serve to death of the youth. The only difference is that, in Owen's case, the perpetrators or beneficiaries of these deaths will be pointed —the rich and their travelling companion, the Anglican Church— in a sequence with curious analogies.

6. ARGUMENT BY ANALOGY

Analogies are comparisons between things, states of affairs, etc. that highlights aspects in which they are thought to be similar. According to van Eemeren and Grootendorst, the comparison relation is presented "as if there were a resemblance, an agreement, a likeness, a parallel, a correspondence or some other kind of similarity" (1992, p. 97). Analogies can help make an unfamiliar concept easier to understand by comparing it to a familiar one. The *argument by analogy* asserts that if the facts relating to A and the facts relating to B are alike in certain known respects, they will be alike in another respect.

The argumentation process that facilitates analogy is based on the abstraction of a generic property from two comparable events, whose connection must be defended. As a contextually essential trait of those events, that property is relevant to the achievement of the communicative intention, to the force of the analogy and the efficacy of the argument by analogy.

We can use an analogy to explain and make vivid the features of an object or a state of affairs. We must notice that the content of an analogy is in some way visual and not completely propositional. In fact, varieties of analogy are essentially visual. The analogy carries sufficient insight to "see" what types of adjustments must be carried out between cases where there are no clear similarities in order to achieve a certain harmony, but it also presupposes a criterion to discover the relevant similarity in the disparate things. The argument is effective with the audience when the object we are making the analogy to is one that has a significant impact and that adds some emotional resonance.

We can also adduce it in order to support a claim with the purpose to persuade and induce beliefs and attitudes in an audience. In our case, the filmmaker passes on the viewers the task of interpreting a sequence and working out which are the ways in which both situations are alike. The filmmaker illustrates by means of a comparison and, by reflecting on it and upon accepting the analogy, viewers may discover new knowledge on both situations, mainly on the real one. This is the genuine value of the analogy: the audience is oriented on the real situation (cf. Ribeiro (Ed.), 2014).

In *War Requiem*, a much more concentrated focus of interpretation is occasioned by the figure of Abraham as an Anglican bishop. If we value the artistic expression and privilege it as a kind of argumentative discourse from which we gain something unavailable to other modes of discourse, the interpretation may become essential for the better argument to be found. So, while our thought might analogize the sequence taking the clue from Jarman's background, the film includes a rhetorical strategy that points contextually to the analogy and assigns the property from one context to another: what war was for Owen, AIDS was for Jarman. Thus, it certifies the film as an *argument by analogy*, multiplying the argumentativeness of the filmic text.

In this way, multimodal argumentation is lodged inside the needs of the filmic art and of the history or the real events. And so it is, because analogies function in a certain context and with certain purpose. Both situations, fictitious or real, resemble each other. Jarman makes vivid the most relevant features of the event. Starting with the knowledge based on the consequences of war to the youth/Owen, the audience is able to find resemblances in the AIDS crisis/Jarman. This allows us to not only identify each implicated group, but also to assign each one a meaning within the broader context and to reach a conclusion in relation to the crisis, changing eventually our beliefs and attitudes.

7. THE PARABLE OF THE OLD MAN AND THE YOUNG

In the reformulation of the story of Abraham and Isaac made by Owen in "The parable of the Old Man and the Young", Abraham stands for the rulers of Europe and Isaac is a typical soldier, representative of all the young men slaughtered. The phrase "one by one" from the poem sung by tenor and baritone emphasizes how mass slaughter is made up of individual deaths. That's why Abraham instead of offering the "Ram of Pride" "slew his son" accentuating his dull perverseness. The deviation from the biblical story is more significant than the similarity: the knowledge of it elevates audience expectations, furnishing a moral setting for the characters to be examined, and fixing the analogy with the AIDS crisis.

Jarman represents Owen's death in the trenches at the hands of Abraham as an Anglican bishop. Simultaneously, we see how happily the supercharged bankers and industrial capitalists who sent their sons to war are applauding. The sequence takes place on two levels: Owen writing the story and experiencing his own death. After this, he is transported by the Unknown Soldier depicted as Christ. Meanwhile the music creates a strong juxtaposition between the optimism of a quasi-sacrifice that strengthens Abraham's ties with God and the grim horror of the death of a young man at the hands of their spiritual father. Hence, the reappearance of Old Soldier in the *Offertorium* suggests that he could be, not Abraham, but Isaac at the end of his days. In Owen's poem, Isaac dies in line with the pessimism that the poet felt about the future of the war as a total cataclysm. In the film, another Isaac, Jarman's Old Soldier, is still alive. Thus, the filmmaker is closer than the poet to the Scripture, and is more faithful to what really happened in the Great War: many of those who fought survived.

However, the retrospective intertextuality of this exquisite sequence is perfectly coherent with the cause-effect relation that reveals the story, making clear that youth violence exert on each other should be rewritten with the language of sacrifice that disguises what the old men do. Death and destruction are depicted as the necessary price of the triumph and the established order. In this sense, the sequence would point to the external context and would be translated as *a non-verbal argument by analogy*: just as the Church was on the side of those who sent their sons to the war, the Church also had a negative role in the AIDS crisis.

The poem "At a Calvary near the Ancre" shows how Owen (1994, p. 23) and many other soldiers were identified with the suffering figure of a Crucified damaged by bombing. In the film, Owen and the German Soldier are "united in Christ" by the nurse and the Unknown Soldier, carrying crowns of thorns. In the case of Owen and Jarman, the highly emotional identification with Christ meets the deep estrangement from the morality of the Church and its representatives. Owen's ambiguous statement about priests "flesh-marked by the Beast" would have resonated on Jarman, who would have been dismayed by the Church's response to AIDS. In fact, the director wrote on the published script (1989a):

> In my heart, I dedicate my film of *War Requiem* to all those cast out, like myself, from Christendom. To my friends who are dying in a moral climate created by a church with no compassion.

In 1987, as Peake (2011, p. 418) comments, impetus by an AIDS-related public antipathy towards homosexuality was high on the Conservative government's agenda. It was the time for the Clause 28 of Local

Government Act 1988, according to which, authority shall not "intentionally promote homosexuality or publish material with the intention of promoting homosexuality". Again, it was more than a mere coincidence in film and in reality.

8. CONCLUSION

Had to be clear that the filmic text contains visual formulations with metaphorical or symbolic meanings that are perfectly understandable both intertextual and contextually. Its strong impression of emotional and thematic coherence is complemented with music, giving rise to a whole with internal and external harmony. So the respect for the principles of pragma-dialectical communication can make it clear that our explanation and exemplification of non-verbal arguments can be a new approach to pursue studies of argumentation, while can afford to interpret and understand one of the strongest anti-war works of art with multimodal arguments much stronger than their verbal counterparts.

ACKNOWLEDGEMENTS: The author is grateful to the Spanish Ministry of Science and Innovation for supporting this work (Projects FFI2011-23125 & FFI2014-53164), and to M.C. Fuster and other speakers in the congress for their thoughtful suggestions.

REFERENCES

Alcolea-Banegas, J. (2009). Visual arguments in film. *Argumentation, 23*(2), 259-275.

Alcolea-Banegas, J. (2011). Teaching argumentation theory and practice: The case of *12 Angry Men*. In P. Blackburn *et al.* (Eds.), *Tools for Teaching Logics. Third International Congress, TICTTL 2011* (pp. 1-8). Berlin: Springer.

Aristotle (1991). *The complete works of Aristotle*. Princeton: Princeton University Press.

Aristotle (2007). *On rhetoric*. Oxford: Oxford University Press.

Groarke, L. (2002). Toward a pragma-dialectics of visual argument. In F. H. van Eemeren (Ed.), *Advances in pragma-dialectics* (pp. 137-151). Amsterdam: Sic Sat.

Groarke, L. (2003). Are musical arguments possible? In F. H. van Eemeren *et al.* (Eds.), *Proceedings of the Fifth Conference of the International Society for the Study of Argumentation* (pp. 419-422). Amsterdam: Sic Sat.

Groarke, L. (2015). Going multimodal: what is a mode of arguing and why does it matter? *Argumentation, 29*(2), 133-155.

Jarman, D. (1989). *War Requiem*. DVD. Kino.

Jarman, D. (1989a). *War Requiem. The Film*. London: Faber & Faber.

Kjeldsen, J. E. (Ed.) (2015). Special Issue: Visual and multimodal argumentation. *Argumentation, 29*(2).

Kurosawa, A. (1983). *Something like an autobiography*. New York: Vintage.

Owen, W. (1994). *The war poems*. London: Chatto & Windus.

Peake, T. (2011). *Derek Jarman*. Minneapolis: University of Minnesota Press.

Rapoport, A. (1960). *Fights, games and debates*. Ann Arbor: University of Michigan.

Ribeiro, J. J. (Ed.) (2014). *Systematic approaches to argument by analogy*. New York: Springer.

van Eemeren, F. H. & Grootendorst, R. (1992). *Argumentation, communication and fallacies*. Hillsdale: L. Erlbaum.

van Eemeren, F. H. & Grootendorst, R. (2004). *A systematic theory of argumentation*. Cambridge: Cambridge University Press.

Watkins, G. (2003). *Proof through the night*. Berkeley: University of California Press.

Willard, C. A. (1989). *A theory of argumentation*. Tuscaloosa: University of Alabama Press.

4

Bounded Agents and Epistemic Vigilance

J. FRANCISCO ÁLVAREZ
Dpto. Lógica, Historia y Filosofía de la Ciencia, UNED, Spain
jalvarez@fsof.uned.es

A great number of philosophical approaches to language are built on a standard notion of rationality as optimization. Other paths, the bounded rationality ones, could improve our understanding of argumentative process. The argumentative theory of reasoning (Sperber and Mercier) could be improved using an approach related to bounded rationality. In a huge part of economic studies the relevance of bounded rationality has been appearing, it is possible that it also happens in argumentation studies

KEYWORDS: Amartya Sen, argumentative theory of reasoning, A. Rubinstein, bounded rationality, D. Sperber, epistemic vigilance, H. Mercier, H. Simon

1.INTRODUCTION

The argumentative theory of reasoning as presented by Hugo Mercier and Dan Sperber is a major shift when considering the relationship between thinking, reasoning and language. In my opinion, it is in fact a reconsideration of the human model as the argumentative agent. From this perspective I think it could be helpful to review some of the factors on the rationality of agents that have been developed in the field of economic studies, and which have led to recent developments in the field of behavioural economics and economics psychology.

Dan Sperber and Deirdre Wilson (1987; 2000), proposing the principle of relevance as a pragmatic approach instead of Grice´s Maxims, pointed to the importance of addressing the model of human being that supported the respective theories. On my part, fourteen years ago I advanced that it made sense to see the discussion in light of reflections made on the rational agent models that had been discussed in economic theory.

D. Mohammed & M. Lewiński (eds.) (2016). *Argumentation and Reasoned Action: Proceedings of the 1st European Conference on Argumentation, Lisbon, 2015. Vol. II, 195-204.* London: College Publications.

Stated briefly, it is betting against the Olympic idea of rationality (Herbert Simon´s terms), in favour of bounded versions of rationality.

I will show that the argumentative theory of reasoning as presented by Mercier and Sperber in their paper "Why do humans reason? Arguments for an argumentative theory" (2011, p. 57) could be improved by using an approach related to the concepts of bounded rationality because non-optimal paths, satisfaction processes, are quite relevant to understanding, for instance, the epistemic vigilance device they propose.

2. MERCIER AND SPERBER ON EPISTEMIC VIGILANCE

Mercier and Sperber have stressed the need to overcome the idea that "reasoning is ... a means to improve knowledge and make better decisions ... the primary function for which it (reasoning) evolved is the production and evaluation of arguments in communication" (2011, p. 58). Even more, "Reasoning enables people to exchange arguments that, on the whole, make communication more reliable and hence more advantageous. The main function of reasoning, we claim, is *argumentative*" (Ibid., p. 60).

As they said in response to their commentators in "Argumentation: its adaptiveness and efficacy": "Ours is a contribution to the growing body of research showing how, and how much, the human mind is a social mind" (2011, p. 101).

> For communication to be stable, it has to benefit both senders and receivers; otherwise they would stop sending or stop receiving, putting an end to communication itself... But stability is often threatened by dishonest senders who may gain by manipulating receivers and inflicting too high of a cost on them. Is there a way to ensure that communication is honest? ... However, for most of the rich and varied informational contents that humans communicate among themselves, there are no available signals that would be proof of their own honesty. To avoid being victims of misinformation, receivers must therefore exercise some degree of what may be called epistemic vigilance. The task of epistemic vigilance is to evaluate communicator and the content of their messages in order to filter communicated information. (Mercier & Sperber, 2011, p. 60)

Mercier and Sperber (2011) have shown that over the last fifty years a lot of work in the psychology of reasoning has spoken about the limits of our reasoning, that:

> humans reason rather poorly, failing at simple logical tasks, committing egregious mistakes in probabilistic reasoning, and being subject to sundry irrational biases in decision making. This work has led to a rethinking of the mechanisms for reasoning, but not - or at least, not to the same degree - of its assumed function of enhancing human cognition and decision making. (Mercier & Sperber, 2011, p. 58)

3. MARCELO DASCAL AND LANGUAGE AS A COGNITIVE TECHNOLOGY

As I have suggested on other occasions (Álvarez 2002, 2005), the language we speak and use to transmit information or to show our feelings is somewhat more than a mere adequate means to codify and transmit information. A wider notion of our conception of language is very relevant to argumentation theories. Some of the ideas come both from the pragmatic turn in language studies and cognitive technology.

There is an interesting connection between the consideration of language as a cognitive technology and our models of human beings related to their capabilities as agents. That connection is also linked to the formulation of very basic aspects both of the theory of rational choice and of our characterisation as rational agents.

My interest is strictly philosophical; therefore I do not pretend to enter *in extenso* in technical problems of economics or linguistics studies. However, I will try to indicate that certain conceptual specifications, such as those referring to the difference between satisficing, optimality and maximization, can be particularly significant for an argumentative theory of reasoning.

The need to understand language, as a cognitive technology, from a wide concept of distributed cognition that allows attention to be paid to pragmatic problems stemming from its very specificity, have been proposed by some cognitive and computer sciences theorists who have developed the line of distributed cognition (Andy Clark).

All technologies that have been produced by human beings suppose, undoubtedly, important cognitive work. Although as Marcelo Dascal (2002, p. 35) has said, "the cognitive effort does not make it, a particular technology, a *cognitive technology*".

Dascal proposes a characterization of 'cognitive technology' (CT) as every systematic means - material or mental - created by humans that are significantly and routinely used for the performance of cognitive aims. By 'cognitive aims', he means either mental states of a cognitive nature (e.g. knowledge, opinion, belief, intention, expectation, decision, plan of action) or cognitive processes (e.g. perception, memorization, conceptualization, classification, learning, anticipation, the formulation of hypotheses, demonstration, deliberation, evaluation, criticism, persuasion, discovery) that lead to cognitive states or help to reach them.

The main objective of his proposal is to criticize the very idea of communication as a primary function of language. The old idea that language *serves to convey thought or other forms of cognitive content, but need not play any role in the formation of the thoughts it conveys.*

A related idea appears in the first chapter of Ariel Rubinstein´s *Economics and Language* (2000), "Choosing the semantic properties of language", when he proposes that "binary relations fulfil certain functions in everyday life" (p. 11).

Rubinstein shows how one of these functionalities, indication-friendliness (which enables the user to indicate any element in any subset of the grand set) appears if and only if it is a linear ordering. So, linear orderings are the most efficient binary relations for indicating every element in every subset.

The same takes place in the case of splitting a set: To introduce one new term so that the speaker can refer to a new set in as accurate a manner as possible (an average).

4. AMARTYA SEN: SATISFICING, MAXIMIZATION AND THE ACT OF CHOICE

The distinction made by Amartya Sen on several occasions, between maximizing behaviour and nonvolitional maximization due to the fundamental relevance of the act of choice, must be placed in a central position in analyzing maximizing behaviour (Sen, 1997, p. 745). It is particularly pertinent when we consider that the user selects information, takes part in conversation and inevitably adopts a decision: "A person's preferences over comprehensive outcomes (including the choice process) have to be distinguished from the conditional preferences over culmination outcomes given the acts of choice" (Sen, 1997, p. 745).

In natural sciences maximization occurs without a deliberate "maximizer", but when the choice is associated with some kind of responsibility, our ranking of outcomes can be changed.

> Choice functions and preference relations may be parametrically influenced by specific features of the *act* of choice (including the *identity* of the chooser, the *menu* over which choice is being made, and the relation of the particular

> *act* to behavioural social norms that constrain particular social actions. (Ibid., p. 746)

Sen warns us that "Whenever the act of choice has significance" the comprehensive analysis of outcomes can have very extensive relevance to problems of economic, political and social behaviour.

I claim that the consideration of language as a cognitive device (instrument) compels us to consider that the act of choice has decisive significance. For instance, research related to metacognition increasingly impels us to see more clearly that self/reference, the possibility of referring to yourself at the very heart of language, is an essential property of language.

The problem is not only reduced to the importance of introducing the act of choice, the process of choice in what is chosen, but moreover it is necessary to consider the act of choice as an inescapable act.

> A chooser, who may have to balance conflicting considerations to arrive at a reflected judgement, may not, in many cases, be able to converge on a complete ordering when the point of decision comes. If there is no escape from choosing, a choice decision will have to be made even with incompleteness in ranking (p. 746).

That connection between Rubinstein´s ideas and the proposal of Mercier and Sperber finds a substantial fulcrum in some methodological and epistemological elements that have appeared in Amartya Sen's works, particularly in Sen (1993; 1997; and 1999). Sen has repeatedly explained the importance of taking into account chooser dependence and menu dependence in preference relations. From this point of view, I mean that argumentative activities are dependent on arguers and we need an analysis of argumentative actions, with parametric dependence of speakers.

It is possible to wonder if the binary relations in which Rubinstein is interested are precisely the same, or a subset of, the permissible preference relations. Any of them must be reflexive, that is to say, that every alternative is seen to be as good as itself. Thus, according to Sen it can possible to establish some very interesting consequences. First, a best alternative must also be maximal, but a maximal alternative need not be best. In particular, this can occur when the set of best or optimal choices is empty but the maximal set, however, is not empty. A classic example, described by Sen, is given by one interpretation of the story of Buridan´s ass, but in a very interesting interpretation. The ass could not rank the two haystacks and had an incomplete preference over this pair. It did not, therefore, have any optimal alternative. Both x and y were maximal – neither known to be worse than any of the other alternatives. In fact, since each was also decidedly better for the donkey than its dying of starvation z, the case for a maximal choice is strong. As Amartya Sen says: "Optimization being impossible here, I suppose we could "sell" the act of choice of maximization with two slogans: (i) maximization can save your life, and (ii) only an ass will wait for optimization" (1997, p. 765).

I mean, that if we connect explicitly the binary relation with the function of choice and its binariness, the contextual dependence of menu becomes pertinent. And it seems that in the case of language one of the essential elements is precisely this kind of menu dependence. Therefore, these questions arise in optimality theories if they do not deal with the importance of distinguishing between optimization and volitional maximization.

Even more, the sequential order and its uses in solving problems is a well-known device in bounded rationality. So, the sequential selection, the definability, and the "language" that a decision maker uses to verbalize his preferences restricts the sets of preferences he may hold (Rubinstein, chapter 4, p. 55), proof that the act of choice must be included in the set of alternatives.

The philosophical interest in this issue appears when we try to study the conceptual relationship between maximization, optimization and satisfying. It is something similar to the approach seen in Rubinstein's Chapter 5, "On the Rhetoric of the game theory" centred on the analysis of the language used by game theorists:

> Words are a crucial part of any economic model. An economic model differs substantially from a purely mathematical model in that it is a combination of mathematical structure and interpretation. The names of the mathematical objects are an integral part of an economic model (Rubinstein, 2000, p. 72).

In this order, we can accept that something similar happens with our models of language. If we do not forget that "the language a decision maker uses to verbalize his preferences restricts the set of preferences he may hold" (Rubinstein, 2000, p. 55).

Rubinstein cautiously says he tries to take "a first step in a much more ambitious research program to study the interaction between economics agents with language as a constraint on agents′ behaviour, institution, communication, etc. However, dreams aside, the goal of this chapter is quite modest" (p. 57).

Simon (1990, p. 5) explained such a link, as follows: "Because of the limits on their computing speeds and power, intelligent systems must use approximate methods to handle most tasks. Optimality is beyond their capabilities; their rationality is bounded." As Winograd and Flores maintain, Simon does not contest the "rationalistic tradition", but only the version that implies perfect knowledge, perfect foresight, and optimizational criteria (Winograd and Flores, 1986, p. 22, see also O'Neill, 2005, p. 296).

The results developed by Sen show that we must pay attention to maximality because it has a wider scope than optimality, and the difference between maximal choice and optimal choice could be substantial whether or not there are choices in the optimal set.

Possibly, what is more important to debate here could be that although "maximization can be matched by an as if optimization exercise, this does not reduce the importance of broadening the focus from optimization to maximization" (A. Sen, 1990, pp. 767-778) .

It can be extremely enlightening to see how Amartya Sen closely links the notion of maximization to the "important and influential concept of satisficing developed by Herbert Simon, which has often been seen as nonmaximizing behaviour (...) The discussion of satisficing versus maximizing has been somewhat deflected by the tendency to identity maximization with optimization" (Ibid., p. 768).

Fourteen years ago I presented a paper, "Bounded Rationality in dialogic games", in Lugano, and I had already taken into account there Rubinstein's works that appear in *Economics and Language* and the comments that van Benthem, Tilman Börgers and Bruce Lipman made about Rubinstein in the final pages of the same book. Here I have tried to develop some of the ideas that were put forth there, and that I have continued working on, because the majority of approaches to dialogic interaction have been built on a very special model of human being, i. e., the rational optimizing decision maker. This is a very special agent that has at least three unbounded capabilities: this agent has, at any time, all possible information and computational abilities; has no limitations; and is able to achieve an optimal degree of communication with the constraints on and means (language) for a feasible set of actions.

Obviously, this is the situation even in Grice´s Cooperative Principle and in his four maxims. Although there are some scholars who use Grice´s view, Cooperative Principle and the four maxims, as an approach that goes beyond the standard vision of rationality, they are usually closely related to some kind of *substantive rationality* rather than to *procedural rationality* (using concepts coined by Herbert Simon):

> The former is concerned only with finding what action maximizes utility in the given situation, hence is concerned with analyzing the situation but not the decision maker ... Procedural rationality is concerned with how the decision maker generates alternatives of action and compares them. It necessarily rests on a theory of human cognition (Simon 1997: 18).

5. A NON-OPTIMAL EPISTEMIC VIGILANCE COULD BE BETTER

D. Wilson and D. Sperber, for example, insist on distinguishing their approach from other related views linked to norms or maxims. But when they proposed the first principle of relevance, the cognitive principle, they said (Sperber and Wilson 2000: 232):

The human cognitive system tends toward processing the *most* relevant inputs available" (emphasis is mine), and "what we do, essentially, is assume that she [the agent] will pay attention to the potentially *most* relevant stimulus, and process it so as to *maximize* its relevance (Ibid.: 232).

Perhaps, if we both remove "most" from this sentence and replace maximize with satisfy, this new set will maintain the mutual predictability that is necessary to achieve a real act of communication. However, we are not, at the same time, able to obtain a single, deterministic output from each dialogic interaction. But why is obtaining these unique results a real or an important goal?

In the first principle of relevance, the problem would appear to be a simple, careless formulation, but the difficulties are more acute when the authors formulate their second principle of relevance, which they name the *optimal relevance* of an utterance: "The Second, or Communicative, Principle of Relevance: Every utterance conveys a presumption of its own *optimal* relevance" (Ibid.: 234). However, if we speak about degrees of relevance, or levels of satisfaction with relevance, we will get quite a different picture of conversational or dialogic interaction. To start with, we need some pragmatic tools. We may understand this to be a simple nuance, but it is a decisive one. For instance, with the idea of optimal relevance, we would not be able to understand nonlinear order in relevance. However, with a quite different notion, the maximal level of relevance, we would obtain several possible equilibria in the action game of language. This plurality would be blurred if we used an optimal notion. Maximal relevance is opposed to optimal relevance. Even if all the maximals were identical, it would be very important to start with a pluralist approach to degrees of relevance. In other cases, the majority of the explanations that have been built on the relevance principle appear to be *ad hoc* reconstructions. For example, they are unable to incorporate either the bargaining process or the roles that the participants are playing. Therefore, in these linguistic reconstructions, the agent is like Laplace´s demon, who is able to know everything about the communicative action, even what the optimal relevance is.

As I said in that paper, a great number of philosophical approaches to language are also built on a standard notion of rationality that shares some kind of optimization idea and some kind of generic principle that speakers try to optimize. The idea is very similar to utility in neoclassic economic theory. If we try to understand the dialogic process only as a means to obtain an optimum of communication, whatever this is, we lost the main function of social interaction. I think the epistemic vigilance can be better understood as a procedural device, a rule of thumb that the participants in a dialogic interaction usually satisfy to some degree. With these ideas I have developed long ago, we are now making some digital tools to organize and improve our argumentative capabilities. For instance, we are building a free plugin that allows to incorporate Oxford-Style debates on your website from https://wordpress.org/plugins/oxford-debate/.

We are showing that the argumentative acts (as individual and social acts) are intertwined with heuristic tools (among them, the argumentative tools) and some meta-devices as epistemic vigilance could help to understand better our argumentative capabilities that conform human traits mainly if we use a non optimizational approach to argumentative procedures.

ACKNOWLEDGEMENTS: I want express that this work had not been possible without two research projects granted by Spanish research authorities: "La construcción de agentes argumentativos en las prácticas del discurso público" FFI2014-53164-P and FFI2011-23125, "La argumentación en la esfera pública: el paradigma de la deliberación". I have presented some of this ideas in several meetings and seminars. I want express particularly gratitude at participants in Conference of IASC, "Paradoxes of conflicts", Lecce (Italy) December, 2014, and Congress of Spanish Society of Logic, Methodology and Philosophy of Science, Barcelona, July 2015.

REFERENCES

Álvarez, J. F. (2002). El tejido de la racionalidad acotada y expresiva (The fabric of bounded and expresive rationality). In M. B. Wrigley (Ed.), *Dialogue, Language, Rationality: A Festchrift for Marcelo Dascal* (pp. 11-29). Campinas (Brasil): University of Campinas.

Álvarez, J.F. (2005). Bounded Rationality in Dialogic Interactions. *Studies in Communication Sciences: Argumentation in Dialogic Interaction* (special issue), 119-130.

Dascal, M. (2002) Language as a cognitive technology. *International Journal of Cognition and Technology* , 1 (1), 35-61. Accessed 28 September 2015

Mercier, H. & Sperber, D. (2011). Why Do Human Reason? Arguments for an Argumentative Theory. *Behavioral and Brain Sciences,* 34, 57-74.

O'Neill M. (2005), The Biology of Irrationality: Crime and the Contingency of Deterrence. In Parisi F. & Smith V.L. (Eds.), *The Law and Economics of Irrational Behavior* (pp. 287-313). Stanford:Stanford University Press. pp. 287-313.

Rubinstein, A. (2000). *Economics and Language* .. Cambridge: Cambridge University Press.

Sen, A. (1993). Positional Objectivity. *Philosophy & Public Affairs, 22*(2), 126-145.,

Sen, A. (1997). Maximization and the Act of Choice. *Econometrica: Journal of the Econometric Society 65*(4), 745-779

Sen, A. (1999). The Possibility of Social Choice. *American Economic Review, 89*(3), 349-37

Simon, H. A. (1990). Invariants of Human Behavior. *Ann. Rev. Psychology*, 41, 1-19

Simon, H. A. (1997). *An Empirically Based Microeconomics.* Raffaele Mattioli Lectures. Cambridge; New York and Melbourne: Cambridge University Press, 1997.

Sperber, D. & Wilson, D. (2000). Truthfulness and Relevance., http://www.phon.ucl.ac.uk/publications/WPL/00papers/wilson_sperber.pdf Accesed 27 september 2015

Winograd, T. & Flores, F. (1986). *Understanding computers and cognition.* Ablex Publishing Corporation: Norwood

5

Impassioning Reason: On the Role of Habit in Argumentation

MICHAEL J. ARDOLINE
Kingston University, UK
Ardoline@lvc.edu

Reason and argument must be understood in their relation to habit for a full account of decision-making. While reason attempts disinterestedness, argument is bound up in interest and passions. Argument, therefore, cannot be separated from habit. As all decision-making requires interest, an understanding of "reasoning well" as an ongoing process in which an agent must continually work to turn reasoned thought into habit through activity of argumentation is required.

KEYWORDS: Aristotle, habit, Hume, interestedness, Malabou, passions

1. REASONING WELL

David Hume famously remarked that "reason is and ought only be the slave of the passions," thereby reversing the historic Platonic formulation. In this paper, I propose that both of these well-known exhortations occlude the distinction between the faculty of reason and the activity of argument. First, I will separate what is meant by "reasoning" versus "arguing", where the former is understood as disinterested in the specific content examined and the latter as a personal or interpersonal commitment to the content of the statements advanced. Second, I will show the inherent role played by the passions in argumentation. This examination will require an understanding of the passions not as givens of a subject, as was a common modern view, but as a system in circuit with the formation and alteration of habits. This circuit will then be expanded in order to understand the role played by disinterested reason in the formation of habits through argumentation. Finally, I will argue that in order to understand how argumentation

D. Mohammed & M. Lewiński (eds.) (2016). *Argumentation and Reasoned Action: Proceedings of the 1st European Conference on Argumentation, Lisbon, 2015. Vol. II, 205-213.* London: College Publications.

forms our decisions, we require an understanding of habit formation akin to that of virtue ethics.

This analogy to virtue ethics is not meant to bring the baggage of an ethical theory or produce a normative account of decision making. Instead, the goal is to produce an understanding of "reasoning well" as an ongoing process in which an agent must continually work to turn reasoned thought into habit through activity, including that of argumentation. This is not to propose argument as the only mediation between reason and habit, nor is it to give pride of place to reason over habit in decision making and action, as I maintain that argument is as much a product of habit as it is of reason (and that reason itself is only possible through habits). Rather, the claim is that by connecting the interestedness inherent in habits with the disinterested reflection of reason through argument formation, the agent becomes more adept at the decision making process.

2. REASONING VS. ARGUING

In contemporary contexts, reasoning (or rationality) and argument are conceived of as connected or necessarily intertwined. For instance, Habermas's theory of communicative rationality connects reason to norms which would make it shareable with or defensible to others (Habermas, 1987, p. 86). This connects with a strong intuition that successful arguments are communicable - in other words they can be understood and be convincing to someone other than the producer of the argument - and (it amounts to the same thing for Habermas) rational. It is my contention that this obscures the nature of argument by presupposing its success conditions (communicatibility and rationality), and thereby leaves the phenomena of reasoning equally murky. Instead, we should start with asking after the nature of argument and reason, not under what conditions would either be successful. A misused reason or a misunderstood argument is no less a use of reason or an argument. To formalize this insight, we have to ask how reason and argument differ. We can see this difference in the language we employ: we "use" reason and we "make" arguments. This leads us to the first distinction between reasoning and arguing: reason is an ability of the mind to make connections (a power we use), and arguing is something we do in order produce some effects in the world on ourselves or others (an activity).

This distinction between a power and an activity is as old as Aristotle's theory of potentiality and actuality, but all that is necessary to our investigation here is the way in which a power and an activity differ. An activity has a telos; it has a purpose for which we undertake it.

A power, on the other hand, is an unused but actualizable pool of potential. It can be employed for some purpose, but it has no necessary purpose and can be exercised without a purpose. I have the power of running; I may run because I am late or because I enjoy running, but I may also burst into a sprint for no reason. That I can run either for punctuality's sake or for enjoyment shows that it has no necessary purpose in itself; that I can run without a justification at all shows that, rare as it may be, a power may be actualized with no purpose. This is not to say that it is uncaused, but to say that it has no final cause or *telos*. It may very well be that some efficient cause (or whatever other causes your ontology may include) has sent me off running (perhaps a sudden chemical rush in my brain which I have no access to). We will see that this being pushed to actualization without a purpose is particularly important to reason.

Therefore, I propose to define reasoning as the disinterested exploration of logical, rhetorical, or practical associations, and arguing as the use of the same sorts of methods for a chosen end in which the arguer is interested, whether explicitly stated or not. These definitions are quite broad and should include sub-optimal and non-rigorous examples of each. For example, my definition of reasoning would include both the usage of invalid connections and the results of free association. To be a product of reason is not to necessarily be reasonable. Similarly, "argument" used here will also include fallacious and unsound arguments, right down to appeals to force and ad hominem. This must be the case or else we will be smuggling in a conception of what it means to reason or argue well and risk begging the question. Worse yet, by only including examples which are successful attempts at reasoning or arguing, we lose the great insights that come from failure.

Disinterest must be understood as disinterested in a specific conclusion or outcome, not in disinterest in the project of reasoning itself. If the latter were the case, it is unlikely one would reason at all. Reason, then is a space of exploration; it is the act in which we play with ideas honestly, in which we can think our dangerous thoughts earnestly. Reasoning makes it possible for us to criticize standards and practices without becoming Protagoras; and to challenge morality without being Thrasymachus. The interest in reasoning lies in the approach to the problem, not in the content of the outcome. One has a problem which has set on the project of reasoning, and they have some criteria for what a solution may look like, but one does not reach a conclusion until reasoning has run its course.

Argument, on the other hand, furthers a purpose, stance, or end of some kind. Because of this purpose, one knows the conclusion they are attempting to reach ahead of time in an argument. From a communal aspect, arguments are made to convince others of some conclusion, value, etc. that we hold. However, under the term argument, I also include a much wider set of situations, specifically those in which one argues with themselves such as self-reflection or self-criticism. Argument is then an intentional, goal directed operation. I imagine everyone reading this has found themselves arguing when they would prefer to be reasoning, perhaps when you find yourself extending an abstract into a paper days before a conference. Both deadline and thesis statement join to form an oppressive force preventing that free exploration of thought which could radically recast your project, or make it impossible. This shows us that reasoning is not an unqualified good, often for pragmatic reasons.

This distinction between reasoning and arguing is not a harmless, mutually exclusive, or harmonious one. We cannot assume that reasoning is simply a power or mode of thought we can switch into at a whim. Most situations are a mix of the two operations, though the cynics among us would have a point if they snarkily responded that argument takes the lion's share of that mix. There are those glorious moments within an argument where true thought takes places and one changes their stance or produces a new one on the spot. More often than not though, argument takes on the guise of reasoning. "Convictions are more dangerous enemies of truth than lies," warns Nietzsche (Nietzsche, 1996, p. 179).

3. ARGUMENT AND PASSION

As argument is tied to goals, interests, convictions, etc., it is then inseparable from what Modern philosophers would refer to as the "passions".[1] Passions denoting desires, faith, and generally anything which could not be arrived at by reasoning alone, both bodily impulses and habits of the mind (which we will see are not so far apart). In this sense, argument is tied to what Hume called "moral reasoning" after his haunting is-ought distinction (Hume, 1739, p. 245). It is only by submitting reason to the passions that we can actually decide on a preferable state of affairs or actions to undertake (Hume, 1739, p. 238).

Perhaps a more modern example is needed lest this historical account seem like necromancy rather than argument (as I am certainly not reasoning here). In her work on the intersection between

[1] Most commonly found in the British Moderns, especially Hume, but receiving its pride of place earlier in Descartes (For a full treatise, see Descartes, *Passions of the Soul*).

neuroscience, philosophy, psychiatry, and psychoanalysis, Catherine Malabou terms a certain type of subject "the new wounded," which is explicated in the eponymously titled work (Malabou, 2012). These are the radical cases of destructive brain lesions, of the fundamental altering of personality by physical or psychical trauma. The sorts of cases British Neurologist Oliver Sacks has made a career by making understandable to those of us lucky enough to not experience this horrifying transformation first-hand. The defining traits of the new wounded is that they lack a connection with their own history by way of a disconnect to their memories, former personality, personal affects, and, in the worst cases, they are left unable to make decisions. They are not comatose or unconscious; their ability to reason is often left intact. Instead, they are left in a Humean hell where they literally cannot take a stance on whether they prefer a paper cut to the destruction of the world.

This breakdown of the machine that we are, this operation of brain minus some of its everyday capacities is illuminating for many reasons, but for our purpose here, it points out the separation between reason and the passions that can take place. A consequence of the state of the new wounded is the utter inability for any sort of talk therapy to help this state (Malabou, 2012, p. 10). This does not mean that the patient is unable to follow, express, or play with ideas, to go along with the psychiatric or psychoanalytic practices, but that these exercises have little to no effect on them. In a strong sense, we can say that they could follow our reason (reason used here in the broadest sense possible); they can understand, but they cannot be persuaded by argument (again, in the loose sense I defined, which would include psychotherapy).

What this shows is that when we argue, it is not the reason of another that we wish to affect. Our great hubris is to assume that our opponents have a deficiency of reason. Reason is, at best, a way to influence an agent, but not what we aim to influence. Rather, if argument is necessary for decision and intentional action, and argument is that hybrid beast made by mating reason and the passions, then it must be the case that we seek to change the passions of our interlocutors via argument. This is perhaps much more obvious in the case where our interlocutors are ourselves. If I am trying to convince myself that that fifth candy bar is not in my best interest, I am clearly not trying to force myself to reason that such action is unhealthy.[2] Clearly, I've already undertaken that step in order to be attempting to convince myself of it in the first place. Rather, I am trying to overcome

---

[2] cf. Aristotle on Akrasia, *Nicomachean Ethics*, Book VII.

my desire to never not be eating chocolate. My reason lacks nothing that argument could give it here other than the efficacy to directly change my passions.

4. PASSION AND HABIT

Returning to the new wounded for a moment, the loss of history and the inability to decide are not isolated symptoms. Rather, they point to a deeper seat of the passions, or at least of the passions we could ever hope to change by argument. We have an everyday term for this concretion through personal history that produces automatic decisions and interestedness in an agent, we call them habits. To define habit, I turn to the work of Bill Pollard and his valiant attempts to reintroduce and defend the notion of habit in the philosophy of action. The definition he gives of habit is action which is automatic, repeated, and that the agent could rationally intervene to prevent from taking place (Pollard, 2002, p. 67). "Could rationally intervene" should be understood in the sense that one could intervene in their automatic reaction of, for example, outrage to bigoted comments if the comment comes from that Aunt no one listens to anyway and therefore the agent rationally decides it's not worth the effort of outrage, but could not, say after rationally adapting a nihilistic worldview, decide it's not worth the effort and intervene in the beating of their heart, at least not by mental control alone.

By this definition, we can argue that passions either are a subset, or are the result of, habits. First, one's passions are automatic; we do not choose to engage passions at will. We may choose to develop a passion, but the passion is only developed after repeated experience or as a want for a repetition of that experience. One usually comes to love exercise well after paying for their first gym membership. Here we have both the automatic condition and the repeat condition. Furthermore, passions can be rationally intervened upon. In fact, most of our institutions attempt to do just this, especially legal systems and religion.

Pollard focuses on the philosophy of action, so he understandably focuses on the role of habit in action (Pollard, 2002, p. 87-89), whereas we are primarily concerned with the role habit plays in thought. However, Pollard's anti-Cartesian stance allows us to equally project into the mind based on his elaboration of habit. In fact, this would be possible even if Pollard did not push an anti-Cartesian thesis. There is a beautiful nexus between the Cartesian and Humean accounts: Descartes labored over habit as a possible bridge between his dualism (Descartes, 1989), whereas Hume's account begins with habit and builds body and mind out of it. While the Humean account gives a much more felicitous starting point for such a project, I take it as a positive sign that these generally diametrically opposed views agree on this point. For either the Humean or Cartesian stance, habit should play an equally explanatory role for thought as it does for action. Our above investigation, however, immediately brings us back to the Humean camp as habits are taken as logically prior to passions and thought (Hume, 1740, p. 29). In other words, if habits are expressed as much bodily as mental, then they constrain our capacity for thought as much as our capacity for action. Here we can reform the definition of reason above: reasoning is non-habitual mental activity. This makes reasoning an especially rare beast.

This rarity of reason is equal parts blessing and curse as the new wounded show us. We may feel as a hostage to those passions which we wish to change, but being free of our passions is an infinitely worse state. These brief moments of disinterestedness made possible in reason are to be cherished and sought after, but we cannot survive there. Rather, the "I" cannot survive there.

5. FROM ARGUMENT TO ACTION

This analysis so far has many moving parts (reason, argument, passions, habit, decision, and action), and there is no simple order of operations to clarify this overall circuit easily. Rather, each term has an effect on and is affected by each other. Furthermore, some increase or decrease the efficacy of others: habit changes what I am likely to reason about and how, argument may change my habits, passions work against arguments towards decisions, and habits may even contradict each other thereby throwing us back upon reason, just to name a few possible transits. In order to make this circuit less unwieldy without adding excessive detail, it suffices here to catalogue the difficulties argument faces in producing its desired action. We have shown that argument's true goal is to change the passions of the interlocutor in order have an effect on their decision making. However, passions cannot be so easily overcome. Rather, argument, in order to affect decision making, must produce repeated action which can become habit. This new habit will produce different passions, preferably those in line with the argument. Furthermore, the arguer always faces the risk that they are merely speaking out of habit, merely expressing their passions in the guise of expressions of pure reason. Ultimately, it is the mediated connection between reason and habit that must be investigated and developed in order to avoid simple rhetoric, or the reactionary arguments which come from one's personal privileged position of their own habits. However, habit is not simply an obstacle to reason as it is one of its logical conditions of possibility, and one can develop habits of reasoning well by repeated action.

As noted above, this is far from the only way to traverse this circuit. The challenges outlined above only hold for those times in which we attempt to start from argument. We could just as easily, and perhaps more easily, began as Aristotle does with action. Action has the benefit of being closer to habit formation; repeated actions become habits. In fact, Aristotle bars reason and argument from the early parts of ethical development. Instead, one becomes ethical by following virtuous citizens and doing as they do; understanding comes later. It is not so farfetched to believe the same about reasoning and decision making. Perhaps without the interest to build the habit of the disinterested play of thought we call reason, thought would be little more than whim and justification of the passion.

For virtue ethics, the grand question is not "what is the good" nor even "what is virtue," but "how do we cultivate a person capable of acting virtuously." In analogy with Aristotle then, our grand problem surrounds the role of argumentation in decision making is not "what is reason" or "what is a good argument," but "how do we cultivate a person who is capable of being affected by good argumentation?" To understand just how dire and difficult this is even for the best of thinkers, look no further than academic intradepartmental politics.

We should pay special attention to the word 'capable' in this question. This is no light reference; habit, a la Hume, must be understood not simply as repeated behavior but as a condition for the possibility of certain behaviors. Here I find issue with Pollard's definition of habit in that it lacks an account of capacity. By only focusing on habits as repeated, automatic behavior under certain conditions, then one loses the productive power of habits. Specifically, habits make possible new ways of thinking or acting. Any athlete can tell you this. As scholars, you know this. Remember the first time you added footnotes to your dinner conversations?

6. CONCLUSION: HABIT'S PRIORITY OVER REASON

We say "the human is the rational animal", and that our habits are "second nature". Could these two phrases be any more mixed up? In what sense could rationality be of a deeper nature than habits? Of course, rationality is more connected to the human in general as a distinguishing feature from other animals, especially more than any specific habit. We learn rationality not rationally, not from reason alone, but from acting it first, by building the habits that make reason possible. But, in lieu of fully showing that rationality is one habit or set of habits among many, this investigation has shown that if reason can play more than a cursory role in decision making and action, then it does not do so directly. As creatures of habit, our only recourse is to understand how it is that argument can allow us to build new habits which produce the passions and abilities to reason which allow for better decision making. Despite the humbling destruction of centuries of pretension to the power of reason and will, there is a respite to be found in this. Through the very operations of argument, reasoning, deciding, and acting, we are engaged in a process of habit formation that can make these operations better. Indeed, we are always participating in our own cultivation, whether successfully or not. In spite of what we've developed here, the question remains: how should we cultivate ourselves? What habits foster better decision making? What can we become?

REFERENCES

Descartes, R. (1989). *Passions of the Soul.* Translated by Stephen Voss. Indianapolis: Hackett Publishing.

Habermas, J. (1987). *The Theory of Communicative Action: Volume II*, Translated by Thomas McCarthy. Boston: Beacon Press.

Hume, D. (1740). *An Enquiry Concerning Human Understanding.*

Hume, D. (1896). *A Treatise of Human Nature.*

Malabou, C. (2012). *The New Wounded.* Translated by Steven Miller. New York: Fordham University Press.

Nietzsche, F. (1996). *Human, All Too Human.* Translated by R. J. Hollingdale. Cambridge: Cambridge University Press.

Pollard, W. J. (2002). *Habits in action: A corrective to the neglect of habits in contemporary philosophy of action.* Durham theses: Durham University, http://etheses.dur.ac.uk/3973/.

6

Computational Modelling of Practical Reasoning Using Transition Diagrams

KATIE ATKINSON
Department of Computer Sciences, University of Liverpool, UK
katie@liverpool.ac.uk

TREVOR BENCH-CAPON
Department of Computer Sciences, University of Liverpool, UK
tbc@liverpool.ac.uk

LATIFA AL-ABDULKARIM
Department of Computer Sciences, University of Liverpool, UK
latifak@liverpool.ac.uk

In practical reasoning an agent chooses an action based on the goals this action is expected to achieve and the values achieving the goals will promote. Argumentation schemes using transition diagrams have been proposed to support practical reasoning, but the schemes proposed previously are limited in terms of expressivity of goals and extent of look ahead. Here we explain how to overcome these limitations, using a set of linked argumentation schemes and associated critical questions, and outline how this can be formalized in ASPIC+.

KEYWORDS: argumentation schemes, diagrams, goals, practical reasoning, transition

1. INTRODUCTION

Practical reasoning is often distinguished from theoretical reasoning in terms of direction of fit (Searle, 2001). Whereas in theoretical reasoning the idea is that agents will conform their beliefs to fit the world as it is, in practical reasoning agents will choose how they wish the world to be and will seeks ways in which they can make the world fit their desires. An obvious consequence of this is that the outcome of theoretical

D. Mohammed & M. Lewiński (eds.) (2016). *Argumentation and Reasoned Action: Proceedings of the 1st European Conference on Argumentation, Lisbon, 2015. Vol. II, 215-225.* London: College Publications.

reasoning is intended to be objective, but the outcome of practical reasoning will often be subjective. All agents are trying to make their beliefs fit the same world but they will differ in their aims, values and aspirations, and hence will make different choices about the way the world should be, and differ also in their abilities and preferences for actions, and so may make different choices about how they can, or can try to, realise their desires, even when these coincide.

But the current state of the world also matters for practical reasoning: the world and its current state determine what we can do, and influences the effects our actions will have. In order to perform an action, its preconditions must be satisfied, and the consequences of an action will typically depend in some way on the current state. These consequences are not, however, always determinate: sometimes the effects of an action may be better considered as a probability distribution (e.g. tossing a coin, cutting a deck of cards), and sometimes also may depend on the choices of other agents able to affect the situation.

Practical reasoning therefore needs to bring together a range of different knowledge types including:

1. the current state of the world;
2. causal relationships relating to the effects of actions;
3. what we can do in the current situation;
4. the actions of others;
5. goals we want to achieve;
6. values promoted and demoted by various actions and realisation of goals;
7. which values are preferred.

These various knowledge types have different statuses. The current state of the world, and what we can do in that state are both objective, although our beliefs may, of course, be mistaken both about what is the case, and what we are capable of doing. The effects of actions are objective in the sense that that we have beliefs about them rather than control over them, but they are often uncertain and need to be described using probabilities, or may even require an understanding of what else will happen simultaneous with our action. The actions of others can be reasoned about, and so we can arrive at more or less certain beliefs about they will do, but they are ultimately outside of our control. With goals and values we move into the realm of the subjective: we are permitted to adopt whatever goal we wish; and to declare values to be promoted or demoted by particular actions, and the realisation of particular goals, so that the acceptability of arguments becomes relative to the audience to which they are addressed (Perelman & Olbrechts-Tyteca, 1969). But others may disagree with us: whether someone enjoys swimming or not depends on the individual: whether the value of fairness is promoted by equality of outcome or equality of opportunity is a matter for debate, and whether it is more important to promote, for example, equality or enterprise is a matter of pure choice, dependent on the preferences of individual agents.

2. MODELLING PRACTICAL REASONING

In multi-agent systems the commonest approach to modelling practical reasoning is perhaps the belief-desire-intention model based on Bratman (1987). Alternative argumentation based approaches have been explored, however, including those based on state transition diagrams (STDs) (e.g. Atkinson & Bench-Capon, 2007), in which possible states of the world are related by the actions which lead from one to another, and agents reason about which transition to follow. An argumentation scheme for practical reasoning was proposed in Atkinson *et al.* (2006), and expressed as an STD in Atkinson and Bench-Capon (2007):

> PRAS: In the current circumstances R
> I should perform action A
> Which will reach new circumstances S
> Realising goal G and
> Promoting value V.

Relating this to the different knowledge types from the previous section, we find that (1) is expressed in the first line, (3) in the second, (2) in the third, (5) in the fourth and (6) in the fifth. Moreover in Atkinson and Bench-Capon (2007) the transitions are *joint* actions, and so represent the effect of the agent's action given the set of actions performed by other relevant agents. Action A may therefore occur in several transitions, and so the particular new circumstances in the third line may rely on knowledge of what other agents will do: knowledge type (4) above. This joint action mechanism can also accommodate probabilities, through the use of a special "agent" (usually called "nature") the action of which will determine which state will be reached. Finally the "should" in the second line appeals to the particular preferences of the agent or audience concerned, which is determined by the value preferences found in knowledge of type (7). Thus PRAS brings together all seven of the knowledge types identified earlier.

Following Walton (1996), in Atkinson and Bench-Capon (2007) challenges to arguments take the form of so-called critical questions. Seventeen critical questions were identified and divided according to three stages of the reasoning: problem formulation, epistemological reasoning and option selection. Much of the knowledge (types 2, 3 and 6) is hard-coded into the transition diagram, and this has to be critiqued at the problem formulation stage. Once an appropriate transition diagram has been agreed, in the epistemic stage assumptions are made to capture beliefs about the current state and what the other agents will do, i.e. knowledge of types (1) and (4). Most of the argument in Atkinson and Bench-Capon (2007) centres on disagreement as to preferences between the values, knowledge of type (7), representing different audiences (Perelman & Olbrechts-Tyteca, 1969). Arguments are generated by instantiating PRAS and its critical questions from the transition diagram and the status of these arguments is evaluated separately according to the preferences of particular audiences using a value-based argumentation framework (Bench-Capon, 2003). Although this model has proved useful in a variety of applications including law (Atkinson & Bench-Capon, 2005), medicine (Atkinson *et al.,* 2006) and e-participation (Bench-Capon *et al.,* 2015), it has some distinct limitations. First, much of the potentially questionable information is implicit in the transition diagram and so argument about what form the transition diagram should take precedes the generation of arguments and so is beyond dispute during the practical reasoning itself. A second limitation is that the goals of Atkinson and Bench-Capon (2007) are very inexpressive – essentially only subsets of states, restricting the representation of type (5) knowledge. A third limitation of Atkinson and Bench-Capon (2007) is that it allows only the immediately next state to be considered.

To address the first of these limitations we have to provide a way of justifying (i.e. arguing for) various elements of the transition diagram, while the second limitation requires a richer notion of goal, and the third a notion of the future. The first two of these problems were addressed in Atkinson and Bench-Capon (2014). In that paper the transition diagram was augmented by several logic programs. The first of these was *GProg* which intensionally defined a set of goals in terms of the basic propositions forming the states of the diagram. These goal propositions either do, or do not, hold, in a given state according to whether they or not they can be derived from the program using the basic propositions of the state as facts. This gives rise to four varieties of goal:

- achievement goals: desirable things which do not hold in the current state, but will be realised by the action;
- remedy goals: undesirable things which hold in the current state, but will be terminated by the action;

- maintenance goals: desirable things which hold in the current state, but which would cease to do so unless the action were taken;
- avoidance goals: undesirable things which do not hold in the current state, but which would occur unless prevented by some action.

Note that the above means that goals depend both on the current state and the next state. A second logic program (*VProg*) maps these goals to the promotion and demotion of values. Together these programs are able to justify claims about which goals will be achieved by a transition, and which values will be promoted by realising these goals. Similarly knowledge of type (2) can be encapsulated in a third logic program (*CProg*) which, given an assignment of truth values to the set of basic propositions representing a state, and a set of actions, one per relevant agent, will determine the assignment of truth values in the next state. Thus this causal model allows explanation of which aspects of the current state and which actions of which agents led to particular aspects of the new state, and so can form the basis of argumentation about these issues.

The third limitation, concerning the extent of look ahead, was addressed in Atkinson and Bench-Capon (2014b). In that paper consideration was extended beyond the next state, to the subsequent states. This gives rise to a further four types of goal:

- enabling goals: desirable things which do not hold in the current state, cannot be realised from the current state, but can be realised from the target state.
- risks: undesirable things which do not hold in the current state, and cannot occur in the next state but which may occur subsequently if a particular action is taken:
- assurance goals: desirable things which hold in the current state, and which will continue to do so if the action is taken;
- prevention goals: undesirable things which do not hold in the current state, and which can never occur if the action is performed.

That paper also recognised that the performance of an action may in itself promote of demote a value (e.g. we may swim for the pleasure of swimming, not to get anywhere). This gives a ninth kind of goal:

- performance goal: the action is in itself desirable and performed for its own sake.

These additional goal types allow the specification of nine kinds of reason in Atkinson and Bench-Capon (2014b), some of which have both positive and negative (arguing for the non-performance of an action) variants.

3. PRACTICAL REASONING AS PROCESS

It was further argued in Atkinson and Bench-Capon (2014b) that practical reasoning is best seen as a *process* (cf. Prakken, 2010b), as suggested by the three stages identified in Atkinson and Bench-Capon (2007). PRAS should thus be viewed as a highly compressed version of this process and its various parts can be fruitfully separated and encapsulated as a cascade of argumentation schemes each with its own characteristic critical questions. The process is shown in Figure 1.

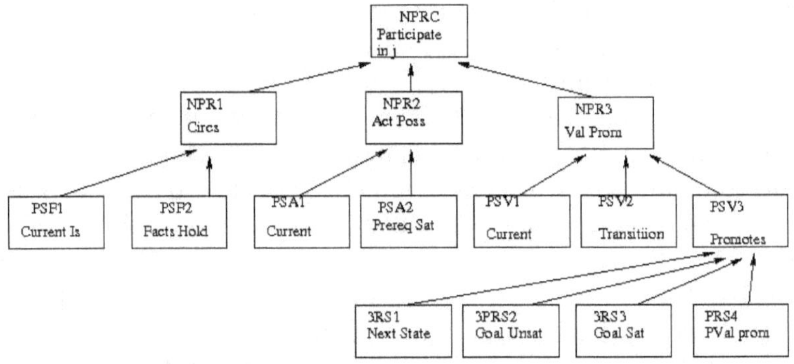

Figure1: Argumentation Schemes from Atkinson and Bench-Capon (2014b)

Thus the New Practical Reasoning Scheme (NPR) concludes that a particular joint action should be participated in on the basis of three premises: NPR1, which states the current circumstances, NPR2 which states that the action required of the agent is possible, and NPR3 which indicates the value that will be promoted by participating.

NPR1 is established on the basis that the circumstances are true of the current state, which can be established using whatever form of theoretical reasoning is appropriate to the particular propositions being questioned.

NPR2 is established by showing that the preconditions for the required action are satisfied by the circumstances.

NPR3 is established by showing the value will be promoted by moving from the current state to the state that will result from participation in the advocated action. Note that this relies on both CProg to determine the transition and its effect and VProg to demonstrate that the value is indeed promoted. The promotion of the value is itself can be established in a number of ways (using GProg) corresponding to the nine goal types listed above (one of which, the simple achievement of a goal in the immediately next state) is shown in the third layer in Figure 1.

NPR can be challenged in several ways:

- It can be **rebutted**, either by offering an argument not to participate in the action, based on one of the negative variants mentioned above, or by offering an argument to participate in some other action, itself based on NPR. In either case the rebutting argument will also be based on a particular value, and acceptability will be decided by the audience according to its value preferences (Bench-Capon, 2003).
- In some situations it may be **undercut**. As Prakken (2010a) argues, undercutters cannot be simply disregarded on the basis of value preferences: if an argument is shown to be inapplicable, then it cannot be used. Thus for undercutters, the undercutting argument must be defeated or the undercut argument withdrawn. This provides a mechanism, if desired, for some values to be given priority, so that all audiences must rank them more highly than other values. In many cases there will be no such values and so undercutters will not be used.
- Finally the argument may be **undermined**, by showing that one of its premises is false. Thus for example an argument based on some proposition *P* being currently true, would be undermined by an argument showing that –*P*. Such arguments are not themselves based on practical reasoning and values, and so must be resolved as appropriate for the argument of the type used in the particular case. Very often such arguments will also not rely on value preferences, but will represent constraints coming from the way the world is. More detail on particular ways of undermining the various different premises is given in Atkinson and Bench-Capon (2014b).

We have presented extensions to the simple argumentation scheme of Atkinson and Bench-Capon (2007). These extensions have then been presented as a process involving several argumentation schemes, which has the effect of separating out the various types of knowledge identified in the introduction. This more complete and better articulated view allows us to choose what will be considered in the argumentation and what will be taken as agreed according to the current dialectical setting. It also enables richer and deeper disagreements, going beyond differences in preferences and priorities to fundamental conceptualisations; it enables the source of disagreement to be located quite precisely, and allows attacking arguments to be expressed more clearly using appropriate argumentation schemes.

4. FORMALISING PRACTICAL REASONING

In the work so far, the approach has been described only in semi-formal terms. Although the transition diagram itself is formally represented, and permits formal statements of argumentation schemes and critical questions, there is a need to represent the machinery uniformly in a formal framework, which will allow the formal proof of properties of the practical reasoning process advocated, such as rationality postulates (Caminada & Amgoud, 2007). A sensible choice of framework would be ASPIC+ (Modgil & Prakken, 2013), a main objective of which is "to identify conditions under which instantiations of the framework satisfy logical consistency and closure properties." (Modgil & Prakken, 2014).

There are a number of similarities between ASPIC+ and our presentation of practical reasoning which make it a good choice as our formal framework. ASPIC+ structures arguments using an argument-subargument structure which corresponds readily to the process of practical reasoning shown in Figure 1. ASPIC+ uses strict and defeasible rules of inference, and these rules correspond to the argumentation schemes sketched here and described in detail in Atkinson and Bench-Capon (2014b). ASPIC+ generates arguments from a set of knowledge bases: our programs GProg, VProg and CProg, together with the facts taken from the current state, will readily instantiate these knowledge bases.

Formally representing the process of practical reasoning in ASPIC+ will allow the demonstration that it satisfies desirable properties and will be an invaluable step towards a full implementation.

5. FUTURE WORK

In the compendium of argumentation schemes presented in Walton *et al.* (2008), there are a number of schemes which are used in practical reasoning, including a scheme from values, a scheme from positive consequences, a scheme from negative consequences, argument from goals, and argument from ends and means. These argumentation schemes are presented individually, and so it is difficult to discern their relationships to one another. Each of these schemes can be related to parts of the process we describe: We thus are able to bring together the piecemeal presentation of Walton *et al.* (2008) into a coherent whole, and better explain why particular schemes appear in various different situations. In real dialogues, there will be a measure of agreement between the parties and so the argumentation will focus on the areas of disagreement. Thus, for example, if the emphasis of the disagreement is on the causal model, we would expect to see means-ends reasoning used, whereas if the disagreement were about subjective preferences, we would expect to see some form of argument from value. By representing our account in a formal framework, we will also be able to put the informal schemes of Walton *et al* (2008) on a formal basis.

Once the practical reasoning process has been formalisised and its properties demonstrated, we will evaluate it by reworking applications previously tackled using PRAS in these terms. Examples that can be used are:

- Law: The property law line of cases stating with Pierson v Post were represented using PRAS in Atkinson and Bench-Capon (2005). This will allow direct comparison of the effectiveness of the new schemes. Moreover this set of cases has become a de facto benchmark in AI and Law (see, e.g. Atkinson, 2012), so that comparison may also be made with other approaches.
- Medicine: A problem concerning a choice between several different drugs to medicate a heart attack was tackled using PRAS in Atkinson et al. (2006). Again this particular scenario has also been addressed in other approaches (e.g. Modgil & Fox, 2006), allowing wider comparison.
- E-Participation: PRAS was used as the basis for the two e-participation tools described in Bench-Capon et al (2015), one of which allows the user to critique a policy proposal and the other which elicits a policy proposal from the user and then supplies a critique. Both of these tools can be re-implemented to evaluate the new schemes.

This series of projects will give a firm basis for determining the value added by addressing the limitations of PRAS which formed the basis for the original implementation.

6. CONCLUDING REMARKS

Practical reasoning plays a central role in most areas of human activity. The need to choose between actions so as to further one's aims constantly arises. Law, medicine and e-participation are three areas of AI where particular attention has been paid to the topic, although it is a pervasive concern of argument and dialogue systems in general. Even where the focus is apparently on theoretical reasoning, this is often being performed in order to inform practical decision making. While approaching practical reasoning with argumentation schemes has proved fruitful, current schemes have limitations: for example the scheme of Atkinson and Bench-Capon (2007) imposes limits of expressiveness and coverage, and the schemes of Walton *et al.* (2008) are presented as a compendium of individual schemes. In this paper we have proposed ways in which the limitations of Atkinson and Bench-Capon (2007) can be removed, so as to provide a structure which can relate the various different schemes of Walton *et al.* (2008) so as to allow a coherent account of the practical reasoning seen as a whole. Additionally, our proposal lends itself to a formal representation in a framework such as ASPIC+ (Prakken, 2010; Modgil & Prakken, 2013), which will allow formal demonstration of its properties.

REFERENCES

Atkinson, K. (2012). Introduction to special issue on modelling Popov v. Hayashi. *Artificial Intelligence and Law, 20*(1), 1-14.

Atkinson, K., & Bench-Capon, T. (2014).States, goals and values: Revisiting practical reasoning. In *Proceedings of the Eleventh International Workshop on Argumentation in Multi-Agent Systems* (ArgMAS 2014).

Atkinson, K., & Bench-Capon, T. (2014b). Taking the long view: Looking ahead in practical reasoning. In *Computational Models of Argument: Proceedings of COMMA*. IOS Press, 109-120.

Atkinson, K., & Bench-Capon, T. (2007). Practical reasoning as presumptive argumentation using action based alternating transition systems. *Artificial Intelligence, 171*(10-15), 855-874.

Atkinson, K., & Bench-Capon, T. (2005). Legal case-based reasoning as practical reasoning. *Artificial Intelligence and Law, 13*(1), 93-131.

Atkinson, K., Bench-Capon, T., & McBurney, P. (2006). Computational representation of practical argument. *Synthese, 152*(2), 157-206.

Atkinson, K., Bench-Capon, T., & Modgil, S. (2006). Argumentation for decision support. In *Database and expert systems applications.* (pp. 822-83). Berlin / Heidelberg: Springer.

Bench-Capon, T., Atkinson, K., & Wyner, A. (2015). Using Argumentation to Structure E-Participation in Policy Making. In *Transactions on Large-Scale Data-and Knowledge-Centered Systems XVIII* (pp. 1-29). Springer.

Bench-Capon, T. (2003). Persuasion in practical argument using value-based argumentation frameworks. *Journal of Logic and Computation, 13*(3), 429-448.

Bratman, M. (1987). *Intentions, plans and practical reasoning.* Cambridge, MA: Harvard University Press.

Caminada, M., & Amgoud, L. (2007). On the evaluation of argumentation formalisms. *Artificial Intelligence, 171*(5), 286-310.

Modgil, S., & Prakken, H. (2014). The ASPIC+ framework for structured argumentation: a tutorial. *Argument & Computation, 5*(1), 31-62.

Modgil, S., & Prakken, H. (2013). A general account of argumentation with preferences. *Artificial Intelligence, 195*, 361-397.

Modgil, S., & Fox, J. (2009). A guardian agent approach to safety in medical multi-agent systems. In *Safety and security in multiagent systems*, 67-79. Springer LNAI 4324.

Perelman, C., & Olbrechts-Tyteca, L. (1969). *The New Rhetoric*, trans. J. Wilkinson & P. Weaver. Notre Dame: University of Notre Dame Press.

Prakken, H. (2010a). An abstract framework for argumentation with structured arguments. *Argument and Computation, 1*(2), 93-124.

Prakken, H. (2010b). On the nature of argument schemes. In C.A. Reed & C. Tindale (Eds.), *Dialectics, Dialogue and Argumentation. An Examination of Douglas Walton's Theories of Reasoning and Argument*, (pp. 167-185). London: College Publications.

Searle, J. R. (2001). *Rationality in action*. Cambridge, MA: MIT press.

Walton, D. (1996). *Argumentation schemes for presumptive reasoning*. Mahwah, NJ: Lawrence Erlbaum Associates.

Walton, D., Reed, C., & Macagno, F. (2008). *Argumentation schemes*. Cambridge: Cambridge University Press.

7

Uncertainty and Fuzziness from Natural Language to Argumentation Models

PIETRO BARONI
DII - University of Brescia, Italy
pietro.baroni@unibs.it

MASSIMILIANO GIACOMIN
DII - University of Brescia, Italy
massimiliano.giacomin@unibs.it

BEISHUI LIAO
CSLC - Zhejiang University, China
baiseliao@zju.edu.cn

There is a big gap between current argumentation models and the capability of faithfully capturing the information conveyed by natural language texts: in particular, the use of current models implies a drastic simplification in the representation of uncertainty and fuzziness. This contribution discusses the modeling challenges posed by the presence of uncertainty and fuzziness in natural language texts and analyzes some research directions aiming at tackling these challenges in the context of argumentation formalisms.

KEYWORDS: argumentation, fuzziness, natural language, uncertainty

1. INTRODUCTION

Uncertainty and fuzziness pervade natural language both explicitly and implicitly. The explicit presence of uncertainty and fuzziness is exemplified by statements like "I believe that tomorrow will probably be a bit colder than today", where the qualifier "probably" indicates (in a fuzzy way) that the subject's belief is accompanied by a certain degree of uncertainty, while the term "a bit colder" provides a fuzzy specification of tomorrow's expected temperature.

D. Mohammed & M. Lewiński (eds.) (2016). *Argumentation and Reasoned Action: Proceedings of the 1st European Conference on Argumentation, Lisbon, 2015. Vol. II, 227-239.* London: College Publications.

In absence of explicit linguistic references, the presence of implicit uncertainty and/or fuzziness can be referred to background domain knowledge. For instance in the statement "Tomorrow will be rainy because the weather forecast says so", the facts that weather forecasts are uncertain and that the notion of being rainy is not crisp can be regarded as generally accepted background knowledge.

Argumentative structures pervade natural language too: not only because humans continuously argue with each other, but also because argumentation is a fundamental form of human rationality (Pollock, 1992; Mercier & Sperber, 2011). Indeed, this is the reason of the increasing interest in the research field of argumentation mining, which "involves automatically identifying argumentative structures within a document, e.g., the premises, conclusion, and argumentation scheme of each argument, as well as argument-subargument and argument-counterargument relationships between pairs of arguments in the document" as stated in the description of the First International Workshop on Argumentation Mining[1].

Given that uncertainty, fuzziness and argumentation live side by side, or even permeate each other, in daily discourse, one might expect that this close relationship has a formal counterpart in the models adopted in formal argumentation research, thus supporting the activities of identification and representation of arguments featuring uncertainty and fuzziness starting from natural language expressions.

Somewhat surprisingly, however, most current formal or semi-formal argumentation models allow for very limited ways, if any, to express explicit uncertainty and fuzziness within argument structure.

This paper, based on a contribution to the ECA 2015 panel "Arguments in Natural Language: The Long Way to Analyze the Reasons to Believe and the Reasons to Act", aims at evidencing and analyzing this gap, at justifying it by pointing out the modeling challenges posed by the presence of uncertainty and fuzziness in natural language, and at discussing some possible future research directions.

2. FORMAL ARGUMENTATION MODELS

Roughly speaking we can classify the models considered in computational argumentation literature into three main categories: abstract frameworks, structured argumentation systems, and semi-formal/diagrammatical schemes. In the sequel we briefly characterize them and discuss their relationships with argument mining.

---

[1] See http://www.uncg.edu/cmp/ArgMining2014/.

Abstract frameworks regard arguments as abstract entities, whose structure and origin are left unspecified, i.e. abstracted away, in order to focus exclusively on the relationships holding among the argument themselves and on the evaluation of argument acceptability according to these relationships.

For instance, in Dung's abstract argumentation frameworks (Dung, 1995) the attack relation only is considered and argument acceptance is assessed on the basis of an argumentation semantics, namely a criterion to select those sets of arguments that can survive together the conflict represented by the attack relation.

Bipolar argumentation frameworks (Amgoud, Cayrol, Lagasquie-Schiex, & Livet, 2008) extend the same conceptual structure by considering support in addition to attack relation, while Abstract Dialectical Frameworks (Brewka & Woltran, 2010) provide a further generalization by considering arbitrary acceptance conditions associated with the links connecting arguments (or, more generally, statements) together. Abstract frameworks, in any case, are high level representation and computation tools, addressing only the fundamental issue of argument acceptance. They are applicable only after arguments and their relationships have been distilled (or abstracted) from some more concrete level. As a consequence, abstract frameworks seem not suitable for argument mining activities, as indeed they take for granted that arguments and their relationships have already been identified.

Structured argumentation systems provide a model for argument structure: they define what an argument is in terms of its components and typically also define the relationships holding among arguments on the basis of finer grained relationships holding among their components.

For instance, in assumption-based argumentation (Toni, 2014) "arguments are deductions of claims using rules and supported by sets of assumptions" and an argument attacks another one if the claim of the former is the contrary of one of the assumptions in the latter.

ASPIC+ (Modgil & Prakken, 2014) provides a richer model of rule-based argument construction by considering the distinction between strict and defeasible rules and introducing several forms of attack (rebutting, undercutting, and undermining). Other examples of structured argumentation systems are DeLP (Garcia & Simari, 2014) and deductive argumentation (Besnard & Hunter, 2014).

Being meant to capture more closely the actual construction of arguments in argumentation processes, structured argumentation systems are more suitable candidate formalisms for argumentation mining. For instance, in the assumption-based model, argument analysis amounts to the identification, within a given text, of the argument claim, of its supporting assumptions and of the rules used for the claim deduction. Similar "component identification guidelines" could be drawn for other structured systems.

Two difficulties must be acknowledged, however, concerning the use of these formalisms for actual argument mining. First, some of these formalisms are still rather abstract, since, for the sake of generality, they leave unspecified some important aspects (e.g. the actual language adopted) hence they are not applicable without making some further specific choices at the implementation level. Second, and more important, they have typically been conceived to capture argumentation in already formalized settings (e.g. argument-based reasoning on possibly inconsistent knowledge bases) rather than at a natural language level. Indeed, due to the enthymematic nature of most natural arguments, some of the argument components encompassed by the above mentioned models, like the assumptions or the rules used, are left implicit in natural language expressions of arguments. Hence, one might argue that such structured formalisms are a suitable target for a "second level" analysis (and completion) of the natural arguments identified in a text, but, in a sense, can be too demanding as a first target formalism for the argument mining process itself. As a matter of fact, an analysis of the references in the papers presented at the First[2] and Second[3] International Workshop on Argumentation Mining shows that the structured argumentation formalisms recalled above are practically absent from current research on argumentation mining, while much more attention has been reserved to the use of semi-formal/diagrammatical schemes, reviewed next.

Semi-formal schemes, often lending themselves to a diagrammatical representation, provide models of argument structure and/or of inter-argument relationships which are typically focused on a few elements, regarded as crucial for the analysis and comprehension of some key aspects of the argumentation process. As such, these schemes do not provide a complete account nor a formal backing of the argumentation process as a whole, to be covered by other models, but rather can be regarded as shedding light on some central points, beneficial for the development of more complete and more formal models.

Examples are the Toulmin model (Toulmin, 1958), subsequently developed by Freeman (Freeman, 1991; Freeman, 2011), Wigmore diagrams (Wigmore, 1931) and Walton's argumentation schemes

---

[2] http://aclweb.org/anthology/W/W14/#2100

[3] http://aclweb.org/anthology/W/W15/#0500

(Walton, 1996; Walton, Reed, & Macagno, 2008). A discussion of the uses of this kind of models for argumentation mining is provided by (Peldszus & Stede, 2013), while an analysis of the papers presented at the First and Second International Workshop on Argumentation Mining and of some earlier influential work (Mochales & Moens, 2011) shows in particular a prevalence in the use of Walton's argumentation schemes.

Accordingly, with the subject of this paper, this points out the need to bring together semi-formal schemes (in particular Walton's argumentation schemes) with uncertainty and fuzziness, as discussed in the next section.

3. UNCERTAINTY AND FUZZINESS IN ARGUMENTATION MODELS

Argumentation is uncertain by nature: arguments are generally regarded as fallible (hence temporary and defeasible) derivations, subject to the possibility that some counterarguments are produced in a dynamic and, in general, open-ended process. Even when a fixed set of arguments is considered, in abstract frameworks most argumentation semantics allow multiple alternative answers (called extensions in the abstract argumentation jargon) to the question "Which arguments are able to survive the conflict together?", leaving open, hence uncertain, the choice of a specific one. The kind of uncertainty captured by these models is qualitative by nature: it refers basically to "yes or no" questions without encompassing uncertainty degrees. Hence it does not appear to be expressive enough to capture the many forms of uncertainty qualifications occurring in natural language, which need to be simplified, or simply ignored, in such a context.

Whether argumentation is fuzzy by nature is a less obvious question, but certainly one can observe that argumentation models are all natively based on crisp, rather than fuzzy, sets, pointing out another gap with respect to the possibility to encompass the fuzziness pervading natural language.

In fact, in recent years the issue of endowing argumentation models with quantitative uncertainty and/or fuzziness has received a significant deal of attention, exemplified by several forms of probabilistic (Li, Oren, & Norman, 2011; Hunter, 2013) and fuzzy (Janssen, De Cock, & Vermeir, 2008; Tamani & Croitoru, 2014) argumentation either at the level of abstract frameworks or of structured argumentation systems.

It may be pointed out however that, as discussed in the previous section, abstract frameworks and structured argumentation systems appear, up to now, to be far from the needs and from the current applications of argumentation mining. Thus the advancements mentioned above do not provide a direct contribution to the issue addressed in this paper, which instead appears to require, as a first step, a treatment of uncertainty and/or fuzziness at the level of semi-formal schemes. To put it in other words, encompassing uncertainty or fuzziness in the models lying at higher levels of abstraction (like abstract frameworks) corresponds to a sort of top-down approach, which takes for granted that uncertainty and/or fuzziness have been already properly identified and somehow preprocessed at lower levels, to be "passed" then to the upper level representation, while argumentation mining appears to require a bottom-up approach, focusing on the first modeling step from natural language sources to semi-formal schemes.

This modeling step appears to pose a set of largely open and challenging questions, as discussed in the next section, where, to keep the presentation compact, we exemplify the many problems arising when addressing the issue of encompassing uncertainty within argumentation schemes.

4. ENCOMPASSING UNCERTAINTY IN ARGUMENTATION SCHEMES

Argumentation schemes are a semi-formal model, based on a structured natural language description of stereotypical defeasible reasoning patterns connecting some premises to a conclusion. Each scheme is equipped with a set of critical questions pointing out potential weaknesses to be identified within actual instances of the scheme. The set of critical questions can be regarded as a sort of checklist to verify whether an actual argument, adhering to the specified scheme, can withstand the relevant most common possible objections.

As such, and as already mentioned, argumentation schemes implicitly involve a basic form of uncertainty since the conclusion is always assumed to follow "presumably" from the premises and some of the reasons to possibly retract the conclusion are explicitly given.

One can then wonder whether this basic pattern can be extended with a richer explicit uncertainty representation aimed at capturing in a more detailed way the variety and nuances present in natural language. In this respect, we can consider two non-disjoint issues: including some form of explicit uncertainty qualification and distinguishing different types of uncertainty (possibly affecting different parts of the scheme).

Both aspects appear to be problematic with respect to the current state of the art and to require substantial advancements both from a conceptual and a formal viewpoint.

As to including some form of explicit uncertainty qualification, it may be observed that in the definitions provided in (Walton, Reed, & Macagno, 2008) some of the schemes already include some explicit linguistic qualification of uncertainty, while others do not.

For instance the scheme "Argument from cause to effect" is defined (omitting the critical questions) as follows:

> Major Premise: Generally, if *A* occurs, then *B* will (might) occur.
> Minor Premise: In this case, *A* occurs (might occur).
> Conclusion: Therefore, in this case, *B* will (might) occur.

Here the word "Generally" in the major premise and the "might" specifications in the parentheses introduce some explicit linguistic references to the fact that the causal relation is uncertain.

Other schemes do not include such linguistic references. For instance, the "Argument from position to know" is defined as follows:

> Major Premise: Source *a* is in a position to know about things in a certain subject domain *S* containing proposition *A*.
> Minor Premise: *a* asserts that *A* (in domain *S*) is true (false).
> Conclusion: *A* is true (false).

Here no explicit linguistic references to uncertainty are present, but one may wonder whether this reflects a substantial difference between the two schemes or these are just different ways of expressing basically the same thing. Indeed it seems that one may suppress the explicit references ("Generally" and the parentheses) in the first scheme without changing its basic meaning (given that implicitly every scheme is defeasible). Dually, one might add some explicit linguistic reference in the second scheme, e.g. adding "possibly" to the Major premise: "Source *a* is possibly in a position to know ...", or replacing "is true" with "is (might be) true" in the Minor premise and in the Conclusion. Also in this case the meaning does not appear to be affected by the modifications.

This simple example shows that a systematic account of the possible uses of linguistic qualifiers of uncertainty within argument schemes would be needed, if they are meant to be used for argument mining activities and to be able to capture the difference between argumentative sentences like "Dr. Smith almost certainly made the right diagnosis because he is recognized as a leading expert in the field" and "Dr. Smith probably made the right diagnosis because he is believed to be sufficiently competent in the field".

As to our knowledge, no attempts to define such a systematic account are available in the literature: some preliminary considerations have been recently provided in (Baroni, Giacomin, Liao, & van der Torre, 2014). As a matter of fact, if undertaken, this activity promises to be a large effort, involving a reassessment of the structure and definition of the schemes as proposed up to now in the literature and, maybe, an orthogonal classification of paradigmatic "uncertainty schemes" (e.g. questioning the credibility of some premise vs. pointing out the presence of exceptional conditions vs. stressing that the relation between premises and conclusion is generally unreliable) that may affect argumentation schemes. We suggest in particular that a systematic classification of the kinds of critical questions associated with the schemes in the literature may provide a starting point for the definition of a set of reference "uncertainty schemes".

While this idea has some appeal (and critical questions provide a concrete starting point) classifying uncertainty schemes seems to call indeed for a classification of "uncertainty types" at a general level and suggests that one may take some inspiration and guideline looking for some general "uncertainty ontology".

This turns out however to be a challenging, if not frustrating, matter.

As an example, the W3C working group for the definition of an uncertainty ontology for semantic web appears to have closed its activities with quite limited results[4]. Indeed, the available material produced by this working group is quite preliminary, has several questionable aspects (e.g. it includes a binary distinction about whether the uncertainty is an inherent property of the world or is a lack of information, and considers fuzzy sets as an uncertainty model), and, by the way, does not contain any mention of the notion of argument at all.

Even worse, it turns out that other attempts to define an ontology of uncertainty in other contexts adopted a very different perspective and terminology, making the notion of uncertainty ontology quite uncertain itself. For instance, in (Blanchemanche, Rona-Tas, Cornuéjols, Duroy, & Martin, 2013) an ontology of scientific uncertainty with specific reference to food risk assessment is considered: differently from the W3C proposal it focuses on the origin of the uncertainty (e.g. data vs. model) rather than on its form (e.g. probability functions vs. belief functions). This shows that quite different perspectives about uncertainty ontology can be, and actually are being, considered: selecting the most appropriate one(s) and/or integrating them in a holistic view is definitely a long way to go.

---

[4] *http://www.w3.org/2005/Incubator/urw3/wiki/UncertaintyOntology.html*

Alternative approaches to deal with this partly discouraging situation from an argumentation mining perspective are discussed in the next section.

5. MANY ROADS TOWARDS THE SAME DESTINATION

Suppose one hypothetically formulates the following well-ordered stepwise research roadmap for encompassing uncertainty and fuzziness within argumentation mining:

1. define a general classification of uncertainty/fuzziness types
2. provide a characterization of the uncertainty/fuzziness types relevant to each argumentation scheme;
3. define a formalism for the representation of uncertainty/fuzziness assessments (of various types) in actual arguments, i.e. in instances of argument schemes;
4. define mechanism(s) to derive an uncertainty/fuzziness assessment for the conclusion of an argument from the assessments concerning the premises and the relevant argumentation scheme;
5. use all the ingredients above for a rich argument mining activity capturing the uncertainty and fuzziness embedded in natural language arguments.

From the discussion in the previous section it turns out that completing this roadmap would be hopeless for years to come, given the huge unresolved challenges inherent to the very first ontological step and to the problems evidenced for step 2 too.

Several strategies can be considered in front of these difficulties.

The simplest and most obvious one consists in ignoring problems that are too difficult to solve: this amounts to overlook the possible presence of uncertainty and fuzziness in natural language arguments and go on with mining activities whose outcomes use formalisms where these aspects are not explicitly encompassed.

Rough as it may seem, this approach basically corresponds to the state of the art, not without some good reasons: mining uncertainty/fuzziness-free arguments is already an extremely difficult task, lying at the frontier of current research. Hence it does not seem wise to consider further complications before having achieved some consolidated solutions to the simpler version of the problem. One could consider the integration of uncertainty and fuzziness within these solutions in a subsequent stage, after the most fundamental aspects of the mining problem have been worked out. A possible objection to this approach would be that uncertainty and fuzziness are among the most fundamental aspects of the mining problem and ignoring them at the beginning is like ignoring them forever. To put it in other words, one might argue that models encompassing them could not be obtained by adapting the uncertainty/fuzziness-free models but would need to be reconsidered from scratch.

Another way to limit complexity is to consider specific application domains where both the variety of arguments to be mined and of the corresponding linguistic expressions is restricted so that satisfactory results can be obtained by some tailored methods trading generality with simplicity and (domain dependent) effectiveness. This kind of specialization has been already considered in several (uncertainty/fuzziness-free) literature approaches. For instance in (Mochales & Moens, 2011) a corpus of well structured legal texts of the European Court of Human Rights is considered, while the work in (Wyner, Schneider, Atkinson, & Bench-Capon, 2012) deals with the domain of product reviews. Adding some limited consideration of uncertainty and/or fuzziness in these domains might turn out to be feasible and provide hints for more general developments in further steps. A pessimistic objection would be that such domain specific achievements have most probably an ad-hoc nature and hence could not be generalized outside their original application domain.

Opposed to the two kinds of "keep it simple first" approaches mentioned above, one can consider at least two varieties of a more principled "start from the foundations" attitude.

On the one hand, a systematic reassessment of semi-formal argumentation schemes, and in particular of argumentation schemes, with explicit consideration of uncertainty and/or fuzziness could be undertaken. This could be done incrementally, by using as a basis the uncertainty related elements already present in the current proposals, e.g. the critical questions in argument schemes. One may object however that this kind of attempt will easily get lost in the plethora of possible different schemes, questions and uncertainty forms without ever achieving a sufficient generality.

In fact, more radically, one may use the difficulties encountered by the attempts to define an ontology of uncertainty as a basis to investigate a new foundational approach able to relate uncertainty, fuzziness, argumentation and their presence in natural language within a unitary and coherent view, ideally based on a few key concepts.

Such a radical foundational approach can be regarded, depending on reader's attitude, either as the only way to go in order to achieve significant long-term results or as a hopeless Holy Grail quest.

Thus, in a typical argumentation style, we have sketched above both some justifications and some rebuttals for the alternative investigation lines one might pursue without expressing any definite judgment about them. Indeed, we don't see any of them as more appropriate or more promising: rather we suggest that while they are alternative at an individual level, in the sense that it is reasonable to assume that they are developed separately by different researchers or research groups, they can be complementary at a community level, given that the results obtained in one direction can be profitably reused in another one.

Indeed, all these different efforts may contribute, through cross-fertilization, to the achievement of their shared long-term goal: e.g. a new kind of argument scheme may find a profitable application in a specific domain, and, in turn, even some ad-hoc results may possibly provide the starting point for new theoretical developments.

This is emphasised by the interdisciplinary nature of argumentation mining, involving a broad range of competences from linguistics and (formal and informal) argumentation to machine learning and uncertainty models, which calls for openness in the relationships among the parts of the community and in the mutual evaluation of their different efforts.

In this sense we believe that, in the long run, openness and availability to cross-fertilization, rather than the identification of the "right" investigation line to follow, will be the key to successfully tackle the research challenges discussed in this paper.

6. CONCLUSION

In this discussion paper, based on a contribution to the ECA 2015 panel "Arguments in Natural Language: The Long Way to Analyze the Reasons to Believe and the Reasons to Act", we have examined some of the main issues and current research challenges concerning the representation of uncertainty and fuzziness in argumentation models suitable for argumentation mining tasks.

It turns out that in the semi-formal argumentation schemes, which, as to recent research trends, appear to be best suited for initial argumentation mining activities, the treatment of uncertainty and fuzziness has received poor attention with respect to more abstract (hence more distant from mining) formalisms.

Filling this gap appears far from being a "conventional" research exercise, as it touches some unresolved fundamental issues, concerning in particular the elusive very nature of uncertainty itself.

A spectrum of investigation approaches, from temporarily ignoring the problem because too difficult to carrying out deep and detailed conceptual analyses, has been discussed, suggesting that a broad community effort, open to explore different directions and to allow their cross-fertilization, appears the best way to go.

ACKNOWLEDGEMENTS: This work has been partially developed during a visit of Pietro Baroni to Zhejiang University funded by the National Science Foundation of China (No.61175058) and Zhejiang Provincial Natural Science Foundation of China (No. LY14F030014). Beishui Liao has been partially supported by the Inter Mobility Project "Formal Models for Uncertain Argumentation from Text" funded by the National Research Fund Luxembourg (FNR).

REFERENCES

Amgoud, L., Cayrol, C., Lagasquie-Schiex, M. C., & Livet, P. (2008). On Bipolarity in Argumentation Frameworks. *International Journal of Intelligent Systems, 23*(10), 1062-1093.

Baroni, P., Giacomin, M., Liao, B., & van der Torre, L. W. N. (2014). Encompassing uncertainty in argumentation schemes. In E. Cabrio, S. Villata, A. Wyner (Eds.) *Proc. of the Workshop on Frontiers and Connections between Argumentation Theory and Natural Language Processing* (pp. 79-85). CEUR Workshop Proceedings, Vol. 1341.

Besnard, P., & Hunter, A. (2014). Constructing argument graphs with deductive arguments: a tutorial. *Argument & Computation, 5*(1), 5-30.

Blanchemanche, S., Rona-Tas, A., Cornuéjols, A., Duroy, A., & Martin, C. (2013). An Ontology of Scientific Uncertainty: Methodological Lessons from Analyzing Expressions of Uncertainty in Food Risk Assessment. *Proc. of the ISA/ESA Mid-term conference Risk and Uncertainty: ontologies and methods.*

Brewka, G., & Woltran, S. (2010). Abstract Dialectical Frameworks. In F. Lin, U. Sattler & M. Truszczynski (Eds.), *Proc. of KR 2010, 12th Int. Conf. on Principles of Knowledge Representation and Reasoning* (pp. 102-111). Palo Alto: AAAI Press.

Dung, P. M. (1995). On the acceptability of arguments and its fundamental role in nonmonotonic reasoning, logic programming, and n-person games. *Artificial Intelligence, 77*(2), 321–357.

Freeman, J. B. (1991). *Dialectics and the macrostructure of arguments.* Berlin: Foris.

Freeman, J. B. (2011). *Argument structure: Representation and theory.* Berlin: Springer

Garcia, A. J., & Simari, G. R. (2014). Defeasible logic programming: DeLPservers, contextual queries, and explanations for answers. *Argument & Computation, 5*(1), 63-88.

Hunter, A. (2013). A probabilistic approach to modelling uncertain logical arguments. *Int. Journal of Approximate Reasoning, 54*(1), 47–81.

Janssen, J., De Cock, M., & Vermeir, D. (2008). Fuzzy argumentation frameworks. In L. Magdalena, M. Ojeda-Aciego & J. L. Verdegay (Eds.), *Proc. of IPMU 08, 12th Int. Conf. on Information Processing and Management of Uncertainty in Knowledge-Based Systems* (pp. 513–520).

Li, H., Oren, N., & Norman, T. J. (2011). Probabilistic argumentation frameworks. In S. Modgil, N. Oren, F. Toni (Eds.), *Theory and Applications of Formal Argumentation - TAFA 2011 Revised Selected Papers* (pp. 1–16). Berlin: Springer.

Mercier, H., & Sperber, D. (2011). Why do humans reason? Arguments for an argumentative theory. *Behavioral and Brain Sciences, 34,* 57-111.

Mochales, R. & Moens, M.-F. (2011). Argumentation mining. *Artificial Intelligence and Law, 19,* 1-22.

Modgil, S., & Prakken, H. (2014). The ASPIC+ framework for structured argumentation: a tutorial. *Argument & Computation, 5*(1), 31-62.

Peldszus, A., & Stede, M. (2013). From argument diagrams to argumentation mining in texts: a survey. *Int. Journal of Cognitive Informatics and Natural Intelligence, 7*(1), 1-31.

Pollock, J. L. (1992). How to reason defeasibly. *Artificial Intelligence, 57*(1), 1-42.

Tamani, N., & Croitoru, M. (2014). Fuzzy Argumentation System for Decision Support. In A. Laurent, O. Strauss, B. Bouchon-Meunier & R. R. Yager (Eds.), *Proc. of IPMU 2014, 15th Int. Conf. on Information Processing and Management of Uncertainty in Knowledge-Based Systems* (pp. 77–86). Cham: Springer International Publishing.

Toni, F. (2014). A tutorial on assumption-based argumentation. *Argument & Computation, 5*(1), 89-117.

Toulmin, S. E. (1958). *The uses of argument.* Cambridge: Cambridge University Press.

Walton, D. (1996). *Argumentation schemes for presumptive reasoning.* Mahwah: Erlbaum.

Walton, D., Reed, C., & Macagno, F. (2008). *Argumentation schemes.* Cambridge: Cambridge University Press.

Wigmore, J. H. (1931). *The principles of judicial proof.* Boston: Little, Brown & Co.

Wyner, A., Schneider, J., Atkinson, K., & Bench-Capon, T. (2012). Semi-automated argumentative analysis of online product reviews. In B. Verheij, S, Szeider & S. Woltran (Eds.) *Proc. of COMMA 2012, 4th Int. Conf. on Computational Models of Argument* (pp. 43-50). Amsterdam: IOS Press.

8

From Beliefs to Truth via Argumentation: Intentionality, Multi-Agent Systems and Community Agreement

ANDRÉ BAZZONI
University of California, Berkeley, USA
bazzoni@berkeley.edu

This paper offers an alternative perspective on the relationship between truth, knowledge and belief based on a novel treatment of the semantics of belief reports. Instead of seeking knowledge through the given notion of truth, we shall *build up* truth through belief and argumentation. The interaction of beliefs inside a multi-agent system is crucial to the constitution (via argumentation) of community agreement, which is in turn construed as the building block of the concept of truth.

KEYWORDS: actual world, argumentation, belief, collective intentionality, community agreement, intentional worlds, possible-world semantics, truth

1. INTRODUCTION

One of the oldest topics in philosophy is the nature of the relationship between truth, knowledge and belief. The typical way of dealing with the issue in the philosophical tradition may be illustrated by the so-called tripartite analysis, according to which the notion of justified true belief specifies the necessary and sufficient conditions for knowledge.

The main goal of this paper is to suggest an alternative perspective based on a novel treatment of the semantics of belief report (and by extension, any propositional attitude, including knowledge), while drawing inspiration from Hintikka's (1962) classical system of epistemic logic, to be presented and dealt with in an informal setting.

Instead of seeking knowledge through the given notion of truth, we shall *build up* truth through the notions of belief and argumentation. More specifically, the interaction of beliefs inside a multi-agent system will be crucial to the constitution (via argumentation) of community agreement, which in turn will be construed as the building block of the

D. Mohammed & M. Lewiński (eds.) (2016). *Argumentation and Reasoned Action: Proceedings of the 1st European Conference on Argumentation, Lisbon, 2015. Vol. II, 241-250.* London: College Publications.

notion of truth. The latter will then be defined not with respect to some independently existent material world, but rather in relation to the logical world consisting of an ever-changing repository of updated knowledge.

This paper is implicitly divided in two parts. In the first part, I introduce (section 2) Hintikka's semantics for belief statements in terms of standard possible-world semantics; and I offer (section 3) a new semantics based on the notion of a possible intentional world, which may be viewed as a simplification of Hintikka's system.

In the second part, I consider (section 4) two possible and different understandings of the concept of "actual world", and provide reasons for working with one of them as regards general philosophical inquiry—which would lead, as we shall see, to attractive ways of addressing old philosophical issues such as the problem of skepticism and the materialism/idealism debate—; and I show (section 5) how the new semantics for belief reports and the retained conception of actual world can work together in the context of multi-agent systems whose internal argumentative interactions yield community-agreement about truth matters, to build up a more philosophically interesting notion of "truth in the actual world".

2. HINTIKKA'S ANALYSIS OF BELIEF

In this first part of the paper, I briefly present Hintikka's semantics for belief (knowledge is formally treated in a similar way), according to which an agent **A** believes some proposition to be true with respect to some possible world **W** if and only if the proposition is true in all possible worlds compatible with *what **A** believes in **W***. Bearing upon this latter italicized expression, I shall introduce in the next section a new kind of possible worlds, which will play the central role in the new semantics of belief.

Hintikka (1962) offers a semantic analysis of belief reports in terms of possible-world semantics. The characteristic feature of this kind of framework is that it interprets sentences of a language not only with respect to the actual state of affairs, but also with respect to *any possible* state of affairs.

To get a feeling for the semantics, consider the sentence: "The capital of France is in Texas". As we know, this sentence is false. On the other hand, we only evaluate it as such because we implicitly consider the world as it actually is as the point of evaluation of the sentence, that is to say, as the state of affairs against which we confront our sentence in order to evaluate it as true or false. The capital of France is actually in

Ile-de-France, but it could have been in Texas, had the history of France been unfolded otherwise.

Possible-world semantics generalizes the notion of truth, and construes the actual world (or the actualized state of affairs) as only one among all possible worlds. What makes the actual world a special one is that it is the one that happens to be actualized, but the other, non-actualized worlds also have important semantic roles to play. For example, conditional sentences such as, "If the capital of France were in Texas, people in Austin would speak French" are elegantly interpreted in terms of possible worlds by paraphrasing it as: "In every possible world in which the capital of France is in Texas, people in Austin speak French". General methods of philosophical argumentation also vastly use the idea of a possible world, for instance in thought experiments such as Putnam's Twin-Earth (cf. Putnam, 1975), in which we are invited to conceive a possible world in which the earth has a counterpart planet resembling it in every respect except for the chemical constitution of water.

The interpretation of modal talk in general, which could not be accounted for in extensional (i.e., based on single points of evaluation) frameworks, is now very naturally captured in terms of quantification over possible worlds. A proposition **P** is *necessary* in world **W** if and only if it is true in *all* possible (with respect to **W**) worlds. **P** is *possible* in **W** if and only if it is true in *some* possible (with respect to **W**) world.

The upshot of Hintikka's analysis was to take belief (and knowledge) as a special kind of modal operator. Specifically, belief is treated on a par with necessity, that is, in terms of universal quantification over possible worlds, the difference being that now instead of a possibility relation between worlds, we have a kind of *compatibility* relation. In practice, this means that given the point of evaluation of a belief report such as, "Ralf believes that the capital of France is in Texas", say the actual world **w**, we wish to consider not the possible worlds relative to **w**, but rather the *compatible* worlds with respect to *what Ralf believes in* **w**. As announced, Ralf's belief will be interpreted by universally quantifying over those compatible worlds, in the following way.

We say that "Ralf believes that the capital of France is in Texas" is true in **w** if and only if "The capital of France is in Texas" is true in all possible worlds compatible with what Ralf believes in **w**. This is indeed intuitive, because if Ralf believes that the capital of France is in Texas, then any world in which the capital of France is *not* in Texas must be incompatible with what Ralf believes; and conversely, if the capital of France is in Texas in any world compatible with what Ralf believes, then Ralf must believe that the capital of France is in Texas, for otherwise the fact that the capital of France is *not* in Texas would be compatible with what Ralf believes, that is to say, there would be some world **w'** compatible with what Ralf believes, in which the capital of France is *not* in Texas, contrary to the assumption.

While it is true that Hintikka's apparatus has since then given room to a number of different systems of epistemic logic, the clause for interpreting the belief operator survives as the basic idea on top of which most semantic analyses of belief are carried out.

3. BELIEF REPRESENTATION AND POSSIBLE INTENTIONAL WORLDS

Rather than interpreting "**A** believes **P** in **W**" by examining whether **P** holds in the totality of possible worlds compatible with the beliefs of **A** in **W**, as in Hintikka's system, we shall restrict the analysis to a single possible world constituted by *what* **A** *believes in* **W**. Such a possible world is simply constituted by the propositions that **A** believes to be true in **W**; this will be called the *possible intentional world of* **A** *with respect to* **W**.

Intuitively, an agent's possible intentional world with respect to some world tells us how that agent understands that world. For instance, if the propositions that John believes to be true in the actual world are, "Berkeley is in California" and "Barack Obama is the present King of France", then these two propositions by themselves constitute a possible world (recall that a possible world may be viewed as the set of propositions that are true in that possible world), namely, John's possible intentional world with respect to the actual world. The term "intentional" comes from Searle's (1983) notion of *Intentionality*—which is in fact a notion with an old philosophical tradition going back to the Scholastics of the Middle Ages; cf. also Brentano, 1873). In short, Searle considers belief to be a particular kind of Intentional state.

Two important characteristics of possible intentional worlds should be emphasized. First, they are partial (unlike "ordinary" possible worlds) as not every proposition necessarily has a truth-value in a given possible intentional world. Second, the notion of possible intentional world is objective, not subjective, in the sense of Frege (1892). Just as understanding the *Sinn* of some proposition is an objective activity shared by the members of a linguistic community, evaluating a proposition is also an objective activity partaken by the agents.

It is curious to find in the work of Quine (1969, p. 309), who rejected any non-extensionalist[1] framework, a paragraph describing

---

[1] One should not at this point confound, however, Intentionality (with a "t") with intensionality (with an "s"). The latter is the antonymous of

almost literally what I have just presented as a possible intentional world (although he calls it a "man's theory"):

> a man's theory on a given subject may be conceived, nearly enough, as the class of all those sentences [...] that he believes to be true.

If the "given subject" is the actual world, for example, then Quine's statement describes roughly but clearly a man's possible intentional world with respect to the actual world. Quine also expresses in the same paragraph his intuition about the objectivity of possible intentional worlds, remarking that "[n]ext we may picture a theory, more generally, as an imaginary's man theory, even if held by nobody". The "impersonal" general theory is simply obtained by abstracting away from the particular man and keeping the set of sentences used to characterize that particular man's theory. The subjective figure of Quine's man is only a dispensable support for the abstract and objective set of sentences that the man believes to be true in the actual world.

We are now in a position to state the new semantics for belief as follows:

> **A** believes **P** in **W** *if and only if* **P** is true in **A**'s possible intentional world with respect to **W**.

This saves all reference to accessible worlds and quantification over them contained in the classical analysis by Hintikka, without losing any intuition at the truth-conditional level (in other words, both interpretations are semantically equivalent).

Formally, a possible intentional world is as uncontroversial as any partial function.[2] Conceptually, it is as plausible as any ordinary possible world, for it is also described as a certain set of propositions of the underlying language.

Chisholm (1963) complained about Hintikka's semantics that it treats knowledge and belief in the same terms as necessity, which fact is, if not entirely unintuitive, at least unmotivated. The semantics I am

---

"extensionality", and the notion of a possible intentional world presupposes the existence of non-extensional contexts. However, as we shall see next in our semantic treatment of belief reports, it is possible to have an extensionalist analysis of language within an intensionalist framework (allowing for various possible worlds). I shall not discuss this issue in any detail here.

[2] There actually is, to be true, a formal difficulty related to the compositional treatment of belief disjunctions such as, "Ralf believes that Mary or Jane won the prize". I cannot treat this matter here, though.

proposing here meets Chisholm's concerns, for it saves all reference to accessible worlds and quantification over them (and hence any analogy with necessity) contained in the classical analysis by Hintikka, without losing any intuition at the truth-conditional level. Moreover, also Hintikka's referentialist inclinations (see his 1969) are fully met by the new semantics (and not by his own framework, due precisely to his use of quantification over possible worlds), since we can now speak of denotation in a single possible intentional world, instead of having to inspect denotation in a whole set of compatible (i.e., accessible) worlds.

4. TRUTH IN THE ACTUAL WORLD

The second part of the paper will focus on the notion of *truth in the actual world*, which corresponds to our by-default (pre-modal) understanding of the concept of truth. This section investigates the notion of actual world, while the next section connects it with our previously defined idea of a possible intentional world.

What is the actual world in the first place? Surely, we may assume the existence of an external material world. Yet even if we do, we may still ask whether it is the locus of evaluation of certain propositions.

The crucial point is that we should draw a line between the material world and the *logical* world. The former is the target of our investigations, while the latter is our point of evaluation. The former exists (if its existence is not accepted, focus is more directly given to the logical world) independently of whether we think or talk about it. However, the latter is the repository of updated information that we acquire *about* the former (or about our perceptions and the like, if the material world is to be rejected).

For instance, when we say that the earth revolves around the sun, we rely on a scientific discovery that reversed the formerly true Ptolemaic system. This is perhaps a case that we feel could not easily be revised by new scientific research. But this is clearly a matter of degree: before the advent of the Relativity Theory, it was commonly thought that science could hardly ever challenge the absoluteness of time. Similarly, suppose physicists come up with an answer to the currently open question of whether the universe will expand forever. This is a yes-or-no question, so suppose that the scientific answer is "yes", hence the sentence, "The universe will expand forever" is true in the actual world. Does it mean that it is true in the external material world? Not necessarily, as it happens, because scientists might one day discover a mistake in the calculations, and revise the answer to "no". And even if the original answer is correct, how could we be ever sure that is so?

Instead of treating the actual world as the external world ("the world out there"), which is a problematic object that gives rise to endless pseudo-philosophical debates such as the materialism/idealism dispute, we may now shift our investigation to the *logical* actual world. Indeed, the external world is the subject of our conceptual apparatus, but its object is the logical actual world, which is a cognitive construction, not an external datum. Human knowledge is the enterprise of *constructing* a logical actual world *on the model of* the external actual world.

This logical world is the result of the interactions between the relevant intentional worlds (or to use Putnam's words, between the relevant body of "experts"—cf. Putnam, 1975; more on relevant worlds below), depending on the nature of the subject or proposition to be assessed.

The logical actual world is according to the new setting a partial world built up from possible intentional worlds, not a metaphysically autonomous total world. The collapse of the presupposition that the actual world is assimilated to the external world thus places the actual world within the class of partial worlds—recall Quine's picture about abstracting away from the imaginary man.

Moreover, the distinction could shed new light on other important philosophical notions apart from truth. To cite but two canonical examples, consider the old cases of the debate between materialism and idealism, and the problem of skepticism.

In our context, materialism may be described as the view that among the total (i.e., "ordinary") possible worlds as defined by standard possible-world semantics, there is a particular world, which is the external world. Idealism, on the other hand, is the theory that no such total world is in fact actualized. The reason why this is, to my view, a pseudo-philosophical debate is that human knowledge is concerned with the construction of a partial logical world based on interactions with the environment (whether human, material, perceptive, conceptual, and so on), not with some metaphysically (isolated, as it were) total world.

This also seems to be the reason why skepticism is at the same time so resistant and so obviously beside the point: on the one hand, a total world is indeed inaccessible to knowledge (for one reason, because there would be no way of knowing that we have achieved knowledge about such an absolute totality; i.e., no way of knowing that we know); on the other hand, however, such a total world (be it made of physical particles or ideas, or whatsoever) is only the subject, not the object of human knowledge. Once we realize that what we purport to know is the content of a partial logical world, we need only set aside the skeptic agent locked into her total world, and get busy building up the real object of our knowledge. We simply claim that we know based on what we believe—or as Dretske (1983) puts it, we cannot believe what we cannot know—, without thereby claiming that what we know is the content of the total world where our skeptic colleague lives her unknowable life.

Recall Quine's imaginary man again: the actual world that human knowledge is concerned with is the partial logical world built up by collective Intentionality and abstracted away from the respective agents. This world is immune to skeptical arguments, for the simple reason that, if skeptical arguments were to threaten the possibility of knowledge, then they would harm the possibility of belief as well, and this would disarm the very basis of any skeptical argument, as stressed by Dretske.

The point is thus: *given* the existence of the material world, it is still not the case that the logical actual world constitutes a perfect match of the external material world. Many truths belonging to the logical world are not truths in the material world (witness the never-ending process of scientific revisions); and most truths in the material world are not even formulated as thoughts and statements, let alone shown to be true in the logical actual world. For example, it must be true or false in the material world whether or not the first asteroid to hit the moon was the size of a watermelon. However, not only do we not know whether it is true or false, but most probably I was also the first person to have ever formulated the question.

5. TRUTH VIA COMMUNITY-AGREEMENT

This last section is devoted to the connection between possible intentional worlds and the logical actual world. The upshot of our discussion so far is that the general expression "is true in the actual world" should in fact be understood as "is true-*for-all-we-know* in the actual world". Then since this for-all-we-know qualification introduces an epistemic element into the notion of truth, we may rephrase the expression as "is true in the logical actual world", which is the kind of actual world that takes into account epistemic factors.

Now what is this "we" in the expression "for-all-we-know"? This is of course an essential part of the whole idea. What I wish to suggest here is that "we" is a kind of "we-intentionality", or *collective intentionality* (cf. Schweikard & Schmid, 2013), and it is constituted by community agreement. Thus the main point that I wish to emphasize in this respect is that even if each agent is associated with a possible intentional world (with respect to the world of evaluation, in our case the logical actual world), not all of those intentional worlds will be relevant to the evaluation of a sentence in that world. For example, I have no role to play in the evaluation of a sentence as, "The universe will expand forever", because I do not belong to the community of relevant agents that are in a position to determine the truth of that sentence. The logical actual world is a kind of compilation of data gathered from *relevant intentional worlds*, or if we follow the main idea in Putnam's (1975) concept of the *division of linguistic labor*—now in connection with the notion of truth instead of meaning as in Putnam's original context—, from intentional worlds associated with the relevant (broadly conceived notion of) body of experts.[3]

Therefore, through individual or collective processes single agents come to form their own possible intentional worlds (with respect to the actual world), but any single sentence *S* singles out among them the ones that are relevant to the evaluation of *S* in the actual world. In other words, *S* determines the set of relevant intentional worlds that get to interact through argumentative processes (e.g., debate, reasoning, persuasion, experimentation) that may in turn lead to the constitution of community agreement, thus determining whether *S* is true in the actual world. The logical actual world is then supplied with the statements judged true by the relevant bodies of experts associated with those sentences.

A final note on the term "argumentation": it is intended here to remain as neutral as possible as to its various characterizations in the literature (whether attached to the notion of rationality, logicality, coalescence, and so forth). Argumentation is thus construed as a type of "glue" for assembling the two fundamental notions dealt with in the preceding sections of this paper, namely (multi-agent) belief and truth.

6. CONCLUSION

This article is an attempt to vindicate the idea that the most interesting conception of truth as regards philosophical analysis is a complex construction grounded on (multi-) agents' epistemic states, not any given absolute disposition of facts. We have seen that the way in which this construction is attained may be described through the mediation of argumentative processes in various degrees depending on what is being

---

[3] In a more fine-grained analysis, we should distinguish the degrees of relevance in the activity of constituting community agreement. For instance, there is a minimal degree of relevance associated with a statement such as, "I'm hungry", whose evaluation is in normal situations exclusively attributed to the utterer (i.e., to its possible intentional world).

evaluated (the extreme case being the one in which argumentation would collapse into the constitution of a single intentional world), which leads to community agreement in multi-agent systems, and then finally to truth in the logical actual world, as opposed to the absolute external world. To achieve this, we crucially used the notion of a relevant intentional world, which comes from our notion of a possible intentional world, introduced in our new semantics of belief reports derived from Hintikka's original system.

ACKNOWLEDGEMENTS: I thank the audience of *ECA Lisbon 2015* for insightful remarks and suggestions on the content of the present paper. This work is part of a research project supported by FAPESP.

REFERENCES

Brentano, F. (1873). *Psychology from an empirical standpoint.* London: Routledge and Kegan Paul.

Chisholm, R. (1963). The logic of knowing. *The Journal of Philosophy, 60*(25), 773-795.

Dretske, F. (1983). The epistemology of belief. *Synthese, 55*(1), 3-19.

Frege, G. (1892). Über Sinn und Bedeutung. *Zeitschrift für Philosophie und philosophische Kritik NF, 100*, 25-50.

Hintikka, J. (1962). *Knowledge and Belief.* Cornell: Cornell University Press.

Hintikka, J. (1969). Semantics for propositional attitudes. In J.W. Davis, D.J. Hockney & W.K. Wilson (Eds.), *Philosophical logic* (pp. 21-45). Dordrecht: Reidel.

Putnam, H. (1975). The meaning of 'meaning'. *Minnesota Studies in the Philosophy of Science, 7*, 131-193.

Quine, W.V. (1969). Reply to Chomsky. In D. Davidson & J. Hintikka (Eds.), *Words and objections* (pp. 302-311). Dordrecht: Reidel.

Schweikard, D.P., & Schmid, H.B. (Summer 2013 Edition). Collective intentionality. In E.N. Zalta (Ed.), *The Stanford Encyclopedia of Philosophy.* URL =<http://plato.stanford.edu/archives/sum2013/entries/collective-intentionality/>.

Searle, J. (1983). *Intentionality.* Cambridge: Cambridge University Press.

9

Criteria for the Reconstruction and Analysis of Doctors' Argumentation in the Context of Chronic Care

SARAH BIGI
Università Cattolica del Sacro Cuore, Italy
sarah.bigi@unicatt.it

NANON LABRIE
University of Lugano, Switzerland
nanon.labrie@usi.ch

The study of medical argumentation typically draws on empirical data. This poses a challenge, as in medical practice, utterances that serve an argumentative purpose are intertwined with utterances that serve different dialogical - as well as clinical - goals. Comprehensive understanding of the communicative context is thus imperative. This contribution seeks to provide a blueprint for the analysis of medical argumentation in context. Doing so, chronic care consultation is used as a particular example of doctor-patient interaction.

KEYWORDS: activity type, argumentative analysis and reconstruction, chronic care consultation, empirical data, medical argumentation

1. INTRODUCTION

The past years have seen an increase in the study of medical argumentation. Several scholars have studied empirical data using both qualitative and quantitative approaches (Bigi, 2014a; Labrie, 2012; 2015; Snoeck Henkemans & Wagemans, 2013). Also in the field of health communication, several authors have recognized and advocated the relevance of studies taking an argumentation theoretical approach (e.g., Labrie, 2015; Labrie & Schulz, 2015; Rubinelli, 2013; Salmon, 2015). It has been argued that the use of argumentation in medical consultation contributes to our understanding of treatment decision-making processes and that, moreover, doctors' provision of

D. Mohammed & M. Lewiński (eds.) (2016). *Argumentation and Reasoned Action: Proceedings of the 1st European Conference on Argumentation, Lisbon, 2015. Vol. II, 251-265.* London: College Publications.

argumentation positively contributes to outcomes of the consultation (Labrie & Schulz, 2014; 2015).

The study of medical argumentation typically relies on empirical data, i.e. video and audio recordings or transcriptions of real life interactions. The analysis of such data provides an opportunity to test and expand theoretical models. Moreover, it allows the analyst to observe and assess the validity of argumentation in practice. The analysis of empirical data is a complex task. Theoretical models of argumentation usually describe an ideal. Yet, in reality argumentation may be difficult to recognize, e.g., for practical reasons, such as cluttering, mumbling, and language ambiguities. Moreover, in medical consultation argumentation occurs along with other forms of discourse and is not nearly always the most common form of interaction, because argumentative goals are intertwined with clinical goals. Consequently, it may be difficult to distinguish between argumentation and speech acts that serve a different dialogical goal. It is, therefore, crucial that argumentation scholars have solid knowledge of the medical context in order to allow for systematic, comprehensive, and meaningful argumentative analyses.

In this contribution, we propose a blueprint for the reconstruction and interpretation of real-life medical argumentation, starting from the analysis and evaluation of chronic care consultation – a specific subtype of the argumentative activity type "medical consultation". Argumentation plays a prominent role in chronic care consultation due to its inherent goals, i.e. patient motivation, behavior change and autonomy (Katon, Von Korff, Lin, & Simon, 2001; Wagner, Austin, Davis, Hindmarsh, Schaefer, & Bonomi, 2001). Moreover, to date, the majority of studies on medical argumentation have focused on general practice, rather than on any other subtypes of consultation. In section 2, the institutional preconditions for medical argumentation in the context of chronic care will be discussed. Section 3 offers the analysis of a real life case and demonstrates the relevance of this blueprint.

2. THE INSTITUTIONAL PRECONDITIONS FOR ARGUMENTATIVE DISCUSSION IN MEDICAL ENCOUNTERS

Over the years, medical consultation has been described as a communicative activity type in which argumentation plays an important role (van Eemeren, 2010; Bigi, 2012; Labrie, 2012; 2013; Pilgram, 2009). Schulz and Rubinelli, for example, refer to medical consultation as an info-suasive dialogue, i.e. "a dialogue blending information and persuasion in an inextricable manner" (2008, p. 426). Yet, applying this definition without further specifications to just any kind of doctor-patient encounter seems overly simplistic and the label "medical consultation" as a blanket term for the communicative activity type requires specification.

First, a distinction should be made between encounters in the context of acute and chronic care. This distinction is based on the nature of the health problem. Acute conditions (e.g., a cold or a broken bone) are diseases that come abruptly and run a short and rapid course. The kind of care implemented for these conditions is generally reactive and episodic (Nuño, Coleman, Bengoa, & Sauto, 2012). Chronic conditions (e.g., diabetes or epilepsy) typically last for an extended period of time and are recurrent in nature. They have a slow onset, a progressive development, and require complex treatment (Nuño et al., 2012).

Second, chronic care consultation should be distinguished from general practice consultation. General practitioners (GPs) typically play the role of gatekeeper. GPs decide to refer a patient to a specialist when the situation is unclear or needs specific attention or treatment. They may also actively collaborate with specialists when patients need support to follow up on the specialist's recommendations.

Based on these distinctions, we argue that Schulz and Rubinelli's (2008) general description of medical consultation as an info-suasive dialogue must be specified. In what follows, we propose a series of methodological steps for doing so. In the next section, first, the institutional preconditions for argumentative discussions in chronic care consultation will be outlined: The initial situation, the starting points, means of argumentation and criticism, and the possible outcomes of an argumentative discussion (van Eemeren & Houtlosser, 2005).

2.1 The initial situation

In medical consultation, argumentative discussions arise when one of the parties disagrees with, or has doubts about, the other party's views or ideas. Most often, although not necessarily, the doctor acts as the protagonist of the discussion. Medical discussions typically concern the interpretation of symptoms, the recommended behaviors and therapeutic actions, and the handling of administrative tasks, such as appointment making, financial matters, and issues concerning the insurance. Often, due to traditional authority roles, patients' disagreement remains implicit. Reversely, doctors may avoid verbal expressions of disagreement with their patients for reasons of politeness. This may make it difficult for the analyst to distinguish between an argumentative discussion and a mere misunderstanding.

This is relevant, from an analytical point of view, because the analyst has to interpret all subsequent dialogical moves depending on the initial situation.

In chronic care, doctors and patients have long-term relationships. This implies that discussions may be ongoing, over the course of several encounters. The analyst should, therefore, ideally aim to collect longitudinal data. Also the consultation phase in which the observed exchange occurs should be considered. For example, the following fragment occurs during the history-taking phase of a diabetes consultation.[1] The doctor observes a very high glucose value in the patient's journal and asks the patient whether she remembers what might have been the cause of the anomaly:

> (1) Patient: I can tell you what happened: I came home after running errands all morning, and I was feeling so weak... I had lunch and then sat in my armchair and fell asleep. I am sure this blocked my digestion and made my blood sugar go up!
> Doctor: Well, no, I don't think this is exactly what happened. If you say you had been out all morning and that you were feeling very weak, perhaps you tried to compensate by eating a little more than usual and this is what made the levels of your blood sugar rise.

This passage, considered out of context, could be interpreted as a difference of opinion between the doctor and patient. However, given the particular consultation phase in which the fragment appears and, additionally, the clinical goal of educating patients about their conditions in view of self-management abilities, it seems more accurate to interpret this exchange as a case of explanation. The doctor explains the patient's symptoms and seeks to modify the patient's beliefs by making certain causal relations explicit – rather than aiming to resolve a disagreement. A different case is (2), which is taken from the treatment decision making phase of another diabetes consultation:

> (2) Doctor: You should try to get some more physical activity
> Patient: I know but I have this leg here that hurts when I walk...
> Doctor: All right, but a fifteen minutes' walk in the park could be enough, and you can stop here and there if you need to

---

[1] All the examples are taken from a corpus of videorecordings of consultations in diabetes care (see, Bigi, 2014a). The original data is in Italian and has been translated by the authors for the purposes of the present contribution.

Here, doctor and patient discuss the patient's need to lose weight. They have different opinions about the attainable goals. This discussion takes place during the so-called patient education and counseling phase. In this phase, due to the clinical goals of the consultation, conflicts of opinion are more likely to occur.

2.2 The starting points

Theories of argumentation generally assume that, in order for a discussion to develop effectively, the parties will have to agree on a certain set of procedural and material starting points (e.g., van Eemeren & Grootendorst, 2004). As is often the case, in chronic care consultation, procedural starting points are mostly left implicit and largely depend on the cultural, administrative, and bureaucratic organization of the health care system.

Material starting points are more often discussed explicitly. Doctor and patient may discuss them within a single consultation or throughout a series of encounters. They are made explicit in an unsystematic way and this process largely depends on the doctors' ability to elicit the relevant information from patients. It is therefore crucial to study the history-taking phase when analyzing argumentation in chronic care consultations. During this phase, many of the medical facts, issues, and conditions emerge that are subsequently used by doctors and patients while arguing in favor or against certain behaviors and therapeutic actions.

2.3 Means of argumentation and criticism

In chronic care consultations there are no predefined, formal rules as to what kind of argumentation and criticism is allowed. The patient-centered approach to care recommends that patients' needs should be taken into consideration and a participatory style of care should be preferred (Stewart et al., 2000). This is particularly true in the context of chronic care, where self-management and adherence are the main goals to be achieved. Descriptions of patient-centered models of doctor-patient interaction at least implicitly suggest that, for example, *ad hominem* or *ad verecundiam* strategies should be avoided in this context. Apart from being considered fallacious moves of argumentation (van Eemeren & Grootendorst, 2004), these strategies should be considered ineffective in terms of the achievement of clinical goals as they may evoke negative emotions such as fear and shame.

There are also means of argumentation that occur particularly frequently in the context of chronic care consultation. The argument from consequences, also known as pragmatic argumentation, is often used in the treatment decision-making phase of the consultation. The effectiveness of this type of argumentation highly depends on the value hierarchies it is based on: Indeed, doctors and patients do not always evaluate consequences by using the same assessment criteria (Walton, 1996; Macagno, forthcoming; Bigi, 2014b). Pragmatic argumentation pointing out the negative consequences of a behavior, in particular, should be used at the right time: Although it is important for patients to know the risks connected to their condition, it may not be helpful to describe those risks immediately, in the early stages of the doctor-patient relationship. As such, a paternalistic argumentative style is not always inappropriate or manipulative (Graffigna, Barello, & Triberti, 2015). Another form of argumentation that is sometimes used in chronic care settings is analogy argumentation. Think of example (4):

> (4) Doctor: You weigh 74 kilos and you should lose 3 or 4 within the next months, but I am not asking you something extraordinary, when you first arrived here you weighed 70 kilos!

This kind of reasoning may become more effective within long-term relationships, in which both parties have had many opportunities to share information about personal facts, preferences, values, and beliefs. Lastly, *non causa ut causa* is sometimes used by patients to interpret their symptoms and argue in favor or against a certain therapy or lifestyle. For example: "since I started taking insulin, I lost 10 kilos". Doctors' replies to these arguments often take the form of dissociations, in which the problem is further specified in order to accomplish not only the goal of winning the discussion, but also the one of patient education.

2.4 Possible outcome

In theory, a discussion can either be concluded by means of a resolution or a settlement (e.g., van Eemeren & Grootendorst, 2004). In reality, discussions in the context of chronic care consultation may end in resolution, settlement, or remain open until the next encounter. The analyst should therefore refrain from looking at single consultations when assessing the outcome of chronic care discussions. Given the long-term relationship between doctor and patient, it may be the case that the parties (implicitly) postpone continuation of the discussion until the next encounter. Other elements that should be taken into consideration are, amongst others: the topic of discussion, the urgency of the problem, the health status of the patient, and the frequency of the encounters, because these could all affect the decision on how to conclude the discussion.

2.5 A blueprint for the analysis of empirical data in the medical setting

So far, we have briefly discussed the institutional preconditions for argumentative practices in the context of chronic care. Chronic care consultation can be seen as an activity type that has a potential for argumentation, but that is neither necessarily nor inherently argumentative in nature. This is a first step that should be considered when reconstructing argumentative discourse within this particular context. In a second step, all of the contextual elements that may affect the quality and effectiveness of argumentative discourse in the medical setting should be systematically taken into account. Drawing on our own empirical work and on insights from previous studies, we propose in Table 1 a description of such contextual elements. We consider this Table by no means definitive or complete, rather we think it should be used as an initial blueprint for the analysis of empirical, argumentative data in chronic care consultation and the medical context in general.

Table 1 describes contextual, textual, and non-verbal elements that can impact on the occurrence, reasonableness, and effectiveness of argumentation in the medical encounter. The analyst may not always have direct access to all elements described in the Table. However, even when only little is known about the context of the argumentative interactions under scrutiny, having the blueprint at hand will prevent the analyst from jumping to conclusions and making hasty assessments. When forced to work with incomplete information, analysts should strive to gain access to professionals in the relevant field of research to discuss the analysis and to interview regarding at least some of the elements described in the Table. In the following section, we provide an example analysis to demonstrate the practical applicability of this blueprint for the reconstruction and analysis of argumentation in empirical data.

Category	Example	Elements that impact on argumentation practices
Contextual elements		
Contextual dimensions (Street, Elwyn & Epstein, 2012)	social, cultural, media, economic	participants' preferences (beliefs, values, prior knowledge)
Structure of the health care system	universal / insurance based / ...	easy/difficult access to health care / quality of care / more or less conflicts of interest / ...
Kind of care	acute / chronic	kind of illness / therapeutic goals / clinical protocols
Specialty	general practice / diabetes / hypertension / cancer / ...	
Location	hospital / outpatient clinic / private clinic / ER / ...	time for communicative exchange / frequency of exchanges / condition of the interlocutors (e.g. in the ER, health condition of the patient, stress of the clinician) / ...
Participants	MD / patient(s) / nurse / other staff / other persons accompanying the patient	different roles / hierarchical configuration / social norms for asymmetric interaction / degree of acquaintance / degree of trust / ...
"Conditions" of care (if available)	time constraints / bureaucracy / waiting room or corridor / average number of patients per hour / first visit or follow-up / ...	psychological and emotional state of participants / degree of burnout of professionals
Structure of the consultation (clinical goals)	"canonical" phases (opening, history, physical examination, patient education and counseling) (Roter & Hall, 2006, p. 112) / other structure	specific dialogical goals of different phases

Textual elements	
Micro-context (van Eemeren, 2011)	the text immediately preceding or following the extract at issue
Linguistic structures	lexical choice / indicators of argumentation / adverbs / adjacency pairs / speech acts / ...
Specific pragmatic function of words	For example: Italian "allora" used for topic shifts / other
Nonverbal communication	
Gestures	Different pragmatic values associated to gestures by different cultures, e.g. the Dutch gesture for "tasty" vs the same gesture in Italian, meaning "beware!"
Signals of empathy	direction of gaze / posture / tone of voice / ...

Table 1: Contextual elements that impact on argumentation practices in the medical context.

3. RECONSTRUCTING AND ANALYZING ARGUMENTATIVE DISCOURSE IN CHRONIC CARE CONSULTATION: AN EXAMPLE FROM DIABETES CARE

In what follows, a passage from a consultation in diabetes care is used to show the usefulness and implications of the kind of analysis proposed in the previous section. The consultation is set within the context of an Italian diabetes outpatient clinic. The participants are the doctor, the patient, and the patient's daughter.

> (5) Doctor: It's a pity because last time, with the nurse, you decided on an action program that has not been fully realized, has it?
> Patient: No no
> Doctor: Because you were coming close to the limit and so you had agreed to work a little more on the measurements after dinner
> Patient: Yes yes, but you know, when you are not feeling very well, like in this period...

Using Table 1 as a guideline, we start by reconstructing the argumentation and by considering all textual elements first, which helps us locate the passage within the history taking phase (micro-context). In this phase, collecting information is the most important dialogical process (structure of the consultation), but the interpretation of the collected information may also occur and trigger argumentative interactions. In this case, the passage occurs directly after doctor and patient have observed a slight worsening of the patient's HbA1c[2]. They are assessing this new information and setting it within the recent complicated period in which the patient has broken her arm and had to undergo surgery and rehabilitation. On the one hand, the doctor needs to make the patient aware of the fact that her situation is worsening as far as her diabetes is concerned; on the other hand, she wants to avoid using shame as a trigger for future behavior change and therefore avoids reprimanding the patient in a direct and harsh way.

The doctor, acting as the protagonist, assumes the standpoint that the patient should have realized the action program. She argues that the patient has previously agreed on this with the nurse. Implicitly, the patient counters the doctor's view, arguing that she has not been feeling well. The doctor uses mitigation strategies (Caffi, 2007) such as "not fully realized" rather than "not realized", "close to the limit" (less threatening), "work a little" to weaken the face threat of the standpoint. If considered in isolation, this extract could be interpreted as an attempt on the part of the doctor to somehow "use" the fact that the patient has not fulfilled her previous commitment as a trigger for future behavior change. If seen this way, the doctor's use of the argument from waste (Walton, 1996, p. 80-83) – "it's a pity..." – could be understood as a means to generate shame in the patient and make her change her behavior. However, a closer consideration of a number of contextual elements allows for a more refined analysis and assessment.

When considered within the context of the overall consultation, we observe that this passage occurs after the issue of the patient's previous commitment with the nurse has been addressed already several times, in different ways. In these previous discussions, the doctor has provided new explanations of why the patient should have adhered to the program suggested by the nurse. In between, she spends some time asking the patient about the accident with her arm and how she is doing. The text immediately preceding the passage reveals that the extract follows directly a shared assessment of the new HbA1c values and constitutes a moment of evaluation of the reasons that led to the worsening of the glucose levels. Therefore, the doctor is not trying to trigger the patient's shame, rather she is suggesting that the patient could have done better for her own good. The patient, on the other hand, uses a strategy of dissociation to provide a reason for not adhering to the program she had agreed to. She does not disagree with the reasonableness of the program she had made with the nurse, but makes a distinction between "adhering to the program in general" and

---

[2] Glycated hemoglobin, or HbA1c, is one of the parameters indicating the gravity of diabetes.

"adhering to the program when one is not feeling well". In this way, she can account for the fact that she has neglected her diabetes and her glucose levels have risen. In fact, the extended discussion on this topic that develops throughout the consultation has an additional function, i.e. to uncover the reasonableness of the patient's behavior in spite of its apparent irrationality. Indeed, one would think that if someone has a health condition, she will strive as best as she can to improve it. This point of view is voiced by the patient's daughter, who is very stern towards her mother and reprimands her harshly for not following the doctor's dietary indications. The doctor, on the other hand, takes up the patient's line of reasoning based on the fact that coping with two serious health conditions at the same time is not easy and in the continuation of the discussion argues in favor of a less strict approach to the problem. Finally, doctor and patient, in collaboration with the daughter, resolve the dispute deciding that the patient will go to the gym twice a week and eat less sweets, but keep her usual amount of fruit.

4. DISCUSSION

The analysis conducted in the previous section aimed at briefly showing the usefulness of taking into account in a systematic way all the relevant contextual elements when reconstructing and interpreting the role of argumentation in the analysis of empirical data, set in the medical context. When conducting analyses aimed at understanding the functions and forms of argumentation in real life contexts, we believe it is risky to rely merely on short extracts of interactions, taken out of context. It is crucial to set the analyzed passages within their wider textual and contextual setting to allow for appropriate interpretations.

Furthermore, we have argued that chronic care is a *potentially* argumentative activity type, as argumentative discourse may occur but is not a prerequisite for effective communication. There are many factors at play that influence the argumentative discourse in chronic care consultation, such as the psychological condition of the patient, including the degree of disease acceptance, the level of engagement, a need for more paternalistic guidance, the competences of the doctor, time constraints, the kind of goals to achieve during the specific consultation. These factors need to be identified and taken into account in a systematic way during the analysis.

For this reason we proposed a list of all the relevant contextual elements that may impact on the occurrence and development of argumentation in the medical setting. This general overview of the features of (chronic care) consultations provides an initial and rough outline of the kind of characteristics that must be kept in mind when analyzing argumentative excerpts of real life data. Thereby, argumentation scholars can give a crucial contribution by providing the theoretical framework and procedural indications to turn argumentation into an actual therapeutic tool for the achievement of therapeutic goals such as patient motivation and adherence (Bigi, 2015; Labrie & Schulz, 2014). They can also provide criteria to outline the scope for argumentation. Finally, a sound theoretical framework will contribute to define pivotal concepts, such as patient preferences, commitment, shared knowledge, and the role of presupposed knowledge. The accuracy of argumentative analyses is crucial, not only because theory should provide an ideal and standardized reflection of empirical reality, but also to allow for methodological rigor (e.g., assessment of inter-rater reliability), and ultimately to ensure the generalizability and representativeness of results.

5. CONCLUSION

Through our example, we have argued for the importance of reconstructing and analyzing empirical data linking it to the relevant contextual and textual elements. Specifically, we believe that future research on medical argumentation should strive to:

1. Work systematically with complete and first hand empirical data;
2. Explicitly acknowledge contextual and textual elements that are relevant for the reconstruction of argumentation within this context;
3. Identify – qualitatively and quantitatively – the most typical functions of argumentation in each sub-field of the medical context in order to be able to plan pertinent and effective interventions to improve the clinical practice.

ACKNOWLEDGEMENTS: Sarah Bigi's research on communicative interactions in chronic care is funded by a grant from the Italian Ministry for University and Research (MIUR – "FIR 2013", Grant: RBFR13FQ5J).

REFERENCES

Barello, S., Graffigna, G., Vegni, E., Savarese, M., Lombardi, F., & Bosio, A.C. (2015). "Engage me in taking care of my heart": a grounded theory study on patient-cardiologist relationship in the hospital management of heart failure. *BMJ Open*, 5, e005582. doi:10.1136/bmjopen-2014-005582

Barry, C.A., Stevenson, F.A., Britten, N., Barber, N., & Bradley, C.P. (2001). Giving voice to the lifeworld. More humane, more effective medical care? A qualitative study of doctor–patient communication in general practice. *Social Science and Medicine, 53*(4), 487-505.

Bigi, S. (2012). Contextual constraints on argumentation. The case of the medical encounter. In F. van Eemeren & B. Garssen (Eds.), *Exploring argumentative contexts* (pp. 289-304). Amsterdam: John Benjamins.

Bigi, S. (2014a). Healthy Reasoning: The Role of Effective Argumentation for Enhancing Elderly Patients' Selfmanagement Abilities in Chronic Care. In G. Riva et al. (Eds.), *Active Ageing and Healthy Living* (pp. 193-203). Amsterdam: IOS Press.

Bigi, S. (2014b). Key components of effective collaborative goal setting in the chronic care encounter. *Communication and Medicine, 11*(2), 103-115.

Bigi, S. (2015). Can argumentation skills become a therapeutic resource? Results from an observational study in diabetes care. In F. van Eemeren & B. Garssen (Eds.), *Scrutinizing Argumentation in Practice* (pp. 281-294). Amsterdam: John Benjamins.

Caffi, C. (2007). *Mitigation.* Amsterdam/London: Elsevier.

Donovan, J.L., & Blake, D.R. (1992). Patient non-compliance: Deviance or reasoned decision-making?. *Social Science and Medicine, 34*(5), 507-513.

Elwyn, G., & Miron Shatz, T. (2009). Deliberation before determination: the definition and evaluation of good decision making. *Health Expectations,* 1-9.

Epstein, R., & Street, R. (2011). Shared Mind: Communication, Decision Making, and Autonomy in Serious Illness. *Annals of Family Medicine, 9*(5), 454-461.

Graffigna, G., Barello, S., & Triberti, S. (2015). *Patient Engagement: A consumer-centered model to innovate healthcare.* Warsaw: DeGruyter Open.

Katon, W., Von Korff, M., Lin, E., & Simon, G. (2001). Rethinking practitioner roles in chronic illness: the specialist, primary care physician, and the practice nurse. *General Hospital Psychiatry, 23,* 138-144.

Labrie, N.H.M. (2012). Strategic maneuvering in treatment decision-making discussions: Two cases in point. *Argumentation, 26*(2), 171-199.

Labrie, N.H.M. (2013). Strategically eliciting concessions from patients in treatment decision-making discussions. *Journal of Argumentation in Context, 2*(3), 322-341.

Labrie, N.H.M. (2015). The promise and prospects of argumentation for public health communication. *Journal of Public Health Research, 4*(1), 547-549.

Labrie, N.H.M., & Schulz, P. J. (2014). The effects of general practitioners' use of argumentation to support their treatment advice: Results of an experimental study using video-vignettes. *Health Communication, 30*(10), 951-961.

Labrie, N.H.M., & Schulz, P. J. (2015). Exploring the relationships between participatory decision-making, visit duration, and general practitioners' provision of argumentation to support their medical advice: Results from a content analysis. *Patient Education and Counseling, 98*(5), 572-577.

Levenstein, J.H., McCracken, E.C., McWhinney, I.R., Stewart, M.A., & Brown, J.B. (1986). The Patient-centered Clinical Method. 1. A Model for the Doctor-Patient Interaction in Family Medicine. *Family Practice, 3*(1), 24-30.

Macagno, F. (forthcoming). Framing decisions – Practical reasoning and values. In A. Rocci (Ed.), *Practical Reasoning and Argumentation*. Berlin: Springer.

Nuño, R., Coleman, K., Bengoa, R. & Sauto, R. (2012). Integrated care for chronic conditions: The contribution of the ICCC Framework. *Health Policy, 105*, 55-64.

Pilgram, R. (2009). Argumentation in doctor-patient interaction: Medical consultation as a pragma-dialectical communicative activity type. *Studies in Communication Sciences, 9*(2), 153-169.

Pomerantz, A. (1984), Agreeing and disagreeing with assessments: some features of preferred/dispreferred turn shapes. In Atkinson J.M. & Heritage J. (Eds.), *Structures of Social Action* (pp. 57-101). Cambridge: Cambridge University Press.

Roter, D. & Hall, J. (2006). *Doctors talking with patients/Patients talking with doctors*. Westport/London: Praeger.

Rubinelli, S. (2013). Rational versus unreasonable persuasion in doctor–patient communication: A normative account. *Patient Education and Counseling, 92*(3), 296-301.

Salmon, P. (2015). Argumentation and persuasion in patient-centered communication. *Patient Education and Counseling, 98*, 543-544.

Schulz, P. J., & Rubinelli, S. (2008). Arguing 'for' the patient: Informed consent and strategic maneuvering in doctor–patient interaction. *Argumentation, 22*(3), 423-432.

Snoeck Henkemans, A. F., & Wagemans, J. H. M. (2012). The Reasonableness of Argumentation from Expert Opinion in Medical Discussions: Institutional Safeguards for the Quality of Shared Decision Making. In J. Goodwin (Ed.), *Between Scientists & Citizens*: Proceedings of a Conference at Iowa State University, June 1-2, 2012 (pp. 345-354). GPSSA.

Stewart, M., Brown, J.B., Donner, A., McWhinney, I.R., Oates, J., Weston, W.W., & Jordan, J. (2000). The impact of patient-centered care on outcomes. *Journal of Family Practice, 49*, 796-804.

Van Eemeren, F. H. (2010). *Strategic maneuvering in argumentative discourse: Extending the pragma-dialectical theory of argumentation* (Vol. 2). Amsterdam: John Benjamins Publishing.

Van Eemeren, F. H., & Grootendorst, R. (2004). *A systematic theory of argumentation: The pragma-dialectical approach* (Vol. 14). Cambridge: Cambridge University Press.

Van Eemeren, F. H., & Houtlosser, P. (2005). Theoretical construction and argumentative reality: An analytic model of critical discussion and conventionalised types of argumentative activity. Retrieved via: OSSA Conference Archive. Paper 9. http://scholar.uwindsor.ca/ossaarchive/OSSA6/papers/9

Wagner, E.H., Austin, B.T., Davis, C., Hindmarsh, M., Schaefer, J., & Bonomi, A. (2001). Improving Chronic Illness Care: Translating Evidence Into Action. *Health Affairs, 20*(6), 64-78.

Walton, D. (1996). *Argumentation schemes for presumptive reasoning*. Mahwah NJ: Lawrence Erlbaum.

10

Argument and Context

JOHN BIRO
Department of Philosophy, University of Florida, USA
jbiro@ufl.edu

HARVEY SIEGEL
Department of Philosophy, University of Miami, USA
hsiegel@miami.edu

Is there ever a straight, unqualified answer to whether an argument is a good argument? Or does it always depend on the context in which the argument is advanced that determines how it is to be assessed? In this paper we argue for the first alternative. While context is often relevant to evaluating various other aspects of argumentation, it does not bear on the assessment of the quality of the arguments used.

KEYWORDS: argument, argumentation, argument purpose, argument quality, context, legal arguments, norms of assessment, persuasion, quarrels, scientific arguments

1. INTRODUCTION: ARGUMENT QUALITY AND THE PURPOSES OF ARGUING

In this paper, we question a view that is we think is surprisingly common concerning the right way to evaluate arguments. The view in question, sometimes called "contextualism",[1] has it that we do not, and should not, hold arguments and arguers to a uniform standard but should take into account the subject matter involved and the purposes, strategies and tactics appropriate to it. We will argue that on a proper understanding of what arguments are, their evaluation is, and should be, uniform in just the way that contextualism questions.

It was once widely believed by philosophers – or so it has been alleged – that whether an argument was a good one depended solely on

[1] For discussion and references see Siegel, 2015.

D. Mohammed & M. Lewiński (eds.) (2016). *Argumentation and Reasoned Action: Proceedings of the 1st European Conference on Argumentation, Lisbon, 2015. Vol. II, 267-279.* London: College Publications.

its logical properties and that the only reason one could have for criticizing an argument was its invalidity. This is held to explain the once-common characterization of fallacies in logic textbooks as arguments that appear to be valid but are not (See Hamblin, 1970; Hansen & Pinto, 1995). It was recognized, of course, that there is another, obvious, reason for rejecting an argument's conclusion: thinking that its premises are false. But that should not be taken to be a shortcoming of the argument *qua* argument, even if it was something for which its user could be faulted.

Not everyone with an interest in arguments shared this view, of course. Rhetoricians, thinking of arguments as instruments of persuasion, have always emphasized the dimension of effectiveness and have tied their standards of evaluation to that. Psychologists, from a slightly different perspective, have also been interested in the role arguments play in belief-formation. In neither case, though, was the epistemic merit of the argument the focus of interest: in the first, because effectiveness and epistemic merit obviously do not track each other; in the second, because a descriptive theory of inference need lay no claim to licensing normative judgments. But it was supposedly common ground among philosophers that in evaluating arguments we needed to pay attention only to their logical properties.[2]

We endorse this philosophical view of argument evaluation when suitably broadened. Evaluating arguments in terms of their logical properties is too narrow, since arguments can be and often are good even though they fall short of deductive validity. Strong inductive and abductive arguments are cases in point. Argument evaluation is not then a matter of logical properties but rather *epistemic* ones: arguments are good when their reasons/premises increase the knowability or rational believability of their conclusions.

2. ARGUMENTS VERSUS QUARRELS

Such a view is, of course, compatible with recognizing that sometimes we use "argument" to mean something to which the above does not apply. Sometimes when people disagree, we say that they are having an argument. When they do, they often use arguments, though not always. Sometimes they just make contrary assertions and bang the table. They can be said to be engaged in a dispute, leading to a conflict and, on occasion, even to a fight, without arguments in the sense of interest here

[2] Whether this picture is accurate or is more of a caricature is open to question. It certainly does not fit what Aristotle has to say about fallacies. (See Biro, 1977)

figuring at all.[3] Let us lump disputes, conflicts, fights and the like together under the label "quarrel", reserving "argument" for the kind of thing about whose evaluation philosophers have taken themselves to be talking about. Quarrels may be events, as with fights, or states, as with disagreements. Either way, they have a temporal location and duration. The arguments we sometimes use are, however, obviously a different kind of thing: they are abstract objects consisting of the propositions expressed by the sentences we utter in speaking or writing. Let us characterize arguments in this sense, for present purposes, simply as structured sets of propositions that may be deployed both in the acquisition of knowledge and in the justification of claims to it. (Nothing in this talk hangs on the precise details of how this characterization is spelled out.[4]) They are, unlike quarrels, repeatable or, if you prefer, multiply instantiable, as events are not. Further, they need no actual deployment for their existence. (I may kick myself for not having used a good argument I should have in our quarrel yesterday.) Even when they are deployed, that need not be to a second party. I can put one to myself, whereas I cannot quarrel alone.[5]

No doubt, we use different measures in different contexts for what counts as success in a quarrel. Sometimes we want to resolve a dispute, sometimes we have reason to prolong it. Sometimes we want to settle a conflict, sometimes we want to sharpen it. Sometimes we fight to the bitter end, sometimes we compromise to make peace. We judge the conduct of the parties to a quarrel by the general criterion of how well it serves their declared or presumed aims, aims that may be served by quarrelling, win, lose or draw. Not so with arguments. Their function is exclusively an epistemic one. Whether given to oneself or to another, our interest in them lies exclusively in whether they give us reason to believe their conclusion.

Those who say that we sometimes – or always – judge arguments simply on how persuasive they are – whether they make us believe their conclusion – may be taken to be saying one of two things. They may just mean that even a good argument, that is, one that does give reason for believing something, may fall on deaf ears, and if it does, it may be judged a failure. While this is obviously possible – indeed,

[3] Last night I was kept awake by an argument people in the hotel room next to mine were having. I could not make out their arguments, though.

[4] For example, on whether the target aimed at should be thought of as knowledge or reasonable belief, actual or possible. Such questions are debated in Biro, 1977 and 1984 and Sanford, 1981 and 1988.

[5] I can have a struggle with myself, being of two minds. Then it is my two minds that are quarrelling; it takes two to quarrel, no less than to tango.

sadly common – we still have to ask, what kind of failure? To say that a good argument may fail to persuade is already to concede that that failure is not relevant to our judging it good. No doubt, persuading is sometimes our aim in giving arguments. But whether the argument we give is a good one must be decided on other grounds. There is nothing in any of this that the defender of an epistemic account needs to worry about.

Advocates of the rhetorical conception may, however, be advancing a more radical claim. They may be saying that persuading is the intrinsic goal of arguing and that therefore the goodness of arguments consists solely, or mainly, in their persuasiveness. One problem for such a view is how to make sense of the possibility just mentioned and its reverse, a bad argument falling on all-too-receptive ears. Another is how to understand single-person uses of arguments. When I ask myself what follows from some premises each of which I have reason to believe but have not connected previously, am I trying to persuade myself of anything?

Similar considerations may be brought against various other conceptions – pragmatic, dialectical, pragma-dialectical – of the role arguments play in the quarrels in which they are deployed and corresponding proposals that they should be evaluated in terms of some other non-epistemic goal, such as the resolution of disputes.[6] True, these considerations are by no means universally seen as decisive, to say the least, and champions of these proposals will find ways to resist. And, of course, there *is* a sense in which it must be uncontroversial that one's view of the proper criteria for assessing how successful one is in deploying an argument does depend on what one takes one's purpose to be. If one's goal is to persuade, it is trivial that one must be judged better or worse in deploying the arguments one does according to how well one does with respect to that goal. But such judgments concern not arguments (in our sense) but deployments of them, not the arguments we use but the use we make of them. If it makes sense to say, as it clearly does, that one can persuade etc. with a bad argument and fail to do so with a good one, that must be because the goodness or badness of an argument is an intrinsic property of it and cannot depend on how well its use serves other goals or purposes.

3. LEGAL AND SCIENTIFIC ARGUMENTS: DIFFERENT NORMS OF ASSESSMENT?

---

[6] There are other problems with such conceptions. They are laid out, among other places, in Siegel and Biro, 2008. We shall not rehearse them here.

So much for the idea that the criteria for assessing arguments should vary with the arguer's purpose. What about the view that they should be subject-sensitive, varying with the kind of thing the parties are quarrelling about? To make that view plausible, people often point to differences between legal reasoning and scientific reasoning. No one can deny that there are differences here, and important ones. Certainly, if we include in what we mean by "reasoning" the way in which arguments are presented, these will loom large. Aside from the conventions and etiquette of arguing, which are obviously different in the two fields but are presumably irrelevant to our question, in what different ways can a given argument be presented? Here are some obvious candidates: confidently or tentatively, aggressively or diplomatically, rigorously or sloppily, in detail or sketchily. Most such things have to do with the arguer's manner. Some may sometimes be indicators of the arguer's assessment of the strength of the argument she is putting forward. But they are all capable of varying independently of whether the argument is a good one: confidence in bad arguments is all too common, timidity in putting forward good ones only a little less so. In our terms, such things are properties of quarrels, not of arguments.

Another dimension of variation may be the order in which different arguments, or even the components of a single argument, are presented. This may have to do with different conventions or with different argumentative strategies. Sometimes one is expected to start with a general premise, sometimes with a particular one, sometimes it is more effective to do one rather than the other. While there may be differences among fields in typical expectations, and in which strategy is more common, it is clear that variations along these lines are possible within each of them. It should also be clear that they do not bear on how *good*, as opposed to how effective, the arguments used are.

In our terms, all such things are properties of quarrels or of their participants, not of arguments. Our question, though, is whether any differences among fields remain that would and should influence our assessment of the arguments presented, rather than the ways in which they are presented. Let us approach the question by looking at these two kinds of reasoning, legal and scientific, in turn, to see how we assess arguments in each. If, on a comparison, these turn out to be the same, contextualism is in trouble.

What are the supposedly distinctive features of legal reasoning? One thing that may be cited is that the premises of the arguments given are different from those in other contexts. In legal systems based on statutes, they will be these. In those based on case law, it will be precedent.[7] With the former, it is easy to see that the neither the semantic status nor the logical role of the statute being appealed is different from that of a fact-stating premise in any other context. There is a statutory speed limit; Speedy drove faster; therefore Speedy broke the law. The properly executed contract calls on Greedy to pay Gullible by now; Greedy has not paid; therefore Greedy is guilty of breach of contract. Slipping arsenic in another's drink is attempting to commit murder; Lefty slipped arsenic into Righty's drink; therefore Lefty is guilty of attempted murder. With arguments that appeal to precedent, the situation is a bit more complicated, though not by much. Their general form is:

> 1) Case A (the present one) is similar (in relevant respects) to an earlier case B
> 2) Case B was decided in way W
> Therefore
> 3) Case A should be decided in way W[8]

With both kinds of argument, our standards of evaluation are the familiar ones we use outside of the legal context. Are the premises true? Is the argument valid or cogent (good, though non-deductive, as with inductive and abductive arguments)? Does it commit any fallacy? Do the premises provide reason for regarding the conclusion as true/probably true/justified?

But, it may be said, this misrepresents real-world legal argumentation. Lawyers use all kinds of arguments to win their case, and not all of these have the neat and tidy form of a deductive argument.[9] Especially in criminal cases, a good attorney will use any argument allowed by the judge to secure a conviction or an acquittal. Suppose I am defending you and think that an appeal to pity, of the sort Socrates refused to resort to at his trial, will help to get you off. Even if I believe you to be guilty, should I not make use of it? Is that not my

---

[7] Few systems are purely of one or the other sort. This does not affect the present point.

[8] There is a background premise with this kind of argument ("Cases relevantly similar to ones already decided should be decided in the same way as those"), from which the conclusion derives its normative force.

[9] It is interesting to note, though, that while we have suggested that the canonical form of an argument appealing to case law is deductive, it may be seen, at another level, as an inductive inference: "Past Fs have been Gs; This is an F; therefore, this is a G." So seen, it may be evaluated in terms of the usual canons of induction.

responsibility as an attorney? We can allow all that and still maintain that none of it is relevant to the evaluation of the argument used. Some arguments may be permissible in one context but not in another. And, arguably, in a legal context the permissible ones include fallacious ones. This is a difference in attitude to argumentation in different contexts. It says nothing about arguments as such. If I resort to a fallacious argument as a matter of tactics, I do not thereby make my argument non-fallacious.

This is not to deny that with some arguments, and with the *ad* variety in particular, whether a given use is fallacious can depend on the circumstances in which it is used. Take *ad hominem* arguments. There are situations in which judging the trustworthiness of a source or the reliability of a witness is the best, perhaps, the only, way in which we can come to a conclusion about some claim. In these cases, a responsibly mounted challenge to character or record is entirely acceptable. Elections amount to a kind of *ad populum* argument and, as long as they are properly conducted, their results can be seen as the conclusion of a sound argument.[10] (At a certain level of generality, jury trials may be seen as another example.) None of this affects how we classify an argument, independently of our judging it in the circumstances fallacious or not. While our classification and our verdict both depend on the particular circumstances, the principles we use to evaluate it do not vary with the subject matter of the argument. But then we do not have the kind of relativity the contextualist claims we do.

A brief look at scientific reasoning yields the same conclusion. Of course, the conventions governing the presentation of arguments, whether in writing or at scholarly gatherings, differ considerably from those that prevail in the law (or in philosophy, for that matter). But here, too, we evaluate the arguments presented independently of these conventions, even if it takes effort to disentangle the former from the latter.

Once we have extracted the argument, we ignore the manner of its presentation and focus on its epistemic properties. Since it is obvious that the function of the arguments scientists give in favor of a theory or claim is to provide reason for believing these, they are judged to be bad arguments if they fail to do so. We criticize them in the same ways as we do any other argument. We may find a logical fallacy (deductive or inductive); we may accuse it of being irrelevant or question-begging; we may, that is, judge that the premises fail to provide independent support for the conclusion. As in other contexts, we accept some appeals to

---

[10] Here, too, there is a background premise, of course, one that an opponent of democracy will reject.

authority and some *ad* arguments and reject others. When we dismiss claims out of hand as pseudo-scientific (from cold fusion to ESP to UFOs to spoon-bending), we implicitly rely on arguments of this sort.

Just as the assessment of scientific arguments is independent of the conventions governing scientific argumentation, so it is independent of considerations of persuasiveness. It is a commonplace of the history of science that revolutionary theories find it hard to gain acceptance. (This is the other side of the coin from the legitimate appeals to normal science and to scientific consensus of the sort just mentioned.) What can someone who insists on the relevance of persuasiveness say about this? Suppose his claim is that an argument's being persuasive is a necessary and sufficient condition of its being good. There are two ways to interpret "being persuasive". We can say that an argument is persuasive only if it actually persuades, that is, if it has been accepted (by all, by the majority, by the experts, or whatever). Or we can say that an argument may be persuasive and yet fail to actually persuade, at least for a time. On the first interpretation, we end up putting the cart before the horse: arguments turn into good ones as a result of being accepted, rather than being accepted because they are recognized (perhaps slowly) as being good.[11] On the second, we have to explain what it is for an argument to be persuasive whether or not it has been accepted. Saying that it is having the potential to be accepted is of no help: the question is precisely what property of the argument gives it that potential. The answer the epistemic account gives is that it is the fact that it can help a rational person to extend his knowledge (or set of justified beliefs). One who rejects that answer must offer another. Whatever that is, it must allow for the possibility of a bad (by epistemic standards) argument's being persuasive. But that is tantamount to conceding the independence claim.

4. SUPPORT: HOW MUCH IS ENOUGH?

A further worry is that the degree of support an argument needs to be judged good itself varies by context. Richard Rudner offered a classic statement of this worry in a quite different philosophical context:

> Now I take it that no analysis of what constitutes the method of science would be satisfactory unless it comprised some assertion to the effect that the scientist as scientist accepts or rejects hypotheses.

---

[11] Here we are echoing Socrates' complaint about Euthyphro's definition of piety as that which is dear to the gods.

> But if this is so then clearly the scientist as scientist does make value judgments. For, since no scientific hypothesis is ever completely verified, in accepting a hypothesis the scientist must make the decision that the evidence is *sufficiently* strong or that the probability is *sufficiently* high to warrant the acceptance of the hypothesis. Obviously our decision regarding the evidence and respecting how strong is 'strong enough', is going to be a function of the *importance*, in the typically ethical sense, of making a mistake in accepting or rejecting the hypothesis. (Rudner, 1953, p. 2)

Rudner's argument is made in the context of a philosophical dispute concerning scientific methodology and the metaphysical status of the so-called "fact/value" dichotomy. His claim is that *the degree to which a body of evidence must support a hypothesis in order for that hypothesis' acceptance to be warranted is relative to context*: whether or not the evidence is strong enough to warrant acceptance of the hypothesis will vary contextually. He offers several examples, one of which involves the toxicity of a drug: we demand that the relevant body of evidence confer a very high degree of support before we accept that a drug is safe, because the cost of getting it wrong is so high (people could die if we do). That is, we regard the evidence as sufficiently strong only if the risk of mistake is very low. The acceptance of other, less dramatic, hypotheses (e.g., that a new acne cream will not cause skin dryness) will require a relatively less powerful case – that is, a weaker evidential case – because the cost of being wrong is smaller.

Rudner's worry concerns the degree of evidential support required for the acceptance of a scientific hypothesis. A parallel claim about argument goodness has been made by Geoff Goddu, who argues, compellingly, we think, that "the correct evaluation of an argument is context dependent" (Goddu, 2003, p. 381), because

> when evaluating an argument...we must take into account not only the actual support that the premises provide, but the degree of support the premises *need* to provide as well. We need to know if the actual degree of support is *enough* and what support is enough will change from context to context. (Goddu, 2004, p. 30, emphases in original, note deleted; cf. also p. 33)

Goddu illustrates his claim with several suggestive examples. The most straightforward is that of the same argument, drawing on the same evidence, put forward by the prosecution in a civil trial and in a criminal one. In the former, the argument is adequate if it establishes the defendant's guilt by a preponderance of evidence; in the latter, the evidence must establish guilt beyond a reasonable doubt. If the argument establishes that the probability of the defendant's guilt is .6, it is strong enough to warrant a guilty verdict in the context of the civil trial but not in that of the criminal trial.

We think that Rudner and Goddu are right: how strong the evidence for a hypothesis must be to warrant its acceptance, how high the degree of support the premises must provide to warrant acceptance of the conclusion, can vary from context to context. This is especially so when a substantial risk is involved in making a mistake in accepting a hypothesis or conclusion. But we do not think this upends our main point. While the answer to "How much support is enough?" may vary with context, how much support the evidence/premises actually provide for the hypothesis/conclusion is determined in the normal way, on the basis of the usual inductive and deductive canons and in accordance with the usual epistemic criteria. A look at Goddu's example makes this clear. While establishing that the probability of the defendant's guilt is .6 is sufficient to warrant conviction (i.e., finding the defendant guilty) in the civil trial but not in the criminal one, that the evidence establishes that the probability of guilt is .6 is not itself context-dependent. The degree to which the evidence establishes the probability of the defendant's guilt is strictly a function of the epistemic relationship between the body of evidence and the proposition that the defendant did it, that is, a function of the relation of support that obtains between the evidence and the proposition.

5. WHAT COUNTS AS EVIDENCE?

A further apparently contextually sensitive criterion of argument quality, especially in legal contexts, concerns *evidence admissibility*. Suppose that during a trial a recording emerges in which the defendant apparently admits to the crime: "I did it, and I enjoyed watching the blood spurt after cutting off his arm!" Expert forensic testimony establishes that the voice on the tape is indeed that of the defendant. Moreover, the fact that the victim's arm was severed was not made public; only the perpetrator (and perhaps his accomplices) and the investigating officers were aware of this grisly aspect of the crime. These facts together constitute strong evidence of the defendant's guilt; they establish that the defendant probably did it. However, after this evidence is presented, the defense establishes that the recording was obtained by way of an illegal wiretap and is consequently ruled inadmissible. Here we seem to have a contextually sensitive criterion of argument evaluation: in the courtroom the argument/evidence does nothing to establish the defendant's guilt, while outside the courtroom it provides considerable support for that conclusion.[12]

We certainly agree that the rules of evidence are of crucial legal importance, and that those rules can render strong cases outside the courtroom inadmissible inside it. Nevertheless, we think it is a mistake to regard them as constituting a contextual criterion of argument evaluation. Rather, the rules reflect one of the key features of the law: an overriding concern for the fair treatment of defendants, who must be regarded as innocent until proven guilty and who must be protected from entrapment and other unsavory tactics of law enforcement and prosecutors. Given the purposes of the law, some good arguments must be excluded from legal proceedings. But this does not make them bad arguments, rather than merely inadmissible ones. The evidence in question does not fail to support the defendant's guilt; it supports it. That it does is compatible with its inadmissibility. As with our earlier examples, we must distinguish the quality of the argument *qua* argument from the purposes of arguers in advancing it. The argument advanced by the prosecution in this example does not accomplish the prosecutors' purpose, to gain a conviction. But the evidence strongly supports the defendant's guilt, nonetheless. The legal system willingly runs the risk of letting the guilty go free as a consequence of restrictions on the admissibility of evidence in order to reduce the probability of the innocent being wrongly convicted. This may well be the right thing to do, both legally and morally. Doing so does not make otherwise strong arguments weak.

6. CONCLUSION: EPISTEMIC, RHETORICAL, AND OTHER DIMENSIONS OF ARGUMENT EVALUATION

If we are right, what looks like contextual variation in norms of argument quality is actually something else: variation in purpose, variation in manner of presentation, variation in circumstances of use, variation in degree of support required for the inference to be good, etc. If so, we have no reason to think that the norms of argument quality are anything other than those allowed by the epistemic view.

Let us suppose that we are right in claiming that the intrinsic function of arguments is the same whatever the context and that the standards we use to evaluate them are so as well. None of this is to deny that arguments have rhetorical, dialectical and other dimensions. Nor is it to say that understanding those dimensions is not just as important in

[12] Of course it does not prove it: for example, the defendant may have merely witnessed the crime (and so was aware of the severed arm), not committed it.

a theory of argumentation as understanding the epistemic one we have emphasized. Insisting that it is the epistemic dimension that is relevant to understanding and evaluating arguments should not be seen as slighting the importance of those other dimensions. Conventions governing quarrels in different domains are of great sociological and practical interest. The strategies and tactics that best yield persuasion are the proper business of rhetoric, as those that best yield consensus are the proper business of pragma-dialectics. We have insisted only that the assessment of arguments, as distinct from that of argumentation, is the province of epistemology. Understanding argumentation in all its aspects requires a division of labour, and we should therefore adopt an ecumenical attitude in the theory of argumentation.[13] But adopting such an attitude is not the same thing as mistaking one aspect of the phenomenon for another, or wanting to reduce one to another. Ironically, contextualism, while claiming the opposite, does just this by refusing to recognize the common epistemic standard for the evaluation of arguments in all contexts, no matter how diverse.

REFERENCES

Biro, J. (1977). Rescuing 'begging the question'. *Metaphilosophy, 8*(4), 257-271.

Biro, J. (1984). Knowability, believability, and begging the question: A reply to Sanford. *Metaphilosophy, 15*(3-4), 239-247.

Biro, J., & Siegel, H. (2006). Pragma-dialectical versus epistemic theories of arguing and arguments: Rivals or partners? In P. Houtlosser & A. van Rees (Eds.), *Considering pragma-dialectics: A festschrift for Frans H. van Eemeren on the occasion of his 60th birthday* (pp. 1-10). Mahwah, NJ: Erlbaum.

Goddu, G. C (2003). The context of an argument. In J. A. Blair, et. al. (Eds.), *Informal logic at 25: Proceedings of the Windsor conference* (pp. 1-14). Windsor, ON: OSSA.

Goddu, G. C. (2004). Cogency and the validation of induction. *Argumentation, 18*(1), 25-41.

Hamblin, C. L. (1970). *Fallacies.* London: Methuen.

Hansen, H. V., & R. C. Pinto (Eds.), (1995). *Fallacies: Classical and contemporary readings.* University Park, PA: Penn State University Press.

Rudner, R. (1953). The scientist *qua* scientist makes value judgments. *Philosophy of Science, 20*(1), 1-6.

Sanford, D. H. (1981). Superfluous information, epistemic conditions and begging the question. *Metaphilosophy, 12*(2), 145-158.

Sanford, D. H. (1988). Begging the question as involving actual belief and inconceivable without it. *Metaphilosophy, 19*(1), 32-37.

[13] We urged just this sort of ecumenicism in Biro and Siegel, 2006.

Siegel, H. (2015). Argumentative norms: How contextualist can they be? A cautionary tale. In F. H. van Eemeren & B. Garssen (Eds.) *ISSA 2014: Selected Essays*, Dordrecht: Springer. In press.

Siegel, H., & Biro, J. (2008). Rationality, reasonableness, and critical rationalism: Problems with the pragma-dialectical view. *Argumentation, 22*(2), 191-203.

11

Argument from Analogy and its Interpretation: A Problem of Evaluation

ANGELINA BOBROVA
Russian State University for the Humanities, Moscow, Russia
angelina.bobrova@gmail.com

The paper contributes to the debates about argument from analogy and its evaluation. In the first part I'll present a three-level procedure that includes syntactic, sematic, and pragmatic analysis. The second part will be focused on the syntactic analysis of arguments from analogy. I will focus on their core structure that shows why, from a structural standpoint, these arguments can never be treated as deductive.

KEYWORDS: analogical argumentation, argument from analogy, argument scheme, core scheme, evaluation procedure

1. INTRODUCTION

Argument from analogy is a well-known type of basic argumentative reasoning. It "begins by using one case (usually agreed on and relatively easy to understand) to illuminate or clarify another (usually less clear)" (Govier, 2013, p. 319). Such arguments are quite popular, yet their popularity is not a safeguard against difficulties and ambiguities caused by analysis or evaluation of arguments from analogy (analogical arguments).

It has to be noted that I am not drawing a distinction between the analysis of an argument and its evaluation. Evaluation usually requires an argument schemes analysis. What is unclear, though, is the matter of this analysis. There is no agreement about the structural foundation of these arguments (the schemes appearance) or about their nature (i.e., the possibility of their deductive understanding).

This paper proposes a method of argument schemes elaboration. In the third section of the paper I will present a three-level procedure that includes syntactic, semantic, and pragmatic analysis. I will argue that this division makes the analysis more sequential and

D. Mohammed & M. Lewiński (eds.) (2016). *Argumentation and Reasoned Action: Proceedings of the 1st European Conference on Argumentation, Lisbon, 2015. Vol. II, 281-290.* London: College Publications.

dynamic. The fourth section will be focused on the syntactic analysis. I will bring evidence in favor of a core analogical arguments scheme, and point at its functions. Such scheme itself will demonstrate why these arguments can hardly be understood as deductive. However, the second section is a brief overview of the history of the problem.

2. TYPES OF ARGUMENTS FROM ANALOGY AND THEIR SHEMES

There is a pile of literature on argument from analogy. As D. Walton points out, it is always risky to write something "on the subject because so many scholars in so many fields had already written so much about it" (Walton, 2014, p. 23). Providing an overview of different types and classifications of argument from analogy, A. Juthe states "that there is consensus neither on what types of analogical argument there are, nor on how to classify them" (Juthe, 2014, pp. 124-125).

In this paper I will follow the line of debates triggered by T. Govier's distinction between *a priori* and inductive analogical arguments (Govier, 1985, 1987). The main differences between them can be summarized as follows:

- *a priori* analogies need not be a real case, they can be hypothetical, whereas inductive analogies cannot start from hypothetical cases;
- *a priori* analogies cannot be empirically testified, while inductive analogies do;
- *a priori* analogies are not predictions, whereas inductive analogies has to be so (Govier, 1989, p. 143).

The schemes of these arguments (*a priori* and inductive analogies respectively) represent these differences structurally (see 1 and 2).

(1) 1. A has x, y, z.
2. B has x, y, z.
3. A is W.
4'. It is in virtue of x, y, z, that A is W.
5. Therefore, B is W (Govier, 1989, p. 144).

(2) 1. A has features x, y, z.
2. B has features x, y, z.
3. A has feature f.
4*. Most things which have features x, y, z, have feature f.
5. Thus, probably, B has feature f (Govier, 1989, p. 141).

Thus, the words "features" (in the premises) and "probable" (in the conclusion) in the scheme of an inductive analogy (2) point to the comparison of real cases, while the absence of these words in the scheme of an *a priori* analogy (1) indicates hypothetical cases.

The idea of *a priori* analogies that Govier illustrated with Thomson's famous argument in defense of abortion caused long disputes. A remarkable statement was made by B. Waller; in his 2001 paper he proposed to mark out the so-called *a priori* normative analogies (see 3).

(3) 1. We both agree with case a.
2. The most plausible reason for believing a is the acceptance of principle C.
3. C implies b (b is a case that fits under principle C).
4. Therefore, consistency requires the acceptance of b (Waller, 2001, p. 201).

Waller states that we come across these arguments when we deal with universal principles (norms, rules, etc.). These principles appear in analogical arguments as a general premise that turns such arguments into deductive. F. Shecaira supported Waller's position but slightly amended the scheme claiming that it "is an accurate representation, not of analogical arguments generally, but of analogical arguments as they are characteristically formulated by professional moral philosophers and judges" (Shecaira, 2013, p. 430).

Eventually, the proposal was rejected because "by adding universal principles the analogies in analogical arguments are turned redundant" (Bermejo-Luque, 2014a, p. 335). However, the issue of analogical argument schemes variety remained. It is still unclear whether "the hypothesis that argument from analogy has two separate schemes is not such a bad one" (Walton, 2014, p. 23).

Today Govier's line has been continued by D. Walton. In a recent paper (Walton, 2014) he distinguishes two schemes that only slightly differ from Govier's. The first scheme deals with two cases and refers to the concept of similarity (see 4), while the second scheme reflects numbers of common or different features (attributes), in which the things compared are supposed to be analogous (see 5).

(4) Similarity Premise: Generally, case *C1* is similar to case *C2*.
Base Premise: *A* is true (false) in case *C1*.
Conclusion: *A* is true (false) in case *C2* (Walton, 2014, p. 23).

(5) Entities *a*, *b*, *c*, *d* all have the attributes *P* and *Q*.
a, *b*, *c* all have the attribute *R*.
Therefore *d* probably has the attribute R (Walton, 2014, p. 29).

Although Walton's proposal looks convincing, a hope to get a unitary scheme of arguments from analogy is still alive "because normally we would just like to have one scheme representing such a basic and distinctive type of argument" (Walton, 2014, p. 23).

In 2004 a core scheme was offered by M. Guarnini (see 6).

> (6) Premise 1: *a* has features *f1*, *f2*, . . . , *fn*.
> Premise 2: *b* has features *f1*, *f2*, . . . , *fn*.
> Conclusion: *a* and *b* should be treated or classified in the same way with respect to *f1*, *f2*, . . . , *fn* (Guarnini, 2004, p. 161).

It was not generally accepted, the biggest criticism being that the scheme dealt with individuals. At the same time, I would not call Guarnini's attempt fruitless. Today, a unitary scheme approach is supported by L. Bermejo-Luque (2012, 2014a, 2014b). Her understanding of analogical arguments as complex second-order speech acts helps her explain how such arguments can be both deductive and defeasible. However, Bermejo-Luque, in fact, avoids structural analysis of analogical arguments but pays attention to their procedural side. She states that analogical arguments can operate as acts of supporting or acts of adducing. In Toulmin's model, to which Bermejo-Luque appeals, they will stand as:

> (1) *data* (i.e., the evidence that is pointed out in support of a conclusion),
> Or
> (2) *backings* (i.e., the fact that stands behind the warrant of the argument and justifies the step from reason to conclusion).

This focus on the procedural, or functional, side of analogical arguments is really valuable, but it should not exclude the scrutiny of the schemes. Otherwise, the analysis will appear to be devoid of its foundation. We now have to find the way to combine the structural and the functional analyses. In my view, it will improve the procedure of evaluation as it will make the process more diversified. In the next section I propose a version of such combination.

3. LEVELS OF ANALYSIS

It is universally understood that an argument from analogy is a weak context-depended argument. It means that its cogency depends not only on structural peculiarities, but also on the context. I will treat the context as a combination of the argument content (a linguistic component) and the speaker's, or utterer's, presuppositions.

First of all, I would like to refer to cognitive science, namely to its three criteria for analogy evaluation (Holyoak & Thagard, 1989; Kokinov, 1996):

(1) structural correspondence between the two situations;
(2) semantic similarity between the elements of both situations;
(3) correspondence established between the most important aspects of both situations with respect to the speaker's goal.

This set of criteria can be successfully adapted to argumentation needs. Similarly, the analysis and evaluation of an argument from analogy can be described as following three levels, or steps:

(1) a level of inference, or Syntax level;
(2) a level of linguistic analysis, or Semantics level;
(3) a level of cognition, or Pragmatics level.

Indeed, when we first come across an argument, we want to be sure that we really deal with an argument from analogy. It means that we are looking for its specific structure: if A is similar to B, and A has X, then B has X as well. The nature of such comparison can vary but these variations are not critical to the general idea of structure identification. So, the level of inference will be the first level of our analysis.

After that, the newly identified argument scheme has to be specified according to the features in which the things compared are supposed to be analogous. This can be done with the content study because here, linguistic peculiarities can serve as markers. Thus, it will be a level of linguistic analysis.

The final level refer to pragmatic aspects of arguments from analogy analysis. We reveal the agents' (both utterer's and listener's) presuppositions that point at the reason for an argument. These presuppositions help to evaluate its relevance to the goal (why the argument has been brought up in the first place). This last step I propose to call the level of cognition.

Although the evaluation procedure demands taking all the levels into account, it is better to examine each step separately. For this reason, I'll concentrate below on the first level (Syntax level), which is the very level of argument schemes analysis.

4. THE FIRST LEVEL OF ANALYSIS

Argument schemes are conceived as a principle that allows us to study the inferential cogency of arguments. The word "inferential" is a keyword here as it means that an argument scheme can clarify the inferential side of arguments. E. Rigotti and S. Greco Morasso offer the following definition: "argument schemes are somewhat abstract structures or forms to which the actual arguments can be ascribed" (Rigotti & Greco Morasso, 2010, p. 491). Walton, Reed and Macagno treat argument schemes as "forms of argument (structures of inference) that represent structures of common types of arguments used in everyday discourse" (Walton, Reed, & Macagno, 2008, p. 1).

So what is an analogical argument scheme, then? Do we need one or several schemes? The debates are still in progress. If we take a unitary scheme approach, it is not clear enough how the scheme can be adjusted to various situations. A several schemes approach, on the other hand, represents a problem of choice, as the borders between different schemes might be blurry when it comes to analyzing the real arguments.

However, if we accept the three-level analysis, it may be not a mistake to deliberate a unitary or a core scheme of analogical arguments that can be used at the inferential level as a marker of analogical reasoning in argumentation:

(9) 1. A has x, y, z.
2. B has x, y, z.
3. A has w.
4. Therefore, B has w.

Indeed, as it has been already mentioned, all types of analogical arguments structurally follow the same line. Namely, being based on some resemblance between A and B, they prescribe feature X to B because A possesses it. Consider two fictional argumentations:

> (7) *Mary's life is like a zebra because zebra's coloration is black and white (it has stripes); Mary's life is full of bad and good events, and black and white stripes are similar to bad and good events respectively.*

> (8) *Mary will (probably) like her new bag because Mary's old bag was black, medium-sized, business-like, and she liked it; Mary's new bag is black, medium-sized, and business-like, and Mary's new bag is similar to the old one.*

In the terms of Govier's classification the first instance looks like an *a priori* analogy, whereas the second one looks like an inductive one. These arguments are certainly different (and Govier's dichotomy accentuates it), but it is not a structural difference. At the most general level of structure these examples are quite similar to each other. An analysis based on Walton's taxonomy will yield the same result.

A unitary scheme perfectly copes with the first step of analysis, which is a general identification of an argument type. Along with that, it does not abandon the advantages of a several schemes approach. All I want to say is that Govier's and Walton's achievements can be completely revealed at the level of Semantics. They do not dramatically influence the basic idea of analogical arguments structure. Thus it would be better to move this part of argument analysis to the next level, which is the Semantics level. In my view, it helps to minimize "the problems concern the way the inferential configuration of arguments is constructed and the degree of specification of the analysis" (Rigotti & Greco Morasso, 2010, p. 491).

Inference analysis emphasizes another peculiar property of arguments from analogy. Analogical arguments almost never occur as is in an argumentation. They are usually accompanied by arguments with deductive, abductive, or inductive inferential backgrounds. Moreover, since arguments from analogy are weak arguments, they are deeply involved in the synthesis of arguments. It is really hard to separate one inference from the other, which can cause confusion with the schemes (especially in the case of a multi schemes approach).

Walton's turn to Shank's and Abelson's script conception demonstrates a perfect example of such synthesis. Walton appeals to this concept of story constructions when he tries to explain the idea of similarity, i.e., the meaning of "case C1 is similar to case C2" in his first scheme (see 4). He understands scripts as "sequences of actions and events of kinds we are all familiar with in everyday life.[1] <...> Scripts are based on common knowledge about the way things are normally done or the way things normally happen in situations that we can be expected to be familiar with" (Walton, 2014, p. 34). However, it is not difficult to see that scripts should be treated as justifications that precede analogical arguments. They should not be understood as parts of arguments from analogy.

---

[1] 1. John went into a restaurant. 2. John sat at a table. 3. A waiter gave John a menu. 4. John ordered a steak and salad dish. 5. The waiter served the steak and salad dish to John. 6. John ate the steak and salad. 7. The waiter gave the bill to John. 8. John paid the bill. 9. John left the restaurant (Walton, 2014, p. 34).

We see the same story in my first example (see 7). We accept the conclusion *"Mary's life is like a zebra"* if and only if we know in advance (before an analogical argument has started) that a white stripe can be associated with a good event, while a black one can be associated with a bad one. Such knowledge precedes analogical arguments and it does not matter whether we treat it as a hypothesis or as a common.

At the same time, the synthesis of arguments does not imply that an analogical argument should be understood as a complex compound argument as Manfred Kraus proposed at the 8th ISSA conference (Kraus, 2015). On the contrary, the unique structure and special functions of these arguments point to their originality. Here, we have a distinctive type of argument that cannot be reduced to a consequence of deduction, induction, and abduction.

Moreover, the idea of synthesis structurally demonstrates how analogical arguments can be deductive yet defeasible (Bermejo-Luque's conception). An argument from analogy may seem deductive if an argument which precedes an analogical one is deductive. We have to take it as a given that general rules and universal propositions or U-claims (the likes of those Waller points to, for example) do not belong to analogical arguments. They impart moral obligations, common rules, etc., which have to be justified before an argument form analogy takes place.

5. CONCLUSION

The procedure of analogical arguments evaluation is not a trivial matter. To make it clearer I propose to diversify an argument scheme analysis. For this reason, I have introduced a three-step procedure of analysis that includes syntactic, semantic, and pragmatic levels. In my view, following these steps will give to the process of evaluation a necessary sequence. First, we deal with the structure of an analogical argument. Then we examine its linguistic peculiarities. Lastly, we look at the agent's presuppositions that help to evaluate how an argument relates to its goal.

I admit that the proposal is still quite a rough one, and it has to be further developed and fine-tuned. In the current paper I focused my attention the first step of analysis (the level of Syntax). I argued that analogical argument identification needs a unitary argument scheme. Such scheme reveals the originality of arguments from analogy, and helps to determine their place in the processes of argumentation. Analogical arguments are weak, which complicates their separation from other arguments. The latter may entail an effect of deductive understanding of an argument from analogy. It happens when structurally analogical arguments are accompanied by the arguments with deductive foundations. In any case, analogical arguments are always defeasible, which is shown with the application of a unitary scheme.

REFERENCES

Bermejo-Luque, L. (2012). A unitary schema for arguments by analogy. *Informal Logic, 32*(1), 1-24.

Bermejo-Luque, L. (2014a). Deduction without dogmas: The case of moral analogical argumentation. *Informal Logic. 34*(3), 313-336.

Bermejo-Luque, L. (2014b). The uses of analogies. In H. J. Ribeiro (Eds.), *Systematic Approaches to Argument by Analogy* (pp. 57-72). Dordrecht: Springer.

Brown, W. R. (1989). Two traditions of analogy. *Informal Logic, 11*(3), 161-172.

Bryushinkin, V.N. (2012). Cognitive approach to argumentation. In *Proceedings on the 5th international conference on cognitive science* (pp. 39-40). Kaliningrad.

Burke, M. (1985). Unstated premises. *Informal Logic Newsletter*, 7, 107-118.

Freeman, J. (2013). Govier's distinguishing *a priori* from inductive arguments by analogy: Implications for a general theory of ground adequacy. *Informal Logic, 32*(2), 175-194.

Govier, T. (1985). Logical analogies. *Informal Logic, 7*(1), 27-33.

Govier, T. (1987). *Problems in Argument Analysis and Evaluation.* Dordrecht: Foris.

Govier, T. (1989). Analogies and missing premises. *Informal Logic, 11*(3), 141-152.

Govier, T. (2002). Should *a priori* analogies be regarded as deductive arguments? *Informal Logic, 22*(2), 155-157.

Govier, T. (2013). *A Practical Study of Argument* (Enhanced seventh edition). Wadsworth: Boston.

Guarini, M. (2004). A defence of non-deductive reconstructions of analogical arguments. *Informal Logic, 24*(2), 153-168.

Holyoak, K., & Thagard, P. (1989). Analogical mapping by constraint satisfaction. *Cognitive Science, 13*, 295-355.

Juthe, A. (2014). A systematic review of classifications of arguments by analogy. In H. J. Ribeiro (Eds.), *Systematic Approaches to Argument by Analogy* (pp. 109-126). Dordrecht: Springer.

Kokinov, B. (1996). Analogy-Making: Psychological data and computational models. In B. Kokinov (Eds.), *Perspectives on Cognitive Science*, vol. 2, NBU Press.

Kraus, M. (2015). Arguments by analogy (and what we can learn about them from Aristotle). In B. Garssen, D. Godden, G. Mitchell & A. F. Snoeck Henkemans (Eds.), *Proceedings of the 8th Conference of the International Society for the Study of Argumentation* (pp. 805-814). Amsterdam: SicSat.

Rigotti, E., & Greco Morasso, S. (2010). Comparing the Argumentum Model of Topics to other contemporary approaches to argument schemes: The procedural and material components. *Argumentation, 24*(4), 489-512.

Shecaira, F. (2013). Analogical arguments in ethics and law: A defence of a deductivist analysis. *Informal Logic, 33*(3), 406-437.

Thompson, J. (1971). In defence of abortion. *Philosophy and Public Affairs, 1*(1), 47-66.

Toulmin, S. (1958). *The uses of argument.* Cambridge. University Press: Cambridge.

Waller, B. (2001). Classifying and analyzing analogies. *Informal Logic, 21*(3), 199-218.

Walton, D., Reed, C., & Macagno, F. (2008). *Argumentation schemes.* Cambridge: Cambridge University Press.

Walton, D. (2012). Similarity in arguments from analogy. *Informal Logic, 32*(2), 190-218.

Walton, D. (2014). Argumentation schemes for argument from analogy. In H. J. Ribeiro (Eds.), *Systematic Approaches to Argument by Analogy* (pp. 23-40). Dordrecht: Springer.

12

Analytical Sociology, Argumentation and Rhetoric: Large Scale Social Phenomena Significantly Influenced by Apparently Innocuous Rhetorical Devices

ALBAN BOUVIER
Institut Jean Nicod Institute, Ecole Normale Supérieure, France
bouvier.alban@ehess.fr

Some rhetorical devices, understood as kinds of small scale "social mechanisms" - whose investigation is the topic of "analytical sociology", a new domain - may play a significant role in the unintentional emergence of social phenomena on a much larger scale (e.g. a few rhetorical tricks used in the famous *Lincoln-Douglas Debates* (1858) might have played a crucial role in the emergence of the American Civil War).

KEYWORDS: American civil war, analytical sociology, argumentation, debates, difference of scales, Lincoln-Douglas, social commitments, social epistemology, social mechanisms, strategic maneuvering, triggering events

1. INTRODUCTION

In this paper, I would like to try to shed light on a kind of small scale mechanisms - namely rhetorical devices – and on the extent to which they can play, along with other small scale mechanisms, a significant role in the emergence of social phenomena on a much larger scale. I hope that these conceptual analyses will help improve our understanding of certain aspects of what is or what could be *explanation* within the context of analytical sociology, a new trend in the social sciences. These conceptual analyses will be based on a case study, which outlines an empirical analysis of a set of argumentation procedures and rhetorical devices used by Abraham Lincoln and Stephen Douglas in the famous Lincoln-Debates held in Illinois during the 1858 senatorial election process. One could argue that apparently innocuous strategic rhetorical devices used in these initially very local political speeches (they were held in small towns) nonetheless played a significant role in

D. Mohammed & M. Lewiński (eds.) (2016). *Argumentation and Reasoned Action: Proceedings of the 1st European Conference on Argumentation, Lisbon, 2015. Vol. II, 291-299.* London: College Publications.

the emergence of the American Civil War. Beginning in April 1861, this conflict caused hundreds of thousands of deaths and led to a huge transformation of the United States, with deep and long impacts on the whole world. Historians generally argue that this war, prepared by many complex events was *triggered* – this specific wording is often used - by the election of Lincoln as the President of the United States in 1860, two years after the famous Lincoln-Douglas Debates. In the empirical section of my paper, I would like to elucidate as much as possible the causal link between the rhetorical strategies used in Illinois 1858, taken as *crucial* events, and the fact that the election of Lincoln as a President acted as a *triggering* event in the Civil War.

I have chosen to examine the Lincoln-Douglas debates because they constitute an exceptional case study for argumentation theory and argumentation theory, from my viewpoint, deals, in its *descriptive* dimension, with analytical sociology and, in its *normative* dimension, with social epistemology. Analytical sociology, or, more specifically, analytical cognitive sociology, investigates how people try to persuade audiences using rhetorical devices, while social epistemology investigates how they *should* persuade audiences using logically valid argumentation procedures. However, in this paper I will address only issues of analytical sociology and not those of social epistemology.

The Lincoln-Douglas debates are exceptional because of their moral, political and historical content – under debate was the legitimacy of the extension of slavery to new states (in particular Nebraska and Kansas) - but also because of their procedural framework. In the seven debates that were held over the course of a few months in seven towns in Illinois, each debater alternated speaking first: the first speaker spoke for one hour, the respondent one hour and a half, and then the initial speaker concluded the debate with a final half-hour speech. Rhetoricians have analyzed the Lincoln-Douglas debates in the past. Gross and Dearin (2003) recently chose the Lincoln-Debates to pedagogically illustrate a few conceptual distinctions introduced by Perelman and Olbrechts-Tyteca in their *Treatise on Argumentation* (1969). Other rhetoricians, in particular David Zarefsky (1990, 2010), have also examined these debates in detail.[1] But none tried to understand rhetorical strategy as part of social mechanisms, nor did they tackle the issue of explanation in social sciences.

I will begin by addressing some elementary issues in analytical sociology. The second - very short - section of my paper will be devoted to the analysis of the concept of scale in relation to triggering and crucial events, with regard to the American Civil War example. Finally,

---

[1] See also Belz (1992) for a general overview.

in the third and main section of my discussion, I will concentrate on the possible role of rhetorical strategic maneuverings, taken as components of social mechanisms, in the emergence of large scale phenomena, and, in particular, in the emergence of the American Civil War.

2. ELEMENTS AND COMPONENTS OF SOCIAL MECHANISMS.

The general framework of this analysis is analytical sociology, of which the main goal is to discover mechanisms generating social facts. The focus on the discovery of mechanisms themselves fits in with the conception of science put forward by Nancy Cartwright (1983) from the standpoint of the philosophy of physics, and by Jon Elster (1998, 1999) from the standpoint of the philosophy of social sciences. The main inspiration of both these authors may have in fact originated in the philosophy of biology. Physiologists are interested in setting up relevant statistical correlations between illnesses, such as lung cancer, and other phenomena, such as tobacco consumption. However, physiologists are also interested in discovering the processes that could explain - at the cellular level - how certain components of tobacco can turn healthy cells in cancerous cells. This productive process is called a mechanism. Peter Hedström, following Jon Elster's seminal ideas, has labeled "analytical sociology" this program of explanation in sociology.

The definition of "mechanism" has been called controversial, but I view Machamer, Darden and Craver's definition - initially forged to apply to *biological* mechanisms - as general enough to apply for sociological mechanisms also: "Mechanisms are entities and activities organized such that they are productive of regular changes from start to finish" (Machamer, Darden, & Craver, 2000).[2] Besides, in certain definitions, the role of "triggering events" at the start of the process is emphasized, as in Hedström & Swedberg (1998) or in Elster (1999), who speaks of "triggered" causal patterns.

The first task, according to the analytical sociology program, should be to analyze mechanisms in their components and also the probable and intricate relationships between these components. The outline of this analysis described in Bearman and Hedström's *Handbook of Analytical Sociology* put forward the role of beliefs, preferences, and emotions. A more thorough reading of the *Handbook* and further related publications reveal the role of other individual mental states, such as perceptions and sensations, as well as the role of cognitive faculties, such as attention, memory and reasoning; together these mental states and cognitive faculties suggest a kind of *cognitive sociology* subprogram

[2] See also Hedström & Bearman, 2009, p. 6.

in analytical sociology. Similarly, the role of certain relations, such as authority relations and trust relations, is obvious in certain investigations of analytical sociologists (Coleman, 1990). However, a more general perusal of literature in the social sciences suggests that other mechanisms may have important roles, such as social commitments, in principle based on mutual trust. Subtler mechanisms, such as those described as "joint commitments" by Margaret Gilbert, have not yet been examined in detail in analytical sociology. Even more in-depth readings in the social sciences show the specific roles played by discourse, rhetoric and argumentation in mental states and certain relationships, but these subjects have not been examined either with the framework of analytical sociology.

3. SCALE-CHANGE EVENTS AS CRUCIAL EVENTS IN THE DYNAMIC OF SOCIAL MECHANISMS.

At this point, the notions of scale and crucial event require a brief additional clarification in relation to the dynamic of social mechanisms.

In the American Civil war case, the election of Lincoln as a President of the United States initiated, or more precisely, *triggered* a specific chain of almost unavoidable events. Immediately after Lincoln's election, and even before his inauguration, South Carolina seceded, followed by six other slave states which formed the Confederation, and other secessions soon followed. Hostilities began when Fort Sumter - where loyal forces were stationed despite South Carolina being a confederate state - was attacked by the Confederate states. Many scholars believe that, following Lincoln's election, this conflict was somewhat *unavoidable*. This chain of almost inevitable events, initiated by a clearly identifiable triggering event, fits in well with the concept of mechanism.

However, what I would like to emphasize here, following Thomas Schelling (1978) and Timur Kuran (1995), is that small scale events can, in and of themselves, produce much larger scale events. These events, which consist in a change of scale – up or down – in a historical process, whether or not they actually trigger a chain of events, I call "crucial events". There were further crucial events in the process leading to the Civil War: the presidential election, of course (both a crucial and a triggering event), South Carolina's secession, and the secession of six other slave states.

My specific claim on this issue regarding the American Civil War process is that the Lincoln-Douglas Debates constituted not only a *preliminary* event to Lincoln's election within the long sequence of events running to the end of the Civil War, but also a *crucial* event in the specific sense here formulated. It was a crucial event because it constituted a scale-up change which gave a national dimension to Lincoln, only a local politician up to that point.

4. SOCIAL COMMITMENTS AND STRATEGIC RHETORICAL MANEUVERINGS AS SPECIFIC MECHANISMS

In this last and longest section of my paper, I would like to focus more precisely on the mechanisms involved in the Lincoln-Douglas Debates. Zarefsky, writing on the subject, stated: "People who are involved in argumentation usually wish not only to resolve a disagreement, but also to resolve it "in their own favor"' (Zarefsky, 2006). Van Eemeren and Houtlosser (1999) have made the additional argument that: "In order to reach this goal, interlocutors may use "strategic maneuvering." I do not contend, as revisionist historians argued in the thirties, that Lincoln's and Douglas' only goal was to be elected and that they "maneuvered" in order to fulfill this goal.[3] To the contrary, my assumption is that Lincoln sincerely desired both the preservation of the Union and the progressive abolition of slavery, and that he viewed his election as a senator as means to achieve these ends.

However, I also assume that Lincoln, along with Douglas, used several rhetorical strategic maneuverings in order to be elected (as a means to fulfill other purposes). Each one, for example, tried to show that their opponent's political program regarding the Nebraska-Kansas Constitution (dealing with their right to decide the legality of slavery) would lead to a fatal *slippery slope*, either towards the abolition or the extension of slavery across the United States. They each also attempted to discursively trap each other. In the first debate in Ottawa, Illinois, Douglas began by asking Lincoln a series of embarrassing questions concerning his previous political commitments regarding abolitionism. Although this strategy disconcerted Lincoln at first, he soon adapted, and, following his friends' advice, adopted a symmetrical strategy. Both tried to lead the other debater to publicly adopt extreme positions, since each believed that the most moderate candidate would win the most votes – but in fact, both of them were moderate in comparison to other political leaders.

The initial focus of both Lincoln's and Douglas' arguments was not the content of their opponent's political programs, but rather their authenticity, as each one accused the other's trustworthiness. Lincoln tried to show, and most likely sincerely (but probably wrongly thought), that Douglas was plotting with other members of Congress to impose

---

[3] See also Belz (1992) on this point.

slavery step by step, law by law, over all of the United States. Douglas, on the other hand, used another strategy as already said: he tried to entrap Lincoln in his own previous commitments or alleged commitments to extreme positions. In this case, trust, either epistemic or moral, was not actually the main issue. Although Lincoln managed to avoid these traps, it was only with difficulty, unlike Douglas; indeed, Douglas' victory at the senatorial election in 1858 as the most moderate candidate was probably in part due to his successful navigation of these debates. Of course, this success was short-lived, because the relative moderation he emphasized in 1858 made him lose the votes of the extreme pro-slavery activists in 1860, during the presidential election. Indeed, these activists instead voted for another candidate, Breckenridge, thereby splitting the pro-slavery camp in two parts and leading to Lincoln's victory.

My claim is that the crucial scale up event in the process leading to the election of Lincoln – and then the Civil War - were the Lincoln-Douglas debates – although they were not themselves triggering events. But there was another event which was, with regard to the LD debates, very important (although not a "crucial" event in the sense I take this word since it did not produce a change of scale): the Republican platform voted in 1854, four years before Lincoln's nomination as representative of the Republican Party.

The New Republican Party Platform was adopted in 1854 at Springfield, Illinois. In this meeting, this Party voted a platform that quite explicitly supported the *abolition* of every law and legal decision in favor of slavery. Although Lincoln did not participate in this meeting, or formally agree to the platform, he was publicly listed as one of the meeting's organizers. As a result, most people believed Lincoln was fully committed to the Party platform ("silence gives consent" entrapment). Then, Lincoln spoke as a political leader accepting to be his party's official candidate. Therefore, he should have felt committed to its platform, even if he had not participated in the 1854 convention. Reciprocally, the members of his party should have felt committed to Lincoln's speech.

Douglas was the first to take the floor at the first meeting. He reminded Lincoln and the audience in straightforward terms of the Republican Platform allegedly signed by Lincoln, pressing Lincoln to explicitly say whether or not he endorsed the stances of the "abolitionist" party. According to Douglas, Lincoln was committed and, on insisting on that point, Douglas was clearly trying to entrap Lincoln in his own previous (real or alleged) commitment.

Lincoln had great difficulty responding to Douglas' strategy, because the Republican Platform was quite explicit in its support of the *immediate abolition of slavery*. He refused to answer Douglas' questions and even left the floor before his allocated time ran out. Nonetheless, Lincoln did attempt to avoid this trap by claiming that he was being misrepresented, since he had in fact refused to take part in the 1854 meeting. Unfortunately, many people did not believe this account, which, although probably true, seemed implausible. Lincoln attempted to frame the debate in much more moderate terms, replacing the opposition slavery-no slavery with a choice between extending slavery or not extending slavery.

During the second debate at Freeport, Lincoln modified his rhetorical strategy in accordance with the advice of certain members of his party, and decided to express his own views on the issues at hand. Consequently, he publicly admitted that he did not agree with many points of the "Abolitionist" Republican convention. In particular, he stated that he was in favor of repealing the Fugitive Slave Law, which punished slaves escaping from Southern states to Northern states. In a way, in responding to Stephen Douglas' rhetorical strategy, Lincoln was forced to violate his joint commitment with the other members of his party, and ran the additional risk of appearing opportunistic. In the end, however, Lincoln managed to avoid the potentially negative consequences of his admission, most likely because the majority of the Republican Party believed that Lincoln could win the senatorial elections with this more moderate program on abolitionism. And rather than being reproached for his disregard for the 1854 platform, Lincoln succeeded in imposing a new platform on the Republican Party.

Finally, I believe that Douglas' strategic rhetorical maneuverings were probably more efficient than Lincoln's in the Illinois senatorial elections. In other words, Douglas succeeded in making Lincoln look like a radical anti-slavery candidate, ready to impose his views by war, rather than a candidate above all else attached to Union. It is for this reason that Lincoln's subsequent election acted as a triggering event of the Civil War.

5. CONCLUSION

In this paper, I have tried to set forth a number of elements, which, I have argued, act as components of social mechanisms. Trust is surely one of these components as are joint commitments. Events effecting changes of scale (often called "threshold phenomena"), are other components of social mechanisms.

However, my main goal has been to emphasize the role of rhetoric maneuverings in the framing or reframing of a particular process, by mixing with other social mechanisms, such as joint commitments. These maneuverings can be especially significant, causally speaking, when they occur within the context of a scale change event

REFERENCES

Belz, H. (1992). Rhetoric and Deliberation in the Debate over Slavery. Lincoln, Douglas, and Slavery: In the Crucible of Public Debate by David Zarefsky. *The Review of Politics, 54*(2), 338-340.

Biggs, M. (2009). Self-fulfilling prophecies. In Hedström and Bearman, 2009 a.

Bouvier A. (2002). An Epistemological Plea for Methodological Individualism and Rational Choice Theory in Cognitive Rhetoric. *Philosophy of the Social Sciences, 32*(1), 51–70.

Bouvier, A. (2007). An Argumentativist Point of View in Cognitive Sociology. In P. Strydom, (Ed.), 'New Trends in Cognitive Sociology', special issue, *European Journal of Sociological Theory, 10*, 465-480

Bouvier, A. (2010). (working paper). Unintentional and intentional social mechanisms. Commitments and entrapments in social life. *International Network for Analytical Sociology*, Paris.

Cartwright, N. (1983). *How the Laws of Physics Lie.* Oxford: Oxford University Press.

Eemeren, F. H. van, & Houtlosser, P. (1999). Strategic maneuvering in argumentative discourse. *Discourse Studies, 1*(4), 479-497.

Elster, J. (1998). A plea for mechanisms. In P. Hedström & R. Swedberg (Eds.), *Social Mechanisms. An Analytical Approach to Social Theory* (pp. 45-73). Cambridge: Cambridge University Press.

Elster, J. (1999). *Alchemies of the Mind: Rationality and the Emotions.* Cambridge: Cambridge University Press.

Feherenbacher, Don E. (1960). The Origins and Purpose of Lincoln's 'House Divided Speech' (pp. 615-643). *The Mississipi Valley Historical Review*

Gross, A., & Dearin, R., (2003). *Chaïm Perelman.* Albany: State University of New York Press.

Hedström, P., & Swedberg, R. (1998). Social Mechanisms: An Introductory Essay. In P. Hedström & R. Swedberg (Eds.), *Social Mechanisms: An Analytical Approach to Social Theory* (pp. 1-31). Cambridge: Cambridge University Press.

Hedström, P., & Bearman, P. (2009a). *Handbook of Analytical Sociology.* Oxford: Oxford University Press.

Hedström P., & Bearman, P. (2009b). What is Analytical Sociology all about? An Introductory Essay. In P. Hedström & P. Bearman (Eds.), *Handbook of Analytical Sociology* (pp. 3-24). Oxford: Oxford University Press.

Hempel, C. (1962). Explanation in Science and in History. In R.G. Colodny (Ed.), *Frontiers of Science and Philosophy* (pp. 7-33). Pittsburgh: University of Pittsburgh Press.

Kuran, T. (1995). *Private Truths, Public Lies: The Social Consequences of Preference Falsification*. Cambridge, Mass.: Harvard University Press.

Machamer, P., Darden, L, & Craver, C.F. (2000). Thinking about Mechanisms. *Philosophy of Science*, *67*, 1-25.

Merton, R., (1967). On Sociological Theories of the Middle-Range. In *On Theoretical Sociology*. New York: The Free Press.

Perelman, Ch., & Olbrecht-Tyteca, L. (1969). *The New Rhetoric: A Treatise on Argumentation*. Notre Dame/ London: University of Notre Dame Press.

Schelling, Th. (1978). *Micromotives and microbehaviors.* Norton.

Zarefsky, D. (1990). *Lincoln, Douglas, and Slavery: in the Crucible of Public Debate*. Chicago: University of Chicago Press.

Zarefsky, D. (2010). Public Sentiment Is Everything: Lincoln's View of Political Persuasion. *Journal of the Abraham Lincoln Association*, *15*(2), 23-40.

13

Enquiring Responsibly in Context: Role Relativity and the Intellectual Virtues

TRACY BOWELL
Philosophy Programme, University of Waikato, New Zealand
taboo@waikato.ac.nz

JUSTINE KINGSBURY
Philosophy Programme, University of Waikato, New Zealand
Justinek@waikato.ac.nz

In previous work we have outlined a distinction between three kinds of intellectual virtues: cognitive, regulatory and motivational. In the first part of this paper we outline this distinction. Using it as a framework for analysis, we develop some case studies through which we consider which of those characteristics are most crucial to inquiring responsibly when occupying particular roles in professional and personal lives. We then consider possible impediments to acquiring and exercising those intellectual virtues.

KEYWORDS: intellectual virtues, virtue and argumentation, role relativity

1. INTRODUCTION

There is a recent trend towards virtue-theoretic approaches to argumentation (see for example, Aberdein, 2010, 2014; Battaly, 2010; Cohen, 2009; Paglieri, 2014.) Virtue argumentation theorists define good reasoning in terms of the intellectual virtues of the reasoner, rather than in terms of how well-supported her premises are and how well-connected they are to the conclusion that is being drawn from them. We have argued elsewhere (Bowell & Kingsbury, 2013) that it is not possible to give a complete analysis of what it is for an argument to be good in terms of the virtues exercised by the arguer in putting it forward: there is an important sense in which a good argument does not become less good when put forward by someone who is not exercising

D. Mohammed & M. Lewiński (eds.) (2016). *Argumentation and Reasoned Action: Proceedings of the 1st European Conference on Argumentation, Lisbon, 2015. Vol. II, 301-309.* London: College Publications.

the relevant virtues. Our own approach is both truth-oriented and character-connected: we argue that the more standard view of good reasoning (what it is and how to develop the capacity for it) is *enhanced* rather than displaced by an understanding of the virtues of the good enquirer.

The virtues that need to be exercised in a cognitive and communicative situation may be influenced by the role one is playing. There is currently little work in this area in the field of argumentation and reasoning, and this is clearly a gap: a better understanding of role relativity would contribute to a more complete picture of what it takes to inquire responsibly, while at the same time acknowledging the fact of and consequences of contextual differences. Pedagogically this understanding might influence both how CT is taught and the preparation people are given for different professional and societal roles: it could contribute to their development as responsible epistemic citizens. The question of role relativity is also relevant to epistemic justice. A better understanding of the ways in which occupying different roles may impede one's ability to exercise the virtues of a good reasoner may go some way towards enabling everyone to participate in proper reasoned enquiry and on equal terms. In what follows we first consider the notion of roles in the context of reasoning and enquiry. We go on to introduce a tripartite distinction between types of virtues of inquiry, which we will use as a framework for our analysis of three cases we consider next. Having considered these three cases and the ways in which they display the role relativity of the exercise of virtues, we end the paper by considering possible role-relative impediments to the exercise of the virtues.

2. ROLES

Interest in role relativity is more common and more developed in ethical theory, and discussion of the question can be found in all of the three common varieties – consequentialist, deontic and virtue-oriented. The latter, though, is most relevant to our discussion. The central idea is that there are virtues that an agent has or might be expected to acquire and exercise by virtue of some particular role she occupies. Such roles seem to fall into two types. Firstly, there are social or institutional roles such as teacher, student, employer, employee, juror, parent, friend. These are roles that one occupies either because of some social relationship or because of the part one plays in some institution or other, for example, when one is a member of a jury. The second type of role that we have in mind is the role that one adopts within an argument exchange itself, such as arguer, respondent/audience and counter-arguer. These are the types of roles one might occupy in the course of a cognitive/communicative exchange involving reasoning. The first type of roles tend to be more stable and of longer duration and one might occupy several at once, for instance, teacher, parent, employee, friend. By contrast, the second type of role is more transient. Clearly, one is unlikely always to be the arguer. Moreover, one's role within an exchange can shift within the context of the exchange. A respondent who offers a counterargument, for example, becomes the arguer and her responsibilities to exercise certain habits of good argument may shift as her role within the exchange shifts.

3. THREE TYPES OF INTELLECTUAL VIRTUE

In previous work we have outlined a distinction between three different kinds of intellectual virtues (Bowell & Kingsbury, 2015). The cognitive virtues are the skills of perception, memory and introspection, the ability to reason deductively and inductively. These are roughly the reliabilist virtues, capacities with which we are naturally endowed, yet which need to be honed and developed. Moreover, we need to be motivated to use them (we discuss this shortly). What we call the regulatory virtues (following Lepock, 2011) are roughly the responsibilist virtues. They regulate the use of our cognitive abilities, and they tend to be closely related to moral virtues. They often serve to constrain our cognitive and communicative exchanges with others, enabling us to treat other enquirers properly and to take proper account of the contributions that they make to enquiry. The following passage from Zagzebski offers a list of these virtues:

> Eg. the ability to recognize the salient facts [to which we would add "and properly to weigh them up as evidence"]; sensitivity to detail; open-mindedness; fairness; epistemic humility; perseverance; diligence, care and thoroughness; the ability to recognise reliable authority; intellectual candour; intellectual courage, autonomy, boldness, creativity and inventiveness (Zagzebski, 2006, p. 114)

Departing from the more usual bipartite account, we introduce a third type of virtue of enquiry generated by the recognition that the acquisition of the virtues requires practice and practice requires motivation. Indeed, one may have fully developed the regulatory virtues yet still not be motivated to exercise them at all the relevant moments: one might be reluctant to become involved in enquiry for various reasons (tiredness, laziness, fear of causing offence, fear of consequences such as job loss, punishment, and so on[1]) Thus the motivational virtues as we conceive them involve valuing knowing how things really are and being bothered to try and find out.

4. CASE 1: THE CRITICAL THINKING TEACHER AND STUDENT

To illustrate this tripartite model and the way in which it provides a framework for thinking about the role relativity of the intellectual virtues, we consider two cases. Take first of all the Critical Thinking teacher teaching the standard 1st year CT course. She needs to exercise the cognitive virtues and regulate them using the virtues generally associated with being a good teacher. She needs to exercise a degree of epistemic humility, but appropriately constrained so that she does not show unwarranted respect to students' wayward contributions. She may give consideration to such contributions, but need not take into account a contribution that is unhelpful or just plain wrong. She needs to exercise a degree of fairness in evaluating the contributions put forward by her students; she should not dismiss out of hand contributions that express positions contrary to her own or from students whom she dislikes or finds irritating. Coupled with that she should remain open-minded - willing to countenance a variety of positions on a topic and contributions from a diverse range of voices. She should also be willing to exercise intellectual adaptability where necessary – for instance when students express some position that she's never thought of before or where a point is expressed in such a way that its meaning is difficult to parse. Such an instance may also call for perseverance, diligence, care and thoroughness. She also needs to be motivated to exercise these virtues in her role as teacher and motivated to model virtuous enquiry. However much a student irritates her, she is sufficiently motivated as a responsible enquirer to be open to learning from the points they raise; she gives them attention equal to that she gives to those students whom she finds less irritating. She puts her own commitments and prejudices to one side when considering students' arguments that are contrary to her views. For instance, a teacher who is a committed vegan yet motivated towards responsible enquiry should approach arguments that support meat eating or argue against vegetarianism with fairness, open-mindedness, and so on.

---

[1] We say more about this later in the paper when we consider barriers to acquiring and exercising the various virtues.

Now let us imagine her students and their response. The students need to exercise the cognitive virtues. They need to bring to bear the regulatory virtues of open-mindedness, the ability to recognise intellectual authority, intellectual courage (if they are to speak up in class, for example), intellectual adaptability to deal with new concepts and to deal properly with challenges to their existing beliefs, perseverance, diligence, care and thoroughness. Finally, they need to be motivated sufficiently in pursuit of knowledge to want to acquire and exercise these virtues and motivated to practice in order to develop them.

5. CASE 2: THE WHISTLE-BLOWER

Our second case concerns a whistle-blower: Suppose that a relatively junior employee of a large organization strongly suspects that a senior manager has behaved unprofessionally and unethically. After reflection, and ensuring he is confident with respect to the evidence available to him, he decides that the correct thing to do, both intellectually and morally, is to report the matter to his manager – to become a "whistle-blower". If he has enquired well up to this point, in addition to exercising cognitive virtues - his perceptual capacities, his powers of memory, introspection and investigation and his ability to reason inductively, he will also have regulated the deployment of those capacities through bringing to bear his ability to recognize salient facts and his sensitivity to detail. He will also have employed intellectual courage, perseverance, diligence and open-mindedness, holding off from jumping to unwarranted conclusions about the senior manager's behaviour and being prepared to revise or relinquish his beliefs in the light of new evidence. In turn, if all the enquirers are responsible, the manager to whom our whistle-blower reports will herself need to marshal intellectual courage in being prepared to follow up the matter; she will also bring to bear perseverance, diligence and open-mindedness. As in the case of the teacher responding to the student, she would draw upon epistemic humility in being prepared to take seriously a matter being raised by a junior member of staff. She too will need to remain open-minded and display intellectual fairness both with respect to the whistle-blower and to the alleged miscreant. Both the manager and the employee (whistle blower) would need the motivation to pursue the truth of the matter exercising the relevant virtues of enquiry as they do so, they would also need to remain motivated through the course of enquiry, in the face of obstacles and challenges to their credibility as enquirers. Notice that the relevant roles in this case are institutional roles, determined by the positions the individuals occupy within an organisation and within the relationship of employee and manager.

6. THE ARGUER AND THE RESPONDENT

Now consider the case of an arguer and a respondent. In cases of argument the relationships between participants and roles are much more fluid than those that ensue from a social or institutional relationship. Having put an argument, one will likely find oneself in the position of respondent once one's interlocutor makes their contribution to the exchange. At the point of putting forward an argument, the onus is on us to exercise those virtues that will contribute towards a good argument – the cognitive virtues, the regulatory virtues of recognising the salient facts, sensitivity to detail, intellectual courage (perhaps you know the argument will be unpopular), being communicative, knowing one's audience and how they are likely to respond, and the virtue of being and remaining motivated to pursue the truth of the matter responsibly. The cognitive virtues also come into play at the point of receiving an argument, but the regulatory virtues that come to the fore may differ from those that bear the greater weight for the person in the arguer's position. Here fairness and open-mindedness in reflecting upon and evaluating the argument are important, as are the ability to recognise reliable authority (where this is relevant to the argument or arguers) and intellectual adaptability and candour (in responding to the argument.) If the respondent provides a counter-argument, she will need to do so with epistemic humility, fairly (responding to the argument given, for example, not some weaker facsimile) and with some degree of courage in pursuing her line of argument. Both enquirers need to be and remain motivated to pursue the truth of the matter.

7. IMPEDIMENTS TO REASONING RESPONSIBLY

In this next section we consider impediments to acquiring and exercising virtues of responsible enquiry. Many of those impediments seem to ensue from our position within relationships and the power dynamics within them, both personal and institutional, and from our social roles. These impediments may be social (such as power relations that result from differences in social position) psychological, intellectual or emotional. It is important to recognise that enquirers don't all start from the same place. Some are advantaged qua enquirers. This could just be because they're more skilful at enquiry, although that itself may be aided by their social situation. But it could also be because of their relatively privileged position in a particular situation. Social marginalization and disadvantage through gender, power race/ethnicity, class, and so on may impede one's ability to acquire the intellectual virtues by limiting one's opportunities to do so. They may also impede one's ability to exercise the intellectual virtues one *has* acquired, for instance when one is in a situation in which one feels intimidated by the social power of other enquirers, or when one's life experience is such that it is hard to exercise open-mindedness. We tend to find fault when the regulatory virtues are not exercised: indeed, the notion of our being responsible for exercising them, of having a choice whether or not to exercise them, is built into our understanding of these types of virtues. However we suggest there are situations in which the enquirer herself may not be wholly responsible for her failure to exercise certain of the intellectual virtues. An illustration of such cases is provided by the case of a parent and child, which provides a simple and clear illustration of how power can come into play in enquiry.

Consider the familiar case where a parent asks a child to do something and the child demands reasons why she should act/refrain from acting. The parent who resorts to "Because I said so" fails to respond responsibly; she favours shutting down the exchange rather than enabling the child to pursue her enquiry as to the reasons why she should do/not do X. She uses her power over her child to avoid further enquiry rather than providing reasons, risking that they be challenged and becoming engaged in an argument. The (implicit?) appeal to parental authority that's made also places a burden on the child to respect that apparent authority and not to further challenge the parent. The child may be fearful of the consequences of doing so. Moreover, the parent's response is also unfortunate in that it undermines the child's attempt to reason and to employ any virtues of enquiry that she may have acquired or be in the process of enquiring. Further, discouraging her efforts, it undermines her motivation to pursue the truth about how things are or should be.

Returning to our whistle-blower: in our narrative, the whistle-blower does manage to exercise the intellectual virtues appropriate to his situation. But it is not difficult to imagine similar cases in which the impediments to doing so are too difficult to overcome. For instance, it may be difficult to be intellectually courageous when one is in a relationship with one's managers that is such that one makes oneself vulnerable if one speaks out. Anyone from a socially marginalized or disadvantaged group may find it harder to summon up intellectual courage, adaptability, fairness, open-mindedness and perseverance if they are accustomed to being marginalized within cognitive exchanges and enquiry, if they are used to their own views being ignored or not being sought at all, or if they have become used to others speaking on their behalf and assuming they have a good understanding of their lives. By contrast, someone who is accustomed to being in positions of power and influence may find it difficult (or even inconceivable) to exercise epistemic humility, to be open-minded towards views that are not aligned with their own. They may find it difficult to recognize reliable authority other than their own.

8. CONCLUSION

We have distinguished between cognitive, regulatory and motivational virtues, and suggested that which regulatory virtues it is appropriate to exercise on a particular occasion is role-dependent: what is required of the whistle-blower is not what is required of the person to whom he reports, and what is required of the teacher is not what is required of the student. These kinds of roles and their demands are relatively stable. Other roles such as the roles of arguer and respondent are more transient, and so it is important to keep track of where you are in an argumentative exchange – it makes a difference to what is required of you. We suggest also that impediments to the exercise of both regulatory and motivational virtue can be role-dependent, taking the roles in question now to include social position more broadly construed. Intellectual courage, for example, may be both more necessary and more difficult to achieve from a position of disadvantage; the motivation towards truth may be overwhelmed by the risk that speaking it will cost you your job. Attempts to develop the regulatory and motivational virtues, both in ourselves and others, will go better if we are sensitive to such issues.

REFERENCES

Aberdein, A. (2010). Virtue in argument. *Argumentation, 24,* 165-179.

Aberdein, A. (2014). In defence of virtue: The legitimacy of agent-based argument appraisal. *Informal Logic, 34,* 77-93.

Battaly, H. (2010). Attacking character: Ad hominem argument and virtue epistemology. *Informal Logic, 30,* 361-390.

Bowell, T., & Kingsbury, J. (2013). Virtue and argument: Taking character into account. *Informal Logic, 33*(1), 22-32.

Bowell, T., & Kingsbury, J. (2015). Virtue and enquiry: Bridging the transfer gap. In M. Davies & R. Barnett, (Eds.), *The Palgrave handbook of critical thinking in higher education* (pp. 233-246). New York: Palgrave Macmillan.

Cohen, D. H. (2009). Keeping an open mind and having a sense of proportion as virtues in argumentation. *Cogency, 1,* 49-64.

Paglieri, F. (2015). Bogency and goodacies: On argument quality in virtue argumentation theory. *Informal Logic, 35*(1), 65-87.

Zagzebski, L. (2006). *Virtues of the mind: An enquiry into the nature of virtue and the ethical foundations of knowledge.* Cambridge: Cambridge University Press.

14

Automatically Identifying Transitions between Locutions in Dialogue

KATARZYNA BUDZYNSKA
Polish Academy of Sciences, Poland / University of Dundee, UK
k.budzynska@dundee.ac.uk

MATHILDE JANIER
Centre for Argument Technology, University of Dundee, UK
m.janier@dundee.ac.uk

JUYEON KANG
Prometil, France
j.kang@prometil.com

BARBARA KONAT
Centre for Argument Technology, University of Dundee, UK
bkonat@dundee.ac.uk

CHRIS REED
Centre for Argument Technology, University of Dundee, UK
c.a.reed@dundee.ac.uk

PATRICK SAINT-DIZIER
IRIT-CNRS, France
stdizier@irit.fr

MANFRED STEDE
University of Potsdam, Germany
stede@uni-potsdam.de

OLENA YASKORSKA
Polish Academy of Sciences, Poland
OYaskorska@gmail.com

D. Mohammed & M. Lewiński (eds.) (2016). *Argumentation and Reasoned Action: Proceedings of the 1st European Conference on Argumentation, Lisbon, 2015. Vol. II, 311-327.* London: College Publications.

The contribution of this paper is theoretical foundations for dialogical argument mining, as well as initial implementation in software for dialogue processing. Automatically identifying the structure of reasoning from natural language is extremely demanding. Our hypothesis is that the structure of dialogue can yield additional clues as to argument structures that are created and cocreated. Our work has been performed using the MM2012 corpus in OVA+.

KEYWORDS: argument mining, dialogue structure, Inference Anchoring Theory

1. INTRODUCTION

Argumentative exchanges expressed through dialogical interaction can involve many more variables and subtleties than do arguments expressed as monologues. In trying to build algorithms that might automatically detect the presence and structure of argument, it might be expected that research should begin with the simpler monological cases before moving on to generalise techniques for dialogue. It turns out, however, that the very complexity that makes dialogue so challenging also offers rich sources of additional information that can be used to guide the automatic recognition process. This paper demonstrates how working with broadcast debate using a relatively new approach to the analysis of dialogical argumentation can offer insight into the dialogue games that participants are playing, and that those dialogue games give detailed grist to the algorithmic mill.

2. INFERENCE ANCHORING THEORY

Inference Anchoring Theory – IAT (Budzynska & Reed, 2011) provides the framework for connecting dialogical structures with argumentative structures thus allowing for the analysis of natural dialogical interactions. IAT is not a general-purpose discourse analysis technique. It is tailored specifically to handle discourse that involves argumentation, i.e., the giving of reasons in support of claims in order to affect an audience. Examples of such discourses are mediation (Janier et al., this issue, vol. I) and debate (Janier & Yaskorska, this issue, vol. I). Let us present the example of a simple dialogue:

(1) a. Bob: *p* is the case
b. Wilma: Why *p*?
c. Bob: *q*

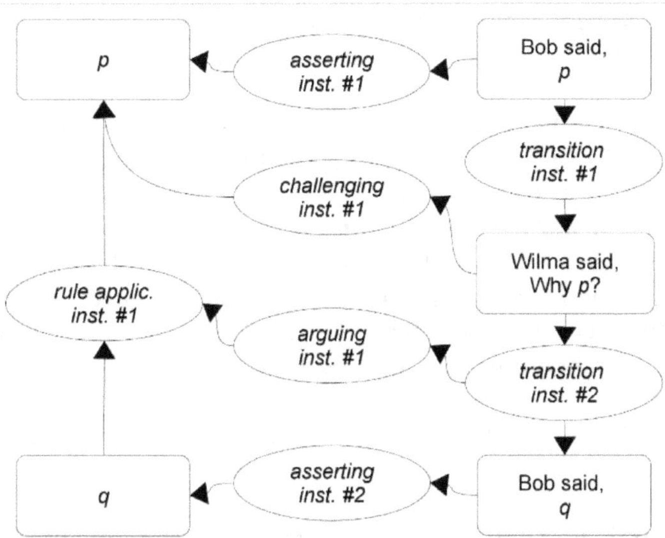

Figure 1. An IAT analysis of the dialogical and argumentative instances.

Figure 1 presents the diagram with IAT analysis of the dialogue from example (1). Right-hand side of the diagram consists of the propositional reports on locutions (such as "Bob said, *p*" etc.). Left-hand side of the diagram consists of nodes with propositional content (in this case: *p* and *q*). Three different types of relation are expressed in Figure 1:

i) relations between locutions in dialogues (transition instance #1 and #2);
ii) relations between propositions (rule application instance #1); and
iii) illocutionary connections that link locutions with their contents (asserting instance #1 and #2, challenging instance #1, arguing instance #1).

The first type of relations refers to rules of protocol which speakers follow to perform locutions during a dialogue game. For example, locution (1-b) is a legal response to (1-a) which means that they are related via some specific protocol rule of the game. Application of those rules creates instances of transitions (transition instance #1 and #2). Relations of type (ii) are typically studied in logic and argumentation theory. In figure 1, we have only one relation of this type: that is relation of inference (rule application instance #1). However, IAT allows for expressing also other relations, such as conflict or rephrasing. Relations of type (iii) are illocutionary connections with which a given locution is performed. In this work, illocutionary connections are intuitively related to various illocutionary forces (i.e.

the speaker's communicative intentions (Searle, 1969)).

Representation of discourse structures involving argumentation requires some way of representing relations connecting those two elements. In IAT this is achieved by the concept of anchoring illocutionary connection. Two general types of anchoring are possible. Example of the first type, presented in the figure 1 is the challenging instance #1, as it is anchored in the propositional report on locution (on the right-hand side) and targets propositional content (on the left-hand side). Illocutionary connections anchored in transition and targeting rule application instances are the second type. In figure 1 an example of such a connection is arguing instance #1. Illocutionary connections anchored in transitions are the focus of this paper.

3. THEORETICAL FOUNDATIONS FOR MINING ARGUMENTS IN A DIALOGUE

The research presented in this paper aims to automatically extract inferential structures (arguments pro-) and conflict structures (argument con-) using as cues their dialogical context (see Budzynska et al., 2014a; Budzynska et al., 2014b). In order to efficiently recognise arguments in a dialogue, we need a solid theoretical foundation which will represent not only elements of arguments and elements of dialogue, but also explain how these two types of structures are connected with each other. Inference Anchoring Theory, described in the previous section, is a good candidate for such a task, however, the application of IAT to the realm of the complex human communicative interactions presents some initial challenges and the theory needs to be adapted so that it is robust enough to describe how people create arguments during the dialogue.

We selected the genre of the radio debate and worked with the BBC Radio Moral Maze programme (the corpus is available at: http://corpora.aifdb.org/, see also Sect. 5). During each 45 minute programme, participants discuss moral aspects of important social and political issues in Great Britain. The programme is chaired by Michael Buerk who leads the discussion between four panellists – public people with a background in social activism (writers, journalists, lecturers, public commentators etc.). Moreover, so-called witnesses are invited who are experts on a given topic and who describe a situation in more detail. Such structured radio debate ensures that participants create many well-formed arguments which can be studied.

Consider the example from the programme on problems of families where the participants discuss whether or not the State should intervene into the problem families in order to decrease the poverty in the country. Here Anne and Ruth were trying to establish whether a number of families who might need an intervention is sufficient for making a moral issue about this. Ruth provides the percentage of badly parented children (according to the Government's criteria) in the troubled families of different boroughs of London.

(2) a. Anne McElvoy: Isn't it a rather specific example?
b. Ruth Levitas: No. Birmingham was given a target of 4180, they estimate they can find about 7% on the troubled families' criteria.

What we would like to capture here in this example is that Levitas introduces the conflict with McElvoy (argument con-) and provides an inference (argument pro-) to support her standpoint. Such a communication dynamic is relatively simple for a human to recognise, but it poses more difficulties for automatic identification of dialogical argumentation[1].

First, McElvoy's standpoint is not asserted explicitly qua an affirmative sentence but as an interrogative one. Yet questions do not aim to provide opinions – they are seeking for them, so the problem is: how Levitas can introduce the conflict if there was no opinion provided in (2-a)? For example, if I say "Do you like apples?", I am not giving my opinion with which the hearer can conflict. I rather want the hearer to provide me with her opinion in the matter that I am asking about. Thus, developing an algorithm which associate a sequence: a question followed by an answer "No", with a conflict structure is not a good solution, because the automatic system would deliver a lot of errors as an output. We need to look for a more fine-grained theoretical grounding here.

A second challenge is related to indexicality which is particularly common in dialogical communication, because people typically do not repeat a material that was already introduced by their opponents in the previous move(s). Imagine I enter the room exactly when Levitas begins her turn. Without knowing what happened before (2-b), it is impossible to reconstruct the propositional content of her first locution. In other words, if I just hear that someone says "No", it is not possible to understand whether she meant "It is not a rather specific example", "I

---

[1] We assume that dialogical argumentation (or: quasi-dialogical argumentation) does not necessarily require two or more speakers. It is sufficient for one speaker to introduce quasi-dialogue, when he cites or refers to opinions of his opponent(s). In such cases, the speaker can provide con-arguments against his opponent(s)' standpoints, and justify his own opinions with pro-arguments.

don't like apples" or anything else. From the point of view of the automatic recognition, we need a good way of instructing an algorithm where to look for the content in such cases and how to extract it[2].

The first challenge was addressed by extending the list of the illocutionary connections with such ones that are typical for dialogical interactions. In the example (2), such a specific connection is assertive questioning which has a dual function of both asserting and questioning (see Budzynska et al. 2014b for other dialogical connections). More specifically, in (2-a) McElvoy does not only seek Levitas' opinion whether it was a rather specific example, but also implicitly conveys her own opinion that it was a rather specific example (see figure 2). In contrast to a pure question which only seeks for the hearer's standpoint, here the speaker (McElvoy) gives her own opinion, and as a result when the hearer (Levitas) responds to such an assertive question, the respondent (Levitas) provides her opinion as well, and as a result she either agrees or disagrees with the previous speaker (in this case – Levitas disagrees with McElvoy introducing conflict between their opinions).

The second challenge, the challenge of indexicality, is addressed by anchoring illocutionary connections in transitions. In figure 2, there are three such cases: disagreeing, asserting and arguing. If annotators analyse the example (2) as it is presented in figure 2 and the data is used to develop an algorithm, then the instruction specifies that it is not enough to look at the single locution ("RL: No" in figure 2) to extract the content of Levitas' assertion, because the asserting connection is not anchored in the locution. What is required is to find the transition, check which locutions is connected ("AM: Isn't it a rather specific example?" with "RL: No"), and then it is possible to recognise what is the propositional content of the second locution. In other words, if Levitas started her turn in (2-b) by saying "It is not a rather specific example" (the response which is fully repeating the content of the question), then the assertion would be anchored in the second locution and there would be no need to look at the history of this locution for the automatic extraction of its propositional content.

[2] Note that the indexicality is not a specific property of argumentation -- in fact it occurs in any natural communication. Nevertheless, while it is rather rarely found in monological argumentation, it becomes an important part of dialogical argumentation.

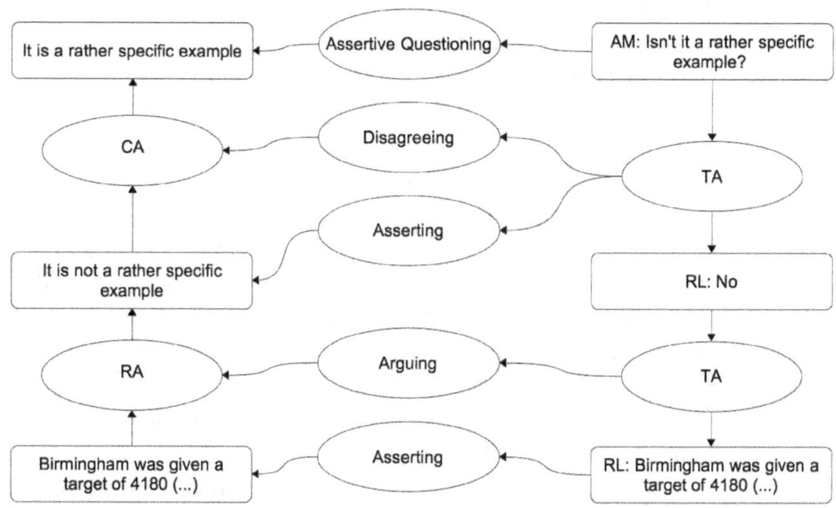

Figure 2. Disagreeing and arguing in a dialogical context

How do these solutions lay foundations for automatically extracting inferential and conflict structures from transitions in a dialogue? We can instruct an algorithm that if it finds a sequence: pure question followed by a word that is equivalent to "No", then there is no conflict introduced, while the conflict structure is created (see CA in figure 2), if the algorithm finds a sequence: assertive question followed by a word that is equivalent to "No". Moreover, since asserting is anchored in the transition between locutions and not in a single locution, in order to extract the content of "RL: No" the algorithm has to inspect the content of the other element of this transition relation.

To automatically recognise the inferential structure (see RA in figure 2), an algorithm has to be instructed that in this type of discourse a sequence: a word that is equivalent to "No" followed by assertion, anchors inference (in fact in many cases some additional cues are needed to increase the efficiency of the instruction). Moreover, the algorithm will extract directly the content of the premise, because the assertion introducing it is anchored in the single locution (see the bottom of the figure 2), however, in order to extract the full inferential structure (both premise and conclusion), it has to search for the transition and the second element of this relation (i.e. the locution "RL: No"). Still this locution does not anchor anything, so that the algorithm should not stop searching upwards into the history of the dialogue. Once it reaches the first locution, it can go back downwards which will allow for the reconstruction of the content of the second locution, which in turn will provide the information what is the conclusion of the inference.

4. RELATED WORK

Automatic argument mining has received a lot of interest during the past years and is now an important application of Computational Linguistics. This line of work started in specific domains, in particular that of legal language, where Mochales-Palau and Moens (2009) identified claims and their justifications in legal texts. Today, the task is sometimes conceived more broadly as finding not just pairs of claim and argument, but more complex structures involving rebuttals, counterrebuttals, etc. Genres that are being studied include scientific papers (e.g., Kirschner et al., 2015) and student essays (e.g. Nguyen & Litman, 2015). Beyond monologue text in "standard" language, lately the research also turned to more dialogical communication. Thus, the Internet Argument Corpus (Walker et al., 2012) is a collection of contributions to internet discussion forums, where users interact with each other to a certain extent. An example for analyzing such discourse automatically is the work on online user comments by Park et al. (2015). Snaith and Reed (this issue, vol. I) propose the automatic method of inducing context-free grammar from a transcript of a dialogue. The extracted grammar describes the formal protocol that governs the interaction of this dialogue. Also, Swanson et al. (2015) propose an operationalization of the notion of "argument clarity" for such comments, where sentences are being rated (via crowdsourcing) in terms of this clarity, and automatically identified on the basis of features that to a good extent are domain-neutral. For transcriptions of oral dialogue, however, we are not aware of any research other than our own.

Mining the structure of argument in text starts with segmentation, i.e., the step of finding the individual spans that correspond to the minimal segments of the argumentative structure. Thereafter, the two computational tasks to be executed on the basis of a text segmentation are (i) to identify the illocutionary connection of individual units, and (ii) to identify the relations between those units. For (i), certain linguistic features of the utterance (e.g., sentence mode, mood, modality, verb class, particles) and the context of recent moves are exploited to compute the most likely speech act. The computational dialogue analysis community has addressed this task for a long time, using both rule-based and statistical approaches; for the latter, see, e.g., (Stolcke et al., 2000). Task (ii) is closely related to efforts in discourse parsing, which, again, usually targets monologue text. One popular framework is Rhetorical Structure Theory (Mann & Thompson, 1988), which posits that a tree structure can be assigned to a text on the basis of recursively linking adjacent segments by means of coherence relations. One well-known approach to automatically compute these relations and the resulting tree structure is that of (Hernault et al., 2010), who divided the task into two separate classifiers for (i) deciding whether to link two segments, and (ii) assigning a relation label to such pairs.

Beyond building a structure description, intellectual argument analysis involves judging the plausibility of instances of argumentation. This is largely beyond the state of the art of automatic analysis, but an important first step into this direction is the automatic classification of arguments in terms of Argumentation Schemes (Walton et al., 2008). Feng and Hirst (2011) showed that this in principle possible, restricting the set of schemes to five frequently-used ones.

5. CORPUS DESCRIPTION

Within the corpus studies, 4 transcripts of the BBC 4 radio program *Moral Maze* were annotated (hereafter MM2012a) according to the IAT framework.[3] MM2012a contains 58000 words and has 284 questions or challenges for about 1417 assertions which are parts of argumentation (from total of 2000 identified sentences). Analyses were carried out using OVA+[4] (Janier et al., 2014) and stored in AIFdb corpus[5] (Reed et al., 2008; Lawrence & Reed, 2014). In the first step, dialogue moves were described with illocutionary connections, distribution of which is presented in table 1.

Annotating corpora is a time-consuming task (particularly with IAT because of the large variety of schemes and categories); moreover, the spoken interactions context makes the task trickier than with monologues, hence the relatively small size of our corpus.

Within 1849 annotated locutions, apart from the most expected illocutionary connections (Assertions: 1417 occurrences), there is also a significant number of illocutionary connections via which participants introduce premises and conclusions for their arguments (AQ, RQ, ACh, RCh, PCon and Con: 256 occurrences). This data illustrates the dynamics of radio debates via the variety of illocutionary connections.

---

[3] Results provided here account for August, 2015; they may differ from previous works given that the corpus was enlarged to add a fourth transcript

[4] http://ova.arg-tech.org

[5] http://corpora.aifdb.org

Illocutionary connections	Occurrences
Assertions (A)	1417
Pure Questions (PQ)	81
Assertive Questions (AQ)	103
Rhetorical Questions (RQ)	70
Total Questions (Q)	254
Pure Challenges (PCh)	7
Assertive Challenges (ACh)	10
Rhetorical Challenges (RCh)	13
Total Challenges (Ch)	30
Popular concessions (PCn)	53
Other	7
Total Concessions	60
Empty (no illocutionary connection in locutions)	88
Total	1849

Table 1. The distribution of illocutionary connections anchored in locutions in MM2012a corpus.

In the next step of the corpus analyses, transitions between locutions were identified and illocutionary connections anchored in those transitions were described. The distribution of the identified illocutionary connections anchored in transitions is presented in table 2.

Illocutionary connections	Occurrences
Arguing	563
Disagreeing	200
Agreeing	101
Restating	78
Asserting	45
Questioning	2
Challenging	3
Other	80
Non-anchoring (no illocutionary connection in transition)	360
Total	1432

Table 2. The distribution of illocutionary connections anchored in transitions in MM2012a corpus.

Within 1432 of all analysed occurrences, 763 (61%) illocutionary connections anchored in transitions are related to the process of argument construction carried out via the illocutionary connections of arguing, disagreeing and agreeing, what illustrates that this type of dialogue is very argumentative. The dialogical dynamics proper to debates were thus identified: Arguing, Disagreeing and Agreeing. Occurrences of other types of argumentative dynamics being far less than 100 each, we decided to group them all together (Other).

The MM2012a corpus has been annotated by two annotators that have the same linguistic training and a good expertise of the IAT theoretical background. Measures of the differences between annotators, calculated before discussion, are summarised in table 3.

Types of annotation	Inter-annotator agreement
segmentation	79%
illocutionary connections (YA)	88%
illocutionary connections (YA) anchored in TA	78%
conflict relation (CA)	76%
inference relation (RA)	86%
transitions (TA)	89%

Table 3. Inter-annotator agreement measures for MM2012a corpus.

We measured the rate of agreement between the two annotators. All in all, the agreement rate is relatively high. For this reason, we consider the framework schemes as stable, easy to identify and accurate. (see more about argument corpus studies with the use of IAT in: Janier & Yaskorska, 2015).

6. AN AUTOMATIC IDENTIFICATION OF ILLOCUTIONARY CONNECTIONS ANCHORED TO TRANSITIONS

Let us now develop the main features of a linguistic model and an implementation that allow to automatically identify illocutionary connections anchored to transitions. In this first experiment, three main illocutionary connections are considered: Disagreeing, Agreeing and Arguing. The other connections such as reframing or conceding are not considered here: they are relatively infrequent. It is interesting to see in this corpus that the level of disagreement is relatively high, this means that dialogues are rather controversial.

This first investigation is based on the linguistic cues found in the units at stake. It is clear that some cases need, in addition, context, knowledge and inference to be identified, however, linguistic analysis is favored because it is simpler, requires less resources and is relatively re-usable, within similar dialogical contexts. Illocutionary connections anchored to transitions are identified on the basis of a pair, adjacent or not, of dialogue units, their contents and the illocutionary connection that is associated to each of them (see section above). It is clear also that in some cases, relatively limited, the taking into account of more than two units would introduce more contextual elements and would help to resolve ambiguities.

For this preliminary mode, our development corpus is composed of 248 already tagged transitions between units. This is not very large, but seems to be sufficient for our current aim. Transitions without any illocutionary connections are not considered in this investigation, they correspond to about 15 to 20% of the situations, and will be investigated in a later stage of the project.

Let us now develop the linguistic model elaborated for each of the illocutionary connections given above. Let us consider a pair of units (U1, U2) and the transition T that occur between them. U1 and U2 respectively have the illocutionary connections Uif1 and Uif2, while T is anchored the illocutionary connection Tif. The model that is developed below considers the linguistic contents of U1 and U2, Uif1 and Uif2 and the fact that these are uttered by the same speaker (S1) or by different speakers (S1 and S2). This analysis has obviously a relational character, between units and speakers.

Since our corpus is quite small, we have first identified linguistic marks, explaining why they contribute to disagreeing, agreeing or arguing and then we have slightly generalized them via synonyms or equivalent expressions. Finally, these marks have been grammaticalized, when relevant, in order to avoid long lists of terms and to favor a "local" grammar approach that characterizes illocutionary connections anchored to transitions.

6.1 A model for Disagreeing

A first parameter to consider are ***the illocutionary connections assigned to U1 and U2***. Some typical cases have been identified such as:

[U1: Assert] & [U2: Assertive Question] & [different speakers]
→ Disagreeing

Where Disagreeing is the illocutionary connection anchored to the transition T. Some typical language expressions may also be involved as constraints in order to confirm disagreement, since there are situations which may be ambiguous with other illocutionary connections.

A second situation is the case where U1 and U2 are uttered by two different speakers, or where there is a reported speech situation, usually in U1, and where ***forms of negation*** are observed in either U1 or U2. Roughly, these forms of negation indicate that the speakers do not share the same point of view. These forms of negation are quite numerous. Let us cite here the main categories:

(1) Variants of negation: *"I do not"*, *"I don't"*, *"I cannot"*, *"can never"*, etc. These forms are essentially observed contrastively in U2
(2) Negatively oriented propositional attitudes: *"I disagree"*, *"I cannot accept"*, etc.
(3) Contrastive connectors between U1 and U2: *however, but*, etc.
(4) Negatively oriented lexical terms: *"sluggish"*, *"bad"*, *"wrong"*, *"aberration"*, *"harmful"*, etc. associated with a judgement in U2 about a fact reported in U1.
(5) Contextually negative terms: *"coercive"*, *"peculiar"*, *"warnings"*, *"dangerous"* found in U2 as a response to U1 and negative judgement terms: *"unwanted"*, *"undesired"*, *"hazardous"* in U2 when a fact or opinion is given in U1
(6) Use of antonyms in U1 and U2, bipolar or continuous: *"expensive / cheap"*, *"moral / immoral"*, or the negation, in any order, e.g.: *"coercive / not coercive"*.

The last main situation to consider occurs when U1 and U2 are produced by the **same speaker**, with two situations:

(a) the use of contrastive expression: P *but* Q or
(b) a reported speech situation, citing someone else or an admitted opinion.

Our lexicon of negative terms contains about 100 terms, which is not very large. This indicates that speakers tend to use terms which can be understood by a large population of listeners. Each of the above situations is expressed by a "local" grammar and may be associated with constraints and restrictions.

6.2 A model for Agreeing

The linguistic model for Agreeing is based on the same philosophy, it may be slightly simpler. It is structured around two main situations:

The Illocutionary connections assigned to U1 and U2 may precisely characterize forms of agreement, e.g.:

[U1: Assert] & [U2: Assertive Question] & [same speaker for U1 and U2] → Agreeing

Typical forms of agreement in U2, with no negation in U1, often define forms of agreement. Among the most typical ones, let us cite:

(1) Typical forms of approving: *yes, OK, I'm happy*, etc.
(2) Forms expressing an opinion that approves U1 contents: *I think, they are right, I agree, I like*, etc.
(3) Typical positive expressions, positive evaluative expressions: *sympathetic, interesting, powerful point*, etc.
(4) Typical positive binders between U1 and U2: *as you say, your own experience, well* not followed by any negative expression, etc.

These forms of agreement are more or less strong (e.g. *I do not disagree* is less strong than *I agree*). Measuring these connections is beyond the present investigation since these are relative to the speaker and to the situation. It is interesting to note that language seems to be richer in negative terms than in positive ones. Our lexicon of positively oriented terms contains at the moment about 60 terms.

6.3 A model for Arguing

Arguing is by large the main situation (70% of the cases in our development corpus, but this may vary from dialogue to dialogue). Arguing can occur between two speakers or a given speaker may be arguing for his/her own views.

In our analysis, Arguing is considered as the by-default option: if no Disagreeing or Agreeing situation has been detected, then by default it is an Arguing illocutionary connection.

However, to confirm our analysis, a few linguistic cues have been identified, which may be used when there are ambiguities with agreeing or disagreeing. These mainly are:

(1) Unit discourse connectors*: but also, if you, indeed, so* (with some constraints), etc.
(2) Connectors such as *but* or *however* introducing U2, without any negative expression in U2.

6.4 Implementation and performances

This relatively simple linguistic model has been implemented on our TextCoop platform using discourse patterns and constraints. The results obtained are reported in table 4, expressed in terms of accuracy (we consider our evaluation as indicative, therefore tests involving precision and recall are not yet relevant):

Illocutionary connection	Correctly recognized	Not correctly recognized
Disagreeing	82%	18%
Agreeing	85%	10%
Arguing	95%	5%

Table 4. Performance results of automated extraction of illocutionary connections.

Results are good in spite of the relative simplicity of the analysis and the complexity of the task. One of the reasons is that speakers of the Moral Maze make their best to use a clear language, well-structured with explicit marks so that they position and argumentation is clear and unambiguous. Results could be different if one considers less controlled dialogues.

An important remark concerns the errors: among the 'not correctly recognized' connections, only about 1/3 of them are due to an incomplete or incorrect linguistic analysis or to language ambiguities, while the other 2/3 would require knowledge and inference to identify the illocutionary connection anchored to the transition. This would be an interesting research direction: the pragmatic forms of disagreement or agreement. In a number of situations, domain knowledge can help identify the correct connection, but this is much more costly in terms of resources than just using linguistic knowledge.

This section has presented a preliminary linguistic and language processing analysis of illocutionary connections anchored to transitions, in addition to the works done on unit delimitation and their illocutionary connection identification, presented above. This work remains largely exploratory and is still preliminary. Results show that linguistic analysis is worth pursuing but that a relatively large number of cases (about 15%) need pragmatic analysis which is not surprising, even in well-formed dialogues. This rate should be higher in less controlled ones.

7. CONCLUSION

This paper has reported on the first few steps of a new methodology for understanding argument structure in dialogue that is predicated on the constraints imposed on interaction by tacit common understanding of the dialogue game that is being played. The work has demonstrated that even quite simple rules of these games – rules that describe some of the ways in which speakers can disagree, agree and argue – can constrain expressions sufficiently to be able to contribute significantly to the extremely demanding AI task of automatically recognising argument structure in free natural language. Though this paper reports early advances, it demonstrates that the approach represents a rich seam of academic investigation.

ACKNOWLEDGEMENTS: We gratefully acknowledge the support of the Polish National Science Center under grant 2011/03/B/HS1/04559, Leverhulme Trust under grant RPG-2013-076 and Innovate UK under grant 101777.

REFERENCES

Budzynska, K., & Reed, C. (2011). Whence Inference? In *University of Dundee Technical Report.*

Budzynska, K., Janier, M., Kang, J., Reed, C., Saint-Dizier, P., Stede, M., & Yaskorska, O. (2014a). Towards Argument Mining from Dialogue. In S. Parsons, N. Oren, C. Reed, & F. Cerutti (Eds.), *Frontiers in Artificial Intelligence and Applications, Proceedings of 5th International Conference on Computational Models of Argument COMMA 2014* (pp. 185-196). IOS Press.

Budzynska, K., Janier, M., Reed, C., Saint-Dizier, P., Stede, M., Yaskorska, O. (2014b). A Model for Processing Illocutionary Structures and Argumentation in Debates. In N. Calzolari et al. (Eds.), *Proceedings of the 9th edition of the Language Resources and Evaluation Conference LREC*, (pp. 917-924).

Feng, V. W., & Hirst, G. (2011). Classifying arguments by scheme. In *Proceedings of the 49th Annual Meeting of the Association for Computational Linguistics: Human Language Technologies-Volume 1* (pp. 987-996). ACL.

Hernault, H., Prendinger, H., duVerle, D. A., & Ishizuka, M. (2010). HILDA: a discourse parser using support vector machine classification. *Dialogue and Discourse, 1*(3), 1-33.

Janier M., Aakhus, M., Budzynska, K. & Reed, C (2016). Modeling argumentative activity in mediation with Inference Anchoring Theory: The case of impasse. In D. Mohammed & M. Lewiński (eds.), *Argumentation and Reasoned Action: Proceedings of the 1st European Conference on Argumentation, Lisbon, 2015. Vol. I, 245-264.* London: College Publications.

Janier, M. & Yaskorska, O. (2016). Applying Inference Anchoring Theory for argumentative structure recognition in the context of debate. In D. Mohammed & M. Lewiński (eds.), *Argumentation and Reasoned Action: Proceedings of the 1st European Conference on Argumentation, Lisbon, 2015. Vol. II, 569-586.* London: College Publications.

Janier, M., Lawrence, J. & Reed C. (2014) OVA+: An argument analysis interface. In In S. Parsons, N. Oren, C. Reed, & F. Cerutti (Eds.), *Frontiers in Artificial Intelligence and Applications, Proceedings of 5th International Conference on Computational Models of Argument COMMA 2014* (pp. 463–464). IOS Press.

Kirschner, C., Eckle-Kohler, J. & Gurevych, I. (2015). Linking the Thoughts: Analysis of Argumentation Structures in Scientific Publications. In *Proceedings of the 2nd Workshop on Argumentation Mining* (pp. 1-11). Denver: ACL.

Lawrence, J. & Reed, C. (2014). AIFdb Corpora. In *Frontiers in Artificial Intelligence and Applications, Proceedings of 5th International Conference on Computational Models of Argument COMMA 2014* (pp. 465–466). IOS Press.

Mann, W. C., & Thompson, S. A. (1988). Rhetorical structure theory: Toward a functional theory of text organization. *Text-Interdisciplinary Journal for the Study of Discourse*, *8*(3), 243-281.

Mochales-Palau, R., Moens, M.F. (2009). Argumentation mining: the detection, classification and structure of arguments in text. In C. D. Hafner (Ed.) *Proceedings of the 12th International Conference on Artificial Intelligence and Law (ICAIL 2009)* (pp. 98-109). Barcelona: ACM Press.

Nguyen, H. & Litman, D. (2015). Extracting Argument and Domain Words for Identifying Argument Components in Texts. In *Proceedings of the 2nd Workshop on Argumentation Mining* (pp. 22-28). Denver: ACL.

Park, J., Katiyar, A., & Yang, B. (2015). Conditional Random Fields for Identifying Appropriate Types of Support for Propositions in Online User Comments. In *Proceedings of the 2nd Workshop on Argumentation Mining* (pp. 39-44). Denver: ACL.

Reed, C. Wells, S., Devereux, J. & Rowe, G. (2008) AIF+: Dialogue in the Argument Interchange Format. In *Proceedings of the 2nd International Conference on Computational Models of Argument (COMMA 2008)* (pp. 172-311).

Searle, J. R. (1969). *Speech acts: An essay in the philosophy of language (Vol. 626)*. Cambridge: Cambridge University Press.

Snaith, M., Reed, C. (2016). Dialogue Grammar Induction. In D. Mohammed & M. Lewiński (eds.), *Argumentation and Reasoned Action: Proceedings of the 1st European Conference on Argumentation, Lisbon, 2015. Vol. I, 667-684.* London: College Publications.

Stolcke, A., Ries, K., Coccaro, N., Shriberg, E., Bates, R., Jurafsky, D., & Meteer, M. (2000). Dialogue act modeling for automatic tagging and recognition of conversational speech. *Computational linguistics*, *26*(3), 339-373.

Swanson, R., Ecker, B., & Walker, M. (2015). Argument Mining: Extracting Arguments from Online Dialogue. In *Proceedings of 16th Annual Meeting of the Special Interest Group on Discourse and Dialogue (SIGDIAL 2015)* (pp. 217-227). Prague: ACL.

Walker M.A., Anand, P., Fox Tree, J. E., Abbott, & R., King, J. (2012). A Corpus for Research on Deliberation and Debate. In N. Calzolari et al. (Eds.), *Proceedings of the 8th Language Resources and Evaluation Conference*

(LREC 2012) (pp. 812-817). Istanbul.

Walton, D., Reed, C., & Macagno, F. (2008). *Argumentation schemes*. Cambridge: Cambridge University Press.

15

Limitations of the Sympathy-based Model of Ethical Deliberation: The Case of Adam Smith and Richard Mervyn Hare

ADAM CEBULA
Cardinal Stefan Wyszyński University, Warsaw, Poland
a.cebula@uksw.edu.pl

The paper examines the key assumptions underlying two prominent theories of moral thinking – Adam Smith's theory of sympathy and R. M. Hare's prescriptivism. Due to some analogous systemic shortcomings inherent in both theories, the specific procedure of ethical deliberation which they both entail is found to be impossible to fulfil. A case is made for the recognition of this as a major deficiency of a prototypal model of moral reasoning devised by Smith and Hare.

KEYWORDS: emotivism, ideal observer theory, preference utilitarianism, prescriptivism, role reversal, sympathy

1. INTRODUCTION[1]

The basic assumption behind Richard Hare's prescriptivism amounts to acknowledging the possibility of rendering any moral judgment as a specific type of precept. Hare's postulate concerning the morphological affinity between descriptive and imperative utterances allows him to lay down a set of rules of the calculus of imperatives, analogous to those of classic syllogistic reasoning. The possibility of applying the principles of non-contradiction or excluded middle to imperative sentences becomes the proof of the universal character of the standards of coherence, invoked typically in the analysis of descriptive assertions.

---

[1] The paper is in large part based on a fragment of the author's book published in Polish (Cebula, A. (2013). *Uczucia moralne. Współczesny emotywizm a filozofia moralna brytyjskiego Oświecenia*. Warszawa: Wydawnictwo Naukowe Semper).

D. Mohammed & M. Lewiński (eds.) (2016). *Argumentation and Reasoned Action: Proceedings of the 1st European Conference on Argumentation, Lisbon, 2015. Vol. II, 329-339.* London: College Publications.

However, Hare's model of a deep structure of language turns out to be insufficient as a basis for the identification of moral judgments. It is impossible to draw a boundary between moral precepts and ordinary imperatives by referring only to the prescriptivist interpretation of evaluative statements. In order to unambiguously identify the former, Hare introduces into his theory its second fundamental constituent – the principle of universalizability of moral judgments. The principle in question consists in the possibility of ascribing to a moral precept a characteristic of immutability in its relation to all the situations - real or hypothetical - which may be considered analogous to the context of its original application.

In order to test the universalizability of a prescription the person issuing it must "assume the roles of" all the persons who may be interested in the direct and indirect consequences of the course of action which it recommends. It is only after the whole series of these virtual "incarnations" of the moral arbiter that the ethical character of the evaluative sentence can ultimately be confirmed.

2. HARE'S UNIVERSALISABLE PREFERENCES AND SMITH'S SYMPATHY

It is easy to see that the universalizability of moral precepts, regarded as the second determinant of the autonomy ethical discourse, differs significantly from the first criterion of that autonomy, expressed by far more modest means. It leads directly to the involvement of Hare's prescriptivism in the whole range of issues related to the concept of ideal moral subjectivity. Among the metaethical theories built around this concept a prominent place is occupied by the ideal observer theory. It is invoked repeatedly by Hare himself: after its appropriate reformulation he considers it to be fully equivalent to the theoretical proposals articulated in "The Language of Morals" and "Freedom and Reason"[2].

The juxtaposition of prescriptivism with the classic type of the theory of ideal observer, presented by Adam Smith in his "Theory of Moral Sentiments," should not, therefore, seem an arbitrary construal. It is Hare himself who suggests that a parallel analysis of both theories might prove a worthy undertaking[3]. The common purpose of both thinkers, which is to overcome the solipsistic threats to ethical sentimentalism by devising procedures of intersubjective coordination of

---

[2] "Of the three theories that I have just shown to be practically equivalent, it is largely a matter of taste which one adopts." (Hare, 1989, p. 158).

[3] "It will be plain that there are affinities, though there are also differences, between this type of theory and my own." (*Hare*, 1963, p. 16).

individual emotions/preferences underlying value judgments, turns out to be a source of the fundamental kinship between the two theories.

The model of moral deliberation developed by Adam Smith provides sufficient evidence to support Paul Ardal's claim that the individual moral consciousness depicted in the "Theory of Moral Sentiments" "can be 'reduced to a species of sympathy'" (Ardal, 1966, p. 133). Smith's sympathy, i.e. a specific identification of a moral arbiter with the person subject to ethical evaluation, can be construed in two ways. Under the weak construal it involves consideration by the moral arbiter of the possibility of finding himself in the same situation as the person whose behavior is being assessed, followed by the examination of his own, hypothetical emotive reactions, which could be the result of this type of placement in the relevant situational context. It is the hypothetical emotional response of the moral arbiter to the circumstances of the moral dilemma that is to constitute the ultimate basis for the final normative verdict ("If I were you, I would do this or that ...").

Under the alternative interpretation, each act of sympathy entails the necessity of a distinct kind of "appropriation" of the identity of the person one sympathizes with. A subject appraising the moral value of a particular act is not supposed to consider the **hypothesis** of his own involvement in the given situation, but to achieve - at the very moment of resolving the moral dilemma - **a state of complete identification** with the agent undertaking the assessed course of action:

> When I condole with you for the loss of your only son, in order to enter into your grief I do not consider what I, a person of such a character and profession, should suffer, if I had a son, and if that son was unfortunately to die: but I consider what I should suffer if I was really you, and I not only change circumstances with you, **but I change persons and characters**. My grief, therefore, is **entirely upon your account, and not in the least upon my own. It is not, therefore, in the least selfish** [emphasis added] (Smith, 2005, pp. 289-290).

As can be seen, according to Smith, a person engaging in ethical deliberation acquires the actual competence to judge another person, only on the condition that at the time of delivering the moral verdict he - in a certain sense - becomes that person (!).

It is the issue of the wobbly identity of the moral arbiter that demonstrates the deep affinity between the metaethical underpinnings of the theories of Adam Smith and Richard Hare. The key thing here is Hare's depiction of the act of ethical deliberation. Just as in the argument of Smith, the issuing of a moral verdict becomes possible, according to Hare, only under the condition of achieving a thorough unification of the perspectives of the observer-arbitrator and the participant/participants of the situation under consideration. This superimposition of completely different perceptual frameworks constitutive of particular instances of moral subjectivity cannot be rendered by use of a standard conditional formula:

> We shall make the nature of the argument clearer if, when we are asking B to imagine himself in the position of his victim, we phrase our question, never in the form `What would you say, or feel, or think, or how would you like it, if you were he?', but always in the form `What do you say (in propria persona) about a hypothetical case **in which you are** [emphasis added] in your victim's position?' (Hare, 1963, p. 108)

The specific appropriation of another person's self-consciousness is supposed to embrace, as Hare writes in another place, the full set of attributes characterizing that person as an autonomous individual:

> B has got, **not to imagine himself** in A's situation **with his own (B's) likes and dislikes, but to imagine himself in A's situation with A's likes and dislikes** [emphasis added.] (Hare, 1963, p. 113).

A common assumption of the theories of Smith and Hare as regards the functional outfit of moral subjectivity capable of carrying out this type of operation is the acknowledgement of existence of a specific psychological capacity – the power of imagination. It is imagination that renders possible the suspension of the boundaries between the "I" of the moral arbiter and the "I" of the person being evaluated. Crossing these boundaries at the moment of solving a moral dilemma, i.e. a temporary "takeover" of another person's "I", involves all the dimensions of individual subjectivity:

> By the imagination we place ourselves in his situation, we conceive ourselves enduring all the same torments, we enter as it were into his body, and become in some measure the same person with him, and thence form some idea of his sensations, and even feel something which, though weaker in degree, is not altogether unlike them (Smith, 2005, pp. 4-5).

It seems, therefore, that the operation of the reversal of roles performed by a moral arbiter is to go far beyond his considering the hypothesis of his own involvement in the situation he deliberates upon. The act of sympathizing with another person entails taking over the very "substrate" of his/her identity - the "I" becomes the "s/he" with all the specific attributes of subjectivity constitutive of that person as an autonomous individual.

3. CIRCULARITY OF SYMPATHY

With such a thorough transformation of the perceptual perspective on each situation, however, it becomes doubtful whether any result of ethical deliberation may ever be obtained. The metempsychotic acquisition of another person's subjectivity, which is repeatedly described by Hare, must in each case result in restoring the overall structure of the moral "I", including the person's ability to deliberate on moral matters. Such a conclusion is confirmed by Hare's analysis of the imaginative reversal of roles between a sadistic torturer and his victim:

> If he [the victim] is suffering like that, he knows that he is, and has the preference that it should stop (a preference of a determinate strength, depending on how severe the suffering is). He thus assents, with a determinate strength of assent, to the prescription that it should stop. This preference and this assent are part of his situation, and therefore part of what I have to imagine myself experiencing, were I to be transferred forthwith into it (Hare, 1989, p. 184).

It is notable that the prescriptivist model of moral deliberation is applied by Hare with unwavering rigidity: subjected to severe physical torture the individual fully retains the ability to articulate his ethical position in terms of universalizable prescriptions. Under such an assumption, however, Hare's postulate of a total identification between the torturer-moralist with his victim must entail the necessity of re-starting the process of moral deliberation - this time from the perspective of the victim, who is supposed to perform the act of sympathizing with her own executioner (!).

The "fast track" solution to the dilemma proposed by the Hare seems to completely ignore the recurrence potential inherent in the act of role reversal between the two equal subjects. The desire to eliminate the suffering expressed by the victim is the most dominant component of his sensations and - for obvious reasons - becomes an essential element of his subjective experience, which is, in turn, "appropriated" by the individual engaging in moral deliberation. Hare's insistence on the most basic nature of the act of identification between the tormentor-moralist and his victim results in a situation in which the former, in order to articulate a fully-fledged, universalizable prescription, laying down his obligation to stop inflicting pain on his defenseless prisoner, must, in the newly adopted role, carry out the whole procedure of moral reasoning from the very beginning. Despite Hare's assurances to the contrary, a direct result of this deliberation does not systematically confirm the self-evident precept applicable in the given case – Stop inflicting pain on your victim! – but brings about a serious procedural complication, assuming the form of the classic "begging the question" fallacy. Paradoxically, the tormented prisoner - or rather the tormentor assuming in thought the "role" of his victim – is not ethically entitled to request directly for the cessation of his torture. By initiating the procedure of moral reflection, he can only reenact the maneuver of shifting the decision-making center back into the imagined (at the second, higher level of "imagined imagination") outlook of the ruthless torturer. At this point, the whole procedure begins anew. The ultimate result of this infinitely multiplied "metempsychosis" founded in Hare's model of ethical deliberation, is thus a kind of "labyrinth of mirrors" - making it impossible to generate a real solution to a moral dilemma.

It is a significant fact that the risk of circular reasoning appears already in the theory of sympathy proposed by Adam Smith. Postulating the need for restraining one's natural affective impulses, Smith introduces into his theory a distinct principle of a kind of "secondary" mediation of emotional experience:

> As [the observers] are constantly considering what they themselves would feel, if they actually were the sufferers, so he is as constantly led to imagine in what manner he would be affected if he was only one of the spectators of his own situation. As their sympathy makes them look at it, in some measure, with his eyes, so his sympathy makes him look at it, in some measure, with theirs, especially when in their presence and acting under their observation: and as the reflected passion, which he thus conceives, is much weaker than the original one, it necessarily abates the violence of what he felt before he came into their presence, before he began to recollect in what manner they would be affected by it, and to view his situation in this candid and impartial light (Smith, 2005, p. 17).

What is significant in this description, just as in Hare's analysis of the emotive experiences of a psychopathic torturer and his victim, is the emphasis on the reciprocal character of sympathy. It is that reciprocity of sympathy which makes the suffering person - the object of the compassion of others - distance herself from her own sensations and look at her situation through the "eyes" of the external observers. The main content of her subjective experience is thus expanded by this "gap" and only in this expanded form can be "adopted" by the sympathetic observers.

Correspondingly to Hare's assumption concerning the full symmetry of the act of "role reversal", Smith's postulate of the reciprocity of sympathy leads to an inevitable circularity of his own formula of moral deliberation. Due to the recursive nature of mutual fellow-feeling, the continuous distancing oneself from the actual content of one's own emotive experience/the emotive experience of others - person A sympathizes with the feelings of person B, striving, in turn, to sympathize with the feelings of person A, who sympathizes with person B, etc. - must result in the endless escape of the final settlement of the moral dilemma. Smith's conviction of the ability of a moral arbiter to break the "vicious circle" and obtain in particular cases an explicit normative judgment appears to go beyond the limitations of his theory.

4. DISPERSION OF MORAL SUBJECTIVITY

Both in "The Theory of Moral Sentiments," as well as in Hare's prescriptivism an attempt is made to put forward an alternative construal of the act of sympathy. "The violator of the more sacred laws of justice" depicted by Smith must experience - sooner or later - a sense of alienation from the motives prompting his vicious conduct:

> They appear now as detestable to him as they did always to other people. By sympathizing with the hatred and abhorrence which other men must entertain for him, he becomes in some measure the object of his own hatred and abhorrence (Smith, 2005, p. 76).

A similar act of "joining in" with the feelings of all the observers of one's actions - real or hypothetical - turns out to be - according to Smith – a necessary condition for the possibility of overcoming one's selfish passions:

> Respect for what are, or for what ought to be, or for what upon a certain condition would be, the sentiments of other people, is the sole principle which, upon most occasions, overawes all those mutinous and turbulent passions into that tone and temper which the impartial spectator can enter into and sympathize with (Smith, 2005, p. 239).

In both cases, assuming this multiplied perspective on the evaluated situation is dependent on the success of a truly impressive operation engaging individual moral subjectivity. It is supposed to consist in the simultaneous (or temporarily unspecified) "locating of" of one's moral "I" in all the positions occupied by people who are in any way interested in a specific outcome of the course of events.

The specific requirements for this kind of "multi-location" of the moral arbiter are outlined by Hare in an explicit manner:

> B has got, **not to imagine himself in A's situation with his own (B's) likes and dislikes, but to imagine himself in A's situation with A's likes and dislikes. But the moral judgement which he has to make about this situation has to remain B's own, as has any other prescriptive judgement that he makes, if it is to have a bearing on the argument** [emphasis added] (Hare, 1963, p. 113)

The superimposition of two completely distinct instantiations of subjective self-awareness is supposed to enable the moral arbiter to enunciate normative prescriptions from two "places" at a time. A moral precept, proclaimed by the moral arbiter in the state of full consciousness of the "appropriated" experience of the other participants of the assessed situation (imagining himself ***"in A's situation with A's likes and dislikes".***), is to remain, at the same time, his own, fully private acknowledgment of the solution to the moral dilemma.

A remarkable complexity of such a position of the moral arbiter may be revealed in a deeper analysis of the jurisdictive/lawgiving state of mind he is supposed to attain. It turns out that the person engaging in moral deliberation, while transferring himself in imagination to the place occupied by (or, to be exact, turning into) the other participants of the assessed situation, is supposed to be experiencing **at the same time** their own - real and present ("I want it here and now), and not hypothetically "simulated" ("I would like that in that situation") – volitional states of mind, resulting from the adequate recognition of the situational determinants of other people's preferences.

It might seem that this elaborate matrix of a relatively stable – i.e. limited to the moment of normative decision-making - fusion of all the relevant perspectives of looking at the situation of a moral dilemma will be a sufficient safeguard against the danger of falling into the trap of the said circularity of moral reasoning. However, subject to a rigorous analysis, the "bi-locating" moral subjectivity turns out to be a grotesque hybrid, shielding a truly fundamental problem of the inter-subjectivist version of emotivism devised by Smith and Hare. It amounts to the question concerning the mode of a subjective anchoring of normative prescriptions articulated by a moral arbiter sympathizing with other people.

The question of the specific "locus subiecti" (the position of the subject) - a clear and at least temporarily stabilized embedment of a moral arbiter's subjective perspective within the structure of the relations between autonomous individuals - becomes even more urgent in view of a possible numerical complication of a situation posing a moral dilemma. It is a perfectly natural thing to envisage a course of events, in which the direct or indirect involvement can be attributed to a larger number of people (perhaps infinitely many people?).

Such a situation, writes the author of "Freedom and Reason", should be categorized in the same manner as the simplified case, representing a binary relation between a person engaged in moral deliberation and a person he reverses roles with. The hypothesis of a sequential ("in random order") consideration of particular episodes of the specific "metempsychosis" carried out by the moral arbiter does not essentially affect the basic logical structure of the original construal depicting the target jurisdictive/lawgiving state of moral self-awareness.

A crucial challenge for Hare's prescriptivism is thus the development of a concept of a subjective super-structure in which the procedure of comparing/compiling the preferences of separate individuals - preceded by their accumulation in that super-subject's memory - can be effectively carried out. Due to the stringent requirements imposed by Hare on the act of role reversal in order to preserve its truly inter-subjective nature, it is impossible to accept James Griffin's proposal of a: "reduction of interpersonal comparisons to intrapersonal ones by appeal to the judger's own preferences as to possible states of himself (Griffin, 1988, p. 76).

It is no wonder, therefore, that, in spite of its radicalism, the postulate of a specific dispersion of consciousness of the moral arbiter, remains the most adequate interpretation of the act of "taking into account" all the individual preferences which are at stake in a given situation. It is explicitly confirmed by the author of "Moral Thinking" himself in his polemic against the argument of Zeno Vendler, questioning the legitimacy of the "multi-locational" division of the moral "I":

> I can **imagine being you but not me being you, or you being me.** For these `you's' and `me's' denote distinct individuals, which cannot be mixed or exchanged.[...] [I]magining being in exactly the same qualitative conditions as another person is

> the same thing as imagining being that person [emphasis added] (Vendler, 1988, p. 183).

Hare's response to this critique does not leave any doubt as to his presumption concerning the potential "stretch" of moral subjectivity:

> [I]n the moral argument we have to appeal to the concern which those people have for themselves in that situation[...]. In so doing, we are taking into account the concerns of two different people, both of whom refer to themselves as `I'. So in that sense **there are two Is and not one** [emphasis added] (Hare, 1988, p. 286).

As it is not difficult to guess, such a conception does not in any way facilitate finding a solution to the fundamental problem of the theories of Smith and Hare, i.e. a profound dichotomy between the natural - psychological and cognitive - constraints of a human individual, and the truly excessive requirements relating to the acts of sympathy he is supposed to undertake. The question of "locus subiecti" - the question of which particular place a subject-anchored universalizable prescription is to be enunciated from – thus remains unanswered.

5. CONCLUSION

The systemic shortcomings which, in a symmetrical pattern, become evident in the course of analysis of the ethical theories of Smith and Hare point out severe limitations of the intersubjective version of emotivism proposed by both thinkers. Surprisingly enough, both theories' failure to overcome the solipsistic impediment to moral sentimentalism, i.e. to devise adequate procedures of transferring acts of ethical evaluation beyond the privacy of an individual moral "I", may be said to have been predicted by the great predecessor of Adam Smith:

> No force of imagination can convert us into another person, and make us fancy, that we, being that person, reap benefit from those valuable qualities, which belong to him. Or if it did, no celerity of imagination could immediately transport us back, into ourselves, and make us love and esteem the person, as different from us (Hume, 1946, p. 69).

Apparently, the emotionality-centered approach to ethics initiated by David Hume almost 300 years ago seems to be doomed from the beginning to stumble into insoluble paradoxes.

REFERENCES

Ardal, P. S. (1966). *Passion and Value in Hume's Treatise.* Edinburgh: Edinburgh University Press.

Griffin, J. (1988). Well-being and its Interpersonal Comparability. In D. Seanor, N. Fotion & R. M. Hare (Eds.), *Hare and Critics. Essays on Moral Thinking* (pp. 73-88). Oxford: Oxford University Press.

Hare, R. M. (1963). *Freedom and Reason.* Oxford: Clarendon Press.

Hare, R. M. (1988). Comments on Z. Vendler. In D. Seanor, N. Fotion & R. M. Hare (Eds.), *Hare and Critics. Essays on Moral Thinking* (p. 286). Oxford: Oxford University Press.

Hare, R. M. (1989). *Essays in Ethical Theory.* Oxford: Oxford University Press.

Hume, D. (1946). *An Enquiry Concerning the Principles of Morals.* LaSalle, Ill.: The Open court publishing co.

Smith, A. (2005). *The Theory of Moral Sentiments. Sao Paulo: MetaLibri.*

Vendler, Z. (1988). Changing Places? In D. Seanor, N. Fotion & R. M. Hare (Eds.), *Hare and Critics. Essays on Moral Thinking* (pp. 171-184). Oxford: Oxford University Press.

16

Shallow Techniques for Argument Mining

JÉRÉMIE CLOS
Robert Gordon University, UK
j.clos@rgu.ac.uk

NIRMALIE WIRATUNGA
Robert Gordon University, UK
n.wiratunga@rgu.ac.uk

STEWART MASSIE
Robert Gordon University, UK
s.massie@rgu.ac.uk

GUILLAUME CABANAC
Université de Toulouse, France
guillaume.cabanac@univ-tlse3.fr

Argument mining has recently emerged as a promising field at the frontiers of the argumentation and text mining communities. However, most techniques developed within that field do not scale to larger amounts of data, depriving us for example of valuable insights in large-scale discussion forums. On two social media datasets, we study different lightweight scalable text mining techniques used within the sentiment analysis community and their applicability to the argument mining problem.

KEYWORDS: argument mining, sentiment analysis, text mining

1. INTRODUCTION

The advent of the Web 2.0 has seen a massive increase in user-generated data in the form of comments and messages such as the ones displayed in Figure 1. Increasingly it is the platform of choice for public debate and conversation, but its traditional "document-centric" focus and associated search and browse methods are less fit for purpose. For

D. Mohammed & M. Lewiński (eds.) (2016). *Argumentation and Reasoned Action: Proceedings of the 1st European Conference on Argumentation, Lisbon, 2015. Vol. II, 341-356.* London: College Publications.

example, in Figure 1 it is not sufficient to only search for keywords but instead be influenced by the tree-like thread structure enforced by users responding to other user comments.

Figure 1: Excerpt from a comment tree where users discuss UK politics.

New research fields such as sentiment analysis and topic modelling have thus emerged in order to fill this need for better and more intuitive ways to help users browse through large amounts of data. While valuable, these techniques inevitably fall short of meeting the representational requirements when dealing with conversational data. For instance, sentiment analysis is only concerned with projecting documents on a negative to positive opinion dimension, and topic modelling is focused on analysing a corpus and identifying central topics. Neither are interested in the conversational dynamics.

Argument mining is able to discover knowledge that would allow us to detect justifications for common opinions, generate fine-grained debate graphs for complex political issues or refine common opinion mining algorithms. There are however many challenges in adapting argument mining algorithms to the scale of the Social Web. Current approaches either rely on computationally expensive NLP techniques or on human annotations, neither of which are transferable to a real time analysis setting where large volumes of data, absence of reliable knowledge sources and informal language are the norm.

In this paper we propose to relax the requirements of an argument mining algorithm by restricting the argument mining task to a target (another argument/expression of opinion) detection and stance (whether it supports or attacks the target) classification task, leveraging existing literature in the sentiment analysis and opinion mining. Our contribution is threefold: firstly, we build a novel dataset based on online comments from the Reddit[1] social website and a noisy labelling process. Secondly, we experiment using three standard unsupervised sentiment analysis approaches in order to measure how well they can approximate the stance classification part of the argument mining process. Thirdly and finally, we improve a PMI-based classifier by incorporating contextual clues in a simple but intuitive way into the classification process.

In section 2 we detail the relationship between argumentation, argument mining and information retrieval, thus justifying and contextualizing our approach. In section 3 we explain the classification approaches that are being compared, as well as the approach we are proposing as an incremental improvement over a naive technique based on strength of association. In section 4 the experimental methodology is presented, together with details of a new dataset, generated for the purpose of this research. Finally, before concluding in section 6, section 5 will analyse the results from the comparative study.

2. BACKGROUND AND RELATED WORKS

Our approach to argument mining is inspired by text analysis and representation schemes which are commonly used in information retrieval. Rather than linguistic rigor, we aim to use knowledge-light representations that can still provide insight about the discussion. As explained in the previous section, we seek to rebuild the argumentation graph underlying a discussion by making the following simplifying assumptions: all comments have an argumentative value, and the target of a comment is always the comment to which it is replying. For the purpose of building a bipolar argumentation graph, we loosely assign to the "attack" relationship defined by Dung (Dung, 1995) the semantics of *overall disagreement*, and to the "support" relationship defined by Cayrol and Lagasque-Schiex (Cayrol & Lagasquie-Schiex, 2005) the semantics of *overall agreement*.

The need to scale classification to large amounts of data requires a simple conceptual representation of arguments, such as the one proposed in Pragmatic Argumentation Theory (PAT) (Van Eemeren et al., 1996; Hutchby, 2013). Fitting a complex model of argument would be computationally expensive and not fit the colloquial nature of social media content and it is thus deemed preferable to use a more accurate and simpler model.

---

[1] http://www.reddit.com

PAT defines an argument as an opinionated piece of text which can arise in the presence of two elements: (1) a **target**, being some other action by another actor which has been called out ; (2) a **stance**, i.e. whether it is supporting or attacking the target. We bypass the target detection step and focus on stance classification, which allows us to use techniques from text mining and sentiment analysis (Pang & Lee, 2008), considering stance of an argument analogous to the sentiment of an opinionated text.

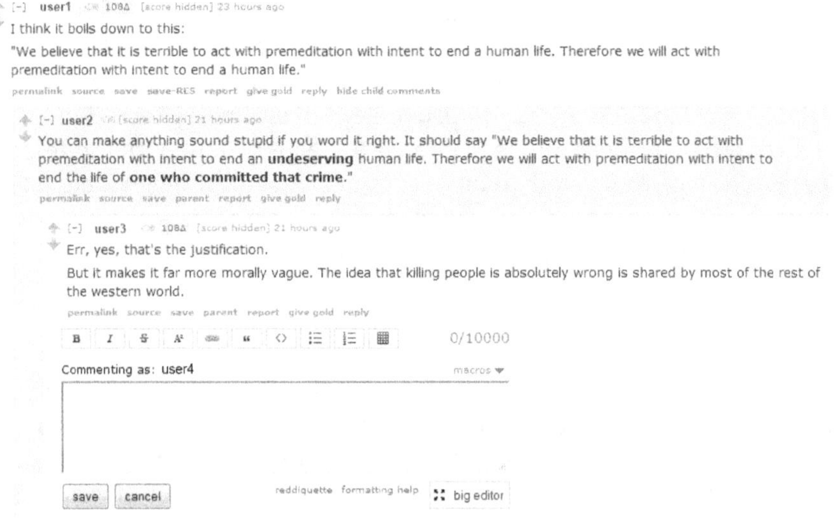

Figure 2: Illustration of how the Reddit commenting system forces the user to place their comment under the contribution

Figure 2 illustrates the way the system incites users to insert their comment under the relevant section of the discussion by presenting it in a threaded structure, allowing us to treat the argument mining problem as a classification task. For example, the comment posted by User 3 in Figure 2 is in agreement with the comment posted by User 2, which makes it a supporting statement.

2.1 Argument mining

Argument mining has been approached in the literature as the study of methods and techniques to detect argumentative discourse units, their role in the argumentation process and how they relate to other argumentative discourse units (Peldszus & Stede, 2013). Early work focused on representing arguments in a restricted manner (Cohen,

1987), but the first steps towards an automated treatment of argument mining (Palau & Moens, 2009; Mochales & Moens, 2011) aimed to mine legal text using supervised learning techniques. More particularly, Palau and Moens (Palau & Moens, 2009) performed a three-step argument analysis by firstly detecting argumentative sentences, secondly identifying whether they were part of a conclusion or a premise, and thirdly classifying the relationships between these sentences, thus trying to mine argumentative structure in these legal texts. Cabrio and Villata (Cabrio & Villata, 2012) on the other hand made use of textual entailment and semantic similarity features in order to train a classifier to recognize attacking arguments, in line with Dung's abstract argumentation framework (Dung, 1995). Similar work along these lines studied the use of context-free grammars (Wyner et al., 2010) to extract arguments but did not handle the detection of their relationships with other arguments, since legal texts mainly deal with cases of monological argumentation. However, these approaches do not transfer well to social media because of their reliance on idiosyncratic features and complex learning methods such as support vector machines (SVM) (Bishop, 2006).

Other works from the argument mining community focused on bridging it to the field of opinion mining (Villalba & Saint-Dizier, 2012) and studied the use of reasoning patterns in user-contributed reviews. They did not however direct their study towards the automated detection of arguments themselves within these textual reviews and instead focused on a descriptive analysis, making them not directly relevant to our work.

2.2 Stance classification

Stance classification becomes relevant to argument mining because of its binary classification nature. However most techniques used in the literature base their work on training a complex classifier using a large number of computationally expensive features (Boltuzic & Snajder, 2014; Anand et al., 2011; Abbott et al., 2011; Somasundaran & Wiebe, 2010; Walker et al., 2012a) or are performed on automatically transcribed text (Wang et al., 2011) or non-conversational content (Bousmalis et al., 2013) with limited applicability to Web2.0.

Much closer to our approach, Yin et al. (Yin et al., 2012) used a logistic regression classifier trained on somewhat complex features and related the notion of local and global stances, building the global stance of a post by computing a sequence of local stances between that post and the first appearance of the topic of discussion

Cardie and Wang (Wang & Cardie, 2014) took a different approach to the stance classification problem in that they used an isotonic conditional random field-based technique to detect local stance. However, they still required a significant training phase and need a large collection of idiosyncratic features, which negates the portability of their approach.

2.3 Sentiment analysis and argument mining

The field of sentiment analysis (Pang & Lee, 2008) also treats the binary classification of large text corpora along the axis of positivity/negativity. Two main families of methodologies emerge:
- **Supervised sentiment analysis** involves the use of supervised machine learning techniques in order to perform sentiment classification. Traditional algorithms known for their versatility and performance on text classification tasks are MaxEnt, SVM, and Naive Bayes (Pang et al., 2002). These algorithms are trained on training data in order to produce a model able to classify future test data.
- **Lexicon-based sentiment analysis** involves the learning/building of lexicons, which are look-up tables of terms with strengths of association scores for different classes, as well as combination rules (Taboada et al., 2011). Test data is directly fed to the lexicon-based classifier which uses the combination rules in order to compute the scores of each class (positive and negative) based on the presence of terms from the lexicon.

Our work can be put in context between the argument mining and the stance detection communities, as we aim for the detection of relationships between textual entities with the bipolar semantics of Cayrol and Lagasquie-Schiex (Cayrol & Lagasquie-Schiex, 2005) but do so by attributing different semantics to their relations.

3. SHALLOW TECHNIQUES FOR ARGUMENT MINING

We refer to the relationship between a comment and its direct parent as the conversational context of that comment. Our goal is to take into account, for all comments, a progressively deeper level of their conversational context and study its effect on the overall classification accuracy. As such, we will review the approaches according to their level of context-informedness as well as their degree of supervision. Here context-informedness refers to the extent to which the approach uses information that is external to the comment, and supervision refers to the extent to which the algorithm requires a human-labelled dataset in order to work.

3.1 Argument mining as lexicon-based classification

In its most basic form, argument mining can be seen as a classification task, where the classes are either support or attack. This basic form entails that any piece of text can be classified by itself without taking into account any notion of conversational context, by simply applying standard text representation techniques and mapping this representation into the class codomain. contains simple lexicon-based classification (referred to as PMILex).

A simple way to reliably classify instances is the use of a lexicon. Because there does not exist a manually built lexicon of local stance, we compute it on a distant-labelled dataset using normalized pointwise mutual information (NPMI) as a measure of strength of association between terms and their class.

$$NPMI(x,y) = \frac{PMI(x,y)}{-\log[p(x,y)]}$$

$$PMI(x,y) = \frac{\log\,(p(x,y))}{p(x)p(y)}$$

This classification rule classifies a user comment x on the basis of a maximized sum of associations between each of its terms t and each class c. Notice that the term class associations can now be exploited as a lexicon.

While this approach is sensible to create a general purpose lexicon, it suffers some flaws in the following cases: (1) if none of the terms used in the child post has an argumentative value or is present within the lexicon, no classification is possible, and (2) some terms might end up with an undeserved score because they accidentally appear more frequently within comments of one class. For example if non-argumentative terms such as "Monday" accidentally co-occur too often within one class, they will be misconstrued as being indicative of that class.

3.2. Context-aware methods for argument classification

The previous method provided a simple mapping from a set of terms to a class label, we now explore different ways to consider context during the classification. This context can originate from either the sentiment contained within the comment, or the conversation that contains the comment. Sentiment-guided methods deal with the sentiment context, i.e. the sentiment that is expressed within the terminology and grammatical structure used in the text. Within the context of a discussion, this sentiment is akin to a global stance taken by the author of that text with respect to the topic at hand. Conversational context-aware methods on the other hand are focused on representing the relationship between a comment and its conversational context, in our situation its parent post. They do so either by detecting attack or support against the author of the parent comment (Wang & Cardie, 2014) or by modifying the importance of some terms based the parent comment (the proposed approaches).

3.2.1 Sentiment-guided methods

Sentiment methods provide a means to use emotive context to infer argument stance, by assuming the stance of a comment as equivalent to its sentiment orientation. We employ a simple sentiment analysis algorithm based on a lexicon (Esuli & Sebastiani, 2006) to which we will refer to as SENTLEX. It operates by looking up positive and negative values of terms present in the comment and summing them separately into a positive and a negative score. The classification rule is based on a simple comparison: a higher positive strength implies a supportive comment, and higher negative strength implies an attacking comment. We use the SMARTSA algorithm (SSA) developed by Muhammad et al. (Muhammad et al., 2013) as an extension of traditional lexicon-based sentiment analysis techniques taking into account additional linguistic factors such as the presence of special terms called modifiers that alter the class values of terms in their vicinity by exaggerating them (amplifiers), reducing them (diminishers) or inversing their polarity (negators).

Sentiment-guided methods assume that global stance of the comment, i.e. how its author feels about the discussion topic, can be used in place of its local stance, i.e. how its author feels about the parent comment. As such they present some flaws whenever (1) those two stances do not align, (2) the stance is expressed in a sentiment-neutral WAY, or (3) the overall sentiment of the sentence ends up being balanced. The following examples illustrate these flaws:

(1) "*I completely disagree with you, this movie was very good and I enjoyed every minute of it.*" Here we can see that the author expresses a positive opinion by disagreeing with the author of the parent comment, thus making an attacking statement.

(2) "*There is nothing in the world that will make me see the situation your way.*" In this comment there is no positive or negative terminology used, while the sentence is written with an attacking stance.

(3) "*I agree that the acting was good, but I am still disappointed that the dialogues were so poorly written.*" Here negative and positive sentiments are equally used, but the stance should be supporting.

3.2.2 Conversational context-aware methods

We explore conversational context-aware methods using the Unsupervised Sentiment Surface algorithm (USS). USS is extended from Cardie and Wang (Wang & Cardie, 2014) and uses a shallow linguistic analysis to compute the average distance between second person pronouns and positive or negative terms. The classification compares those average distances, classifying the comment as attacking if negative terms are on average closer to second person pronouns and supporting otherwise. USS works on the intuition that the stance of the comment is contained within explicit references to the parent posts: such references can be analyzed by detecting second person pronouns (e.g. "you", "your", etc.) and their polarity by searching their grammatical neighborhood for sentiment-bearing terminology.

For example, "*I don't agree with you and I think your opinion is wrong*" would be interpreted as an attacking statement because of the overwhelming proximity of negative terms ("*don't*", "*wrong*") near second person pronouns ("*you*", "*your*"). This approach can also give flawed results when there is ambiguity contained in the text in the following: (1) when a positive (respectively negative) term is accidentally closed to a second person pronoun which is semantically linked to a negative (respectively positive) term (2) when a more complex sentence structure is used where the polarity of a term is implicitly negated, or (3) whenever no second person pronouns or sentiment-bearing terms are used. The following examples illustrate the first two cases:

(1) "I like you, but you are wrong." Here we can see that d(like, you) < d(you, wrong) where d is a distance function, which would classify this instance as a supporting statement.

(2) *"I can barely tolerate that you believe yourself to be right."* Here the sentence structure puts *yourself* very close to *right*, which will classify the sentence as a supporting statement. However, it is clear under a human eye that the sentence has a disapproving tone.

The USS approach is based on the assumption that over a significant number of sentences these errors would cancel each other out and result in a good overall classification accuracy.

3.2.3 Context-informed feature vector enrichment

Term vector representation is a convenient way to work with text data because of its simplicity and versatility (more details can be found in (Manning et al., 2008)). It is compatible with lexicon-based classification as well as more standard supervised classifiers and extremely common in text classification. In this section we use conversational context to alter that feature vector of a comment based on its parent comment. We experiment using two basic variations of this alteration. The classification rule we use is to assign the class label c that maximizes the association score between an instance x and c. The association score is computed differently according to the enrichment scheme.

$$Classification(x) = ArgMax_c[AssociationScore(x, c)]$$

Intersection-based vector enrichment. We used a lexicon computed similarly to the simple PMI lexicon discussed in a previous section, and compute the association score between an instance x and a class c as the sum of NPMI scores between all terms t with c where t is present in both the x and its parent p(x).

$$AssociationScore_1(x, c) = \sum_{t \in x \cap p(x)} NPMI(t, c)$$

Union-based vector enrichment. We modify the classification rule and compute the association score as the sum of NPMI scores between all terms t with a given class c where t is present in either the instance x and its parent p(x).

$$AssociationScore_2(x, c) = \sum_{t \in x \cup p(x)} NPMI(t, c)$$

The algorithm described in Figure 3 represents the general classification algorithm of our approaches, where the changing part is the way the instance I is created from the comments P and C. The classification is done by a simple summation of terms, which is a highly scalable operation with a negligible computational cost.

```
Data: Child comment C and parent comment P
Result: A class label L
Create instance I from P and C;
AL ← 0;
DL ← 0;
For Term T in I do
  AL ← AL + agreementValue(T);
  DL ← DL + disagreementValue(T);
End
If DL ≤ AL then
  Return agreement ;
Else
  Return disagreement ;
End
```

Figure 3: General classification algorithm

4. EXPERIMENTAL DESIGN

In this section we detail our experimental design, firstly by going over our datasets and the way they were collected, secondly moving on to the pre-processing steps that were performed in order to sanitize the data, thirdly to the metrics used in order to compare approaches, and finally to our experimental methodology.

4.1 Datasets

We performed our experiments on two social media datasets: the Internet Argument Corpus (referred to as IAC) and the Reddit Noisy-Labelled Corpus (referred to as RNLC). Statistics on the corpora can be found in Table 1.

The Internet Argument Corpus (IAC). The IAC (Walker et al., 2012b) is a corpus of forum comments manually labelled by 5 annotators on a degree of agreement/disagreement with their parent comment on a scale of -5 to 5. A subset of this dataset was used for our experiment, by selecting the comments that ensured disjoint class membership (meaning filtering out comments with an average score close to 0).

Dataset	*IAC*	*RNLC*
Number of comments	1856	3086
Average terms/sentence	40.3	35
Average sentences/comment	2.9	7.8
Instances of agreement	928	1543
Instances of disagreement	928	1543
Common vocabulary size	6036	
Total vocabulary size	25004	

Table 1: Descriptive statistics on the dataset

The Reddit Noisy-Labelled Corpus (RNLC). The RNLC is a new corpus of comments extracted from the Reddit and automatically labelled with a binary class using evidence contained within the comments. Explicit expressions such as "*I [positive adverb] agree*" and "*I [positive adverb] disagree*" variations were used to detect evidence of a comment belonging to a class. In the case of the presence of conflicting evidence, i.e. expressions acting as strong evidence towards both classes, the comments were not considered. Remaining comments were automatically assigned to their respective class and the corresponding sentences were deleted from the comments in order to avoid a class bias advantage. That labelling process is inspired from distant supervision learning (Mintz et al., 2009) whereby highly discriminative expressions are used as class label proxies.

Both datasets were pre-processed by removing comments that were deemed as non-constructive because of their limited length. A threshold was empirically chosen based on a human observation of the data and all comments composed of less than 20 words and/or with a vocabulary of less than 10 words were considered as noise and removed from the data.

No stemming was applied due to the unreliable vocabulary used in social media, meaning that a rule-based procedure for would reunify terms that are semantically distant and thus remove information from the datasets. For the same reason no lemmatization was applied, since dictionary-based lemmatizers would at best be ineffective and at worst detrimental to our approach.

Finally, in the absence of information about the real class distribution, we artificially enforced a uniform class distribution by subsampling the majority class.

4.2 Evaluation metrics and experimental protocol

We chose classification accuracy as our evaluation metric because balanced data renders other threshold metrics (such as F_1-Score) less meaningful and used the standard 10-Fold cross-validation protocol for machine learning experiments (Bishop, 2006) for both our distantly learned approaches and our unsupervised approaches in order to preserve as fair a comparison as possible.

5. RESULT AND DISCUSSION

The results shown in Table 2 show that while a standard lexicon built on a background corpus with a simple bag-of-words representation does not significantly outperform standard sentiment analysis or stance classification techniques, changing the representation of the instances by adding some form of context improves our classification accuracy again by a significant margin (+5.7% compared to a similar approach without context and +7.8% compared to the best baseline).

Dataset/Method	*SentLex*	*SentLex*	*SmartSA*	*USS*	PMILex	PMILex+inter	PMILex+union
IAC	0.5042	0.5260	0.5147	0.5061	0.5362	**0.5899**	0.5696
RNLC	0.4670	0.4664	0.4522	0.4718	0.4843	**0.5304**	0.5043

Table 2: Accuracy of the compared approaches.

We note that between the two approaches to introduce context in the bag of words representation, the best accuracy was given by the approach using the intersection of the bag of words representations of child and parent comments rather than the union (which preserves more information). A possible reason for this is that the intersection of bags of words preserve topically relevant terms which can then be used in the classification process. In the context of classifying argumentative stance, such result implies that evidence of this stance is contained within common information present in both child and parent comment.

However, this does not account for the fact that the union of bags of words outperforms the standard methods while potentially adding noise into the representation. This leads us to think that further refinement of the model could be done by using a term weighting scheme as a middle ground between adding information using the union operator and filtering information using the intersection operator.

The accuracy scores however while being an improvement over the baselines are too low for an operational context. More work is required in improving the naive way in which we added context at classification time.

6. CONCLUSION

Argument mining in the context of classification has much to gain from shallow techniques borrowed from information retrieval, text mining and sentiment analysis research. A comparative analysis of representation techniques for classifying the parent-child relationships in threaded posts show that sentiment analysis approaches can be successfully adopted for argument mining provided that the conversational-context is captured in representation schemes. We show that the simple bag-of-words representation contextualised by parent-child vocabulary intersection leads to significant improvements over comparable baseline approaches. Following on from these results we aim to explore conversational-contextual enrichments that can further improve representation for argument classification. For this in addition to the parent-child single level relationship, we intend exploring further levels of context such as from siblings to ancestors.

REFERENCES

Abbott, R., Walker, M., Anand, P., Fox Tree, J. E., Bowmani, R., & King, J. (2011, June). How can you say such things?!?: Recognizing disagreement in informal political argument. In *Proc of the Workshop on Languages in Social Media* (pp. 2-11). Association for Computational Linguistics.

Anand, P., Walker, M., Abbott, R., Tree, J. E. F., Bowmani, R., & Minor, M. (2011, June). Cats rule and dogs drool!: Classifying stance in online debate. In *Proc 2nd workshop on computational approaches to subjectivity and sentiment analysis* (pp. 1-9). Association for Computational Linguistics.

Bishop, C. M. (2006). *Pattern recognition and machine learning*. Springer.

Boltuzic, F., & Šnajder, J. (2014, June). Back up your stance: Recognizing arguments in online discussions. In *Proceedings of the First Workshop on Argumentation Mining* (pp. 49-58).

Bousmalis, K., Mehu, M., & Pantic, M. (2013). Towards the automatic detection of spontaneous agreement and disagreement based on nonverbal behaviour: A survey of related cues, databases, and tools. *Image and Vision Computing*, *31*(2), 203-221.

Cabrio, E., & Villata, S. (2012, July). Combining textual entailment and argumentation theory for supporting online debates interactions. In *Proc of the 50th Annual Meeting of the ACL: Short Papers-Volume 2* (pp. 208-212). Association for Computational Linguistics.

Cayrol, C., & Lagasquie-Schiex, M. C. (2005). On the acceptability of arguments in bipolar argumentation frameworks. In *Symbolic and quantitative approaches to reasoning with uncertainty* (pp. 378-389). Berlin / Heidelberg: Springer.

Cohen, R. (1987). Analyzing the structure of argumentative discourse. *Computational Linguistics, 13*(1-2), 11-24.

Dung, P. M. (1995). On the acceptability of arguments and its fundamental role in nonmonotonic reasoning, logic programming and n-person games. *Artificial Intelligence, 77*(2), 321-357.

Esuli, A., & Sebastiani, F. (2006, May). Sentiwordnet: A publicly available lexical resource for opinion mining. In *Proc of LREC* (Vol. 6, pp. 417-422).

Hutchby, I. (2013). *Confrontation talk: Arguments, asymmetries, and power on talk radio.* Routledge.

Manning, C. D., Raghavan, P., & Schütze, H. (2008). *Introduction to information retrieval* (Vol. 1, p. 496). Cambridge: Cambridge University Press.

Mintz, M., Bills, S., Snow, R., & Jurafsky, D. (2009, August). Distant supervision for relation extraction without labelled data. In *Proc of the Joint Conf of the 47th Annual Meeting of the ACL and the 4th Intl Joint Conf on NLP of the AFNLP: Volume 2-Volume 2* (pp. 1003-1011). ACL.

Mochales, R., & Moens, M. F. (2011). Argumentation mining. *Artificial Intelligence and Law, 19*(1), 1-22.

Muhammad, A., Wiratunga, N., Lothian, R., & Glassey, R. (2013). Domain-Based Lexicon Enhancement for Sentiment Analysis. In *SMA@ BCS-SGAI* (pp. 7-18).

Palau, R. M., & Moens, M. F. (2009, June). Argumentation mining: the detection, classification and structure of arguments in text. In *Proc of the 12th intl conf on AI and law* (pp. 98-107). ACM.

Pang, B., & Lee, L. (2008). Opinion mining and sentiment analysis. *Foundations and trends in information retrieval, 2*(1-2), 1-135.

Pang, B., Lee, L., & Vaithyanathan, S. (2002, July). Thumbs up?: sentiment classification using machine learning techniques. In *Proc of the ACL-02 conference on Empirical methods in NLP Volume 10* (pp. 79-86). ACL.

Peldszus, A., & Stede, M. (2013). From argument diagrams to argumentation mining in texts: A survey. *Int Journal of Cognitive Informatics and Natural Intelligence (IJCINI), 7*(1), 1-31.

Somasundaran, S., & Wiebe, J. (2010, June). Recognizing stances in ideological on-line debates. In *Proc of the NAACL HLT 2010 Workshop on Computational Approaches to Analysis and Generation of Emotion in Text* (pp. 116-124). ACL.

Taboada, M., Brooke, J., Tofiloski, M., Voll, K., & Stede, M. (2011). Lexicon-based methods for sentiment analysis. *Computational Linguistics, 37*(2), 267-307.

Van Eemeren, F. H., Grootendorst, R., Jackson, S., & Jacobs, S. (1993). *Reconstructing argumentative discourse.* Tuscaloosa, AL: University of Alabama Press.

Villalba, M. P. G., & Saint-Dizier, P. (2012, September). Some Facets of Argument Mining for Opinion Analysis. In *COMMA* (pp. 23-34).

Walker, M. A., Anand, P., Abbott, R., & Grant, R. (2012, June). Stance classification using dialogic properties of persuasion. In *Proc of the North American Chapter of the ACL: Human Language Technologies* (pp. 592-596). Association for Computational Linguistics.

Walker, M. A., Tree, J. E. F., Anand, P., Abbott, R., & King, J. (2012). A Corpus for Research on Deliberation and Debate. In *LREC* (pp. 812-817).

Wang, L., & Cardie, C. (2014). Improving agreement and disagreement identification in online discussions with a socially-tuned sentiment lexicon. *ACL 2014*, 97.

Wang, W., Yaman, S., Precoda, K., Richey, C., & Raymond, G. (2011, June). Detection of agreement and disagreement in broadcast conversations. In *Proc of the 49th Annual Meeting of the ACL: Human Language Technologies: short papers-Volume 2* (pp. 374-378). ACL.

Wyner, A., Mochales-Palau, R., Moens, M. F., & Milward, D. (2010). *Approaches to text mining arguments from legal cases* (pp. 60-79). Berlin / Heidelberg: Springer

Yin, J., Thomas, P., Narang, N., & Paris, C. (2012, July). Unifying local and global agreement and disagreement classification in online debates. In *Proc of the 3rd Workshop in Computational Approaches to Subjectivity and Sentiment Analysis* (pp. 61-69). ACL.

17

Reasonable Agents and Reasonable Arguers: Rationalization, Justification, and Argumentation

DANIEL H. COHEN
Colby College, USA
dhcohen@colby.edu

Data from neuroscience suggest that, contrary to the conference theme, argumentation and reasoning are not the main vehicles for our decisions and actions. They are "fifth wheels" on those vehicles: ornate but ineffective appendages whose maintenance costs exceed their contributions. Although the data, their interpretations, and their putative implications all deserve challenge, this paper explores how to accept and incorporate these findings into a coherent view of what we do when we reason.

KEYWORDS: argumentation, deliberation, justification, rationalization, virtue argumentation

1. INTRODUCTION

Providing arguments to justify our actions is routine, but psychological data show that much of our apparent reasoning is actually rationalization: our reasoning has little or no effect on the decisions we reach. Furthermore, despite our self-proclaimed status as reasons-responsive agents, it is *not* routine for us to respond very well to reasoning. The give and take of argumentation does not usually end with anyone's surrender, rarely elicits the desired concessions, and hardly ever succeeds in producing resolutions that are satisfactory all around. It seems that our reasoning is a charade and our arguments make no difference. We have been wasting our time: argumentation theory is much ado about nothing!

The situation is not that bleak, of course. First, there are some ambiguities in the neurological and psychological data: there is room to question the interpretations that lead to the conclusion that our reasoning has little or no effect on our decision-making. Second, there are occasions when we really do deliberate and argumentation really

D. Mohammed & M. Lewiński (eds.) (2016). *Argumentation and Reasoned Action: Proceedings of the 1st European Conference on Argumentation, Lisbon, 2015. Vol. II, 357-366.* London: College Publications.

does cause changes in how others think and act. The conclusions that we rationalize *more* than we reason and that argumentation is *mostly* idle leave room for those exceptions – and open the door for strategies to expand the range of effective reasoning and magnify its causal footprint. What I want to focus on, however, is the "third rail" here: defending the coherence, cogency, importance, and necessity of argumentation as a practice even if and when it is *post facto*, causally inefficacious rationalization.

The key ingredients in the defense are the difference between the causal and logical orders, how we individuate arguments, the different purposes implicit in the different demarcations, and the causal role of character in argumentation. The result is greater clarity about what it is to be a good arguer, what it means to be an effective arguer, and how those two concepts relate. What connects them is the rare argumentative virtue exhibited by those arguers who manage to bring out the best in their fellow arguers.

2. THE MAIN VEHICLES OF OUR DECISIONS?

The stated premise of this conference is that "argumentation and reasoning are the main vehicles for our decisions and actions." It is an optimistic, even idealistic, statement about human rationality, but unfortunately as it stands it is quite false.

In the first place, the phrase "*our* decisions" is ambiguous. It can be interpreted as referring either *distributively* to the decisions we make as individuals or *collectively* to the decisions we come to in groups. If the claim targets individuals, then we have to explain away the empirical evidence that the reasons we offer to explain and justify our decisions often have no causal role at all. Our stated reasons tend to be after-the-fact rationalizations rather than genuine deliberations leading to the act.[1] There is even evidence for thinking that our conscious decisions to act generally occur after the motor neurons have already initiated the processes leading to actions.[2] At the individual level, then, argumentation and reasoning are at least sometimes "fifth wheels" when it comes to the essential elements of human agency: acquiring beliefs, making decisions, and initiating actions.

---

[1] Kornblith, 1999 provides a philosophical account of rationalizations, but see Haidt, 2001 and Mercier and Sperber, 2011 for the empirical data.

[2] Wagner and Wheatley, 1999 is both a source and a summary of evidence for the claim that conscious decisions generally *follow* the initiation of motor actions.

The phrase "our decisions" is still problematic when applied to groups. Argumentation may indeed be the "main vehicle" when it comes to, say, a small circle of friends deciding where to go to dinner, but it would be hasty to generalize from that example. We cannot ignore the effects of group dynamics, social forces, and other economic, historical, institutional, cultural, and psychological contingencies on the entire gamut of group decision-making. Argumentation might play a decisive role in a medium-sized committee accepting a report, but its efficacy in rallying an entire polity around its government's war effort is harder to credit. I do believe that reasoning and argumentation sometimes do take lead roles, and that there are institutions and procedures we can put into place to help bring about those occasions, but those claims need supporting evidence and argument; argumentation theorists should not just take them on faith.

It is, then, at least arguable that argumentation and reasoning cannot be considered the main vehicles for *our* actions in either the distributive or the collective sense.

The metaphorical characterization of reasoning and argumentation as "the main vehicles" for our agency is also dangerously ambiguous. If that phrase is meant to refer to the causal mechanisms that bring about our decisions and actions, then once again we would need empirical data to support the claim, especially given the substantial body of empirical data suggesting otherwise. If, on the other hand, the claim is that reasoning and argumentation are the *vehicles* by which we navigate around our actions and decisions, i.e., the conceptual tools we use to understand, explain, justify, evaluate, motivate, and otherwise process our own agency, then it is unobjectionable, but approaching the tautologically vacuous: the way to justify an action is to provide a justification.

There is an important point here about explanation that almost always needs emphasis: even though we refer to premises and to causes alike as reasons, the logical relation of premises to conclusions is different than the physical relation of causes to effects. One relates propositions; the other connects events. Thus, argumentation and reasoning could be the main vehicles in justifying explanations of actions even if they are not the main vehicles in causal explanations of actions. Even if the reasons we offer to justify an act have no role in causing it – in which case they might be fairly called rationalizing – their status as justifiers are not necessarily undermined. Justifications can be causally impotent but logically powerful at the same time.

There is another kind of causal impotence to acknowledge. Besides the questions about the causal role of the reasons we give, there are questions about the causal efficacy of the reasons we hear. Even very good reason-givers are not necessarily very reasons-responsive. We are skillful at constructing arguments, rationalizations or not, and part of that skillfulness requires providing premises that are at least apparently relevant, sufficient, and acceptable (thus, passing the "RSA Test" introduced in Johnson & Blair, 1977). That is, good arguers – i.e., skillful reasons-givers – have to be able to identify good reasons in order to use them. Thus, the people who are the most adept at constructing arguments should also be the best at recognizing and appreciating the arguments of others. Ironically, the opposite seems to be the case: arguers who excel at being proponents are often the most reason-resistant (Kornblith, 1999, p 184). In part, this is because the skill-set used by proponents to construct arguments is equally useful for opponents in constructing counter-arguments. Further, elements from that skill-set are also applicable in raising objections. The more skillful arguers are, the less likely one is to hear them utter the words, "I guess you were right, after all," even when they are presented with very good reasons in the most cogent of arguments. The most skillful proponents in arguments are not necessarily either the most effective or the best arguers.

When it comes to decisions and actions, then, it is entirely possible for us to go through the motions of argumentation and reasoning skillfully but with no real effect. For all we can tell, any argument we present might be idle.

3. REASONS FOR REASONING

Suppose then, for the sake of argument, that human reasoning is often rationalizing and that, appearances to the contrary, arguments rarely, if ever, cause changes in belief. Let us also grant that even skillful arguers seldom elicit concessions of defeat from their interlocutors – as seldom as those skillful arguers offer their own concessions. Nevertheless, even in that worst-case scenario, reasoning and argumentation are integral to decision and action. To get to the conclusion that reasoning is a sham and argumentation is useless, we would still have to make a fallacious inference and accept some dubious assumptions.

The flawed inference is from the fact that for all we can tell *any* argument might be post facto rationalization to the conclusion that *all* of them might be. The scope of the quantification invalidly changes. Even the most pessimistic neuroscientists acknowledge the possibility of exceptions: the generalizations are just *generally* the case, rather than universal necessities. Thus, there is traction for what I earlier called the project of "expanding the causal footprint' of reasoning (a project developed in Cohen, 2016).

The dubious assumptions concern the boundaries of reasoning and argumentation. We assume that reasoning ends when it reaches its conclusion, and it is that assumption that justifies dismissing rationalizations as mere charades because their conclusions are present from the outset. There is a parallel assumption about argumentation: arguments end when the arguers reach resolution, settlement, or simply disengage. That means an argument is effective only if it results in changes in the arguers' beliefs in the course of arguing and as a result of the argumentation. Delayed effects that occur after an argument's "end" are not counted. But they do indeed count!

3.1 Justification

The first inference relies on common, but skewed notions of how the structures and purposes of reasoning relate, namely that since the logical order begins with the premises and ends with the conclusion, the end of the argument, meaning its *goal*, must be the same as the end of the argument, in the sense of its *terminus*. Every time the rhetorical order of presentation follows that pattern, the idea is reinforced, suggesting that the conceptual order also follows suit. This is evident in how we talk about reasoning: The language of inference, thinking, and presentation all follow the language of time. The premises are the *starting points*, intermediate steps *follow from* them, until at last we *arrive at the conclusion*. And, of course, the fact that we call the conclusion "*the conclusion*" further cements its position as the terminal point of the process.

While that may be how calculation works, it serves poorly as a template for other kinds of reasoning. For example, in most legal reasoning, the cognitive order starts with the inferential conclusion: prosecutors and defense attorneys know the conclusions for which they will be arguing right from the start. Formal debates, political argumentation, and personal confrontations exhibit the same pattern. Were we so inclined, we might even posit that there is a single universal template for all arguments, that the conclusion of every argument is an instance of the single conclusion, "I am right and you are wrong," and that this conclusion is always present from the start. Arguers just need to fill in the details.

I am being facetious, of course. Reasoning does not fall into the single pattern in which the conclusion comes first. But neither does it fall into the neat pattern with the conclusion last – rhetorically, cognitively, or temporally. Nor does the resolution of argumentation always involve categorical winning and losing. We should be wary of the imperative to generalize. Forced generalizations trivialize theories by turning them into Procrustean Beds.

The reason this matters here is that two very different patterns of reasoning are integral to our decisions and actions – *deliberation* and *justification* – and they exemplify opposite patterns.[3] In one, the conclusion comes at the end of the cognitive process; in the other, it is there from the start. Even if much of the reasoning we offer in the service of justifying our action is rationalization, to call it a sham simply because the conclusion came first would reveal a fundamental misunderstanding of the mechanics of justification-reasoning.

A justification is not compromised by being the product of *post facto* rationalization rather than antecedent deliberation. If an action can be justified, it does not matter whether the justification comes before or after: *post facto* justification is still justification. Habitually or spontaneously generous acts performed without deliberation are still acts of generosity.[4] Even acts of generosity performed for selfish motives can be justified, although it remains open to us to make a different evaluation of the agent in such cases – we can distinguish when the *act* is justified for the agent from the not-always congruent cases in which the *agent* is justified in performing the act.[5] After all, some actions that are perfectly justifiable are never justified: they could be given justifications even if none is ever provided.

The extension and application of this distinction regarding *justifiable acts* to *reasonable agents* yields a very Aristotelian virtue ethics conclusion: what makes agents *reasonable* is whether they act

---

[3] Walton and Krabbe, 1995 use "deliberation" to refer to dialogues whose goal is deciding on a course of action. I think it is consonant with that usage to include reasoning intent on deciding what to believe (rather than, say, "inquiry" which they reserve for proving or disproving hypotheses already in hand, or "discovery," which has the proper sequencing but lacks the connotations connecting it to motivation and practical reason). "Justification" as it is used here maps most directly onto persuasion dialogues, but would also encompass part of what goes on in negotiations since negotiators know in advance what they would like to get out of the negotiation. The mapping is inexact because reasoning can occur outside of dialogues as well as within them.

[4] After-the fact justification is not the same as "retrospective" justification, although both might be might be possible. See Calhoun, 1992 for an analysis of the latter concept.

[5] Greco, 1999, building on Sosa, 1991, constructs a parallel distinction for epistemology between when a proposition is (objectively) justified for a believer and the believer is (subjectively) justified in believing that proposition, as part of his argument for reliabilism in virtue epistemology.

reasonably, more than whether there are (in some Platonist sense) reasons for their acts. If character is a greater determinant of behavior than reasoning, then inculcating the proper habits – the virtues – will help ensure that we will be reasonable and our behavior will be justifiable, even if we do not reason our way to our actions and cannot justify them. *Acting reasonably does not actually require reasons!*

We can extend this distinction to argumentation in order to appreciate argumentation's effectiveness.

3.1. Effective argumentation.

The flawed assumptions concern argumentation and epistemology, namely, that argumentation is useless if arguers emerge from arguments with all their prior beliefs unchanged. This embodies a profound misunderstanding of human epistemology and a failure to fully appreciate what argumentation can do and how it can do it.

The implicit epistemological picture is that we are defined by the sets of propositions we believe. An argument for a proposition, *p*, is successful when that proposition is added to the interlocutor's belief set. First, a model that recognizes only belief and non-belief as relevant propositional attitudes is a very impoverished one. Believing comes in kinds and degrees. Arguments affect beliefs in many ways. They strengthen and weaken them; they clarify and articulate them; and they help us understand and appreciate them by locating them relative to other beliefs. Second, this picture treats all beliefs as discrete and proposition-like. If that were so, the Cartesian project of doubting everything and starting over with a single, first belief might actually make sense, but the idea of an epistemological field with a solitary propositional belief is deeply incoherent. Rather, to use Wittgenstein's image (Wittgenstein, 1969 §140), "When we first begin to *believe* anything, what we believe is not a single proposition, it is a whole system of propositions. (Light dawns gradually over the whole.)" Third, the narrow focus on belief acquisition is deforming because it excludes emotions, attitudes, achievements, abilities, foibles, and the rest of our cognitive lives – all of which are appropriate targets for argumentation. Together, these assumptions severely constrict the admissible range of possible effects from argumentation.

Another set of common but un-noted assumptions concerns the boundaries of arguments: When does an argument begin, when does it end, and who counts as a participant? These matter when we treat arguments as dynamic interactions between agents, rather than as abstract, static inferential structures of propositions. In particular, we need to push back on the notion that arguments always have clear endpoints. Does our argument end when we disengage? What if we go our separate ways but continue inner dialogues between ourselves and imagined others? Does my belated admission to myself that you had the better of it – but only after you left, after my successful rhetorical and dialectical maneuvering, and after my adamant refusals to give so much as an inch in your presence – mean that the argument was a draw and you did not persuade me? But you *did* convince me; it just took longer than the actual dialogue. If the argument were over when we stopped, then we would be in the oxymoronic situation of having to say both that the argument ended without you convincing me of your position and yet that your argument really did convince me!

It may be just semantics as to where we mark an argument's end, but the long-term effects of argumentation should be taken into account. Some of the apparent ineffectiveness of argumentation is an artifact of where we draw the boundaries and how we tally up effects. Even if we rarely hear the words, "*I guess you were right and I was wrong*," that does not mean arguments never bring about situations in which those words could be said. Of course, *skillful* arguers can avoid ever having to say those words, but *good* arguers will recognize that they could be said – and *very good* arguers are happy to say them.[6] Arguers who are both good and *effective* create those situations for other good arguers.

Since the cognitive effects of one argument might not be visible until the next one, the conclusion that there are no such effects is unwarranted.

4. CONCLUSION: VIRTUE AND CHARACTER

Let me briefly summarize these conclusions regarding the inefficacy of argumentation and reasoning, with the hopes that they may serve as signposts for further development.

A problem arises from the ontological difference between lines of reasoning and causal chains. The gap between them (along with assumptions about causation's role in explanations) creates space for skepticism about reasoning and argumentation: perhaps they are explanatory "fifth wheels".

By itself, the causal inefficacy of argumentation does not make arguments useless epiphenomenal danglers because its social functions remain. Argumentation provides justifications for our actions, and the cogency of our reasons here is independent of whether those reasons

[6] This line of reasoning, along with the distinction between good arguers and effective arguers, is developed in Cohen, 2016.

are also causes. Presumably, every action has a causal explanation but not every action is justified. Causal stories are not enough. For justification we need argumentation – so justification justifies argumentation.

The apparent inefficacy of arguments is partly a matter of where the boundaries of argumentation and reasoning have been drawn and how effects have been counted.

Moreover, even if reasoning is completely outside the causal nexus, so premises only have status as reasons rather than causes, arguments also have status as events with their own effects. There is good arguing and there is effective arguing. They are different.

Arguments as events have effects while they are being conducted – physical, cognitive, emotional, etc. – but also they may also have long-term and delayed effects on their arguers that are manifest beyond the argument's narrowly defined spatio-temporal boundaries.

Finally, the ability to give reasons and the propensity to be reasons-responsiveness are both constitutive of what it is to be reasonable, so constructing good arguments is not the only or the best measure of reasonableness. The skillful hypocrite ("bullshitter"), a sophistic reasons-giver who constructs persuasive arguments but does not buy into them even when they are cogent, is not a better arguer than someone who recognizes and accepts the cogent ones.

Therefore, the most *effective* arguers – those with the most effect – are those who don't just produce good arguments, but who also bring out the best in their opponents, i.e., help make them able to recognize and *be affected by* reasons. If you want your arguments to have more effect on me, the best thing you can do is to make a better arguer out of me.

REFERENCES

Calhoun, C. (1992). Changing one's heart. *Ethics, 103*, 76-96.

Cohen, D. (2016). Argumentative virtues as conduit's for reason's causal efficacy. *Topoi* (forthcoming).

Greco, J. (1999). Agent reliabilism. In J. Tomberlin (Ed.), *Philosophical perspectives 13: Epistemology* (pp. 273-296). Atascadero, CA: Ridgeview Press.

Haidt, J. (2001). The emotional dog and its rational tail: A social intuitionist approach to moral judgment. *Psychological review, 108*(4), 814-834.

Johnson, R., & Blair, J.A. (1997). *Logical Self-Defense*. Toronto: McGraw-Hill.

Kornblith, H. (1999). Distrusting reason. *Midwest Studies in Philosophy, XXIII*, 181-196.

Mercier, H., & Sperber, D. (2011). Why do humans reason? Arguments for an argumentative theory. *Behavioral and Brain Sciences, 34*(2), 57-111.

Sosa, E. (1991). *Knowledge in Perspective.* Cambridge: Cambridge University Press.

Walton, D., & Krabbe, E.C.W. (1995). *Commitment in dialogue.* Albany: SUNY Press.

Wegner, D.M., & Wheatley, T. (1999). Apparent mental causation: Sources of the experience of will. *American Psychologist, 54*(7), 480-492.

Wittgenstein, L. (1969). *On Certainty.* G.E.M. Anscombe & G.H. von Wright (Eds.), D. Paul & G.E.M. Anscombe (trs.). New York: Harper Torchbooks.

18

Arguments and Decisions in Contexts of Uncertainty

VASCO CORREIA
ArgLab, Universidade Nova de Lisboa, Portugal
vasco.correia75@gmail.com

This article argues that debiasing techniques meant to reduce biases in argumentation and decision-making are more effective if they rely on environmental constraints, rather than on cognitive improvements. I identify the four main factors that account for the inefficiency of critical thinking with regard to debiasing and claim that extra-psychic strategies are more reliable tools for counteracting biases in contexts of uncertainty. Finally, I examine several examples of debiasing strategies that involve contextual change.

KEYWORDS: accountability, biases, contextualism, critical thinking, debiasing, incentives, rationality, selective exposure

1. INTRODUCTION

There is growing consensus among philosophers and social scientists that critical thinking is by and large ineffective in preventing biases, and that debiasing endeavours are bound to fail (Arkes, 1986; Croskerry et al., 2013a; Fischhoff, 1982; Mercier & Sperber, 2011; Paluk & Green, 2009; Willingham, 2007). This claim is seemingly supported by multiple empirical studies indicating that people's cognitive and motivational biases are overall "immune" to critical thinking. Some authors have therefore concluded that critical thinking "does not seem to yield very good results" (Mercier & Sperber, 2011, p. 65), or even that it "has proven to be absolutely worthless" (Arkes, 1981, p. 326). What is worse, it appears that some debiasing techniques can backfire and end up amplifying, rather than reducing, people's biases and prejudices (Galinsky et al., 2000; Sanna et al., 2002). In light of these results, it may be tempting to jump to the conclusion that arguers and decision-makers are better off without debiasing programs. Fishhoff (1982, p. 431), for one, explicitly makes that suggestion: "A debiasing procedure may be

D. Mohammed & M. Lewiński (eds.) (2016). *Argumentation and Reasoned Action: Proceedings of the 1st European Conference on Argumentation, Lisbon, 2015. Vol. II, 367-378.* London: College Publications.

more trouble than it is worth if it increases people's faith in their judgmental abilities more than it improves the abilities themselves".

In this article, I argue that this claim is flawed and contend that there are valid reasons to entertain a certain degree of optimism regarding the efficacy of (some) debiasing strategies. I begin by identifying the four main factors that seemingly explain the inefficiency of critical thinking in preventing biases, namely: (a) unawareness of biases, (b) cognitive limitations (c) lack of motivation, and (d) inadequate correction. At the same time, this diagnosis calls for a theory of debiasing capable of surmounting each of those hindrances. This will lead me to develop, in section 3, what I call a *contextualist approach to debiasing*, which predicts that debiasing strategies tend to be more effective when they rely on environmental constraints and social structures, rather than on cognitive improvements and critical thinking. Finally, in section 4, I examine several debiasing strategies that involve contextual constraints. Crucially, my purpose is to show that, unlike intra-psychic (or cognitive) debiasing strategies, and indeed critical thinking, extra-psychic (or contextual) debiasing strategies reliably succeed in reducing a certain number of biases.

2. THE LIMITS OF DEBIASING

There is an ongoing debate over whether biases and heuristics are maladaptive mechanisms that systematically lead to irrational thinking and poor decision-making (Dunning, 2009; Elster, 2007; Kahneman, 2011; Stanovich, 2011), or, on the contrary, adaptive effects that help us make decisions under conditions of uncertainty (Gigerenzer & Todd, 2000; Gigerenzer, 2008; Stich, 1990). In my view, the two claims are not fully incompatible and it seems reasonable to conciliate them by suggesting that biases can indeed be adaptive in certain contexts, particularly under constraints of time and information, although they can *also* prove costly or even disastrous in other contexts, particularly when overconfidence and unrealistic optimism lead people to mismanage risks, ignore red flags, procrastinate, and neglect useful information[1]. To that extent, it is not inconsistent to maintain that biases should in principle be reduced, but not always. In a recent paper, Kenyon & Beaulac (2014, p. 344) put it perhaps more elegantly when they write: "Biases should be mitigated *when* they are problematic, and not because they are per definition problematic".

---

[1] Taylor (1989, p. 237) acknowledges this aspect: "Unrealistic optimism might lead people to ignore legitimate risks in their environment and to fail to take measures to offset those risks".

Be that as it may, if we accept the idea that it would be beneficial and desirable to counteract a certain number of biases, the question to be asked is of course "*Can* biases be debiased?" In other words: "Do debiasing strategies effectively work?" Some authors seem confident that cognitive illusions can be mitigated by raising awareness of the existence of biases and by teaching people critical thinking. Thagard (2011, p. 160), for one, suggests that "critical thinking can be improved, one hopes, by increasing awareness of the emotional roots of many inferences". Jonhson & Blair (2006, p. 201), in a similar vein, stress the importance of knowing and practicing the correct forms of argumentation as a means to prevent irrational thinking: "Logic alone is not enough, but awareness of the criteria of good argument, plus practice, plus self-knowledge and knowledge pertinent to the issue - all of these must be integrated into the evaluation of argumentation".

To be sure, training in argumentation and reasoning skills alone can prove insufficient, to the extent that arguers can be fully logical in their reasoning, and biased nonetheless. It is not strictly speaking *illogical*, for example, to search for evidence that supports what one already believes in, as it happens with the confirmation bias (or myside bias). More generically speaking, as Paul (1986, p. 379) points out:

> It is possible to develop extensive skills in argument analysis and construction without ever seriously applying those skills in a self-critical way to one's own deepest beliefs, values, and convictions.

This is why virtue-based approaches to argumentation suggest that training in formal reasoning must be supplemented by the acquisition of good habits of thinking capable of ensuring the rationality of people's reasoning even when they are not being vigilant against biases. The hope is that a reinforcement of the arguer's epistemic virtues and skills can forestall, or at least mitigate, his or her irrational tendencies. In the best-case scenario, the virtues of one's "analytical thinking" (the so-called System 2) would be progressively incorporated into one's "intuitive thinking" (or System 1) and produce almost effortlessly a rational and fair judgment.

Such optimism about critical thinking has been challenged by a number of authors. In fact, there seems to be growing consensus that critical thinking is ineffective in preventing cognitive and affective biases. While some authors cautiously warn that critical thinking "is not as effective as one might hope" (Kenyon & Beaulac, 2014, p. 343), or that it can only claim "modest benefits" (Willingham, 2007, p. 12), others contend more bluntly that it "does not seem to yield very good results" (Mercier & Sperber, 2011, p. 65), or that it "has proven to be absolutely worthless" (Arkes, 1981, p. 326).

Several factors seem to explain the ineffectiveness of critical thinking as a means to reduce biases. First, people are remarkably unaware of their own biases, even after being informed about the phenomenon. Wilson et al. (2002, p. 185) stress this difficulty:

> In order to avoid [mental] contamination, people must first detect that it exists. This is often quite difficult, because people have poor access to the processes by which they form their judgments.

This difficulty is perhaps explained by what Pronin et al. (2002) call the "bias blind spot", which is the tendency to believe that other people are more biased than we are. Perhaps ironically, most people seem to be biased even with respect to how biased they are, which presumably hinders their ability to detect and correct potential biases.

Second, it has been hypothesized that biases are difficult to suppress insofar as they serve the role of protecting our belief systems from doubts and challenges that would otherwise destabilize our ability to reason and decide swiftly in conditions of uncertainty. Snelson (1993, p. 48) goes as far as to suggest that biases as a whole form an "ideological immune system" meant to protect one's pre-existing beliefs from foreign ideas and hypotheses. This could explain why, according to the author, "Educated, intelligent, successful adults rarely change their most fundamental premises" (*ibid.*). This view is consistent with a phenomenon Stanovich (2005, p. 163) calls "dysrationalia", which he defines as "the inability to think and behave rationally despite adequate intelligence". In more colloquial terms, Stanovich's view is that *intelligence* (or what is perceived as such) does not entail *rationality*. In fact, it has been argued that intelligent and knowledgeable individuals can be all the more prone to biases, inasmuch as they possess the information and the argumentative skills necessary to rebut potential challenges. Taber & Lodge (2006, p. 757) dubbed this phenomenon the "sophistication effect" and were able to demonstrate it experimentally by showing that politically knowledgeable individuals are more susceptible to affective biases than their unsophisticated peers, presumably "because they possess greater ammunition with which to counterargue incongruent facts, figures, and arguments".

Third, even when people acknowledge their propensity to be biased, they may not be motivated to engage in debiasing efforts. As Lilienfeld et al. (2009, p. 394) suggest, maybe "people do not perceive these efforts as relevant to their personal welfare". After all, biases are often considered adaptive—rightly or wrongly—insofar as they sometimes boost people's self-confidence, motivation and mood (Taylor and Brown, 1988). According to Mercier & Sperber (2011) some biases would even prove beneficial when it comes to common welfare. The confirmation bias, for example, seems to encourage the different interlocutors to come up with the best available arguments in support of their standpoints, thereby "contribut[ing] to an efficient form of division of cognitive labor" (Mercier & Sperber, 2011, p. 65). This suggestion can be challenged by the argument that the confirmation bias *also* has undesirable effects, insofar as it involves a neglect of disconfirming evidence that contributes to perpetuate false beliefs, prejudices and stereotypes. However, individuals may be unaware of these maladaptive effects—or even believe that such effects serve their interest of being able to convince others—and therefore lack the motivation to do whatever is necessary to counteract them.

Finally, even if arguers are both aware of their biases and motivated to correct them, their debiasing efforts may nevertheless be inappropriate, or even backfire. According to Wilson et al. (2002, p. 191) there are three types of errors in debiasing attempts: "*insufficient correction* (debiasing in the direction of accuracy that does not go far enough), *unnecessary correction* (debiasing when there was no bias to start with), and *overcorrection* (too much debiasing, such that judgments end up biased in the opposite direction". While cases of insufficient correction are easily understandable in light of the above-mentioned factors, cases of unnecessary correction and overcorrection are perhaps more puzzling. Yet, several studies confirm that certain biases can be amplified by debiasing attempts. For example, Sanna et al. (2002) found that the hindsight bias is attenuated when the subjects have to come up with a few counterfactual thoughts, but increased when they try to come up with many counterfactual thoughts. Likewise, when subjects are instructed to avoid stereotypes about a given group, such instructions appear to remind them of those stereotypes and paradoxically increase their implicit biases (Galinsky & Moskowitz, 2000). The point to be made is that debiasing can backfire, either because the wrong strategy is applied or because an appropriate strategy is poorly executed.

To sum up, the inefficiency of critical thinking in preventing biases can be linked to four main factors: (1) people's unawareness of their own biases, (2) the entrenchment of biases in our cognitive structures (3) the lack of motivation to debias, and (4) inadequate correction. If this diagnosis is correct, critical thinking alone seems indeed insufficient to counteract biased reasoning and irrational thinking. As Wilson et al. (2002, p. 192) observe, "just because people attempt to correct a judgment they perceive to be biased is no guarantee that their result will be a more accurate judgment". However, that does not entail that debiasing in general is doomed to fail or that we should remain skeptical regarding people's ability to improve the rationality of their beliefs, inferences and choices. As we shall see in the next section, this simply means that debiasing efforts need to be based on more sophisticated strategies that take into account people's cognitive and motivational limitations.

3. THE CONTEXTUALIST APPROACH TO DEBIASING

Lest one draw the conclusion that debiasing is ineffective insofar as critical thinking is ineffective, it is worth stressing that *debiasing cannot be reduced to critical thinking*. According to Willingham's (2007, p. 8) definition, "critical thinking consists of seeing both sides of an issue, being open to new evidence that disconfirms your ideas, reasoning dispassionately, demanding that claims be backed by evidence, deducing and inferring conclusions from available facts, solving problems, and so forth". Other definitions of critical thinking also highlight what seem to be its most essential aspects, namely: (1) open-mindedness, (2) rational thinking, and (3) self-critical evaluation of one's thinking skills[2]. *Debiasing*, on the other hand, demands much less from individuals, and is generally understood as a strategy (or set of strategies) that is designed to suppress/mitigate biases, or at least to suppress/mitigate their effects.

Bearing this distinction in mind, it is easy to grasp that debiasing can be achieved without critical thinking. The peer review system, for example, is a notoriously effective debiasing strategy that does not require any cognitive change akin to critical thinking. Even if the peer reviewer is racist or misogynous, such lack of impartiality will not have an effect on his or her evaluation of the author's work, since it remains anonymous. In such cases, debiasing strategies seek to suppress *the effects* of a given bias, leaving the bias itself intact. Thus, for example, when the journal *Behavioural Ecology* decided to adopt a peer review process, they found that it led to a 33 per cent increase of representation of female authors (Budden et al., 2008). Although reviewers did not become less biased thanks to this debiasing strategy, at least their biases did not have an impact on their assessment of the quality of women's work. In other cases, as we will see, debiasing strategies are effective in suppressing the biases themselves, and not just their effects.

[2] Cf. Fisher (2011, p. 4), Lau (2011, p. 2), Siegel (1988, p. 32).

This demarcation between debiasing and critical thinking is paramount to understanding why there are reasons to remain optimistic regarding our ability to counteract implicit biases. While critical thinking requires a number of cognitive virtues and skills at an individual level (open-mindedness, two-sidedness, critical awareness, impartiality, etc.), the claim I wish to make is that the most effective debiasing strategies are those that rely instead on non-cognitive (or extra-psychic) structures. This is what I propose to call the *contextualist approach to debiasing*, i.e. the prediction that debiasing strategies have a better chance of becoming effective if they rely on extra-psychic, environmental and social structures, rather than on cognitive improvements at an individual level. This model is in line with what Thaler & Sunstein (2008) call a "choice architecture", which seeks to promote rationality, not by reforming people, but by reforming the contexts in which people reason and make decisions.

This approach was firstly conceptualized by Larrick (2004) under the label of the "Technologist" approach. The author points out that:

> Debates about rationality have focused on purely cognitive strategies, obscuring the possibility that the ultimate standard of rationality might be the decision to make use of superior tools (Larrick, 2004, p. 318).

Interestingly, the "tools" metaphor is also used by other advocates of the contextualist approach: Soll et al. (2013) refer to "debiasing tools", Elster (1989) to a "toolbox of mechanisms", and Gigerenzer (2002) to an "adaptive toolbox". The reason for this is presumably that tools are per definition exterior to the subject and meant to increase one's natural capacities. In a similar fashion, Kenyon & Beaulac (2014) describe debiasing strategies as elements of "contextual engineering", in line with Thaler and Sunstein's (2008) notion of "choice architecture". Finally, Stanovitch (2011, p. 8) also argues that it is worth resorting to *environmental changes* when purely *cognitive changes* seem insufficient: "in cases where teaching people the correct reasoning strategies is difficult, it may well be easier to change the environment so that decision making errors are less likely to occur". The point to be made is that the contextualist approach is already partly endorsed by a number of authors, although none of them focuses exclusively on this aspect.

That being said, the contextualist approach does not make the claim that intra-psychic debiasing strategies and critical thinking are bound to fail. It merely makes the prediction that extra-psychic (or contextual) strategies are both more effective and more reliable in preventing biases. The reason for this is that extra-psychic devices are specifically designed to address the types of problems that critical thinking alone is unable to overcome; namely: (1) unawareness of biases, (2) cognitive limitations, (3) lack of motivation, and (4) inadequate correction. To make this point clearer it is perhaps useful to examine in greater detail some those strategies.

4. EXTRA-PSYCHIC DEVICES

The main advantage of extra-psychic strategies is that they take into account the individual's cognitive and motivational limitations and often impose indirect methods of control. Thus, for example, instead of relying on people's willingness to debias themselves, contextual debiasing requires the creation of external constraints meant to enhance people's motivation to correct their biases (or to sanction their inability to do so). A compelling example of this type of strategy is *accountability*. Tetlock (2002, p. 590), who conducted empirical research on accountability for decades, maintains that holding people accountable for the rationality of their thinking is one of the most effective ways to prevent biases: "There is indeed a substantial list of biases that are attenuated, if not eliminated, by certain forms of accountability". This list includes the biases of overconfidence, primacy, overattribution, illusory correlation, and the fundamental attribution error.

The reason why accountability is such an effective debiasing strategy is that it counteracts each of the four above-mentioned hindrances. First, when subjects know their arguments will be scrutinized, they are understandably more motivated to make sure that those arguments are sound and rational. In one study, for example, it appeared that subjects accountable to unknown audiences tended to engage more in self-criticism and to raise potential objections to their own views, not because they cared more about the truth than those who were not accountable, but presumably because they feared that others might find serious flaws in their position (Tetlock & Boettger, 1989). Second, this motivation to engage in self-scrutiny also works as an incentive to increase the subjects' awareness of potential biases and to overcome their resistance towards challenging ideas and counterarguments. And finally, accountability mitigates the problem of inadequate correction, to the extent that the interlocutors' feedback contributes to hold in check both unnecessary correction and overcorrection. There are, however, limitations to this strategy, given that accountability may incidentally amplify other types of biases (group polarization, compromise effect, etc.). A person who is accountable to a group with known views, for example, may end up saying precisely what (he or she believes) the others wish to hear.

Another extra-psychic debiasing tool is the *control of the source of information*. According to Wilson et al. (2002, p. 192) this is "is the most effective defence against [mental] contamination: A stimulus that never enters our minds cannot bias our judgment or feelings". We have seen that the system of peer review is a good example of how effective it can be to limit exposure to information susceptible to influence one's judgment. Although biases are not suppressed by such a procedure, their effects are neutralized. Yet in other cases selective exposure can effectively prevent the formation of biases. Croskerry et al. (2013b, p. 66) give a plausible example of this when they note that "some emergency physicians avoid reading nurse's notes until after they have assessed the patient". By choosing not to expose themselves to some of the information, those physicians avoid potential framing effects in their diagnosis. That being said, selective exposure may also amplify certain biases, and particularly the confirmation bias. It is reported, for example, that Dick Cheney always requires the television set to be tuned on Fox News before entering a hotel room in order to avoid being exposed to "left-wing propaganda".

A third debiasing strategy that relies on contextual constraints is *group interaction*. As Larrick (2004, p. 326) points out, "groups serve as an error-checking system during interaction". After all, as the proverb as it, "Two heads are better than one". In addition, groups tend to bring forward more diverse perspectives than lone individuals, which in turn works as a demonstrably effective debiasing tool, bearing in mind that several studies indicate that subjects who consider a range of alternative perspectives tend to be less prone to one-sided thinking (Wilson et al., 2002; Pronin et al., 2002). Furthermore, Mercier & Sperber (2011, p. 65) point out, groups can contribute to hold the confirmation bias in check, given that each element of the group tends to focus primarily on the arguments that best support his or her view, thereby exposing their contradictors to the best available counterarguments. The drawback of this strategy, however, is that it only works effectively if the members of the group are individuals who hold different views on the topic at stake, for otherwise they may end up reinforcing their shared beliefs in virtue of a group polarization effect.

Finally, the use of *incentives* can also contribute to mitigate certain biases, insofar as they are meant to increase people's motivation to be vigilant and rational in their thinking. Thaler & Sunstein (2008) promote this type of approach in their book *Nudge*. Instead of enforcing the desired outcomes by resorting to sanctions or accountability, they argue, it is more fruitful to implement what they call *nudges*, i.e.

incentives that promote rational thinking and deciding by using aspects of the environment. As Lilienfeld et al. (2009, p. 394) observe, "research suggests that at least some cognitive biases may be reduced by enhancing participants' motivation to examine evidence thoughtfully". However, it must be stressed that incentives have no effect on certain biases (e.g. overconfidence, hindsight, framing effects), and can even amplify other biases (e.g. representativeness). In fact, while incentives seemingly counteract the problem of lack of motivation to debias, they can exacerbate the problem of inadequate correction. As Pelham & Neter (1995, p. 591) put it: "If the only tool at a person's disposal is a hammer, convincing the person to work harder can only lead to more vigorous hammering".

5. CONCLUSION

If the Contextualist approach to debiasing is correct, we have a better chance to promote the rationality of our arguments and decisions by deliberately submitting them to external constraints and social contexts. This is not to say that critical thinking and other intra-psychic debiasing strategies are worthless, but simply that they are not as efficacious in dealing with the four main factors that explain the persistence of biases. Unlike cognitive change, contextual constraints reliably enhance one's (1) awareness of biases, as well as one's (2) motivation to debias; and also help (3) overcoming cognitive limitations, and (4) avoiding inadequate correction. Granted, we have seen that almost every debiasing technique has its shortcomings. Furthermore, as Wilson et al. (2002, p. 200) point out, "these techniques are in their infancy and their reliability, validity, and controllability are still an open question". The hope, however, is that such techniques may produce significant improvements *as a whole*. To that extent, the question is not whether debiasing works or not (in absolute terms), but rather which specific debiasing technique works better in which type of context, and what type of side-effects it might have.

REFERENCES

Arkes, H. (1981). Impediments to accurate clinical judgment and possible ways to minimize their impact. *Journal of Consulting and Clinical Psychology, 49*, 323–330.

Budden, A. E., T. Tregenza, L. W. Aarssen, J. Koricheva, R. Leimu, & C. J. Lortie (2008). Double-blind review favours increased representation of female authors. *Trends in Ecology & Evolution, 231*, 4–6.

Croskerry, P., Singhal, G., & Mamede, S. (2013a). Cognitive debiasing 1: origins of bias and theory of debiasing. *Quality & Safety, 22*(2), 58–64.

Croskerry, P., Singhal, G., & S. Mamede, S. (2013b). Cognitive Debiasing 2: Impediments to and Strategies for Change. *Quality & Safety, 22*(2), 65–72.

Dunning, D. (2009). Disbelief and the neglect of environmental context. *Behavioral and Brain Sciences, 32*, 517-518.

Elster, J. (1989). *Nuts and bolts for the social sciences.* Cambridge: Cambridge University Press.

Elster, J. (2007). *Explaining social behavior.* Cambridge: Cambridge University Press.

Evans, J. (2007). *Hypothetical thinking.* Hove, New York: Psychology Press.

Fischhoff, B. (1982). Debiasing. In D. Kahneman, P. Slovic, & A. Tversky (Eds.), *Judgment Under Uncertainty* (pp. 422-444). Cambridge: Cambridge University Press.

Fisher, A. (2011). *Critical thinking.* Cambridge: Cambridge University Press.

Galinsky A., Moskowitz G., & Gordon, B. (2000). Perspective taking. *Journal of Personality and Social Psychology, 784*, 708–24.

Gigerenzer, G. (2008). *Rationality for mortals.* Oxford: Oxford University Press.

Gigerenzer, G., & Todd, P. (2000). Précis of *Simple heuristics that make us smart. Behavioral and brain sciences, 23*, 727-780.

Johnson, R., & Blair, A. (2006). *Logical self-defense.* New York: International Debate Association.

Kahneman, D. (2011). *Thinking, fast and slow.* New York: Farrar, Straus and Giroux.

Kenyon, T., & Beaulac, G. (2014). Critical thinking education and debiasing. *Informal logic, 34*(4), 341-363.

Larrick R. (2004). Debiasing. In D. Koehler & N. Harvey (Eds.), *The Blackwell Handbook of Judgment and Decision Making* (pp. 316-337). Oxford: Blackwell Publishing.

Lau, J. (2011). *An introduction to critical thinking and creativity.* New Jersey: Wiley & Sons.

Lilienfeld, S., Ammirati, R., & Landfield, K. (2009). Giving debiasing away. *Perspectives on Psychological Science, 4*(4), 390-8.

Mercier, H., & Sperber, D. (2011). Why do humans reason? *Behavioral and Brain Sciences, 34*, 57-111.

Paul, W. (1986). Critical thinking in the strong and the role of argumentation in everyday life. In F. Eemeren, R. Grootendorst, A. Blair & C. Willard (Eds.), *Argumentation* (pp. 379-382). Dordrecht: Foris Publications.

Pelham, B., & Neter, E. (1995). The effect of motivation of judgment depends on the difficulty of the judgment. *Journal of personality and social psychology, 68*(4), 581-594.

Pronin, E., Lin, D., & Ross, L. (2002). The bias blind spot: Perceptions of bias in self versus others. *Personality and Social Psychology Bulletin, 28*, 369–381.

Sanna, L.J., Schwarz, N., & Stocker, S.L. (2002). When debiasing backfires. *Journal of Experimental Psychology, 28*, 497-502.

Siegel, H. (1988). *Educating reason*. London, New York: Routledge.

Snelson, J. (1993). The ideological immune system. *Skeptic Magazine, 1*(4), 44-55.

Soll, J. Milkman, K., & Payne J. (2014). A user's guide to debiasing. *http://opim.wharton.upenn.edu/~kmilkman/Soll_et_al_2013.pdf*.

Stanovich, K. (2005). *The robot's rebellion*, London: The University of Chicago Press.

Stanovich, K. (2011). *Rationality and the Reflective Mind*. New York: Oxford University Press.

Taber, C., & Lodge, M. (2006). Motivated skepticism in the evaluation of political beliefs. *American Journal of Political Science, 50*(3), 755-769.

Tetlock, P. (2002). Intuitive politicians, Theologians, and Prosecutors. In T. Gilovich, D. Griffin & D. Kahneman (Eds.), *Heuristics and biases* (pp. 582-599). Cambridge: Cambridge University Press.

Tetlock, P. (2005). *Expert political judgment*. Princeton, NJ: Princeton University Press.

Tetlock, P., & Boettger, R. (1989). Accountability. *Journal of Personality and Social Psychology, 57*, 388-398.

Thagard, P. (2011). Critical thinking and informal logic. *Informal logic, 31*(3), 152-170.

Thaler, R., & Sunstein, C. (2008). *Nudge*. New Haven and London: Yale University Press.

Willingham, D. (2007). Critical thinking: Why is it so hard to teach? *American Educator, 31*(2), 8-19.

Wilson, T. D., Centerbar, D. B., & Brekke, N. (2002). Mental Contamination and the Debiasing Problem. In T. Gilovich, D. Griffin, & D. Kahneman (Eds.) *Heuristic and Biases* (pp. 185-200). Cambridge: Cambridge University Press.

19

Instrumental Rationality as a Component of Epistemic Vigilance in a Persuasion Dialogue

KAMILA DEBOWSKA-KOZLOWSKA
Faculty of English, Department of Pragmatics of English, Adam Mickiewicz University in Poznan, Poland
kamila@wa.amu.edu.pl

In this paper, I argue that exercising epistemic vigilance by a hearer relies on both epistemic and instrumental rationality. I propose a cognitive framework for instrumental rationality. Using the concept of the Beneficial Cognitive Model, I show that hearer's instrumental rationality relates to evaluating speaker's arguments on the basis of whether they activate the (mental) beneficial topics in the mind of the hearer which are within the area of hearer's interest of persuasion.

KEYWORDS: beneficial, cognitive, epistemic, instrumental, persuasion dialogue, pragmatics, rationality, vigilance

1. INTRODUCTION

This paper presents a cognitive pragmatics account of hearer's processing of a fallacious argument. The cognitive pragmatics approach adopted here relies on the joint and complementary forces of linguistic pragmatics represented here by both Relevance Theory (Sperber & Wilson, 1986) and the Argumentative Theory of Reasoning (Mercier & Sperber, 2011a) and psychology of reasoning (e.g. Evans, 2006; Stanovich, 2011). Such an approach to the study of language processing is methodologically united with the postulates offered in Noveck and Sperber (2004, pp. 7-13).

I especially deal here with the cases when the hearer's verbal behaviour signals the acceptance of a fallacious argument. For instance, when the hearer rephrases the opponent's fallacy in his own argumentative utterance in the way that is beneficial to him. I limit the account of a fallacy processing to a persuasion dialogue. I argue that processing of a fallacious counter-argument by a hearer in a persuasion

D. Mohammed & M. Lewiński (eds.) (2016). *Argumentation and Reasoned Action: Proceedings of the 1st European Conference on Argumentation, Lisbon, 2015. Vol. II, 379-391.* London: College Publications.

dialogue is similar to processing a non-fallacious counter-argument since in both cases the hearer will try to use the counter-argument to realise his persuasive goal.

I use the definition of a persuasion dialogue proposed in Debowska-Kozlowska (2014a). For a conversational exchange to be called a persuasion dialogue, two conditions need to obtain. First, there needs to be a conflict of opinion at the beginning of the exchange and the participants need to have the shared, global goal to resolve the disagreement. Second, at least one participant needs to have a persuasive aim to convince his opponent to accept an opposing point of view (cf. Walton, 1995; Walton & Krabbe, 1995). In this paper, the description of the cognitive processing of a counter-argument is restricted only to the participant with the persuasive aim.

My claim is that the traditional view of fallacies relates to the epistemic type of rationality which assumes that people's verbal behaviour is rational when they follow the principles of logic or rules specified by a given normative system. I argue that exercising epistemic vigilance in a persuasion dialogue in terms of argument's consistency and supplementing it with further reflective inferencing focused on argument's reasonableness is not, however, primarily based on epistemic rationality. I claim that the process of evaluation relies far and foremost on instrumental rationality (cf. Stanovich, 2011). The conception of epistemic vigilance is treated by Sperber (2001) and Mercier and Sperber (2011a) as a cognitive device of a hearer which helps him trace inconsistencies in the content and structure of speaker's argumentation and evaluate any aspect of untrustworthiness of the speaker. In their account, Sperber and Mercier do not, however, consider epistemic vigilance with reference to a specific type of a dialogue.

Relying on the discussion of the results of experimental studies on reasoning presented by Stanovich (2011, pp. 63-71), I propose a cognitive framework for instrumental rationality of a hearer in a persuasion dialogue. I call the framework the Beneficial Cognitive Model (hence BCM). I indicate that hearer's instrumental rationality relates to evaluating speaker's arguments on the basis of whether they activate the (mental) beneficial topics in the mind of the hearer which are within the area of hearer's interest of persuasion. To delineate the relation between the use of instrumental rationality and the BCM, I use the conception of the massive modularity framework of the mind. In RT and the ATR, two modules are of special interest, namely the comprehension module and the argumentation module. In this paper, however, I apply the terminology used in Stanovich (2011) to emphasise that the argumentation module has the reflective and algorithmic mode. Both the reflective and algorithmic mind need to be taken into consideration to account for the hearer's processing of a counter-argument in a persuasion dialogue. The paper shows that a mental beneficial topic gets activated in the mind of a hearer in the process of intuitive inferencing that takes place in the comprehension module of the mind, known also as the autonomous mind (cf. Stanovich, 2011). However, the further use of the beneficial topic for the evaluation of a counter-argument is carried out through reflective inferencing that occurs in the reflective mind. The paper shows also the role of the algorithmic mind in the hearer's evaluation of a counter-argument and explains in what way cognitive decupling and serial associative cognition enable the use of a beneficial topic for the assessment.

Section 2 discusses intuitive and reflective processing and introduces the concept of epistemic vigilance. Also, Oswald and Hart's (2013) perspective on cognitive processing of effective source-related fallacies (i.e. ad hominem, ad verecundiam, ad populum) is presented. Their focus is on the situation when a fallacy is not noticed by an opponent since he does not use his cognitive resources to evaluate the argument (e.g. he is not good at proper reasoning or focuses on figuring out the meaning of the message). In this paper, the focus is on both situations, i.e. when a hearer does not notice that a fallacy has been committed and when the hearer is aware that the fallacy has occurred. Thus the shift of cognitive resources of a hearer is considered only in relation to instrumental rationality. I show that in both cases (noticing and not noticing a fallacy) it is the activation of a mental beneficial topic that is the deciding factor for the evaluation of the argument in terms of how it contributes to the successfulness of the hearer in terms of his persuasive goal.

Section 3 is devoted to the conceptualisation of instrumental and epistemic rationality. It explains also in what way the reflective and algorithmic mind coordinate the use of these two types of rationality. Relying on the information from section 3, section 4 elaborates on why the BCM should be seen as a cognitive framework for instrumental rationality in a persuasion dialogue. It explains the process of a counter-argument evaluation by a hearer. It shows that the activated mental topic from the BCM of a hearer is subsequently used in serial associative cognition or cognitive decupling. After the reflective mind sends out a call to start one of these operations, the algorithmic mind determines the hearer's ability to maximally use the activated beneficial mental topic.

2. INTUITIVE AND REFLECTIVE PROCESSING OF ARGUMENTATION

As explained in Debowska-Kozlowska (2014b), a hearer of a counter-argument uses both types of processing, i.e. intuitive and reflective. Despite the fact that scholars apply various terminology to talk about these dual processes, they generally agree on their characterisation (see Mercier & Sperber, 2009; Reber, 1993; Bickerton, 1995; Evans, 2006, 2009; Stanovich, 1999).

The intuitive type of processing a message is the main interest of Relevance Theory proposed by Sperber and Wilson (1986). The scholars focus on the way a hearer understands an upcoming stimulus, that is an utterance. In the psychology of reasoning (see Evans, 2009; Stanovich, 2011), the intuitive process of metarepresenting the meaning of an utterance is described as fast, unconscious, associative and contextualized. Relevance-theorists believe that the intuitive processing occurs in different metarepresentational modules in the human mind which interact with each other. Of main interest to the relevance-theorists is the comprehension module in which inferences are drawn about the representation of the utterance (Wilson, 2000).

In a persuasion dialogue, both participants act as a speaker and a hearer. Thus, both of them will interpret each other's public representations, i.e. utterances. Mental and public representations may become, as Sperber (2000, p. 3) indicates, "objects of second-order representations or metarepresentations". Metarepresentations are grouped into four types: (1) "mental representations of mental representations", (2) "mental representations of public representations", (3) "public representations of mental representations", (4) "public representations of public representations". The first type refers to the situation in which a hearer mentally represents the beliefs of his opponent. The second type pertains to the typical relevance-theoretic situation of processing the speaker's meaning in the mind of the hearer. The third type describes a situation in which a participant verbally represents the beliefs of his opponent. The fourth type presents a situation in which a participant verbally represents the interpretation of his opponent's utterance. The second and fourth type of representation are of interest to this paper. Processing the meaning of the opponent's argumentative utterance is treated as a pre-step for its evaluation. In a persuasion dialogue, the intuitive inferencing is broken off by reflective reasoning when the meaning of a counter-argument is already represented in the mind of the hearer and the evaluation process of the content of the counter-argument is initiated. The fourth type of the representation is assumed to be the result of hearer's evaluative procedure.

The reflective type of processing a message has been extensively discussed by Sperber and Mercier in their Argumentative Theory of

Reasoning (Mercier & Sperber, 2009; Mercier & Sperber, 2011a, 2011b). The function of reflective inferencing is described as evaluative since it enables a hearer to consciously compare and assess a counter-argument. Controlled, slow and sequential nature of reflective processing makes it possible for the hearer not only to evaluate arguments but also to accept or reject the conclusions coming from those arguments. In the ATR a lot of attention is given to the concept of epistemic vigilance (e.g. Sperber et al., 2010; Mercier, 2011). I claim that exercising epistemic vigilance by a hearer with a persuasive view in a persuasion dialogue is a specific process that needs to take into account hearer's persuasive wants and desires. Below, I will present Mercier and Sperber's conceptualisation of epistemic vigilance and, later on, explain why it is not directly applicable to the case of a persuasion dialogue.

As Mercier and Sperber (2011a) indicate, if the hearers do not want to be deceived or misguided, they need to be vigilant. In other words, they need to carefully and actively judge the speaker and his speech. Exercising epistemic vigilance is defined as the use of a cognitive skill that allows to examine the information through performing three actions. The first and the second action, i.e. "coherence checking" and "assessing relevance", are, in fact, treated as the same strategy of argument's content and structure evaluation involving two complementary steps. The third action, "trust calibration" refers to the verification of speaker's dependability (Sperber, 2001; cf. Santibáñez Yáñez, 2012; Debowska-Kozlowska, 2014b). The exercising of epistemic vigilance in terms of an argument's content is mainly related to paying attention to "logical relationships" and "evidential relationships" of justificatory potential of speaker's argumentation. One way of checking the coherence of the content, as Sperber (2001) indicates, may be exercised by observing how the logical terms (e.g. "if", "and", "or", "unless") and other sentence connectors (e.g. "therefore", "since", "but", "nevertheless") are used. The application of epistemic vigilance in terms of the structure of argumentation is related to the assessment of the relations between arguments. For instance, speaker's compound argumentation might be evaluated through checking whether he properly uses argumentative indicators. If the use of epistemic vigilance by a hearer suggests the rejection of an argument, the further processes of reflective reasoning can help the hearer to re-evaluate the argument as valuable or retain the decision about the argument's rejection. For instance, if in the process of coherence checking it is confirmed that logical terms and words indicating inferential relationship are used incorrectly, further reflective reasoning can help see whether the propositional content expressed after those connecters is fallacious or not in traditional terms.

Mercier and Sperber's (2011a, cf. Sperber et al., 2010) concept of epistemic vigilance, as shown above, is based on only one type of rationality, i.e. epistemic rationality. Exercising vigilance in a persuasion dialogue is, however, related to what is salient, essential and advantageous for a hearer in terms of his persuasive aim. Thus I claim that in the processes of the evaluation of a counter-argument a hearer often uses some cognitive shortcuts. In other words, I argue that the hearer performs coherence checking, relevance assessment and trust calibration only if these actions help him achieve what he wants. In a persuasion dialogue, epistemic assessment that aims at acquiring genuine knowledge is thus not only related to epistemic rationality. Searching for genuine knowledge about a counter-argument in a persuasion dialogue relates to scrutinising its advantageousness.

The above claims about the use of cognitive shortcuts in the epistemic assessment are motivated by the cases presented in Oswald and Hart (2013) and Oswald (forthcoming). Oswald and Hart (2013) discuss the examples of cognitive shortcuts applied by a hearer while evaluating source-related fallacies (i.e. ad verecundiam, ad populum, and ad hominem). For instance, they focus on the cases when a hearer accepts an argument even though not a proper authority has been mentioned to prop up the speaker's conclusion. Oswald and Hart (2013, p. 4) explain that the reason for the acceptance is the fact that human cognitive system uses some cognitive shortcuts/heuristics instead of consciously evaluating speaker's "competence, benevolence, expertise, reliability, credibility" with reference to the use of a contextually appropriate source. The reason behind the use of cognitive shortcuts might be that the cognitive competences of a hearer are not sophisticated enough to recognise a fallacy (cf. Oswald, 2011). In the case of a persuasion dialogue, a hearer will often use the shortest processing route to satisfy his persuasive need to win the discussion.

The idea that epistemic assessment is not only confined to epistemic rationality is elaborated on in Oswald (forthcoming). He gives support for the claim that epistemic assessment might happen not only in the argumentation module but also in the comprehension module of the mind. In fact, Sperber et al. (2010, p. 374; cf. Oswald, forthcoming) also notice that the comprehension module is equipped with an evaluative function which allows for "a cost-effective epistemic assessment". I argue that the use of an instrumental rationality for evaluating a counter-argument of an opponent is also based on cognitive heuristics. Contrary to Oswald (forthcoming), who exemplifies the use of the cognitive heuristics in the autonomous mind, I want to focus on the case of using the cognitive shortcuts in the argumentation module. Both instrumental and epistemic rationality are related to the analytic system of the argumentation module. Nevertheless, only the reflective processing relying on instrumental rationality is directly goal-oriented.

The next section offers the description of two types of rationality and explains in detail why the use of instrumental rationality is based on using cognitive shortcuts. Additionally, I show that the argumentation module discussed by Mercier and Sperber, needs to be considered as located in, what Stanovich (2011) calls, reflective and algorithmic mind. These two structures coordinate the use of instrumental rationality.

3. INSTRUMENTAL AND EPISTEMIC RATIONALITY TYPES

Evans and Over (1996) describe instrumental and epistemic rationality as rationality 1 and rationality 2 respectively. Their definitions are as follows:

> Rationality 1: Thinking, speaking, reasoning, making a decision or acting in a way that is generally reliable and efficient for achieving one's goals.
>
> Rationality 2: Thinking, speaking, reasoning, making a decision or acting when one has a reason for what one does sanctioned by a normative theory. (Evans & Over, 1996, p. 8)

In general, if one's activities are aimed at the achievement of his goal then they are based on instrumental rationality (Over, 2004; Stanovich, 2011). Epistemic rationality, also referred to as evidential rationality or rationality of action, relates to following the principles of logic or rules specified by a normative theory, model or condition. According to epistemic rationality, people's actions must be based on "good" reasons. Whether the reasons are good or not depends on the normative theory, model or condition chosen as a point of reference.

Two groups of scholars, i.e. Meliorists and Panglossians, focus on the measuring of the rationality of a given behaviour in experimental conditions (Stanovich, 2011, pp. 8-10; see also Stanovich, 1999). Their interpretations of the results, however, differ significantly. It is due to the fact that they assume opposing positions on whether a researcher has the right to provide criteria for rational behaviour before an experiment starts. Meliorists believe that rules of normative systems assumed in advance allow to identify a thinking error of a participant of an experiment. In the Panglossian tradition, instrumental rationality explains human performance in experiments through maximization of participant's goal fulfilment. Data gathered through psychological experiments which confirm that there are individual differences in the reflective reasoning suggest that humans often base their rational judgements on "the expected utility". As Stanovich (2011, p. 6) indicates, "a person chooses based on which option has the largest expected utility" for the person. The expected utility is a motive for a cognitive shortcut which facilitates the processing of an argumentative message.

Different interpretations of the results of the experimental studies by Meliorists and Panglossians motivated Stanovich (2011, p. 33) to design "the tripartite structure and locus of individual differences" which aims to explain the variations. The structure which he designed using the results of empirical studies draws the distinction between individual cognitive styles and individual cognitive abilities as responsible for different patterns of responses and thus results interpretation. Individual's acting in accordance with his instrumental rationality is dependent on his cognitive styles and abilities. Stanovich (2011) emphasises that unique cognitive styles of an individual result from his cognitive abilities. Therefore, except for acknowledging the importance of the autonomous mind, he proposes to consider the partition of analytic system. He discusses "individual differences in rational thinking dispositions" as being monitored by individuals' reflective minds and "individual differences in fluid intelligence" as being controlled by their algorithmic minds.

Thinking dispositions (i.e. cognitive styles) pertain to the belief structure of an individual and to his attitudes. As Stanovich (2011) shows, thinking dispositions are also based on individual's goals and the way he perceives those goals in terms of their importance. Thinking dispositions pertain thus to the belief structure of an individual which is created as a result of the interaction with the norms of a given society and unique personal evaluation of those norms. Additionally, reasoning dispositions are influenced by the individuals' attempts to realise their private goals in a given situation (e.g. in a persuasion dialogue). Clearly, we can thus observe two types of rationality being engaged in the functioning of the reflective mind. At the level concerned with the adjustment of individual's beliefs and attitudes to the requirements of a given society we observe epistemic rationality. At the level concerned with the efficient pursuing of one's goals we observe instrumental rationality. To properly manage private goals, i.e. in agreement with his beliefs structure (see Burgoon et al., 1994), an individual relies on his cognitive ability. Different cognitive abilities of individuals refer to their algorithmic minds which control and enable the proper operations to be taken. The proper operations in the case of a persuasion dialogue refer to the maximised utilisation of instrumental rationality.

4. BENEFICIAL COGNITIVE MODEL AND INSTRUMENTAL RATIONALITY

In this section, I will introduce the concept of the Beneficial Cognitive Model (Debowska-Kozlowska, 2014a) to show that in a persuasion dialogue hearer's rationality relates to evaluating speaker's arguments on the basis of whether they activate the (mental) beneficial topics in the mind of the hearer which are within the area of hearer's interest of persuasion.

In Debowska-Kozlowska (2014a), I define the concept of the Beneficial Cognitive Model as a set of mental beneficial topics which help a hearer resolve the conflict of opinion but only in his favour. The term "topic" is used to mean a mental conception in the mind of a hearer which might be transformed into an utterance when it is publicly expressed. Thus a question, a standpoint or an argument, etc. might be treated as a public externalisation of a mental topic. With reference to the mental conception of a topic, I differentiate between prototype and radial topics. Prototype topics are satisfying, salient essential and advantageous points for a hearer in terms of his persuasive wants and desires while radial topics do not relate to hearer's persuasive wants and desires. However, radial topics might be helpful and profitable for a hearer in terms of his other wants and desires, e.g. collaborative ones.

I argue that the BCM is a part of the cognitive environment of a hearer. In the relevance theoretic terms, the cognitive environment is the set of old facts and assumptions that construct the integrated context that helps an individual better understand an upcoming information. If any of the topics from the BCM is activated, e.g. through opponent's uttering the content synonymous with a given prototype topic of a recipient, then the human mind of the recipient will maximise the relevance of the stimulus for him. This process will take place intuitively. The activation of a prototype topic in a persuasion dialogue is, however, also supplemented by the reflective inferencing based on instrumental rationality. The hearer will use reflective inferencing based on instrumental rationality to maximise his profit after hearing the utterance.

Reflective reasoning based on instrumental rationality is coordinated by the reflective and algorithmic mind. Reflective reasoning of a hearer is thus dependent on his individual cognitive styles and abilities. The cognitive styles of the hearer determine the way in which an activated prototype topic is used by him. His cognitive competencies determine his ability to optimally use the activated prototype topic.

In a persuasion dialogue, a hearer uses the activated prototype topic through the process of reflective reasoning. The use of the prototype topic in the analytical processing by the hearer is related to making the counter-argument of an opponent advantageous for the hearer. For instance, the hearer might rephrase a fallacious counter-argument in the way that will allow him to realise his persuasive needs.

Relying on RT, the ATR and the dual-process literature (e.g. Evans, 2003, 2006; Stanovich, 1999, 2011), I claim that the use of a prototype topic in reflective reasoning involves either cognitive decoupling (hence CD) or serial associative cognition (hence SAC). My claim is that the serial associative cognition and cognitive decoupling facilitate the process of evaluation of a fallacious argument in terms of persuasive gains. These operations make it possible for the counter-argument to be turned into a move that is beneficial for the hearer. Since SAC relies on a shorter processing route than CD, it also involves less cognitive effort than CD. Humans with high cognitive competencies are more prone to engage in CD than humans with low cognitive competencies when they know it will bring them some immediate persuasive gains. It thus the reflective inferencing based on the operations of SAC that is the least effortful cognitive model for instrumental rationality. Below the two operations are explained.

When a participant of a persuasion dialogue hearers an utterance, he processes its meaning automatically. The intuitive processing cannot, however, handle the generation of a higher a response. The autonomous mind responsible for intuitive processing initiates some "preattentive processes" which stimulate the processing through the reflective mind (Evans, 2006, 2009). The reflective mind subsequently sends the signal to "decouple" the metarepresentation of a speaker (i.e. his "public representation of his mental representations") with the further conceptualization of the public representation by a hearer. As Stanovich (2011) indicates, the algorithmic mind allows to uphold the decoupled representations. In the process of cognitive decoupling, hearer's cognitive ability enables him to "distance oneself from representations of the world so that they can be reflected upon and improved" (Stanovich, 2011, p. 51).

Hearer's mental representation of the speaker's public representation and its evaluation with relation to hearer's individual goal is possible due to cognitive simulation. After a prototype topic is activated in the mind of a hearer, he uses it in hypothetical thinking to simulate a situation or an event similar to this expressed in the argumentative content. The hypothetical scenario is, however, advantageous for the hearer. A good example of putting the hypothetical thought into words is a conditional utterance expressing a favourable scenario for the participant of a persuasion dialogue.

Serial associative cognition does not involve the manipulation of alternative worlds. In the case of a persuasion dialogue, serial associative cognition relies just on the propositional content of a counter-argument. After the propositional content of the counter-argument activates a prototype topic in the mind of the hearer, the most accessible structure of the world activated by the prototype topic is used in reflective inferencing. This type of processing is a way of displaying a satisficing bias (Evans, 2006; Debowska-Kozlowska, 2014b). The evaluation of an argumentative content is based on the assessment which is most satisfying for the hearer at a given moment (even if the evaluation is based on false beliefs). The notion of satisficing bias appears to successfully explain the human mind's need to ease the cognitive load and process only the most reachable information. Stanovich (2011) argues, however, that the term "focal bias" is more appropriate as far as the associative mode of thinking characterising serial associative cognition is concerned. The reflective reasoning is in this case based on a "single focal model" that generates related thoughts. The umbrella term "focal bias" subsumes such notions as e.g. Johnson-Laird's (2006) principle of truth, the conception of focusing (Legrenzi et al., 1993), the focalism (Wilson et al., 2000). The important feature of focal bias is the propensity for sustaining serial associative cognition and thus not providing for the operations of decupling.

5. CONCLUSION

Epistemic vigilance in a persuasion dialogue is primarily based on instrumental rationality. In the process of evaluating, a hearer in a persuasion dialogue relies on the expected utility of a counter-argument. His persuasive goal determines the way of assessing the counter-argument. Reflective processing of the counter-argument is based on the operations of SAC or CD. Both of these operations make use of a prototype topic from the BCM. Since both of these operations are goal-oriented, they use cognitive shortcuts. However, SAC uses a shorter processing route as it does not involve hypothetical thinking in language processing. No matter what the cognitive abilities and styles of an individual hearer are, he is expected to assess the counter-argument relying on a prototype topic. However, the specific operations a particular hearer applies are highly dependent on his individual cognitive abilities (i.e. fluid intelligence). Thus, the conception of epistemic vigilance in a persuasion dialogue combines the interaction between the cognitive abilities and cognitive styles of a hearer. Not only the reflective mind but also the algorithmic mind takes an active role in the evaluation of a counter-argument.

REFERENCES

Bickerton, D. (1995). *Language and human behavior.* Seattle: University of Washington Press.

Burgoon, M., Hunsaker, F. G., & Dawson E. J. (1994). *Human Communication.* Thousand Oaks: Sage Publications.

Dębowska-Kozłowska K. (2014a). Processing topics from the Beneficial Cognitive Model in partially and over-successful persuasion dialogues. *Argumentation,* 28(3), 325-339.

Dębowska-Kozłowska K. (2014b). Intuitive and reflective inferencing in counter-argument processing. In I. Witczak-Plisiecka (Ed.), *Cognitive and Pragmatic Aspects of Speech Actions* (pp. 239-260). Frankfurt am Main: Peter Lang.

Evans J. St. B. T. (2003). In two minds: Dual-process accounts of reasoning. *Trends in Cognitive Sciences,* 7, 454–459.

Evans, J. St. B. T. (2006). The heuristic-analytic theory of reasoning: Extension and evaluation. *Psychonomic Bulletin & Review, 13*(3), 378–395.

Evans J. St. B. T. (2009). How many dual process theories do we need: One, two or many? In J. Evans & K. Frankish (Eds.), *In two minds: Dual processes and beyond* (pp. 33-54). Oxford: Oxford University Press.

Evans, J. St. B. T., & Over, D. E. (1996). *Rationality and reasoning.* Hove, U.K.: Psychology Press.

Johnson-Laird, P.N. (2006). *How we reason.* Oxford: Oxford University Press.

Legrenzi, P., Girotto, V., & Johnson-Laird, P.N. (1993). Focusing in reasoning and decision making. *Cognition, 49,* 37-66.

Mercier, H. (2011). When experts argue: Explaining the best and the worst of reasoning. *Argumentation, 23*(3), 313-327.

Mercier, H., & Sperber, D. (2009). Intuitive and reflective inferences. In J. St. B. T. Evans & K. Frankish (Eds.), *In two minds: Dual processes and beyond* (pp. 149-170). New York: Oxford University Press.

Mercier, H. & Sperber, D. (2011a). Why do humans reason? Arguments for an argumentative theory. *Behavioral and Brain Sciences, 34*(2), 57-74.

Mercier, H., & Sperber, D. (2011b). Argumentation: its adaptiveness and efficacy. Response to commentaries on 'Why do human reason?'. *Behavioral and Brain Sciences, 34*(2), 94-111.

Noveck, I. A., & Sperber, D. (Eds.). (2004). *Experimental Pragmatics.* Basingstoke: Palgrave Macmillan.

Oswald, S. (2011). From interpretation to consent: Arguments, beliefs and meaning. *Discourse Studies, 13*(6), 806-814.

Oswald, S. (forthcoming). Rhetoric and cognition: Pragmatic constraints on argument processing. *Proceedings of the Fifth Conference of Intercultural and Social Pragmatics (EPICS V).*

Oswald, S., & Hart, Ch. (2013). Trust based on bias: cognitive constraints on source-related fallacies. In D. Mohammed & M. Lewiński (Eds.), *Virtues of Argumentation. Proceedings of the 10th International Conference of the Ontario Society for the Study of Argumentation* (pp. 1-13). Windsor, ON: OSSA.

Over, D.E. (2004). Rationality and the Normative/Descriptive Distinction. In D. J. Koehler & N. Harvey (Eds.), *Blackwell Handbook of Judgment and Decision Making* (pp. 3-18). Malden, MA: Blackwell Publishing.

Reber, A. S. (1993). *Implicit learning and tacit knowledge.* New York: Oxford University Press.

Santibáñez Yáñez, C. (2012). Mercier and Sperber's Argumentative Theory of Reasoning: From the Psychology of Reasoning to Argumentation Studies. *Informal Logic, 32*(1), 132-159.

Sperber, D. (2000). Introduction. In D. Sperber (Ed.), *Metarepresentations: A Multidisciplinary Perspective* (pp. 3-16). Oxford: Oxford University Press.

Sperber, D. (2001). An evolutionary perspective on testimony and argumentation. *Philosophical Topics, 29*, 401-413.

Sperber, D. & Wilson, D. (1986). *Relevance: Communication and cognition.* Oxford: Blackwell.

Sperber, D., Clément, F., Heintz, Ch., Mascaro, O., Mercier, H., Origgi G., & Wilson, D. (2010). Epistemic Vigilance. *Mind & Language, 25*(4), 359–393.

Stanovich, K. E. (1999). *Who is rational? Studies of individual differences in reasoning.* Mahwah: Lawrence Erlbaum Associates.

Stanovich, K. E. (2011). *Rationality and the reflective mind.* New York: Oxford University Press.

Walton, D. (1995). *A pragmatic theory of fallacy.* Tuscaloosa: University of Alabama Press.

Walton, D., & Krabbe, E. (1995). *Commitment in dialogue: Basic concepts of interpersonal reasoning.* State University of N.Y. Press.

Wilson, D. (2000). Metarepresentation in linguistic communication. In D. Sperber (Ed.), *Metarepresentations: A Multidisciplinary Perspective* (pp. 411-448). Oxford: Oxford University Press.

Wilson T. D., Wheatley Th., Meyers J. M., Gilbert, D. T., Axsom, D. (2000). Focalism: A source of durability bias in affective forecasting. *Journal of Personality and Social Psychology, 78*(5), 821-836.

20

Strategic Maneuvering to Diminish Political Responsibility in a Press Conference

YELIZ DEMIR
Department of English Linguistics, Hacettepe University, Turkey
yelizd@hacettepe.edu.tr

KEREM YAZICI
Preparatory School of English, Ufuk University, Turkey
kerem.yazici@ufuk.edu.tr

It is an essential requirement of democracy that politicians provide account of their words and actions to the public. This paper aims to show how a politician carries out the accountability procedure in a press conference by exploiting the three aspects of strategic maneuvering. The paper draws its data from the political press conference held by the former Turkish Prime Minister Erdoğan, following the mine accident that took place in Soma, Turkey, in 2014.

KEYWORDS: accountability, critical event, political responsibility, pragma-dialectics, press conference, strategic maneuvering

1. INTRODUCTION

Political press conferences (also called news conferences) are one of the prominent activity types in the political domain which instantiates the accountability practice. "Accountability" refers to "the principle that governmental decision-makers in a democracy ought to be answerable to the people for their words and actions" (Besette, 2001, p. 38). It involves politicians' informing the public about the policies to be adopted, reasons why certain decisions have been made or policies have been implemented (Andone, 2015).

Although in some western nations political press conferences are held on regular basis during political campaigning processes in order to inform the public about party programs or decisions, (see

D. Mohammed & M. Lewiński (eds.) (2016). *Argumentation and Reasoned Action: Proceedings of the 1st European Conference on Argumentation, Lisbon, 2015. Vol. II, 393-406.* London: College Publications.

Bhatia, 2006; Clayman & Heritage, 2002; Eshbaugh-Soha, 2012), they are predominantly held on case-specific circumstances, especially when there is a critical happening that is of public concern. For whatever reason a press conference is held, the politician holds the responsibility to account about topics that are within the scope of his political conduct. However, being able to account is especially important when a politician or the party he/she is representing is assumed responsible for a critical event that has undesirable consequences for the public. And if this party is in charge of the government in a country, then the pressure for accountability becomes more dominant. Under such a condition, political press conferences serve as an instrument for a politician to justify the position of the government by means of argumentation and convince the public about the acceptability of the standpoint he/she has adopted concerning its role in this critical event. By adopting the pragma-dialectical framework, this paper sets out to explain how a politician manoeuvres strategically in a press conference for the purpose of diminishing political responsibility when his party which is in charge of the government is assumed responsible for a critical event.

2. STRATEGIC MANEUVERING IN THE CONTEXT OF A POLITICAL PRESS CONFERENCE ADDRESSING A CRITICAL EVENT

The notion of "strategic maneuvering" was introduced by van Eemeren and Houtlosser (1997) in their extended pragma-dialectical program in order to embrace the two simultaneously pursued goals in argumentation: the dialectical goal of reasonableness and the rhetorical goal of effectiveness. According to pragma-dialectical approach (van Eemeren and Houtlosser, 1997; 2002; 2005; van Eemeren, 2010), arguers resort to "strategic maneuvers" in order to diminish the tension between these two goals.

Van Eemeren and Houtlosser (2002) distinguished between three interrelated aspects of strategic maneuvering: (a) selecting from the topical potential, (b) meeting the audience demand, and (c) exploiting presentational devices. In strategically maneuvering between their dialectical and rhetorical aims, parties opt for the topics that they find easiest to discuss, they consider the audience expectations in formulating their standpoints and converge to the points they think the audience will agree with, and they use the most effective presentational devices to convince the opposing party of the acceptability of the standpoint they adopted.

In a press conference addressing a critical event about which the government is held responsible, the politician representing the government will have to maneuver strategically in order to resolve the difference of opinion in his own advantage. The primary difference of opinion in the context of such a political press conference is between the politician (and/or the government he is representing) and the public who questions and disapproves his actions.

However, because the politician and the public cannot interact directly to realize this difference of opinion in a press conference, the antagonism that the public might want to voice is embodied in the voice of the journalists who have the responsibility to hold the politician to account by asking not only information-seeking but also challenging questions to serve the public interest. The institutional context of a political press conference offers certain constraints for a political figure to maneuver strategically to steer the discussion to his favor.

A political press conference is institutionally organized in four phases. Bhatia (2006, p. 179) categorizes these phases as "the opening sequence, the individual voices, the interactional sequence and the closing sequence". Her characterization of the political press conference, however, is based upon a two-party conference, held by a host politician in a country and a guest politician invited to that country to deliberate on topics that are of concern to both nations. Because this format is somewhat different from the one we are dealing with, we adopt the following characterization of the phases of a press conference held by a single politician to address a critical event: an opening phase, a monological statement phase, an interactional phase, and a closing phase. The opening and closing phases are almost always very short and include a conventional format of introduction and end by the politician. These phases exhibit almost no argumentative quality compared to the body part of the press conference which includes the monological statement phase and the interactional phase. Therefore, we will restrict our attention to the body part of the press conference and address some possible strategic maneuvers the politician can employ in order to steer the direction of the discussion to his/her advantage.

The monological statement phase is the phase in which the politician provides individual evaluation of the critical issue under discussion. At this phase the politician does not take any questions from the journalists; therefore, he can take advantage of being the only voice to formulate advantageous starting points for the discussion. In a press conference that is held by a politician to address a critical event, the politician can be expected to choose from the topics that help him foreground the positive responsibility that was undertaken by the government and background the topics that may cause a negative evaluation of their responsibility in the event. It is also likely that the politician will predominantly address the expectations of the ones who suffer the consequences of the critical event directly. When the event brings about far-reaching effects in the society, the politician may opt for a strategy to unify the audience and call for a need to act as allies in the process of overcoming the negative consequences of the event, therefore emphasizing the inappropriateness of contravening the government in this critical issue. In an attempt to decrease the opposition in the audience, the politician may also make use of presentational devices that are more likely to draw wider acceptance from the audience. Zarefsky (2008) mentions the tendency of politicians to use condensation symbols as a strategic device to attract more positive response from the audience. Due to their widely agreed positive connotation, the politician addressing the critical event in a press conference may use national and religious condensation symbols as a strategy to call for unification in reacting to this event.

Although we mention these maneuvers in relation to the monological statement phase, they are not necessarily observed only at this phase. Similar strategies can be observed in the interactive stage as well. However, by nature, interactive stage offers certain other opportunities for a politician to maneuver strategically to attain an advantageous position in the discussion.

In the interactional phase the politician starts taking questions from the journalists. At this phase it is the journalists who decide about which topics to foreground in the discussion. Journalists are institutionally obliged to pursue the public interest. Andone (2013) points out that journalists abide by the norm of "due impartiality", which refers to allowing a variety of views to be heard and not giving prominence to one view over another. And when public interest is at issue, journalists can voice relevant antagonism against these different views in an equal way.

At the interactive stage of a press conference, the politician may come up with strategies to cope with the critical questions of the journalists. Evasive strategies (see Bhatia, 2006; Clayman & Heritage, 2002) can be among the most widely adopted strategies by a politician at this phase to avoid discussing the topics that would give an impression that the government did not fulfil its responsibilities in dealing with the event.

The conventional direction of the discussion in the interactive phase is from the journalists to the politician and not vice versa. This constraint attributes the burden of proof to the politician unidirectionally as he is the one who has to give an account. However, the politician may strategically shift the roles and critically evaluate the journalists' questions to gain advantage in the discussion.

In this section, we have suggested how in the context of a press conference a politician can resort to strategic maneuvers for the purpose of diminishing the political responsibility attributed to the government concerning a critical event. In order to illustrate such an endeavor of a politician, we will draw, in the following section, some examples from the political press conference held by the former Turkish Prime Minister Erdoğan on May 14, 2014 following the Soma mine accident.

3. A CASE IN POINT: THE SOMA PRESS CONFERENCE BY ERDOĞAN

The Soma press conference was held following the disastrous mine accident that took place on May 13, 2014 in Soma, Turkey, claiming the lives of 301 miners. The Soma disaster was the most fatal mine disaster in Turkey`s history and one of the most fatal accidents in the history of mining all over the world, according to a report of the Chamber of the Mining Engineers of Turkey ("TMMOB Soma maden kazası raporunu açıkladı", 2014).

Following this disastrous event, there emerged a number of questions in the public (voiced in various newspapers) regarding the responsibility of the government in preventing the accident. One was about whether proper inspections of the mine were carried out by the experts from the Ministry of Energy and Natural Sources and the Ministry of Labor and Social Security concerning the operation of the mine and the safety conditions of the workers. Another concern was whether the government put the required sanctions on those firms which turn out not to comply with the regulations for the operation of these mines. In the two main phases of the press conference (i.e., the monological statement phase and the interactional phase), Erdoğan maneuvers strategically with the topical potential, audience demand, and presentational devices to diminish the political responsibility attributed to the government concerning its role in the causes of the event.

3.1 Strategic maneuvering in the monological statement phase

In the monological statement phase of the press conference, Erdoğan maneuvers strategically with the topical potential by foregrounding the activities undertaken by the government following the mine accident rather than focusing on the reasons why the accident had occurred. The topics chosen at this phase help Erdoğan formulate an advantageous starting point for the discussion: the starting point that the government takes its responsibility and initiates all the necessary actions in the event of a critical incident. This starting point carefully disguises the responsibility attributed to the government in relation to the causes of the incident. The following two statements suggest how Erdoğan makes use of the available topics:

> (1) Immediately after hearing about the accident we put every governmental means available into use instantly from the most direct to the most indirect ones. We cancelled both our national and international programs. [...] Under the coordinatorship of Taner Yıldız (Minister of Energy and Natural Sources), our Ministry of Health, AFAD (The Disaster and Emergency Management Presidency of Turkey), Kızılay (Turkish Red Crescent), [...] Turkish Coal Mining Enterprises, search-and-rescue teams from all neighboring cities, medical care teams, experts, and technicians arrived in the region. We have spent all the effort in order to reach our brothers in the mine as soon as possible with the most cutting-edge equipment. We have put into use every governmental means on facing this large-scale and sorrowful event. [Translation from Turkish by the authors, YD & KY: the original forms of the statements are given in the footnotes.[1]]

The first extract suggests that the government acts swiftly to remedy the needs immediately after the mine accident. As the most urgent concern at that moment was search-and-rescue operations, Erdoğan focuses highly on this topic. In order to eliminate possible accusations concerning the pace and the technical capability of the government in reacting to the disaster, he stresses that as a government they did their best to reach the miners as quickly as possible, and they did so with the most developed technological equipment.

Erdoğan also stresses the government's role as an authority in revealing the causes of the accident and initiating legal actions against those who are found responsible for the accident. The statement (2) indicates that Erdoğan tries to assure the public that the government will be acting responsibly and stern in clearing up the case and identifying the responsible agents.

---

[1] Kaza haberini duyar duymaz, en yakından en uzağa kadar, kademe kademe tüm imkanlarımızı çok hızlı bir şekilde seferber ettik. Gerek ulusal gerekse uluslararası programlarımızı iptal ettik. [...] Taner Yıldız beyin koordinatörlüğünde, Sağlık Bakanlığımız, AFAD, Kızılay, [...] Türkiye Taş Kömürü İşletmeleri, bütün komşu illerden arama-kurtarma ekipleri, sağlık ekipleri, uzmanlar, teknisyenler bölgeye intikal ettiler. Mümkün olan en hızlı şekilde, en modern teçhizatla madendeki kardeşlerimize ulaşmanın gayreti içerisinde olduk. Böyle büyük çaplı ve acı bir hadise karşısında devletin gereken tüm imkanlarını devreye aldık.

> (2) My fellows, I want everyone to be confident about this: This accident will be investigated and is already being investigated extensively, right down to the last detail. We don't let any probabilities overlooked, and we won't do so. The case will be cleared up and steps will be taken to satisfy both the families of the victims and the public. Everyone should rest assured about this.[2]

The topical choice focusing on the post-accident roles of the government helps Erdoğan diminish the political responsibility that can be attributed to the government in relation to the causes of the accident.

In this phase, Erdoğan also maneuvers strategically with the audience demand by addressing the expectations of especially those who suffer the consequences of the accident more directly, that is, the victims' families. The following quote illustrate this:

> (3) And from this stage on we will do whatever is necessary [for the victims' families]. I would like to lay it on the line: no matter what is needed, financially or morally, we as the government of Turkey, the Ministry of Labor and Social Security and the enterprise itself will take all the necessary steps.[3]

Also notable at the point is the Ex-Prime Minister's aligning the government with the "concerned" citizens, whom he addresses as a targeted audience for his words. The following statement suggests this:

> (4) I thank especially all my citizens who have preserved their composure in this tragic incident and who pray for the salvation of our miners in every corner of our country.[4]

---

[2] Arkadaşlar şundan herkesin emin olmasını istiyorum. Bu kaza en ince ayrıntılarına kadar, en küçük detaylarına kadar araştırılacak ve araştırılıyor. Hiçbir ihmalin göz ardı edilmesine izin vermeyiz, vermeyeceğiz. Olay aydınlatılacak, hem yakınlarının, hem kamuoyunun tatmin olacağı adımlar da atılacaktır. Herkes bu konuda müsterih olsun.

[3] Bu aşamadan itibaren de gereken her şeyi yapacağız. Ve şunu çok açık net söylüyorum; madden ve manen ne gerekiyorsa bizler Türkiye Cumhuriyeti devleti olarak, kendi zaten Çalışma ve Sosyal Güvenlik Bakanlığı olarak, bunun yanında işletme olarak da atılması gereken adımlar nelerse bunların hepsi de yapılacaktır.

[4] Bu acı hadisede metanetini muhafaza eden, ülkenin her köşesinde madencilerimizin selameti için dualar eden milletime de özellikle teşekkür ediyorum.

The appreciation expressed for the "citizens who have preserved their composure" indicates that Erdoğan distances this audience from others who have a critical attitude against the government. As extract (5) indicates, Erdoğan warns the audience that these are malevolent opportunists who want to take advantage of the situation. Framing the audience this way, Erdoğan chooses to respond to the expectations of only those who act concordantly with the government and ignores others who are critical about the government's role in this issue.

> (5) But I would like to declare one more thing to my citizens here in front of all the media, and it is this: There are some extremists, some groups who would like to take advantage of such critical times and draw opportunities for themselves. I would also like to remind my citizens not to give any value to those groups for the peace of our country and for our unity and solidarity. [5]

Erdoğan also makes use of certain presentational devices to strengthen the effect of his argument at this phase of the conference. A notable presentational characteristic of his speech is his reference to religious themes or expressions.

Religion can be regarded as one of the condensation symbols for a nation (see Zarefsky, 2008 for using condensation symbols in political argumentation) and can be expected to attract more positive responses than negative ones in the public. For this reason, and also because he is a conservative politician, Erdoğan provides a religious evaluation of this mine accident at various times in order to soften or modify the perceptions about this hard event. In the following quote, he tries to soften the perception of death while doing the difficult job of mining.

> (6) May Allah surround our brothers who lost their lives with his mercy. Our god states in Koran that "man can have nothing but what he strives for." May Allah accept our brothers to the heaven, who have lost their lives in the pursuit of earning their keep and providing a living for their children.[6]

---

[5] Fakat burada, tüm medyanın huzurunda milletime bir şeyi daha duyurmak istiyorum; bu tür havaları fırsat bilip, bunları istismar etmek isteyen bazı aşırı uçlar var, gruplar var. Bunlara da değer verilmemesinin ülkemizin huzuru için, ülkemizin birliği, beraberliği için çok çok önemli olduğunu hatırlatmak istiyorum.

[6] Rabbim, hayatını kaybeden kardeşlerimizi rahmeti ile kuşatsın. Kur'an-ı Kerim'de 'Hakikaten insanoğlu için emeğinden başkası yoktur' buyuruyor

As (6) illustrates, by referring to Koran, Erdoğan argues that the ones who die while they are striving for making a living for their family are appreciated by the god and are likely to be awarded with a place in heaven. This way of representing the death of the miners may serve to soothe the perceptions about this tragic event and encourage a submissive rather than a critical attitude in the audience.

All in all, by manipulating the topics in an advantageous way, selectively addressing the audience expectations, and using religious-orientated presentational devices, the objective of Erdoğan's strategic maneuvering in this phase is to portray the government as a protective power that remedies the needs, in either economical or in moral terms, a fellow that stands next to the victims' families, and an authority that undertakes necessary actions following the disaster. Foregrounding the post-disaster actions taken by the government helps Erdoğan in strategically diminishing the political responsibility attributed to the government regarding its role in the causes of the mine accident.

3.2 Strategic maneuvering in the interactional phase

In the interactional phase of the press conference, Erdoğan takes the questions of the journalists regarding the mine accident. Some of these questions are directed at eliciting information about the event and some at critically evaluating the government`s role in the event. More important for the purpose of our study, however, are Erdoğan's responses to the critical questions because he exploits these critical questions in his strategic maneuvering with an attempt to diminish political responsibility. We will specifically concentrate on his replies to one of the critical questions.

This critical question is one which directly addresses the government's responsibility for the causes of the mine accident. The question runs as follows:

> (7) Journalist: How could this enterprise, which conducts such a dangerous business and which also has the potential to bring with it a huge disaster like this, carry on doing business without being prepared for an accident like this one? Who has the responsibility in this? Is there any responsibility for your government or your ministries?[7]

Rabbimiz. Rabbim, ekmeğinin peşinde, çoluk çocuğunun rızkının peşinde hayatını yitiren bu kardeşlerimizin mekanlarını cennet eylesin.

[7] Bu kadar tehlikeli bir iş yapıp da böyle bir kazaya hazırlıklı olmayan, hatta bu kadar büyük bir felaketi beraberinde getirme potansiyeli olan bir işletme nasıl

On receiving this question, Erdoğan replies with the following words:

> (8) Dear friends, I believe you don't follow closely, as a reporter, how coal mines work around the world. This can be due to the fact that there are not many coal mines in Qatar [referring to the journalist who is from Al Jazeera Turk based in Qatar]; there is natural gas. I am going to give you some numbers here. It's crucial in order for you to understand what is going on. [8]

In (8) we witness that Erdoğan strategically changes the conventional role of a politician in a press conference; it is a shift from having to give an account in response to the critical question of a journalist to critically evaluating the journalist`s question. Erdoğan establishes such a shift for the purpose of evading to discuss the government's responsibility. His reply indicates that it is not the government who should be blamed regarding this issue; it is the fault of the journalist who asks such a question without investigating how mines operate all over the world. At this point, the topic Erdoğan foregrounds in the discussion is "mine accidents all over the world", which helps him to draw the conclusion that accidents happen naturally in mines:

> (9) Look, in England, I'm going back in time a bit, 1862, collapse in a mine, 204 people died. [...] In Belgium 1887, methane explosion, 120 [casualties]. [...] Coming to France, 1906, the second most fatal coal mine accident in the world history; 1099 casualties. If we look at more recent times, [...] China, in 1942, it is thought that gas and coal dust was the reason for the mining accident resulting in the most fatal mine accident in the world. Do you know the number of casualties? 1549... These accidents are usual happenings in mines. Take America, with all its technology and everything. [...] In 1907, gas and coal dust explosion in two seperate mines, 361 [casualties]. Friends we, firstly, should not interpret these

---

olup da faaliyetlerine devam edebildi? Burada sorumluluk kime ait acaba? Hükümetinize, ilgili bakanlıklara, Çalışma ya da Enerji Bakanlığı`a düşen bir sorumluluk var mıdır?

[8] Değerli arkadaşlar bir gazeteci olarak zannediyorum dünyada kömür madenlerinin nasıl çalıştığını pek yakından takip etmiyorsunuz. Bu belki de şundan kaynaklanıyor olabilir; Katar'da pek kömür madeni yok, orada doğalgaz var. Ben size şimdi şurada birkaç tane rakam vereceğim, neyin ne olduğunu görmeniz bakımından bu çok çok önemli.

> incidents in the [...] coal mines as if these types of accidents never occur in such places. These are ordinary things.[9]

The extract (9) illustrates that Erdoğan tries to portray mine accidents as ordinary events which take place all over the world, regardless of the countries' facilities and technologies. He makes reference to various mine accidents that happened in the 19th and 20th centuries and emphasizes the accidents that caused more casualties than the one in Turkey. He uses the record of mine accidents all over the world as a concrete material starting point which can be assumed to attract little opposition and critique from the audience.

Erdoğan also seems to recognize the necessity to respond to the questions of the critical audience about whether the government took the necessary precautions to prevent the accident. The following words show this:

> (10) Look, as a result of the safety and security checks conducted at the end of March on this mine, [...] it was deemed successful in occupational health and safety issues.[10]

Indirectly, Erdoğan maintains in this extract that the government fulfilled its responsibility in terms of safety and security checks of the mine. As a result of the inspections carried out, the mine company was found to be functioning successfully. When this explanation is combined with his remarks in (9) stressing that mine accidents are common events independent of how successful a mine is operating, it becomes clear that there is no one who can be blamed. The cause of the accident is, then, the mining occupation itself, which is dangerous in nature.

---

[9] Bakınız İngiltere'de şöyle biraz geçmişe gidiyorum, 1862, bu madende göçük 204 kişi ölmüş. [...] Belçika'da '87 metan gazı patlaması 120. [...] Fransa'ya geliyorum, 1906, dünya tarihinin en ölümlü ikinci kömür madeni kazası, ölen 1099. Daha yakın dönemlere şöyle geleyim diyorum. [...] Çin 1942'de dünyanın en çok ölümlü gaz ve kömür madeni kazasına gaz ve kömür tozunun karışmasının neden olduğu sayılıyor ölüm sayısı ne biliyor musunuz? 1549. [...] Bu ocakların bu noktada bu tür kazaları sürekli olan şeyler. Bakın Amerika. Teknolojisiyle her şeyiyle. 1907, iki ayrı madende grizu ve kömür tozu patlaması 361. Arkadaşlar yani biz bir defa bu tür ocaklarda, kömür ocaklarında, madenlerde bu olanları, lütfen buralarda hiç bu tür olaylar olmaz diye yorumlamayalım. Bunlar olağan şeylerdir.

[10] Bakın bu ocakla ilgili Mart ayı sonunda yapılmış olan gerek sağlık gerekse güvenlik kontrolünde bu ocağın, bu noktada işçi sağlığı ve iş güvenliği noktasında başarılı olduğu tespit edilmiştir.

Erdoğan also resorts to certain religious presentational means as in the following statement:

> (11) Look, there is a phenomenon called "occupational accident" in the literature. This is not only bound to the mines. It can also occur in other fields. Occupational accident. This is what happened here too. It is in the *fitrat* of this business. There is no such thing like "no accidents will happen" in mines.[11]

The word *fitrat* refers in Islam to the peculiarities that are "woven into the spirit and soul of human beings" ("God-given nature", 2015). The meaning of the word can be extended to refer to other peculiarities that certain phenomena carry by nature. By emphasizing that accidents are part of the *fitrat* of mining, that is, they are part of the natural peculiarities of this business, Erdoğan also indirectly promotes the idea that "we should accept it and surrender to destiny" as we cannot change it no matter what we do. The strategic appeal to a widely held religious notion in this example aims at attracting consent rather than criticism concerning the event.

In short, Erdoğan's strategic choice of topic, his addressing the audience demand and his appeal in the presentation to religious motives in response to the critical question, which directly addresses the responsibility of the government for the accident, indicate that he tries to diminish or disclaim the political responsibility for the causes of the accident attributed to the government.

4. CONCLUSION

We have shown that argumentation is an instrument in the hands of a politician to direct the discussion to his advantage in a political press conference addressing a critical event. When the critical event is such that political responsibility is attributed to the government, the politician speaking on behalf of the government maneuvers strategically to steer the discussion to his advantage by diminishing the political responsibility concerning the critical event. In order to illustrate the case, our study drew on the press conference held by the former Turkish Prime Minister Erdoğan following the Soma mine disaster.

---

[11] Bakın literatürde iş kazası denilen bir olay vardır. Bu sadece madenlerde olur diye bir şey yok. Başka işlerde de olur. İş kazası. Burada da olur. Bunun yapısında fıtratında bunlar var. Hiç kaza olmayacak diye bir şey madenlerde yok.

It was observed that the politician maneuvers strategically to diminish and disclaim the political responsibility attributed to the government for the conditions affecting the mine accident in Soma. He does so by foregrounding topics that assign advantageous starting points to him, framing the audience to respond to, and using religious presentational means to attract support rather than opposition. During the press conference examined, Erdoğan defends the standpoint that "the government is not responsible for the mine accident in Soma" and the main arguments he presents are the following: (1) The government is doing its utmost to provide technical, financial, and moral support; (2) The government has initiated legal and administrative procedures to reveal the bodies responsible for this catastrophic event; (3) These things happen in mining; it is the fate of the miners. As these arguments suggest, Erdoğan tries to depict mine accidents as ordinary events, and when it comes to accounting for the government's responsibility in the event, he chooses to give an account concerning the post-accident responsibilities of the government which awards him an advantageous position in the discussion, leaving aside the discussion about the roles attributed to the government pertaining to the causes of the event.

ACKNOWLEDGEMENTS: We would like to thank Prof. Dr. Frans H. van Eemeren and Dr. Corina Andone for their helpful comments on this paper.

REFERENCES

Andone, C. (2013). *Argumentation in political interviews: Analyzing and evaluating responses to accusations of inconsistency.* Amsterdam: John Benjamins.

Andone, C. (2015). The burden of proof in dealing with political accountability. In R. Săftoiu, M. Neagu, & S. Măda (Eds.), *Persuasive games in political and professional dialogue* (pp. 19-38). Amsterdam: John Benjamins. DOI: 10.1075/ds.26.02and.

Besette, J. M. (2001). Accountability: Political. In N. J. Smelser and P. Baltes (Eds.), *International encyclopedia of the social & behavioral sciences.* Retrieved from http://www.sciencedirect.com/science/article/pii/B0080430767010913

Bhatia, A. (2006). Critical discourse analysis of political press conferences. *Discourse and Society, 17*(2), 173-203.

Clayman, S., & Heritage, J. (2002). *The news interview: Journalists and public figures on the air.* Cambridge: Cambridge University Press.

Eemeren, F.H. van. (2010). *Strategic maneuvering in argumentative discourse: Rxtending the pragma-dialectical theory of argumentation*. Amsterdam: John Benjamins.

Eemeren, F. H. van, & Houtlosser, P. (1997). Rhetorical rationales for dialectical moves. In J. Klumpp (Ed.), *Proceedings of the Tenth NCA/AFA Conference on Argumentation* (pp. 51-56). Annandale, VA: Speech Communication Association.

Eemeren, F. H. van, & Houtlosser, P. (2002). Strategic maneuvering in argumentative discourse: Maintaining a delicate balance. In F. H. van Eemeren & P. Houtlosser (Eds.), *Dialectic and rhetoric: The warp and woof of argumentation analysis* (pp. 131–159). Dordrecht: Kluwer Academic.

Eemeren, F.H. van & Houtlosser, P. (2005). Theoretical construction and argumentative reality: An analytic model of critical discussion and conventionalised types of argumentative activity. In D. Hitchcock & D. Farr (Eds.), *The Uses of Argument: Proceedings of a Conference at McMaster University* (pp. 75-84). Hamilton, ON: Ontario Society for the Study of Argumentation.

Eshbaugh-Soha, M. (2012). The politics of presidential press conferences. *American Politics Research*, *41*(3), 471-497.

God-given nature (2015, 27 May). Retrieved from http://www.imamreza.net/eng/imamreza.php?id=9604

TMMOB Soma maden kazası raporunu açıkladı (2014, 18 September). In the official website of the Chamber of the Mining Engineers of Turkey. Retrieved from: http://www.maden.org.tr/genel/bizden_detay.php?kod=9432

Zarefsky, D. (2008). Strategic maneuvering in political argumentation. *Argumentation*, *22*, 317-330.

21

Fallacy as Vice and/or Incontinence in Decision-Making

IOVAN DREHE
Romanian Academy – Iași Branch, Romania
drehe_iovan@yahoo.com

In my paper I aim to present a possible approach to the theory of fallacy specific to virtue argumentation theory. This shall be done employing conceptual pairs as virtue/vice or continence/incontinence, and illustrated by means of Aristotelian practical syllogisms. Based on these considerations I will then focus on two topics: 1. the possibility of a causal relation between incontinence and vice; 2. the difference between sophisms and paralogisms from the perspective of virtue argumentation.

KEYWORDS: akratic break, argumentational incontinence, argumentational vice, argumentational virtue, causality, fallacy, interference, practical syllogism, virtue argumentation

1. INTRODUCTION

Since C. L. Hamblin's influential book, *Fallacies*, has been published in 1970, a lot of work and research was done on the subject of argumentation errors following Hamblin's stress on the need of novel approaches to push the field forward, beyond the traditional treatments. Among these we can count informal and formal approaches, epistemic ones, dialectical and dialogical ones, pragma-dialectical ones etc. (for a survey see Hansen, 2015), each being determined by the specific argumentation theories advocated by their proponents.

For the last ten years a new theoretical approach to argumentation theory has been taking shape, the virtue theoretic approach to argumentation, importing from, accommodating and adding to the conceptual content present in virtue theories existent in ethics and epistemology. The main contributions in this direction were made principally by two scholars, Andrew Aberdein and Daniel Cohen (e.g. Cohen, 2009; Aberdein, 2010; etc.), both arguing for the importance and potential of an *aretaic* turn in the field of argumentation studies.

D. Mohammed & M. Lewiński (eds.) (2016). *Argumentation and Reasoned Action: Proceedings of the 1st European Conference on Argumentation, Lisbon, 2015. Vol. II, 407-415.* London: College Publications.

As in the case of the other approaches to argumentation, in virtue argumentation theory attempts are made to construct a specific view on the subject of argumentation errors. Bad arguments, or fallacies, were already discussed by Andrew Aberdein (2014) and Daniel Cohen (2005) in relation to the concept of vice. In what follows I intend to propose a new possible way of dealing with fallacious arguments along the lines of virtue argumentation theory, by trying to underline that another concept might be relevant in addition to the concept of (argumentational) vice: incontinence (gr. *akrasia*) or weakness of will[1].

2. CRITERIA FOR A "GOOD" ARGUMENT

Considering the fact that virtue theory was initially revived in ethics, drawing a good deal from Aristotle's views present in his ethical works, and that there are fallacies that are intentionally conceived in order to deceive, we might be inclined to think that what can be considered a good or a bad argument may have more than an ethical undertone. For example, if we consider what would be a proper criterion to judge an argument and pronounce it "good" or "bad", then surely different answers might be provided, depending on the choice of ethical theory. For example, from an "utilitarian" perspective, if the argument is motivated by the principle of "the greatest good for the greatest number of people", then the argument is "good". If it is motivated by "sinister interest", then it is "bad". From a "pragmatic" point of view, if an argument is "efficient" (i.e. it persuades the audience), then it can be considered "good". If it is "inefficient" and it fails to persuade, then it can be labelled as "bad". If we consider this from a "normative" perspective, then an argument is "good" if it respects the principles of valid reasoning and "bad" if it fails for some reason to do so. Since this way of seeing arguments as "good" or "bad" resemble the ways "human action" is considered from an ethical perspective, I think one more perspective can be accommodated: that of "virtue theory".

Thereby, from a "virtue theoretic" perspective, an argument can be considered "good" if the arguer argues (or the audience assesses) as a virtuous arguer would do in a specific situation. If the arguer "fails" to argue in this manner, then his argument is "bad". Some problems I intend to address in this paper are related to this "failure" on behalf of the arguer, failures which make his arguments "bad". One of these problems is the following: why does the failure occur?

---

[1] I already discussed this in (Drehe 2016, sect. 4). In the present paper I will attempt to develop it a bit further.

3. CONTINENCE AND INCONTINENCE

As expected, the starting point will be Aristotle's discussion about what it means for a man to be continent or incontinent in respect to his actions. Firstly, what does it mean for a man to be continent, or to be rationally in control of his own actions? For Aristotle, the continent man is the one who knows about the defective quality of his appetites and refuses rationally, after some deliberation/calculation, to follow on their path (*Nicomachean Ethics* VII, 1, 1145b10-11; 1145b13-14; etc.).

As an example of such continent behavior I will use the illustrative tool provided by Aristotle, the practical syllogism. The agent who has practical wisdom deliberates prior to any action and the result of such deliberation is a decision regarding the course of action thought best. This deliberation is done using practical syllogisms (see e.g. *On the Soul* III, 11, 434a16-19; *Movement of Animals* 7, 701a11-20; *Nicomachean Ethics* VI, 5, 1140b4-5; VII, 3, 1147a25-31; VII, 4. etc.). An example of practical syllogism conducive to continent behavior is the following:

> 1. Major premise: "Sweet things should be avoided in case of the agent with diabetes";
> 2. minor premise: "This particular agent has diabetes";
> 3. Conclusion: "Therefore the agent should refrain from sweet things."

However, it is difficult for the agent to remain continent as long as the appetites interfere. For example, the propensity for sweet things interferes and modifies the practical syllogism thus:

> 1. M.P.: "Sweet things should be tasted";
> 2. m.p.: "This particular thing is sweet";
> 3. C.: "Therefore it should be tasted".

This way, because of the appetites, the agent becomes incontinent or akratic.

What about continence and incontinence with regard to argumentative acts? In this case we can also illustrate a continent behavior via a practical syllogism[2].Practical syllogism in the case of

---

[2] In (Drehe 2016) I called this a "deliberational syllogism" in order to distinguish it from the "practical syllogism" specific to moral contexts. However, in order to avoid adding more terminology at this point, I will use the phrase "practical syllogism".

argumentative acts should be preliminary to acts of arguing/assessing arguments in the same way as in moral contexts is preliminary to action. What results is a decision regarding the course considered right in the process of arguing/assessing. In the case of an arguer the practical syllogism might look like this:

> 1. M. P.: "You should construct your argument based on rule X of valid reasoning in order to persuade a listener in an argumentational situation of type A";
> 2. m. p.: "You are in a situation of type A";
> 3. C.: "For a good outcome, you should apply rule X."

And in the case of the listener/audience an example of practical syllogism for good outcome in the evaluation of arguments:

> 1. M. P.: "You should evaluate the arguer's argument using the rule X relevant to an argumentational situation of type A";
> 2. m. p.: "You are in a situation of type A";
> 3. C.: "For a good outcome, you should evaluate based on rule X."

An example of practical syllogism usable to avoid illicit ad hominem arguments can be the following:

> 1. M. P.: "Arguments of agents should not be evaluated on the basis of what we know about the agent's deeds or beliefs."
> 2. m. p: "I have knowledge of this agent's deeds and believes"
> 3. C.: "I should refrain from evaluating this agent's arguments based on my knowledge of him."

As in the case of actions, mentioned above, there are interfering entities both in argument construction and assessment. These entities may be of different kinds: appetites, passions, dispositions, emotions, urges, desires, biases, beliefs etc. A human agent virtually cannot think without interferences of this kind in his reasoning process (even though he is resisting and in the end manages to bring about his conclusion rationally). If an interference is successful, then the agent might become incontinent. If he resists and manages to keep his reasoning process on the "right track", then he remains continent.

For example, in the case of the agent who formulates the practical syllogism to help him avoid fallacious ad hominem argument, his temper might interfere and get the better of him and, as a result, modify his syllogism in the following manner:

1. M. P.: "Morally dubious persons should refrain from formulating arguments on certain topics."
2. m. p.: "This particular person in front of me is morally dubious in my view."
3. C.: "Therefore I should not care to listen to his argument."

So, in brief, argumentative incontinence for the arguer appears when the arguer considers the X line of argument as best suited in a situation, but takes Y line of argument, because of some kind of interference. As for the listener: argumentative incontinence appears when considers that an argument is properly evaluated in X manner, yet he does not assess the argument in this manner because of some kind of interference.

4. CAUSALITY AND AKRATIC BREAKS

Interference may bring about akratic breaks and akratic breaks modify the practical syllogism at some point, which in turn leads to incontinence. Amelie Oksenberg Rorty (1980) discusses four types of akratic breaks: 1. The agent knows the right aims, but fails to commit to them: "akrasia of direction or aim".2. The agent knows which aims are right, is committed to them, but still fails in identifying the proper context to act (in this case: argue): "akrasia of interpretation"; 3. The agent knows the right aims, commits to them, knows the right context to act/argue, but fails to reason towards a conclusion or to make a decision towards acting or arguing: "akrasia of irrationality"; 4. The agent knows about the right aims, is committed to them, interprets the context properly, validly deduces a conclusion or takes a decision to act or argue, but actually fails to do so: "akrasia of character".

So, it seems that there are levels of interference, as is the case of akratic breaks, but there are also interfering entities. In short, an akratic break is caused by an interference. But what kind of interferences are there and what is the manner of their interfering action? As a sketch we may distinguish between interference from the outside of the practical syllogism and interference from the inside.

Firstly, let us consider the possible interferences from the outside. For example, virtues can interfere with the practical syllogism (to keep the argument/action on the "right track"); vices, on the other hand, might derail the argument; in addition, one can discern many interfering entities: dispositions, beliefs etc. as already mentioned in the previous section. One interesting question at this point would be the following: are vices different from dispositions or beliefs? Or these too can be considered vices insofar they harm the argument (in the same way they could be considered virtues if they keep the argument valid/sound)? Anyway, interferences from the outside seem to causally determine argumentational incontinence because they cause akratic breaks.

Secondly, there might be interferences from the inside of the practical syllogism. For example, one kind of akratic break can change the composition of the practical syllogism. Also, one might wonder whether an akratic break can cause another one as a direct result of it?

These are questions that involve too many ramifications to be tackled in the body of the present text. However, I will attempt to provide several examples of akratic breaks and their respective causing interfering entities.

> 1. Akrasia of irrationality caused by dispositions of the body: when an arguer or listener cannot reason properly due to some temporary or permanent physical impediment (sickness, madness etc.)
> 2. Akrasia of interpretation caused by interference of intellectual disposition: when the agent confuses a situation where the other party actually tries to have a deliberational dialogue with the agent, but because of an intellectual disposition such as excessive and unwarranted suspiciousness he interprets it as an adversative or eristic dialogue.
> 3. Akrasia of aim caused by the interference of a temperamental disposition: a person who delivers a speech is heckled from the audience. The person loses his or her composure because of a temperamental disposition such as irritability and gets aggressive towards the audience.

These are only several imaginable instances of interference causing akratic breaks at different levels. If these could be represented in a table, that table would look like Table 1 below.

Of course, there are many more instances, real or imaginary, that can be used to fill in the table (represented in table 1 by *x*'s). However, this would involve more research and adjustments of the model in order to make it coherent and plausible. At this time at least the following points should be considered as important: 1. the need of a taxonomy of interfering entities; 2. an analysis of the possibility of akratic breaks and their explanatory usefulness and efficiency[3]; 3. the

---

[3] There might be some objections against the suitability and explanatory power of the practical syllogism/akratic breaks model in this case. Of course the new model can always be considered and filled in with the relevant details. For

possibility that certain types of entities cause only certain types of akratic breaks; 4. identifying eventual new kinds of akratic breaks in cases where the practical syllogism is not appropriate anymore.

Akrasia type / Interfering entity type	Akrasia of Aim	Akrasia of interpretation	Akrasia of irrationality	Akrasia of character
Dispositions of the Body	x	x	Example 1 from above	x
Temperamental dispositions	Example 3 from above	x	x	x
Intellectual dispositions	x	Example 2 from above	x	x
Beliefs	x	x	x	x
Etc.	x	x	x	x

Table 1. Interfering entities and akratic breaks caused by them

It seems obvious at this point that vices (considered as a type of interfering entities along the lines discussed above) can bring about fallacious argumentative behavior by causing incontinence in argumentation.

5. DEFINING A FALLACY

In light of the things considered above we can attempt to give a definition to fallacies. In broad lines a fallacy can be considered to be an error in argument (in the formulation or in the assessment of an argument) arising from an akratic break of a certain type caused by an interference by one or more interfering entities.

If we consider things at a different level, where we want to distinguish sophisms (=intentional fallacies) from paralogisms (=non-intentional fallacies), then we might say that: 1. A sophism is an error in argument arising from an akrasia of aim (resulting from an akratic break at the level of the Major Premise), caused by a certain interfering entity that makes the agent intentionally use another line of argument/evaluation; 2. A paralogism is an error in argument arising

---

example, in the eventuality of building a new such model one needs to distinguish between the actual act of arguing or assessing an argument and the preliminary fore-thinking process of which the result is the decision to argue/assess in one way or another.

from an akrasia of interpretation or irrationality (caused by their respective akratic breaks), caused by certain interfering entities.

Another question seems to come about at this point: what about akrasia of character, i.e. the choice to argue or not, the choice to evaluate or not – is this a fallacy or not?

6. CONCLUDING REMARKS

Obviously, this way of considering fallacies (i.e. from the perspective of vice/incontinence) does not exclude by necessity or supersedes the classical treatments. On the contrary, it can act as a complementary approach and it has the potential to offer alternative insights on the way valid arguments or fallacies appear.

ACKNOWLEDGEMENTS: This paper is supported by the Sectoral Operational Programme Human Resources Development (SOP HRD), financed from the European Social Fund and by the Romanian Government under the contract number POSDRU/159/1.5/133675.

REFERENCES:

Aberdein, A. (2010). Virtue in argument. *Argumentation, 24*(2), 165-179.

Abderdein, A. (2014). Fallacy and argumentational vice. In D. Mohammed & M. Lewinski (Eds.), *Virtues in Argumentation*: *Proceedings of the 10th International Conference of Ontario Society for the Study of Argumentation (OSSA), May 22-25, 2013.* Windsor: OSSA.

Aristotle. *Movement of Animals.* Translated by A. S. L. Farquharson. In (Barnes 1984).

Aristotle. *Nicomachean Ethics.* Translated by W. D. Ross, revised by J. O. Urmson. In (Barnes 1984).

Aristotle. *On the Soul.* Translated by J. A. Smith. In (Barnes 1984).

Barnes, J. (Ed.) (1984). *The Complete Works of Aristotle. The Revised Oxford Translation.* Princeton: Princeton University Press.

Cohen, D. H. (2005). Arguments that backfire. In D. Hitchcock & D. Farr (Eds.), *The Uses of Argument,* (pp. 58-65). OSSA2005.

Cohen, D. H. (2009). Keeping an open mind and having a sense of proportion as virtues in argumentation. *Cogency, 1*(2), 49-64.

Drehe, I. (2016). Argumentational virtues and incontinent arguers. *Topoi.* Forthcoming.

Hamblin, C. L. (1970). *Fallacies.* London: Methuen.

Hansen, H. (2015). Fallacies. In *The Stanford Encyclopedia of Philosophy* (Summer 2015 Edition), Edward N. Zalta (Ed.), URL = <http://plato.stanford.edu/archives/sum2015/entries/fallacies/>.

Rorty, A. O. (1980). Where does the akratic break take place? *The Australasian Journal of Philosophy*, *58*(94), 333-346.

22

Arguments for an Informational Layer in Theories of Argumentation

SJUR KRISTOFFER DYRKOLBOTN
Utrecht University, The Netherlands
s.k.dyrkolbotn@uu.nl

When people argue, about what do they disagree? Perhaps nothing at all, except the nature of the disagreement. At any rate, arguers often disagree about the meaning and relevance of arguments, in ways that invariably influence their opinions about argument strength. The prevalence of such higher-order dynamics is an argument for an informational layer in models of argumentation. In my paper, I elaborate on this claim and argue that it is relevant, even for logicians.

KEYWORDS: deep disagreement, dialogue type, divergence, fallacy, formal logic, informational layer, interpretation

1. INTRODUCTION

It is beyond doubt that humans like to argue, but it also seems that they have a propensity for committing fallacies, both logical and communicative. Hence, for those interested in developing a descriptive theory of argument and reason, it makes sense to look for models of argumentation that capture essential features of argumentative agents, agents whose primary aim is not to argue correctly to arrive at the truth, but to argue convincingly, in order to win the debate.

Such models need to take into account the fact that humans are inherently social agents, and that argumentation has meaning because it brings arguers into contact with each other, in a way that allows them to exert mutual influence on each other's subjective viewpoints. This vision of argumentation is gaining ground in cognitive science, with recent research suggesting that reasoning as such evolved to help us become better arguers, a theory that has been dubbed the argumentative theory of reason (Mercier & Sperber, 2011).

Importantly, Mercier and Sperber argue convincingly that "bad" argumentation can sometimes be beneficial in a normative sense, not

D. Mohammed & M. Lewiński (eds.) (2016). *Argumentation and Reasoned Action: Proceedings of the 1st European Conference on Argumentation, Lisbon, 2015. Vol. II, 417-429.* London: College Publications.

only to egocentric arguers who wish to win debates, but also to social groups that might in the end have more to gain by approaching "truth" or "rationality". However, in order for debate to be beneficial in this sense, it requires social conditions and protocols capable of transforming individual biases and argumentative transgressions into collective insights. Success in this regard becomes the mark of a "good" process of argumentative interaction.

Unfortunately, existing models of both rhetoric and argumentation tend to focus on the prescriptive aspects of argumentative reasoning, by formulating standards of "good" argumentation and "effective" rhetoric that arguers should follow. The argumentative theory of reason indicates that this approach is incomplete, sometimes even inappropriate, when considering argumentation in the context of multi-party deliberation and interaction.

To address this, existing theories of argumentation should be supplemented by an explicit account of the multiple, often questionable, viewpoints that agents are likely to endorse regarding the content and meaning of their own and each other's arguments. The goal should be to formulate theories that can accommodate multiple viewpoints that are irreducible; according to the argumentative theory of reason, maintaining semantic diversity must necessarily become a key overarching constraint on normative theories.

Hence, the ideals of good argumentation need to be recast as social standards that pertain to sets of diverging interpretations of argumentative meaning. In the following, I will explore this idea in some more depth, by arguing that argumentation theorists need to devote more attention to what is sometimes called the "informational" layer of reasoning (Stenning & van Lambalgen, 2008, p. 348).

2. ARGUMENTS AND MULTIPLE INTERPRETATIONS

Despite the fact that argumentation theorists often focus on persuasion dialogues, it is clear that argumentation is a multi-faceted phenomenon; sometimes, people argue to persuade, but at other times, they argue to explore hypothetical scenarios, decide on a course of action, or simply to gain more information about a state of affairs (Walton & Krabbe, 1995; Walton, 2007). It is clearly useful to keep in mind these distinctions, to be more precise when addressing various phenomena associated with argumentation. It can also be appropriate, as dialogue theorists do, to divide discussions into segments where each is labelled by a specific dialogue type and analysed accordingly.

Indeed, this methodological approach might represent a move forward compared to a narrow argumentation-as-persuasion perspective, which threatens to block out an important part of the larger picture associated with typical argumentative exchanges in social life.

However, any purported categorisation of discussions into dialogue forms will invariably leave us with an incomplete and contestable representation of the meaning of arguments and argumentative behaviour. To suggest the plausibility of this claim, it should be enough to recall that most arguments encountered in practice are enthymematic, or even recalcitrant (Paglieri & Woods, 2011). Hence, it must be expected that participants, audiences, and theorists, will all entertain different interpretations of discussions and the meaning of their various argumentative components.

Moreover, while it is helpful to have an analytic language at one's disposal for expressing such interpretations, no theory can be expected to resolve all interpretative divergences among those who partake in and observe arguments. Some have thought of this as marking the inherent subjectivity of argumentation (Kock, 2007), but others have stressed how it gives rise to important mechanism whereby people come to change their positions and collectively develop novel views on the matter under consideration (Wohlrapp, 1998).

The latter mechanism is particularly crucial, also in relation to the insights reported in (Mercier & Sperber, 2011). Indeed, if we take the argumentative theory of reason to heart, the indeterminateness of argumentation structures should be approached as a potential strength, not a weakness. Moreover, the plurality of interpretations encountered among those who engage with arguments in real life should be recognised as an interesting object of study, not as an obstacle to analysis that should be removed by "analytic reconstruction" (van Eemeren & Grootendorst, 2004) or the like.

It should be noted that embracing multiple interpretations at the informational level does not imply scepticism with respect to the analysis that takes place once an interpretation is fixed. This is an important point to note for anyone interested in reconciling the argumentative theory of reason with normative argumentation theories. If the argumentative theory is understood as emphasising the value of plurality in interpretation, it does not undermine the relevance of traditional argumentation theories, but merely suggests a new avenue of research.

So far, this avenue appears to be under-explored; most theories of argumentation have little to say about the interplay between argument moves and diverging interpretations of meaning. This is unfortunate, since argumentation can clearly be beneficial also when arguers appear to "talk past each other". The valuable social functions of argumentation can even be enhanced in these cases, because it might increase understanding and stimulate critical thinking, even if it does not reduce the degree of disagreement.

For instance, if two people attempt to persuade each other, and both agree that they are engaging in argumentation for that purpose, the function of this argumentation at the societal level can still be mainly connected with the flow of information it engenders, not the flow of acceptable conclusions one might hope (often in vain) to distil from it. Such "mixed dialogues", as they are called in (Walton & Krabbe, 1995), are particularly likely to involve subtle interplays between different interpretations of meaning.

A typical example would be a debate with participants that are in so-called "deep disagreement", meaning that they cannot reasonably be expected to make much progress towards resolution (Fogelin, 1985). An instance of this could be, for instance, a debate on abortion involving passionate pro-life and pro-choice participants. From the internal workings of the argument moves, such a dialogue would probably come to be classified as persuasive, if not also eristic. At the same time, it would cloud the meaning of such an exchange to regard it as being about who argues most convincingly, or directed towards "resolving a difference of opinion" (van Eemeren & Grootendorst, 2004, p. 57).

These are clearly not the most important aspects of debate when the participants disagree as fundamentally as they often do on the issue of abortion. Rather, the potentially positive functions at the social level tend to be almost purely informational and emotive: to build some mutual understanding, to rid the world of some unnecessary prejudice, while informing us of pressure points that should be dealt with to avoid a complete breakdown of communication accompanied by heightened tensions.

Of course, communication as such might fail, resulting in more mutual suspicion and prejudice. But when is this likely to be the result, and when is confrontation likely to be beneficial? This is a question that a theory of argumentation should address. To do so, it needs to pay more attention to the informational layer and the dynamics of interaction that can result when opposing ways of looking at the world collide in incommensurable arguments.

To apply the pragma-dialectical theory or the like to analyse such exchanges, resulting no doubt in more blame than praise all around, does not seem like the appropriate way to proceed. Specifically, the worry is that such theories would distract us from many important social functions of argumentation because these functions pertain specifically to the functional contributions made by instances of bad argumentation.

If bad arguments are not properly understood, but painted as enemies of fruitful discussion, argumentation theories might even come to discourage debate on issues that really matter, because resolution appears impossible. In this way, the theory of argument itself might end up fuelling the misguided idea that there is "no point arguing" for people who disagree fundamentally about how to interpret key semantic building blocks of the discussion.

This further suggests that a new analytic approach is called for, which sets out to study, rather than conflate, the multiple interpretations of argument structures and their constituents that typically arise in real-life argumentation scenarios. For an interesting demonstration of this, presented as a defence of argumentation theory that succeeds by implicitly widening the definition of its purpose, I point to (Freeman, 2012).

This article starts out by noting how Stanley Fish's work (especially (Fish, 1996)), presents a challenge to argumentation theories such as the pragma-dialectical theory, before presenting an argument leading to the conclusion that "Fish's scepticism of argumentation [is] not justified on any level" (Freeman, 2012, p. 74). However, at this point in the article, argumentation is not understood as an activity that aims to resolve differences of opinion, but rather as a form of interaction that is purposeful whenever it "may lead to a deeper understanding of one's world view and a more mature commitment to it" (Freeman, 2012, p. 73).

Hence, it would appear that Fish's scepticism is at least sufficiently justified to cause an interpretative shift in the mind of a critic, significantly broadening the meaning of argumentation. This is a positive development, which might in turn suggest new strategies for analysing arguments and new normative principles, for instance to ensure that argumentation remains a force for good even when there is deep disagreement, as discussed in (Zarefsky, 2012, pp. 80-85).

Although this points beyond the current boundaries of argumentation theory, an increased sensitivity to interpretation issues also seems to have implications for some core topics that already enjoy pride of place, including the study of fallacies.

3. THE INFORMATIONAL TAKE ON FALLACIES

Whether or not an argument can be described as fallacious is clearly dependent on how the argument is interpreted. Hence, a judgement to the effect that an argument is a fallacy is in general no more secure than the semantic interpretation that gives rise to this judgement. Unfortunately, this contingent nature of the designation of fallacies is not at the centre of attention in most argumentation theories.

The underlying sensitivity to interpretation might be acknowledged, but then in a way that does not explore the informational layer of argumentation in sufficient depth. For example, consider the so-called Mexican War Argument, originally due to (Rescher, 1964, p. 82), as discussed in (Walton, 2007, p. 225; Walton et al., 2010, p. 220):

> The United States had justice on its side in waging the Mexican war of 1848. To question this is unpatriotic, and would give comfort to our enemies by promoting the cause of defeatism.

According to Rescher's original analysis, this argument is fallacious – an illicit instance of argument from consequences. Specifically, the reasons given in the second sentence are not regarded as relevant to the issue presented in the first (whether or not the US had justice on its side). But arguably, this "fallacy" is in the eye of the beholder. In fact, the discussion of the example offered by Walton takes us halfway to this conclusion, without making the final step.

Specifically, Walton analyses the argument as one involving a tension between different dialogue modes, where the argument as stated is fallacious with respect to the persuasive dialogue regarding the justness of the war, but would be acceptable in a dialogue regarding whether it is appropriate to question the justness of that war in the first place (Walton, 2007, p. 225).

Hence, Walton acknowledges that no fallacy can be found in the argument structure as such, since the designation of fallacy depends on the interpretation of the mode of the argument. But, according to Walton, the persuasive interpretation is "correct" and the fallacy reemerges.

Arguably, this re-invention of the fallacy reflects that the interpretation layer has been dealt with too superficially. Specifically, no clear argument is provided to clarify why we should be committed to the interpretation of the argument that Walton endorses. Moreover, the Mexican War Argument would appear to leave room for several alternative interpretations, also as a persuasive argument in favour of the justness of war.

First, the argument might be understood as an enthymeme, where the implicit premise is that there has already been a lot of discussion and convincing evidence presented in support of the claim that the US had justice on its side. In this case, the argument is relevant, because it acts as a reference to other unstated arguments that the arguer expects us to reflect on and accept, for the reasons stated. The argument might then be classified as a meta-argument, but not as a fallacy.

To bring out the point more clearly, we could substitute a more recent war for the Mexican War; say the war in Afghanistan. If someone used the same argument form in this context, it would be hard not to think of 9/11 when evaluating the argument, even for those disagreeing with the war and/or rejecting the argument. This, moreover, is something that the arguer might well anticipate, and also be entitled to anticipate, since the standard justification for the war in Afghanistan is shared background information to the argument.

In any case, it seems wrong to be too confident in one's interpretations, especially if this results in an uncharitable reading of someone else's argument as a fallacy. The point is not that Walton's interpretation of the Mexican War Argument is necessarily wrong, just that it is not the only reasonable candidate. Indeed, it is possible to interpret the argument in a yet another way, as making reference to a default rule whereby patriots should normally believe that a war is just whenever their country engages in it. Again, the background information must be taken into account, giving rise to the interpretation whereby the intended message is that since the US did in fact engage in the war, the default rule must be applied, so the war must be considered just.

Again, a variant of the argument brings out the point more clearly, namely the following Canadian War Argument:

> The United States will have justice on its side and should wage war against Canada in 2016. To question this is unpatriotic, and would give comfort to our enemies by promoting the cause of defeatism.

Here the default rule cannot be applied, and most people would immediately recognise the argument as pure nonsense. Arguably, we are still not in the presence of any fallacy, but we are certainly in the presence of an argument that looks *much more* unreasonable than the original variant. If the informational layer of argumentation is not considered relevant, this difference risks disappearing, as the argumentation theory is likely to classify this argument as being fallacious in the same way, and for the same reason, as the Mexican War Argument.

Now, while the similarities between them are interesting, it is not appropriate to disregard the fact that there are also important differences, arising from how people are likely to interpret these arguments in light of shared background information and their attitude towards the persons making them. To conclude, the designation of the Mexican War Argument as a fallacy seems unhelpful as a general proposition.

However, the underlying observation that led to this designation can be helpful in a completely different way, namely as a reason for revising one's interpretation. Indeed, if Walton argues that the argument is a fallacy in the persuasive mode, this in itself is arguably a normative reason to prefer a different interpretation.

At the same time, an arguer that makes arguments that are hard to interpret in good faith without revealing "fallacies" is likely to lose the argument. Plainly, such an arguer will tend to ask too much of us. This social observation, then, can be expected as an important remainder of the traditional account of fallacies even after the informational level has been incorporated into the theory.

The general mechanism is also on display in argument patterns based on formal logic. For a concrete example, consider the following argument, where Paul is apparently committing the fallacy of affirming the consequent.

> Teacher: Miriam hates to study, but if she has an essay due she will be in the library.
> Paul: I saw Miriam in the library, so from what you said we should conclude that she has an essay due.
> Teacher: No, that would be a fallacy.

Paul is being reasonable, but the teacher, having studied some logic and argumentation theory, insists that he argues fallaciously. The teacher thinks he has good reason to do so, since Paul appears to be inferring that Miriam has an essay due from the fact that she is in the library, even though the teacher only said that she would be in the library if she had an essay due.

However, the teacher also said that Miriam hates to study. Hence, a reasonable interpretation of his statement is that unless she has an essay due, she would be unlikely to be in the library. In fact, this interpretation of the semantic content of the teacher's claim is much more reasonable then the literal interpretation. Clearly, having an essay due cannot possibly imply that Miriam will be in the library; Miriam can have an essay due for a week at least, and during that time, she is likely to sleep, eat, drink, and relax, mostly outside of the library.

The formal interpretation is therefore senseless, especially under the classical logic mode of thinking that the teacher appears to insist on. Paul's interpretation, on the other hand, is quite sensible; giving rise also to a reasonable default inference that Miriam has an essay due.

This example demonstrates the danger of focusing on the clarity of ideal forms when it is highly unclear whether people (should) interpret information in a way that conforms to them. Many examples of this danger can be found, including in psychological research that purports to show that humans do not reason very well, or do not reason according to logical principles.

Here the researchers often appear to reason just like the teacher in the example above, passing judgement and drawing inferences about test subjects based on interpretations that are assumed conclusive but turn out to be contestable. This has been demonstrated very effectively by the work of (Stenning & van Lambalgen, 2008), who highlight how reasoning to an interpretation can give rise to diverging interpretations of logical tasks.

Interestingly, the use of formal logic itself serves as an important guide towards this conclusion, since modern logic often permits formal modelling of diverging interpretations of the information presented to test subjects, thereby demonstrating how the "wrong" answer might very well also be logically correct, even if it does not conform to the researcher's expectations (Stenning & van Lambalgen, 2005).

The broader point is that how people (choose to) interpret semantic information is crucial also to how they reason and argue. Hence, declaring a form of argument as a fallacy, without considering alternative interpretations that render it innocuous, is not good practice. Arguably, when argumentation theorists reconstruct arguments in a given form, thereby conflating different possible interpretations to fit their own models, they also create fallacies where none previously existed.

To counter this mechanism, it is of little help to change one's mode of reasoning about arguments, by insisting that such reasoning should be "informal". Indeed, given the progress in formal logic in recent decades, producing a range of non-classical systems for formalising the "informal", there is reason to question whether the traditional conceptual distinction between argumentation and formal logic is meaningful at all. Rather, it seems that logicians and argumentation theorists should both embrace multiple interpretations of meaning and study the interactions between them.

Making progress here might sound like a daunting task. However, just as Stenning and van Lambalgen made use of formal logic to make their point about the plurality of interpretation in the context of empirical work in psychology, formal logic can arguably be a great help also in making room for a distinct informational layer in theories of argumentation.

4. TOWARDS FORMALISATION OF THE INFORMATIONAL LAYER IN ARGUMENTATION

Formal logic is a useful tool for describing reasoning processes compactly, allowing for greater precision and more efficient management of complexity. As such, formal logic has little or no intrinsic normative content, but can be an invaluable help when formulating normative theories and reasoning based on exogenous parameters.

This is how formal methods are now being used in many areas of computer science, including with respect to the (computational) study of argumentation, as discussed for instance in (Prakken, 2006; Rahwan & Simari, 2009). Importantly, the formal turn in argumentation theory makes it very natural to explore more precisely what an informational layer in argumentation could look like.

Interestingly, work on the (modal) logical foundations of argumentation theory has already begun, with some interesting results obtained in (Grossi, 2010), based on the argument graphs from (Dung, 1995). So far, however, most of this work has been concerned with the standard reading of such graphs as an "objective" representation of the argumentation structure under consideration. The shift to an informational treatment, where an argument graph would only be one among a whole set of subjective interpretations, has not yet received much attention.

However, some formal work on this has been carried out based on the perspective of social choice theory. Specifically, multiple interpretations of argumentation structures have been thought to give rise to an aggregation problem, where the goal is to merge different representations (Coste-Marquis et al., 2007; Gabbay & Rodrigues, 2013; Dunne et al., 2012). The main focus so far has been on imposing principles of synthesis that appear "fair" and "rational" in lieu of the standard understanding in social choice theory.

Arguably, however, this way of thinking does not do justice to the processes of divergence and reconciliation that take place when agents negotiate about meaning through arguments. In order to improve on current models, the modal logic approach might be better suited, especially if argument graphs can be encapsulated in branching-time temporal logics like those of (Ben-Ari et al., 1983).

Such logics are well known from computer science, and they are also used for the study of actions and capabilities, both from a practical and a philosophical viewpoint, see, e.g., (Belnap et al., 2001; Alur et al., 2002; Broersen, 2011). To develop a formal theory of the informational layer in argumentation, a promising direction would be to merge these techniques with established techniques in formal argumentation. In this way, one could hope to use existing computational models in the style of Dung to represent the cognitive states of the arguers.

Then, argumentative deliberation could be modelled as a computational process, meaning that computational logic could be used to describe how the arguers' cognitive states, as well as the global state of the dialogue, develops in response to individual or collective actions (argument moves).

With this framework in place, insights from computational logic could be applied to describe and reason about various normative principles in procedural terms, by expressing constraints using temporal formulas that ensure certain desirable invariants, such as the absence of "vicious" cycles, prevention of deadlock, commitment to the agenda, faithfulness to previous assent etc. This would allow us to specify different constraints depending on the nature of the debate, and to reason formally about what different constraints imply and how they are related.

Importantly, receiving input from traditional (informal) argumentation theories could be of great assistance. It seems clear, in particular, that alongside the development of descriptive tools in formal logic, research should be done on how these tools can be used to facilitate normative analysis. If formal logicians and argumentation theorists jointly pursue the challenge of establishing a secure basis for studying the informational layer in argumentation, it is my belief that interesting insights would result, amenable to fruitful expression in both formal and informal languages.

5. CONCLUSION

This paper has argued that an explicit informational layer is needed in order to put normative theories of argumentation on a more secure descriptive footing. By giving the informational layer pride of place, argumentation theory should also become more relevant to related theories about reasoning in argumentative settings, especially the argumentative theory of reason formulated by Mercier and Sperber.

Importantly, this theory is pluralistic in the sense that it acknowledges the important synergies that can arise from different intuitions about what constitutes "good" or "rational" argumentation. Moreover, it emphasises how instances of "bad" argumentation can also have valuable social functions.

Arguably, the most interesting form of plurality is the one that arises from the informational layer, where people invariably interpret data differently in ways that influence how they reason and argue. Importantly, embracing this sort of plurality is not tantamount to endorsing relativism, nor should it fuel criticism of argumentation theory as such. However, it does lead to a call for a more nuanced take on some core notions, especially that of a fallacy.

The paper concluded by arguing in favour of a re-appreciation of formal logic, and by giving a brief sketch of how (computational) modal logics can be used to incorporate the informational layer in formal theories. Hopefully, informal logicians and argumentation theorists will take an interest in this work, to help ensure that it remains relevant to its purported object of study.

REFERENCES

Alur, R., Henzinger, T. A., & Kupferman, O. (2002). Alternating time temporal logic. *Journal of the ACM, 49*(5), 672–713.

Belnap, N., Perloff, M., & Xu, M. (2001). *Facing the Future: Agents and Choices in Our Indeterminist World.* Oxford: Oxford University Press.

Ben-Ari, M., Pnueli, A., & Manna, Z. (1983). The temporal logic of branching time. *Acta Informatica, 20*(3), 207–226.

Broersen, J. (2011). Making a start with the *stit* logic analysis of intentional action. *Journal of Philosophical Logic, 40*(4), 499–530.

Coste-Marquis, S., Devred, C., Konieczny, S., Lagasquie-Schiex, M. C., & Marquis, P. (2007). On the merging of Dung's argumentation systems. *Artificial Intelligence, 171*(10-15), 730–753.

Dung, P. M. (1995). On the acceptability of arguments and its fundamental role in nonmonotonic reasoning, logic programming and n-person games. *Artificial Intelligence, 77*, 321–357.

Dunne, P. E., Marquis, P., & Wooldridge, M. (2012). Argument aggregation: Basic axioms and complexity results. In *COMMA*, 129–140.

Fagin, R., Halpern, J. Y., Moses, Y., & Vardi, M. Y. (1995). *Reasoning About Knowledge.* Cambridge, Massachusetts: MIT Press.

Fish, S. (1996). Why we can't all just get along. *First Things, 60*, 18–26.

Fogelin, R. (1985). The Logic of Deep Disagreements. *Informal Logic, 7*(1), 3–11.

Freeman, J. B. (2012). Can argumentation always deal with dissensus? In F. H van Eemeren & B. Garssen (Eds.), *Topical Themes in Argumentation Theory: Twenty Exploratory Studies.* Berlin: Springer.

Gabbay, D. M., & Rodrigues, O. (2013). An equational approach to the merging of argumentation networks. *Journal of Logic and Computation, 24*(6), 1253-1277.

Grossi, D. (2010). On the logic of argumentation theory. In W. van der Hoek, G. A. Kaminka, Y. Lesperance, M. Luck, & S. Sen (Eds.), *AAMAS*, 409–416.

Kock, C. (2007). Norms of legitimate dissensus. *Informal Logic, 27*(2), 179–196.

Mercier, H., & Sperber, D. (2011). Why do humans reason? Arguments for an argumentative theory. *Behavioral and Brain Sciences, 34*, 57–74.

Paglieri, F., & Woods, J. (2011). Enthymemes: From reconstruction to understanding. *Argumentation, 25*(2), 127–139.

Prakken, H. (2006). Formal Systems for Persuasion Dialogue. *The Knowledge Engineering Review, 21*, 163-188.

Rahwan, I., & Simari, G., (Eds.) (2009). *Argumentation in artificial intelligence.* Springer.

Rescher, N. (1964). *Introduction to Logic.* New York: St. Martin's Press.

Stenning, K., & van Lambalgen, M. (2005). Semantic interpretation as computation in nonmonotonic logic: The real meaning of the suppression task. *Cognitive Science, 29*(6), 919–960.

Stenning, K., & van Lambalgen, M. (2008). *Human Reasoning and Cognitive Science.* Cambridge, Massachusetts: MIT Press.

van Eemeren, F. H., & Grootendorst, R. (2004). *A Systematic Theory of Argumentation: The Pragma-dialectical Approach.* Cambridge: Cambridge University Press.

Walton, D. (2007). *Dialog Theory for Critical Argumentation.* Amsterdam: John Benjamins.

Walton, D., Atkinson, K., Bench-Capon, T., Wyner, A., & Cartwright, D. (2010). Argumentation in the framework of deliberation dialogue. In C. Bjola, & M. Kornprobst, (Eds.), *Arguing Global Governance: Agency, Lifeworld and Shared Reasoning.* Taylor & Francis.

Walton, D., & Krabbe, E. C. W. (1995). *Commitment in Dialogue: Basic Concepts of Interpersonal Reasoning.* New York: State University of New York Press.

Wohlrapp, H. (1998). A new light on non-deductive argumentation schemes. *Argumentation, 12*(3), 341–350.

23

Familiars: Culture, Grice and Super-Duper Maxims

MICHAEL A. GILBERT
York University, Canada
Gilbert@YorkU.ca

Gilbert has introduced and expanded on the concept of "familiars". This talk argues that the concept is central to the idea of everyday argumentation. Using Grice's ideas on cooperation it is argued that cultures and fields may have differing rule sets dictated by meta-maxims or Super-Duper maxims. These must be considered for successful argumentation.

KEYWORDS: argumentation theory, cross-cultural argument, ethos, familiars, Gilbert, Grice, maxims, source credibility

1. INTRODUCTION

In 2007 I introduced the notion of "familiars" (2007b) and more recently (2014a, 2014b) I expanded on that concept. In this talk I want to continue that expansion and further argue that the concept is central to the idea of everyday argumentation, and especially to the ideas of audience and context.

Let me begin by fleshing out what I mean by the term "familiars." In its simplest sense familiars are the people we know and are familiar with. This, however, comes in a wide range of degrees. For some people, those who are closest to us, the degree of familiarity is extreme and interaction is frequent, as for example, family, friends, colleagues, and, for some, neighbours. Others are familiar but the relationship is less intimate. This group might include professionals such as your doctor and dentist, and tradespeople such as your auto mechanic, plumber, cleaners, and such like. In addition, there are a host of others who are also acquaintances: friends of friends, old mates, school chums, and so on. With all of these the familiarity means that the opening stage of argument, as described in pragma-dialectics (Eemeren & Grootendorst, 2004), has typically been previously settled.

D. Mohammed & M. Lewiński (eds.) (2016). *Argumentation and Reasoned Action: Proceedings of the 1st European Conference on Argumentation, Lisbon, 2015. Vol. II, 431-438.* London: College Publications.

It is important to note that having a stage settled does not mean it cannot be re-opened. Andrew and Simon might, for example, sit down in the pub and immediately begin their usual argument about hockey. They need neither the confrontation stage nor opening stage and just begin. But if Andrew suddenly becomes heated Simon might interject with something like, "Hey, man, chill! What's going on? This is hockey we're talking about." So on the one hand the appeal might be to re-open the confrontation stage and see if somehow the topic was changed, or, on the other, to go to the opening and see if the rules have been altered. The heightened emotion might, for example, indicate a complete change of subject (Gilbert, 1997).

When an individual is not in a close circle the argument partner still has information available to her. It is rare that we know nothing at all about someone, and even rarer that we undertake an argument with such an individual. Figure 1 demonstrates the various circles of familiarity.

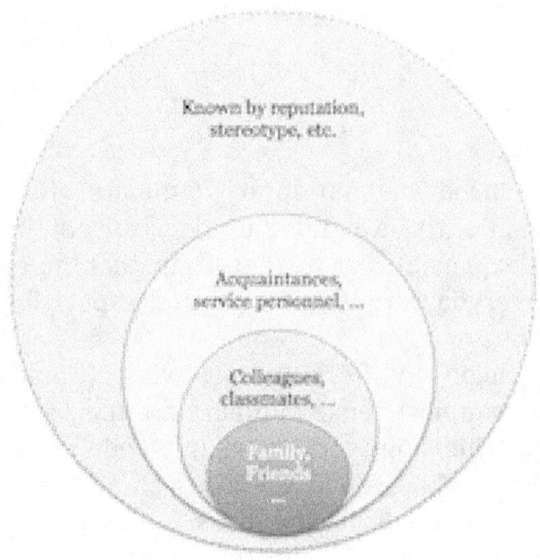

Figure 1 – Circles of familiarity.

The outer circle is quite large and has a dotted line rather than a solid line as its border to indicate it lack of strict limits. It is within this circle that one is more likely to spend time in the confrontation and opening stages of argument. However, even here there is information that can be brought to bear to orientate oneself to the dissensual situation. First of all, how do you know this person? Is he a friend of a friend? Does he work in your field? What is his background? Does he come from an argumentative culture or an avoidance culture? We know, or imagine we know, different things about different cultures.[1] All of these, while not obviating the need for the first two stages will certainly inform them.

A familiar with whom one is arguing is an audience. Perelman (1982) points out that a protagonist and her audience must have certain elements in common. These elements include beliefs and values, facts and loci, goals and objectives, and without some minimal foundation an argument cannot get off the ground. We make assumptions regarding our audience and use them to initiate the encounter. I quote myself:

> Whenever you speak with someone you adopt a particular approach based on who you believe they are and what you imagine their goals to be. If you encounter someone who sounds like a foreigner and has a bewildered look, you will assume your "help the tourist" mantle: you'll put on a smile and try to look unthreatening and helpful. If your manager is talking to you, you will look interested and keen so as to portray your usefulness to her. (Gilbert, 2014, p. 78)

Perelman's point is also emphasized by Toulmin (1958) when he talks about fields. Familiars, even simple cultural familiars, make assumptions about what is and is not acceptable behaviour, what language may be used, and who may argue with whom.

Underlying all this and, perhaps, providing the outside limit, is a Gricean perspective on the fundamentals of communication. That is, so long as (1) the Gricean Maxims are being followed; and (2) The Maxims are recognizable to both communicants, then the conversation can proceed with at least a minimum degree of familiarity. Requirement (1) is familiar, but I am making it into a necessary but not sufficient condition. I do this because Grice's maxims must be adapted to field and cultural norms and expectations. Gricean Maxims as expressed in his essay (Grice, 1975) are geared to British middle-class values and expectations. I will refer to these maxims as GM-B. Other cultures and, possibly, fields require adjustment. Reasoning, and especially the selection of an appropriate reasoned response in a given situation depends on knowledge of the participants, their particular interrelationship, and the ways in which responses, though reasonable, may be socially or even morally enjoined. In argument we expect certain

---

[1] This is not to endorse stereotyping. The fact is that we use whatever information is available to us in order to argue well and successfully. However, as interaction commences the individual characteristics will supersede whatever presuppositions we might have had for a good arguer.

responses and make certain assumptions based on our knowledge of our interlocutor. This comes down to what might be considered a Meta Cooperative Principle which, while it can go awry, usually does not. The point with respect to familiars, however, is that the way we reason and the argument choices we made are frequently governed by the social interactive assumptions guiding our reasoning.

To underline the importance of cultural variation in Gricean Maxims, I provide examples. The first concerns the Supermaxim of Quality: "Supermaxim: Try to make your contribution one that is true" (Grice, 1975, p. 27). I spend a good deal of time in Mexico, and on several occasions have asked for directions. Until I learned the subtleties of Mexican culture, I frequently followed directions that were, simply, made up - invented out of whole cloth. Once, when searching for a small museum I asked someone at a cultural centre for its location. He was about to respond when a colleague entered. "Oh," he said, "good, Pedro will know. I was going to lie." Another instance had me in a taxi in a small town when the driver, who was not local, asked someone for the location of a specific street. When the woman finished her answer, the driver said, "She doesn't know," and headed out in a different direction. I learned that while the Supermaxim of GM-B was always tell the truth, the Super-Duper maxim of GM-M, if you will, was never be rude to a stranger, and not offering assistance is rude. (The trick, by the way is simple: if someone thinks before answering, say thanks and ask someone else.)[2]

Other examples involve the role of disagreement in Jewish culture. Schriffin (1984) investigated and "proved" what everyone raised Jewish knows, viz., that arguing is not merely a way to resolve a disagreement, but a method of being sociable. This flies in the face of maxims regarding saying the minimum necessary and remaining relevant. Volatility and drama are important parts of Italian arguing, and the rules for arguing in Japan are highly structured depending on age and status. My Asian students often explain that they will never disagree with an elder even if they know the older person is wrong. Again Super-Duper maxims are in play. Directness also varies greatly depending upon culture. In many Canadian First Nations a question may be answered by a story rather than directly. For that matter, directness in answering is often considered a masculine response as opposed to a more encompassing and relational feminine response. "Getting to the point," may, in some contexts be inappropriate or rude. It pays to

[2] I should also add that this is by no means universal, and there may well be any number of Mexicans who will say that they regretfully cannot help you.

remember that Grice was an upper middle class white male, and being succinct and direct are important values.

The outside circle of familiars is the one in which we know the least about our individual interlocutor and, hence, must rely on less precise information. Jill may know someone's background culture, field of work or business, gender, age and so on, and these bits of information may help her create an initial expectation, but in these cases the opening stage of the interaction will play a more significant role than in situations involving those in closer circles. If Jill knows she is having a meeting with someone in an outer circle she may well do research, ask people, and/or, at the very least, be open to rules and values she might not normally expect.

There are many other aspects relevant here, but time does not permit their exploration. For example, work spearheaded by McCroskey and his associates (McCroskey & Dunham, 1966; McCroskey & Young, 1981) concerning ethos, now usually referred to as "source credibility," provides common ways for evaluating familiars of all degrees. More recently, some of this can be seen in work of O'Keefe (O'Keefe, 2002, 2009). The idea of source credibility can then be slotted into the Gricean framework discussed above to provide a loose model of an extension of source credibility as applied in dynamic and actual situations. How we evaluate sources will also be dependent, to one degree or another, on cultural and contextual schemes, thereby making meta adjustments to the Gricean Maxims.

Among the approaches to Argumentation Theory most currently available, none should have an issue with the idea of familiars. Obviously, a rhetorical approach not only uses it, but also embraces it insofar as the concept of audience is central. The situation is not so straightforward with respect to the dialectical approach. This approach has a set of rules designed to promote *good and reasonable* argument as its foundation. The difficulty, of course, is the definition of "good and reasonable." The rules of both Informal Logic [IL] and Pragma-dialectics [PD] are intended to foster the proper conduct of arguments and aid in their leading to proper conclusions. The way they are structured is analogous to Grice's GM-B, that is, they are largely formed for and by white western males who have a particular approach and attitude toward argumentation. This is fine and well so long as the participants agree that the rules put forward are the ones to play by.

Cultures, fields and audiences differ one from another in many ways, and among those ways are some that work against a "standard" set of rules upon which all can agree. Even within some fields certain basal assumptions are rejected by other fields. Willard (1983) uses the example of cost-benefit analysis, CBA, which has a ground assumption that everything can be converted into a monetary value. When Jack's dog Topsy is run over and killed he is extremely upset and angry. The insurance company of the driver is not; they simply take the dog's age and breed into consideration and come up with a figure. As Topsy was mature and a mixed breed, Jack feels cheated. But the ground assumption of CBA is what runs the insurance world, and he has no chance. Note that this argument between Jack and the insurance company has no opening stage: it is all pre-set by law and contract. This means that the interaction is not a critical discussion, and, presumably, falls outside the realm of pragma-dialectics.

In PD and IL the assumption is that one will accept the ground rules, but that does not always happen. It will happen with familiars in all but the most outer circles, and that is why I say that familiars are not anathema to dialectical theories. When it comes to cultural assumptions, however, the situation becomes murkier. I have argued (Gilbert, 2007a) that some cultures reject even basic principles such as the Law of Non-contradiction. Though this has been opposed by some, (Siegel, 2007), it nonetheless poses difficulties for dialectical approaches for the simple reason that it actually happens. To resolve this, there needs to be a consideration of meta issues regarding rule sets and an awareness of what I have referred to as Super-Duper maxims.

2. CONCLUSION

In conclusion, there really is no need for tension between the dialectal approaches and the idea of familiars. If PD and/or IL take the position that the rules as stated are impervious to change, then the many situations in which protagonists insist that their rules are different means that those contexts are not arguments. A la positivism's rejection of non-testable hypotheses when Jill and Jack argue in a culture that defies or rejects classical rules, they are, perhaps, ruminating or noodling or chatting or what have you. This seems like a strange and certainly Draconian approach, but it is one solution nonetheless. Alternatively, the dialecticians might consider the opening stage to include such basic ideas as the cultural or contextual milieu so as to set up the parameters of which rules, fallacies and techniques are and are not acceptable. Perhaps it is best to consider the stage as a meta-opening stage or a preamble stage in order to determine whether or not it is possible to proceed.

Brockriede's sixth characteristic expounded in "Where is Argument?" states that "a frame of reference shared optimally" is required for argument to occur (Brockriede, 1990). I'm not sure about "optimally," but we must have some familiarity with the argumentative values, rules and presumptions made by our interlocutors. We must become familiar with unfamiliars, understand their maxims, background and culture in order to have a valuable dissensual interaction.

REFERENCES

Brockriede, W. (1990). Where is argument? In R. Trapp & J. E. Schuetz (Eds.), *Perspectives on argumentation: Essays in honor of Wayne Brockriede* (pp. 4-8). Prospect Heights, Ill.: Waveland Press.

Eemeren, F. v., & Grootendorst, R. (2004). *A systematic theory of argumentation: The pragma-dialectical approach.* New York: Cambridge University Press.

Gilbert, M. A. (1997). *Prolegomenon to a pragmatics of emotion.* Paper presented at the 1st International Conference of the Ontario Society for the Study of Argumentation, St. Catherine's, ON CA.

Gilbert, M. A. (2007a). Informal logic and intersectionality. In H. V. Hansen & R. C. Pinto (Eds.), *Reason Reclaimed: Essays in Honor of J. Anthony Blair and Ralph H. Johnson*: ValePress.

Gilbert, M. A. (2007b). Natural normativity: Argumentation theory as an engaged discipline. *Informal Logic, 27*(2), 149-161.

Gilbert , M. A. (2014a). *Arguing with people.* Calgary, Alberta: Broadview Press.

Gilbert , M. A. (2014b). *Rules is rules: Ethos and situational normativity.* Paper presented at the International Society for the Study of Argumentation 2014, Amsterdam.

Grice, P. H. (1975). Logic & Conversation. *Studies in the way of words.* Cambridge, Mass.: Harvard University Press.

McCroskey, J. C., & Dunham, R. E. (1966). Ethos: A confounding element in communication research. *Speech monographs, 33*(4), 456-463. doi: 10.1080/03637756609375512

McCroskey, J. C., & Young, T. J. (1981). Ethos and credibility: The construct and its measurement after three decades. *Central States Speech Journal, 32*(1), 24-34. doi: 10.1080/10510978109368075

O'Keefe, D. J. (2002). Communicator Factors. In D. J. O'Keefe (Ed.), *Persuasion: Theory & research* (2nd ed., pp. xvi, 365 p.). Thousand Oaks, CA: Sage Publications.

O'Keefe, D. J. (2009). Persuasive effects of strategic maneuvering: Some findings from meta-analyses of experimental persuasion effects research. In F. H. Eemeren (Ed.), *Examining Argumentation in Context: Fifteen studies on strategic maneuvering.*

Perelman, C. (1982). *The realm of rhetoric.* Notre Dame, Ind.: University of Notre Dame Press.

Schiffrin, D. (1984). Jewish argument as sociability. *Language in Society, 13*(3), 311-335. doi: 10.2307/4167542

Siegel, H. (2007). Multiculturalism and argumentative norms. In H. V. Hansen & R. C. Pinto (Eds.), *Reason Reclaimed: Essays in Honor of J. Anthony Blair and Ralph H. Johnson*: ValePress.

Toulmin, S. (1958). *The uses of argument.* Cambridge: Cambridge University Press.

Willard, C. A. (1983). *Argumentation and the social grounds of knowledge.* University, Ala.: University of Alabama Press.

24

What (the Hell) is Virtue Argumentation?

G.C. GODDU
University of Richmond, USA
ggoddu@richmond.edu

The purpose of this paper is (i) to determine the nature of virtue argumentation—to determine what aspect of argumentation the theory is trying to explain and (ii) to pose some challenges that such a theory needs to overcome.

KEYWORDS: virtue argumentation, virtue theory

1. INTRODUCTION

Virtue argumentation has become a hot topic in argumentation theory of late. The theme of the 2013 OSSA Conference was Virtues of Argumentation and Andrew Aberdein's virtue argumentation bibliography[1] continues to grow quite quickly. Given that argumentation is an intellectual activity, at least some of the intellectual virtues should apply. Yet one might still wonder what the relevant virtues (or vices) are, to what degree various argumentative phenomena can be explained by appeal to intellectual or argumentative virtues (or vices), or whether argumentation has any virtues (or vices) all its own.[2]

As Aberdein's bibliography attests, there is significant work on some of these questions. But in this paper, I shall try to answer a much more fundamental, perhaps unvirtuously phrased, question—what the hell is virtue argumentation?

2. VIRTUE THEORIES

[1] See *http://my.fit.edu/~aberdein/VirtueBiblio.pdf*.

[2] That last question arose in a brief conversation I had with Andrew Aberdein while listening to another talk on virtue argumentation and was at least the catalyst for this paper.

D. Mohammed & M. Lewiński (eds.) (2016). *Argumentation and Reasoned Action: Proceedings of the 1st European Conference on Argumentation, Lisbon, 2015. Vol. II, 439-448.* London: College Publications.

To begin to answer that question, I start with some virtue theories I do understand.

First, Virtue Ethics: Ethics is generally concerned with what we ought to do. Different ethical theories provide different answers to the question—"what ought we to do?" According to utilitarians we ought to do that which maximizes, in some relevant sense, overall utility. According to deontologists we ought to do that which is in accord with universal moral laws. According to virtue ethicists we ought to do that which is in accord with exemplifying the relevant virtue or virtues. So when we ask why the good Samaritan ought to stop and help the traveller, the utilitarian will say because doing so maximizes overall utility, the deontologist will say because doing so is in accord with a moral law or maxim such as "treat others as you would wish to be treated", while the virtue theorist will say because doing so exemplifies the virtues of charity and kindness.

But the virtue theorist might go on to say that this characterization misses a major strength of virtue ethics—the focus on the agent rather than the act. For example, Aberdein, (2007, p. 17) writes: "A substantial innovation of most virtue theories is that they are explicitly agent-based rather than act based." Interestingly, this "strength" has also been pointed to as a weakness—the virtue ethicist being charged with changing the subject. Both sides are mistaken here. Firstly, all three theories can give coherent answers to the questions: "What ought we do?" and "What is a good person?" Unsurprisingly the utilitarian will answer the latter by saying the good person maximizes overall utility, while the deontologist says the good person acts in accord with the universal moral laws, while the virtue theorist says the good person exemplifies the relevant virtues. Secondly, we fully expect the goodness of acts and the goodness of persons to be intertwined—good people generally do good acts and doing good acts is generally part and parcel of being a good person.

So whence the notion that virtue theories are agent based rather than act based (or at least not agent based)? There is a difference amongst the three theories—virtue ethics accounts for the goodness of acts and the goodness of persons in terms of the character of persons; utilitarians in terms of the utility state of the whole universe; and deontologists in terms of conformity or non-conformity to universal moral laws. So virtue theories, unlike most versions of the other two theories, account for the target phenomena—the goodness of acts or persons, in terms of something generally internal to agents, whereas the latter account for the target phenonema in terms of external features or standards.[3] But without a good argument that the explanation of the target phenomena (good acts or good persons or both) *ought* to be either internal or external, this difference speaks neither in favour nor against any of the theories.

Next, Virtue Epistemology: Epistemology is generally concerned with what we ought to believe or know. Different epistemological theories provide different answers to the questions—what ought we believe or what do we know? According to the evidentialist, say, we ought to believe that which is supported by the evidence, where most often evidential support is a matter external to the agent. According to the virtue epistemologist we ought to believe that which is acquired in an intellectually virtuous way. Again the two theories differ in how to account for the target phenomena, justified believing, relative to whether the explanation is internal or external to the agent. And again without a good argument that the explanation ought to be external or internal the difference speaks neither in favour nor against either theory.[4]

Similar remarks apply to virtue jurisprudence[5] and other virtue theories, but I turn now to virtue argumentation.

3. VIRTUE ARGUMENTION?

So how then to understand virtue argumentation? Argumentation is generally concerned with arguments. Different theories provide different answers to the question—what is a good argument? But now we have a problem that does not immediately emerge for either of our previous virtue theories—what exactly is the target phenomena? "Argument" is ambiguous[6] between, at least, things such as

---

[3] The external/internal distinction here is rough at best. For example, a Kantian might say that the source of universal moral law is the self-legislation by autonomous moral agents and so something internal to the agent after all. My thanks to colleague Brannon McDaniel for pointing out this possibility.

[4] The internalism/externalism debate in epistemology, i.e. whether an agent needs to be appropriately mentally connected to whatever makes a belief that P justified in order for the agent to be justified in believing that P cuts across the two theories sketched above. For example, an evidentialist-internalist would say that the agent must be appropriately aware that the evidence in fact supports the belief in order to be justified, in which case the justification for believing is partly external (the evidence supporting the belief) to the agent and partly internal (the awareness that the evidence supports the belief).

[5] See, for example, Solum (2003).

[6] But see Simard-Smith and Moldovan (2011) for a discussion that "argument" is not ambiguous.

reason/conclusion complexes and acts such as, say, trying to convince someone there is a problem determining the target phenomena.[7]

I doubt that the target phenomena is *good argument* in the sense of good reason/conclusion complexes; that is, I doubt that virtue argumentation is meant as a competitor to such claims as:

> x is a good argument iff x is sound
> or
> x is a good argument iff x's premises are acceptable, relevant, and independently support the conclusion
> or
> x is a good argument iff x instantiates no fallacy.

What would such a competitor be like? Perhaps:

> x is a good reason/claim complex iff x is put forward by a virtuous arguer? (or would be put forward in situation y by the ideal virtuous arguer?) (or the presenter of x has argued virtuously?)

None of these strike me as very plausible and, with perhaps the exception of some of Aberdein's (2014) comments concerning *ad hominem*, I see little evidence that virtue theorists are trying to analyse good reason/claim complexes.

Regardless, here are some reasons to think virtue theorists ought not be in the business of explaining the goodness of reason/claim complexes. Firstly, if the explanation for why a virtuous arguer never knowingly puts forward fallacious arguments is that such arguments violate conditions on truth, or relevance, or sufficient support, then the goodness of the argument is not really being explained by the virtue of the arguer, but rather the virtue of the arguer is in part being explained by conformity to these external criteria on arguments. Indeed, I suspect that an arguer who comes to know that a proposition P is false and yet still uses it as a premise is not being a virtuous arguer, nor is one who knows that the premises do not sufficiently support the conclusion and

---

[7] Belief is often considered ambiguous between the content of the belief and the act of believing—so the content being justified might be different than the act being justified. (See also Note 4). Hence, epistemologists need to take care that they are talking about the same kind of justification. Presumably the virtue epistemologists are interested in justifying the agent's believing, even if part of that justification involves the content's logical relationship, with other things the agent knows.

yet presents that argument anyway. But it is the failing to respect these external criteria that is making the arguer non-virtuous. In other words, the satisfaction of, or conformity with, the external criteria is explaining, at least in part, both the goodness of the argument and the virtue of the arguer.

Secondly, the same reason/claim complex put forward to relevantly similar audiences in relevantly similar circumstances by a virtuous arguer on the one hand and a lucky, non-virtuous arguer on the other, will not affect the goodness of the reason/claim complex. The virtuous arguer has, we can suppose done his or her homework and knows that the premises are true and can be adequately defended for the given audience and that the premises adequately support the conclusion to the satisfaction of the audience in the given situation. The lucky arguer has had the same argument pop into his or her head as a result of staring too long into the fire after too much to drink, but does nothing to actually check the truth of the premises or the adequacy of their support and yet, fortuitously, offers that argument to the appropriate audience in the appropriate circumstances. The non-virtuousness of the lucky arguer will not make his or her arguing fail in the circumstance precisely because the goodness of the reason/claim complex itself is not being affected by the arguer (and the arguer was lucky enough to present the argument to the right audience in the right circumstances.)[8]

Finally consider Linda Zagzebski's (1996, p. 79) claim about virtue theories:

> By a pure virtue theory I mean a theory that makes the concept of a right act derivative from the concept of a virtue or some inner state of a person that is a component of virtue. This is a point both about conceptual priority and about moral ontology. In a pure virtue theory the concept of a right act is defined in terms of the concept of a virtue or a component of virtue such as motivation. Furthermore, the property of rightness is something that emerges from the inner traits of persons.

Virtue theories are supposed to explain some normative property of acts in terms of the virtues of the actors. Agents are morally justified in doing x or epistemically justified in believing x iff the doing or believing emerges from the relevant virtues of the agent. Hence, to keep the parallel, the normative property explained should be of some kind of act, and the only relevant act is the arguing. Arguings, then, are good if they emerge from the argumentative virtues of arguers.

---

[8] See Bowell and Kingsbury (2013) for a similar argument.

Indeed, I cannot make sense of most of Daniel Cohen's (2009, 2013a, 2013b) uses of the word 'argument' unless I read it as "arguing". For example, Cohen's (2013b) suggestion that the concept of argument be expanded to include the context is perfectly reasonable if we are talking about arguings, since arguings are events that take place at certain times in certain places—times and places are part of the identity conditions of that act or event.[9]

Note that if the target is good arguings, virtue argumentation is going to be irrelevant to those interested in determining the nature of good arguments rather than arguings—it just isn't a competitor for answering the former question. Nor should those interested in good arguments automatically assume that the answer to the good argument question answers the good arguings question. Bowell and Kingsbury, (2013, p. 23) for example write: "we think that what makes it the case that an arguer has argued well is that they have presented an argument that is good." But consider the following:

> *Tetsugakusya wa kenmeidesu. Miyamoto Musashi wa tetsugakusya. Dakara, Miyamoto Musashi wa kenmeidesu.*

Even supposing the argument expressed is a good argument, the arguing was, for most readers, I suspect, horrible. Nor would the presentation of a perfectly non-trivial sound argument that was interspersed with vitriol and unflattering remarks on the intellectual capacity of the audience be a good arguing. At the very least there is more to good arguing than just presenting a good argument, just as there is more to a good poetry reading than merely presenting a good poem—good poems read in dull monotones are not good poetry readings.

Even if theorists concerned with good arguments are not, directly at least, competitors with those interested in the goodness or

---

[9] Note that his concern about the "virtues of venues" doing irreparable damage to the concept of virtue as an inculcated habit of character is misplaced since the goodness of the arguing is what is being analysed in terms of the virtues of the arguers—*recognizing* the appropriateness of the venue is presumably a virtue of the arguer (as is recognizing the truth of premises, etc.) Hence, the fact that the virtuous arguer needs to be attuned to normative properties such as the goodness of arguments or the appropriateness of venues, does not require extending the concept of virtues to either venues or arguments. But see the second worry in section 4 for a new problem that might emerge if one takes this route.

badness of arguings, I suspect there are some genuine competitors. For example, pragma-dialecticians might hold that an arguing is good if all participants do not violate the pragma dialectical rules. Others might hold that an arguing is good if it convinces (or perhaps rationally convinces) the target audience of the conclusion. I shall not try to adjudicate amongst these competitors here. Instead I will conclude with two potential worries for virtue argumentation construed in terms of explaining the goodness of arguings.

4. TWO WORRIES

Worry One: What exactly is the goodness the virtue theorist is trying to analyse? Cohen (2013a, 2013b) talks of "fully satisfying" arguments. Mathematicians talk of "beautiful" or "elegant" proofs. But satisfaction is a psychological property and beauty and elegance may be in the eye of the beholder. Mathematicians, for example, often think logic proofs are inelegant, whereas logicians tend to think mathematical proofs are gappy. Hence, whatever this goodness is, one might worry it has little to do with the goodness or badness of the reasoning going on and ultimately it has to be the reasoning that is the base concern in argumentation theory.

Here is the worry put another way. Good arguments, while perhaps neither necessary nor sufficient for good arguings, at least constrain good arguings. Knowingly presenting a bad argument seems, prima facie, to generate a bad, disingenuous, unvirtuous arguing. It may be that all parties, acting in good faith, with the best evidence available to them at the time, in fact use bad arguments. But the parties are not blameworthy for such uses and we might still say the arguing is the best it could be in the circumstances. But suppose the arguer does give a good argument. The worry is that outside of personal preference or satisfaction or judgments of beauty or elegance concerning the "giving", there is nothing over and above the good argument to the good arguing. In the case in which the arguer gives a bad argument, but it was the best that could be done in the circumstances and the arguer presents well, we might say at least that the arguer argued as virtuously as was possible and that the arguing was as good as it could be given the circumstances. But in either case we judge that the arguer/audience/judges performed or presented well—but why, as an argumentation theorist, should I care about the goodness of the performance?

I grant that the performance has to be at least understandable to hope to achieve even the minimal goal of truth-propagation. But that is true of communication in general—to achieve the desired goal, most of the time, the communication needs to be understood. Hence, my Japanese syllogism above fails to meet even this minimal communication requirement. But once we have met this minimal communication requirement, what other than "performance" satisfaction are we talking about when we talk about good arguings?

Worry Two: The second, initially less important, worry is that the virtues of the virtuous arguer are just the standard intellectual virtues of open-mindedness, the right balance between credulity and skepticism, perseverance, intellectual courage, confidence in reason, conscientiousness, understanding, wisdom, honesty...etc. (See Aberdein, 2010). But since arguing is an intellectual exercise it is no surprise that arguing is under the purview of the intellectual virtues. Of course we want to instill open-mindedness, etc. in arguers since those intellectual virtues seem conducive to the general intellectual goals of acquiring truth, understanding, and wisdom. But these virtues apply to other intellectual activities such as experimenting or analysing. So beyond noting the fact that as an intellectual activity, arguing will exhibit the general qualities of intellectual activities, why care about the virtues in argumentation theory?

I finish by considering some possible answers. Firstly, isn't the fact that good arguing, as an intellectual activity, exhibits the intellectual virtues, enough to merit the desire to understand how good arguing exemplifies those virtues? Secondly, perhaps the intellectual virtues combine or weight differently in the case of argumentation versus experimentation or analysis. Perhaps experimenters need more patience than arguers. Thirdly, perhaps particular virtues manifest themselves differently in different intellectual contexts. Cohen (2009) for example argues that epistemic open-mindedness is different than critical thinking open-mindedness. I am sceptical of Cohen's particular example since the virtue could be described as "appropriate open-mindedness" and that virtue seems applicable in both cases—the epistemologist who lets his or her open-mindedness hinder the acquisition of knowledge is not being appropriately open-minded.

But "appropriateness" raises a new worry. If being appropriately open-minded is recognizing when to be more open minded (and when to stop) or less open-minded (and when to stop) and this recognition requires being attuned to certain qualities in the world, i.e. when you question this, but not that you get it right, isn't it ultimately the rightness achieved that is explaining both the goodness of the process and the goodness of processor? But then it is not the virtues themselves that are of interest, but rather the right outcomes in the world that those virtues track. Here is the problem put another way.[10] Consider an omniscient and omnipotent arguer who has to argue with us mere mortals. Presumably such an arguer can argue perfectly, and yet exhibit no virtues, no inculcated habits of mind. So whatever explains the goodness of the being's arguings, it is not the argumentative virtues.

Perhaps whatever is explaining the arguing's goodness in the omniscience case, whatever it is that the omniscient/omnipotent being is tracking, is also explaining the goodness in more everyday cases. Unfortunately, as mere mortals, we need to work to inculcate habits of mind that increase the chances we will track these right outcomes. Without doubt inculcating the intellectual virtues in our students is a worthwhile endeavour. But I still wonder whether it is truly the virtuousness of the arguer that is explaining the goodness of the arguing.

5. CONCLUSION

Virtue argumentation, to be a kind of virtue theory like virtue ethics or virtue epistemology, is concerned with the goodness of acts of arguing and not the goodness of arguments (reason/claim complexes). But now the virtue argumentation theorist has several questions to answer. For example, what is this goodness of arguings and is it anything interesting over and above the goodness of arguments? Is it really the case that the goodness of arguings is explained by the virtuousness of arguers? Perhaps virtue argumentation theorists can answer these questions, but until they do I see no reason to shift argumentation theory's focus on good arguments.

ACKNOWLEDGEMENTS: An earlier version of this paper was presented at the European Conference on Argumentation (2015). I thank the audience of that presentation for helpful discussion and questions. I also thank my colleague, Brannon McDaniel, for useful comments on earlier versions.

---

[10] The problem, a version of the Euthyphro problem, is a problem for all virtue based accounts of good x. Essentially is x good because the virtuous person does it or does the virtuous person do it because x is good.

REFERENCES

Aberdein, A. (2007). Virtue argumentation. In F.H. Van Eemeren, et al. (Eds.), *Proceedings of the Sixth Conference of the International Society for the Study of Argumentation*, vol. 1, (pp. 15-19). Amsterdam: Sic Sat.

Aberdien, A. (2010). Virtue in argument. *Argumentation, 24*(2), 165-179.

Aberdein, A. (2014). In defence of virtue: The legitimacy of agent-based argument appraisal. *Informal Logic, 34*(1), 77-93.

Bowell, T., & Kingsbury, J. (2013). Virtue and argument: Taking character into account. *Informal Logic, 33*(1), 22-32.

Cohen, D. (2009). Keeping an open mind and having a sense of proportion as virtues in argumentation. *Cogency, 1*(2), 49-64.

Cohen, D. (2013a). Skepticism and argumentative virtues. *Cogency, 5*(1), 9-31.

Cohen, D. (2013b). Virtue, in context. *Informal Logic, 33*(4), 471-485.

Simard-Smith, P., & Moldovan, A. (2011). Arguments as abstract objects. *Informal Logic, 31*(3), 230-261.

Solum, L.B. (2003). Virtue jurisprudence: A virtue-centered theory of judging. *Metaphilosophy, 34*(1-2), 178-213.

Zagzebski, L. (1996). *Virtues of the mind*. Cambridge: Cambridge University Press.

25

The Pragmatic Force of Making Reasons Apparent

JEAN GOODWIN
Iowa State University, USA
goodwin@iastate.edu

BETH INNOCENTI
University of Kansas, USA
bimanole@ku.edu

Making arguments makes reasons apparent. Sometimes those reasons may affect audiences. But over-emphasis on effects distracts from other things that making arguments accomplishes and thus fails to account for its pragmatic force. We advance the normative pragmatic program on argumentation through case studies of how early advocates for women's suffrage in the US made arguments to demonstrate that they were persons capable of making reasons apparent and that their actions were reasonable.

KEYWORDS: Argumentation, argument, enactment, function of argument, normative pragmatics, women's suffrage

There are more things in heaven and earth, Horatio, than are dreamt of in your philosophy.

1. INTRODUCTION

Over the past 20 years there has been increasing focus on pragmatic theories—those which, in the words of the conference theme, approach argumentation "as a mode of action." In Figure 1 we attempt to represent in a neutral way the assumptions shared by many pragmatic theories. At its most basic, a pragmatic theory asks us to understand what O'Keefe (1982) termed argument_1s—the premise/conclusion units people exchange with each other—by placing them in the context of the argument_2s in which they occur—the transactions between speakers and audiences.

D. Mohammed & M. Lewiński (eds.) (2016). *Argumentation and Reasoned Action: Proceedings of the 1st European Conference on Argumentation, Lisbon, 2015. Vol. II, 449-462.* London: College Publications.

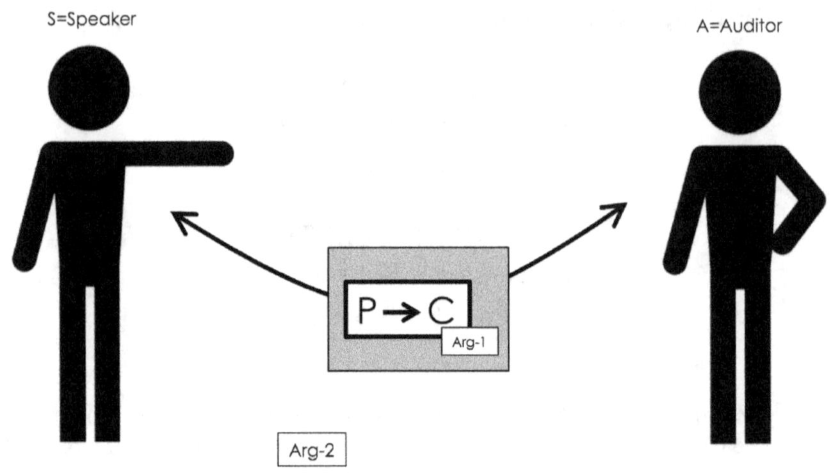

Figure 2 - Elements of a pragmatic theory of argumentation

Functionalist theories of argumentation are a subset of pragmatic theories—probably the dominant one. As Dima Mohammed has pointed out in her recent, incisive article (2015), functionalist theories pick out some argumentative activity—either the collective interaction or some individual activity going on within it—and assert that it has an "intrinsic goal." As Mohammed further points out, the intrinsic goals that have been put forward are quite varied. For example, perhaps S's argumentative activity is to:

- justify C to A (Bermejo-Luque, 2010, perhaps)
- invite A to infer C (Pinto, 2001, since partially reconsidered in Pinto, 2009)
- increase A's adherence to C (Perelman & Olbrechts-Tyteca, 1969; the view often ascribed to rhetoric)
- rationally persuade A that C (Johnson, 2000)
- critically test C, in order to induce A to accept C as a standpoint, thus rationally resolving the disagreement between A and S (pragma-dialectical theories generally)

There are important (and unresolved) differences among these views. But all share a focus on A, the audience addressed. S's argumentative activity is aimed to change either A's cognitive attitude towards C (e.g., infer, adhere, be persuaded) or to change his "external" activities with respect to C (e.g., openly accept as a standpoint). In this view, a pragmatic theory of argumentation investigates the interpersonal conditions and activities that have to take place in order for the specified function to be realized; these lead to norms for argumentation.

We have critiqued many aspects of functionalist theorizing (Goodwin, 2001a, 2001b, 2007a, 2007b; Innocenti, 2005, 2006, 2011a, 2011b). Here we want to continue to urge an enlargement of view by shifting attention from what making an argument_1 does to an audience to an even more basic "mode of action." In this paper, we will present two case studies demonstrating how making argument_1s accomplishes important tasks that have nothing to do with changing an audience's relationship to C. Instead, it is simply S's putting P→C out there in the world that has force; she does something worth doing by making P→C apparent, by showing P→C, by making P→C manifest, by making P→C explicit—that is, in the shorthand we will use in this paper, by *making a reason apparent.* In each case study we will see:

1. S makes a reason apparent.
2. S cannot expect A to accept (etc.) C.
3. S does something by making a reason apparent.

We will draw the two case studies from the early women's suffrage movement in the US: one from 1848, one from 1869-1875. While women in the mid-19th century were achieving important gains, little advance was made or reasonably could be made on obtaining the right to vote. As we will show, throughout the period deeply entrenched beliefs worked against them. One representative anecdote gives the flavour of the dominant beliefs suffrage advocates faced. As late as 1915, a *New York Times* editorial called for readers to vote "no" in a referendum on women's suffrage in part because women "lack the genius for politics" and, while men "vote according to judgments founded on observation and knowledge acquired in the pursuit of their daily business," women "would inevitably attempt to decide such matters empirically or emotionally" ("The woman suffrage crisis," 1915).

In these circumstances, it was not reasonable to expect that making argument_1s would change audiences' relationships to C, the claim that women should be able to vote. Nevertheless, advocates used argument_1s, and used them well. We close this paper by drawing out the implications of these cases for pragmatic theories of argumentation.

2. CASE 1: 1848: THE FIRST WOMEN'S RIGHTS CONVENTION AND ITS AFTERMATH

The first women's rights convention met in Seneca Falls, where Elizabeth Cady Stanton lived and worked for reform movements including women's rights and the abolition of slavery while caring for her five children. Organizers put a brief notice in the local newspaper that a women's rights convention would be held in a local church, and about 300 people attended, including at least 40 men (Flexner & Fitzpatrick, 1996). Among the charges against men in the famous "Declaration of Sentiments" that emerged from the event was a complaint about suffrage: "He has never permitted her to exercise her inalienable right to the elective franchise."

Public opinion, even among some convention organizers (Wellman, 2004), viewed women's suffrage as preposterous, however. Leading activist Lucretia Mott reportedly said upon hearing Stanton read the complaint about suffrage in the "Declaration," "Oh Lizzie! If thou demands that, thou will make us ridiculous! We must go slowly" (qtd. in Lutz, 1940). In the first half of the 19th century it was "obvious" that women ought to be excluded from the public sphere, including both voting and public speaking. One of the main grounds for this exclusion was that women lacked the intellectual capacity to participate in public life; they were ruled by their bodies and emotions and had visions limited to the domestic sphere (Isenberg, 1998). Consider some examples:

Women's supposed irrationality and seductiveness could be invoked to preclude their participation in public advocacy (Welter, 1966; Zaeske, 1995), and their so-called superior moral virtue could be raised as a reason to block their participation in raucous partisan debates (Blackwell 2004). Some myths were sanctioned by science. In 1873 Edward H. Clarke, a medical doctor, wrote about educating women in a book that went to 17 editions in 13 years (Cayleff, 1992). He insisted that a girl ought not "work her brain over mathematics, botany, chemistry, German, and the like" because it is not possible to "safely divert blood from the reproductive apparatus to the head" (Clarke, 1873, p.126). Women did not have the intellectual ability to vote and, even if they did, they ought not use it to think about political matters lest their reproductive organs atrophy and home and society suffer. In his lecture on "Woman" delivered in Philadelphia in 1849, Richard Henry Dana made the commonplace assertion that there is a divine, natural order, and woman's place in that order is in the home where she may exercise influence in the world "mediately [. . .] by permeating the masculine actor with the feminine of her own nature" (qtd. in Henry,

1995, p. 13). Woman and man ought to act together; "the heart is then working with the head; there is union between the affections and faculties" (qtd. in Henry, 1995, p. 13). One need not read long or deeply in texts of the time to see the "pervasiveness of that philosophy in the larger socio-political culture" (Henry, 1995, p. 12).

In short, as one historian has put it, "the notion of political equality for women was so radical that for a long time it was virtually impossible even to imagine woman suffrage" (DuBois, 1987, p. 839). Although these sentiments were commonplace, Stanton herself seems to have been surprised by the negative reaction to the event. McMillan, in her comprehensive study of the Seneca Falls Convention, reports that Stanton

> later recorded in her memoir, "No words could express our astonishment on finding, a few days afterward, that what seemed to us so timely, so rational, and so sacred, should be a subject for sarcasm and ridicule to the entire press of the nation." Those who attended the Convention were derided as 'women out of their latitude' and encouraged to spend their time more productively by tending to their domestic duties.

The convention's demands were characterized as "'impracticable, absurd,...ridiculous...excessively silly...[and] unnatural'" (McMillen, 2008, p. 99).

The speech Stanton gave several times after the convention (available at Stanton, 2010) can be seen as an attempt to overturn some of these stereotypes. The speech is highly argumentative. We encourage you to read it. Less important than what the argument$_1$s were, however, was the fact that *Elizabeth Cady Stanton was making them*. By standing up and putting argument$_1$s out there—by making reasons apparent—she was showing that she was the kind of person who could make reasons apparent. She was demonstrating that she had the capacity to participate in public life, that she had reason as well as emotion, that she had a vision of affairs beyond the domestic sphere.

Contemporary rhetorical scholarship calls the technique Stanton was employing *enactment*. In enactment, a claim is supported by the *activity* of making the claim.[1] Campbell and Jamieson have described enactment as a form "in which the speaker incarnates the argument, *is* the proof of the truth of what is said" (Campbell & Jamieson, 1978, p. 9; see also Campbell, 1988; Crenshaw, 1997; Lewis, 2011). In Michael

---

[1] The technique is thus the inverse of the performative contradiction, where the speaker's making of a claim serves to undermine it; it is the "I am alive" in contrast to the "I am dead."

Mendelson's words, the speaker "embodies" the claim being made; "the subject of one's discourse is rendered in the very form of that discourse" (1998, p. 38). Similarly, Palczewski has remarked, "the power of the (presentational) proof exceeds the (discursive) words" (2002, p. 7). Stanton was not alone among the early suffrage advocates in using enactment; many women defended their ability to be in public, to speak, and to reason by in fact being in public, speaking, and reasoning (e.g., Daughton, 1995; Huxman, 2000; Linkugel, 1993).

When Stanton engaged in argumentative activities as a form of enactment, she was performing a specific and forceful "mode of action":

1. In her speech, Stanton made many reasons apparent in support of her claims for women's rights, and for suffrage in particular.
2. In light of the opposition the movement was already experiencing, it was unlikely that her audiences would seriously consider, much less be moved by, her demands for the vote.
3. Nevertheless, by making reasons apparent, Stanton accomplished something: she showed to her audiences that she, a woman, was a person capable of making argument_1s.

3. CASE 2: 1869-1875: THE "NEW DEPARTURE"

Step forward two decades. In the post-Civil War period, many suffrage activists expected women to be enfranchised alongside formerly enslaved men. When those expectations proved false, the women's movement fell into disarray; the organization splintered, and the leadership had no clear strategy for moving ahead. This changed in 1869 with the emergence of a new strategy, sometimes called the "New Departure," resting on an argument_1 that women *already* had the right to vote (Balkin, 2005; DuBois, 1987). In its basic form, the argument_1 went like this:

> P1. Women are citizens of the United States.
> P2. Voting is a privilege or immunity of citizenship.
> P3. The (new) 14th Amendment to the Constitution provides that "no State shall make or enforce any law which shall abridge the privileges or immunities of citizens of the United States."
> C. Therefore, no state can abridge women's right to vote.

Based on this reasoning, women should not have to plead for the vote to be granted. They already had the right, and should be able to claim the support of the courts if it was denied. Hundreds of women across the country therefore presented themselves at the polls, requested ballots, and in some cases managed to cast them (Ray, 2007).

In addition to motivating women to vote, the core argument$_1$ was also presented to audiences outside the movement. Most famously, Susan B. Anthony and her lawyers made it in defending her against charges of illegal voting (Richards, 2007). It was made before the US Supreme Court in the case of *Minor v. Happersett*, brought to force local officials to register Virginia Minor (Ray & Richards, 2007). Victoria Woodhull made it when giving the first speech by a woman to a Congressional committee (DuBois, 1987; Jones, 2009). And it was made by many less famous women when they showed up at voting places.

The argument$_1$ was met with scorn in almost every case. For example, the state didn't even bother to send a lawyer to the Supreme Court hearing in the *Happersett* case, counting (correctly) on the Court to decide unanimously against the woman's claim. The lower court had written that the right of states to limit the suffrage to men

> cannot at this day be questioned. The (I may say) universal construction of the Constitution of the United States on this subject, and the almost universal practice of all of the States in reference to this subject, from the adoption of the Constitution to the present time, ought to be sufficient to prevent the necessity of an investigation of this subject now. There are certainly some questions that the courts of the country have a right to consider as settled, and that question I think is one of them ("Minor v. Happersett," 1873).

Note that the court here is not engaging the argument$_1$; it is dismissing it as one that does not even deserve "investigation," since women's position is "settled" and "cannot...be questioned." Nowadays, we may have a hard time recovering the "off the wall" (Balkin, 2005) nature of the New Departure reasoning, since we find the question "settled" on the other side. To get a sense of how bizarre women may have sounded, try substituting the words "seven year olds" for "women" in the core argument$_1$ identified above.

Nevertheless, women continued to make the argument$_1$, and that making did have force. Consider the following dialogue, reported by Ray (2007, p. 12) from a contemporary woman's periodical:

> **Miss B.** Here is my vote, sir, (handing in her ticket).
> **Judge.** What is the name?
> **Miss B.** Carrie S. Burnham.
> **Judge.** Where do you reside?
> **Miss B.** No. 1329 Vine street.

> The parties in charge of the window books promptly consulted their lists and found that these answers were correct.
> **Judge.** I am sorry, Miss Burnham, but I have instructions not to receive your vote.
> **Miss B.** *Why not, sir? I am a citizen. I pay taxes. I am governed, and I have a right to vote.*
> **Judge.** I cannot receive your vote.
> **Damon [Kilgore].** What reason do you assign?
> **Judge.** This is not the place to argue the matter. I cannot take the vote.
> **Damon.** Let us not proceed too hastily. Allow me to present to your consideration the result of laborious research in this matter.
> Mr. Killgore then drew forth document after document, in order to convince the judge of the election that a lady had the right to vote. The decision of the judge could not be changed, and Mr. Killgore and his lady friend re-entered their carriage.

Miss B. shows up at a voting place and enacts the ritual of voting. As Ray has said, this is the "the appropriation of the cultural performance of a ritual for rhetorical ends." Her performance is rejected; at that point, she makes the (emphasized) core argument$_1$. What does this making *do*? It has no chance of persuading—as the official points out, this is not the "place" for considering the argument$_1$. Nevertheless, making the argument$_1$ does an important job. As Ray also points out, an out-of-place rhetorical performance can be interpreted as a parody; it is so off-the-wall that it must be some sort of game, or practical joke, or, as we would say nowadays, "performance art." In that case, the woman's attempt to vote would be taken as outrageous or silly. Making the argument, however, demonstrates that the activity (a woman voting) is an activity that is supportable by reasons. Thus it demonstrates that Miss B. is *serious*. She has a reason for thinking that she can vote—a reason that she shows to the world in making the argument$_1$.

This demonstration can have impact on the world; for example, it makes it harder for the judge to just push Miss B. out of the polling place or otherwise treat her with disrespect.[2] The demonstration can also have an impact on the demonstrator herself. Ray notes that "some women [attempting to vote] reported despondence and a motivating anger, whereas others expressed joy or renewed self-respect" (Ray, 2007, p. 16). As Stillion Southard has said about the New Departure

---

[2] It should be noticed that Mr. Killgore, stepping in to speak for his companion, threatens to reduce the event to farce; the gentleman doth argue *too much,* methinks.

strategy more broadly, "it pressed the issue of woman suffrage into the privileged spaces of national politics and created the opportunity for women to enter these spaces and enact the citizenship rights they sought. Most significantly, the New Departure prompted women to leave their homes, walk to voting polls, and cast ballots" (2011, p. 43). This effect of speaking on the speaker herself Dale Hample has termed "arguing to display identity" (Hample & Irions, 2015); Richard Gregg (1971) once called it the "ego-function of rhetoric." What is involved, Gregg explains, is not so much "self-persuasion" about the particular "claims or the sense and probity of appeals and arguments," but instead *self-constitution:* "establishing, defining, and affirming one's self-hood as one engages in a rhetorical act" (p. 74).

Here again we find women engaging in argumentative activities to perform a specific and forceful "mode of action":

1. In making the New Departure argument$_1$, women like Miss B made apparent a reason why the US Constitution guaranteed women's suffrage.
2. In light of long-established precedent and hostile public opinion, it was unlikely that their audiences would seriously consider, much less be moved by, the argument$_1$.
3. Nevertheless, by making a reason apparent, the women accomplished something: they showed to their audiences—and to themselves—that their attempts to vote were reasoned; they were not joking, but serious.

4. IMPLICATIONS

To summarize, these two case studies show:

1. S's making argument$_1$s can have the "mode of action" of making reasons apparent.
2. Even when S's making a reason apparent is unlikely to have any impact on A's relationship to C,
3. nevertheless, S's making a reason apparent can accomplish many tasks of individual and social importance.

Are these uses of argument$_1$s weird, atypical, rare? No. Many individuals and groups advocating causes outside the mainstream need to demonstrate that they aren't nuts in order to have a voice in the public sphere and be taken seriously. Making reasons apparent is a strategy for accomplishing this. Various forms of enactment—including the enactment of reason-giving—were typical of the women's movement from its inception to the granting of suffrage by the 21st Amendment in 1920, and beyond. And there are further potential uses; for example, as pointed out by Scott Jacobs in the discussion of this paper, making reasons for C apparent can break the taken-for-granted appearance that ~C is obvious. Making reasons apparent can make it apparent that there is a disputable issue, with reasons on both sides.

What is the relationship between making reasons apparent and the functions of argumentation asserted by functionalist theories? We have critiqued functionalist theories for not defending why the specific function each asserts is *the* function of argumentation, with all other argumentative activities and purposes merely parasitic on it (Goodwin, 2007a). Here we provide an argument that many asserted functions are parasitic on *making reasons apparent*. In order to affect an audience in any way (to persuade them, to induce them to alter their standpoint, etc.) a speaker first has to make a reason apparent. It is not possible to change A's relationship to C without making a reason apparent, although it is possible (as we have shown) for S to make a reason apparent without trying to change A's relationship to C. Making reasons apparent is thus a task pragmatically necessary for any audience effect. Therefore, if there is a function of argumentation, making reasons apparent is more likely it.[3]

Do we *want* to take making reasons apparent to be the function of argumentation? We don't see why we would. If making reasons apparent is indeed the function of argumentation, it is going to be hard to derive substantial norms from it; it is too thin. What interpersonal or discursive conditions need to be in place for making reasons apparent to get its job done? The leading candidates might be:

1. What is made apparent has to be a reason. Note that it doesn't have to be a very good reason; it just has to be a P→C unit: a premise, in some sort of support relationship with a claim/conclusion. Interestingly, this suggests that argumentative activities rely on a

---

[3] As an aside: in the discussion after this paper was presented, several commentators including Geoffrey Goddu suggested that there can be an implied audience targeted for persuasion even when the immediate audience cannot be moved. We certainly agree that argumentative activities commonly have multiple goals, both short and long-term; that is a basic assumption in rhetorical studies. For example, suffrage advocates were looking forward to shifting public opinion in the long run and claiming recognition from an audience of posterity. But it is hard to build a *pragmatic* theory of argumentation around an imagined audience. The aim of pragmatic theories has been to gain better understandings of argument$_1$s by embedding them in the immediate context of the argument$_2$s in which they occur. Invoking more diffuse contexts like social controversies or a universal audience will not give the same sort of traction for theory-building.

conception of "→"—i.e., that a pragmatic theory of argumentation needs support from a theory of reasons.

2. The discourse has to make the reason apparent. This suggests a norm of clarity in making $argument_1$s.

In our experience, both of these norms are indeed important for teaching, where we find ourselves frequently asking our students to make $argument_1$s (not tell stories) and to use lots of indicator words to make the structure of their $argument_1$s clear. However, these two items hardly exhaust the norms needed to capture the goodness of $argument_1$s or $argument_2$s. In particular, since A, the audience of the argumentative activity, isn't relevant to this function of making reasons apparent, it is unlikely that this function will provide a basis for audience-regarding norms like fairness to A or a responsibility to listen to A's counter-arguments.

If it seems strained or unimpressive to talk about *making a reason apparent* as the "function of argumentation" why not drop the function talk and just say that to make an $argument_1$, a speaker has to make a reason apparent? Making a reason apparent is what making an $argument_1$ *is* (Jacobs, 2000; O'Keefe, 1982). Adopting this approach, argumentation theories with a pragmatic bent remain interested in what people can do individually and collectively by making $argument_1$s. But instead of anointing one or more of these doings as *functions*, all such "modes of action" are embraced as *uses*.[4] Pragmatic theories of argumentation are responsible for understanding argumentative activities in all their cunning, their richness and diversity, including accounting for how arguers constitute or make contextually determinate the norms governing their argumentative activities. The normative pragmatic approach to argumentation theory has been doing just that.

ACKNOWLEDGEMENTS: We give hearty thanks to the conference organizers for arranging this inaugural event, and to the audience for their useful feedback.

REFERENCES

Balkin, J. M. (2005). How social movements change (or fail to change) the Constitution: The case of the New Departure. *Suffolk University Law Review, 39*, 27-65.

---

[4] Or to use Mohammed's proposed terminology, *uses* or *purposes*.

Bermejo-Luque, L. (2010). Intrinsic versus instrumental values of argumentation: The rhetorical dimension of argumentation. *Argumentation, 24*(4), 453-474.

Blackwell, M. S. (2004). Meddling in politics: Clarina Howard Nichols and antebellum political culture. *Journal of the Early Republic, 24*(1), 27-63.

Campbell, K. (1988). Enactment as rhetorical strategy in the year of living dangerously. *Central States Speech Journal, 39*(3-4), 258-268.

Campbell, K. K., & Jamieson, K. H. (1978). Form and genre in rhetorical criticism: An Introduction. In K. K. Campbell & K. H. Jamieson (Eds.), *Form and genre: Shaping rhetorical action* (pp. 9-32). Falls Church: Speech Communication Association.

Cayleff, S. E. (1992). She was rendered incapacitated by menstrual difficulties: Historical perspectives on perceived intellectual and physiological impairment among menstruating women. In A. Dan and L Lewis (Eds.), *Menstrual Health in Women's Lives* (pp. 229-35). Urbana: University of Illinois Press.

Clarke, E. H. (1873). *Sex in Education; or, Fair chance for the girls.* Boston.

Crenshaw, C. (1997). Resisting whiteness' rhetorical silence. *Western Journal of Communication, 61*(3), 253-278.

Daughton, S. M. (1995). The fine texture of enactment: Iconicity as empowerment in Angelina Grimké's Pennsylvania Hall Address. *Women's Studies in Communication, 18*(1), 19-43.

DuBois, E. C. (1987). Outgrowing the compact of the Fathers: Equal rights, woman suffrage, and the United States Constitution, 1820-1878. *The Journal of American History, 74*(3), 836-869.

Flexner, E., & Fitzpatrick, E. (1996). *Century of struggle: The woman's rights movement in the United States.* Cambridge: Belknap.

Goodwin, J. (2001a). The noncooperative pragmatics of arguing. In E. T. Nemeth (Ed.), *Pragmatics in 2000: Selected papers from the 7th International Pragmatics Conference* (Vol. 2, pp. 263-277). Antwerp: International Pragmatics Association.

Goodwin, J. (2001b). One question, two answers. In H. Hansen, C. W. Tindale, J. A. Blair, R. H. Johnson, & R. C. Pinto (Eds.), *Argumentation and its Applications.* Windsor, ONT: Ontario Society for the Study of Argumentation.

Goodwin, J. (2007a). Argument has no function. *Informal Logic, 27,* 69-90.

Goodwin, J. (2007b). Theoretical pieties, Johnstone's impiety, and ordinary views of argumentation. *Philosophy & Rhetoric, 40,* 36-50.

Gregg, R. B. (1971). The ego-function of the rhetoric of protest. *Philosophy and Rhetoric, 4,* 71-91.

Hample, D., & Irions, A. (2015). Arguing to display identity. *Argumentation, 29*(4), 389-416.

Henry, D. (1995). Text and context: Lucretia Coffin Mott's "Discourse on Woman." *Rhetoric Society Quarterly, 25,* 11-19.

Huxman, S. S. (2000). Perfecting the rhetorical vision of woman's rights: Elizabeth Cady Stanton, Anna Howard Shaw, and Carrie Chapman Catt. *Women's Studies in Communication, 23*(3), 307-336.

Innocenti, B. (2005). Norms of presentational force. *Argumentation and Advocacy, 41,* 139-51.

Innocenti, B. (2006). A normative pragmatic perspective on appealing to emotions in argumentation. *Argumentation, 20,* 327-343.

Innocenti, B. (2011a). Countering questionable tactics by crying foul. *Argumentation and Advocacy, 47,* 178-188.

Innocenti, B. (2011b). A normative pragmatic model of making fear appeals. *Philosophy and Rhetoric, 44,* 273-290.

Isenberg, N. (1998). *Sex and citizenship in antebellum America.* Chapel Hill: University of North Carolina Press.

Jacobs, S. (2000). Rhetoric and dialectic from the standpoint of normative pragmatics. *Argumentation, 14,* 261-286.

Johnson, R. H. (2000). *Manifest rationality: A pragmatic theory of argument.* Mahwah, NJ: Lawrence Erlbaum Associates.

Jones, J. (2009). Breathing life into a public woman: Victoria Woodhull's defense of woman's suffrage. *Rhetoric Review, 28*(4), 352-369.

Lewis, T. (2011). Winning woman suffrage in the masculine west: Abigail Scott Duniway's frontier myth. *Western Journal of Communication, 75*(2), 127-147.

McMillen, S. G. (2008). *Seneca Falls and the origins of the women's rights movement.* Oxford: Oxford University Press.

Mendelson, M. (1998). The rhetoric of embodiment. *Rhetoric Society Quarterly, 28*(4), 29-50.

Minor v. Happersett, 55 58 (Supreme Court of Missouri 1873).

Mohammed, D. (2015). Goals in argumentation: A proposal for the analysis and evaluation of public political arguments. *Argumentation.* (Online first).

Linkugel, W. (1993). Anna Howard Shaw: A case study in rhetorical enactment. In K. K. Campbell (Ed.), *Women public speakers in the United States, 1800-1925* (pp. 409-420). Westport: Greenwood Press.

Lutz, A. (1940). *Created equal: A biography of Elizabeth Cady Stanton, 1815-1902.* New York: John Day.

O'Keefe, D. J. (1982). The concepts of argument and arguing. In J. R. Cox & C. A. Willard (Eds.), *Advances in argumentation theory and research* (pp. 3-23). Carbondale: Southern Illinois University Press.

Palczewski, C. (2002). Argument in an off-key. In G. T. Goodnight (Ed.), *Arguing communication and culture: selected papers from the Twelfth NCA AFA Conference on Argumentation,* vol. 1 (pp. 1-23). Washington DC: National Communication Association.

Perelman, C., & Olbrechts-Tyteca, L. (1969). *The new rhetoric: A treatise on argumentation* (J. Wilkinson & P. Weaver, Trans.). Notre Dame: University of Notre Dame Press.

Pinto, R. C. (2001). The relation of argument to inference. In *Argument, Inference and Dialectic* (pp. 32-45). Dordrecht: Kluwer Academic Publishers.

Pinto, R. C. (2009). Argumentation and the force of reasons. *Informal Logic, 29*(3), 268-295.

Ray, A. G. (2007). The Rhetorical Ritual of Citizenship: Women's voting as public performance, 1868–1875. *Quarterly Journal of Speech, 93*(1), 1-26.

Ray, A. G., & Richards, C. K. (2007). Inventing citizens, imagining gender justice: The suffrage rhetoric of Virginia and Francis Minor. *Quarterly Journal of Speech, 93*(4), 375-402.

Richards, C. K. (2007). Susan B. Anthony: "Is it a crime for a U.S. citizen to vote?" (3 April 1873). *Voices of Democracy, 2*, 189-209.

Stanton, E. C. (2010). Address by Elizabeth Cady Stanton on woman's rights, September 1848. *The Elizabeth Cady Stanton & Susan B. Anthony Papers Project.* Retrieved from http://ecssba.rutgers.edu/docs/ecswoman1.html

Stillion Southard, B. A. (2011). *Militant citizenship: Rhetorical strategies of the National Woman's Party, 1913-1920.* College Station: Texas A&M University Press.

Wellman, J. (2004). *The road to Seneca Falls: Elizabeth Cady Stanton and the first woman's rights convention.* Urbana: University of Illinois Press.

Welter, B. (1966). The cult of true womanhood: 1820-1860. *American Quarterly, 18*(2), 151-174.

"The woman suffrage crisis." (1915, Feb. 7). *New York Times*, C2.

Zaeske, S. (1995). The 'promiscuous audience' controversy and the emergence of the early woman's rights movement. *Quarterly Journal of Speech, 81*, 191-207.

26

Getting Involved in an Argumentation in Class as a Pragmatic Move: Social Conditions and Affordances

SARA GRECO
IALS, Università della Svizzera italiana, Switzerland
sara.greco@usi.ch

TEUTA MEHMETI
University of Neuchâtel, Switzerland
teuta.mehmeti@unine.ch

ANNE-NELLY PERRET-CLERMONT
University of Neuchâtel, Switzerland
anne-nelly.perret-clermont@unine.ch

This paper investigates argumentative discussions in a school activity involving Albanian-speaking pupils in Switzerland. Our aim is to understand how pupils respond to the issue proposed by the teacher for their discussion: do they deal with it? If they introduce new issues, how does the teacher react? We explore social conditions and affordances on the pupils' argumentative interventions, focusing on how issues emerge and how younger interlocutors feel entitled to take part in a discussion.

KEYWORDS: critical discussion, issue, migrant children, opening stage, students interactions

1. POSITION OF THE PROBLEM

In the literature on argumentation and education, there seems to be consensus on the fact that it is difficult to find argumentation by children in the classroom (Kuhn, 1991; Garcia-Mila & Andersen, 2008, Andriessen & Schwarz, 2009; Erduran & Jiménez-Aleixandre, 2008; Schwarz, 2009). In particular, a "deficit approach" is often used to explain migrant children's school failure. Migrants are expected to fail

D. Mohammed & M. Lewiński (eds.) (2016). *Argumentation and Reasoned Action: Proceedings of the 1st European Conference on Argumentation, Lisbon, 2015. Vol. II, 463-478.* London: College Publications.

because of their *cultural distance* with the school (e.g. de Haan & Elbers, 2004).

In this paper, we analyse a case in which a pedagogical activity was designed in order to foster migrant children's participation (Mehmeti & Perret-Clermont, 2016). The focus was on a group supposed to be at risk of school failure, namely Albanian-speaking students living in Switzerland (see for instance Becker, Jäpel, & Beck, 2011; Burri-Sharani, Efionayi-Mäder, Hammer, Pecoraro et al., 2010; Kronig, 2003; Müller, 2001). This paper is part of an ongoing research project on children's argumentation that has been developed at the University of Neuchâtel and at the Università della Svizzera Italiana since 2008-2009[1].

In our approach to this topic, we feel that, before assessing children's argumentative skills or designing argumentation activities for the school, we need to observe in detail how argumentation develops in adult-children discussions. In this paper, our general question is whether all the possibilities of argumentation are fully exploited in this specific setting, i.e. if the discussion is argumentative in a proper sense, what the children's and the teacher's contributions are, how the teacher deals with the children's arguments and vice versa. Such research is relevant for education, as it might invite to think about conditions that could favour a "thinking space" (Perret-Clermont, 2015).

2. THEORETICAL FRAMEWORK FOR THE ANALYSIS

Because our aim is to analyse what happens within a classroom discussion in argumentative terms, we adopt the pragma-dialectical model (van Eemeren & Grootendorst, 2004) in order to reconstruct argumentation in our data. This model foresees four stages of a "critical discussion", in which participants solve their difference of opinion on the merits. In the *confrontation stage*, a difference of opinion emerges between the arguers. In the *opening stage*, the arguers try to establish "how much relevant common ground they share" (van Eemeren & Grootendorst, 2004, p. 60). In the *argumentation stage* of a critical discussion, arguments in support or against a standpoint are advanced and critically tested (ibid.). The concept of critical testing is crucial to argumentation: in fact, the arguers are not simply trying to win their cause but they want to do this by remaining within the boundaries of

[1] A recent development is a grant obtained by the Swiss National Science Foundation (No100019_156690, "Analysing children's implicit argumentation: Reconstruction of procedural and material premises", applicants: A.-N. Perret-Clermont, S. Greco, A. Iannaccone, A. Rocci).

reasonableness. A *concluding stage* occurs if the difference of opinion is resolved on the merits.

Because the discussion we are considering is multiparty and happens in a face-to-face setting, in which all participants are contributing without pre-determined positions and roles, it is particularly important for us to understand how each participant contributes to raise issues, advance standpoints and present arguments.

As van Eemeren and Grootendorst (2004, p. 118) put it, the *analytic overview* helps bring to light which points are at dispute, which parties are involved in the difference of opinion, what their procedural and material premises are, which argumentation is put forward by each of the parties, how their discourses are organised, and how each individual argument is connected with the standpoint that it is supposed to justify or refute.

From a methodological viewpoint, this model serves as a grid for the analysis of argumentation in real communicative interactions. It has a *heuristic* function, insofar as it helps elicit argumentative discussions from conversations that might be non-completely argumentative. It also has an *evaluative* function, as it gives a "normative standard" of how a discussion should proceed in order to resolve the difference of opinion in a reasonable fashion.

In addition to the model of a critical discussion, when analysing "how each individual argument is connected with the standpoint that it is supposed to justify or refute" (ibid.), we will introduce the Argumentum Model of Topics (Rigotti & Greco Morasso, 2010) for the reconstruction of argument schemes. The AMT allows systematically distinguishing premises that are *procedural* in nature, i.e. inferential connections, from premises that are *material* in nature, including cultural or contextual general assumptions as well as factual data.

3. DATA

We rely on a case study, already discussed in Mehmeti & Perret-Clermont (2016) relative to a class within a school of Albanian language and culture, including 8 to 13 years old children. This class is organized and sustained by the association of teachers and parents "Lidhja e Arsimtarëve dhe Prindërve Shqiptarë" (LAPSH). All these students attend a Swiss public school for their regular schooling.

The two teachers (who had no previous training on argumentation) accepted to follow an activity designed by one of the researchers (Teuta Mehmeti). It was expected that a deep attention paid to the psychosocial conditions, favouring the development of a "thinking space" (Perret-Clermont, 2015), could lead children to enter argumentation and show cognitive and social skills. In that activity, students are invited to actively discuss an issue of world importance; the teacher has a more peripheral role, being supposed to provoke discussion and the development of students' thinking. This activity avoids the pressure of normative assessments, which can hinder children's competencies (Butera, Buchs, & Darnon, 2011). It echoes for some points the principles for experiment design for argumentation (Jimenez-Aleixandre, 2009).

Each teacher was given the following protocol describing the activity (Mehmeti & Perret-Clermont, 2014):

> *1)* The teacher presents the researcher: a friend who studies psychology and education and is interested in what children do during classroom activities. She explains that this lesson is different from usual: children have to play an important role conducting the discussion; the teacher will be confined to a more passive role.
> *2)* She says that she expects the students to work in dyads first. She organizes these dyads, and informs them that she will give two photographs (figure 1) to each dyad. She writes three questions on the blackboard:
>
> 1. *What do you see in these pictures? Describe.*
> 2. *Where could these two pictures have been taken?*
>   2.1 *What are the characteristics of this country?*
> 3. *What creates pollution?*

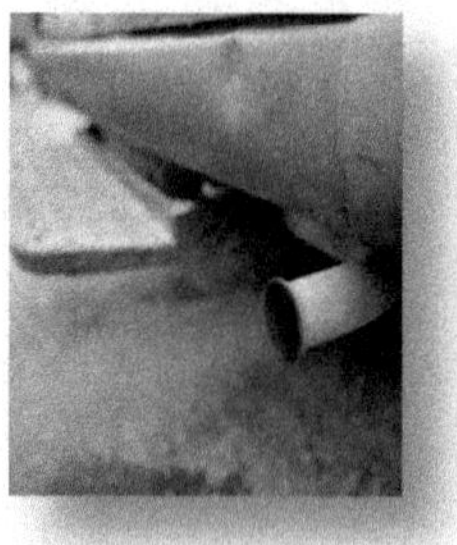

Figure 1 – Photographs given by the teacher

She tells the children that they have to discuss these questions in their dyad. When they reach an agreement, they will go and write their answer on the blackboard.

> *3)* The teacher asks the dyads to choose which member of the dyad will write the answers on the blackboard; then, s/he

does so. Then the teacher explains that one dyad will discuss the answers of another dyad.
4) The teacher draws attention to some of the answers written on the blackboard and opens the discussion to the whole class.

4. ANALYSIS

We will analyse Extract 1, located in the third step of the activity (when all the pupils have written their answers on the blackboard and the teacher designates a group X to discuss the answers of a group Y); this extract has already been discussed from other viewpoints by Mehmeti and Perret-Clermont (2016). We will (a) reconstruct the different argumentative discussions that are present; (b) analyse to what extent these discussions are developed.

The original language is French. Albeit pupils are normally expected to speak Albanian in this special school, the teacher allowed them to use French, because it turned out that their command of Albanian was not sufficient to speak about pollution.

Contrary to other pupils who have answered by mentioning a city either from Albania or Kosovo in question 2, the dyad formed by Burim and Arlind has written "We don't know". Discussions are started by the children and the teacher around the different answers emerged. At a certain moment, Burim intervenes to defend his dyad's answer (turn 1).

1	Burim	*J'ai écrit « on ne sait pas » [à la question 2] mais pour dire que je ne suis pas d'accord avec les autres parce que [la pollution] c'est un problème qui est présent partout.*	I wrote « we don't know » [to the question 2] but to say that I don't agree with the others because it [pollution] is a problem that is present everywhere
2	Teacher	*Et ça veut dire qu'en Suisse aussi ?*	And does it mean that in Switzerland too?
3	Burim	*Oui*	Yes
4	Teacher	*Ah oui, et où par exemple ?*	Oh yes, and where for example?
5	Burim	*Ben j'ai déjà vu mais aussi parce qu'il y a plein de grandes entreprises et industries qui produisent des choses, ça aussi ça pollue*	Well, I have already seen it but also because there are lot of big companies and industries that produce things, this also pollutes
6	Valon	*Oui mais quand même en Suisse y'a beaucoup moins parce que par exemple y'a pas ces déchets comme ca partout*	But still in Switzerland there is much less because for example there is not so such waste everywhere like that
7	Teacher	*Et comment ça se fait?*	And how does it come?
8	Shpresa	*Ben parce que la Suisse c'est pas un pays pauvre*	Well because Switzerland is not a poor country.
9	Teacher	*Et donc?*	So what?
10	Valon	*On peut payer pour enlever les déchets*	The removal of waste can be paid for
11	Teacher	*Où paye-t-on pour ça, comment ça se passe*	Where is it paid for? How does that work?
12	Burim	*Les impôts*	The taxes

Participants: three pupils (Burim, Valon, Shpresa) and the teacher

Table 1 – Extract 1

4.1 Analytic overview of argumentation in Extract 1

As a first step of our analysis, we will propose an analytic overview of argumentation in Extract 1. Because this will be very important for the following discussion, we will not only mention standpoints and arguments, but we will also make *issues* explicit on which argumentation develops. Introducing a new issue, in fact, means introducing a new argumentative discussion; and, as it will be shown, there are three potential argumentative discussions emerging in Extract 1. We will equally specify the initiators of these issues. These changes require us to make a few amendments on how the analytical overview is presented in van Eemeren, Grootendorst and Snoeck-Henkemans (2002), although the main concepts remain unvaried.

A first important change (see Table 2) is that we mark the *issue* on which participants are discussing with a letter (A, B, C). Issues are listed in connection with the bit of the discussion (turns) in which they emerge. A standpoint which is advanced on a given issue is marked with the traditional pragma-dialectical notation: A1 will be a standpoint on issue A, B1 a standpoint on issue 2, and so on. We also insert protagonist(s) and antagonist(s).

As it emerges in Table 2, the initial issue for the discussion corresponds to the second question proposed to the students: "Where could these pictures have been taken?". When Burim starts talking at turn 1, he is giving an argument to explain why he wrote "We don't know" on the blackboard.

From a linguistic viewpoint, Burim's standpoint B1 is outside of the paradigm of expected answers for the question "Where could these two pictures have been taken?". Burim is raising a meta-issue (issue B), which could be formulated as: "Can we answer the question "where have these pictures been taken"?". His standpoint on this issue is "No", and he gives an argument for it: pollution is everywhere. However, we do not mean that Burim necessarily has a polemical objective in mind. His answer could be his way of discussing the teacher's question and it is difficult to say, on the basis of the data at our disposal, how much he feels the need to oppose his fellow students.

To this, the teacher immediately responds by asking questions that challenge argument B1.1 (turns 2 and 4). By this doing, she also shifts the issue of the discussion, which becomes: "Is there pollution in Switzerland? (and where?)".

The discussion is then moved on this new issue (issue C) and the first to take a position is, again, Burim. This is not surprising: having said that pollution is everywhere, he now only needs to specify his position speaking about Switzerland. At turn 5, Burim repeats his standpoint "Yes" and gives as an argument the fact that there are big companies and industries in Switzerland; and this also pollutes.

Turns	**Issue**	**Standpoint and arguments (protagonists)**	**Antagonists**
Preceding discussion	A (Teacher) Where have these pictures been taken?	Other pupils: A1 In Kosovo or Albania (various answers written on the blackboard)	
1-2	B (Burim) Can we answer the question "where have these pictures been taken"?	Burim: B1 We cannot answer B1.1 Because pollution is everywhere	Teacher: ? Challenges B1.1 and thus opens issue C
2-12	C (Teacher) Is there pollution in Switzerland (and where)?	Burim: C1 (Yes) there is pollution in Switzerland C 1.1 there are lot of companies and industries which produce things, this also pollutes	Valon: C1" No there is much less pollution in Switzerland C1.1" There is not so such waste everywhere like that Valon + teacher + Shpresa + Burim: C.1.1.1" Because the Swiss pay to remove waste (via taxation) C.1.1.1.1" Because Switzerland is not a poor country

Table 2 - Analytic overview of argumentation in Extract 1

Valon, another pupil, assumes a contrary standpoint, saying that in Switzerland pollution is "much less"; he presents an argument for his standpoint: there is not so much waste (turn 6). At this point, Valon's argumentation is further developed with a series of subordinative arguments produced by Shpresa (turn 8), Valon (turn 10) and Burim himself (turn 12). Each of these arguments (represented in table 2 as C1.1.1" and C1.1.1.1") is solicited by a question asked by the teacher.

4.2 Discussion

The analytic overview in section 4.1 is similar to a tree in which some branches are not fully flourishing, while others are. Some branches have been cut as soon as they started to grow. The gardener, in this metaphor, is the teacher, who controls the development of the discussion. Concerning this metaphor, we think that it is important to point out some hypotheses on these processes: what seems to motivate students' arguments, what they interpret about the issues that are addressed in the activity or in teacher's interventions, how they follow or interrupt their line of reasoning.

4.2.1 Legitimate issues and meaningful issues

A first observation concerns the issue set by the teacher in her question: "Where have these two pictures been taken?" Our first hypothesis is that, asked to this group of pupils, this might have been perceived as a rhetorical question. In fact, it is evident for most pupils that the two pictures have been taken in Kosovo or Albania; all answers are consonant except for Burim and Arlind's.

If this hypothesis is true, i.e. if the answer is obvious, why asking this question? Generally speaking, an informative question must ask for something that is not known, while rhetorical questions ask for something that the answerer should know already (Gobber, 1999). In argumentative terms, if there is no actual difference of opinion (because the "issue" is not a real issue) no discussion needs to be opened. Independently from the original intentions with which this question had been formulated, it is possible that Burim thinks that somehow the picture in question is leading pupils to give answers that support a stereotyped and prejudiced view of Kosovo and Albania.

A second observation also concerns the emergence of issues. When, at turn 1, Burim sets a new issue (B), he makes a very courageous move, as he questions his teacher's question, by saying that it is not possible to answer it. In ordinary school situations, teachers' questions are reputed to be part of a "didactic contract" (Brousseau, 1980; Schubauer-Leoni, 1986; Sensevy & Mercier, 2007). As such, they are perceived as meaningful and not discussed. However, following the design proposed by the researcher, the teacher has announced that this lesson will be different. It could be that, therefore, Burim dares to question what is normally taken for granted.

Burim provides an argument for his standpoint (see table 2):

B1 We cannot answer
B1.1 Because pollution is everywhere

The teacher reacts by challenging B1.1 repeatedly at turns 2 and 4. We cannot know if this is done on purpose or inadvertently, but what happens here is that an issue introduced by a young pupil is immediately abandoned as an effect of the teacher's questioning. So, the argumentative discussion on issue B is left unaccomplished, as *we do not get to any concluding stage*. This might induce to think that the pupil's issue (B) was not legitimate.

4.2.2 Developing different lines of argument?

Burim reacts to issue C on which the teacher has steered the discussion:

C1 Yes there is (pollution in Switzerland)
C 1.1 there are lot of companies and industries which produce things, this also pollutes

To this, Valon replies with another standpoint and argument opposing Burim's (see table 2):

C1" No there is much less pollution in Switzerland
C1.1" There is not so much waste everywhere like that

So, even though the teacher has not accepted Burim's issue (see section 4.1.1), she still finds herself confronted with a good example of argumentation developed by her pupils. Specifically, the scenario is that of a *mixed dispute* with a protagonist and an antagonist (Burim & Valon) who both have advanced arguments in favour of their respective standpoints.

The teacher intervenes in this discussion asking questions (turns 7-9-11) in such a way that she seems to be willing to develop children's argumentation. We found a similar role of adults' questioning in previous research (Greco Morasso, Miserez Caperos, & Perret-Clermont, 2015). However, in this case, the teacher only talks to Valon, while she does not interact with Burim, thus only developing one side of the mixed dispute. Different persons contribute to this development: Shpresa (turn 8), Valon (turn 10), and even Burim (turn 12), who has been able to follow this line of argument, while de facto abandoning his own.

Thus, a mixed dispute that was balanced up to turn 6 becomes unbalanced after the teacher's interventions. Because the goal of the argumentation stage of a critical discussion is submitting the parties' standpoints and arguments to critical scrutiny in order to resolve a difference of opinion on the merits (see section 2), *the argumentation stage as it is developed here is questionable at the least.* In fact, one side is completely abandoned while the opposing side is developed; they are not really confronted. As a consequence, there is no proper concluding stage of this discussion. Whether the students have reached agreement or not is not clear because no space is given for exploring this aspect.

Moreover, the teacher's questions induce her students to provide *subordinative* argumentation in support of Valon's argument. On the opposite, there is no exploration (*inventio*) of further arguments in support of Valon's (or Burim's) standpoints. Our hypothesis is that the teacher's questioning about Valon's argument creates new expectations and interpretations of the activity: it could be that students feel that they do not need to critically think about arguments but rather to answer the teacher's questions.

Notably, fostering a process of *inventio* on all sides of an argumentative dispute is not strictly required by a standard critical discussion, unless it is explicitly functional to the goal of resolving the difference of opinion. Thus, one could object that it might have been unnecessary to raise further arguments in this case. However, the context in which this discussion takes place must be taken into account in order to fully appreciate the potential value of assigning a broader space to *inventio*. In fact, because the aim of the activity was to foster students' argumentation, it might have been functional to give a broader space to the exploration of other arguments.

4.2.3 What is pollution? A problem with the opening stage

The difference of opinion between Burim and the others could be resolved, at least in part, by tackling the meaning that they attribute to the term "pollution". In fact, while the photos obviously point to two specific forms of pollution (waste and cars' smoke), the linguistic term "pollution" per se covers a wider area of phenomena. When Burim alludes to industrial pollution, he relies on this broader interpretation.

Now, if this broader meaning of pollution is adopted, Burim's claim that pollution is everywhere is difficult to contradict. The AMT representation of his argument (standpoint C1 and argument C1.1) is represented in figure 1. Burim adopts a *locus from cause to effect* to show that there are different independent causes for pollution and some of them (big companies and industries that produce things) are present in Switzerland; therefore, pollution is necessarily present in Switzerland.

However, if one takes the specific meaning of pollution that is suggested by the photos, then Burim's claim is not acceptable because his main contextual or cultural premise, i.e. the endoxon, would fail to be true.

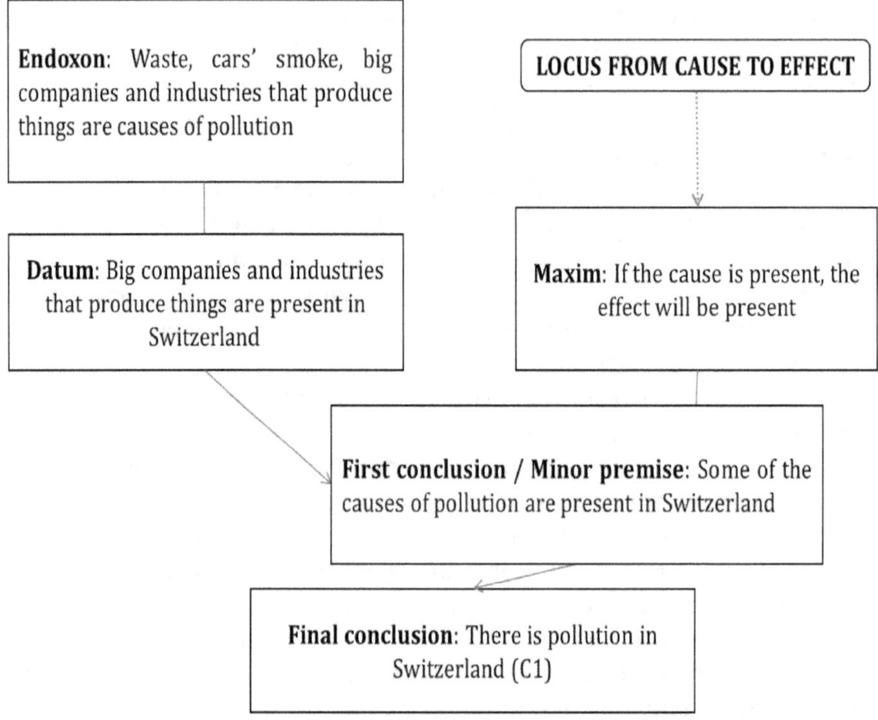

Figure 2 - AMT analysis of Burim's argument (standpoint C1 and argument C1.1)

In this passage, we have a typical case of ambiguity raised by a natural language term (i.e. *pollution*). As a consequence, there is a problem with the opening stage, as participants do not agree on the meaning of a term that is central to the discussion. Problems with the opening stage, especially linked to ambiguities, may generate misunderstandings and conflicts (Dascal, 2003).

Thus, this ambiguity should be resolved for the discussion to proceed in a reasonable way. Clarifying this term could bring to a resolution of the difference of opinion on issue (C), as participants might agree that there is *industrial* pollution in Switzerland, while other forms of pollution (such as waste abandoned in the streets) would be less frequently seen in Swiss cities. However, the discussion proceeds without solving this problem, thus leaving the difference of opinion unresolved.

5. CONCLUSIONS AND OPENINGS

This paper has analysed a case of argumentation in the classroom in which a group of pupils discuss on an issue related to pollution. This discussion testifies to the presence of argumentation in a classroom in which minority children are involved. In this case, and contrary to previous studies, we found that children do not lack basic skills of argumentation: they are able to spontaneously open new issues for a discussion; they advance standpoints and arguments. Moreover, they can follow the teacher when she opens new paths for their discussion.

However, we have found that the three argumentative discussions opened by the teacher and her pupils are not corresponding to an ideal critical discussion because some elements are missing. In section 4.2.1 and 4.2.2, we have found that the *concluding stage* was missing; in section 4.2.2, this was linked to a problem with the critical testing of arguments in the *argumentation stage*. In section 4.2.3, we discussed a problem with the *opening stage*, based on the ambiguity of the term "pollution".

We assume that a teacher has a special role, to some extent exceeding the "normal" or "canonical" role of a participant to an argumentative discussion: she is in control of turn taking, she manages issues and helps develop arguments. Ideally, as we think, she is also in charge of the fact that each argumentative discussion in the classroom might become an occasion for young pupils (and herself) to learn more about argumentation as a form of reasonable resolution of disagreement.

Following this line of reasoning, we have two aspects on our agenda for future work.

1. The "special role" of the teacher, who participates in the pupils' argumentative discussion without being a canonical participant, is similar to the role of a *dispute mediator* (Greco Morasso, 2011). The latter is responsible for the creation of an argumentative space in which parties, who enter the discussion as conflicting disputants, can resolve their conflict via argumentation. We therefore assume as a working hypothesis for future research that, if a teacher (or adult) wants to

foster children's argumentation, it might be useful to model his or her role upon the mediator's.

2. We need to better clarify the argumentation context in which this discussion takes place in terms of an *argumentative activity type* (van Eemeren, 2010). This is, however, a delicate item on our research agenda. In fact, one cannot simply say that there is one activity type corresponding to "discussions in the classroom". Many things happen in the classroom. Not all discussions are argumentative; and not all of them have the same goals and characteristics. Therefore, we should carefully consider what activity types are happening in a classroom, considering whether students and teacher(s) are involved in a formal or informal discussion; what subject is being taught; what the goal of that segment of interaction is, and so on. Moreover, as we think, classroom activities must not be confined into rigid boundaries, so one should not think to establish activity types whose boundaries are too rigid. As Danish and Enyedy (2015) show, teachers may capitalize on discussions that arise from students even though they are outside the boundaries of a pre-defined activity type.

ACKNOWLEDGEMENTS: We are grateful to the Swiss National Science Foundation for its support (contract n°100019-156690/1 (Perret-Clermont/Greco/Iannaccone/Rocci) which is allowing to bring forward our research on children's argumentation.

REFERENCES

Andriessen, J., & Schwarz, B. (2009). Argumentative design. In N. Muller Mirza & A.-N. Perret-Clermont (Eds.), *Argumentation and education: Theoretical foundations and practices* (pp. 145-174). New York: Springer.

Becker, R., Jäpel, F., & Beck, M. (2011). *Statistische und institutionelle Diskriminierung von Migranten im Schweizer Schulsystem. Oder: Werden Migranten oder bestimmte Migrantengruppen in der Schule benachteiligt?.* Universität Bern: Institut für Erziehungswissenschaft. Abteilung Bildungssoziologie.

Brousseau, G. (1980). L'échec et le contrat. *Recherches, 41*, 177-182.

Burri-Sharani, B., et al. (2010). *La population kosovare en Suisse.* Berne: Office fédéral des migrations (ODM).

Butera, F., Buchs, C., & Darnon, C. (Eds), (2011). *L'évaluation, une menace ?* Paris: PUF.

Danish, J. A., & Enyedy, N. (2015). Latour goes to kindergarten: Children marshalling allies in a spontaneous argument about what counts as science. *Learning, Culture and Social Interaction, 5*, 5-19.

Dascal, M. (2003). Understanding misunderstanding. In M. Dascal, *Interpretation and Understanding* (pp. 293-321). Amsterdam: John Benjamins.

de Haan, M., & Elbers, E. (2004). Minority status and culture: local constructions of diversity in a classroom in the Netherlands. *Intercultural Education, 15*(4), 451-453.

Eemeren, F.H. van (2010). *Strategic manoeuvring in argumentative discourse: Extending the pragma-dialectical theory of argumentation*. Amsterdam: John Benjamins.

Eemeren, F. H. van, & Grootendorst, R. (2004). *A systematic theory of argumentation: The pragma-dialectical approach*. Cambridge: CUP.

Eemeren, F. H. van, Grootendorst, R., & Snoeck-Henkemans, A. F. (2002). *Argumentation: Analysis, evaluation, presentation.* Mahwah, NJ: Lawrence Erlbaum Associates.

Erduran, S., & M.P. Jiménez-Aleixandre. (2008). *Argumentation in Science Education: Perspectives from classroom-based research.* New York: Springer.

Gobber, G. (1999). *Pragmatica delle frasi interrogative. Con applicazioni al tedesco, al polacco e al russo*. Milano: ISU.

Greco Morasso, S. (2011). *Argumentation in dispute mediation: A reasonable way to handle conflict.* Amsterdam: John Benjamins.

Greco Morasso, S., Perret-Clermont, A.-N., & Miserez, C. (2015). L'argumentation à visée cognitive chez les enfants. In N. Muller Mirza and C. Buty (Eds.), *L'argumentation dans les contextes de l'éducation* (pp. 39-82). Bern: Peter Lang.

Jiménez-Aleixandre, M. P. (2008). Designing argumentation learning environments. In. S. Erduar, & M.P. Jiménez-Aleixandre (Eds.), *Argumentation in Science Education: Perspectives from classroom-based research* (pp. 91-115). New York: Springer.

Kronig, W. (2003). Eléments d'interprétation du faible taux de réussite scolaire des enfants immigrés dans le degré primaire. In CDIP (Ed.), *Le parcours scolaire et de formation des élèves immigrés à «faibles» performances scolaires. CONVEGNO 2002: Rapport final* (pp. 24-33). Berne: Conférence suisse des directeurs cantonaux de l'instruction publique (CDIP).

Kuhn, D. (1991). *The Skills of Argument.* Cambridge: CUP.

Mehmeti, T. & Perret-Clermont, A.-N. (2014, August). *Seeking success of migrant pupils through designed tasks: A case study with Albanian pupils in Switzerland.* Paper presented at EARLI SIG 10, 21, 25 joint meeting: Open spaces for interaction and learning diversities, Padua, Italy.

T. Mehmeti, & A.-N. Perret-Clermont (2016). Seeking Success of Migrant Students through Designed Tasks: A Case Study with Albanian Students in Switzerland. In A. Surian (Ed.), *Open Spaces for Interactions and Learning Diversities* (pp. 137-150). Rotterdam: Sense Publishers.

Müller, R. (2001). Die Situation der ausländischen Jugendlichen auf der Sekundarstufe II in der Schweizer Schule – Integration oder Benachteiligung? *Schweizerische Gesellschaft für Bildungsforschung, 23*(2), 265-298.

Muller Mirza, N., et al. (2009). Psychosocial processes in argumentation. In N. Muller Mirza & A.-N. Perret-Clermont (Eds.), *Argumentation and education: Theoretical foundations and practices* (pp. 67-90). New York: Springer.

Perret-Clermont, A.-N. (2015). The Architecture of Social Relationships and Thinking Spaces for Growth. In C. Psaltis, A. Gillespie & A.-N. Perret-Clermont (Eds.), Social Relations in Human and Societal Development (pp. 51-70). Basingstokes (Hampshire, UK): Palgrave Macmillan.

Rigotti, E., & Greco Morasso, S. (2010). Comparing the Argumentum Model of Topics to other contemporary approaches to argument schemes: The procedural and material components. *Argumentation, 24*(4), 489-512.

Schubauer-Leoni, M.L. (1986). *Maître-élève-savoir : analyse psychosociale du jeu et des enjeux de la relation didactique.* Thèse de doctorat en Sciences de l'éducation, Université de Genève.

Schwarz, B. (2009). Argumentation and learning. In N. Muller Mirza & A.-N. Perret-Clermont (Eds.), *Argumentation and education: Theoretical foundations and practices* (pp. 91-126). New York: Springer.

Sensevy G. & Mercier A. (dir.). (2007). *Agir ensemble: l'action didactique conjointe du professeur et des élèves.* Rennes: Presses universitaires de Rennes.

Images:
1) https://kasaselimi.files.wordpress.com/2010/03/mbetruina.jpg
2) thinkstockphotos/stockbyte:
http://cache4.asset-cache.net/xr/56530327.jpg?v=1&c=IWSAsset&k=3&d=8A33AE939F2E01FF5442AB8FC2AF2ED849B62DCF13617E5E26F109DF68AEEBDABCC685C059D63657

27

Analysing Arguments in Decision Making Discourse

KIRA GUDKOVA
Philological Department, Saint Petersburg State University, Russia
gudkovakira@bk.ru

The paper deals with the analysis of argumentation from the perspective of practical reasoning. I conduct a comparative analysis of the university students' dialogues in which they advocate the proposition of value and the proposition of policy. The argument analysis focuses on the structure of the reasoning and the relevance of the argument. The results obtained reveal the differences and similarities in arguments in two types of persuasive dialogues.

KEYWORDS: argument, decision making dialogue, educational dialogue, value stating dialogue

1. INTRODUCTION

The paper deals with the analysis of argumentation from the perspective of practical reasoning. I focus mainly on the problems of argumentation literacy in educational sphere. Since the introduction of the Testing System in Saint Petersburg State University, there has been a renewed interest in argumentation from a pedagogical perspective. Such renewal in argumentation results from the realization that argumentation skills are a necessary part of education at all levels. Consequently, the number of papers and research projects on the problems of argumentation has increased greatly. Yet one cannot help but wonder how many misconceptions, among teachers and students, exist about what a good argument is, whether an argument is reasonable or fallacious.

Saint Petersburg State University has established B2 level of English as a graduation requirement. Thus, all students of all faculties sit a final English test aimed at assessing productive and receptive skills of the graduates. The speaking part of the test contains two parts: a monologue and a dialogue in which students have to defend a standpoint with the help of argumentation. There are two types of

D. Mohammed & M. Lewiński (eds.) (2016). *Argumentation and Reasoned Action: Proceedings of the 1st European Conference on Argumentation, Lisbon, 2015. Vol. II, 479-492.* London: College Publications.

disputes: either disputes asserting something to be good or bad, wright or wrong, effective or ineffective or disputes over what should or should not be done. Thus in two types of tasks students are supposed to advocate either proposition of value or proposition of policy. The proposition of value establishes a judgmental standard. Any attempt to demonstrate something to be good, right or effective depends on the criteria for goodness, rightness, or effectiveness. The advocate of a proposition of value normally applies his/her own criteria. Arguing a policy proposition calls for a more sophisticated and developed series of arguments.

The methodological frameworks I refer to is an approach suggested by K. Rybacki & D. Rybacki for argumentative reconstruction (Rybacki & Rybacki, 1991) and dialogue typology (Walton, 1998; 2006) for dialogue analysis. As a case study, I conduct an analysis of the recorded students' dialogues in which they advocate either the proposition of value or the proposition of policy. The dialogues are authentic conversational exchanges that took place at the final test that students sat after finishing their English course. The dialogues present the material for identification, analysis and evaluation of argumentation. The results obtained show the strategies used by participants and the peculiarities of arguments put forward in two types of persuasive dialogues. The choice of recorded dialogues as a source of data is motivated by their argumentative nature.

2. THE CHARACTERISTICS OF EDUCATIONAL DIALOGUE

Different communicative practices have developed within various domains of communicative activity. These practices show themselves in various speech events.

> Communicative activity types are conventionalized practices whose conventionalization serves through the implementation of certain 'genres' of communicative activity the institutional needs prevailing in a certain domain of communicative activity (van Eemeren, 2010, p. 139).

The communicative activity types in the educational communication include, among many others, lectures, seminars, presentations, dialogues. Not all of them make use of argumentation; however, educational dialogues always contain argumentation, as it is a kind of obligation imposed on the participants. They have to conduct the conversational exchange within the frame of civilized argumentative behavior.

A dialogue is a type of goal-directed conversation in which two participants are participating by taking turns.

> A dialogue is defined as a normative framework in which there is an exchange of arguments between two speech partners reasoning together in turn-taking sequence aimed at a collective goal (Walton, 1998, p. 30)

At each move, one party responds to the previous move of the other party. Each dialogue is a connected sequence of moves that has a direction of flow. In my research, I adopt this general definition as the dialogues I am investigating fit it: there are two participants who take turns, the participants respond to the previous moves of each other.

Besides these general characteristics that all types of dialogue possess the educational dialogue has some peculiarities that are specific to this particular communicative practice. Let us consider these features in detail. The main characteristic of educational dialogue is that it is specific rule-governed communication. One set of rules is the grammar of the English language that governs all communication in English. In educational dialogue, the rules of grammar have become of great importance as the accuracy in choosing grammatical structures directly affects the outcome of communication, in other words the score obtained. The participants (especially the student) are well aware of that. It is not an easy task to produce good arguments in your native tongue; it can be a challenge to do that in a foreign language. The students concentrate chiefly on the accuracy and this leads to the similarity of the arguments.

In addition to the rules of grammar, educational dialogues have their own particular rules that govern communication and that are specified through formal instructions. One of the most important characteristics of educational dialogue is civility. As D. Walton notes civility means that the participants take turns making various moves (Walton, 2006, p. 172). In our case, civility is defined by the environment in which the dialogues take place. The positions that the participants occupy are strictly set: the one party is always an interlocutor and the other is a student. Thus, there is always an age gap between the parties and, in addition, the parties occupy different social positions, with one of them always being superior. These conditions influence the argumentation and pose some challenges for the students, for though they can produce good arguments, while doing so they are restricted by the civility rules. The participants take turns and neither party dominates the dialogue. Hence, some good arguments may be left behind as the proposition or the wording are considered not appropriate for the situation of argumentative exchange with an adult.

The time allocated to the task is strictly limited and this as well has a certain influence on the argumentation of the participants. They should keep to the point and try not to produce long chains of arguments.

These characteristics of educational dialogue affect the arguments that are put forward. Although I expect arguments that can be fallacious from theoretical point of view as I deal with practical reasoning, yet there are fallacious arguments that are not possible in educational dialogue due to the civility rules of the exchange. Namely, I do not expect such arguments as ad hominem and ad baculum. They are impossible in this type of communicative activity due to the mentioned above specific conditions in which the communication exchange takes place (age gap, social positions, and social roles).

D. Walton distinguishes five types of dialogues in which argumentation can occur: persuasion, dialogue, inquiry, negotiation, deliberation, information-seeking dialogue and eristic dialogue. Each dialogue possesses its own characteristics and goals. In addition, while each type of dialogue has its collective goal, each participant has an individual goal. For instance, the goal of persuasion dialogue is resolving or clarifying an issue, whereas the goal of a participant is persuading other party. Similarly, the goal of deliberation dialogue is deciding best available course of action, while the goal of a participant is coordinating goals and actions (Walton, 1998). Argumentation in a deliberation dialogue usually takes the form of practical reasoning, chaining together goals and possible courses of action. The conclusion states that some particular course of action is chosen.

I distinguish two types or genres of educational dialogue. Both of them possess the general characteristics and have their goals: value estimating dialogue and decision making dialogue. An individual goal of the first type is similar to the goal of persuasion dialogue and the goal of the second type is close to the goal of deliberation dialogue as D. Walton describes them. I am interested mainly in these two types as they represent the dialogues in which students take part.

2.1 Examples of value stating and decision making dialogue

The task students are supposed to do is written in their cards. Here is the example of the assignment in which students produce a decision making dialogue.

> (1) You have to explain which graduation scenario you would like to try. You are supposed to choose from the three options given:
> find a job,
> teach a subject you majored in,
> get a master's degree.

Thus conversing with the interlocutor students have to make a decision and choose one scenario from those provided in the card. The decision should be supported by arguments.

Let us look at the examples of the assignment for a value stating dialogue.

> (2) You are preparing for a debate: "Is advertising a necessary evil?" You are supposed to focus on the following areas where agencies and companies use advertising:
> education,
> travel,
> music and cinema,
> food.

In this type of dialogue students are supposed to decide whether something (in our case advertising) is good or bad, necessary or not.

Some decision-making dialogues include values and participants are deciding on a course of actions that is best, the most important etc. Here are the examples of the assignments:

> (3) You are to choose the best way to study a foreign language from three possible options given:
> language school in your country,
> studying abroad
> studying on your own.

One more example:

> (4) You have to discuss which one programme of the City Development plan would be the most important for young people today. You have three options to choose from:
> free Wi Fi in public places,
> volunteer litter clean-up events,
> communal public bikes.

These are authentic assignments for students to produce dialogues that were recorded and that present practical material for analysing and evaluating argumentation.

3. ANALYSING STRATEGIES IN VALUE STATING DIALOGUE

Values are deeply rooted mental states and they are formed early in life. They predispose us to categorize something as existing somewhere along a mental continuum ranging from highly positive to highly negative. Values do not exist independent of each other; they exist in a hierarchy, with some values being deemed more important than others in a given set of circumstances (Rybacki & Rybacki, 1991, p. 42).

Value argumentation communicates our feelings about something and the standards of judgment from which those feelings derive. The purpose of argumentation is how we should judge something. Value propositions identify a value object and a value judgment, a general evaluation – good- bad. According to K. Rybacki & D. Rybacki, there are several steps involved in advocating the proposition of value: 1. the first step is defining the value object. At this step, the definition should be given to clarify the value object 2. The second step is identifying the hierarchy. The arguer identifies the field from which value standards are taken. 3. The next step is specifying the criteria for evaluation. The participants use a generally understood and accepted framework of values. If the criteria are not readily accepted, the arguer may combine value standards from different fields and establish the relevance between the value object and the criteria. 4. The forth step includes measuring the value object against the value judgment criteria that demonstrate that the value object fits the criteria on the basis of the following: a. What element of the society is influenced by the value object? (argument on effect); b. To what degree or in what amount does the effect occur? (argument on extent; c. What is the cause that produces the effect and extent of the value object? (argument on inherence) (Rybacki & Rybacki, 1991, pp. 72-73).

3.1 Dialogue analysis

3.1.1 Examples from recorded dialogues

Below are excerpts from the dialogues in which students judge advertising according to commonly accepted criteria and put it low in the hierarchy of societal values.

> (5) A lot of people are annoyed by advertising, they think that there is too much advertising everywhere: on the TV or the streets, you just can't get, on the radio, you just can't get away from it.

> (6) Personally, as a person who doesn't watch much TV, I don't encounter advertising as often in my life, but sometimes encountering it I do think that it is sometimes too aggressive or doesn't give enough information, not informative enough, too aggressive so, advertising

> (7) The advertising like McDonalds, KFC and others they make people go to this place, to this fast food restaurant, especially children who see this advertising and ask their parents – Oh, I want to go, and they don't know that it can be very harmful because there is no information about the drawbacks of this food advertising.

> (8) Advertising of products and sometimes they are not, not very good for the society. For example, beer or alcohol, generally, is, it doesn't give any benefits to the society and so tobacco or certain lifestyle, for example, I don't know, nightclubs, etc.

In the following excerpts, advertising is placed on the positive side of the scale.

> (9) I think advertising is necessary evil, because now people, very often people don't know what to choose, because we have so many, so many things that, so many things which they can choose. That's why it's necessary.

> (10) social advertising is very good, its usually comes up with the education at schools or you can see it all over the city in the billboards, at the radio, on the television, and it really helps the society to know what's wrong and what's right.

> (11) I think the best advertising nowadays is advertising about food. Because all people need food every day. And it's most useful thing. Because there are a lot of advertisements and you can compare them. For example, on this advertisement this butter costs this number and on this advertisement it costs so. And then you decide where you should go.

> (12) But I can say that our modern world will not be good without advertising cause our modern lifestyle is very fast. And it's good way to take information very fast.

Let us look at how the above-mentioned steps are taken in the educational discourse about the role of advertising in different spheres.

3.1.2 Defining the value object

In the examples provided, the value object (advertising) is instantly recognizable and does not require a lengthy definition (all students share a common knowledge about advertising they know what advertising is and what its functions are).

3.1.3 Identifying the hierarchy

The arguer identifies a particular value standard that serves as a potential source of criteria for judging the value object. He identifies the field from which value standards are drawn and locates the value object in a value hierarchy. The advocate has a range of fields of argument – legal, moral, ethical, political, economic, and scientific – from which to choose. The advocate decides to combine value standards from different fields. Having chosen the field, he or she puts forward arguments that demonstrate the position of the value object in a value hierarchy.

In the given examples the fields chosen are ethical – not very good for society, can be very harmful; moral_- the value of privacy is highlighted aggressive, too much advertising; economic - not informative, doesn't give any benefits to society, good way to take information.

3.1.4 Specifying the criteria for evaluation

At this step, the advocate states what criteria will be used. Those who oppose the positive role of advertising in different spheres of our life apply the following criteria, as you can see in examples 13-18: advertising is not very good for society.

> (13) Advertising of products and sometimes they are not, not very good for the society.

It does not give many benefits to society

> (14) For example, beer or alcohol, generally, is, it doesn't give any benefits to the society and so tobacco or certain lifestyle, for example, I don't know, nightclubs, etc.

It can be harmful

> (15) The advertising like McDonalds, KFC and others they make people go to this place... it can be very harmful because there is no information about the drawbacks of this food advertising.

It is everywhere

> (16) You just can't get away from it.

It is not informative

> (17) It doesn't give enough information

It is annoying

> (18) A lot of people are annoyed by advertising.

Those students who argue that the role of advertising in our society is positive apply the following criteria, as you can see in examples 19-22: advertising helps us to choose from many options

> (19) you can compare... For example, on this advertisement this butter costs this number and on this advertisement it costs so. And then you decide where you should go.

> (20) very often people don't know what to choose, because we have so many, so many things that, so many things which they can choose. That's why it's necessary.

It really helps the society to know what is wrong and what is right.

> (21) social advertising is very good it really helps society to know what is wrong and what is right.

It saves our time

> (22) it's good way to take information very fast.

3.1.5 Measuring the value object against the value judgment criteria

Measuring the value object with the criteria is the fourth step in value advocacy. Three sub issues should be addressed:

a. What element of the society is influenced by the value object? Examples taken from the recorded dialogues show the following:

> (23) society – in case of social advertising
> children - fast food restaurants

b. To what degree or in what amount does the effect occur?

> (24) lot of people are annoyed by advertising, all people need food
> our modern world will not be good without advertising

Effect and extent arguments are necessary elements. If a value object exists extensively, but not have much effect on society, concern may not be justified. Equally, if the effect of the value object is very serious but does not extend to a great number of individuals, concern may not be justified.

c. What is the cause that produces the effect and extent of the value object? Are the effect and its extent the result of something intrinsic to the value object? Advertising companies do not care about people being influenced by what they see and hear and this attitude may cause a lack of concern about the amount of advertising. Advertising is an undeniable part of sales strategies. Helps to promote values. Inherency arguments prove that the effect and the extent attributed to a value object are central to the value system of society or some elements of it.

4. ANALYSING STRATEGIES IN DECISION MAKING DIALOGUE

There are several patterns of organization for policy advocacy: traditional organization – is used when the reason for change involves righting past wrongs and showing the subsidiary benefits of the proposed policy. The comparative advantage structure – is used if the reason for change relates to the attainment of a more desirable future state. It compares the proposed policy to existing policy and argues that the proposed policy is more advantageous. The organization of comparative advantage advocacy begins with the presentation of the policy proposal. The advocate then indicates one or more advantages to be achieved by adopting this proposal. Each advantage should be unique. Only the proposed policy is capable of achieving it. In addition to demonstrating uniqueness, the advocate establishes a quantitative and qualitative measure of each advantage's value. The third type of organization is goals-criteria advocacy. It begins by examining what society values are and the goals it has set to achieve these values. The proposed policy is examined in terms of value criteria that measure its ability to obtain the desired goal (Rybacki & Rybacki, 1991, pp. 82-85).

While advocating the proposition of policy, several issues should be addressed. The first issue is identification of the disparity. Answering the question is important because if no reason for change exists change is unwarranted. The consequences of the disparity also should be taken into account. The second issue is the advocating a way to remedy the disparity. The remedy is a new policy by which a preferred state is reached. The third issue is the consequences of the proposed policy.

4.1 Dialogue analysis

In decision-making dialogue students take different roles: that of a proponent and the role of an opponent. That is why it happens that students not only argue in favour of a certain policy but also oppose some other policy. That happens because of the role that the interlocutor plays in the dialogue. As there is always a choice presented in the task he or she puts forward arguments in favour of other policies that students chose to advocate or refute.

In the examples given below the students advocate the best programme for young people and put forward arguments for installing free Wi-Fi in public spaces.

> (25) And I think that we really need it in public places because we have a lot of tourists, we have ... we have special gadgets with Wi-Fi connection and we can't ... we can't connect every time when we go out from home and we should have these places with free Wi-Fi that we can save our moneys.

> (26) Because many devices use Internet very often: telephones, laptops and so on and many people need, need to communicate and use Internet to call anybody from a lot of places...People can use, for example, Skype unlike standard calls of telephone and do it for free.

> (27) in some places with a free Internet can help us... help us to use Wi-Fi ... to use Internet when we, for example, can't have money on the phone, when we can't put money on our gadgets.

The following examples show the strategies used to oppose the policy proposition. The opponent of the policy tries to demonstrate that good and efficient reasons exist to consider the policy unacceptable.

> (28) At the same time I don't agree with that idea that it would be useful and worthwhile for our society to extend such places
>
> (29) because our people, particularly young people, are too lazy and Wi-Fi is quite aid for them and it's not so good idea to use Wi-Fi in public places.
>
> (30) So a lot of young people don't have enough money to use not free Internet access so free Wi-Fi is quite attractive item of public places which attract a lot of people to go there, to sit there, sitting all time there and spending a lot of time doing nothing, only surfing the Internet.
>
> (31) parents buy themselves childrens , for example, iPads, laptops with Internet connection and it is ... it could be really bad for their psychology ... psychologies because there are many websites with bad information for them with really... it could be really avoid them.

Let us see how all three issues are addressed in the argumentation exchange in which the student opposes the installation of free Wi-Fi in public places. The opponent of the policy tries to demonstrate that good and efficient reasons exist to consider the policy unacceptable. First, the arguer identifies the disparity. The reason for change exists. Moreover, if the proposed policy is accepted the reason for change will be resolved. There will be free Wi-Fi in public places. As you can see in example 32,

> (32) So, firstly, I'd like to speak about free Wi-Fi- in public places. So unfortunately nowadays there are not so many places where free Wi-Fi is provided.

The arguer choses to oppose the proposed policy and refutes the reason for change, as you can see in example 33,

> (33) At the same time I don't agree with that idea that it would be useful and worthwhile for our society to extend such places because our people, particularly young people, are too lazy and Wi-Fi is quite aid for them and it's not so good idea to use Wi-Fi in public places.

The strategy that the opponent choses is the strategy of refuting the change. He performs a "worst case" analysis of the situation and portrays the consequences as bad for society, especially for its most vulnerable part – young people. As you can see in example 34,

> (34) So a lot of young people don't have enough money to use not free Internet access so free Wi-Fi is quite attractive item of public places which attract a lot of people to go there, to sit there, sitting all time there and spending a lot of time doing nothing, only surfing the Internet.

The reason for the proposed policy exists, as there are not enough places with free Wi Fi so the change is warranted. However, the consequences of the proposed policy are portrayed as bad for society.

5. ARGUMENT ANALYSIS

Argumentation in educational dialogue is in the form of practical reasoning and the arguments put forward are often fallacious from theoretical point of view. As I mentioned above the setting of argumentative exchange is formalized, and specific time limit and responsibilities are imposed on the participants (age gap, social positions). That is why there are no ad hominem and ad baculum arguments. They are impossible in educational dialogue. The arguments that occur: hasty generalization, arguments from analogy, arguments from consequences. Shifting grounds often occur and is understandable because the nature of the dialogue predisposes these fallacies. Below are some examples of the arguments put forward in the dialogues: hasty generalization,

> (35) Our people, particular young people are too lazy I think in our time every person can afford some gadgets with the access to Wi-Fi and ...

Argument from negative consequences,

> (36) it could be really bad for their psychology ... psychologies because there are many websites with bad information for them with really... it could be really avoid them.

Argument from analogy,

> (37) I think it (amusement park) would be safe and children and adults will enjoy it because you know about Disneyland and some other parks.

6. CONCLUSION

Educational dialogue is a genre of communicative activity that has developed within educational domain and that has both a normative and a descriptive status. This dual nature of the dialogue influences the argumentation that takes place. We deal with argumentation (not persuasion) in educational dialogue as its normative nature predisposes that students are not going to elicit an emotional reaction from the examiner. Rather they will try to operate on the rational level. They focus on proof and reasoning.

The descriptive nature of educational dialogue predisposes that argumentation takes the form of practical reasoning. Advocating or opposing value propositions students combine value standards from different fields; for the most part, they employ moral, ethical arguments. Advocating or opposing policy propositions students use mainly comparative advantage structure. As naïve arguers, students communicate their own ideas about an object or policy in question; they base their reasoning mostly on their personal life experience.

The dual nature of educational dialogue influences the arguments that are put forward. The normative aspect makes it impossible the use of certain arguments (ad hominem, ad baculum). The descriptive aspect manifests itself in the arguments that are for the most part fallacious from theoretical point of view, the most common being hasty generalization, shifting ground.

REFERENCES

Eemeren, F. H., van. (2010). *Strategic maneuvering in argumentative discourse.* Amsterdam: John Benjamins.

Rybacki, K., & Rybacki, D. (1991). *Advocacy and opposition. An introduction to argumentation.* Englewood Cliffs, NJ: Prentice Hall.

Walton, D. (1998). *The new dialectic. Conversational contexts of argument.* Toronto: University of Toronto Press.

Walton, D. (2006). *Fundamentals of critical argumentation.* Cambridge: Cambridge University Press.

28

Automatic Exploration of Argument and Ideology in Political Texts

GRAEME HIRST
Department of Computer Science, University of Toronto
gh@cs.toronto.edu

VANESSA WEI FENG
Department of Computer Science, University of Toronto[1]
vanessa.w.feng@gmail.com

The underlying argumentation of politically-opinionated texts tends to be informal and enthymematic, and commingled with non-argumentative text. It usually assumes an ideological framework of goals, values, and accepted facts and arguments. Our long-term goal is to create computational tools for exploring this kind of argumentation and ideology in large historical and contemporary corpora of political text. Overcoming the limitations of contemporary lexical methods will require incorporating lexical, syntactic, semantic, and discourse-pragmatic features into the analysis.

KEYWORDS: argumentation schemes, discourse parsing, framing, ideology, political argumentation, natural language processing, rhetorical structure theory, shibboleth, vocabulary

1. INTRODUCTION

Politically opinionated texts, whether written or spoken, are naturally occurring argumentation. They include oral speeches by members of a legislature and written opinion pieces in news publications. The goal of a politically opinionated text is to persuade the hearer or reader that a particular political position is correct, thereby changing or reinforcing the present beliefs of the hearer or reader. However, the underlying argumentation tends to be informal and enthymematic, and, especially

[1] Author's present affiliation: Workopolis Partnership, Toronto, Canada.

D. Mohammed & M. Lewiński (eds.) (2016). *Argumentation and Reasoned Action: Proceedings of the 1st European Conference on Argumentation, Lisbon, 2015. Vol. II, 493-504.* London: College Publications.

in oral speeches, commingled with non-argumentative text. In particular, it tends to assume an ideological framework of goals, values, and accepted facts and arguments.

Converse (1964) defines an *ideology* as a system of beliefs that is "bound together by ... constraint or functional interdependence" — that is, an individual's political beliefs are not chosen at random; rather, they fit together into a broader system. The most fundamental and enduring dimension of variation in ideology is *left versus right*, a divide that is pervasive in politics (Cochrane 2013, 2015). People of differing ideological positions will often *frame* matters differently in argumentation on any particular issue, where the *framing* of an issue is an ideological viewpoint or perspective on that issue: that is, a set of beliefs, assumptions, and pre-compiled arguments. Entman (1993, p. 52) describes framing as a matter of "selection and salience. To frame is to select some aspects of a perceived reality and make them more salient in a communicating text, in such a way as to promote a particular problem definition, causal interpretation, moral evaluation, and/or treatment recommendation for the item described." For example, on the issue of how much immigration should be allowed into their country, one person might frame the argument as one of economic benefit or detriment, whereas a second person might frame it as one of the benefits or problems of multiculturalism, and a third might frame it as an imperative, or not, of social justice.

And this leads to the idea of computational methods that could look at a political discourse and identify the *ideological framework* that the speaker or writer is implicitly using — in practice, some kind of quantifiable semantic reflections of ideologically charged ideas or beliefs. The work that we will present below is directed towards this long-term goal, putting an emphasis on automatically finding the relations between clausal units and on finding the unspoken, and possibly ideological, premises in an enthymematic argument. This would include the creation of computational tools for finding and analyzing argumentation in large corpora of political texts, both historical and contemporary. For example, these tools might answer, or help us answer, questions such as *Find arguments that support the Antwerp debt-reduction plan*, *How do opponents of the Cabbage Abatement Act justify their positions?*, and *What ideological frameworks were used to argue against immigration in 1905?* We envision users of such a system to include political historians, journalists, and ordinary citizens. (In our own work, we are focusing on the digitized archives of the parliamentary proceedings of Canada, the U.K., and the Netherlands.)

Building a system such as this is an ambitious, long-term project for the research field. Its components include automatically discriminating the argumentative portions of the texts from the non-argumentative and metadiscursive portions. In the former, we want to then automatically find the structure of the arguments, distinguishing the premises from the conclusion, identifying the argumentation scheme, and, where possible, the unstated argumentative and ideological elements. Prior research with similar goals has taken approaches largely based on text-classification methods with primarily lexical features, achieving only modest success.

Automatically identifying implicit assumptions and conclusions remains a distant goal, but it is one that can be aided by simpler methods for the identification of ideological positions and background knowledge about particular ideologies, such as those to be discussed in the next two sections. If, in the course of automatic political argument analysis, a known ideological assumption can be fitted to the hypothesized argument, then confidence increases in both the identification of the argument and the identification of its underlying ideology.

2. THE ROLE OF VOCABULARY

In a political debate, or some other expression of a position on some specific topic, where exactly in the language that speakers and writers use does their ideology become apparent? We might expect that it's not in the words themselves, because the words relate to aspects of the topic of the debate regardless of which side the speaker[2] is on, and that it's only at higher levels — sentences and text — that the ideology becomes apparent. But in fact, what we find is that different ideological frameworks lead to different word usage even for the same topic.

So then perhaps we can identify the ideology of a speaker just from the vocabulary that they use in their argument — so-called *bags of words* with a weighted frequency count of each word spoken. And researchers in Natural Language Processing have tried to do exactly this. Overall, the results have been mixed. For example, Diermeier, Godbout, Yu, and Kaufmann (2007) tried to classify U.S. senators as ideologically liberal or conservative just by looking at the words they used. They found that this was easy to do for senators who were at the extremes of the ideological spectrum; but they couldn't do it for

---

[2] We will generally use the word *speaker* to subsume both speakers and writers. In the experiments described below in this section, the data is written transcripts of political speeches.

senators who were in the middle of the spectrum, for whom they obtained essentially chance accuracy. Nonetheless, when they looked at the vocabulary that discriminated the extreme senators, they found a few easy shibboleths: For example, if an extreme senator says the word *gay*, they're liberal, and if they say *homosexual* they're conservative. And that one word (if they use it at all) is sufficient to accurately classify them. But usually, it isn't that easy.

We followed up on Diermeier et al.'s work (Hirst et al., 2014) by looking at speeches in the European Parliament, where there is a multi-party spectrum of ideology that is much broader than in the U.S. Congress and in which a left–right ideological division is dominant (Hix et al., 2007). We took the English version of the proceedings of the European Parliament from 2000 to 2010, and asked whether we could classify each speaker, using only their vocabulary, as left-wing or right-wing, and *a fortiori* classify them by party membership. Figure 1 shows the ideological spectrum of the parties in the period that we studied. For ideology classification, to create a left–right split, we removed the ALDE in the centre, and grouped the other parties as either left-wing or right-wing. For the party-membership classification, we removed the small right-wing parties from the data and classified only members of the five largest parties. The classification algorithm that we used was a support-vector machine with 5-fold cross-validation.

		← Left	**-Centre-**	**Right →**	
European United Left / Nordic Green Left (GUE/NGL)	Progressive Alliance of Socialists and Democrats (PES)	The Greens / European Free Alliance (EFA)	Alliance of Liberals and Democrats (ALDE)	European People's Party (EPP)	Small right-wing groups (ECR, EDD, UEN, EFD, ITS)

Figure 1 – The ideological spectrum of parties in the European Parliament in 2000 to 2010.

We found that we could distinguish between speakers from left-wing parties and those from right-wing parties with an accuracy of 78.5%; this was 28 points above the baseline of just choosing the most frequent class, which was 50.5%. Further, we could distinguish which of the five major parties a speaker belonged to with an accuracy of 61.8%, which was 23 points above the most-frequent-class baseline. And again there were a few easy shibboleths: for example, the words *profits* and *militarization* indicate a speaker from the hard left, and the words *subsidiarity* and *competitiveness* indicate a speaker from the hard right.

However, there are serious limits to this approach. For example, we found that it utterly failed on speeches in the Canadian Parliament (Hirst et al., 2014). The method was able to distinguish the language of a governing party from that of an opposition party with very high accuracy — 84 to 97% depending on the exact conditions — but what the classifier had actually learned was the language of political attack and defence, with little or no expression of ideological positions at all. This finding reflects the adversarial nature of Canadian politics.

In addition to finding ideological positions across topics, vocabulary-based methods are also used for *stance detection*, the task of determining the speaker's position, pro or con, on a specific known issue. This is typified by the work of Anand et al. (2011) and Somasundaran and Wiebe (2010).

Now, a critic of all this work might say, with some justification, that it completely evades most of the problem. We don't want to know only what a speaker's ideology or position is; we also want to know *how* they argue for or justify that position. Yet all these methods do is use the speaker's vocabulary, without even any consideration of the order in which the words were uttered, let alone any thought about meaning or content or structure of the argumentation itself! Nonetheless, they demonstrate that textual analysis does not always need to use structure or meaning or deep semantic analysis to succeed in its aims. But surely we can do even better if, yes, we start using a more linguistically informed analysis and incorporating syntactic, semantic, and discourse-pragmatic features into the analysis, as we will now discuss.

3. SHALLOW LINGUISTIC ANALYSIS TO RECOGNIZE ARGUMENTATION SCHEMES[3]

Argumentation schemes are the templates or structures from which ordinary textual arguments are built — common forms of argument that are more usually presumptive and defeasible than deductive. Walton, Reed, and Macagno (2008) have catalogued 65 distinct schemes, and each scheme has an associated set of critical questions that challenge arguments in the scheme and their implicit premises. Many of these schemes are quite rare, so we concentrated on the five schemes that are

[3] This section is based on work that was first presented by Feng and Hirst (2011).

most frequent in the Araucaria database, a corpus of annotated arguments produced at the University of Dundee[4]:

- *Argument from example*, and *argument from cause to effect*, whose meanings are clear from their names.
- *Practical reasoning,* which is an argument that a certain pre-condition should be brought about in order to achieve a goal.
- *Argument from consequences,* which is an argument that something should be done because the consequences will be good, or should not be done because the consequences will be bad.
- *Argument from verbal classification*, which is a quasi-syllogistic — "quasi-" because it depends on defeasible classifications.

It should be understood that in this work we are *not* recognizing the presence in the text of the arguments themselves or their elements — their premises and conclusions. This is a task that has been studied, with some moderate success, by other researchers (e.g., Mochales and Moens, 2008, 2009a, 2009b; Stab and Gurevych, 2014; Nguyen and Litman, 2015), and we see argument-scheme recognition as being "downstream" in the analysis pipeline from this, assuming its eventual success.

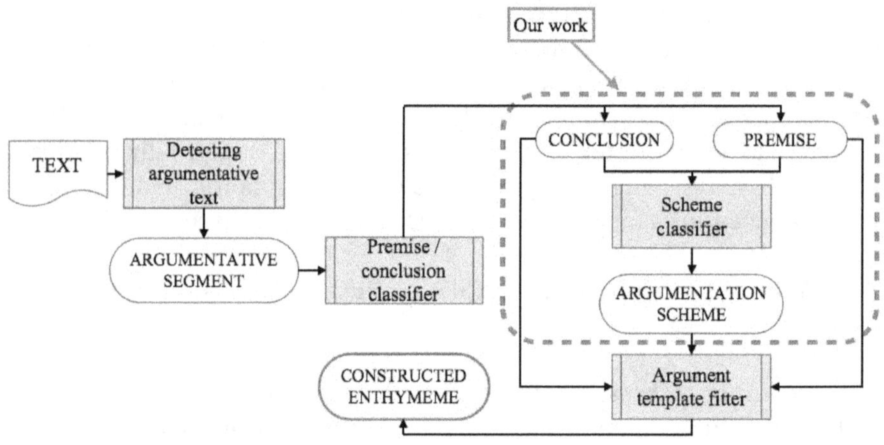

Figure 2 – The classification of argumentation schemes within an argument analysis system.

Hence, in a completed system (figure 2), prior processes would pick out argumentative segments of the input text, and try to identify the premises and conclusions in each. Given that, we then classify the

[4] http://araucaria.computing.dundee.ac.uk/doku.php

argumentation scheme that is being used, which, in turn, would be used by further processes to reconstruct the full enthymematic argument, and beyond that (not shown in the figure) to start to identify the ideological framing of the argument. Thus we are crucially assuming that current research on other aspects of argument analysis will be successful: detection and classification of the components, and determining whether an argument is linked or convergent — whether it requires a conjunction or merely a disjunction of its premises. We cast the problem as one of text classification.

Before any analysis of the text for argumentation, we first analyze it syntactically— specifically, a dependency analysis — using the Stanford parser (de Marneffe, MacCartney, and Manning, 2006)[5], which is a standard in the field. Then, to recognize argumentation schemes, we use a number of features that apply to all five schemes and some that are specific to each argumentation scheme. The features that apply to all schemes mostly concern the textual structure of the argument or whether it is a linked or convergent argument; the features are listed in table 1.

The location of the conclusion in the text.
The location of the first premise.
Whether the conclusion appears before the first premise.
The interval between the conclusion and the first premise.
The ratio of the length of the premise(s) to that of the conclusion.
The number of explicit premises in the argument.
Type of argumentation structure: linked or convergent.

Table 1 - Features used in classification for all five argumentation schemes.

The scheme-specific features are words and semantic patterns. For example, for argument from example, we look for the words *for example*, among others. For argument from cause to effect, we look for verbs that indicate cause, and we also use a number of textual patterns that indicate causal relationships (Gîrju, 2003); and analogously for practical reasoning. For argument from consequences, we look for propositions that are positive and negative, which we determine from the sentiment rating of the words in the *General Inquirer*, a computational lexicon (Stone, Dunphy, Smith, and Ogilvie, 1966)[6]. And

[5] http://nlp.stanford.edu/software/lex-parser.shtml

[6] http://www.wjh.harvard.edu/~inquirer

for argument from verbal classification, we look for textual similarities between the premises and the conclusions, and for appropriate dependency relations in both premises and conclusions: copulas, expletives, and negative modifiers. Details of these features are given by Feng and Hirst (2011).

Our classification algorithm was C4.5 (Quinlan, 1993), which builds a decision-tree for classification from a set of features, and we trained it on the Araucaria corpus of arguments from Dundee, introduced above, which is annotated with argumentation schemes. We created ten random pools of data in which baseline guessing was always 50%, and we then did 10-fold cross-validation on this data. We used two methods of evaluation: a one-versus-others classification in which we try to discriminate one scheme from all the others, and a pairwise classification for each of the ten possible pairings of the five schemes. Our evaluation metric was average accuracy.

The results for one-against-others classification are shown in table 2. We were able to distinguish argument from example and practical reasoning from all the others with accuracies above 90%. For argument from cause to effect, we achieved a more modest 70%, and accuracies were around 63% for the other two — which is still well above the baseline of 50%. The low accuracy for these last two schemes is probably due at least partly to the fact that they don't have the obvious cue phrases or patterns that the other three schemes have; and it is also perhaps because they were the schemes for which we had the least available training data. Table 3 shows the results for pairwise classification. For some pairs, we get near-perfect accuracy: practical reasoning versus argument from consequences and versus verbal classification; and practical reasoning versus argument from example and versus argument from cause to effect both achieve 93–94%. Many other pairs achieve accuracies around 86%. The result that stands out as poorest, at 64%, is between verbal classification and argument from consequences, which were also our two poorest categories for one-against-others classification.

Scheme	**Accuracy (%)**
Argument from example	90.6
Argument from cause to effect	70.4
Practical reasoning	90.8
Argument from consequences	62.9
Argument from verbal classification	63.2

Table 2 – Results of one-against-others argument scheme classification.

	Accuracy (%)			
	Example	**Cause**	**Reasoning**	**Consequences**
Cause	80.6			
Reasoning	93.1	94.2		
Consequences	86.9	86.7	97.2	
Classification	86.0	85.6	98.3	64.2

Table 3 – Results of pairwise argument scheme classification.

These results, then, can be the basis for future work to recover the missing premises of arguments. Some of these premises will be implied by the structure of the argumentation scheme itself, in conjunction with its critical questions. Others, we hope, will be found in a large set of what we are calling *"axioms"* of political argumentation and ideology, which we are presently working to derive from large corpora. Additional such "axioms" may be generated from the text itself, by textual entailment, implicature, or logical necessity. And some will come from searching more-general background knowledge, which is another present research area.

4. RECOGNIZING DISCOURSE STRUCTURE

Last, we briefly discuss the role of discourse parsing in the recognition of argumentation schemes — that is, determining the structure of an argumentative text in terms of *Rhetorical Structure Theory* (RST) (Mann & Thompson 1988). RST builds trees of relationships between the units of a discourse — the so-called *elementary discourse units* (EDUs), which are usually clauses or clause-like constituents of the text. There are 16 classes of relations possible between these units or between groups of units, which are listed in table 4. Some of the names, such as CAUSE and ENABLEMENT, already hint at the relationship between RST and argumentation. And in most of these relationships, a distinction is made in which one EDU is more prominent in the discourse than the other; the prominent unit is called the *nucleus* and the other is called the *satellite*.

ATTRIBUTION	CONDITION	EVALUATION	SUMMARY
BACKGROUND	CONTRAST	EXPLANATION	TEMPORAL
CAUSE	ELABORATION	JOINT	TOPIC-CHANGE
COMPARISON	ENABLEMENT	MANNER-MEANS	TOPIC-COMMENT

Table 4 – Classes of relationships in Rhetorical Structure Theory.

A number of the discourse relationships of Rhetorical Structure Theory (RST) have clear counterparts as argumentative relationships, and in the case of arguments, the RST structure of the text will mirror, at least to some extent, the structure of the argument; therefore, an RST analysis of a text will be an important component of the structural analysis of argumentation in a text. In one experiment using five common argumentation schemes, Cabrio, Tonelli, and Villata (2013) showed that an RST relation did indeed match the cognate argumentation scheme in about two-thirds of the cases where annotation was possible at all. Hence, RST relationships will be an important feature both in the analysis of arguments, and in recognizing argumentative text in the first place; and hence discourse parsing becomes part of this work. Discourse parsing is not a new topic, but recent work both by us (Feng and Hirst, 1012, 2014) and by others (Joty, Carenini, Ng, and Mehdad, 2013) has aimed to improve it substantially. This includes improving the initial segmentation into units, improving the parsing itself by using more linguistic knowledge and by building a smarter parser that works differently between sentences than within sentences. Feng and Hirst's parser also includes a post-editing adjustment process.

The results are generally increased accuracies in all aspects of the procedure compared to earlier work, and getting closer to human levels. Even though the role of some RST relations in the structure of argument is unclear or uncertain, we hypothesize that as features for classification, they will nonetheless have a positive effect on the structural analysis. RST structure, then, becomes another important feature for our recognition of arguments and argumentation schemes, and forms part of our current research.

5. CONCLUSION

Research in the automatic (or semi-automatic) analysis of political and opinionated text has begun to look more deeply at arguments, opinions, and ideologies. Shallow, word-based methods often suffice for simple analyses, but more-linguistically informed methods are necessary to get at the actual structure of arguments. This is a useful end in itself, but beyond that, we expect that this research will also become part of the general idea of semantic search, so that, with future developments, argumentation, opinion, and ideology can be used as facets in Web searches, or searches of large document collections; in automatically answering questions; and in automatically creating summaries and syntheses of large numbers of documents.

ACKNOWLEDGEMENTS: This work was financially supported by the Natural Sciences and Engineering Research Council of Canada.

REFERENCES

Anand, P., et al. (2011). Cats rule and dogs drool!: Classifying stance in online debate. *Proceedings, 2nd Workshop on Computational Approaches to Subjectivity and Sentiment Analysis,* Portland, Oregon, pp. 1–9.

Cabrio, E., Tonelli, S., & Villata, S. (2013). A natural language account of argumentation schemes. *Proceedings, XIII International Conference of the Italian Association for Artificial Intelligence* (LNAI 8249), Turin, pp. 181–192.

Cochrane, C. (2013). The asymmetrical structure of left / right disagreement: Left-wing coherence and right-wing fragmentation in comparative party policy. *Party Politics,* 19(1), 104–121.

Cochrane, C. (2015). *Left and Right: The Small World of Political Ideas.* Kingston, Ont.: McGill–Queen's University Press.

Converse, P.E. (1964). The nature of belief systems in mass publics. In D.E. Apter (Ed.), *Ideology and Discontent* (pp 206–261), London, UK: Collier-MacMillan.

de Marneffe, M.-C., MacCartney, B. & Manning, C.D. (2006). Generating typed dependency parses from phrase structure parses. *Proceedings, 5th International Conference on Language Resources and Evaluation,* Genoa, 449–454.

Diermeier, D.; Godbout, J.-F.; Yu, B.; & Kaufmann, S. (2007). Language and ideology in Congress. *Annual Meeting of the Midwest Political Science Association.*

Entman, R.M. (1993). Framing: Toward clarification of a fractured paradigm. *Journal of Communication,* 43(4), 51–58.

Feng, V.W. & Hirst, G. (2011). Classifying arguments by scheme. *Proceedings, 49th Annual Meeting of the Association for Computational Linguistics,* Portland, Oregon, pp. 987–996.

Feng, V.W. and Hirst, G. (2012). Text-level discourse parsing with rich linguistic features. *Proceedings, 50th Annual Meeting of the Association for Computational Linguistics,* Jeju, Korea, pp. 60–68.

Feng, V.W. and Hirst, G. (2014). A linear-time bottom-up discourse parser with constraints and post-editing. *Proceedings, 52nd Annual Meeting of the Association for Computational Linguistics,* Baltimore, pp. 511–521.

Gîrju, R.. (2003). Automatic detection of causal relations for question answering. *Proceedings of the ACL 2003 Workshop on Multilingual Summarization and Question Answering,* pp. 76–83.

Hirst, G., Riabinin, Y., Graham, J., Boizot-Roche, M., & Morris, C. (2014). Text to ideology or text to party status? In B. Kaal; E.I. Maks; & A.M.E van Elfrinkhof (Eds.), *From text to political positions: Text analysis across disciplines* (pp. 93–115), Amsterdam: John Benjamins.

Hix, S., Noury, A.G., & Roland, G. (2007). *Democratic Politics in the European Parliament.* Cambridge: Cambridge University Press.

Joty, S., Carenini, G., Ng, R.T., & Mehdad, Y. (2013). Combining intra- and multi-sentential rhetorical parsing for document-level discourse analysis. *Proceedings, 51st Annual Meeting of the Association for Computational Linguistics,* Sofia, 486–496.

Mann, W. & Thompson, S. (1988). Rhetorical structure theory: Toward a functional theory of text organization. *Text,* 8(3):243–281.

Mochales, R. & Moens, M.-F. (2008). Study on the structure of argumentation in case law. *Proceedings of the 2008 Conference on Legal Knowledge and Information Systems,* pp. 11–20.

Mochales, R. & Moens, M.-F. (2009a). Argumentation mining: the detection, classification and structure of arguments in text. *Proceedings, 12th International Conference on Artificial Intelligence and Law,* pp. 98–107.

Mochales, R. & Moens, M.-F. (2009b). Automatic argumentation detection and its role in law and the Semantic Web. *Proceedings of the 2009 Conference on Law, Ontologies and the Semantic Web,* pp. 115–129.

Nguyen, H.V. & Litman, D.J. (2015). Extracting argument and domain words for identifying argument components in texts. *Proceedings of the 2nd Workshop on Argumentation Mining,* Denver, 22–28.

Quinlan, J.R. (1993). C4.5: Programs for machine learning. *Machine Learning,* 16(3):235–240.

Somasundaran, S. & Wiebe, J. (2010). Recognizing stances in ideological on-line debates. *Proceedings, Workshop on Computational Approaches to Analysis and Generation of Emotion in Text,* Los Angeles, 116–124.

Stab, C. & Gurevych, I. (2014). Annotating argument components and relations in persuasive essays. *Proceedings, 25th International Conference on Computational Linguistics (COLING 2014),* Dublin, 1501–1510.

Stone, P. J., Dunphy, D. C., Smith, M. S., & Ogilvie, D. M. (1966). *General Inquirer: A computer approach to content analysis.* Cambridge, MA: MIT Press.

Walton, D., Reed, C., and Macagno, F. (2008). *Argumentation schemes.* Cambridge University Press.

29

Persuasion, Authority, and the (Common Law) Foundations of Transnational Legal Decision-Making

GRAHAM HUDSON
Ryerson University, Canada
graham.hudson@crim.ryerson.ca

This paper outlines a model of argumentation that formulates the processes by which international and comparative human rights law influence the reasoning of domestic judges. I argue that the persuasive influence of such law flows from, and is justifiable by reference to, a distinctive mode of rational argumentation centred around precedent and analogy. If sound, this model helps explain how persuasive influence may be distinguished from political or ideological power (i.e. authority) and how decisions to use such law are constrained by formal and informal institutions of interpretation i.e. are justifiable in jurisprudential terms.

KEYWORDS: analogical legal reasoning, international/ comparative human rights, persuasive authority, transnational law

1. INTRODUCTION

Domestic courts increasingly cite (and presumably rely) on international and comparative human rights law when making decisions. This practice may be described as part of what has been termed the "migration of constitutional ideas" -- a phenomenon by which judges, lawyers, legislators, human rights advocates, and other legal actors carry constitutional values, norms, and concepts across institutional divides (e.g. public/private law) and jurisdictional boundaries (Choudhry, 2011). Cast in its best light, the migration of constitutional ideas facilitates the development and pan-constitutional influence of the human rights enterprise, leading to the more complete protection and promotion of human dignity across national and other normative borders. The practice does not sit well, however, with the tenets of positivism or liberal constitutionalism, whereby decisions of

D. Mohammed & M. Lewiński (eds.) (2016). *Argumentation and Reasoned Action: Proceedings of the 1st European Conference on Argumentation, Lisbon, 2015. Vol. II, 505-518.* London: College Publications.

domestic judges must at all times be premised on domestic legal norms ultimately sourced within a state's constitution. As norms produced by outside political and ideological communities, international and comparative human rights lack this pedigree.

So it is that judicial reliance on international and comparative human rights has yet to be justified in jurisprudential terms. Some attempts have been made by way of reference to poorly defined (and arguably oxy-moronic) concepts, such as "persuasive authority" (Moran, 2005, pp. 158-159; Moran, 2003). This term suggests that international and comparative human rights law consists in external norms that judges may, but are not required to, use as reasons for a decision. But little effort has been invested in outlining precisely how such norms influence judicial reasoning, how persuasive influence may be distinguished from political or ideological power (i.e. authority), or how decisions to use such law are constrained by formal and informal institutions of interpretation. While there is considerable literature on persuasion in the context of domestic, common law settings (e.g. MacCormick, 1978, 2005; Schauer, 2008; Walton, 2002; Hage, Span & Lodder, 1994; Prakken, 1993), this literature presumes agreement about underlying criteria or "rules of recognition" that do not extend so far as to account for legal norms sourced in outside legal orders.

In this paper, I will use varying conceptions of transnational law to situate the migration of constitutional ideas along a continuum that reflects varying orders of disillusionment or disengagement with traditional accounts of judicial reasoning. I will relate these theoretical perspectives to strands of argumentation theory, and claim that the persuasive influence of international and comparative human rights such law flows from, and is justifiable by reference to, a distinctive mode of rational argumentation centred around precedent and analogy. While schematic, this model accounts for how the distorting effects of political and ideological power are constrained by a host of procedural rules that govern argumentation, orienting it towards the generation of shared understandings among diversely situated participants. If sound, this model will show that transnational legal decision-making is justifiable by reference to rather ordinary principles of common law reasoning and, ultimately, liberal constitutionalism. This comes at the cost, however, of sidelining the full, critical potential of more radical, but theoretically under-developed, conceptions transnational law.

2. TRANSNATIONAL LEGAL REASONING

Legal reasoning is, among other things, concerned with the sorts of considerations that contribute to the determination of a legal issue. Traditionally, much as today, philosophers have been pre-occupied with how jurists reason with state law. Accounts of common law reasoning, for example, tend to be embedded in the tenets of liberal constitutionalism, which supports a highly compartmentalized conception of law as a closed system of rules, delineated in terms of subject matter and jurisdiction. Philosophers have accordingly been concerned with (the boundaries) of positive law (e.g. statutory provisions, constitutional norms, doctrine), and only secondarily with those extra-legal materials that supplement but do not contest state authority.

Transnational legal reasoning differs from common law reasoning in that it is concerned with problems and/or solutions that transcend the jurisdictional boundaries of any given domestic legal system. We may begin to understand the nature of such reasoning by unpacking the concept of transnational law. According to Phillip Jessup, transnational law includes "all law which regulates actions or events that transcend national frontiers" (Jessup, 1956, p. 136); to actions and events, I would add relationships. At root, transnational law is part of a tradition of breaking faith with state law or, to put it another way, decoupling law from national political structures (Scott, 2009; Zumbansen, 2012). Depending on how far one wishes to push this tradition, transnational legal reasoning may describe three possible types of reasoning: transnationalized legal traditionalism, transnational legal decisionism, and transnational socio-legal pluralism (Scott, 2009).

Transnationalized legal traditionalism is the mode of transnational reasoning most in keeping with liberal constitutionalism, describing reasoning about positive law in transnational *contexts*. Here, state law is the exclusive authoritative base of decisions about actions, events and relationships that, as an empirical matter, transcend national borders. The power of judges to make decisions about transnational issues rests on authority that the state delegates, or tolerates, and at any time may revoke. For example, the state may empower a domestic court to exercise universal jurisdiction over international crime, to enforce the decisions of a regional or international court or tribunal, or interpret some piece of legislation consistently with international human rights. In kind, it may revoke this power through statute.

Transnational legal decisionism begins the process of decoupling law from national political structure. Whereas transnational legal traditionalism posits that it is simply the *context* of state law that is transnational, transnational legal decisionism posits that law itself becomes transnationalized, *operating* in a semi-autonomous transnational normative space. This body of law emerges through discursive interactions among jurisdictionally diverse courts and other interpretive communities. Communication arises horizontally (e.g. between courts of the same status, whether national or regional), vertically (e.g. between a national and "supranational" court), and through a mixture of both (Slaughter, 1994). These discursive exchanges result in the construction of doctrine, or a set of decisions, principles, and rationales that judges use to solve legal problems that occur/recur in multiple jurisdictions e.g. (inter)national security, transnational/international crime, environmental issues. While this body of law is irreducible to any given national political structure, it doctrine remains tethered to the state insofar as it is the product of judicial practice that the state recognizes, authorizes, or tolerates i.e. could terminate at any point; it is one possible outcome of transnationalized legal traditionalism.

Finally, there is transnational socio-legal pluralism, where law is entirely decoupled from national political structures. The legal pluralist element recognizes that law is not conceptually tied to the state. With the exception of critical legal pluralists (Kleinhans & MacDonald, 1997), those in this tradition see non-state normative orders as potentially having systematic qualities, which is to say, among other things, that a community has established an institutionalized, structured approach to the creation, development, interpretation and application of its law; there is a language or code that structures decisions about what is law/not-law. This perspective on transnational legal reasoning supports, on the one hand, a potentially complete decoupling of law from national political structures while, on the other hand, a positivist orientation towards sources and systems (Teubner, 1997).

3. PERSUASION, AUTHORITY, AND LAW

Transnational legal traditionalism and transnational legal decisionism are the two most obvious candidates for conceptualizing the migration of constitutional ideas. The comparative attractiveness of each depends in large part on their ability to address three outstanding questions: 1) whether, how, and why international and comparative human rights exert a distinctive persuasive influence on judicial reasoning, 2) how this influence fits into and reconstitutes interrelationships among diverse discursive communities located across jurisdictional and institutional divides, 3) how the use of extrinsic legal norms as bases of decision-making may be justified in jurisprudential terms.

The concept of persuasion has some history in domestically-oriented legal philosophy. The term "persuasive authority", for example, is used –perhaps oxy-moronically- to describe positive law that lacks binding or precedential weight, including *obiter dicta*, dissents, and extra-doctrinal case law (Lamond, 2012; Schauer, 2008). By contrast, there is virtually no detailed literature on the question of whether international and comparative law may serve as persuasive authorities in domestic law.

There is, however, some work on how international law may exert a distinctive persuasive influence in the sphere of international relations. Jutta Brunnée and Stephen Toope's "interactional" account of international law, for example, outlines a hypothetically distinctive logical form and rational force of international legal argument (Brunnée & Toope, 2000). Brunnée and Toope outline: 1) conceptual and functional differences between "material capabilities and interests, and normativity" (i.e. political authority and persuasion, respectively), and 2) the "constraining and facilitating (formal and informal) institutions of legal interpretation" that insulate participants to legal interaction from coercion (Brunnée & Toope, 2000, p. 25). In my view, this theory can be readily adapted to explain transnational legal interactions that occur (in) between international and domestic law.

Influenced by dialogical argumentation theory, Brunnée and Toope begin by insisting that legal interaction is ultimately oriented towards the generation of shared understanding i.e. inter-subjective structures and institutions "that help in specifying the interests that motivate action: norms, identity, knowledge, and culture" (Brunnée & Toope, 2000, p. 29). They outline a loose framework of legal interaction that ensures that each participant to an argument has an equal opportunity to initiate and perpetuate discourse, forward challenges, criticisms, interpretations and explanations, express their internal values, attitudes, feelings and other features of their personal identities, and share in the invocation and application of the regulatory rules that structure power relations (Burleson & Kline, 1979, pp. 421-423). Notice that these four conditions of equality are essentially procedural in orientation emphasizing, not what is argued, but *how* it is argued. These conditions, and the relatively stable, enduring rules that are used to secure them, frame an ideal speech situation within which personalized arguments are both permissible and encouraged, but which are nonetheless appraised relative to standards of public reason and their capacity to resist challenge.

The source of law's persuasive influence is to be found, not so much in general values and principles of a substantive character, but in the dialogical processes through which authoritative decision-makers address legal subjects with the reasons for a specific rule. Decisions that abide by rules of rational argumentation are most likely to rest on reasons that draw from the values, beliefs, interests, and expectations of those who will be made subject to obligation. In this way, legal subjects can endorse a rule or obligation as in keeping with core elements of their identities i.e. as persuasive.

Crucially, this process results in the dialectical reconstitution of both law and subject. Legal subjects change by being open to the culture, knowledge, identities, and values of other disputants, which is expressed in terms of law and public reason; they are changed by the process of identifying and helping construct shared understandings. Law also changes, as its content is indelibly marked by exogenous values, cultures, epistemologies, and norms. This process does not simply happen organically, but is guided also by what Lon Fuller describe as a "morality of aspiration"- a moral imperative that enjoins state officials to produce, interpret, and apply positive law in such a way as to protect and promote individuals' capacity to form and realize personal life projects. A morality of aspiration is concerned with the growth and development of positive law towards the actualization of social justice and collective well-being. As an empirical matter, law need not and may not work in this way, but if it is to be truly persuasive and aspirational, diverse discursive communities must be included in the production, interpretation, and application of legal obligation. This interactive diversity of knowledge and rigorously structured mode of discursive interaction helps ensure that law is developed in ways that are responsive, not just to the internal logic of formal doctrine, but also to the extrinsic identities, interests, knowledge, and aspirations of all who are subject to legal obligation.

Although schematically relayed here, we can see how Brunnée and Toope's theory clarifies the conceptual and functional difference between persuasion and authority in at least international contexts. This is both promising and challenging when applied to domestic contexts, where we have seen persuasion and authority are intimately connected. I believe, however, that Brunnée and Toope's may be adapted to explain how international and comparative human rights exert a distinctive persuasive influence on domestic decision-making. The answer lies in the nature of analogical legal reasoning.

4. ANOLOGICAL LEGAL REASONING

For our purposes, analogical reasoning occurs whenever a court "draws on similarities or dissimilarities between the present case and previous cases which have no precedential weight in the case at hand (Toulmin,

1958, p. 202; Perelman & Olbrechts-Tyteca, 1969, pp. 372-374). In the context of common law reasoning, analogies are used to demonstrate that the relationship between the facts and law in one case or situation is similar to the relationship between the facts and law in another. From these recognized or readily observable similarities, judges infer that there is a further hidden, typically more abstract similarity. Analogies accordingly guide the discovery of new ideas, relationships, and properties in ways that may be intuitive, logical, or both (Brewer, 1996).

The role of analogies in legal reasoning is contested in both analytical and normative terms. Analytically, there is disagreement about whether analogical legal reasoning has a distinctive logical form and what this means for its rational or persuasive force. These debates have normative implications, as many think that analogical reasoning is capricious, arbitrary, or unbounded by formal institutions of interpretation. Grant Lamond sheds some light on these issues by highlighting three types of analogical reasoning, which he positions along a continuum ranging from most to least consistent with common law precepts. The three types of analogies are classificatory, close, and distant.

Classificatory analogies are most consistent with common law precepts, being concerned with whether or not a new case belongs to an "existing category" i.e. is governed by a given line of precedent. (Lamond, 2014, p. 8). Judges here use analogies to decide whether an immediate case is governed by a common law rule. This process requires determining whether the facts of an immediate case are such as to function as the factual predicate of the rule i.e. sufficiently similar to those classes of facts to which a legal rule is supposed to apply. This process is problematic when factual similarities are not self-evident as well as when the content (and reasons) for a rule are ambiguous - a rather common occurrence. Lamond argues that there are a range of logical and professional techniques jurists use in order to guide analogical reasoning of this sort. This includes identifying the reasons for a rule, its relationship to principles, and its place within the larger doctrinal, institutional, and structural context (Lamond, 2014, pp. 6-9).

Close and distant analogies are different from classificatory analogies insofar as they are not oriented towards the enforcement of rules -- they guide deliberation precisely when there is no rule to enforce i.e. a case does not fit within an existing category governed by precedent. Since a decision has to be made, judges will have to forge ahead in the face of legal gaps. Rather than do so arbitrarily, they may (and should) rely in part on persuasive authorities that deal with similar factual situations or legal issues. Embedded within these non-binding (elements of) cases will be a line of thought or rationale as to how the rule in those cases coheres with the rules, principles, and objectives of overarching doctrine. Even though a case is non-binding, it provides insights into how a postulated new legal rule may be tailored to advance the aims and spirit of doctrine.

Close and distant analogies in this way simultaneously facilitate and constrain the development of doctrine; they frame the discovery of new laws while ensuring that new rules cohere with principles and values of escalating generality. Lamond for this reason argues that that "neither the doctrine of precedent itself nor close analogies can be properly understood except in relation to each other". (Lamond, 2014, p. 11). Close and distant analogies differ in the extent to which they (reliably) supplement the doctrine of precedent. Close analogies describe instances when judges draw from cases that are part of the same doctrinal field as the novel case (e.g. torts, property, contracts); distant analogies draw cases from outside doctrines. The "closer" the non-binding case is to the (sub-)doctrine that frames the legal issues of the immediate case, the more useful it is and, hence, the greater the persuasive force of the analogy. It follows that distant analogies are less persuasive or justifiable than close analogies. This is because they do not reliably advance the aims of precedent (i.e. internal consistency and structured development of doctrine) and are more "haphazard" or less constrained by (in)formal institutions of interpretation (Lamond, 2014, p. 15).

Lamond's theory blurs the boundaries between, if not outright conflates, persuasion and authority; non-binding laws are not authoritative in the sense that they provide complete reasons for applying a rule, but they are authoritative in the sense that they are, and can only be, persuasive because they are part of the corpus of positive law. Lamond does not, for example, discuss how non-state law or normative orders might be persuasive authorities if and when they perform the function of facilitating and constraining the development of doctrine. This is surprising; I would expect any norm that furthers the internally consistent development of doctrine to be a legitimate consideration for precisely the reasons Lamond provides. Why should it matter whether a norm that facilitates the development of doctrine is state or non-state in origin? It would seem that Lamond prioritizes form over function, where the relative persuasiveness of a legal norm – at least to those committed to positivist theory-- is always conditional on its proximity to political authority.

In global and multicultural contexts, the development of some kinds of doctrine may well depend on analogical reliance on informal or non-state law. Lamond surely is right to insist that there is great virtue in honouring and respecting the enduring values that are germane to legal doctrine; there should of course be constraints on decision-making, and pre-existing law is one of them. However, I do not share Lamond's faith in the possibility, much less desirability, of conclusively demarcating doctrine from outside normative systems and orders. We know that state law is, as a matter of fact, saturated with non-law, rife with internal contradictions and paradoxes, and composed of a blend of concepts cobbled together from across a plurality of doctrines and non-state sources (Waddams, 2003). Law has never been, and will never be, otherwise. Reverence for state law in the context of analogical reasoning interferes with a morality of aspiration and the generative energy necessary to advance the responsiveness of state law to ambient values, principles and expectations. It does so by maintaining rote hierarchical distinctions between law and not-law, privileging the rules and rationales of state actors above those of non-state interpretive communities that share in the construction of meaning. This, in turn, precludes judicial reliance on a wide range of normative materials that are essential to the persuasiveness -- the very meaning -- of law to persons other than (although including some) judges and members of the legal profession.

5. ANALOGICAL REASONING AND THE MIGRATION OF CONSTITUTIONAL IDEAS

My principal claim is not that we should reject Lamond's account of the function of analogy. To the contrary, I claim that the performance of this function is not hindered by, and actually depends upon, analogical reasoning with exogenous law. This is precisely what is going on with the migration of constitutional ideas. Consider the "relevant and persuasive" doctrine developed by the Supreme Court of Canada (SCC) (Hudson, 2008). According to this doctrine, judges should interpret the *Charter of Rights and Freedoms* in relation to binding and non-binding international human rights as well as comparative human rights. Non-binding international law includes treaties not signed by Canada, decisions of international and regional courts not binding on Canada (e.g. the European Court of Human Rights), and the views of interpretive communities that have no authority to produce international law e.g. treaty monitoring and reporting bodies. It turns out this is actually the bulk of international human rights law, as most treaty provisions are so vague as to be directly inapplicable to a legal issue absent supplemental interpretation. This is similar to constitutional rights texts, except questions about the meaning and application of the latter are authoritatively settled by courts. It bears repeating that the specification of treaties is performed largely by discursive communities that have no authority to impose (general) legal obligations; at best, regional or international courts may impose judgments on states parties to a proceeding only. These judgments are no more (and no less) than "persuasive" to those judges who wish to rely on it.

The SCC's inclusion of non-binding international law and comparative law as legitimate bases of judgment indicates both that compliance is not the exclusive objective of the doctrine, and, that legal norms need not be authoritative to be considered persuasive. The purpose of this doctrine is threefold: 1) to protect and promote human rights *per se*, 2) to improve the quality of judgment by exposing jurists to the assumptions, biases, and inequalities built into domestic structures and institutions 3) to establish transjudicial linkages that help regulate the full geographic scope of transnational social interactions and relationships. Taken as a whole, these functions are oriented towards engendering congruence between state law(s) and background social values, beliefs, and practices that have strong global and multicultural dimensions. The doctrine reflects the view that state law is/should be produced, interpreted, and applied by both state and non-state actors who interact in relative autonomy from political and ideological power. Canadian courts are just one of these fora, but they remain an institution distinctively well-suited to securing the recognition of rights and the protection of high priority values when groups have been excluded from state political processes.

Coming full circle, we can see how the migration of constitutional ideas is justifiable in jurisprudential terms. On the one hand, we may adopt what Mattias Kumm calls a "cosmopolitan" paradigm of constitutionalism, where the basis of legal authority lies in "a complex standard of public reason" as well as the "legitimate concerns of outsiders" (Kumm, 2009, p. 286). So-called "small 'c'" constitutions, such as the *Charter* or English constitution, are mere iterations of a holistic, "large 'C'" constitutional order that has no political boundaries. International law – another parochial iteration of this cosmopolitan constitutional order-- may then take priority over the democratic will of domestic political communities under certain circumstances. Depending on the institution that has produced a norm, that institution's recognized authority over a given issue, and the material role of a political community in making a problem better or worse, a domestic court could conceivably be authorized to prioritize an international legal norm over their home state's constitution. Analogy on this reading would function more or less along the lines proposed by Lamond (i.e. as classificatory, close, and distant), except we would expand our conception of authority to include laws produced by transnational and international political communities.

On the other hand, we may retain a "statist" conception of constitutionalism, where authority flows from the will of "the people"; here, the "constitution is seen as the legal framework through which a political community governs itself as a sovereign nation" (Kumm, 2009, p. 286). However, we could see analogy as a mechanism for strengthening the aspirational qualities of constitutional doctrine, the development of which remains an end-in-itself. The objective, in other words, always is in improving state law, rather than the grander project of tending to an autonomous global rule of law. We can imagine, quite easily, that the integrity and development of constitutional doctrine that operates in transnational contexts depends on healthy, mutually constructive relationships with any legal order that is similarly committed to the protection and promotion of human rights. This is so not simply because of the transnational social, political, cultural, and economic character of many contemporary problems. It is because all human rights law, whether constitutional, international or comparative, are cognate – they share a common historical and conceptual origin. International and foreign courts, along with non-state discursive communities, work with similar if not identical principles (e.g. equality, freedom of expression, privacy, liberty), and have produced a wide range of rationales for how these principles may be actualized in the concrete context of recurring legal problems. When faced with similar legal problems, reliance on these rationales can help judges better align possible decisions with the principles and objectives of their own constitutional rights doctrine.

In this way, international and comparative human rights are not persuasive because of their proximity to political authority, but because they provide judges with insight into the practical and normative (de)merits of possible decisions that are ultimately sourced in the domestic law. Contrary to the ethos of cosmopolitan constitutionalism, this sort of analogical reasoning would be constrained by precedent in two ways. First, exogenous rationales could be used if, and only to the extent that, an issue is ungoverned by the doctrine of precedent. Second, they would be used only to the extent that they help judges develop doctrine consistently with its internal principles, rules, and objectives. Transnational legal decision-making in this way simply is an instance of close or distant analogical reasoning, depending on whether or not one respectively sees international and comparative human rights jurisprudence as being in some sense a "part" of constitutional rights doctrine (and *vice versa*) or as a set of external but nonetheless cognate doctrines.

6. CONCLUSION

This paper has set out to highlight the common law foundations of transnational legal reasoning. It is, of course, highly schematic and exploratory. However, I believe that transnational legal reasoning, like all common law reasoning, can be usefully approached as a distinctive mode of rational argumentation centred around precedent and analogy. One may take this conception as far as she would like. On the far end of the spectrum, we may adopt a cosmopolitan approach. Analogy here serves as one of the mechanisms through which decision-makers construct a body of law that is either autonomous from state law altogether or is tether to state law *per se*, but irreducible to the law of any one state. This approach has affinities with transnational socio-legal pluralism and transnational legal decisionism, respectively.

While I see the normative appeal of this conception, much theoretical and empirical work remains to be done to adequately outline how this body of law arises, develops into a system, and is sustained by actors situated across multiple jurisdictions. While I am hopeful this can be done, we are also free to conceive of transnational legal reasoning as supplemental to state law. Viewing the migration of constitutional ideas through the prism of transnational legal traditionalism, we need not say anything more than that transnational legal reasoning helps facilitate and constrain the development of doctrine in response to transnational environments and towards values of social justice and human dignity. This practice is not only justifiable -- it is necessary to judicial decision-making in global and multicultural contexts.

REFERENCES

Brewer, S. (1996). Exemplary reasoning: Semantics, pragmatics, and the rational force of legal argument by analogy. *Harvard Law Review, 109*, 923-1038.

Brunnée, J., & Toope, S., (2000). International law and constructivism: Elements of an interactional theory of international law. *Columbia Journal of Transnational Law, 39*(19), 19-74.

Burleson, B. B., & Kline, S. L. (1979). Habermas' theory of communication: A critical explication. *The Quarterly Journal of Speech, 65*(4), 412-428.

Hage, J.C., Span, G.P.J., & Lodder, A.R. (1992). A dialogical model of legal reasoning, In C.A.F.M. Grütters et al. (Eds.). *Legal knowledge based systems, information technology and law. JURIX '92*, (pp. 135-146). Lelystad: Koninklijke Vermande.

Hudson, G. (2008). Neither here nor there: The (non-)impact of international law on judicial reasoning in Canada and South Africa. *Canadian Journal of Law and Jurisprudence, 21*(2), 321-354.

Jessup, P. (1956). *Transnational law.* Yale University Press.

Kleinhans, M.-M., & MacDonald, R. A. (1997). What is critical legal pluralism? *Canadian Journal of Law & Society, 12*(2), 25-46

Kumm, K. (2009). The cosmopolitan turn in constitutionalism: On the relationship between constitutionalism in and beyond the state. In Jeffrey L. Dunoff & Joel P. Trachtman (Eds.), *Ruling the world? Constitutionalism, international law, and global governance.* Cambridge: Cambridge University Press

Lamond, G. (2012). Persuasive authority in the law. *The Harvard Review of Philosophy*http://philpapers.org/asearch.pl?pub=3115, *17*(1), 16-35.

Lamond, G. (2014). Analogical reasoning in the common law. *Oxford Journal of Legal Studies, 34*(3), 567-588.

MacCormick, N. (1978). *Legal reasoning and legal theory.* Oxford: Oxford University Press.

MacCormick, N. (2005). *Rhetoric and the rule of law. A theory of legal reasoning.* Oxford: Oxford University Press.

Moran, M. (2005). Influential authority and the estoppel-like effect of international law. In Hilary Charlesworth, et al. (Eds.). *The fluid state: international naw and national legal systems* (pp. 156-187). The Federation Press.

Moran, M. (2003). Authority, influence and persuasion: *Baker*, Charter values and the puzzle of method. In D. Dyzenhaus (Ed.), *The unity of public law* (p. 389). Oxford: Hart Publishing.

Perelman, C., & Olbrechts-Tyteca, L. (1969). *The new rhetoric: A treatise on argumentation.* University of Notre Dame Press.

Prakken, H. (1993). *Logical tools for modelling legal argument.* Dissertation Amsterdam. Amsterdam.

Schauer, F. (2008). Authority and authorities. *Virginia Law Review, 94*, 1931-1961.

Scott, C. (2009). Transnational law as proto-concept. *German Law Journal, 10*(7), 859-876.

Slaughter, S. –M. (1994). A typology of transjudicial communication. *University of Richmond Law Review, 29*, 99-137.

Teubner, G. (1997). 'Global bukovina': Legal pluralism in the world society. In Gunther Teubner (Ed.), *Global law without the state* (pp. 3-28). Dartmouth: Aldershot.

Toulmin, S. (1958). *The uses of argument.* Cambridge: Cambridge University Press.

Waddams, S. (2003). *Dimensions of private law: Categories and concepts in anglo-american legal reasoning.* Cambridge: Cambridge University Press.

Walton, D. (2002). *Legal argumentation and evidence.* Penn State University Press.

Zumbansen, P. (2012) Transnational law, evolving. In J. Smits (Ed.). *Encyclopedia of Comparative Law [2nd Ed.]* (pp. 899-925). Edward Elgar.

30

Pragmatic Argumentation in the Law-Making Process

CONSTANZA IHNEN JORY
Institute of Argumentation, Law Faculty, Universidad de Chile, Chile
cihnen@derecho.uchile.cl

Pragmatic argumentation -argumentation for or against a course of action based on its desirable or undesirable consequences- plays a central role in the law-making process. With a view to contributing to the public accessibility and scrutiny of parliaments' legislative work, this paper proposes instruments for the analysis and evaluation of pragmatic argumentation brought forward by governments to justify a government bill in the context of British second reading debates.

KEYWORDS: argumentative pattern, British Parliament, law-making debates, pragmatic argumentation, second reading

1. INTRODUCTION

Bills are often justified before a legislature by reference to their desirable effects on society. Thus, to mention only a few examples, Tony Blair's government sought to justify the *Terrorism Bill* (2005-06) as a means "to contest and then to defeat" the "terrorist threat" facing the UK in the aftermath of the London bombings (HC Deb, 26 October 2005, c326); the Conservative-Liberal Democrat coalition led by David Cameron argued for the *Welfare Reform Bill* (2011-12) as a means to reach out to people who have become "trapped in a permanent state of worklessness and dependency" (HC Deb, 9 March 2011, c919); and Obama's administration introduced the *Affordable Health Care Bill* (2009-10) "to provide affordable, quality health care for all Americans and reduce the growth in health care spending" (H.R. 3962, 29 October 2009).

Argumentation that seeks to justify (or refute) the acceptability of a proposed course of action by reference to the action's desirable (or undesirable) consequences is generally referred to as "pragmatic argumentation" (Perelman & Olbrechts-Tyteca, 1958/2000; Schellens,

D. Mohammed & M. Lewiński (eds.) (2016). *Argumentation and Reasoned Action: Proceedings of the 1st European Conference on Argumentation, Lisbon, 2015. Vol. II, 519-538.* London: College Publications.

1987; van Eemeren & Grootendorst, 1992).[1] With a view to contributing to the public accessibility and scrutiny of parliaments' legislative work, this paper proposes instruments for the analysis and evaluation of pragmatic argumentation brought forward in favour of a government bill. To design these instruments, I will take into consideration the institutional constraints affecting the use of pragmatic argumentation. Since institutional constraints vary considerably from legislature to legislature, and even within the same legislature, from one legislative stage to the next, I will focus on a well-demarcated context, namely British debates on public bills that take place at "second reading." Second reading is the first substantial stage of the British legislative process.[2] My theoretical starting point is the pragma-dialectical theory of argumentation (van Eemeren & Grootendorst, 1984, 1992, 2004; van Eemeren, 2010).

2. THE ARGUMENTATIVE PATTERN OF THE GOVERNMENT

Bills must go through several stages before they become an Act of Parliament: first reading, second reading, committee stage, report stage, and third reading. Each stage is governed by specific discussion procedures to ensure that the constituent parts of a bill -policies, policy objectives, clauses and schedules- are discussed in an orderly, efficient, and piecemeal fashion.[3]

---

[1] Pragmatic argumentation has a positive and negative variant. In the positive variant, a recommendation to carry out an action is justified by reference to its desirable consequences; in the negative variant, a recommendation not to carry out an action is justified by reference to its undesirable consequences. Since the paper deals with pragmatic argumentation presented *in favour* of a bill, I shall henceforth refer only to the positive variant.

[2] The UK Parliament is at the centre of my analysis because the British legislative system has served as a model for several countries, in particular those within the Commonwealth. Thus, the instruments developed can be extended, with some adjustments, to a number of other legislatures. My analysis centres on the second reading of a bill because debates at this stage generate material starting points for those taking place in subsequent legislative stages. The analysis of second reading is therefore a necessary point of departure for the analysis of the remaining stages of the law-making process. Similarly, my decision to focus on public bills (bills which change the law as it applies to the general population) is that such bills are by far the most common type of bill introduced in Britain. The majority of public bills are government bills.

[3] I use "discussion procedures" as an umbrella term to cover a broad set of regulations that can be found in *The Standing Orders of the House of Commons*;

Discussion procedures define the motion that should be debated at each legislative stage and the *prima facie* case the government needs to develop to justify their standpoint towards each motion. Since, as a rule, the government conforms to these institutional exigencies, their discourse is characterised by the presence of a distinct argumentative pattern at each stage.

An "argumentative pattern" (van Eemeren & Garssen, 2013, p. 7) consists of a specific combination of argument schemes, within a particular type of argumentation structure, used to defend a specific type of standpoint. Figure 1 represents the argumentative pattern used by the government at second reading. (I use pragma-dialectical notational conventions to represent the relation between arguments).[4]

1 The bill should be read a second time
- 1.1 The principles of the bill are acceptable
  - 1.1.1a We should legislate on domain D
    - 1.1.1a.1a Legislation on domain D leads to consequence Y_0
    - 1.1.1a.1b Consequence Y_0 is desirable
  - 1.1.1b The policies informing the bill P_1, P_2, etc. are acceptable on balance
    - 1.1.1b.1a Policy P_1 is acceptable
      - 1.1.1b.1a.1a Policy P_1 leads to consequence Y_1
      - 1.1.1b.1a.1b Consequence Y_1 is desirable
        - 1.1.1b.1a.1b.1a Consequence Y_1 leads to consequence Y_0
        - 1.1.1b.1a.1b.1b Consequence Y_0 is desirable
      - 1.1.1b.1b Policy P_2 is acceptable
        - 1.1.1b.1b.1a Policy P_2 leads to consequence Y_2
        - 1.1.1b.1b.1b Consequence Y_2 is desirable
    - 1.1.1b.1c [Etc., etc., etc.]

Figure 1 – Argumentative pattern of the government at second reading

The argumentative pattern used by the government at second reading can be established on the basis of an examination of the discussion procedures governing the conduct of business at this stage, as well as

---

The Standing Orders of the House of Lords; the *Guide to Making Legislation* (2014) published by the Cabinet Office; the *Legislation Fact Sheets Series* prepared by the House of Commons and the House of Lords Information Office; and personal communication with the House of Commons Information Office and with several MPs from parliamentary session 2009-10.

[4] This argumentative pattern can vary to some degree in practice. The government may expand the structure by responding to real or anticipated criticisms against the acceptability, relevance or sufficiency of the arguments they have advanced to ground the acceptability of their standpoint. Moreover, the government may just as well decide not to mention or not to defend some of the propositions included in the structure because, for example, the Opposition has previously made it clear that they are uncontroversial. Finally, some elements in the structure (usually, evaluative premises) can remain implicit in the government's discourse.

empirical observations of real-life second reading debates.[5] According to discussion procedures, the goal of second reading is to discuss and vote upon the motion "That the bill be now read a second time." In essence, the motion means 'The bill should proceed to the next legislative stage.' The government has -when the public bill under discussion is a government bill- a positive standpoint towards this proposition (1). Discussion procedures also establish that in order to provide sufficient support for their standpoint, the government must persuade their opponents that the so-called "principles" of the bill are acceptable (1.1). Proving the acceptability of the principles of the bill entails, firstly, justifying the proposal to legislate on a given domain -terrorism, welfare, health, etc.- (1.1.1a) and, secondly, justifying the acceptability of the main policies informing the bill (1.1.1b; 1.1.1b.1a, 1.1.1b.1b, etc.).[6] The government need not demonstrate that every single policy in the bill is acceptable, however. It suffices to show that the policies are acceptable on balance.[7]

There are also conventions regulating the justification of the proposal to legislate in a given domain and the policies of the bill. Thus, the government is expected to justify the proposal to legislate on the basis of some overall purpose Y_0 of the bill (1.1.1a.1a & 1.1.1a.1b) and each policy of the bill on the basis of some particular objective that the policy is meant to achieve (e.g., 1.1.1b.1a.1a & 1.1.1b.1a.1b).

---

[5] Due to space limitation, I can only carry out a partial analysis of second reading procedures and analyse only one token of second reading debates. I am fully aware though that a more extensive analysis of procedures is required and that more debates need to be analysed to further support my claims. See Ihnen Jory, 2012 (chapters 4 and 5 in particular), for a detailed examination of second reading procedures and a justification of the argumentative pattern proposed.

[6] Policies are not identical to the clauses laid down in a legislative text. Clauses implement policies. For example, suppose the government has the policy of creating some criminal offence. The clauses implementing the policy would stipulate, among other things: (1) the definition of the offence; (2) a defence to the offence; (3) the sanction to the offence, etc. At second reading, the government has no obligation to justify the acceptability of the clauses of the bill. For the same reason, no amendments to the clauses are voted upon at this stage. Clauses and amendments are mainly discussed and certainly voted upon at subsequent legislative stages.

[7] From the perspective of the opponent, policies are acceptable 'on balance' if: (a) some of the policies informing the bill are sufficiently desirable to outweigh some other undesirable policies in the same bill; or (b) it is likely that the undesirable policies will not be part of the bill at the end of the legislative process.

Finally, the policies informing the bill ought to be a relatively coherent set. For example, a counter-terrorism bill is expected to comprise a set of counter-terrorism measures, rather than policies trying to deal with health-related issues. In practice, the government achieves this coherence by presenting the policies as indirect means to realising the overall objective of the bill. More specifically, realising the objective of each policy is claimed to be a means to achieving the overall objective of the bill (e.g., 1.1.1b.1a.1b.1a & 1.1.1b.1a.1b.1b).

Let me give an example to illustrate how the argumentative pattern introduced can be actualised in practice (figure 2). During the second reading of the *Terrorism Bill* (2005-6), Charles Clarke, the then Home Secretary, sought to justify the acceptability of the principles of the bill by explicitly arguing that the UK had to strengthen their counter-terrorism legislation ((1.1).1a), and implicitly suggesting that the main policies informing the bill were acceptable ((1.1.1b)). The Minister spent considerable time trying to justify two controversial policies of the bill: the policy of creating a new offence relating to encouragement to commit terrorist acts ((1.1.1b).1a); and the policy of extending the maximum period of detention for terrorist suspects without a trial from 14 to 90 days ((1.1.1b).1b).

In order to justify the proposal to legislate on counter-terrorism, the government argued that legislation on this domain was part of their strategy to contest the unprecedented terrorist threat they faced ((1.1).1a.1a & (1.1).1a.1b). To support the acceptability of the policies of the bill, Mr Clarke mentioned the intentions underlying them. For example, he argued that the intention underlying the new offence of indirect encouragement was to "make it clear that the glorification of terrorism is not a legitimate point of view" (e.g., (1.1.1b).1a.1a & (1.1.1b.1a.1b)).

Finally, the desirability of making clear the illegitimacy of glorifying terrorism was implicitly supported by the argument that making this clear will contribute to contesting terrorism ((1.1.1b.1a.1b).1a & (1.1.1b.1a.1b).1b).

1 The TB should be read a second time (c322)
(1.1) The principles of the TB are acceptable
(1.1).1a We should strengthen our counter-terrorist legislation (c322, c327)
(1.1).1a.1a Strengthening our counter-terrorist legislation is part of our strategy to contest the terrorist threat we face (c327)
(1.1).1a.1b We need to address this threat (c322)
(1.1.1b) The policies informing the TB are acceptable on balance
(1.1.1b).1a We should create an offence of encouragement of terrorist acts (c327, c334)
(1.1.1b).1a.1a Creating an indirect encouragement offence will make it clear that the glorification of terrorism is not a legitimate point of view (c327)
(1.1.1b.1a.1b) Making it clear that the glorification of terrorism is not a legitimate point of view is desirable
(1.1.1b.1a.1b).1a Making it clear that the glorification of terrorism is not a legitimate point of view will help us contest the terrorist threat we face (c327)
(1.1.1b.1a.1b).1b We need to address this threat (c322)
(1.1.1b).1b There is a compelling case to extend the maximum period of detention of terrorist suspects prior to charge from 14 to 90 days (c340, c342)
(1.1.1b).1c [Etc., etc., etc.]

Figure 2 – Mr Clarke's *prima facie* case at the second reading of the *Terrorism Bill*

3. ANALYSING PRAGMATIC ARGUMENTATION OF THE GOVERNMENT

Pragmatic argumentation plays a vital role in the government's discourse at second reading. Within the government's argumentative pattern, pragmatic argumentation is used to justify at least three different standpoints: the proposal to legislate in a given domain (1.1.1a); the acceptability of the policies of the bill (e.g., 1.1.1b.1a); and the desirability of bringing about the objectives underlying each policy (e.g., 1.1.1b.1a.1b). Indeed, the argumentation brought forward by the government to support these standpoints is an instantiation of the argument scheme underlying pragmatic argumentation. The argument scheme of pragmatic argumentation is presented in figure 3.

1 Action X should be carried out
1.1a Action X leads to consequence Y
1.1b Consequence Y is desirable

Figure 3 – Argument scheme of pragmatic argumentation

Pragmatic argumentation can have almost infinite manifestations within the government's discourse. There are, however, at least two manifestations that have become fairly standard. I shall label them "goals case" and "problem-solving case" manifestations of pragmatic argumentation.[8]

In a goals case, an action is presented as a means to attain a desirable goal. The argumentation provided to justify the policy of creating an offence of encouragement to commit terrorist acts is an example: Mr Clarke presented the policy as a means to the objective of "making it clear that the glorification of terrorism is not a legitimate point of view," objective which the government considered desirable (c327). Figure 4 shows the structure that can be used to reconstruct a goals case pragmatic argumentation, followed by a reconstruction of the argumentation used to justify the creation of the encouragement offence.

1 Action A should be carried out
  1.1a Action A is a means to goal G
  1.1b Goal G is desirable

1 We should create an offence of encouragement of terrorist acts (c327, c334)
  1.1a Creating an encouragement offence will make it clear that the glorification of terrorism is not a legitimate point of view (c327)
  (1.1b) Making it clear that the glorification of terrorism is not a legitimate point of view is desirable

Figure 4 – Pragmatic argumentation in a goals case

In a problem-solving case, the action is presented as a means to solve (or prevent) some present (or future) problem. Mr Clarke's argument for the policy consisting of extending the maximum period of detention of terrorist suspects prior to charge is an example: to demonstrate the acceptability of this policy the he argued that the 90 days proposal was necessary because "significant conspiracies to commit terrorist acts have gone unprosecuted as a result of the time limitations placed on the control authorities following arrest" (c340). Figure 5 presents the structure that can be used to reconstruct a problem-solving case, followed by the reconstruction of the argumentation presented by the government to justify the 90 days proposal.

---

[8] The distinction is based on the case construction strategies distinguished in the debate literature (e.g., Freeley & Steinberg, 1961/2009; Jasinski, 2005; Ziegelmueller & Dause, 1975; Inch & Warnick, 1994). Case construction strategies refer to different affirmative case construction strategies. Another case construction strategy often mentioned in the literature -and whose implementation may result in yet another standard manifestation of pragmatic argumentation- is the "comparative advantages case" (Inch & Warnick, 1994; Jasinski, 2001).

1 Action A should be carried out
  1.1a Acton A will stop (or prevent) situation S
  1.1b Stopping (or preventing) situation S is desirable

1 There is a compelling case to extend the maximum period of detention of terrorist suspects prior to charge from 14 to 90 days (c340, c342)
  1.1a Extending the maximum period of pre-charge detention of terrorist suspects will prevent that significant conspiracies to commit terrorist acts go unprosecuted as a result of the time limitations placed on the control authorities following arrest (c340)
  (1.1b) Preventing significant conspiracies to commit terrorist acts from going unprosecuted is desirable

Figure 5 – Pragmatic argumentation in a problem-solving case

At bottom, both manifestations of pragmatic argumentation are equivalent with regard to their propositional content, as they are both attempts at justifying a future course of action on the basis of the action's desirable consequences. Nevertheless, from a rhetorical point of view, it can make a difference for the arguer to develop a problem-solving case rather than a goals case, or vice versa. It all depends on the arguer's rhetorical situation. If the proponent of the pragmatic argument has identified a problem and has successfully attributed the responsibility of the problem to the opposition, then it is in principle more expedient for the proponent to build a problem-solving case, and more advantageous for the opponent to build a goals case. For example, during the British 2010 general election campaign, it was clearly advantageous for the Conservatives to build a problem-solving case, given that the country was in the midst of a global financial crisis and Labour had been in power since 1997. Not surprisingly, then, the Conservatives campaigned under a slogan along the lines of 'We need a Conservative government to get us out of the mess that Labour got us in' (Conservative Party, 2010). By contrast, it was more advantageous for Labour to use a goals case by calling the public to vote Labour in order to "secure economic growth" and "improve living standards for all, not just a few" (Labour Party Manifesto, 2010).

4. EVALUATING PRAGMATIC ARGUMENTATION OF THE GOVERNMENT

From a pragma-dialectical point of view, an argumentation is reasonable if the parties to the discussion agree that this is the case *as the result of* having put the argumentation to an ideal critical test, in light of the commitments the parties have assumed in empirical reality (van Eemeren & Grootendorst, 2004). Accordingly, the evaluative instrument I have built consists of an ideal procedure for critically testing the reasonableness of the government's pragmatic argumentation in the context of second reading discussions. This ideal procedure takes the form of "dialectical profiles," specifying the sequence of questions and answers that the proponent –in this context, the government– and opponent of pragmatic argumentation can or should put forward, in order to establish the acceptability, relevance and sufficiency of the proponent's pragmatic argumentation.[9] The normativity underlying these instruments is both intrinsic and "extrinsic" (van Eemeren & Garssen, 2013, p. 5). It is intrinsic because the questions are partly based on an analysis of the inherent properties of pragmatic argumentation – the types of propositions and the internal structure of this type of argumentation, for example. However, the questions are also based, and to a significant degree, on extrinsic considerations pertaining to the institutional make-up of second reading. In other words, the instruments I will propose are nothing but an attempt at systematising normative intuitions that lawmakers already have in their role as language users and legislators.

The standards of acceptability, relevance and sufficiency of pragmatic argumentation vary, to some extent, depending on the standpoint that the government is trying to justify. Since it is impossible within the scope of this paper to propose instruments for each of the three pragmatic argumentations I have identified in the government's argumentative pattern, I will focus on the assessment of pragmatic argumentation used to defend the policies informing a bill solely. To facilitate the presentation of the evaluative instruments, I have outlined this form of pragmatic argumentation in figure 6.

1 Policy P_1 (or P_2, etc.) is acceptable
 1.1a Policy P_1 (or P_2, etc.) leads to consequence Y_1
 1.1b Consequence Y_1 is desirable

Figure 6 – Pragmatic argumentation to justify the policies of a bill

4.1 Acceptability

Pragmatic argumentation is acceptable if the parties agree that the causal proposition "Policy P_1 leads to consequence Y_1" and the evaluative proposition "Consequence Y_1 is desirable" are acceptable as a result of going through the ideal procedure specified in figure 7. Although the causal proposition is conventionally presented in the

[9] Dialectical profiles "specify the sequential pattern of moves that the parties are allowed to make, or should make, in a particular stage of a critical discussion in order to realize a particular dialectical goal" (van Eemeren & Houtlosser, 2007). Even though dialectical profiled were originally introduced in the pragma-dialectical theory to account for the strategic dimension of argumentative discourse, I believe they can just as well be used for evaluative purposes.

argument scheme first, it makes sense to start examining the evaluative proposition first, as the evaluative premise provides the motivational basis of the argument.[10]

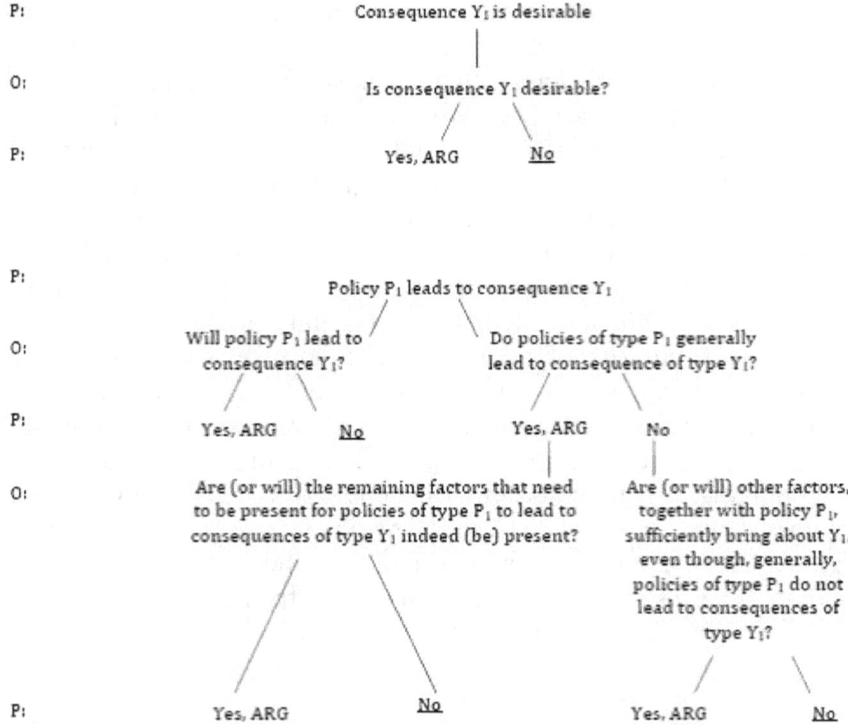

Figure 7 – Dialectical profile to evaluate acceptability

The acceptability of the evaluative proposition "Consequence Y is desirable" is crucial to judge the reasonableness of pragmatic argumentation: if the parties do not share the evaluative premise, then the argumentation provides no support to the standpoint. The test to establish the acceptability of the evaluative premise takes different

[10] In figure 7, 'P' stands for proponent; 'O' for opponent; 'Yes, ARG', for a positive answer to the opponent's question plus argumentation to justify the positive answer; 'No' stands for a negative answer; and 'No', stands also for a negative answer *and* a 'loosing' move of the proponent and the end of the dialogue. For brevity's sake, I am assuming that the opponent in all cases accepts the argument given by the proponent to defend her positive answer 'Yes' immediately. Of course, in reality, this need not be the case. Proponent and opponent often have arguments about the worth of the arguments they bring forward.

forms depending on whether the government has developed a goals case or a problem-solving case. In a goals case, the opponent of the argument needs to ask whether goal G_1 is indeed desirable. In a problem-solving case, the question can take two forms: 'Is stopping (or preventing) situation S_1 indeed desirable?' or 'Is situation S_1 indeed undesirable?'

Pragmatic argumentation will not pass the evaluative test either if the parties do not accept the causal proposition. The causal proposition can be criticised directly by asking whether policy P_1 will indeed lead to consequence Y_1. However, the causal proposition can also be criticised indirectly. The opponent can attack the proposition indirectly in two ways, depending on whether the proponent is committed to the statement 'Generally, policies of type P_1 to lead to consequences of type Y_1'. If the proponent is committed to the acceptability of this causal generalisation, then the opponent can ask the proponent whether the remaining factors that need to be present for policies of type P_1 to bring about consequences of type Y_1 are (or will be) present. An affirmative answer to this question entails a commitment by the proponent not only to the causal generalisation but also to a statement to the effect that the causal generalisation has been applied correctly. If the opponent is satisfied with the proponent's justification provided to support his affirmative answer, then the causal proposition is acceptable. However, if the proponent is not committed to the acceptability of the causal generalisation, the opponent needs to follow a different path of examination. Thus, the relevant question in this case is whether there are other factors, together with policy P_1, which can bring about Y_1, despite the fact that, generally, policies of type P_1 do not lead to consequences of type Y_1. If the proponent answers affirmatively, and gives an argument to that effect that satisfies the opponent, then the causal premise is acceptable.

4.2 Relevance

Pragmatic argumentation can have acceptable premises yet fail the evaluative test because it is irrelevant or insufficient. An argument is relevant if both parties agree that this is the case in the first instance, or the second instance, i.e. after a sub-discussion.

A pragmatic argument needs to be relevant in two respects. First, the pragmatic argument scheme must be a commonly accepted means of defence in the context of the discussion. In other words, both parties need to agree with the pragmatic rule of conduct: in principle, actions that lead to desirable consequences should be carried out (and those leading to undesirable consequences should be avoided). Since, as demonstrated by the analysis carried out in section 3, pragmatic argumentation is a conventionalised means of arguing in favour of a legislative proposal, the relevance of pragmatic argumentation can be, in this respect at least, taken for granted in the context of second reading.

Second, pragmatic argumentation also needs to be relevant in the specific sense of having probative bearing on the immediate conclusion the arguer is trying to justify. This sense of 'relevance,' however, cannot be defined in advance by means of a set of individually necessary and jointly sufficient conditions (or a set of questions and answers in this case). The relevance of pragmatic argumentation understood in this way, needs to be judged case by case.

4.3 Sufficiency

Pragmatic argumentation is sufficient if the opponent is satisfied that the proponent has correctly answered all critical questions relevant to this type of argumentation. In figure 8, I have outlined the set of critical questions relevant to pragmatic argumentation put forward by the government in order to justify the policies of a bill at second reading.

4.3.1 Feasibility

Critical questions 1 to 6 refer to the feasibility of the policy recommended by the government. Any proposal to perform an action entails that the action is feasible. Hence, pragmatic argumentation will fail to provide support to its standpoint if the policy recommended is unfeasible. A policy may be unfeasible because it is non-permissible or because it is unworkable.

In a law-making context, a policy can be non-permissible because it is incompatible with a shared moral principle, the current legal framework, or the Constitution. However, none of these incompatibilities necessarily mean that the policy must be discarded. Consider the case, first, of an incompatibility with some moral principle (question 1). Whether a moral principle should take precedence over the desirable consequences brought about by the policy is not something that is stipulated anywhere in parliamentary procedures and should be left, for that reason, up to the parties of the discussion (question 2). In this way, it is up to them to follow a teleological or a deontological conception of reasonable actions. Of course, if both parties adhere to a deontological conception and agree that the policy is inconsistent with a moral principle, then the pragmatic argument under scrutiny would fail, and the policy would have to be discarded.

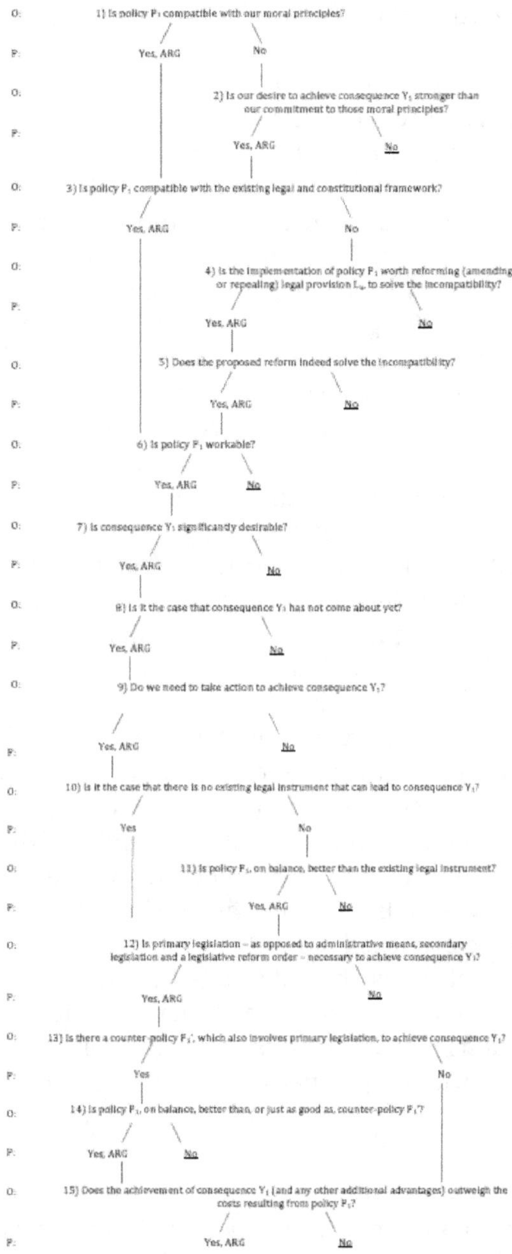

Figure 8 – Dialectical profile to evaluate sufficiency

Consider now the case of an incompatibility with the current legal framework (question 3). Surely, policies must fit harmoniously within the existing body of law. This requisite is not unique to the UK. However, it is just as clear that the demand for coherence should not preclude the possibility to revise and reform existing law. For this reason, the identification of an incompatibility between the government's proposed policy and existing legal provisions leads to a further critical question, namely: "Is the implementation of the proposed policy worth reforming (i.e., amending or repealing) the existing legal provision to solve the incompatibility?" (question 4). In practice, a bill's drafting team will often anticipate criticisms of incompatibility, and make proposals to reform existing legislation within the same bill in which the policy has been introduced. The proposed reform may not solve the incompatibility, however, which is why an additional question needs to be asked: "Does the proposed reform indeed solve the incompatibility (question 5)?"[11]

The same evaluative test applies in case the policy is incompatible with the British "unwritten" Constitution (questions 3, 4 and 5). Due to the principle of parliamentary sovereignty, the UK Parliament has the right to make or unmake any law whatever, including laws of a constitutional nature, by a simple Act of Parliament. There are no entrenched procedures (such as a special power of the House of Lords, or the requirement for a referendum) by which the unwritten constitution may be amended (Blackburn, n.d.). As a result, an inconsistency between constitutional law and a government's policy does not necessarily mean that the policy should be discarded. Just as it happens in case there is an inconsistency with non-constitutional law, constitutional law can be revised by Parliament so as to make it consistent with the proposed policy.[12]

Finally, a policy may be unfeasible because it is unworkable, that is to say, impossible to carry out due to factual limitations (question 6). "Factual" limitations may refer to "brute facts" –such as the capacity of a country to grow certain type of crops or produce some type of energy–

---

[11] If the reform proposed amounts to repealing the conflicting legal provision, then the incompatibility is automatically "solved."

[12] However, in practice, reforming some constitutional documents, such as the *Magna Carta* and the *Human Rights Act* (1998) is unlikely. As Turpin and Tomkins (2007, p.75) point out, "Parliament is limited in its legislative activity by an awareness of what is politically unfeasible or likely to create an adverse public opinion."

but most of the time, they refer to "social facts".[13] An example of a social fact amply used to criticise welfare policies, and social programmes more generally, is the lack of sufficient resources to fund social policies or programmes.

4.3.2 Significance

Critical question 7 tests how significantly desirable the consequence of the proposed policy is. This question manifests in discourse differently depending on whether the government has developed a goals case or a problem-solving case strategy. If the government has decided to develop a goals case, the opponent would have to ask whether the goal that the policy is supposed to achieve is significantly *desirable*. By contrast, if the government has developed a problem-solving case, the question is whether the problem that the policy is said to solve is significantly *undesirable*. Failing to demonstrate that the consequences of the policy are significantly desirable can affect the sufficiency of pragmatic argumentation because every law-making process involves costs of at least two sorts. First, the administrative costs associated with any law-making activity: legislating involves the use of a vast amount of human and material resources. Second, legislating on a given issue, to achieve consequence Y_1, often involves the cost of not achieving some other desirable consequence Y_1'. Achieving desirable consequences Y_1 and Y_1' simultaneously can be impossible either because achieving one is inherently incompatible with realising the other, or because there is lack of time and resources to achieve both consequences simultaneously. Hence, to motivate sufficiently the acceptability of the policy, it has to be shown that achieving consequence Y_1 outweighs (at the very least) the costs involved in the law-making activity. Certainly, the more significantly desirable the intended consequence of the policy is, the better the chances that this desirable consequence will outweigh the costs involved in the law-making process.

4.3.3 Necessary means

Once it is clear that the policy is feasible and that the intended consequence of the policy is significantly desirable, the means-end relation between the policy and the intended consequence needs to be examined carefully.

---

[13] I use the distinction between "brute" and "social" facts introduced by Searle (1995).

Pragmatic argumentation should mention a policy that is a necessary means, or the best means, to achieve the intended consequence. Questions 8 to 13 test whether the policy proposed is necessary. One way of testing whether the policy is necessary is asking whether consequence Y_1 has not come about yet (question 8). Again, this question takes different forms depending on whether the government has developed a goals case or a problem-solving case. In a goals case, the question is whether goal G_1 has not been attained yet. In a problem-solving case, the question is whether undesirable situation S_1 exists or will exist –i.e., whether the presumed problem indeed exists or is likely to develop in the future.

Critical questions 9 to 13 relate to the different types of actions that the British government and Parliament can take to achieve social objectives and solve social problems. First, a policy is unnecessary if it is clear that consequence Y_1 will come about independently of whether some action (of any kind) is taken on the side of the government (question 9). One way of formulating this criticism is to argue that 'the problem will solve itself'.

Next, a policy is unnecessary if there is an existing legal instrument that can be used to achieve intended consequence Y_1 (question 10). If the objective of the proposed policy can be achieved by some existing legal provision, then the measure should be discarded, *unless* the government can prove that the measure is better, on balance, than the existing legal instrument (question 11). The existing legal instrument need not have been designed for the same purpose than the proposed policy –it is enough that the instrument *can* be used for the same purpose.

Another way of testing whether the policy is necessary is to inquire whether primary legislation –as opposed to secondary legislation, a legislative reform order or some administrative means– is necessary to achieve consequence Y_1 (question 12). According to the *Guide to Making Legislation* (2014, p. 23) published by the Cabinet Office, secondary legislation, legislative reform orders, and administrative means are preferable means of achieving governmental objectives, because parliamentary time is scarce.[14] Primary legislation

[14] However, the idea that secondary legislation and legislative reform orders should be preferred by the government over primary legislation is understandable from a strategic perspective, but controversial from a democratic point of view. Secondary legislation greatly increases the power of the executive and is less subject to critical examination than primary legislation. For a critique of the government's use of secondary legislation, see Porter (2009).

refers to an Act of Parliament. An Act is a bill approved by both the House of Commons and the House of Lords and formally agreed to by the reigning monarch (Acts, n.d.). Policies included in a public bill are by definition primary legislation instruments.[15] Hence, if the intended consequence of the policy can be achieved by any of the aforementioned means, the policy would be unnecessary and the government's pragmatic argumentation, at the very least, weak.

Finally, even if it is clear that some action needs to be taken, that there is no legal instrument at hand to achieve the intended consequence, and that primary legislation is necessary, the policy proposed by the government could still be unnecessary. Indeed, the policy would be unnecessary if there is a counter-policy, which also involves some form of primary legislation, to achieve consequence Y_1 (question 13). If there is no counter-policy available, then the argumentation has passed the necessary-means test, and a final critical question, pertaining to the costs of the policy, needs to be considered (question 15). However, if there is a counter-policy available, the policy is not necessary, and the question concerning which policy is the best means needs to be deal with first (question 14).

4.3.4 Best means

Indeed, the government can maintain their pragmatic argumentation, even though the proposed policy is not necessary, as long as the policy is better than, *or at least, just as good as* the counter-policy (question 14). When policy and counter-policy are *equally advantageous* it is indifferent which of the two is chosen. Critical question 14 proposes that when both policies are equally advantageous, then the government's policy should prevail over the counter-policy. The reason for this is that in British second reading debates, the government, as proponent of the policy, has already developed a full plan (a set of clauses and schedules) for the policy's implementation at the moment of defending their policies. In contrast, when the opponent proposes a counter-policy during second reading they do not necessarily have a

[15] Secondary legislation is usually concerned with detailed changes to the law, under an existing Act of Parliament. These changes range from the technical, to fleshing out Acts with greater detail. Often an Act contains only a broad framework of its purpose and more complex content is added through secondary legislation (Delegated legislation, n.d.). A legislative reform order is an order by a Minister that acquires legal status after being approved by Parliament (Legislative Reform Orders, n.d.). Administrative means refer to any instrument available to the executive, which does not require legislation.

full-fledge counterplan for its implementation. Full-fledged counterplans are put forward and discussed later, at committee stage. At least in second reading then, in case policy and counter-policy are equally advantageous, it seems more efficient to decide in favour of a government policy, for which a plan has already been developed.

4.3.5 Cost-benefit

Finally, whether the policy is a necessary means or a best means, examining the costs of the policy –any costs, not only those involved in every law-making process– is an important step in the evaluation of any pragmatic argumentation (question 15).

5. CONCLUSION

In this paper, I have proposed instruments for the analysis and evaluation of pragmatic argumentation brought forward by the government to persuade legislators to vote in favour of a bill at the legislative stage of second reading. To develop these instruments, I have taken into account not only the "intrinsic" properties of pragmatic argumentation, but also some "extrinsic," institutional, constraints affecting the use of pragmatic argumentation.

In my view, another, "extrinsic" constraint that needs to play a role in the evaluation of pragmatic argumentation is a political ideal of legitimate law. Since the normative standpoints under discussion refer to the provisions of a bill, which propose the creation of social rules, additional soundness conditions will have to be taken into account as well. For example, it might be reasonable to include in the test of sufficiency the following critical question: 'Have all the parties that will be affected by the policy taken part, directly or indirectly, in the discussion concerning its acceptability?' Whether this question ought to be part, indeed, of the evaluation of pragmatic argumentation, and to what degree a negative answer would affect the sufficiency of the argumentation are topics to be dealt with in future research.

REFERENCES

Cabinet Office (2014). *Guide to Making Legislation*. Retrieved May 6, 2015, from *http://www.cabinetoffice.gov.uk/resource-library/guide-makinglegislation*

Conservative Party Manifesto (2010). Retrieved from https://www.conservatives.com/~/media/files/activist%20centre/press%20and%20policy/manifestos/manifesto2010

Delegated legislation (n.d.). Retrieved May 15, 2015, from http://www.parliament.uk/about/how/laws/delegated/

Eemeren, F.H. van (2010). *Strategic maneuvering in argumentative discourse. Extending the pragma-dialectical theory of argumentation.* Amsterdam: John Benjamins.

Eemeren, F.H. van, & Garssen, B. (2013). Argumentative patterns in discourse. In D. Mohammed & M. Lewinski (Eds.), *Virtues of argumentation: proceedings of the 10th International Conference of the Ontario Society for the Study of Argumentation (OSSA)*, 22-26 May 2013 (pp. 1-15). Windsor, ON: OSSA.

Eemeren, F.H. van, & Grootendorst, R. (1984). *Speech acts in argumentative discussions. A theoretical model for the analysis of discussions directed towards solving conflicts of opinion.* Berlin: De Gruyter.

Eemeren, F.H. van, & Grootendorst, R. (1992). *Argumentation, communication and fallacies. A pragma-dialectical perspective.* Hillsdale, NJ: Lawrence Erlbaum.

Eemeren, F.H. van, & Grootendorst, R. (2004). *A systematic theory of argumentation. The pragma-dialectical approach.* Cambridge: Cambridge University Press.

Eemeren, F.H. van, & Houtlosser, P. (2007). Seizing the occasion: parameters for analyzing ways of strategic manoeuvring. In F.H. van Eemeren, J.A. Blair, C.A. Willard & B.J. Garssen (Eds.), *Proceedings of the Sixth Conference of the International Society for the Study of Argumentation* (pp. 375-380). Amsterdam: SicSat.

Freeley, A.J., & Steinberg, D.L. (2009). *Argumentation and debate: Critical thinking for reasoned decision-making* (12th ed.). Australia, etc.: Wadsworth Cengage Learning. (Original work published 1961).

Ihnen Jory, C. (2012). *Pragmatic argumentation in law-making debates: Instruments for the analysis and evaluation of pragmatic argumentation at the Second Reading of the British Parliament.* (Doctoral dissertation). University of Amsterdam.

Inch, E.S. & Warnick, B. (1994). *Critical thinking and communication: The use of reason in argument* (2nd ed.). New York: Macmillan.

Jasinski, J. (2001). *Sourcebook on rhetoric. Key concepts in contemporary rhetorical studies.* Thousand Oaks, CA: Sage Publications.

Labour Party Manifesto (2010). Retrieved from http://www2.labour.org.uk/uploads/TheLabourPartyManifesto-2010.pdf

Legislative Reform Orders (n.d.). Retrieved May 15, 2015, from http://www.parliament.uk/business/committees/committees-archive/regulatory-reform-committee/regulatory-reform-orders/

Perelman, C., & Olbrechts-Tyteca, L. (2000). *The new rhetoric. A treatise on argumentation*. Notre Dame: University of Notre Dame Press. (Original work published 1958).

Porter, H. (2009, May 20). A new politics: Restrict the use of secondary legislation. *The Guardian*. Retrieved from http://www.theguardian.com/commentisfree/2009/may/20/parliament-reform-legislation

Schellens, J.P. (1987). Types of argument and the critical reader. In F.H. van Eemeren, R. Grootendorst, J.A. Blair, C.A. Willard (Eds.), *Argumentation: Analysis and Practice* (pp. 34-41). Dordrecht: Foris Publications.

Searle, J.R. (1995). *The construction of social reality*. New York: Free Press.

Turpin, C., & Tomkins, A. (2007). *British Government and the Constitution. Texts and materials*. Cambridge: Cambridge University Press.

Ziegelmueller, G., & Dause, C.A. (1975). *Inquiry and advocacy*. Englewood Cliffs, NJ: Prentice Hall.

31

A Computational Study of the Vaccination Controversy

SALLY JACKSON
University of Illinois at Urbana-Champaign, USA
sallyj@illinois.edu

NATALIE LAMBERT
University of Illinois at Urbana-Champaign, USA
nwhite6@illinois.edu

New communication technologies can alter familiar social practices in ways that are neither intended nor desired. As a new communication ecology forms around digital communication networks, the practice of argumentation may be changing, appearing (at least at first) to be becoming less reasonable. We examine a familiar controversy (over childhood vaccination), using new computational tools to investigate the puzzling argumentative maneuvers of both the anti-vaccination movement and the relevant expert communities.

KEYWORDS: argumentation, macroscopes, media ecology, vaccination controversy

1. INTRODUCTION

Marshall McLuhan famously pointed out that new media enter society "disguised as degradations of older media" (cited in E. McLuhan & Zingrone, 1995, p. 274). So it is with argumentation in the digital age; the sort of discourse made possible by the Worldwide Web and its many social platforms certainly appears to be a degradation of what we remember as high points in human reasonableness. It appears highly anomalous, and deeply suspect, when viewed through the lens of legacy concepts and tools. The rise of social media has been blamed - frequently, and by credible commentators - for amplifying all that is unreasonable in society. But if McLuhan was right that this might be a disguise for some more profound change, we should look especially

D. Mohammed & M. Lewiński (eds.) (2016). *Argumentation and Reasoned Action: Proceedings of the 1st European Conference on Argumentation, Lisbon, 2015. Vol. II, 539-552.* London: College Publications.

closely for what we might not be seeing. We should question our gut reactions.

In roughly 3 decades, humanity has experienced a shift from one media ecology to another: from the electronic age, so-called, to the digital age. In the 1980s, people still got most of their information on public health from newspapers, television, their own physicians, and word-of-mouth. Twenty years later, certainly by 2005, anyone could create their own publications (including blogs), and these would be discoverable by interested readers through computational search methods that draw no distinction between professional journalism and "citizen" journalism. In 2015, it is entirely commonplace for all kinds of people, from the most educated to the least, to rely on their own intentional searches for information rather than on information curated for them by elites.

We have been looking at a controversy (over childhood vaccinations) that spans the decades before and after this shift in communication environment. This time period is of interest because it offers at least the hope of seeing how argumentation changes as we settle into a new media ecology. Our theoretical aim is to try to understand how argumentation practice adapts itself to major changes in communication technologies—especially how it is adapting itself to the global data communication infrastructure and all of the new media that have grown over this infrastructure.

Our object of study is a health controversy occurring mainly in the US and UK. How to characterize this controversy is itself controversial. It has come to be known as the anti-vaccination controversy, but those on the anti-vaxx side object to this. In the US, the headline issue of the controversy is whether parents should be compelled to have their children vaccinated against infectious diseases. In the UK, where vaccinations are not required by law, the headline issue is whether parents should voluntarily have their children vaccinated according to the recommendations of health authorities. In both contexts, a scientific question figures centrally: Does childhood vaccination cause dangerous side effects? One conjecture in particular has come to stand for the controversy as a whole: that the combined measles-mumps-rubella (MMR) vaccine causes autism.

People disagree on exactly what started the controversy, but what is clear from news archives is that public interest in autism was rising throughout the 80s and 90s, and speculation about its causes was common. Figure 1 shows the number of stories on autism located by Lexis-Nexis search of major world news publications between 1985 and 2000. The first scientific report exploring a link between MMR and autism (Wakefield et al., 1998) was anticipated by several prior press mentions of a suspicion of such a link, along with other suspicions of environmental causes for an increase in rates of autism. Although interest was clearly already rising by the time of the Wakefield publication, there is a dramatic uptick around the same time. The suspected link between autism and vaccination became an active matter of discussion.

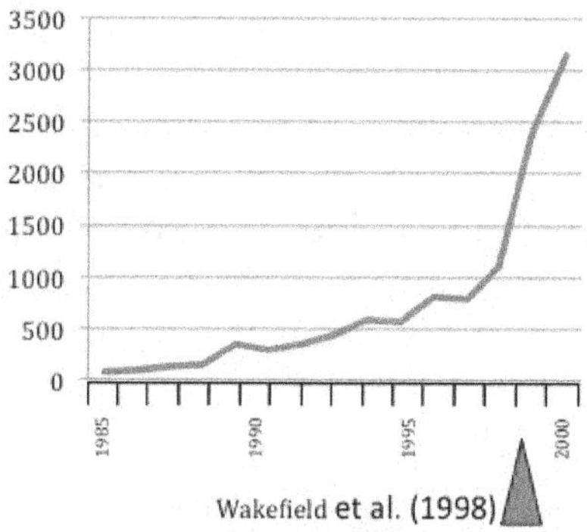

Figure 1. Stories on autism in major news publications between 1985 and 2000, retrieved from Lexis-Nexis.

Even today, most people who are not directly involved are aware of the controversy through news and editorial coverage. In the mid-1990s, though, the first algorithmic search engines appeared, changing the information environment dramatically. What this has meant for the vaccination controversy is that an endless stream of parents join the discussion when it becomes relevant to their concerns, conducting web searches on vaccination both in response to, and in place of, information encountered from other sources. Three potentially important consequences of the changing communication environment are the proliferation of information sources, the rise of algorithmic search methods, and the rise of Open Access as a societal expectation. These have contributed to increasing penetrability of what used to be separate spheres of discourse.

2. THE SHAPE OF THE CONTROVERSY, 2015

Broadly, we see this controversy as ranging across mainstream journalism, citizen journalism, science journalism, scientific research reporting, government agency statements, and more.

News coverage of the controversy by mainstream journalism is still very active. Over the past 20 years, there have been over 15,000 articles in major world news publications that discuss the link between autism and vaccination. Current press coverage often identifies the Wakefield publication as the beginning (and the source) of the controversy, ignoring the speculations that were already being covered by press prior to that study's appearance. From "inside" the controversy, and also from an analytic perspective, it seems more accurate to say that journalists do not simply report the controversy so much as actively participate in it—initially stirring public concern over the safety of vaccination, and more recently fueling public outrage against vaccine resistors (e.g., Salzberg, 2015).

For people deeply invested in the controversy, especially those with an enduring interest in autism, a wealth of unique content occurs not in mainstream journalism but in blogs and other social media. The blogosphere is quite heterogeneous; it is the home to anti-vaccination groups (like Age of Autism, http://www.ageofautism.com/), but also the home to autism advocacy groups whose official position has become pro-vaccination over time (like Autism Speaks; see for example, https://www.autismspeaks.org/science/policy-statements/information-about-vaccines-and-autism). Blogs "cover" authentic news that the mainstream press ignores. Fewer in number, but still of interest, are blogs and other social media sites featuring parents who advocate for vaccination (such as Voices for Vaccines, http://www.voicesforvaccines.org/).

Government agencies participate actively in attempting to persuade as many citizens as possible to cooperate with public health initiatives; they respond continuously to what officials believe to be misinformation delivered through other sources. In both the US and UK, public health officials acknowledge that there is public uncertainty over the safety of vaccination, which they try to address by denying that there is any controversy among medical experts. In other words, agency officials do not stand outside the controversy but actively argue and counter-argue.

Nor does the scientific research community stand apart from the controversy. For scientists interested in this controversy, the scientific question of interest is what causes autism, and the thousands of studies that have been done on genetic bases for autism are assumed to weigh against the public conjecture that autism is caused by vaccination. However, scientists have also attended closely to the public controversy, and both individually and collectively, the health research community has attempted to "settle" the controversy. Significant efforts of this kind are a 2004 review of the vaccination/autism link commissioned by the Institute of Medicine (published as *Immunization Safety Review: Vaccines and Autism*, 2004), a 2007 National Academies workshop on environmental factors in autism (with proceedings published as *Autism and the Environment,* 2008), and a 2011 Institute of Medicine review of suspected adverse effects of vaccination (with results published as Stratton et al., 2012). Social scientists have also been involved, for example in efforts to understand why it is so difficult to persuade people to accept the conclusions of the scientific community (for a few examples using diverse approaches, see Burgess, Burgess, & Leask, 2006; Poltorak et al., 2005; Ruiz & Bell, 2014).

Until the mid-1990s, and even for some time after that, scientific information and official government positions would have been difficult for ordinary citizens to locate and read, and it would have been plausible to treat these as something external that could be referred to in argument rather than as argumentative content--at best, introduced in an appeal to expert authority. We treat scientific research literature symmetrically with other texts relevant to this controversy, not as an external standard against which to evaluate other moves within the controversy. Treating scientific literature as consisting of moves within an argumentative discourse is different from trying to interpret the scientific consensus from the literature so that it can be deployed by one side or another within the controversy. As recently as 20 years ago, it might have been reasonable to bracket the scientific research literature and see it as separate from public discussion. Now, though, with the digitization of former print media, and with the Open Access movement, nearly all of this content is retrievable and reusable by people with and without scientific training. This may not "change everything," but it is a feature of the new media environment that needs examination.

By now, thousands of scientific reports and tens of thousands of other texts make up the content of this controversy that is available to anyone with an Internet connection. These are also our data. No one reads all of this content, nor should researchers expect to be able to do so. But it is all present in the information environment and retrievable by interested individuals with varying motivation and skill. Its existence and availability cannot simply be ignored. This creates a very challenging situation for the analyst trying to characterize the controversy and its various competing points of view.

3. SEEKING A MACROSCOPIC VIEW

We need tools that allow us to handle data of size and complexity too great for unaided human processing. Katy Börner (2011) introduced the term "macroscope" to refer to a class of computational tools designed to aid in examination of phenomena too large or too complex to be seen with unaided human senses. Macroscopes can be designed for any field of study, including argumentation; they may replace conventional methods, but more likely they will cooperate with conventional methods, as we will illustrate later.

Nowadays, everyone who searches the Internet approaches information computationally, relying on computational algorithms to locate relevant content and to provide clues to its value (such as likes and upvotes). A macroscopic approach to argumentation research means depending on algorithms to do many of the things that used to be done by an analyst working with small amounts of data: algorithmic search, but also algorithmic text processing, including sentiment analysis and relationship analysis.[1]

Macroscopes may combine diverse computational tools, passing the results of one analysis as data for the next analysis. So, for example, we can apply natural language processing tools applied to large bodies of blog content to identify promising search terms to use in querying news databases like Lexis-Nexis, or we can use analytics from science databases (such as Web of Science or PubMed) and attempt to connect these with happenings in news or social media. A variety of methods have been proposed for representing the temporal unfolding of a conversation, and the results of content queries can be the data for these temporal reconstructions. The point is that a macroscope usually involves combinations of computational tools.

To adequately represent this particular controversy, we ultimately need investigative tools and analytic methods that will allow us to see how activities in several separate spheres influence activities within the other spheres. We need to see not only what sense the public makes of scientific research and its translation into expert opinion, but also how public pushback exerts influence on what experts need to have an opinion about, and how this in turn exerts influence on the scientific research agenda.

4. EARLY FINDINGS: DENIALISM AND FAILURE TO RESPOND

---

[1] There are challenges to relying on algorithms, of course. Online search algorithms result in different results for individual searchers, because search algorithms are tailored to users' search and email history. These are particularly likely to affect the rank-ordering of results. When searching the Web, we use browser settings that depersonalize our searches, but this does not mean that we can determine how the controversy looks to an arbitrary searcher. This variability is a challenge to taking a macroscopic approach because unless search engine companies like Google share their data, researchers are limited to simulating popular searches or observing the search behaviors of a sample of people out of the vast numbers utilizing algorithms online.

In contrast with argument analysis based on reconstruction and assessment of individual texts, a macroscopic view draws attention to the larger patterns in a body of discourse. We illustrate this with an examination of something widely believed not only by journalists, medical experts, and members of the public, but also by many argumentation theorists: that the anti-vaccination movement is primarily a consequence of science illiteracy or, worse, an expression of science denialism. In blogs like Age of Autism, and especially in comments contributed by readers, many statements can be found that, in or out of context, can be indicted on these grounds. The press has been active in promoting both the scientific illiteracy and the science denialism themes for several years now.

Nevertheless, the common portrayal of people who refuse to vaccinate their children as science illiterates or denialists is difficult to reconcile with two facts we have observed: first, the close and detailed attention the activist community pays to science; and second, the sustained effort the activist community has made to try to change the scientific research agenda.

Although no generalization will be true for all anti-vaccination groups, at least some show a high level of attentiveness to scientific research. For example, after the US Institute of Medicine issued its 2011 report rejecting the link between MMR vaccine and autism, Age of Autism (http://www.ageofautism.com/2011/07/part-3-.html) offered a detailed study-by-study response. While its conclusions may be wrong in whole or in part, the effort put into direct inspection of scientific reports makes clear that this group fully buys into the notion that scientific research is needed and can be informative. Objections to the Institute of Medicine conclusions are expressed in terms of the validity of research design, the strength of evidence provided by individual studies, and the trustworthiness of the individuals who conducted the research.

More telling than the close attention paid to published research and authoritative reports is the sustained effort to participate in shaping the research agenda. This effort has in fact been partly successful. In 2004 the US Congress passed legislation mandating increased research on autism, and an Inter-Agency Coordinating Council was formed to regularly plan and review the research done across federal agencies involved in health science research. In 2007, members of the advocacy community succeeded in getting a workshop on environmental factors in autism sponsored by the National Academies, for the specific purpose of giving activists a chance to speak directly to scientists and to attempt to change scientific opinion on what might be the most promising directions for autism research (resulting, as noted earlier, in *Autism and the Environment*, 2008).

A great deal of attention has already been paid to how unsuccessful the public health establishment and mainstream medical science have been in eradicating the anti-vaccination position (Salzberg, 2015). Science denialism is a handy explanation for this, but as we have tried to show, it is not consistent with the persistence of the activist community in advocating for more research and for a change in research priorities. By treating all texts and participants symmetrically, we can ask a parallel question about how successful the activist community has been in persuading the scientific research community to address its questions.

The scientific community responded immediately and forcefully to the 1998 study that suggested a link between MMR vaccination and autism. That study, led by British physician and researcher Andrew Wakefield, was eventually retracted by the journal's editors and found to have been fraudulent. Other highly speculative research published at about the same time drew no such response. Wakefield's study, announced in a press release and heavily covered by British journalism, attracted so much public interest that it had to be debunked. However, this flurry of work did not really answer a question that had been on people's minds even before Wakefield published his study: why autism was increasing so rapidly.

In the years prior to Wakefield's infamous study, many ideas were circulating in public discussion to try to account for why autism cases were becoming more frequent. Many of these had to do with environmental toxins—air pollution, water pollution, pesticides, and so on, and also vaccinations (Forum on Neuroscience and Nervous System Disorders & Institute of Medicine, 2008). Let's look at what medical science was concerned with during the same period. Figure 2 provides a synoptic view of the rise, and change, in scientific interest in autism. We generated the time lines by query PubMed using carefully chosen query strings and noting how many articles were published each year matching each query string. The blue line is the line that corresponds most closely with the continuing interests of activist groups: the number of articles addressing environmental risk factors in autism. Notice how this is dwarfed by the red line, representing the number of articles addressing genetic factors in autism.

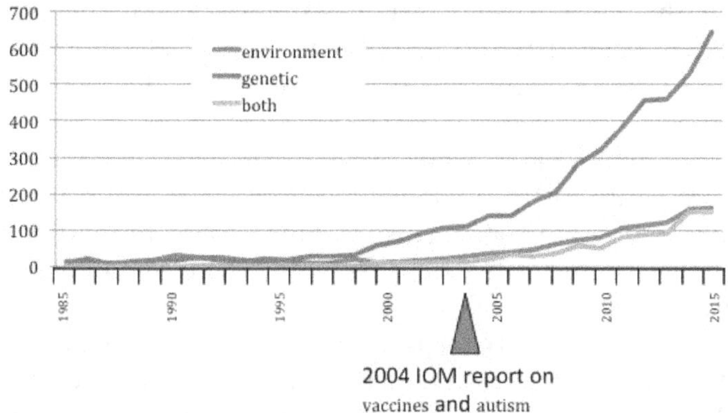

Figure 2. Research reports in PubMed linking autism to environment or genetics or both.

This analysis is (admittedly) quite crude: We did not classify studies ourselves, but used keywords that would allow us to track a contrast that has crystallized relatively recently between genetic and environmental risk factors for autism—notably, in response to public pressure on the scientific community. Our query strings work best for the newest publications, those that tend to use vocabulary that has become standardized only in the last ten years. Until the advent of automated search, journal titles were far more idiosyncratic in phrasing. But if we simply search on autism, by itself, the vast bulk of research up to the 1990s was still concentrated on diagnosis, description, and behavioral symptoms of the disease, not its cause. Only a few articles each year mentioned any potential cause of autism in its title.

So these trend lines are faulty in all kinds of ways, but if we treat them as rough indicators of the relative attention paid to genetic and environmental factors, something interesting becomes obvious. Just as the public health authorities have been unable to eradicate the beliefs that lead to refusal of vaccination, the activist community has been unable to shift scientific priorities toward a search for environmental causes of the increase in prevalence of autism. There appeared to be some softening in 2007, with the National Academies workshop, but not much really changed. What appears in the figure as a rise in research on environmental factors is actually a rise in research that searches for environmental factors associated with gene expression or with non-inherited gene abnormalities. The green curve that tracks so closely with the blue curve is the number of reports that deal with both environmental and genetic factors. It shows that autism research is completely dominated by genetic explanations for the disease--and not really concerned with explaining its increased prevalence. Until very,

very recently, the consistent response to increased numbers of autistic children has been that the increase is most likely due to change in diagnostic criteria, with no real increase in the condition being diagnosed.

Treating scientists and activists symmetrically, the charge that anti-vaxxers are science denialists not only appears inconsistent with their deep interest in the science, but also starts to look like a rather objectionable rhetorical tactic, justifying a failure to respond to the substance of an opposing argument.

Cued by these broad patterns, we wondered whether activists have ever attempted to call scientists on this failure to respond. We found that although activists' arguments make headlines only when they are manifestly unreasonable or when they are voiced by celebrities, there have been sustained efforts to enforce norms of responsiveness on the scientific community. These efforts can often be found in remarks delivered at scientific meetings, in testimony before congressional committees, and so on. Examples (1) and (2) show passages from the transcribed remarks of Mark Blaxill, a citizen participant in the 2007 National Academies workshop. Blaxill is the parent of an autistic child, the author of a book titled *Age of Autism*, and a primary writer for the Age of Autism blog. Although he is not scientifically trained, he is a well-educated citizen who has invested heavily in trying to understand autism and in trying to influence public health research. In his remarks, he is directly addressing the scientific community, and his argument is about whether the scientific community is "being reasonable."

> (1) I want to talk a little bit about the burden of proof on time trends. I would make the suggestion that given the increases that we have seen, the notion that the reported increases are an artifact is a hypothesis, and it is a testable hypothesis. (*Autism and the Environment*, p. 273)

Blaxill continues by enumerating evidence from public data sources to establish doubt that the increase over time in autism diagnoses can be explained away as artifact. But his most important argumentative move is in pointing out that the artifact idea is itself only conjectural, not supported by evidence. He concludes by describing what it will take to get citizens to believe that there really is no autism epidemic:

> (2) Then I would ask the question in terms of studies, I think we should pursue studies to clarify uncertainties, but I would urge us to consider changing the burden of proof. Rather than

> saying the burden of proof is to demonstrate that all this is real, I would say the burden of proof is to demonstrate that it is artifactual.
> If that is the case, we ought to think about changing our official narrative, because the expression of doubt about the increases creates the sense that we have a mystery and a puzzle, and no sense of urgency. The recognition of the reality changes the entire dynamic. I think a lot of us are saying we need to treat autism as an emergency, and that is what all the data points to. (*Autism and the Environment*, p. 274)

With some justice, Blaxill insists that scientists do more than repeat over and over that the apparent increase in autism could be an artifact of changing diagnostic criteria. He himself *extends* the argument that the increase is real, introducing evidence that takes the change in diagnostic criteria into account. Not unreasonably, he expects the other side to extend their own argument as well, by producing evidence to discount how much of the apparent increase can be explained away.

5. CONCLUSION

A standard intratextual analysis of arguments about MMR and autism involves locating a claim and assessing whether the arguments supporting it are any good. We might apply intratextual techniques to diagramming Blaxill's argument. A macroscopic approach opens a whole set of other possibilities.

First, and most obviously in this controversy, a macroscopic approach reveals that what is "at issue" may not be the same for everyone who has an opinion. How relevant is genetics to the controversy? That depends, for the activist community, on whether genetics is on a path to discovering why autism is increasing so rapidly. The scientific community has barely acknowledged the possibility of an authentic increase and sees the search for an environmental cause of the increase as a distraction from the discovery of genetic causes for the various forms of autism.

Second, at least conjecturally, we will find in many controversies that the quality of individual arguments as formulated at any point in time is far less important than the elaboration of a standpoint in response to criticism and counterargument. For our particular controversy, it's interesting that the research community has never really taken up sensible questions originating in the commonsense experience of citizens—but has responded forcefully when it has suited their own purposes.

Third, controversies are very dynamic—and it is never obvious when they have reached their "last word." In the vaccination controversy, all sides have shown greater complexity of argument over time, more nuanced reasoning, and of course change in exactly what is claimed. The argument for resisting vaccination has been built up over time so that it is no longer based on Wakefield's retracted paper; consequently, simply reiterating that the paper has been discredited cannot dissuade people from resisting vaccination.

A macroscopic approach does better than an intratextual approach in seeing that the overall quality of the discourse is not really a property of the defenses mounted for each distinguishable standpoint, but a product of the engagement between and among diverse points of view. In principle, citizens' questions, and even their challenges to expert opinion, can help improve the scientific basis for expert opinion by pointing to possibilities that would never otherwise have been considered. Similarly, direct engagement with scientific work, and direct practice in trying to interact with scientists, can produce stronger citizen advocacy.

As Jackson (2015) argued, "Without some form of accountability to ordinary citizens, experts and expert communities may feel that they deserve deference, but ordinary citizens do not have to agree to this. In such cases, experts must make their way in argumentation just as any other arguer would." One consequence of the controversy has been detailed attention to the safety of MMR, a kind of due diligence that is essential for rational trust in experts. The continuing protests aim for a credible explanation for the increase in incidence of autism. Putting to rest the question of whether *vaccination* is the cause still leaves an open question. The scientific community continues to expect deference, but in the new media ecology, the public is equipped to demand a much more active role in getting science done on questions it considers important.

The new media ecology draws more and more content into a common space that allows new visibility for all kinds of discourse. And it provides unprecedented resources for organizing—what Shirky (2007) described as "ridiculously easy" organizing. In the vaccination controversy, we confront not just a large number of voices, but multiple organized communities, some of which pay close and constant attention to the others, and some of which do not. This means that to understand the controversy, we need to look at how arguments within one sphere influence activities and arguments within other spheres—or how they fail to do so. We see in this controversy that scientific research and expert opinion based on this research do have influence within the public. (The autism activist community is now quite divided on the question of whether vaccination causes autism, as well as over whether autism is a disease that needs to be cured or prevented.) But we also see that, at least in principle, an informed and engaged public can exert influence on what research gets done. That is happening slowly as the activist community gets better at articulating a credible research agenda. But as Jackson (2015) suggests, in a media ecology that prefers a logic of delegation over a logic of deference, treating a failure to defer to expertise as irrational has become obsolete.

ACKNOWLEDGEMENTS: This work was supported by a grant from the University of Illinois Research Board. Scott Jacobs provided useful comments and suggestions for the conference paper on which this contribution is based.

REFERENCES

Börner, K. (2011). Plug-and-play macroscopes. *Communications of the ACM, 54*, 60-69. doi: 10.1145/1897852.1897871

Burgess, D. C., Burgess, M. A., & Leask, J. (2006). The MMR vaccination and autism controversy in United Kingdom 1998–2005: Inevitable community outrage or a failure of risk communication? *Vaccine, 24*(18), 3921-3928.

Forum on Neuroscience and Nervous System Disorders & Institute of Medicine (2008). *Autism and the environment: Challenges and opportunities for research: Workshop proceedings.* Washington, DC: The National Academies Press.

Immunization Safety Review: Vaccines and Autism (2004). Washington, DC: The National Academies Press.

Jackson, S. (2015). Deference, distrust, and delegation: Three design hypotheses. In F. H. van Eemeren & B. Garssen (Eds.), *Reflections on theoretical issues in argumentation theory*, ch. 17. Dordrecht: Springer Argumentation Library, Vol. 28.

McLuhan, E., & Zingrone, F., Eds. (1995). *Essential McLuhan.* New York: BasicBooks.

Poltorak, M., Leach, M., Fairhead, J., & Cassell, J. (2005). 'MMR talk' and vaccination choices: An ethnographic study in Brighton. *Social Science & Medicine, 61*(3), 709-719.

Ruiz, J. B., & Bell, R. A. (2014). Understanding vaccination resistance: Vaccine search term selection bias and the valence of retrieved information. *Vaccine, 32*(44), 5776-5780.

Salzberg, S. (2015, February 1). Anti-vaccine movement causes worst measles epidemic In 20 years. *Forbes.* Retrieved from http://www.forbes.com/sites/stevensalzberg/2015/02/01/anti-vaccine-movement-causes-worst-measles-epidemic-in-20-years/

Shirky, C. (2007). *Here comes everybody: The power of organizing without organizations.* Penguin.

Stratton, K., Ford, A., Rusch, E., & Clayton, E. W. (2012). *Adverse effects of vaccines: Evidence and causality.* Washington, DC: The National Academies Press.

Wakefield, A. J., Murch, S. H., Anthony, A., Linnell, J., Casson, D. M., Malik, M., Berelowitz, M., Dhillon, A. P., Thomson, M. A., Harvey, P., Valentine, A., Davies, S. E., & Walker-Smith, J. A. (1998/Retracted 2010). Ileal-lymphoid-nodular hyperplasia, non-specific colitis, and pervasive developmental disorder in children. *The Lancet, 351*(9103), 637–641. doi:10.1016/S0140-6736(97)11096-0 (Retraction published February 2010, *The Lancet, 375*[9713], p. 445)

32

Verbal Swindles, Frauds, and Other Forms of Deceptive Manipulation in the Bush Administration Case for Invading Iraq: How to Exploit Pragmatic Principles of Communication so as not to Lie

SCOTT JACOBS
University of Illinois at Urbana-Champaign, USA
csjacobs@illinois.edu

The American news media irresponsibly stood silent as Bush Administration spokespersons made a bogus case for invading Iraq. Journalists could see and should have challenged spokespersons for exploiting pragmatic principles of communication to implicate what was false or unfounded without explicitly stating lies. The defectiveness of the case was a matter of obvious inconsistency with facts that were commonly known and easily available to anyone who bothered to note and check what was being implicated.

KEYWORDS: deception, implicature, Iraq War, journalism, manipulation, news coverage, normative pragmatics, political discourse

1. INTRODUCTION

On May 1, 2003, President George W. Bush jumped out of his Lockheed S-3B Viking jet and onto the deck of the USS Abraham Lincoln where, after changing out of his flight suit, he stood beneath a banner announcing, "Mission Accomplished," and he declared: "Major combat operations in Iraq have ended" (Sanger, 2003). But already, the public justification for the invasion of Iraq had begun to unravel. Today, it is clear that the Bush Administration's case for the invasion of Iraq was based almost completely on a tissue of falsehoods, false both in detail and in general rationale.

There were no weapons of mass destruction. No chemical weapons or chemical weapon programs. No biological weapons. No smallpox. No anthrax. No bio-weapon labs sneaking about the desert on

D. Mohammed & M. Lewiński (eds.) (2016). *Argumentation and Reasoned Action: Proceedings of the 1st European Conference on Argumentation, Lisbon, 2015. Vol. II, 553-567.* London: College Publications.

the back of tractor-trailer trucks. There was no nuclear weapons program. The intercepted aluminum tubes that were carted around the hallways of Congress by Administration officials eager for war really were just for battlefield rockets and not for uranium centrifuges. And the only quantities of yellowcake uranium anyone had to worry about were not from Niger, but were the 550 tons of it that Iraq had already stockpiled before the first Gulf War and were safeguarded since that time by the International Atomic Energy Agency. There was no alliance with al-Qaida. There were no al-Qaida terrorists operating in Iraq—at least, not until American and British troops drew them in. There were no cheering crowds standing along the roadsides with flowers to greet the troops as liberators. Nor, as it turned out, would the war pay for itself.

As things began to fall apart in Iraq, a flood of stories poured out reporting on what had gone wrong: the prewar pressures on CIA analysts to come up with favorable intelligence (Risen, 2003), the con-artist defector informants (Miller & Drogin, 2005), the cooked intelligence stove-piped directly up to the White House (Hersh, 2003), the gullible journalists all too eager to get a front-page scoop (Okrent 2004), and the knowledge inside government that war had been decided long ahead of time and that, in the words of the infamous Downing Street Memo, "intelligence and facts were being fixed around the policy" (Vlahos, 2005). In the face of these astonishingly sudden revelations, it was hard not to just shake your head in disgust and think, as Michael Massing (2004) put it, "Now they tell us."

To this day, American journalists, government officials and politicians of all stripes still hide behind the feeble pronouncement of U.S. Weapons Inspector David Kay before the Senate Armed Services Committee: "We were almost *all* wrong. . . . we were *all* wrong" (Kay, 2004).

But not everyone ***was*** wrong about these matters before the war. There were plenty of reasons ***at the time*** to see through or at the very least to challenge Administration claims. The basis for doubt was readily available in the public record.

My purpose in this paper is not to explain why American journalists failed so miserably in carrying out their duty to serve as public watchdogs and as custodians of the democratic deliberation. I simply want to show that they had the opportunity to do so. I will make two points:

(1) Mainstream American journalists should have known or at the very least strongly suspected that they were being played by the Bush Administration because what was being implicated and pragmatically presupposed was manifestly inconsistent with public knowledge.
(2) Mainstream American journalists should have seen through the humbug by taking note of how, in making its case for war, the Bush Administration routinely and gratuitously violated pragmatic principles of communication.

2. PRAGMATIC PRINCIPLES OF COMMUNICATION IN ARGUMENT

The pragmatic principles of communication that I have in mind are exemplified in Grice's (1989) Cooperative Principle[1] and the attendant conversational maxims.[2] But there are many other formulations, and I am not particularly concerned with exactly which formulation of rational standards for communication one prefers to use. All have their advantages, and none are so precise, consistent or complete as to allow exact and unequivocal specification of particular implicatures and so to demand their adoption. All the available formulations depend upon prior and independent natural language user intuitions to drive any analysis.

Grice's system has the nice feature of isolating expectations of truthfulness, informativeness, relevance, and perspicuity, all with respect to purpose. These expectations serve as interpretive standards

---

[1] Grice's Cooperative Principle states:

"Make your conversational contribution such as is required, at the stage at which it occurs, by the accepted purpose or direction of the talk exchange in which you are engaged."

[2] Grice's Conversational Maxims and submaxims:

Quality: Try to make your contribution one that is true

- Do not say what you believe to be false

- Do not say that for which you lack adequate evidence

Quantity: Be informative

- Be as informative as is required (for the current purposes of the exchange)

- Do not be more informative than is required

Relation: Be relevant

Manner: Be perspicuous

- Avoid obscurity of expression

- Avoid ambiguity

- Be brief (avoid unnecessary prolixity)

- Be orderly

that fit well with the interests and norms of argumentation theorists. Apply them to the purpose of justifying a claim, and out falls something very much like, say, Johnson and Blair's (2006) standards of premise acceptability, sufficiency, and relevance. And the presumption that the speaker is observing the Cooperative Principle and Conversational Maxims—or is at least trying to satisfy them as best as feasible—amounts to what argumentation scholars will recognize as the kind of charity principle that guides analytic reconstructions of arguments (Wilson, 1959).

Sperber and Wilson (1995, p. 270) emphasize the tendency in communication toward a minimization of interpretive effort, on the one hand, and a maximization of (positive) interpretive effects, on the other. Their Presumption of Optimal Relevance emphasizes this expectation of efficiency in communication.[3] Interpretation elaborates in a direction of least resistance, an idea that replaces Grice's assumption that literal meaning is the starting point for interpretive elaboration.

Along these lines, Horn's (1984) R- and Q-Principles find in Grice's maxims something like Zipf's (1949) Principle of Least Effort.[4] People tend to follow the line of least resistance that is compatible with achieving a task. Horn's two principles suggest a kind of division of labor by which both hearer and speaker work to minimize effort, given the need to achieve the task of communication.

Levinson's (2000) Q-, M- and I-Principles call attention to the central role of stereotypical knowledge and (ab)normal means of expression in making inferences. Each principle is elaborated by an attendant speaker's maxim and a recipient corollary that is warranted by that maxim.[5] While Levinson (2000) is concerned primarily with

---

[3] Where relevance = positive inferential effects, the Presumption is:
a. The ostensive stimulus is relevant enough for it to be worth the addressee's effort to process it.
b. The ostensive stimulus is the most relevant one compatible with the communicator's abilities and preferences.

[4] Horn's Q[uantity] and R[elation] Principles are:
Q-principle:
Say as much as you can (given the R-principle)
R-principle:
Say no more than you must (given the Q-principle)

[5] ***Q-Principle***: What isn't said, isn't. (What you do not say is not the case.)

Speaker's maxim. Do not provide a statement that is informationally weaker than your knowledge of the world allows.

Recipient corollary. Take it that the speaker made the strongest statement consistent with what he knows.

generalized conversational implicatures, the principles would seem to apply beyond standardized or conventionalized formulas to particularized conversational implicatures as well.

All these standards can be applied in more specific fashion. In the upcoming examples, I will refer to a sort of corollary of Levinson's principles, what I call a Normal Forms assumption (borrowing from Cicourel, 1974, based on Schutz, 1964).

3. GEORGE BUSH'S CINCINNATI SPEECH

Now, the way these principles apply to the kind of snow job that the Bush Administration pulled in making the case for the invasion of Iraq is this: Very little of what anyone claimed or argued amounted to outright lies. Rather, the falsehoods occurred mainly at the level of what was implicated by assertions and what the very acts of assertion pragmatically presupposed. And that made anything that was meant difficult to pin down or to spell out. Any criticism always had the force of a glancing blow because the impact could never be a complete denial or direct contradiction of what was said. It could only have the force of a "yes, but . . ." kind of objection.

Let me illustrate by turning to a passage from a speech that President Bush gave in Cincinnati, Ohio, on October 7, 2002. The speech was delivered a few days before Congress was scheduled to vote on the War Powers Act that would authorize the President to use military force against Iraq. The speech and the upcoming vote occurred less than one month before the upcoming national election for seats in the House of Representatives and the Senate. The speech was promoted beforehand as an answer to widespread calls for evidence and proof justifying Administration claims that Iraq posed a serious threat to national security.

M-Principle: What's said in an abnormal way isn't normal.
Speaker's maxim. Indicate an abnormal, non-stereotypical situation by using marked expressions that contrast with those you would use to describe the corresponding normal, stereotypical situations.
Recipient corollary. What is said in an abnormal way indicates an abnormal situation.
I-Principle: What is expressed simply is stereotypically exemplified.
Speaker's maxim. Produce the minimal linguistic information sufficient to achieve your communicational ends.
Recipient corollary. Amplify the informational content of the speaker's utterance, by finding the most *specific* interpretation consistent with stereotypical knowledge, up to what you judge to be the speaker's point.

The following passage from the speech occurred just after President Bush had laid out his case that Iraq possessed weapons of mass destruction and programs for their development and production and for their delivery.

> 1. And that is the source of our urgent concern about Saddam Hussein's links to international terrorist groups.
> 2. Over the years, Iraq has provided safe haven to terrorists such as Abu Nidal, whose terror organization carried out more than ninety terrorist attacks in twenty countries that killed or injured nearly 900 people, including 12 Americans.
> 3. Iraq has also provided safe haven to Abu Abbas, who was responsible for seizing the Achille Lauro and killing an American passenger.
> 4. And we know that Iraq is continuing to finance terror, and gives assistance to groups that use terrorism to undermine Middle East peace.
> 5. We know that Iraq and the al Qaeda terrorist network share a common enemy -- the United States of America.
> 6. We know that Iraq and al Qaeda have had high-level contacts that go back a decade.
> 7. Some al Qaeda leaders who fled Afghanistan went to Iraq.
> 8. These include one very senior al Qaeda leader who received medical treatment in Baghdad this year, and who has been associated with planning for chemical and biological attacks.
> 9. We have learned that Iraq has trained al Qaeda members in bomb making, poisons, and deadly gases.
> 10. And we know that after September 11, Saddam Hussein's regime gleefully celebrated the terrorist attacks on America.
> 11. Iraq could decide on any given day to provide a biological or chemical weapon to a terrorist group or individual terrorists.
> 12. Alliances with terrorists could allow the Iraqi regime to attack America without leaving any fingerprints (Bush, 2002).

This is the entire case establishing Iraq as "the central front in the war on terrorism" and making the invasion of Iraq out to be an act of pre-emptive self-defense and a justified reaction to, if not outright retaliation for, the al-Qaeda attack of September 11th.

Practically everything said in this paragraph is seriously misleading. While I don't have the space to go through it line by line, let me just point out some of the most egregious cases of misleading implicature. In each case, the implicature is invited on the unwitting presupposition that the speaker is fulfilling a principle that is instead being covertly violated.

Consider the phrasing used in sentences 2 and 3. Iraq is said to have "provided safe haven" to both Abu Nidal and Abu Abbas. By Levinson's I-Principle and M-Principle, a listener could reasonably expect that ordinary meanings apply to both cases. President Bush has not used a "marked" expression as would be called for by the M-Principle in order to flag the listener away from assuming a normal, stereotypical situation related to the circumstances of providing safe haven. He has used a simple, straightforward phrase of the sort that would warrant listeners in assuming that the speaker is invoking the I-Principle. The way the two principles work together can be expressed as a kind of Normal Forms Assumption: What is said represents ordinary meanings and ordinary circumstances unless otherwise indicated. This assumption warrants any hearer filling in those unstated details in the situation being referred to that would be commonly known, would ordinarily obtain, and would be relevant to understanding what was meant by the terms and phrases used.

President Bush used this phrase, "provided safe haven," despite the inconsistency of its implicatures with information that was a matter of public record regarding both the case of Abu Nidal and the case of Abu Abbas. Normally, when someone is provided safe haven, the provider is protecting that someone, and that someone has sought out or at least voluntarily accepted that protection. Yet, a New York Times article in August had reported that Abu Nidal had tried to enter Iraq on a forged passport but had been discovered and placed under house arrest (Schmemann, 2002).[6][7]

Moreover, Nidal was dead. That summer, when Iraqi officials came to arrest him, he had been shot in the head—four times. As a later article in the New York Times dryly put it, "The Iraqis said it was a suicide, but few people outside Iraq believed it" (Melman, 2003). I doubt even the Bush Administration believed it. After all, one bullet to the head could be a suicide. Maybe even two. But ***four***?! These details seem pretty far afield of the sort of stereotypical situation that President Bush's simple phrasing would lead a listener to infer.

---

[6] Quotation of news sources is intended to simply give some sense of what was already a matter of public record before the invasion of Iraq commenced on March 19, 2003. Of course, what is now easily discoverable and retrievable on the worldwide web was nowhere near as accessible at the time to the casual public.

[7] "An Iraqi official quoted by Reuters said Abu Nidal had entered Iraq from Iran early last year on a forged passport but had been discovered and placed under house arrest."

Another kind of problem arises in the case of Abu Abbas. Providing someone with safe haven ordinarily suggests that that someone needs protection from danger of some kind, or could at least be thought to need such protection. In the case of a known "terrorist" one would readily imagine that the danger would involve international efforts to kill or imprison the evil doer. Yet no such danger existed. Abu Abbas had been granted amnesty six years earlier, in 1996, in an international deal following the 1992 Oslo accords between Israel and the Palestine Liberation Organization (Johnston, 2004).[8] In fact, Israeli courts confirmed his immunity in 1999 (Joffe, 2004), and until the Palestinian uprising in 2000 when he left for Iraq, Abbas had been freely living in Gaza and travelling to the West Bank through Israeli checkpoints and Israeli controlled territories (Goldenberg, 2000). And the danger presented by the Palestinian uprising, long since over, had nothing to do with Abbas ever being a terrorist.

Or consider another way in which President Bush's references to Nidal and Abbas were misleading. In both cases, President Bush provides detailed elaboration their terrorist activities. ("Abu Nidal, whose terror organization carried out more than ninety terrorist attacks in twenty countries that killed or injured nearly 900 people, including 12 Americans" and "Abu Abbas, who was responsible for seizing the Achille Lauro and killing an American passenger"). By Gricean logic (Quantity submaxim: "Do not be more informative than is required") or by Horn's R-Principle ("Do not say more than you must"), providing this extra information implicates that it must be required for some purpose. Such elaboration goes beyond just an act of reference fixing. Presumably, the descriptions are also meant to emphasize and call attention to the especially dangerous quality of these "terrorists" ***to Americans***. The implicature would seem to be that this danger to Americans is why Saddam Hussein was interested in providing them "safe haven" in his country. But the public record indicated just the opposite of what was implicated by President Bush. When he died that summer, the New York Times reported that Nidal was believed to have been suffering from heart disease and diabetes and was a heavy drinker. And his network had become largely moribund and there was no evidence that he had conducted any terrorist actions since his move to Baghdad (Schmemann, 2002). This hardly sounds consistent with the

---

[8] "Abbas was said to have renounced terrorism after the 1992 Oslo accords between Israel and the Palestine Liberation Organization. He was granted amnesty under an Oslo-related deal in 1996 and left Baghdad, where he had been living, to return to his home in the Gaza Strip. In 2000, apparently prompted by the Palestinian uprising, he returned to Baghdad."

idea that Saddam Hussein was arranging for his country to serve as base of operations for an active and serious terrorist threat to the United States, something akin to the prominent public memory of how the Taliban allowed Afghanistan to serve as safe harbor and staging area for Osama bin Laden and al-Qaida's 9-11 attackers.

There is a similar discrepancy between implicature and public record with respect to Abu Abbas. Not only had Abbas renounced terrorism (Johnston, 2004), he had publicly "castigated the September 11th attacks and had pointedly damned al-Qaida" (Joffe, 2004).

Or consider the implicatures invited by another turn of phrase. In the sixth sentence, Bush states: "We know that Iraq and al-Qaeda have had high-level contacts that go back a decade." The verb phrase, "go back" marks the whole clause "that go back a decade" as having a special meaning in a way that contrasts with the use of "ago" as in the more ordinary phrase, "a decade ago." Bush's phrasing implicates a continuous series of contacts up to and including the present time. It does so by invoking a perspective that starts in the present time and moves to an endpoint a decade ago. The implicature is that Iraq and al-Qaeda have some kind of operational relationship that is current and has been ongoing since that decade endpoint. Had "a decade ago" been used, the suggestion would have been of a punctuated set of events, of contacts that were not presumed to include the present or near-present time, and that were not presumed to constitute some kind of frequent, active, or ongoing coordination. Of course, if the contacts did not constitute that kind of coordination, there would be little point in mentioning them as they would not provide much evidence of a threatening nexus of danger.

Virtually identical phrasing ("going back") was also used by CIA Director George Tenet (C.I.A. letter, 2002)[9] and by Secretary of Defense Donald Rumsfeld (Schmitt, 2002).[10] This was a practiced usage. And apparently the implicatures were apparent enough and considered misleading enough that several former CIA officers went on the record

---

[9]"We [the CIA] have solid reporting of senior-level contacts between Iraq and al Qaeda going back a decade." CIA Director George Tenet. Letter to Senator Bob Graham, Chairman of the Senate Intelligence Committee, October 7, 2002.

[10] "On Thursday, Mr. Rumsfeld said that contacts between Al Qaeda and Iraq had increased since 1998. 'We do have solid evidence of the presence in Iraq of Al Qaeda members, including some that have been in Baghdad,' he said. 'We have what we consider to be very reliable reporting of senior-level contacts going back a decade, and of possible chemical- and biological-agent training'."

to contradict them. Bob Baer (Borger, 2002)[11], Vince Cannistraro (Slavin & Diamond, 2002)[12], and Daniel Benjamin (Dreyfuss, 2002)[13] all stated that officials of Iraq and al-Qaeda had met in the early 1990s and then again in 1998, but that nothing ever came of the meetings. While a meeting in 1998 is not just "a decade ago," neither is it the ongoing, active coordination that the Bush team so carefully insinuated.

Finally, consider a different sort of problem, one illustrated in sentence 9: "We have learned that Iraq has trained al Qaeda members in bomb making, poisons, and deadly gasses." Like use of the verb *know* in lines 4, 5, 6, and 10, the verb *learned* presupposes a strong factual basis for assertion and overtly calls attention to the epistemic status of the proposition in the *that* clause ("Iraq has trained al Qaeda members in bomb making, poisons, and deadly gasses"). Using such verbs goes beyond a commitment to the truth of the asserted proposition to an

---

[11]"There is already considerable skepticism among US intelligence officials about Mr. Bush's claims of links between Iraq and al-Qaida. Bob Baer, a former CIA agent who tracked al-Qaida's rise, said that there were contacts between Osama bin Laden and the Iraqi government in Sudan in the early 1990s and in 1998: 'But there is no evidence that a strategic partnership came out of it. I'm unaware of any evidence of Saddam pursuing terrorism against the United States.' A source familiar with the September 11 investigation said: 'The FBI has been pounded on to make this link.'"

[12] "Intelligence officials referred inquiries to the White House, then changed their stance later in the day. One official said there is credible information about discussions of safe haven between Iraq and al-Qaeda. Vince Cannistraro, former CIA counterterrorism chief, said the only known discussion of that kind occurred in 1998 when Farouk Hijazi, Iraq's ambassador to Turkey and reputedly a top Iraqi intelligence official, went to Afghanistan after al-Qaeda bombed two U.S. embassies in Africa. Hijazi offered al-Qaeda sanctuary in Iraq, but terrorist leader Osama bin Laden turned it down, Cannistraro says, because he did not want to become a tool of Iraqi leader Saddam Hussein. Cannistraro said it is possible some al-Qaeda members changed their mind after U.S. and Afghan forces overthrew Afghanistan's Taliban regime. But he accused the Bush administration of overstating uncorroborated information from al-Qaeda detainees. 'They're cooking the books,' Cannistraro said."

[13] "Daniel Benjamin, co-author, with Steven Simon, of The Age of Sacred Terror, was director of counterterrorism at the National Security Council (NSC) in the late 1990s, and he oversaw a comprehensive review of Iraq and terrorism that came up empty. 'In 1998, we went through every piece of intelligence we could find to see if there was a link [between] al-Qaeda and Iraq,' says Benjamin. 'We came to the conclusion that our intelligence agencies had it right: There was no noteworthy relationship between al-Qaeda and Iraq. I know that for a fact. No other issue has been as closely scrutinized as this one.' The State Department's annual review of state-sponsored terrorism hasn't mentioned any link, either."

affirmation of the justified basis for that commitment. In Gricean terms, the speaker is not just observing the second sub-maxim of Quality ("Do not say that for which you lack adequate evidence"); the speaker is openly committing to having engaged in an especially stringent standard of due diligence so as to rule out doubts on the matter.

Yet even as President Bush and other members of his administration were repeatedly making this kind of commitment, intelligence analysts were denying that these claims had been certified by the intelligence community (Drew, 2002).[14] It makes one wonder what kind of especially strong standard of due diligence was being alluded to. What is especially bizarre is that Administration officials openly acknowledged that the source for their assertions came almost exclusively from the uncorroborated reports of individual detainees (Pincus, 2003)[15] and of individual defectors. And it was an open secret that those detainees had given the information after having been subjected to "extraordinary renditions" to military states like Syria or Egypt, (Barton, 2001) and then subjected to "enhanced interrogation techniques" when returned to American custody (Bennis, 2003)[16]. With regard to the defectors, virtually all of them had been obtained, managed and promoted by the Iraqi National Congress (Dreyfuss, 2002).[17] Administration officials were quite open about this. Reliance on

---

[14] "Though pressed to do so by administration officials, the CIA has been unable to come up with any confirmed link between Iraq and the attacks on September 11. Intelligence officials said that they couldn't confirm Bush's claim in a speech in Cincinnati that Iraq helped to train al-Qaeda operatives in 'bomb-making and poisons and deadly gases.'"

[15] "'We've learned,' Bush said in his speech, 'that Iraq has trained al Qaeda members in bomb making and poisons and deadly gases.' But the president did not mention that when national security adviser Condoleezza Rice had referred the previous month to such training, she had said the source was al Qaeda captives."

[16] "A key component of the alleged Iraq-al Qaeda link is based on what Powell said 'detainees tell us...'. That claim must be rejected. On December 27 the *Washington Post* reported that U.S. officials had acknowledged detainees being beaten, roughed up, threatened with torture by being turned over to officials of countries known to practice even more severe torture. In such circumstances, nothing "a detainee" says can be taken as evidence of truth given that people being beaten or tortured will say anything to stop the pain."

[17] "The Pentagon's war against the CIA relies heavily on intelligence from the Iraqi National Congress. But most Iraq hands with long experience in dealing with that country's tumultuous politics consider the INC's intelligence-gathering abilities to be nearly nil. Yet, Perle, Woolsey and the Pentagon's policy-makers increasingly use the INC as their primary source of information

this source occurred even though the INC head, Ahmed Chalabi, and the organization was openly dismissed by intelligence analysts as corrupt and unreliable (Strobel & Landay, 2002).[18] [19]

4. CONCLUSION

What can be seen in these few lines from an early speech by President Bush is indicative of the Administration's entire public relations campaign. It was a campaign based on bluff, bluster, and implicature. And it was openly at odds with the public record. Too bad American journalists didn't pick up on it. Mainstream American journalists studiously avoided framing their reports of Administration claims to show how this information called into question both the reliability and the veracity of those claims and arguments. Neither did they confront or challenge Administration officials to explain the discrepancies. Maybe if they had been trained in linguistic pragmatics they would have at least known how to talk about what was going on right before their eyes.

---

about Iraq's weapons programs, its relationship to terrorism and its internal political dynamics. 'A lot of what is useful with respect to what's going on in Iraq is coming from defectors, and furthermore they are defectors who have often come through an organization, namely, the INC, that neither State nor the CIA likes very much,' Woolsey told me."

[18] "A senior U.S. military official, speaking on condition of anonymity, expressed grave fears that civilian officials in the Pentagon may be blindly accepting assertions by Chalabi and his aides. [. . .] 'Our guys working this area for a living all believe Chalabi and all those guys in their Bond Street suits are charlatans. To take them for a source of anything except a fantasy trip would be a real stretch,' one official said. [...] Pentagon officials say the INC has been a valuable conduit of information from Iraqis inside the country and has helped arrange the recent defection of four members of Saddam's regime, including one with intimate knowledge of hidden Iraqi facilities for weapons of mass destruction. [...] The CIA refused to meet with the defectors until ordered to do so and has systematically disparaged the Iraqi opposition, they said. [...] The CIA severed its relationship with the INC after the group was unable to account for millions of dollars in covert aid."

[19] "'The [INC's] intelligence isn't reliable at all,' says Cannistraro. 'Much of it is propaganda. Much of it is telling the Defense Department what they want to hear. And much of it is used to support Chalabi's own presidential ambitions. They make no distinction between intelligence and propaganda, using alleged informants and defectors who say what Chalabi wants them to say, [creating] cooked information that goes right into presidential and vice-presidential speeches'" (Dreyfuss, 2002).

REFERENCES

Barton, G. (2001, Dec. 19). Broad effort launched after '98 attacks. *Washington Post.* Retrieved from http://www.washingtonpost.com/wp-dyn/content/article/2010/03/11/AR2010031102582.html.

Bennis, P. (2003, Feb. 5). Powell's dubious case for war. *Foreign Policy in Focus.* Retrieved from http://fpif.org/powells_dubious_case_for_war/.

Borger, J. (2002, Oct. 8). White House 'exaggerating Iraqi threat' Bush's televised address attacked by US intelligence. *Guardian.* Retrieved from http://www.theguardian.com/world/2002/oct/09/iraq.usa.

Bush, G. W. (2002, Oct. 7). President Bush outlines Iraqi threat. Transcript. *White House Press Release, Office of the Press Secretary.* Retrieved from http://georgewbush-whitehouse.archives.gov/news/releases/2002/10/20021007-8.html.

C.I.A. letter to Senate on Baghdad's intentions. (2002, Oct. 9). *New York Times.* Retrieved from http://www.nytimes.com/2002/10/09/international/09TTEX.html.

Cicourel, A. V. (1974). *Cognitive sociology.* New York: Free Press.

Drew, E. (2002, Dec. 5). War games in the Senate. *New York Review of Books, 49(19).* Retrieved from http://www.nybooks.com/articles/archives/2002/dec/05/war-games-in-the-senate/.

Dreyfuss, R. (2002, Nov. 21). The Pentagon muzzles the CIA. *American Prospect.* Retrieved from http://prospect.org/article/pentagon-muzzles-cia.

Goldenberg, S. (2000, April 28). Israel lets in Achille Lauro hijacker turned peacemaker. *Guardian.* Retrieved from http://www.theguardian.com/world/2000/apr/29/israel.

Grice, P. (1989). *Studies in the way of words.* Cambridge: Harvard University Press.

Hersh, S. M. (2003, Oct. 27). The stovepipe. *New Yorker.* Retrieved from http://www.newyorker.com/magazine/2003/10/27/the-stovepipe.

Horn, L. R. (1984). Toward a new taxonomy for pragmatic inference: Q-based and R-based implicature. In D. Schiffrin (Ed.), *Georgetown University round table on languages and linguistics* (pp. 11–42). Washington, DC: Georgetown University Press.

Joffe, L. (2004, March 11). Abu Abbas. Palestinian guerrilla leader who presided over the notorious Achille Lauro cruise ship hijacking 19 years ago. Obituary. *Guardian.* Retrieved from http://www.theguardian.com/news/2004/mar/11/guardianobituaries.israel

Johnson, R. H., & Blair, J. A. (2006) *Logical self-defense.* New York: International Debate Education Association.

Johnston, D. (2004, March 10). Leader of '85 Achille Lauro attack dies at prison in Iraq. *New York Times.* Retrieved from http://www.nytimes.com/2004/03/10/world/leader-of-85-achille-lauro-attack-dies-at-prison-in-iraq.html.

Kay, D. (2004, Jan. 28). Former top U.S. weapons inspector David Kay testified Wednesday before the Senate Armed Services Committee about efforts to find weapons of mass destruction in Iraq. Transcript. *CNN.com*. Retrieved from http://www.cnn.com/2004/US/01/28/kay.transcript/.

Levinson, S. C. (2000). *Presumptive meanings: The theory of generalized conversational implicatures*. Cambridge: MIT Press.

Massing, M. (2004). *Now they tell us: The American press and Iraq*. New York: New York Review of Books.

Melman, Y. (2003, Feb. 9). Iraq's ties to terror: The threat isn't easy to read. *New York Times. Week in Review.* Retrieved from http://www.nytimes.com/2003/02/09/weekinreview/the-world-iraq-s-ties-to-terror-the-threat-isn-t-easy-to-read.html.

Miller, G., & Drogin, B. (2005, April 1). Intelligence analysts whiffed on a 'Curveball'. *Los Angeles Times.* Retrieved from http://articles.latimes.com/2005/apr/01/nation/na-curveball1.

Okrent, D. (2004, May 30). Weapons of mass destruction? Or mass distraction? *New York Times, Week in Review.* Retrieved from http://www.nytimes.com/2004/05/30/weekinreview/the-public-editor-weapons-of-mass-destruction-or-mass-distraction.html.

Pincus, W. (2003, June 22). Report cast doubt on Iraq-Al Qaeda connection. *Washington Post*, p. A01.

Risen, J. (2003, March 23). C.I.A. aides feel pressure in preparing Iraqi reports. *New York Times.* Retrieved from http://www.nytimes.com/2003/03/23/world/a-nation-at-war-intelligence-cia-aides-feel-pressure-in-preparing-iraqi-reports.html.

Sanger, D. E. (2003, May 2). Bush declares 'One victory in a war on terror.' *New York Times*, pp. A1; A16.

Schmemann, S. (2002, Aug. 21). Iraqi official affirms death of Abu Nidal; Suicide hinted. *New York Times.* Retrieved from http://www.nytimes.com/2002/08/21/world/iraqi-official-affirms-death-of-abu-nidal-suicide-hinted.html.

Schmitt, E. (2002, Sept. 28). Rumsfeld says U.S. has 'bulletproof' evidence of Iraq's links to al Qaeda. *New York Times.* Retrieved from http://www.nytimes.com/2002/09/28/world/threats-responses-intelligence-rumsfeld-says-us-has-bulletproof-evidence-iraq-s.html.

Schutz, A. (1964). *Collected papers II: Studies in social theory* (A. Broderson, Ed.). The Hague: Nijhoff.

Slavin, B., & Diamond, J. (2002, Nov. 18). Experts skeptical of reports on al-Qaeda-Baghdad link. *USA TODAY.* Retrieved from http://usatoday30.usatoday.com/news/world/2002-09-26-iraq-alqaeda_x.htm.

Sperber, D., & Wilson, D. (1995). *Relevance: Communication & cognition*, 2nd ed. Oxford: Blackwell.

Strobel, W. P., & Landay, J. S. (2002, Oct. 24). Infighting among U. S. intelligence agencies fuels dispute over Iraq. *Knight Ridder.* Retrieved from http://www.mcclatchydc.com/news/special-reports/iraq-intelligence/article24433516.html.

Vlahos, K. B. (2005, June 1). Downing Street memo mostly ignored in U.S. *FoxNews.com.* Retrieved from http://www.foxnews.com/story/2005/06/01/downing-street-memo-mostly-ignored-in-us.html.

Wilson, N. L. (1959). Substances without substrata. *Review of Metaphysics, 13*(4), 521-539.

Zipf, G. K. (1949). *Human behaviour and the principle of least effort.* Cambridge: Addison Wesley.

33

Applying Inference Anchoring Theory for Argumentative Structure Recognition in the Context of Debate

MATHILDE JANIER
Centre for Argument Technology, University of Dundee, UK
m.janier@dundee.ac.uk

OLENA YASKORSKA
Polish Academy of Sciences, Poland
OYaskorska@gmail.com

This research is motivated by the growing interest in methods for argument recognition in natural dialogues. Our aim is to describe the argumentative structure of dialogues in debate. To this end, we concentrate on a formal description of the dialogues based on both analytical and quantitative corpus studies. The studies helped us reveal the characteristics of the argumentation in the context of debate, which we are specifying in detail using Inference Anchoring Theory.

KEYWORDS: corpus studies, dialogue structure, Inference Anchoring Theory, natural communication

1. INTRODUCTION

Studies presented in this paper are driven by the currently growing interest in the recognition and description of argumentation means in real-life dialogues. In particular, we are interested in the formal description of the reply structure in different types of conversation. In our studies three types of dialogues are taken into account: debate (Yaskorska, 2014), mediation (Janier, Aakhus, Budzynska, & Reed, 2014b, 2016) and financial discourse (Budzynska, Rocci, & Yaskorska, 2014). Within the analyses of those types of dialogues we aim to define rules according to which disputants pursue their conversation, dialogical situations and factors determining arguments pro- and con-particular standpoints as well as the strategies they use for achieving their goals in the dialogue. Our description is used in studies on

D. Mohammed & M. Lewiński (eds.) (2016). *Argumentation and Reasoned Action: Proceedings of the 1st European Conference on Argumentation, Lisbon, 2015. Vol. II, 569-586.* London: College Publications.

computational techniques for argument structure extraction from resources in natural language (see for example Budzynska, Janier, Kang, Reed, Saint-Dizier, Stede, & Yaskorska, 2014; Budzynska et al., 2016).

In order to reach all those objectives we need a model for a dialogue driven by the empirical studies of particular discourse, such as investigated in (Janier, Aakhus, Budzynska, & Reed, 2014, 2016; Yaskorska, 2014). Our research, then, is based on analytical and quantitative studies of transcripts of real-life dialogues. In this paper we concentrate on the genre of moral debate which is a field investigated jointly by the authors. We aim to present fragments of characteristics for debate reply structure via which participants introduce arguments. In the corpus studies we use Inference Anchoring Theory (IAT) (Budzynska & Reed, 2011), an analytical method for dialogical argument representation.

Studies presented in the paper contribute not only to the computational models for argument extraction as to the field in which they were initiated. We also aim to bring the contribution to the existing formal models of dialogue called dialogical systems. The concept of such models was introduced by Hamblin (1970) as a rule-governed structure of organised conversation. The main goal of such systems is to model contexts for everyday conversation that will allow us to analyse argumentation performed in natural communication (Walton & Krabbe, 1995; Visser, 2013). Currently a lot of systems are built depending on the type of conversation those systems aim to model, e.g. system DC by Mackenzie (1979), system TDG by Bench-Capon (1998), or the work in (Visser, Bex, Reed, & Garssen, 2011). Yet, existing models for conversation are built as abstract systems applicable to most types of conversations. Our focus is to create a dialogue model for a particular discourse basing on the empirical analyses of real-life dialogues.

The paper consists of two parts. In the first part, we introduce the methodology for the description of dialogue reply structures that consists of a theoretical framework for the analysis of argument means in the dialogue as well as corpus studies applied for the description of the structure of this type of conversation. In the second part, we show the process in which formal description for the dialogue type of debate can be created. In particular, we show how Arguing (Section 3.1), Disagreeing (Section 3.2) and Agreeing (Section 3.3) are conducted in a moral debate and how we represent those situations in the IAT framework.

2. EMPIRICAL STUDIES FOR THE FORMAL DIALOGUE DESCRIPTION

2.1 Theoretical framework

The dialogue structure analysis is based on two theoretical models describing dialogical means for argumentation: Inference Anchoring Theory (IAT) (Budzynska & Reed, 2011) and formal dialectics (Hamblin, 1970). On the one hand, IAT allows for the representation and description of the structure of dialogical arguments; on the other hand, formal dialogue systems theory allows us to present and describe structural rules for dialogues. According to this latter framework, formal dialogical systems contain three main types of moves (Prakken, 2006). The first type, called locution rules, determines what type of locutions a player can execute and what type of illocutionary forces can be associated with them during a conversation, e.g. a participant may be allowed to use: *claim p*, for asserting a proposition *p*; *why p?*, for challenging a standpoint *p*, and *retract p*, for withdrawing *p*. The second type of moves, called commitment rules, describes how a particular locution affects the commitment store of a player, e.g. the performance of *claim p* results in adding the proposition *p* into the speaker's commitments. Finally, the key element of a dialogue system is its structural rules describing what kind of locution a player can execute at a particular stage of the dialogue, e.g. after *why p*? the participant can utter: *argue*(*p,q*); or *retract p.*

In accordance to formal dialectics, for the description of the dialogue structure of the discourse of debate we define a dialogue move as Illoc_conn(*p*), where *Illoc_conn* is the illocutionary connection a player can associate to a particular move in the dialogue, and *p* is a the propositional content of the move (Searle, 1969). For the analyses of discourse of debate we use a particular set of illocutionary connections typical of the investigated type of dialogue.

In the proposed method for argument recognition, we apply the normative approach for dialogue structure description: we describe the structural rules of real-life dialogues performed radio debates. To do this, we analyse transcripts of *The Moral Maze* radio program (for a detailed description of corpus studies see Section 2.2). The description of the argument structure is performed through the application of Inference Anchoring Theory, which shows the interrelation between dialogical process and argument structure. The main goal of the theory is to show "how the complex language structures (particularly inference) are linked to communicative structures (such as e.g. speech acts of arguing or disagreeing)" (Budzynska, Janier, Reed, & Saint-Dizier, 2013). The IAT framework assumes that argumentation structures are anchored in the dialogical process via illocutionary connections related to the illocutionary force (see Searle, 1969).

In IAT, dialogues are represented as graphs. Imagine the conversation between two participants, Bob and Alice, about funding science. This dialogue, presented in Example (1), is analysed as in figure 1.

(1)
(a) Bob: *We should increase funding for science.*
(b) Alice: *Why should we increase funding for science?*
(c) Bob: *Science is necessary for successful industry.*

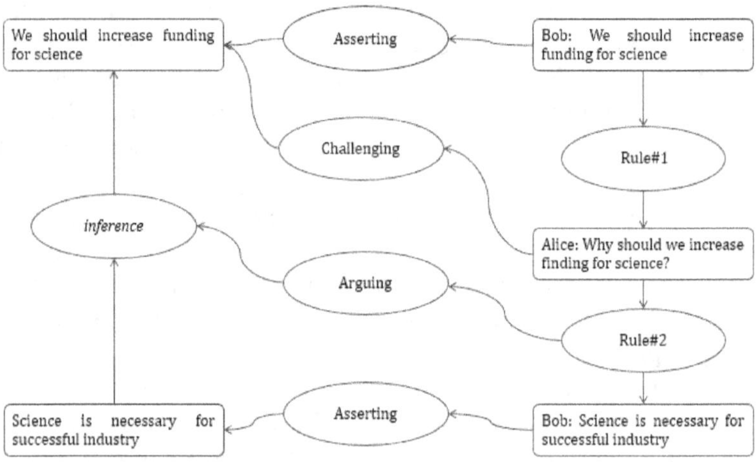

Figure 1. IAT representation of example (1)

On the right hand side of the IAT diagram (figure 1), locutions (i.e. the utterances of the dialogue between Bob and Alice) are represented. They are connected via transition nodes, instantiated Rule#1 and Rule#2 and representing specific dialogue rules according to which the conversation is governed, like e.g. a participant can perform an assertion after a challenging move. On the left hand side of the diagram, the propositional contents of the locutions are represented, and the propositional contents of Bob's locutions are connected by an inference node that shows the relation of inference between them. When a participant performs an argument against a particular statement, the relation between the two conflicting propositional contents is represented by a conflict node. Argument structure and dialogue process are linked via illocutionary connections, represented by nodes that connect them (centre of the graphs). The first, second and fourth nodes in the centre show the illocutionary connections between the locutions and the propositional contents. The third node is the illocutionary connection anchored in the transition Rule#2: it shows that, through this transition scheme, Bob has performed an argumentation.

After the representation of the interrelation between the dialogical and argumentative processes, we can specify the formal description of the dialogue process as follows:

(2)
(L1) Bob: We should increase funding for science.
(IC1) Asserting
(R1) after Asserting participant can perform Challenging
(L1) Alice: Why should we increase funding for science?
(IC2) Challenging
(R2) after Challenging participant can perform Asserting
(L3) Bob: Science is necessary for successful industry.
(IC3) Asserting
(IC4) Arguing

where Ln denotes participants' moves in a dialogue, ICn - the illocutionary connections used in a dialogue, Rn - the sequence of moves in a dialogue that we can identify as rules according to which participants engage in a dialogue.

IAT allowed us to elicit dialogical and argumentative specificities and the relation between them. Using all these elements and combining them in rules from formal systems, we can describe the rule R1:

R1: After Challenge(*p*) a player can introduce an argument - pro *p* via Assertion(*q*).

2.2 Corpus studies for the formal dialogue description

The formal dialogue description for the discourse in debate is based on corpus studies of the MM20120 corpus[1] containing four transcripts of the BBC Radio 4 program *Moral Maze* (for more details see: Budzynska, Janier, Reed, Saint-Dizier, Stede, & Yaskorska, 2014). This program typically involves a moderator, a panel of four persons and several witnesses who discuss on moral aspects of current controversial issues. The total corpus contains 58,000 words and has 297 questions or challenges for about 1500 assertions. The transcripts contain also

[1] The data presented throughout this paper are based on analyses and statistics realized in February 2015: they may differ from our previous or later works on the topic.

discourse regulators (DR), meant to manage the overall discussion, which occur on average every 15 units. This shows the vitality of the discussions and the diversity of the sub-topics addressed.

In this paper, we use the IAT framework as a method for corpus analysis, and show examples of cases that are characteristic for radio debate discourse. The IAT analyses were carried out using OVA+[2] (Janier, Lawrence, & Reed, 2014). The totality of the analyses of the MM20120 corpus can be consulted in the AIFdb Corpora[3], a large repository of argument analyses for users to create and share and reuse argument graphs (Lawrence & Reed, 2014; Lawrence, Janier, & Reed, 2016). The corpus has been annotated by two annotators who have the same linguistic training and a good expertise of the IAT theoretical background. They analysed the transcripts separately and then discussed their analyses in order to provide a single, stable analysis they both agree with. Obviously, consensus was not easy to reach, but disagreement situations were mostly due to relatively complex structures in syntax. Results of the inter-annotator agreement rate, calculated before discussion, are summarized in table 1.

Types of annotation	**Inter-annotator agreement rate**
segmentation	79%
illocutionary connections anchored in locutions	88%
illocutionary connections anchored in transitions	78%
conflict relations	76%
inference relations	86%
transitions	89%

Table 1. Inter-annotator agreement rate for the MM20120 corpus

Results of the inter-annotator agreement rate range from 76% for the annotation of conflict rules to 89% for the identification of transition nodes. The relatively low result regarding the conflict relations was mainly due to the disagreement upon the understanding of the propositional content of assertive questions and challenges: it is not always easy to define if, through assertive questioning, a speaker gives her own opinion or only checks if she understood what her opponent

---

[2] http://ova.arg-tech.org

[3] http://corpora.aifdb.org

said. As an example, see this question: "So when black people say, for example, Jews control the bank, that's not prejudice?". Is the speaker trying to say that *that's prejudice;* or is she asking to her opponent to confirm that he thinks *that's not prejudice*? All in all, the agreement rate is relatively high; in particular for the attribution of illocutionary connections. For this reason, the schemes of illocutionary connections can be considered as stable, easy to identify and accurate. Moreover, the identification of transitions show that the annotators almost always agreed on the rules of the dialogues, which is particularly important for the following of the task.

Corpus studies were realized in two main steps. Firstly, locutions made by participants during the discussion had to be described with the relevant and applicable illocutionary connections. The set of such illocutionary connections include Asserting (A) - when a participant introduces his beliefs; concessions: Conceding (Con) - when a participant admits someone else's words, Popular Conceding (PCn) - when a participant admits a generally accepted knowledge; three types of questions: Pure Questioning (PQ) - when a participant asks his antagonist to provide his opinion, Assertive Questioning (AQ) - when a participant, apart from asking, also conveys his own opinion, and Rhetorical Questioning (RQ) - when a participant introduces his standpoint but chooses the form of a question; three types of challenges, distinguished symmetrically to the typology of questioning: Pure Challenging (PCh), Assertive Challenging (ACh) and Rhetorical Challenging (RCh) (for more details about this typology see: Budzynska, Janier, Reed, & Saint-Dizier, 2013). The agreement-seeking type of dialogue, typical to radio debate, can be characterised by the distribution of the illocutionary connections (table 2).

Within the 1950 annotated locutions, apart from the most expected illocutionary connection, namely Asserting (A, 1502 occurrences), the most frequent is Assertive Questioning (AQ, 107 occurrences). According to this data and the definition of AQs we can assume that in the MM20120 debates participants tend to present their beliefs under the form of questions inviting their antagonists to confirm the statement.

In the second step of the corpus studies, argument structures in dialogical context had to be reconstructed. The analysis of the arguments performed during the dialogue was made on the level of argumentative units, which, according to IAT, include the propositional content of locutions and the relations between them. The analysis of these structures allows for a correct parsing of dialogue units and for establishing TA relations (transitions) between them, as well as identifying the illocutionary connections anchored in these transitions via which participants express arguments (i.e. in this approach: argumentation, agreement and disagreement).

Illocutionary connections	Occurrences
Assertions (A)	**1502**
Pure Questions (PQ)	83
Assertive Questions (AQ)	107
Rhetorical Questions (RQ)	72
Total Questions (Q)	**262**
Pure Challenges (PCh)	10
Assertive Challenges (ACh)	12
Rhetorical Challenges (RCh)	13
Total Challenges (Ch)	**35**
Popular concessions (PCn)	56
Concessions (Con)	7
Total Concessions (Cn)	**63**
Empty (no illocutionary connection in locutions)	**88**
Total	**1950**

Table 2. The distribution of illocutionary forces anchored in locutions in the MM20120 corpus.

For the purpose of the current paper, we take under consideration only the simple types of transition schemes where participants perform arguments with simple structures: one premise and one conclusion. Within the transitions, the MM20120 corpus has 17 schemes of transitions containing Arguing (285 cases), 16 schemes of transitions containing Disagreeing (106 cases) and 7 schemes of transitions containing the Agreeing illocutionary connection (40 cases). Tables 3, 4 and 5 detail the types of transitions (i.e. the sequences of illocutionary connections) for each argumentative action (Arguing, Disagreeing and Agreeing). To describe the types of transitions (TA) we used the notation Illoc_conn ; Illoc_conn-> Illoc_conn, where the two firsts illocutionary connections refer to the sequence of illocutionary connections anchored in two adjacent locutions, and the third illocutionary connection corresponds to the one anchored in the transition node between the first two: *A ; A -> Arguing,* as to be read as follows: an assertion followed by an assertion anchor the illocutionary connection of Arguing.

Arguing	
Type of TA	Occurrences
A; A	213
A; PQ	1
A; AQ	3
A; RQ	5
A; PCn	10
PQ; A	2
PQ; PCn	1
AQ; A	10
AQ; AQ	4
RQ; A	13
RQ; AQ	5
RQ; RQ	4
PCh; A	3
ACh; A	1
PCn; A	8
PCn; PCn	1
RCh; AQ	1
Total:	**285**

Table 3. Distribution of transitions anchoring the illocutionary connection of Arguing in the MM20120 corpus

We can see from the three tables that the predominant way of introducing arguments in the Moral Maze radio debate via the Arguing, Disagreeing and Agreeing illocutionary connections is the transition scheme *A; A* (Asserting followed by Asserting). Using this sequence, participants introduced 73% of all Arguing, 55% of Disagreeing, and 55% of Agreeing. Such a distribution of transitions was expected, given the predominant number of assertions in the whole corpus (see table 2). These statistics illustrate the general dynamics in this type of dialogue: players tend to introduce and argue for and against their beliefs via statements. Yet, this means that there is also a significant percentage of transition schemes through which participants take part in the argumentative process: 27% for Arguing (76 TAs), 45% for Disagreeing (47 TAs) and 45% for Agreeing (14 TAs).

Disagreeing	
Type of TA	Occurrences
A; A	57
A; PQ	1
A; AQ	8
A; RQ	3
A; PCh	3
A; ACh	1
A; RCh	3
A; PCn	1
A; "no"	2
AQ; A	14
AQ; RQ	2
AQ; "no"	2
RQ; A	6
RQ; PCn	1
ACh; A	1
PCn; A	1
Total:	**106**

Table 4. Distribution of transitions anchoring the illocutionary connection of Disagreeing in the MM20120 corpus

Agreeing	
Type of TA	Occurrences
A; A	17
A; "yes"	3
PQ; "yes"	1
AQ; A	6
AQ; "yes"	5
ACh; "yes"	1
PCn; A	7
Total:	**40**

Table 5. Distribution of transitions anchoring the illocutionary connection of Disagreeing in the MM20120 corpus

3. ARGUMENTATIVE DYNAMICS IN DEBATE

The participants' goal in a debate is to introduce their opinion on a topic as well as justify that opinion. The IAT analyses of the dialogues enabled us to show how participants manage to argue in this real-life context. Also, quantitative studies allow us to initiate the formal description of the reply structure of this type of dialogue. In this section we present examples of argumentative dynamics characteristic for the moral debate from the MM20120 corpus and our analyses of such examples. We will also show a first step towards the formal dialogue rule description basing on the statistic data presented in Section 2.2.

3.1 Arguing

In the moral debate participants introduce pro-arguments not only by asserting beliefs. It is important to look at complex situations of argumentation (i.e. other than an assertion followed by another assertion) and check their frequency in the corpus in order to define what other argumentative sequences (in other words, dialogical means of arguing) are used in debates. Players, for example, tend to pose Assertive Questioning through which they ask their opponents to provide their opinion while also conveying their own beliefs. This illocution is often used for argument construction, as we can see from the table 3 (Section 2.2) – where we can find five transition schemes for argument construction containing AQ.

Let's consider a fragment, taken from a transcript of the corpus where the speakers talk about problem families, increasing poverty in the country, and possible ways to deal with this problem. At one moment a participant claimed that problem families are in a circle of poverty and this circle is created by the culture of dependency grounded through generations. Disputants seemed to agree that such situation provokes some kind of disincentives to work in problem families and so, we need to use the carrot and stick approach to force them to find a job. One of the speakers, Romin, did not agree with such a standpoint. She manifested her point of view by asking two questions:

(3)
Romin: *Does it create these disincentives? Will a carrot and a stick approach to welfare state actually solve them?*

In example (3), we see an argument where no linguistic markers such as 'because' explicitly shows the speaker's line of reasoning. What is more, instead of claiming her beliefs, she poses two questions. Yet, we clearly understand Romin's opinion and its justification. This example is analysed in OVA+ as follows:

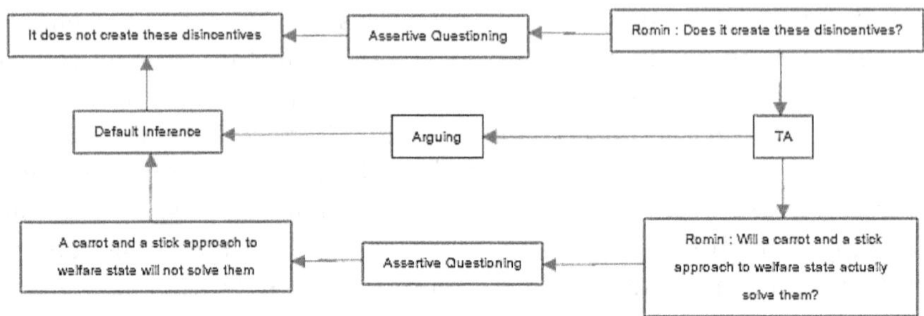

Figure 2. Analysis of example (3)

The analysis represented in figure 2 illustrates that Romin introduces an argument of the form *p then q* performing a sequence of moves consisting of two assertive questions. In the corpus of four trancripts, we found four cases of such dialogue situations, which means that participants of moral debate tend to use such means for arguing in average at least once during the conversation. Basing on this assumption, we can define a dialogue rule describing this way of performing argumentation such that:

> R2: After an Assertive Questioning*(p)* a player can introduce a pro-argument for *p* via move Assertive Questioning*(q)*

3.2 Disagreeing

People engage in a debate when they do not have the same opinion on a specific topic, that is, they disagree with each other. Disagreement can be easily spotted in a discussion e.g. when a speaker says "No" or "I disagree". Yet, in the corpus, there are not many such clear cases. As we can see in table 4 (Section 2.2), there are only two transition schemes representing this type of Disagreeing basing on 4 examples out of 106 cases. The other 16 transition schemes represent more complex dialogical situations where participants disagree. According to the data from table 4, disagreement is most of the time performed by a speaker who asserts a claim after his opponent asserted a claim as well. Leaving aside this type of disagreement, disagreeing via an assertive question following a previous speaker's assertion represents 17% of all disagreement cases (eight occurrences).

Let's regard the case from Example (4). This excerpt is taken from a transcript where speakers wonder if the current British government should be held responsible for the crimes it committed during the imperial era. During the conversation, participants were posing a standpoint about immoral behaviour of British soldiers, and living victims who need to be compensated for this terrible history they went through. This stance initiated a discussion about the inclusivity and exclusivity of black people in western countries. Here, we have Melanie's utterance manifesting her disagreement with her opponent about the stereotypes and prejudices:

> (4)
> Melanie: *What you are saying is people who are powerless cannot be prejudiced. So when black people say, for example, Jews control the bank, that's not prejudice?*

Analysing example (4) is complex for two reasons. First, because Melanie is reporting her opponent's claim ("you are saying"); then, because she uses her opponent's claim to introduce con-argument. The IAT analysis of this example is shown in figure 3.

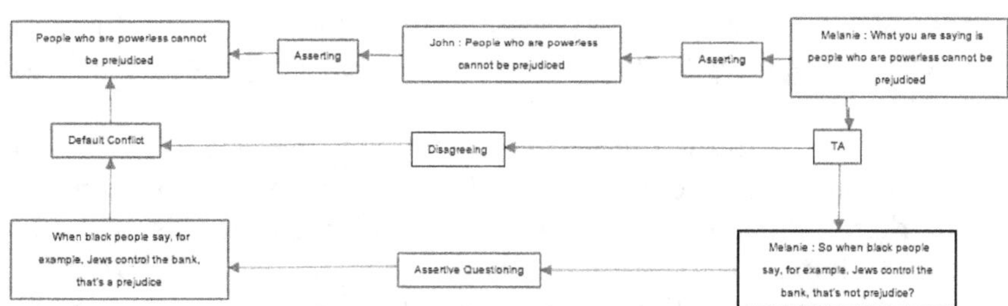

Figure 3. Analysis of example (4)

IAT allows us to analyse reported speech by representing (i) the propositional report of a locution event (right hand side of the graphs), (ii) the propositional report of the speaker's locution which is reported (centre of the graph), (iii) the propositional content of this latest locution (left hand side) and (iv) the illocutionary connections between them. Figure 3 shows that *Melanie says* that *John said* that "people who are powerless cannot be prejudiced". She then asks a question, an assertive one, propositional content of which is in conflict with John's claim. Her question allows her to make sure that John really thinks that powerless people cannot be prejudiced. The question being assertive, she also shows that she disagrees with that but in a softer way than if she herself had asserted this propositional content.

To summarize, Melanie disagrees with John's claim and introduces her disagreement with an assertive question. As an initial step of formal reply structure for debate description, we can denote this situation as a sequence of Assertion followed by Assertive Questioning. In our corpus we found 8 occurrences of such cases. We then have the dialogue rule for con-argument introduction in debate as follows:

> R3: After an Asserting*(p)* a player can introduce a con-argument for *p* via an Assertive-Questioning*(q)*

3.3 Agreeing

In a debate, if people who disagree engage in an argument it is because they want, in the end, that their opponents accept their opinion i.e. they seek for an agreement. This dialogical dynamic is, as for disagreement, sometimes very easy to detect (e.g. when a speaker says "I agree"). According to the data presented in table 5 (Section 2.2), 4 transition schemes, out of the 7 identified, respond to such simple situations. Yet, agreement can be found also within more complex dialogical situations. If we do not take the sequence *assertion-assertion* into account (for, once again, this form represents more than the half of all types of agreement), it appears that participants show their agreement by using an assertion as a reply to their opponent's Assertive Questioning 43% of the time.

Let's take the Example (5), where the speakers discuss moral aspects of getting into debt. Here, Clifford refers to the situation when free-market goes bad and poor people are blamed for this at the first place. He claims that this is a moral problem. Also he wants his opponent to commit to this claim as well. He then performs an Assertive Questioning:

> (5)
> Clifford: *But that's a big moral problem, isn't it?*
> Jamie: *It is a moral problem.*

How can we show that, in this exchange, the two participants agree with each other? Let's have a look at figure 4:

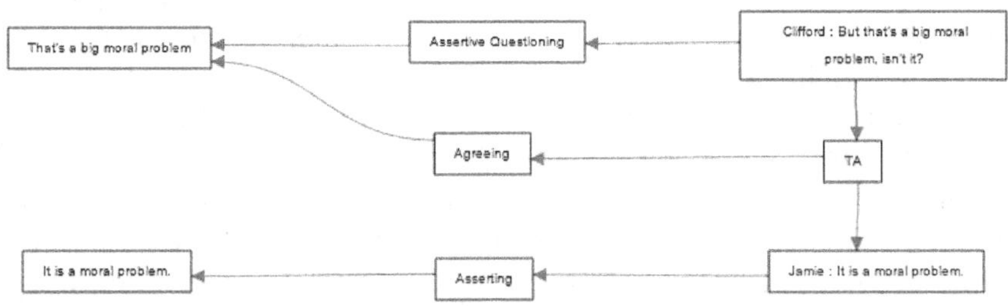

Figure 4. Analysis of example (5)

Clifford's question, once again, is an assertive question; therefore, he conveys his opinion, and the propositional content of Jamie's assertion is really similar to the one of Clifford's and could be seen as a way for Jamie to say "Yes". The fact that this was replied as the answer to an assertive question is the only reason why we can say that Jamie agrees with Clifford. If Clifford's question had been "Is it a moral problem?'"(i.e. if he had used a pure question), Clifford would have only looked for Jamie's opinion and Jamie would have claimed that "It is a moral problem"; therefore, the transition between the two moves would not have anchored any argumentative function. We can now describe a rule for the agreement in moral debate:

> R4: After an Assertive-Questioning*(p)* a player can agree with *p* via an Asserting*(q)*.

4. CONCLUSIONS

In this paper we have proposed a method for the formal dialogue description of the discourse in debate. The analysis of dialogues in this context has allowed us to detect the illocutionary connections used to present arguments -pro and -con particular standpoints, and to define rules according to which disputants pursue their conversation (i.e. their strategies).

The process we have followed for dialogue rules introduction is constituted of three steps: (i) analyses of transcripts of a particular conversation type (here, radio debates) with the identification of the structures and dynamics of the argumentation performed during the dialogue; (ii) statistical analyses of the results of the analyses of transcripts; and (iii) dialogue rule descriptions.

The analysis of the dialogues in the MM20120 corpus revealed the characteristics of the argumentation in the context of radio debate. As an example, we have shown the function of the illocutionary force of assertive question: the interest of speakers in using assertive questions is represented by two types of dialogical strategies speakers can execute during the dialogue. On the one hand, assertive questions (as much as assertions) allow speakers to build arguments. On the other hand, assertive questions are useful in probing the opinion or soliciting agreement of one's interlocutor: posing an assertive question not only commits the speaker to its propositional content (what we can feel in the assertiveness) but also triggers the opponent's agreement. Corpus analyses also revealed that a recurrent feature in debate is reported speech. Indeed, the Moral Maze program starts with the moderator reporting a political or public character's declaration. Then, throughout the debate, witnesses are invited to give their opinion on this, and the panellists often report a speaker's previous claim to initiate a new discussion on the topic at stake. IAT made it possible to represent this dialogical feature and elicited the interest that speakers may have in using it.

The formal description of the dialogical process along with the indication of the argumentation structures will be implemented in algorithms for the automatic recognition of arguments in the dialogues in debate.

ACKNOWLEDGEMENTS: The authors would like to acknowledge that the work reported in this paper has been supported in part by the British Leverhulme Trust under grant RPG-2013-076, in part by the Polish National Science Centre under grant UMO-2014/12/T/HS1/00490.

REFERENCES

Bench-Capon, T. J. (1998). Specification and implementation of Toulmin dialogue game. In *Proceedings of JURIX*, December.

Budzynska, K., & Reed, C. (2011). Whence inference. In *Technical report*, University of Dundee.

Budzynska, K., Janier, M., Reed, C., & Saint-Dizier, P., (2013). Towards Extraction of Dialogical Arguments. In *Proceedings of 13th International Conference on Computational Models of Natural Argument (CMNA13)*.

Budzynska, K., Janier, M., Reed, C., Saint Dizier, P., Stede, M., & Yaskorska. O. (2014). A model for processing illocutionary structures and argumentation in debates. In *Proc. of the 9th edition of the Language Resources and Evaluation Conference (LREC)*.

Budzynska, K., Janier, M., Kang, J., Reed, C., Saint Dizier, P., Stede, M., & Yaskorska. O. (2014). Towards Argument Mining from Dialogue. In *Frontiers in Artificial Intelligence and Applications*, vol. 266, 185–196. *Computational Models of Argument (COMMA14)*, IOS Press, September.

Budzynska, K., Janier, M., Kang, J., Konat, B., Reed, C., Saint-Dizier, P., Stede, M., & Yaskorska, O. (2016). Automatically identifying transitions between locutions in dialogue. In College Publications. *Studies in Logic and Argumentation*. In D. Mohammed & M. Lewiński (eds.), *Argumentation and Reasoned Action: Proceedings of the 1st European Conference on Argumentation, Lisbon, 2015. Vol. II, 311-327.* London: College Publications.

Budzynska, K., Rocci, A., & Yaskorska, O. (2014). Financial Dialogue Games: A Protocol for Earnings Conference Calls, *Frontiers in Artificial Intelligence and Applications. Proc. of 5th International Conference on Computational Models of Argument COMMA 2014, Simon Parsons, Nir Oren, Chris Reed, Federico Cerutti (Eds.)*, IOS Press, (266): 19-30

Hamblin, C. L. (9070). *Fallacies*. Vale Press.

Janier, M., Lawrence, J., & Reed C. (2014). OVA+: An argument analysis interface. In *Computational Models of Argument (COMMA)*, vol. 266, 463–464. IOS Press.

Janier, M., Aakhus, M., Budzynska, K., & Reed, C. (2014). Games Mediators Play: Empirical methods for deriving dialogue structure. In *1st International Workshop for Methodologies for Research on Legal Argumentation (MET-ARG).*

Janier M., Aakhus, M., Budzynska, K., & Reed, C. (2016). Modeling argumentative activity in mediation with Inference Anchoring Theory: The case of impasse. In D. Mohammed & M. Lewiński (eds.), *Argumentation and Reasoned Action: Proceedings of the 1st European Conference on Argumentation, Lisbon, 2015. Vol. I, 245-264.* London: College Publications.

Lawrence, J., & Reed, C. (2014). AIFdb Corpora. In *Computational Models of Argument (COMMA)*, vol. 266, 465–466. IOS Press.

Lawrence, J., Janier, M., & Reed, C. (2016). Working with open argument corpora. In D. Mohammed & M. Lewiński (eds.), *Argumentation and Reasoned Action: Proceedings of the 1st European Conference on Argumentation, Lisbon, 2015. Vol. I, 367-380.* London: College Publications.

Mackenzie, J. D. (1979). How to stop talking to tortoises. In *Notre Dame Journal of Formal Logic*, *20*(4), 705–717.

Prakken, H. (2006). Formal systems for persuasion dialogue. In *The Knowledge Engineering Review*, *21*(02), 163–188.

Reed, C. Wells, S., Devereux, J., & Rowe, G. (2008). AIF+: Dialogue in the Argument Interchange Format. In *Frontiers in artificial intelligence and applications*, 172, 311.

Searle, J. R. (1969). *Speech acts: An essay in the philosophy of language.* Cambridge University Press.

Visser, J. (2013). A formal account of complex argumentation in a critical discussion. In D. Mohammed & M. Lewiński (Eds.), *Virtues of argumentation: proceedings of the 10th International Conference of the Ontario Society for the Study of Argumentation (OSSA).*

Visser, J., Bex, F., Reed, C., & Garssen, B. (2011). Correspondence between the pragma-dialectical discussion model and the Argument Interchange Format. In *Studies in Logic, Grammar and Rhetoric, 23*(26), 189–224.

Walton, D. N., & Krabbe, E. C. W. (1995). *Commitment in dialogue: Basic concepts of interpersonal reasoning*. State University of New York Press.

Yaskorska, O. (2014). Recognising argumentation in dialogical context. In *Proceedings of the 8th Conference of the International Society for the Study of Argumentation* (ISSA).

34

Strategic Maneuvering with *That Says it All* and *That Says Everything*

HENRIKE JANSEN
Leiden University Centre for Linguistics, The Netherlands
h.jansen@hum.leidenuniv.nl

This paper discusses the expressions *that says it all* and *that says everything* when used in presenting an argument, as in: 'It must have been a rich place. The area has many gothic cathedrals, that says it all.' It will be argued that (1) *that says it all* constitutes the argument's (explicitly conveyed) inference license, (2) a protagonist uses this phrase for strategic reasons, and (3) its strategic function may entail some specific fallacies.

KEYWORDS: burden of proof fallacy, derailment, idiomatic expression, inference license, presentational device, strategic maneuvering, unclearness fallacy

1. INTRODUCTION

This paper is about arguments that are presented with the idiomatic expressions *that says it all* and *that says everything*, as in fragment (1) below. Fragment (1) is a comment posted in response to an article in 'MercoPress', a news site calling itself a South Atlantic News Agency. The author of this comment is participating in a discussion about the legitimacy of Argentina's invasion of the Falklands in 1982 and the UK's entitlement to reclaim the island. The particular fragment is a response to another comment in which the word 'Kelpers' is used for people who live on the Falklands. The use of this word is considered to be offensive by the author of (1):

> (1) I politely asked you not to use what many consider an offensive term, typically Think [another commentator, HJ] uses it with abandon with a huge measure of personal abuse thrown in. Used in such a boorish manner, when clearly aware

D. Mohammed & M. Lewiński (eds.) (2016). *Argumentation and Reasoned Action: Proceedings of the 1st European Conference on Argumentation, Lisbon, 2015. Vol. II, 587-599.* London: College Publications.

> of its offensive nature, well that says everything about your attitude towards the Falklanders. (...)
>
> But yes, if Think and his ilk wish to use racist terminology when there is a very ready alternative [that] says it all really. We can pretty much ignore them as a irrelevant racist fuckwits.
>
> 87 JustinKuntz (#), Aug 31st, 2010 - 04:22 pm
> (Comment to 'Malvinas: Argentina working for more than "a gesture of support" from Unasur', article in MercoPress, August 27th, 2010 - 08:25 UTC;
> http://en.mercopress.com/2010/08/27/malvinas-argentina-working-for-more-than-a-gesture-of-support-from-unasur)

In (1) two arguments are presented using the phrases *that says it all* and *that says everything*. In this paper, I will discuss how both expressions are used to fulfil an arguer's rhetorical purpose. I will show that these phrases constitute the explicit inference license of the argument of which they are part. I will also discuss their strategic functions and some fallacies that arise when the strategic use of these maneuvers is derailed. This paper therefore contributes to the pragma-dialectical theory of strategic maneuvering by regarding *that says it all* and *that says everything* as presentational devices, i.e. as means of formulating an argument in an effective way.[1]

Before presenting my analysis, I have to add two preliminary notes. Firstly, this research is occasioned by the Dutch idiomatic expression *dat zegt alles*. This Dutch expression is very frequently used. It can be translated into English by both *that says everything* and *that says it all.*[2] *Dat zegt alles* pre-eminently fulfils an argumentative function

---

[1] Analyses of other presentational devices conducted from a pragma-dialectical perspective can be found in (a.o.): Boogaart (2013), Jansen (2009; 2011), Jansen, Dingemanse & Persoon (2011), Snoeck Henkemans (2005; 2007; 2009; 2011; 2013), Tonnard (2011), Tseronis (2009), van Eemeren & Houtlosser (1999a; 1999b; 2000a; 2000b; 2002), Zarefsky (2006).

[2] In a column in *The New York Times Magazine* (20 May, 1984) William Safire suggests that *that says it all* is stronger than *that says everything*: "What is this it all that everybody is saying? The expression goes beyond the mere 'everything' to encompass 'all that is possible.' For example, 'the man who has everything' is an impoverished oaf compared with 'the man who has it all.' Helen Gurley Brown, the Cosmopolitan editor who discovered 'Sex and the Single Girl,' entitled her memoirs 'Having It All' - that is, living life to the ne plus ultra - and 'it all' meant a combination of love, success, health, money, good looks, fame and contentment sometimes summarized in fast-food restaurants as 'the works.'"

in discourse, and the same holds for related expressions making use of the pattern *that says X, X* being *enough* (in Dutch: *genoeg* or *voldoende*), *much* (in Dutch: *veel*), *a lot* (in Dutch: *het nodige*), *something* (in Dutch: *iets* or *wel wat*), *little* (in Dutch: *weinig*) and *nothing* (in Dutch: *niets*). A theoretical analysis of the English uses of the *that says X*-pattern shows that these can fulfil the same argumentative function as the Dutch ones.[3]

My second note concerns the class of arguments for which a presentation with *that says X* seems typical. A collection of Dutch and English arguments using these expressions shows that they are mainly used in symptomatic argumentation. Symptomatic arguments are those in which the premise refers to something that is symptomatic of the subject or object mentioned in the standpoint (van Eemeren, Houtlosser, & Snoeck Henkemans, 2007, p. 154).[4] Or, in other words, the premise of such an argument includes a sign that is suggested to entail the acceptability of the standpoint (such arguments are therefore also called 'arguments from sign'). This can be demonstrated by fragment (1), where we see two arguments in which the opponent's behaviour is presented as a sign of his character. In the first argument, the behaviour consists of using an offensive term while being aware of its offensiveness; in the second, it consists of using an offensive term when there is a non-offensive alternative. In both arguments the behaviour displayed by the arguer's opponent is presented as a reason supporting a negative standpoint about the opponent's attitude.

2. ARGUMENTATIVE FUNCTION

The arguments in (1) show two ways in which *that says it all* and *that says everything* can be used, namely with or without the addition of an 'about' phrase. For convenience the fragment is repeated below and the two arguments are numbered (a) and (b):

> (1a) I politely asked you not to use what many consider an offensive term, typically Think uses it with abandon with a huge measure of personal abuse thrown in. Used in such a boorish manner, when clearly aware of its offensive nature, well that says everything about your attitude towards the Falklanders.

---

[3] This also holds for the synonymous *that tells you X*-pattern.

[4] This definition applies to the 'prototypical variant' of symptomatic argumentation. Van Eemeren, Houtlosser and Snoeck Henkemans (*Ibidem*) also distinguish an 'inverse variant', in which the symptomatic relationship goes the other way around, i.e. when the standpoint mentions something that is symptomatic for what is expressed in the premise.

> (1b) But yes, if Think and his ilk wish to use racist terminology when there is a very ready alternative [that] says it all really. We can pretty much ignore them as a irrelevant racist fuckwits.

Although in this fragment 'about' is only added to *that says everything*, it could also be added to *that says it all*. If the phrases are extended with 'about', the proposition following 'about' gives an indication of how the standpoint should be interpreted. Often this is just a slight indication, of which a more specific interpretation should be grasped on the basis of the context. In argument (a) the standpoint is a negative evaluation of the opponent's character. It is not obvious precisely how far this negative evaluation goes, but in any case, the arguer seems committed to the standpoint that his opponent's attitude is 'offensive'. In argument (b) the addition 'about' is lacking; instead, the argumentation contains an explicit standpoint, which specifies the negative typification of the opponent's character, namely that this person is a racist fuckwit. This stronger standpoint was perhaps already intended in argument (a), but of course we can't be sure about that. After all, the second argument contains the new premise, arguing that an alternative word for 'Kelpers' is available.

A reconstruction of both arguments shows that *that says everything* and *that says it all* both constitute their arguments' inference license. In argument (a) the word *that* in *that says everything* refers to the sentence part *Used in such a boorish manner, when clearly aware of its offensive nature*. This sentence part functions as a premise in an argument because *says everything* is a metonymical way of indicating that a conclusion can be drawn from it.[5] As I just noted, the words *about your attitude* give the clue that the standpoint must be about the attitude of the person who is addressed, and the context reveals that it must be a negative evaluation of this attitude. This analysis results in the following reconstruction. In this reconstruction, 1 presents the standpoint (here placed between brackets because it was left unexpressed by the arguer), 1.1 the premise and 1.1' the element that usually (but not always, as in this case) remains unexpressed: the inference license.

> (1. Your attitude towards the Falklanders is offensive)
> 1.1 You use an offensive term in a boorish manner while being clearly aware of its offensive nature

[5] Here I mean metonymical in the sense of Lakoff & Johnson (1980) and Lakoff (1987).

1.1' *That says everything* about your attitude
= If people use an offensive term in a boorish manner while being clearly aware of its offensive nature, this indicates that they have an offensive attitude

A largely similar reconstruction can be made of argument (b), the only difference being that (b) contains an explicit standpoint whereas the standpoint of argument (a) was left unexpressed by the arguer. In this argument, the phrase *says it all* connects – again – the explicit premise with the (also explicit) standpoint. The standpoint is constituted by the sentence that follows this expression. This gives the following reconstruction:

1. Think and his ilk are racist fuckwits
1.1 They wish to use racist terminology when there is a very ready alternative
1.1' *That says it all*
= If people use racist terminology when there is a very ready alternative, this indicates that they are racist fuckwits

3. STRATEGIC POTENTIAL

To gain more insight into the strategic potential of *that says it all, that says everything* and their synonyms, it is useful to compare the way in which these expressions function as an inference license with the way this is done by a regular inference licence. A regular inference license consists of a conditional *if...then* sentence, in which the content of the premise is connected to the content of the standpoint. If this element remains unexpressed in the argumentation, which is often the case, the analyst has the task of making it explicit (reconstructing it). This can either be done with an *if...then* sentence that literally repeats what is expressed in the premise and the standpoint, or by creating a more generalised version that abstracts from the particulars mentioned in those elements.[6] Examples of the latter are the *if...then* sentences that were added to the reconstructions of premises (1.1') of arguments (1a) and (1b).[7]

---

[6] The literature on argumentation theory contains different views about whether unexpressed premises should be analysed at all, and if so, whether they should be reconstructed as a generalised conditional or not.

[7] Van Eemeren & Grootendorst (1992, p. 64) take the view that an *if...then* formulation constitutes the starting point for the reconstruction of unexpressed premises: the variant in which the premise and the standpoint are

There are some differences between a 'regular' inference license on the one hand and *that says it all* and *that says everything* on the other. Firstly, *that says it all* and *that says everything* do not repeat the wordings of the premise or the standpoint, in either a literal or a generalised way. They only contain reference to the premise with anaphorical use of *that*, and they have no standpoint content whatsoever. And *that says it all* and *that says everything* are also different for another reason, since these expressions do not simply express a link between the premise and the standpoint in a neutral way: *that says it all* and *that says everything* convey the meaning that the matter is solved now, that the premise given provides sufficient ground to accept the standpoint. These two characteristics of an inference license formulated by *that says it all* and *that says everything* allow for two strategic uses.

One strategic use of *that says it all* and *that says everything* relates to arguments whose standpoint has been left unexpressed. Normally, when an argument contains an unexpressed standpoint, the elements that make up the argument consist of an explicit premise and an explicit inference license, as in (2):

(2) This is a stupid question and those will not be answered.[8]

In this argument, the standpoint, i.e. that this question will not be answered, has remained unexpressed, which is apparent in the following reconstruction:

(1.	This question will not be answered)
1.1	This is a stupid question
1.1'	Stupid questions will not be answered
	(= If a question is stupid, it will not be answered)

---

literally repeated is called the 'logical minimum'. The next step in the reconstruction process is that the logical minimum is generalised to a 'pragmatic optimum' by 'making it as informative as possible without ascribing unwarranted commitments to the speaker and formulating it in a colloquial way that fits with the rest of the argumentative discourse.' In the above case, the pragmatic optimum version of the unexpressed premise would read 'People using an offensive term in a boorish manner, while being clearly aware of its offensive nature, have an offensive attitude'.

[8] The same argument can even be expressed by merely stating the inference license: 'Stupid questions will not be answered.' After all, this conditional implies both the premise and the standpoint.

Leaving a standpoint implicit when the premise and the inference license are already expressed is in line with van Eemeren & Grootendorst's Principle of Communication (1992, p. 50 ff.; 2004, p. 75 ff.). This Principle has a similar status to Grice's Cooperative Principle (Grice, 1975) but is more specific because it incorporates insights from Searle's theory of speech acts and their conditions (Searle, 1979). The Principle of Communication consists of the commandment *Be clear, honest, efficient and to the point.* The efficiency part relates to Grice's Quantity Maxim and the preparatory conditions of a speech act, stating that interlocutors should not say more than necessary. It is specified in two rules of communication, saying that no *redundant* or *unnecessary* speech acts should be performed, and that no *meaningless* speech acts should be performed (van Eemeren & Grootendorst, 1992, p. 62; van Eemeren & Grootendorst, 2004, pp. 77-78). On the basis of this principle, it can be argued that explicitly expressing a standpoint is an infringement of the first of these rules, since it is superfluous to do so once the premise and the inference license are given. After all, these two elements would provide a hearer with enough information to grasp the unexpressed premise's content.

However, the above line of reasoning does not apply to *that says it all* and *that says everything*. While a regular inference license like the one in (2) refers to a standpoint content, *that says it all* and *that says everything* do not. The inference license expressed by *that says it all* and *that says everything* only indicates that a conclusion should be drawn. Although both expressions suggest that it is obvious which standpoint content the arguer is aiming at, it can sometimes be pretty difficult to grasp what exactly this should be taken to mean. This even holds for cases in which a clue is given with the addition *about [...]*. Remember argument (a) in fragment (1):

> (1a) I politely asked you not to use what many consider an offensive term, typically Think uses it with abandon with a huge measure of personal abuse thrown in. Used in such a boorish manner, when clearly aware of its offensive nature, well that says everything about your attitude towards the Falklanders.

It is clear that the standpoint to be reconstructed involves the attitude of the addressed, and that it gives a negative evaluation of this attitude. However, it is not clear how far the negative characterisation of this attitude may go. As I argued above, the arguer seems at least committed to the standpoint 'you have an offensive attitude to Falklanders'. At the same time, there seems to be a slight suggestion that it involves more, but what this more is, is left to the interpretation of the hearer.

One can think of several reasons why a protagonist would want to leave the interpretation of the standpoint to the hearer. One reason is that the standpoint contains a negative characterisation of things or people that it would be offensive to express. For example, the Dutch corpus shows that the Dutch equivalent of *that says it all* and *that says everything* is very often used by soccer players or soccer coaches who wish to imply a negative conclusion about the way their team has played, but also wish to avoid expressing this conclusion's exact wording. Another reason is that the standpoint is about something emotional or just very complicated and hard to put in only one sentence. In such cases, it may be difficult for the arguer to find the proper formulation and it can be an advantageous move to leave the interpretation of the standpoint to the hearer. This is strategic because it is likely that an antagonist would interpret the standpoint in a way that is most logical to him, including the nuances and conditions under which it is acceptable to him. It provides the protagonist with the opportunity to avoid committing himself to a specific content in advance, and it also leaves him space to distance himself from the content that is reconstructed by the antagonist, if it were beneficial for him to do so.

The second strategic use of *that says it all* and *that says everything* has to do with their semantics. These expressions suggest that the proposition functioning as a premise obviously implies the standpoint and is therefore a sufficient reason for accepting it. They convey the impression that no further explanation or support is needed. They are therefore a strategic means to close a discussion or even to prevent it from starting. It can discourage the opponent from expressing criticism because the way the inference license is formulated may make him doubt his own knowledge and good judgement. Moreover, a less critical or less attentive antagonist may be misled if he does not take the trouble to find out what the inference license actually says and whether this is acceptable.

4. DERAILMENTS

It has become clear by now that *that says it all* and *that says everything* are a means to leave the standpoint implicit and to present the inference license as self-evident. These two uses yield two potential fallacies that could be committed by presenting an argument with these phrases.

Firstly, leaving it up to the antagonist how the standpoint should be interpreted, could derail into a violation of the tenth pragma-dialectical discussion rule, the 'language use rule'. This discussion rule reads: "Discussants may not use any formulations that are insufficiently clear or confusingly ambiguous, and they may not deliberately misinterpret the other party's formulations" (van Eemeren & Grootendorst, 2004, p. 195). This discussion rule is violated if an argument in the form of *that says it all* and *that says everything* is put forward and its context does not provide (enough) clues for an interpretation of the standpoint, thus allowing for many possible interpretations. Consider the argument in (3) for an example:

> (3) Jenny Diski: "I am against anti-Semitism and racism in general..."
> Only in general? Why even write "in general" when that word was not needed?
> I think that says it all.
>
> (ItsikDeWembley, April 1, 2014 at 6:18 am; http://cifwatch.com/2014/04/01/guardian-review-of-the-film-noah-culls-parable-about-israeli-land-grabs-in-the-biblical-story/)

This fragment is taken from a website called "UK Media Watch – Promoting fair and accurate coverage of Israel". Its author is responding to a passage in a review by Jenny Diski on the movie Noah. In this passage Diski says: "(...) I find myself in a double difficulty. I am against antisemitism and racism in general, but I am also against the idea of Zionism and dismayed by its consequences."

How should the standpoint in this comment be reconstructed? It seems that more than one fair interpretation is possible – fair with respect to what the author can be said to have actually committed himself to. One such interpretation is 'Diski is not against anti-Semitism and racism'. But the author could surely distance himself from this interpretation were he met with criticism. For example, he could nuance this standpoint to a version reading 'It is suspected that Diski makes an exception to the unacceptability of anti-Semitism and racism'. Or he could say that his standpoint is not about Diski but about all people who make exceptions to the unacceptability of anti-Semitism and racism, and that it is saying that people should not make exceptions to them. That so many options are possible shows that the arguer is evading his responsibility of committing himself to a standpoint. I consider this to be a derailment of the strategic use of *that says it all* and *that says everything*. In cases like this, the arguer violates the tenth pragma-dialectical discussion rule by committing the fallacy of unclearness (van Eemeren & Grootendorst, 1992, p. 197).

A second fallacy that relates to presenting an argument with *that says it all* and *that says everything* is connected with evading the burden of proof. Both expressions can be regarded as a means to immunise the argumentation against critical questions. By using them, the arguer suggests that the argument needs no further back-up. It gives the impression that it is obvious that the standpoint follows from the premise or that it is obvious which implicit standpoint the arguer believes must follow from the premise. Because self-evidence is an aspect of these expressions' meaning, *that says it all* and *that says everything* may – under certain conditions – constitute manifestations of the fallacy of evading the burden of proof. According to pragma-dialectics, a discussant committing such a fallacy violates the second rule of a critical discussion, the 'obligation-to-defend rule'. This rule reads: "Discussants who advance a standpoint may not refuse to defend this standpoint when requested to do so" (van Eemeren & Grootendorst, 2004, p. 191). This means that the fallacy of evading the burden of proof applies to standpoints for which no support at all is given.

However, in contrast to the pragma-dialectical description of the burden of proof-fallacy, the argumentative use of *that says it all* and *that says everything* does not involve the standpoint, but rather the inference license; and moreover, the use of these expressions does indeed involve argumentation, since there is at least one other statement that functions as a premise. Therefore, the unreasonable use of *that says it all* and *that says everything* should be regarded as a specific variant of this fallacy. This variant arises if the premise offered for the conclusion does not live up to the requirements of proof that these expressions impose, i.e. that the premise provides all-embracing support of the standpoint. It is therefore possible to detect such a fallacy by asking the critical questions associated with the argumentation scheme that hosts one of these expressions. Of course, the critical questions relevant to an argumentation scheme are primarily used to detect fallacies that consist of an incorrect application of such a scheme. But they serve equally well to decide whether an arguer evaded his burden of proof when presenting an argument with *that says it all* or *that says everything*. That is, if the arguer is unable to provide a satisfactory answer to the critical questions concerning the argumentation scheme, he is guilty of evading the burden of proof because he used an expression suggesting that no further support was needed, whereas further support was indeed needed.[9]

---

[9] Van Eemeren & Houtlosser (2002, pp. 151-152) describe the derailment of a similar argumentative move: the *conciliatio*. A *conciliatio* consists in optimally adapting to an opponent's starting point (arguing *ex concessis*): the arguer takes the starting point of his opponent to let it function as a premise in his own argumentation. Such a move derails if the proponent formulates his

5. CONCLUSION

In this paper I have argued that the argumentative use of *that says it all* and *that says everything* causes these expressions to function as an argument's inference license. Formulating an argument by using an expression of this kind has two strategic advantages for an arguer, i.e. leaving the standpoint implicit and suggesting that no further argumentative support is needed. Both of these strategic uses can derail into a fallacy, respectively the fallacy of misusing unclearness and a variant of the fallacy of the burden of proof.

REFERENCES

Boogaart, R.J.U. (2013). Strategische manoeuvres met sterke drank: redelijk effectief? [Strategic maneuvers with hard liquor: Reasonably effective?] In T. A. J.M. Janssen & T. van Strien (Eds.), *Neerlandistiek in Beeld* (pp. 283-292). Amsterdam/Münster: Stichting Neerlandistiek VU/Nodus Publikationen.

Eemeren, F.H. van, & Grootendorst, R. (1992). *Argumentation, communication and fallacies. A pragma-dialectical perspective.* Hillsdale, NJ etc.: Erlbaum Publishers.

Eemeren, F.H. van, & Grootendorst, R. (2004). *A systematic theory of argumentation: The pragma-dialectical approach.* Cambridge: Cambridge University Press.

Eemeren, F.H. van, & Houtlosser, P. (1999a). Over zekere waarden. Een analyse van twee objectief waarderende standpunten. [On certain values. An analysis of two objectively evaluative standpoints.] *Taalbeheersing, 21,* 179-186.

Eemeren, F.H. van, & Houtlosser, P. (1999b). William the Silent's argumentative discourse. In F.H. van Eemeren, R. Grootendorst, J.A. Blair & C.A Willard (Eds.), *Proceedings of the fourth conference of the International Society for the Study of Argumentation* (pp. 168-171). Amsterdam: Sic Sat.

Eemeren, F.H. van, & Houtlosser, P. (2000a). Rhetorical analysis within a pragma-dialectical framework. The case of R.J. Reynolds. *Argumentation, 14,* 293-305.

Eemeren, F.H. van, & Houtlosser, P. (2000b). De retorische functie van stijlfiguren in een dialectisch proces: strategisch gebruikte metaforen in Edward Kennedy's Chappaquiddick speech. [The rhetorical function of figures of speech in a dialectical process: Strategically used metaphors in Edward Kennedy's Chappaquiddick speech.] In R.

---

argument in such a way that the opponent's acceptance of this starting point's justificatory power (implying the proponent's standpoint) is presupposed, whereas it is unlikely that the opponent would accept it.

Neutelings, N. Ummelen & A. Maes (Eds.), *Over de grenzen van de taalbeheersing* (pp. 151-162). Den Haag: SDU Uitgevers.

Eemeren, F.H. van, & Houtlosser, P. (2002). Strategic maneuvering in argumentative discourse: A delicate balance. In F.H. van Eemeren & P. Houtlosser (Eds.), *Dialectic and rhetoric: The warp and woof of argumentation analysis* (pp. 131-159). Dordrecht etc.: Kluwer.

Eemeren, F.H. van, Houtlosser, P., & Snoeck Henkemans, A.F. (2007). *Argumentative indicators in discourse. A pragma-dialectical study.* Dordrecht: Springer.

Grice, P. (1975). Logic and conversation. In P. Cole & J. Morgan (Eds.), *Syntax and semantics. Volume 3: Speech acts* (pp. 41-58). New York: Academic Press.

Jansen, H. (2009). Arguing about plausible facts. Why a reductio ad absurdum presentation may be more convincing. In E.T. Feteris, H. Kloosterhuis & H.J. Plug (Eds.), *Argumentation and the application of legal rules* (pp. 140-159). Amsterdam: Sic Sat.

Jansen, H. (2011). "If that were true, I would never have ..." The counterfactual presentation of arguments that appeal to human behaviour. In F.H. van Eemeren, B.J. Garssen, D. Godden & G. Mitchell (Eds.), *Proceedings of the seventh conference of the International Society for the Study of Argumentation* (pp. 881-889 CD). Amsterdam: Sic Sat.

Jansen H., Dingemanse, M., & Persoon, I. (2011). Limits and effects of *reductio ad absurdum* argumentation. In H. Jansen, T. van Haaften, J. de Jong & W. Koetsenruijter (Eds.), *Bending opinion. Essays on persuasion in the public domain* (pp. 143-158). Leiden: Leiden University Press.

Lakoff, G. (1987). *Women, fire, and dangerous things. What categories reveal about the mind.* Chicago/London: University of Chicago Press.

Lakoff, G. & Johnson, M. (1980). *Metaphors we live by.* Chicago/London: University of Chicago Press.

Searle, J.R. (1979). *Expression and meaning. Studies in the theory of speech acts.* Cambridge: Cambridge University Press.

Snoeck Henkemans, A.F. (2005). What's in a name? The use of the stylistic device metonymy as a strategic manoeuvre in the confrontation and argumentation stages of a discussion. In D. Hitchcock (Ed.), *The uses of argument: Proceedings of a conference at McMaster University* (pp. 433-441). Hamilton: Ontario Society for the Study of Argumentation.

Snoeck Henkemans, A.F. (2007). Manoeuvring strategically with rhetorical questions. In F.H. van Eemeren, J.A. Blair, C.A. Willard & B. Garssen (Eds.), *Proceedings of the sixth conference of the International Society for the Study of Argumentation* (pp. 1309-1315). Amsterdam: Sic Sat.

Snoeck Henkemans, A.F. (2009). Manoeuvring strategically with 'praeteritio'. *Argumentation, 23*, 339-350.

Snoeck Henkemans, A.F. (2011). The contribution of praeteritio to arguers' strategic maneuvering in the argumentation stage of a discussion. In T. van Haaften, H. Jansen, J.C. de Jong & W. Koetsenruijter (Eds.), *Bending opinion. Essays on persuasion in the public domain* (pp. 133-143). Leiden: Leiden University Press.

Snoeck Henkemans, A.F. (2013). The use of hyperbole in the argumentation stage. In D. Mohammed & M. Lewiński (Eds.), *Virtues of Argumentation. Proceedings of the 10th International Conference of the Ontario Society for the Study of Argumentation (OSSA), 22-26 May 2013* (pp. 1-9). Windsor, ON: OSSA.

Tonnard, Y. (2011). *Getting an issue on the table: A pragma-dialectical study of presentational choices in confrontational strategic maneuvering in Dutch parliamentary debate.* Dissertation University of Amsterdam. Alblasserdam: Haveka.

Tseronis, A. (2009). *Qualifying standpoints. Stance adverbs as a presentational device for managing the burden proof.* Dissertation Leiden University. Utrecht: LOT.

Zarefsky, D. (2006). Strategic maneuvering through persuasive definitions: Implications for dialectic and rhetoric. *Argumentation, 20*, 399-416.

35

Overcoming Obstacles to the Use of Peer Grading in the Assessment of Written Arguments

DAVID KARY
Law School Admission Council, USA
dkary@lsac.org

I address two obstacles to the use of peer grading in assessing written arguments: (1) peer graders are not motivated to give their best effort and (2) peer graders lack expertise in argument analysis. Regarding (1), I propose a way of motivating peer graders by scoring their efforts. As for (2), I propose a "scaffolded" scoring rubric that is progressively structured to guide the nonexpert grader through the evaluation of a written argument.

KEYWORDS: assessment, peer grading, scoring rubrics, written arguments

1. INTRODUCTION

Peer grading takes place when an instructor or some other test giver has students/examinees evaluate each other's work and assign grades to it. Peer grading schemes typically have examinees begin by themselves completing an assignment and later moving on to evaluate the work of a small number of their peers on that same task. Before they can grade someone else's work, graders might undergo a training or calibration phase (which might include some sort of test to ensure that graders meet a minimal performance threshold). The test giver usually provides benchmarks or a scoring rubric to guide the grading process. While peer grading schemes like these might prove cumbersome to implement using traditional classroom methods, computer-assisted platforms can remove many of the practical obstacles to using peer grading in the classroom.

There are two main benefits of using peer grading. One is a savings in cost and labor that can allow wider use of constructed response assessments. Because it divides the burden of grading among students, peer grading can make constructed response assessments a

D. Mohammed & M. Lewiński (eds.) (2016). *Argumentation and Reasoned Action: Proceedings of the 1st European Conference on Argumentation, Lisbon, 2015. Vol. II, 601-615.* London: College Publications.

viable alternative for instructors who would otherwise find them cost-prohibitive. In this respect, peer grading can be seen as a competitor with automated essay scoring, both of which are means of inexpensively grading constructed response test items. (Balfour, 2013, pp. 40-41)

The second benefit of peer grading lies in the educational opportunities that the task of grading one's peers might offer. Topping (1998) provides a good summary. The task of evaluating the work of others gives student-graders additional time on task "that can help to consolidate, reinforce, and deepen understanding in the assessor." (p. 254) Peer grading can also promote the development of a vocabulary for discussing quality in a particular discipline, and improve students' metacognition by giving them an opportunity to compare their own performance with that of their peers. Moreover, peer grading schemes might result in faster feedback and more of it: "While peer feedback might not be of the high quality expected from a professional staff member, its greater immediacy, frequency and volume compensate for this." (p. 255)

While these potential benefits are attractive to many, peer grading will have limited value as an educational tool if peer grading schemes cannot produce valid assessments of student performance. In this paper, I will discuss the issue of the validity of scores produced through peer grading schemes, but my larger focus will be on ways of improving validity in peer grading. I proceed by addressing two well-known obstacles to achieving valid assessments through peer grading—lack of motivation and lack of appropriate expertise. I will consider possible solutions for each obstacle in turn. The ultimate goal is that of seamlessly integrating these possible solutions within one peer grading scheme. To that end, I will outline a peer grading scheme that takes steps toward overcoming both obstacles. This peer grading scheme is designed for use with the assessment of written arguments.

2. THE QUESTION OF VALIDITY

Topping (1998) surveys 31 studies of the reliability and validity of peer assessment. His findings were somewhat mixed:

> The majority of studies (18) suggest that peer assessment is of adequate reliability and validity in a wide variety of applications. However, a substantial minority (7) found the reliability and validity of peer assessment unacceptably low in particular projects. (p. 258)

Topping's approach of investigating the validity of peer grading through a survey of individual studies is appropriate, given that the validity of peer grading scores will always depend on the individual details of the course subject matter, the peer grading scheme, and how well the grading scheme is implemented. Although his study is now dated, there is little reason to believe that peer grading is implemented in a significantly better way today than it was in the 1990s.

One telling aspect of Topping's conclusion is the statement that peer assessment is of *adequate* reliability and validity. Adequate reliability and validity is not hard to achieve in cases where peer grading is used for assessments with very low stakes—if, for example, the peer-graded portion of one's final course grade is 10 percent or less. Reports of peer grading being used for much more than this are rare. The contribution from peer graded work is often on par with contributions from components like class participation.

A careful validity study conducted by Kulkarni et. al. (2013) is instructive. That study analyzed peer grades given in a course called "Human Computer Interaction," which was delivered online through the online educational platform Coursera. Like most validity studies for peer assessment schemes this one was based on measuring the agreement between peer grades and expert or staff grades. It found that in an initial scoring session, 34.0% of submissions had median peer grades within 5% of staff grades and 56.9% of submissions had median peer grades within 10% of staff grades. In a second scoring session, the agreement improved somewhat: 42.9% of submissions had median peer grades within 5% of staff grades and 65.5% of submissions had median peer grades within 10% of staff grades.

If these peer grades comprise only a small part of the total grade for a course, these correlations are probably adequate. But when the stakes are higher, it's unlikely that we'd be satisfied with these levels of agreement. Even in the more successful second scoring session, 34.5% of the peer grades disagreed with staff grades by 10% or more. Given that a significant percentage of these peer grades may well diverge from staff grades by much more than 10%, this peer grading scheme isn't delivering the kind of fairness that we would rightfully expect if the assessment had greater stakes.

So clearly there is room for improvement in the validity of peer-given scores, and if peer grading is to play a more central role in assessment, there is a need for such improvement.

3. TWO OBSTACLES TO VALID ASSESSMENT THROUGH PEER GRADING

3.1 Lack of motivation

When courses have a peer grading component, it's typical for students to be required to participate in the peer grading. Generally, there is a grade penalty for not participating. But this does not motivate students to devote their best efforts to the task. Intrinsic rewards may well exist in many cases—students can gain a sense of empowerment and purpose when they are put in a position to judge their fellow students' work—but there is reason to be skeptical about the durability of these rewards.

Educational research into the issue of peer graders' motivation is sparse. One study (Wilson et al., 2015) suggests that peer graders can and do become disenchanted and unmotivated, but more and better research is necessary to determine the extent of this reaction.

There are other grounds to be concerned about peer graders' motivation however. One factor is the circumstances in which peer graders find themselves. With no rewards for grading well, a peer grader may or may not be inclined to put in the required effort. Even if you give it your best effort, it's natural to suspect that many of your peers are not putting in a strong effort. And if you should see poor quality peer-grader comments appended to your work, this would seem to confirm your suspicions. The outcome is demotivation—why should I put in the work if others aren't?

A second reason for concern about peer graders' motivation is the nature of the work. Grading essays and other constructed responses requires extended periods of concentration and offers few intellectual rewards. The work can get repetitive very quickly. It stands to reason that motivation can wear thin.

3.2 A possible solution: grade the graders

It shouldn't surprise anyone that one response to the problem of motivation for peer graders is to build an extrinsic reward into the peer grading system. This can be accomplished by assigning grades to peer graders based on the quality of their work, as measured by certain indicators. One peer grading platform, Calibrated Peer Review™ (CPR), has peer graders go through a calibration stage before they are permitted to do peer assessment, and ultimately assigns grades to peer graders on the basis of "their ability to evaluate the calibrations, their ability to evaluate their fellow students' work, and their ability to evaluate their own work." (Robinson, 2001, p. 475) Another platform
Peerceptiv™ (formerly SWoRD: Scaffolded Writing and Rewriting in the Discipline), assigns grades to peer graders on the basis of consistency (i.e., "systematic deviations from the average rating") and the quality of

written feedback. (Cho, Schunn, & Wilson, 2006, pp. 894-895) The companies behind these platforms claim some success in motivating peer graders by giving performance grades. I won't evaluate their approaches to rating grader performance in this paper, but I will outline a distinct approach to rating peer graders in Section 4.

3.3 Questionable competence

An obvious objection to the very notion of peer grading is to contend that student graders—as their institutional role suggests—are not competent to judge the proficiency of other students taking the same course. There are several responses to this objection. One is to point out that a serious peer grading scheme would never have a student's work being assessed by just one peer. Peer grading schemes usually arrive at a grade for a particular assignment by combining several peer scores using a measure of central tendency, such as the median score. Another response is to point out that grades given by instructors and other professional graders are also subject to drawbacks. Professional graders might be called on to assess dozens or even hundreds of papers on the same subject and consequently suffer from boredom and an understandable loss of mental acuity. Peer graders, on the other hand, might be limited to only three papers on a given subject. At the same time, peer graders might have insights into the work of their peers that elude professional graders who have long forgotten what it's like to be a student. (See Cho, Schunn, & Wilson, pp. 891-892.)

But these responses do not adequately address the legitimate concerns that can be raised about the use of peer grading for important assessments. A pool of peer graders could share certain weaknesses (such as being prone to the same mistaken belief or pervasive reasoning error) that are not found among professional graders. Using multiple graders and median scores is of no benefit in this case. More importantly still, these responses don't address the fact that peer graders lack expertise in the discipline and experience in making complex evaluations, both generally and within the discipline. This problem becomes more acute when weaker students in are called on to grade their more advanced classmates.

It's little wonder then that students themselves frequently suspect the competence of peer graders. (Kaufmann & Schunn, 2010, pp. 388-389.) This negative opinion of one's fellow peer graders could very well lead to worse peer grading results in that it can lead to diminished effort on one's own part.

The question remains: How can students who haven't mastered a particular skill be expected to judge other students' proficiency?

3.4 A possible solution: a scaffolded rubric

As mentioned earlier, peer graders are usually given a scoring rubric to guide them in the grading process. This rubric will provide a set of criteria for evaluating the important dimensions of the work, and, for each criterion, it will provide descriptions of performance levels. A research review by Jonsson and Svingby (2007) found strong evidence that the reliability of the scoring of constructed response assessments is enhanced by the use of scoring rubrics (p. 141) but concluded that it was still an open question whether using rubrics improves the accuracy of peer scoring (p. 138).

Judging by the pervasiveness of scoring rubrics in peer grading schemes, it's fair to say that the use of rubrics is generally thought to improve the accuracy/reliability of peer scores. The question of the architecture of these rubrics needs further attention, however. Many, perhaps most, peer-scoring rubrics are designed like simplified versions of rubrics for instructors or other expert graders. These might improve the reliability of peer scores by keeping graders focused on the same set criteria and by helping to align their notions of each performance level, but they might also seem hopelessly abstract or complicated to the peer grader, or just plain boring. There is definitely room for more innovative styles of scoring rubrics that connect better to peer graders.

I'll use the term "scaffolded rubric" for the style of rubric that I propose.[1] In the broadest terms, a scaffolded rubric is one that is designed primarily to guide the nonexpert grader. In designing such a rubric, the goal is arrive at a structure that makes the most of what judgment the nonexpert grader already possesses. A scaffolded rubric will do this by drawing the user's attention to the basic foundations of the work being assessed and then working upward.[2]

A scaffolded rubric will be structured like an analytic rubric in that it will call on graders to rate the component parts of the work. Like

---

[1] There is some danger in using the term "scaffolded" here in that the notion of scaffolding is commonly used, and perhaps overly used, in educational literature. The general notion behind "instructional scaffolding" is that the instructor provides developmentally appropriate supports (scaffolding) to promote learning among students and gradually removes those supports as those students develop competence. I'm using the notion of scaffolding for a different end—not for the sake of promoting learning (though that is always welcome), but for improving the accuracy of peer grades.

[2] The Peerceptiv™ peer assessment platform refers to itself as "scaffolded peer assessment," but the literature for that platform doesn't address scaffolding in scoring rubrics.

an analytic rubric, it elicits ratings for several criteria. The order in which these criteria are presented is crucial, however, because the understanding that peer graders gain when considering earlier criteria improves their performance when they are making evaluations according to later criteria.

In the Appendix, for example, basic criteria like organization and presentation precede higher order criteria like the logical strength of the argument given for the preferred choice and that given against the alternative choice. These criteria, in turn, precede the elicitation of an overall score for the written argument. This is because tracking the organization of the argument and its presentation will be helpful in assessing logical strength. (But the converse does not hold true.) And given the directions for the writing task, it's crucial to assess the strength of the argument for the preferred choice (and that against the alternative choice) before giving the written argument an overall rating.

Besides presenting criteria in a progressive order, a scaffolded rubric should also promote careful re-reading and re-consideration of the short essay. That is, it should raise questions that grader will not feel confident to answer if she has not read the essay carefully, and it should nudge the grader to re-read when necessary. Needless to say, a scaffolded rubric should use vocabulary and concepts that are appropriate to its users rather than the vocabulary and concepts suitable only for expert graders.

The approach that I prefer in the design of a scaffolded rubric is building it as a questionnaire eliciting a progressive series of ratings. A questionnaire format is more familiar to students than that of standard rubrics, and it gives the user a distinct path to follow. At the same time, it can accommodate the essential features of a rubric, such as statements of criteria and descriptions of performance levels.

4. A PROPOSED PEER GRADING SCHEME (ADDRESSING BOTH OBSTACLES)

4.1 The need for a simple, user-friendly design

To better motivate peer graders, we endorse the notion of adopting a peer grading scheme that gives performance grades to peer graders. In theory at least, the use of a well-designed, scaffolded rubric would also enhance motivation by helping to break down a complex task into more manageable bites. However, if scaffolded rubrics and features that rate the performance of peer graders result in a complicated peer grading scheme—e.g., one that leaves graders unsure of what to do next or why they're being asked to do what they're doing—then we run the risk of defeating our purpose by actually de-motivating peer graders.

This is to say that the design and implementation of the peer grading scheme is crucial. The platform should not overtax the user, nor should it run the risk of leaving the user bored or confused. While following good design principles in developing a software platform would be helpful in this regard, it would also be advantageous to closely integrate the means by which we rate peer graders with the scaffolded rubric and the rest of the peer grading scheme.

4.2 The basic outline

The goal then is to develop a peer-grading scheme that uses a highly scaffolded rubric, provides a way to rate peer graders, and doesn't overtax or confuse peer graders. At the same time, the performance grade for peer graders should be valid as a supplementary indicator of the same ability measured by the base assessment. For example, if the base assessment is writing an argumentative essay, then the grade for peer grader performance should also be a good indicator of one's ability to write an argumentative essay.

To meet this goal, I propose a peer grading scheme that works like a double-duty questionnaire. To the peer grader, it would be a questionnaire that implements a scaffolded rubric. At the same time it functions as a platform for rating peer graders. The discussion that follows describes such a peer grading scheme, geared toward grading argumentative essays. The Appendix presents a rough prototype of such a scheme.

In this scheme, peer graders are assigned regular unscored essays to grade and they are also assigned one or two pre-scored essays. There is no way for the peer grader to determine which is which. They grade the unscored and the pre-scored essays in the same fashion. With the unscored essays, the questionnaire-style scaffolded rubric is there to guide them in their grading. With the pre-scored essays, that same rubric functions as a testing instrument: the ratings that they give are used to calculate their score for grading performance.

A questionnaire-style scaffolded rubric has the peer grader give several ratings, which provide several data points for calculating the peer grader's performance score. The peer graders will know ahead of time that they will receive pre-scored essays to rate and that part of their grade on the overall assignment will be determined by how well they grade those pre-scored essays. Hence the graders have an incentive to grade all the essays as carefully as they know how.

Broadly stated, peer graders' performance will be determined on the basis of how much their ratings diverge from the credited "expert" ratings that the essay was given when it was pre-scored. If ratings that the peer grader gives agree with the expert grader's rating at every data point, the peer grader would receive the best possible grader rating. This grader rating diminishes each time the peer grader gives a rating that diverges from the expert grader's rating, based on how much it diverges.

In assessing peer grader performance, the pre-scored essay in conjunction with the questionnaire functions as a set of "rating items". The peer grader's performance on these rating items determines their score for their grading efforts.

A rating item is a slight variation on what Scriven (1991b) called a "multiple-rating item,"[3] which is a variety of test item consciously modeled on the task of evaluating performance.

> The multiple-rating item, unlike any other standard type of test, moves a testee in to the role of instructor/evaluator, since the typical multiple-rating item requires grading (real or hypothetical) student work. This is a good way, and perhaps the best way, to get the learner to the point where they can critique their own work, i.e., internalize the standards of merit in the field. (Fisher & Scriven, 1997, p. 178)

As this quotation suggests, Fisher and Scriven are well aware that rating items can be used to assess peer grader performance, and they make the case that multiple-rating items test the same sorts of skills that contribute to good writing in a discipline. Fisher and Scriven contend that multiple-rating items tap into "...higher order thinking skills running from deep understanding, analysis [and] evaluation, to critical thinking..." (p. 177) If these skills overlap sufficiently with the skills utilized in writing a strong essay, it justifies combining students' scores for essay writing with their scores for grading essays written by their peers.

To recap then, for unscored essays the peer grader completes a questionnaire that functions as a scaffolded rubric and generates a score

---

[3] A multiple-rating item presents one "stem" to the examinee and asks the examinee to rate (on an n-point scale) each of several "targets" that are based on that stem. In what I am calling "rating items," the directions for the essay writing assignment stand in as the "stem" in Scriven's parlance and the pre-scored essay is the only "target." This "target" is subjected to a set of ratings for different criteria, not just a single rating. (See Scriven & Fisher, 1997, pp. 173-180.)

for that essay. For pre-scored essays, the questionnaire functions as a set of rating items that generates a score for the peer grader's performance. (See Figure 1.) The question of how these scores might be generated is addressed in the next section.

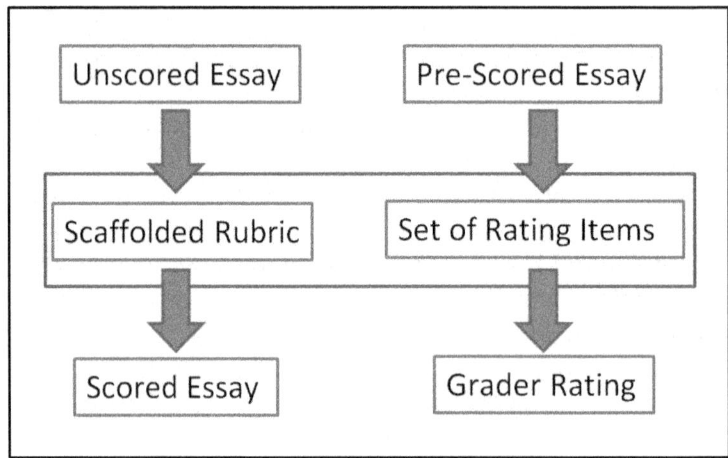

Figure 1. A schematic of the proposed peer grading scheme

4.3 Scoring

There are two steps to computing the scores for the written essays within this peer grading scheme. The first is determining the score that each individual peer grader gives the essay. The second is determining the final essay score from the scores derived from all the peer graders to whom it was assigned.

With a scaffolded rubric like that in the Appendix, arriving at the scores from individual peer graders could be as simple as taking the Overall Score submitted by the grader. In this case, we would be using holistic scoring with all of the preceding ratings in the scaffolded rubric merely functioning as an apparatus to aid the grader in reaching an Overall Score. Note that the preceding ratings in the rubric each draw the grader's attention to the various component features of the essay that the designers of the assessment deem relevant, which should, in theory, stimulate deliberations that enhance holistic scoring.

An alternative to holistic scoring is to implement analytic scoring based on a scoring formula that takes into account all of the ratings elicited by the scaffolded rubric. This could involve taking the mean of all the ratings, or calculating a weighted mean. The details of the scoring formula could be arrived at on the basis of field testing or on the basis of what skills or content are thought to deserve the most consideration.

This demonstrates how the scaffolded rubric that I'm proposing does not fit the usual distinction between analytic rubrics and holistic rubrics. While it's structured like an analytic rubric, a scaffolded rubric does not have to use analytic scoring. The overall score can be generated using a formula that combines component scores, or it can be arrived at holistically (though not without strong guidance from the rubric).

Once the scores from individual raters are determined, there are a couple of options for determining the final essay score. The most straightforward option is to take the median (or some other measure of central tendency) of the scores delivered by some minimum number of peer graders. Another option is to weight the individual peer graders' scores based on the performance score that they receive for their grading.

This brings us to the question of how to calculate peer grader performance. As indicated earlier, peer graders' performance would be determined on the basis of how much their ratings diverge from the credited "expert" ratings that the essay was given when it was pre-scored. Unfortunately, this is as much detail as I'm prepared to suggest at this time. The problem is that most obvious ways of arriving at a score for grading performance (i.e., the simplest scoring models) are vulnerable to a simple strategy on the part of peer graders. If our scoring model is based entirely on tracking the sum of the differences between peer graders' ratings and expert ratings, for example, a crafty peer grader might just pick the middle rating in every instance. For no real effort at all, they stand a good chance of achieving a respectable score. Or a crafty peer grader who wants to achieve more than just a respectable score might employ a more sophisticated strategy of actually reading and rating the essay but "cheating" toward the middle rating as a kind of insurance against missing by a lot.

Needless to say, it's imperative for the success of any peer grading scheme that peer graders don't employ strategies or shortcuts but make sincere attempts to rate the assigned essays using the rubric provided. It seems clear then that a sophisticated scoring model is necessary in the scoring of peer grader performance. Many different more-sophisticated scoring models are possible. However, the choice between them is a technical matter that could very well require complex simulations and careful field testing. This is not to say that an adequate scoring model can't be found, but it is a complex issue that runs beyond the scope of this paper.

5. CONCLUSION

Peer grading may have value as a source of new learning opportunities for students, but the validity of peer grades is generally suspect for anything but a low stakes assessment. In this paper, I outlined a peer grading scheme that might offer a means of improving the validity of peer grades by simultaneously addressing two of the obstacles to accuracy in peer grading: lack of motivation and questionable competence among peer graders.

While this peer grading scheme might hold promise, the design is far from complete. The next step is further research into how to score grader performance, which remains an open question. If that hurdle can be overcome, this peer grading scheme could be refined and developed further. At that stage it would undoubtedly benefit from collaboration with course instructors and testing with prospective users.

REFERENCES

Balfour, S. P. (2013). Assessing writing in MOOCs: Automated essay scoring and calibrated peer review. *Research and Practice in Assessment, 8*, 40-48.

Cho, K., Schunn, C.D., & Wilson, R.W. (2006). Validity and reliability of scaffolded peer assessment of writing from instructor and student perspectives. *Journal of Educational Psychology, 98*(4), 891-901.

Cho, K., & Schunn, C.D. (2007). Scaffolded writing and rewriting in the discipline: A Web-Based reciprocal peer review system. *Computers and Education, 48*(3), 409-426.

Fisher, A., & Scriven, M. (1997). *Critical Thinking: Its Definition and Assessment.* Point Reyes, CA: Edgepress.

Jonsson, A., & Svingby G. (2007). The use of scoring rubrics: Reliability, validity, and educational consequences. *Educational Research Review, 2*, 130-144.

Kaufman, J., & Schunn, C.D. (2010). Students' perceptions about peer assessment for writing: Their origin and impact on revision work. *Instructional Science, 39*, 387-406.

Kulkarni, C., et al. (2013). Peer and self assessment in massive online classes. *ACM Transactions on Computer-Human Interaction, 20*(6), 33.

Likkel, L. (2012). Calibrated Peer Review essays increase student confidence in assessing their own writing. *Journal of College Science Teaching, 41*(3), 42-47.

Robinson, R. (2001). An application to increase student reading and writing Skills. *The American Biology Teacher, 63*(7), 474-480.

Scriven, M. (1991b). Multiple-rating items. California. Retrieved from ERIC database. (ED 340 76B)

Topping, K. J. (1998). Peer assessment between students in colleges and universities. *Review of Educational Research, 68*(3), 249-276.

Wilson, M. J., et al (2015). 'I'm not here to learn how to mark someone else's stuff': An investigation of an online peer-to-peer review workshop tool. *Assessment and Evaluation in Higher Education, 40*(1), 15-32.

APPENDIX: A Peer Grading Prototype for the Assessment of Written Arguments*

Directions: The scenario presented below describes two choices, either one of which can be supported on the basis of the information given. Your essay should consider both choices and argue for one over the other, based on the two specified criteria and the facts provided. There is no "right" or "wrong" choice: a reasonable argument can be given for either.

Denyse Barnes, a young country music singer who has just released her debut CD, is planning a concert tour to promote it. Her agent has presented her with two options: she can tour as the opening act for Downhome, a famous country band that is mounting a national tour this year, or she can be the solo act in a tour in her home region. Using the facts below, write an essay in which you argue for one option over the other based on the following two criteria:

- Barnes wants to build a large and loyal fan base.
- Barnes wants to begin writing new songs for her next CD.

Downhome is scheduled to perform in over 100 far-flung cities in 8 months, playing in large arenas, including sports stadiums. This ambitious schedule would take Barnes far away from her home recording studio, where she prefers to compose. Downhome's last concert tour was sold out, and the band's latest release is a top seller. Many concertgoers at large arenas skip the opening act. But it is possible that Barnes would be invited by Downhome to play a song or two with them.

The solo tour in her home region would book Barnes in 30 cities over a 4-month period, including community theaters and country-and-blues music clubs, a few of which have reputations for launching new talent. These venues have loyal patrons; most shows are inexpensive and are well-attended, even for new talent. Barnes would have a promotion budget for her solo tour, but it would be far smaller than that for Downhome's tour.

---

* The directions and writing prompt in this appendix are reproduced with the permission of the Law School Admission Council. © 2015 Law School Admission Council Inc.

Grading Instructions:

Several factors can contribute to the quality of a written argument. For each goal described below, rate the writer's argument on a scale ranging from 1 (poor) to 5 (excellent).

Ratings Descriptions

1. Entirely fails to achieve the goal; the response is flawed to the point of being wholly ineffective
2. Demonstrates only minimal progress toward the goal; the response has glaring deficiencies
3. Demonstrates significant progress toward the goal, but the response has significant deficiencies
4. Nearly achieves the goal; the response has a few noticeable deficiencies
5. Achieves the goal entirely

Organization

Goal: A well-organized response will state a clear thesis and clearly convey how the argument for that thesis proceeds. The parts of the argument should be easy to track and be ordered in a natural way.

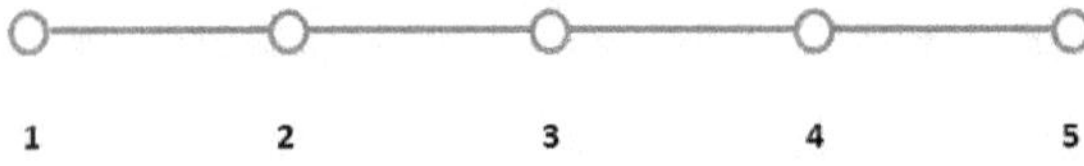

Clear and Succinct Presentation

Goal: Does not raise irrelevant considerations; is not overly repetitive; uses language that is understandable and not vague or convoluted

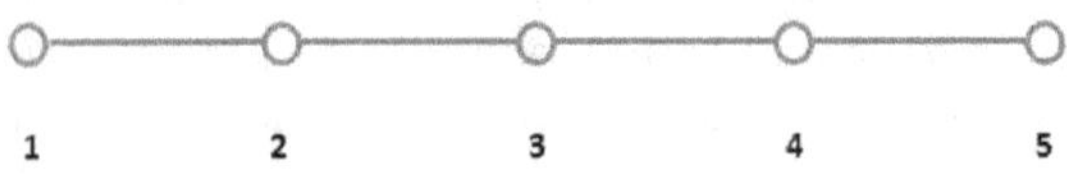

Tone and Credibility

Goal: The essay is written in a way that conveys a positive impression of the writer. The writer reaches conclusions with a level of confidence that is appropriate given the evidence presented. The writer's tone and demeanor is appropriate for the subject matter.

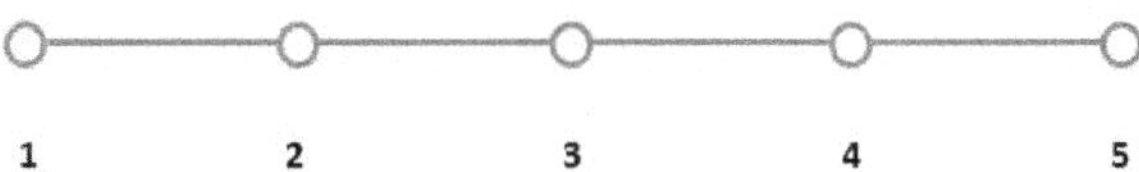

Logical Strength of the Argument Against the Alternative Choice

Goal: Carefully considers arguments that might be presented for the alternative choice and gives effective and well-reasoned rebuttals that are based on the facts and criteria stated in the prompt.

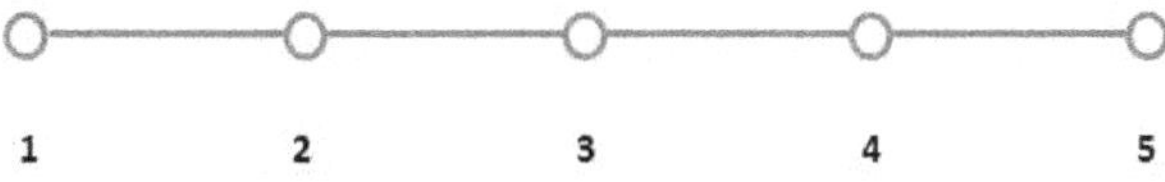

Logical Strength of the Argument For the Preferred Choice

Goal: The writer gives cogent reasons in favor of the choice defended. These will be based on the facts and criteria stated in the prompt. The writer's argument will be free of reasoning errors and resistant to objections and counterexamples.

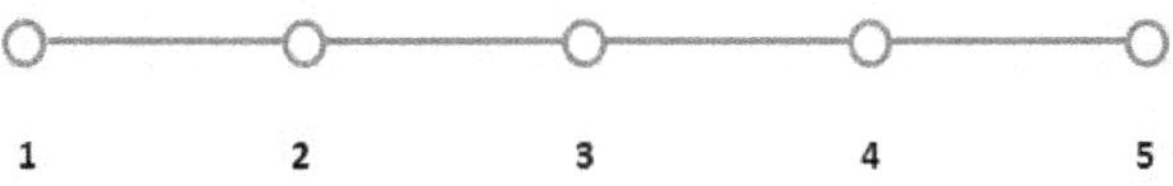

Overall Score

Goal: The essay presents a strong and thoughtful case for the choice that the writer advocates. (Note that this should be your rating of the response as a whole. It is not necessary that your overall score agree with the ratings that you've given for the various components of the writer's argument.)

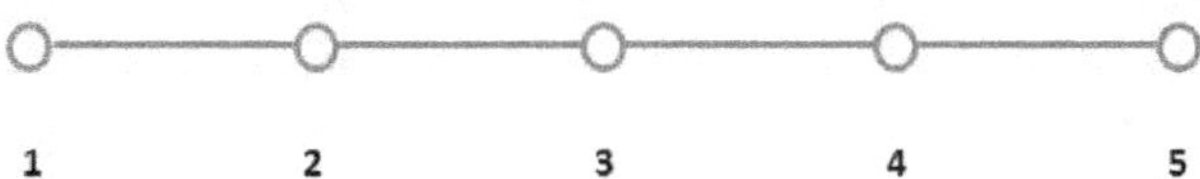

36

Types of Reasoning in Argumentation

IRYNA KHOMENKO
Taras Shevchenko National University of Kyiv, Ukraine
khomenkoi.ukr1@gmail.com

This paper focuses on two types of reasoning in argumentation: object reasoning and meta-reasoning. Both of them are considered from a standpoint of informal logic, a discipline located in the borderland between logic and epistemology.

I look at object reasoning as a subject matter of informal logic, aiming to figure out key features of logical reasoning and real arguments. I consider meta-argumentation as a methodological approach with distinguished tiers (construction, interpretation, and evaluation) connected to three kinds of meta-reasoning.

KEYWORDS: argument, disputing tier, evaluation, interpretation, logical tier, meta-argument, object argument, real argument, reasoning

1. INTRODUCTION

Argumentation theory has a long history. The best way to describe its contemporary developments, in my view, is by considering various theoretical perspectives and approaches. In this regard, it can be presented as different opportunities in argumentation studies: theoretical and empirical; analytical and practical; normative and descriptive; formal and informal, etc.

For my present purposes, it is important to stick to the last point on this list, in particular to the informal approaches. A number of different streams can be identified as informal: American tradition of communication studies and rhetoric, linguistic approaches, pragma-dialectical approach, informal logic, etc. I would like to note that my research in argumentation relates to informal logic.

D. Mohammed & M. Lewiński (eds.) (2016). *Argumentation and Reasoned Action: Proceedings of the 1st European Conference on Argumentation, Lisbon, 2015. Vol. II, 617-627.* London: College Publications.

It should be pointed out that various approaches to informal logic have been offered in literature.[1] Moreover, there are various suggestions on using other labels as a title for this discipline. For example, practical logic, philosophy of argument, theory of argument, applied epistemology, theory of reasoning, theory of critical thinking, etc.

Because of this, one can occur many interpretations of what informal logic is. The closest to my viewpoint would be the definition, established by Blair and Johnson.[2] However, I would like to clarify it by considering informal logic as a normative study of real argument with developed standards, criteria, and procedures for the its interpretation and evaluation.

I use the term "real argument" because informal logicians focus solely on this kind of reasoning. Thus, it can be claimed that such argument is a subject matter of informal logic.

Unfortunately, in spite of numerous papers, books, and textbooks published over the last thirty years, consensus as to what a real argument is has not been achieved so far.

A quick inspection of reasoning types by various theorists shows that the main differentiation is drawn between deductive and inductive reasoning. Several researchers add the third type – abductive reasoning. Thereby the question arises: could these instances of logical reasoning be considered as types of real argument? If we give a positive answer, then how should we relate to the fact the unity of informal logicians around the idea that key feature of a real argument is a kind of reasoning which are not a subject matter of formal logic. This point could be illustrated with Johnson's quote about the formal logic gap.[3] Govier also part companies with Johnson.[4]

---

[1] "The term informal logic does not refer to one well-delineated approach. It rather refers to a collection of attempts to develop and theoretically justify a method for the analysis and evaluation of natural language arguments in different context of use that is an alternative to formal logic" (van Eemeren, Garssen, Krabbe, Snoeck Henkemans, Verheij & Wagemans, 2015, p. 374).

[2] "Informal logic is the best understood as a normative study of argument. It is the area of logic, which seeks to develop standards, criteria and procedures for the interpretation, evaluation, and construction of arguments and argumentation used in natural language" (Blair & Johnson, 1987, p. 148).

[3] "[...] virtual disappearance from the mandate of logic of the focus on real argument" (Johnson, 2000, p. 105)

[4] "[...] what should be obvious: that the understanding of natural arguments requires substantive knowledge and insight not captures in the rules of axiomatized systems" (Govier, 1987, p. 204).

If we have negative reply, then it is unclear what types of reasoning represent a notion "real argument".

In my view, when answering this question it is necessary to differ logical reasoning and real arguments. The distinction is based on the formal and informal understanding of the notion "reasoning".

Thus, the aim of my paper is to present an approach to classification of reasoning types in argumentation within informal logic.

2. REASONING AND REAL ARGUMENT

Let us turn now to the definition of such notions like "reasoning" and "argument".

In general, reasoning is presented as an activity of human mind or as a special kind of thinking, interrelation of thoughts.

Speaking of the formal understanding, reasoning can be defined as a system composed of premises and conclusion. Certain thought (conclusion) is based on others (premises) or derive from others. We can distinguish various types of reasoning taking into account logical forms. In this sense scholars distinguish, for example, such logical reasoning, as deductive, inductive and abductive.

It should be pointed out that the notion "argument" has various labels in informal studies. Among them are real, natural, every day, actual, real-life, ordinary, mundane, marketplace argument. By now we have witnessed many attempts to produce definition of real argument. However, in my view, none of them is clear enough.

For example, according to Blair and Johnson real argument is:

> actual natural language arguments used in public discourse, clothed in their native ambiguity, vagueness and incompleteness.
>
> [...] arguments that have actually been used to try to persuade people, the sorts of arguments the student will encounter outside the classroom" (Johnson & Blair, 1994, p. 6).

Groarke thinks that real arguments are:

> the arguments found in discussion, debate and disagreement as they manifest themselves in daily life (Groarke, 2011).

With regard to clarifying this term I consider it as a complex kind of reasoning, which is used in argumentation as a form of dialogical interaction, where arguers aim is to resolve a conflict of opinions expressed by verbal means.

In my view, based on the analogy from formal logic where object language differs from meta-language, real argument consists of two sub-reasoning: object argument and meta-argument. Let us look closer at both of them.

3. SUB-REASONING OF REAL ARGUMENT

In general, term "object argument" refers to reasoning about such objects as historical events, social events and politics, news in mass media and social networks, advertising, corporate and governmental communications, personal exchange and practical problems. According to informal logicians, these are areas for real argument's applications.

Now let us turn to definition of this type of reasoning in argumentation. I see object argument as set of statements that seeks to justify a conclusion by supporting it with premises; to defend it from objections; or both goals.

With regard to the components of object argument, I believe that we can use the traditional approach: object argument can be considered as a system composed of premises and a conclusion. Conclusion is a statement that is based on other statements, called "premises". Both notions are mutually interdependent and hang upon the context of argumentation.

Thus, it could be possible to define object argument as a conclusion-premises complex. Object arguments represent logical tier of argumentation. On this tier we can think of, for example, deductive or inductive reasoning and use the means of both formal and informal logic for their analysis.

The next item on our agenda is to explain the notion of meta-argument. Here, I use this notion with the following meaning: meta-argument is a reasoning about one or more object-arguments. Object argument in particular discussion is a subject matter of certain meta-argument.

Meta-arguments represent the tier of argumentation, called "disputing tier". I introduce it based on Johnson's treatment of dialectical tier. He defines "dialectical tier" of argument in his book "Manifest Rationality" as follows.

> In addition to this illative core, an argument possesses a dialectical tier in which the arguer discharges his dialectical obligations (Johnson, 2000, p. 168).

It is not difficult to find out various clarifications of Johnson's definition that have been proposed by Johnson himself, Finocchiaro (2013), Govier

(2000), Hichcock (2002), Hansen (2002) and others. However, generally speaking, dialectical tier in their investigations is connected with the key function of argumentation – rational persuasion.

In the present context, the point I would like to stress is that I follow Johnson's idea, however, suggest the following elaboration. First, I consider the disputing tier is a tier of argumentation, which includes the interpretation and evaluation of object argument with standpoint as well as the defence from possible criticism (objections, observations, counterarguments, refutations, etc.).

Let us now focus on the issue of defining interpretation. It can be seen as a description of construction or reconstruction of object argument details in order to ensure their understanding. While we talk about own argument, we concentrate on its construction. In case when we analyze arguments of others, we focus on its reconstruction.

Interpretation of meta-argument relates to the replies on at least the following questions.

(1) How arguments are expressed and stated?
(2) What is their structure? (serial, linked, independent, etc.)
(3) How this structure may be pictured in a structure diagram?
(4) What missing premises can be find in the argument?
(5) What is the type of missing premises?
(6) What missing premises can be included in reconstruction of the argument?

Another aspect of meta-argument is object argument's evaluation, namely the assessment of its merits. Let us turn to criteria approaches.

Here it was suggested by different points for distinction a good argument from a bad one. For instance, speaking of traditional criteria we can talk about "soundness" and "validity". An argument is good if and only if it is formally valid and its premises are true. It should be noted that only validity is a pure logical criterion because we can identify validity of argument by logical methods. Meanwhile, it is not possible to establish whether its premises and conclusion are true or not. The fact that by following these criteria all the good arguments are being reduced to deductive ones proves how strong they are. There are obvious counter-examples to the hypothesis that an argument is good if and only if it is sound in this technical sense. We can see that some arguments which we take to be good are not sound by reflecting on examples of perfectly acceptable arguments whose premises are not all true, or whose inferential step is not deductively valid. In this regard the point of view language arguments is wide spread and recognized within informal logic.

In order to avoid such restriction researchers tried to offer another evaluation approach to argument. For example, Hamblin distinguishes alethic, epistemic, and dialectical criteria for good argument (Hamblin, 1970, pp. 224-252). Unfortunately, it should be noted that although given points were identified clear definition of good argument has not been provided.

Informal logicians have developed other criteria for assessing an argument. For instance, a triad of relevance, acceptability and sufficiency (RAS criteria) is often used as a popular set of criteria in this regard. To be considered as a good argument, its premises must be acceptable, relevant to the conclusion and sufficient to support it.

Blair consider such criteria as follows:

> [...] an argument is a good one if its grounds or premises are singly or in combination relevant as support for the claim in question, individually acceptable, and together (if relevant and acceptable) sufficient to support the claim on behalf of which they were offered (Blair, 2011, p. 87).

From my point of view argument appraisal according to these criteria is much closer to epistemological than to formal logical one.

Firstly, all of these criteria are explicated via epistemological concepts such as belief, knowledge, common knowledge, justification, and others. For example, one of the standard tests for premises acceptability is whether a premise satisfies the common knowledge condition.

Secondly, informal researchers hold that these criteria are best conceived as relative to particular person in particular time and in particular epistemological situation. For example, acceptability is relative to the particular evaluator or to the particular audience that judges availability of the argument. This criterion refers not to the fact that the evaluator or audience accepts the premises, but to the fact that it is reasonable for the evaluator or audience to accept the premises, whether or not they in fact do so. Thus, premises can be acceptable to a particular person, even though the person does not really accept them. Moreover, it can be acceptable even if it is false. It is possible the situation when a false premise is acceptable to someone while that person has good reason to accept it.

Finally, this method is not a properly developed logical one to be used for identifying the goodness of an argument within informal logic. Let us consider relevance criterion. In short, relevance within informal logic is the relation between premises and conclusion of an argument. But here property of the premise to be relevant to its conclusion is called "premissary relevance" or "local probative relevance." This relevance differs from logical relevance, which can be defined as a relationship of entailment between a set proposition and another one.

I suggest that evaluation of meta-argument at least should reply to the following questions:

> (1) Are the premises of certain object-argument relevant to the conclusion?
> (2) Are the premises of certain object-argument acceptable?
> (3) Are the premises of certain object-argument sufficient to support the conclusion?

4. TYPES OF META-ARGUMENT

Moving on, I will try to clarify the notion of meta-argument by distinguishing and describing its various types.[5]

The first step in this direction is to consider monological and dialogical argumentation. In this point, I would like to reference A. I. Goldman, whom I support.

> If a speaker presents an argument to an audience in which he asserts and defends the conclusion by appeal to the premises, I call this activity argumentation. More specifically, this counts as monological argumentation, a stretch of argumentation with a single speaker... I shall also discuss dialogical argumentation in which two or more speakers discourse with the another, taking opposite sides of the issue over the truth of the conclusion (Goldman, 1999, p. 144).

I distinguish two types of monological argumentation: (1) self-monological; (2) extra-monological. Moreover, I consider four types of dialogical one: (1) unmixed dialogical; (2) mixed dialogical; (3) extra-unmixed dialogical; (4) extra-mixed dialogical.

Let us try to describe each of them.

4.1 Types of meta-reasoning in monological argumentation

It or more readers, listeners or observes who not necessarily need should be pointed out that here I take into account the fact that arguing

---

[5] It is pointed out that some researchers have studied meta-argumentation and meta-argument (Wooldridge, McBurney, & Parsons, 2009; Boella, Gabbay, van der Torre, & Villata, 2009; Modgill & Bench-Capon, 2011; Finocchiaro, 2013; Gabbay, 2013)

requires one arguer with the role of proponent. Concerning opponent, it may be one to reply. Nevertheless, proponent should take into account possible objections.

I would like to begin with self-monological argumentation. It occurs when the proponent expresses a point of view on a question and offers as support for this position one or more premises in view of the possible criticism from the opponent.

In other words, it can take place when the subject matter of meta-argument is explicitly proponent's object argument and implicitly the objections of opponent. In this case, we can say that argumentation has a specific feature, called "reflexivity".

Thus, disputing level of such argumentation is represented by the meta-argument, which connected with Interpretation (Pr→Pr) and Evaluation (Pr→Pr/Op).[6]

I call this meta-argument "a self-reflective one".

The second type of monological argumentation is extra-monological one. It takes place when expert critically analyzes self-monological argumentation. It is necessary to keep in mind that expert should take into account objections of opponent in spite of the fact that he has not expressed them explicitly.

In other words, it can occur when the subject matter of expert's meta-argument are on one hand proponent's object argument explicitly and on the other hand the objections of opponent implicitly. In this case, we can say that argumentation has such feature as transitivity.

Here the meta-argument can be represented as Inerpretation (Exp→Pr), and Evaluation (Exp→Pr/Op).

I call such meta-argument "a reflective one".

4.2 Types of meta-reasoning in dialogical argumentation

Let us take a closer look on dialogical argumentation. It should be noted that I consider the cases when arguing requires two arguers: proponent and opponent. Wherein they are a real person or a couple of real people. Moreover, expert is involved in event of extra dialogical argumentation.

Foremost I focus on unmixed dialogical argumentation. It takes place when (1) the proponent proposes the argument to justify his own standpoint and (2) the opponent raises objections, but own standpoint has not been expressed.

In this case, we can say that argumentation has such features as reflexivity and partial symmetry.

---

[6] I use the following symbols for describing of meta-argument; Pr (proponent); Op (opponent); Exp (expert); → orientation of argumentation.

Let us bring attention to the disputing level of argumentation. Here meta-argument could be represented as follows. First, Interpretation (Pr→Pr) and Evaluation (Pr→Pr/Op); Interpretation (Op→Pr) and Evaluation (Op→Pr).

I call this meta-argument "an incomplete one".

Now let us consider a mixed dialogical argumentation. This type takes place when the arguers express to each other divergent opinions on some question and critically analyze them.

In this case, we can say that argumentation has such features as reflexivity and symmetry.

Here the meta-argument is connected with the following types. First, Interpretation (Pr→Pr) and Evaluation (Pr→Pr/Op). Second, Interpretation (Pr→Op) and Evaluation (Pr→Op/Pr). Third, Interpretation (Op→Op) and Evaluation (Op→Op/Pr). Forth, Interpretation (Op→Pr) and Evaluation (Op →Pr/Op).

I call this meta-argument "a complete one".

Another kind of dialogical argumentation is an unmixed extra-dialogical one. It occurs when expert critically analyzes the unmixed dialogical argumentation.

In this case, in my view, argumentation has such features as transitivity and partial symmetry.

According to this situation, disputing level of argumentation may be represented as the following types of meta-argument: Interpretation (Exp→Pr) and Evaluation (Exp→Pr/Op).

I call this meta-argument as "a complex incomplete one".

The last type of dialogical argumentation that I consider in my paper is a mixed extra-dialogical argumentation.

This argumentation takes place when expert makes a critical analysis of mixed dialogical argumentation. In this case, argumentation has such features as transitivity and partial symmetry.

Here the meta-argument may be represented as the following types. First, Interpretation (Exp→Pr) and Evaluation (Exp→Pr/Op). Second, Interpretation (Exp→Op) and Evaluation (Exp→Op/Pr).

I call this meta-argument "a complex complete one".

5. CONCLUSION

In this paper I have presented my reflections on types of reasoning in argumentation within informal logic. In conclusion I would like to summarize the main points of my paper.

First, real argument as a subject matter of informal logic consists of two sub-reasoning: object argument and meta-argument.

Second, object arguments signify logical tier of argumentation. Here we could mentioned logical types of reasoning (deductive, inductive, abductive, etc).

Third, meta-arguments represent disputing tier of argumentation. It includes interpretation and evaluation of object arguments with standpoint as well as defence it from possible criticism (objections, observations, counterarguments, refutations etc.).

Forth, types of meta-reasoning depend on argumentation kinds. I propose to distinguish two types of monological argumentation: self-monological and extra-monological with corresponding meta-reasoning - self-reflective and reflective ones.

Moreover, I consider four types of dialogical argumentation: unmixed dialogical, mixed dialogical, extra-unmixed dialogical, extra-mixed dialogical. According to this classification, I analyse the key features of meta-reasoning, such as incomplete, complete, complex incomplete and complex complete.

REFERENCES

Blair, J. A. (2011). *Groundwork in the theory of argumentation.* New York: Springer.

Blair, J. A., & Johnson, R. H. (1987). The current state of informal logic. *Informal Logic, 9,* 147-151.

Boella, G., Gabbay, D. M., van der Torre, L., & Villata, S. (2009). Meta-argumentation. Modelling I: metodology and techniques. *Studia Logica 93,* 297-355.

Finocchiaro, M. A. (2013). *Meta-argumentation. An approach to logic and argumentation theory.* London: College Publications.

Johnson, R. H., & Blair, J. A. (1994), *Informal logic: past and present. In New essays in informal logic,* (pp. 1-19). Winsdor, Ontario, Canada: Informal Logic.

Eemeren, F. H. van, Garssen, B., Krabbe, E. C. W., Snoeck Henkemans, F. A., Verheij, B., & Wagemans, J. H. M. eds (2015). *Handbook of argumentation theory.* Dordrecht: Springer Reference.

Gabbay, D. M. (2013). *Meta-logical investigations in argumentation networks.* London: College Publications.

Goldman, A. I. (1999). *Knowledge in a social world.* Oxford: Clarendon.

Govier, T. (1987). *Problems in argument analysis and evaluation.* Dordrecht, Holland: Foris Publication.

Govier, T. (2000). Critical review; Johnson's *Manifest Rationality. Informal Logic, 17,* 407-419.

Groarke, L. (2011). Informal logic. In *Stanford Encyclopedia on Philosophy.* http://plato.stanford.edu/ entries/logic-informal/.

Hamblin, Ch. L. (1970). *Fallacies.* New York: Methuen.

Hansen, H. V. (2002). An exploration of Johnson's sense of 'argument'. *Argumentation, 16,* 263-276.

Hichcock, D. (2002). The practice of argumentative discussion. *Argumentation, 16,* 287-298.

Johnson, R. H. (2000). *Manifest Rationality: A Pragmatic Theory of Argument.* Mahwah, NJ: Lawrence Erlbaum and Associates.

Modgill, S., & Bench-Capon, T. G. M. (2011). Metalevel argumentation. *Journal of Logic and Computation, 21,* 959-1003.

Wooldridge, M., & McBurney, P., & Parsons, S. (2005). On the metalogic of argument. In *Proceedings of the Fourth International Joint Conference on Autonomous Adgents and Multi-Agent Systems (AAMAS-05),* Utrecht, July 2005.

37

Prosodic Features in the Analysis of Multimodal Argumentation

GABRIJELA KIŠIČEK
University of Zagreb, Croatia
gkisicek@ffzg.hr

The term "prosodic" refers to such features as pitch, temporal structure, loudness and voice quality, emphasis, accentuation and (non-)fluencies of the speaker. Though 21st century public discourse is demonstrably multimodal, there remains a need to fully recognize more than a merely verbal mode of argument. Based on the analysis of TV commercials, this paper explores the argumentative value of prosodic features in multimodal discourse. We argue that the argumentative reconstruction of multimodal discourse should take prosodic features into account.

KEYWORDS: advertising, argumentation, auditory argument, intonation, multimodal discourse, pitch, prosodic features, visual argumentation, voice

1. INTRODUCTION

Grown-ups love figures... When you tell them you've made a new friend they never ask you any questions about essential matters. They never say to you "What does his voice sound like? ......
Antoine de Saint-Exupery *The Little Prince*

Over the last few decades the realm of argumentation studies has expanded to the analysis of argumentation in multimodal discourse. Unlike mono-modal verbal argumentation, multi-modality implies arguing with visuals (which have been extensively researched), but also with tastes, smells, gestures, music, editing and video framing, as well as with prosodic features such as intonation, voice quality, pitch and pitch-range, emphasis, pauses, tempo, and volume. The past decades have witnessed extensive debate between argumentation scholars (e.g.,

D. Mohammed & M. Lewiński (eds.) (2016). *Argumentation and Reasoned Action: Proceedings of the 1st European Conference on Argumentation, Lisbon, 2015. Vol. II, 629-642.* London: College Publications.

Birdsell & Groarke, 1996; Blair, 1996; Fleming 1996; Birdsell & Groarke 2007) on whether an image can be an argument and whether it is indeed possible to argue without words. Recently, not only visual but also multimodal argumentation became a focal point of interest among those scholars who believe that the more traditional understanding of argumentation as a mono-modal verbal activity should revised to include other modes that contribute to argumentation discourse.

What do these scholars mean by "modes of argumentation" and "multi-modal argumentation"? According to Groarke (2015, p. 149), modes are defined in terms of "the ingredients used in constructing arguments." Multi-modal discourse may thus consist of the verbal part of a message, but it can also include the numerous non-verbal modes to construct an argument.

In his commentary to visual argumentation, Godden (2015, p. 237) emphasizes that, although argument may have different components in different modes, there are essential functional components that make argument an argument.

> The important idea here is that arguments need not be built with words; rather, they can also be built from images, or from a variety of different components. Functionally, though, all arguments are built from the same stuff: claims and reasons. Things lacking these functional components, no matter what they are composed of (whether words, images, or what-have-you), are not arguments.

The aim of this paper is not to determine supremacy of one mode over the other, but to emphasize the importance of prosodic features in the multimodal discourse, and to argue for the need to give an account thereof when dealing with argumentative discourse. A lot of research, analysis and theoretical background has been provided for visual argumentation (Lake & Pickering, 1998; Kjeldsen, 2012; Groarke & Tindale, 2013) which explains and develops tools for the analysis and evaluation of arguments presented through images, photographs and other appeals to eye. However, appeal to the ear appears to be neglected both in argumentation theory and analysis. Therefore, this paper seeks to answer the question whether prosodic features have any role in argumentative discourse, what that role might be, and how it can be best understood. Thus, our goal is to better understand what argumentatively relevant information can be conveyed by means of prosody that can hardly be imagined to be conveyed by a different kind of mono-modal verbal discourse format, i.e., cannot be replaced with a written message or cannot be read to the same effect in a different manner, with different prosodic qualities. As argumentatively relevant information we consider all information that needs to be considered when seriously assessing the acceptability of a standpoint.

2. PROSODIC FEATURES AND SPEECH SOUND

Prosodic features refer to both voice and speech cues of the speaker. They include such features as pitch, temporal structure, loudness and voice quality, emphasis and accentuation, but also (non-)fluencies of the speaker. An extensive literature on nonverbal communication research generally strengthens the view that such features have an important communicative role. For instance, Vroomen, Collier, & Mozziconacci (1993, p. 577) write: "The communicative function of prosody is most readily associated with the expression of emotion and attitude."

Recent reviews have shown that vocal expressions of specific emotions (e.g., anger, fear, happiness, sadness) are generally recognized with above-chance-accuracy, also cross-culturally, and to be associated with relatively distinct acoustic characteristics (Juslin & Laukka, 2003; Laukka, 2008). Besides a correlation between prosody and emotions (Davitz, 1964; Scheerer, 1972; Vroomen, Collier, & Mozziconacci, 1993; Neuman & Strack, 2000), prosodic features are connected to the perception of a speaker's personality, credibility, his *ethos* (Kramer, 1977, 1978; Berry 1990, 1992; Kimble & Seidel, 1991; Zuckerman & Miyake, 1993; Hickson, 2004; Zuckerman & Sinicropi, 2011). Past research has particularly confirmed that, among other elements of nonverbal behavior, prosodic features are associated with the persuasiveness of the speaker and the audience's change of attitudes (Burgoon, Birk, & Pfau, 1990; Knapp, 2002). For instance, fluency, variations in pitch, higher intensity (i.e., louder speech) and faster tempo are positively connected with greater persuasiveness. Based on empirical research (e.g., Smith, Brown, Strong, & Rencher, 1975; Surawski & Ossof, 2006; Bartsch, 2009), one may cautiously conclude that a lower vocal pitch, a faster speech rate, and a relative absence of non-fluencies goes along with higher ratings for speaker's competence and dominance, *ceteris paribus*. Zuckerman & Driver's (1989) research on vocal attractiveness proposed that, similar to attractive faces, attractive voices may elicit a more positive interpersonal impression. They found that professional judges, for instance, were able to agree on whether voices are attractive or not, and that more attractive voices were associated with more a favorable impression of the speaker's personality. As mentioned earlier, attractive voices include lower pitch, absence of nasality or extreme harshness. Subsequent work has largely replicated these results, showing that vocal attractiveness can be compared to effects of physical attractiveness (e.g., Berry 1990, 1992; Zuckerman, Hodgins, & Miyake, 1990). Thus, speakers with more attractive voices are generally more favorably perceived by others. Rezlescu et al. (2015) sought to better determine the correlation between attractive voices and attractive faces and its effect on the perception of a speaker's trustworthiness and dominance. Face and voice are important because they represent two critical cues that used to derive a first impression, and thus present a rich source of socially relevant information. Looking at a face or hearing a voice, humans can reliably infer an individual's sex, age, identity, and emotional state (e.g., Banissy et al. 2010; Meyer et al., 2007; Scott 2008). But faces and voices also prompt spontaneous evaluations related to attractiveness and to character traits such as trustworthiness and dominance (Willis & Todorov, 2006; Vukovic et al., 2011). Rezlescu et al. (2015) experimentally confirmed previous results and concluded that voices, just like faces, can lead to the formation of consistent trait impressions of trustworthiness, attractiveness, and dominance. These insights, of course, are regularly sought to be exploited in the public sphere.

3. PROSODIC FEATURES AND ARGUMENTATION

Based on the findings that voice quality and other prosodic features can be manipulated to deliberately sound in a certain manner, it is easy to see that, among other private or public purposes, it can be used for persuasion and argumentation.

Examples from political discourse readily come to mind, for the main goal of a political campaign is to persuade an audience to vote for a specific candidate. The politician's *ethos* remains is an important means of persuasion, since her qualities as a leader may serve as a premise in constructing the argument.

> (1) Major premise: strong, confident and determent people are the best political leaders.
> Minor premise: candidate C is a confident, strong and determined person.
> Support: prosodic features and its connection to the speaker's personality
> Conclusion: therefore, C is the best choice as a political leader.
> Final claim: vote for C.

Prosodic features thus serve to support the minor premise: her voice quality, volume, tempo etc. can hence replace a verbal message – it need not be stated that she is a strong, determined, trustworthy person;

rather, this can be inferred from the manner of her speech, i.e., its prosodic features.

Another example, one in fact used in judicial discourse in the case of clemency for a death-penalty convict, is the following. Using a multimodal (https://www.youtube.com/watch?v=KhFoeJPP6HE), discourse the convict and his supporters argue that he should not be executed. The supporting argument pivots on his change of character: he is no longer a criminal (gang leader) but a peaceful person who dedicates his life to helping others. The argument may be reconstructed as follows:

> (2) Implied PREMISE: Rehabilitated, changed people should be spared from death sentence.
> Stated PREMISE: Stanley Williams is a changed, different person than he was 20 years ago. He is now a peaceful.
> SUPPORT: Prosodic features of his speech indicate that he is a gentle, peaceful, even a submissive person.
> CONCLUSION: Stanley Williams should not be executed.

Prosodic features that support his change are: higher pitch (gentleness, lack of dominance and masculinity), tempo (slower tempo, lack of confidence), and volume (low self-esteem, lack of determines and confidence, introverted character type). The same argument could be reconstructed by using the Toulmin model:

> (3) Ground: Stanley Williams is changed, rehabilitated.
> Warrant: Rehabilitated people should not be executed.
> Claim: Stanley Williams should not be executed.

In the relevant video, the changed personality of the convict was supported not solely by prosodic features, but also with images and testimonies of his friends, people he had helped, etc. The video combines visual and verbal argumentation, while prosodic features play an important role. The audience invited to evaluate his character based on prosodic features of his speech, in order to conclude: "He doesn't sound like a criminal".

Besides political and judicial discourse, further examples may be found in a wide variety of public discourse events such as arguing about social issues but also marketing and promotion, especially in television commercials.

3.1. Prosodic features in multimodal discourse by way of examples of TV commercials

Arguing in favor of visual argumentation, Groarke & Tindale (2013) make clear that not every image is an argument. They draw a fourfold distinction between *visual flags* (an image that merely draws attention to the argument but does not contribute meaning to it), *visual demonstrations*, *symbols*, and *arguments*. Following in their footsteps, a similar classification can be proposed for arguments that appeal to the ear. Some of these merely draw attention to the verbal part of an argument, and so can be regarded as *auditory flags*.

A good example is a commercial where a baby speaks with an adult voice (https://www.youtube.com/watch?v=ETCoyGloQno). Seeing a baby while hearing an adult male voice results in surprise, humor and interest. Similar effects are achieved in commercials where animals speak with a human voice such as the *Dear kitten* commercial for cat food (https://www.youtube.com/watch?v=aBrSvHPY1NQ), where an older cat (a father) gives advice to his baby. This includes introducing him to the rules of living with people, on how the world functions (in analogy with a human father-son relation); the conclusion is that to have a quality-life one should eat quality-food (a specific brand being introduced). The prosodic features used are tone and quality of voice (male, deeper voice, with little harshness), indicating that it is an older man, tempo (slower tempo with frequent pauses), used as when explaining something slowly and calmly, the intonation (low rising intonation couture) slightly resembling that of speaking with a smile. While the auditory part of the message (the voice, tone and the manner of speaking) is consistent with a father-son conversation, the verbal part is adapted to the cat's life (what to do to obtain a fresh portion of delicious food, how to make people cuddle and carry you, etc.); the visual message-part shows cat and kitten. The contradiction between auditory and visual draws attention to listen to the verbal; the analogy between humans and cats leads to the conclusion that one should treat cats like humans.

According to Groarke & Tindale (2013, p. 145), nonverbal demonstrations also have a function in multimodal argumentative discourse.

> The most basic way in which non-verbal elements function as argument components occurs when music, sounds, images, or even aromas provide evidence for some conclusion.

As for prosody, one can find *auditory demonstrations* in a variety of everyday situations. For instance, when calling in to inform colleagues that one will be absent from a meeting due to severe cold, one can demonstrate one's poor health condition by demonstrating hoarseness, i.e., offer the auditory companion of a sore throat and a heavy cold.

Groarke & Tindale (2013, p. 152) also discuss visual symbols which are "often used either to state a position or to make a case for one" emphasizing that a "context in which an image appears makes it plausible to interpret the image as a visual argument."

Similar things can be said for *auditory symbols*, which are based on the above mentioned empirical research as to how voices are stereotypically perceived. Unlike visual symbols – for good examples of which see Groarke & Tindale (2013, p. 152) – dominantly institutionally learned (through school or driving lessons etc.) auditory symbols seem not to be intuitively perceived, but are rather a result of media and popular culture influence, and largely based on stereotypes. Thus, how does a smart, educated person sound? How does a less smart person sound? Can we tell a dock worker from a lawyer just based on how they sound? Empirical research on various different languages and cultures has answered that question affirmatively (e.g., Labov, 1966; Giles, Bourhis, & Davis, 1975; Eshling, 1978; Honey, 1989; Lippi-Green, 1997; Coupland & Bishop, 2007; Kišiček, 2012), the research findings being presently used in public discourse. All features of voice quality can be regarded as symbols: deep voices with volume (both female and male) for cultivated, educated speakers; faringalised or nasalized voices for less bright people; high pitched voices with palatalized pronunciation for not less bright women particularly, etc.

An example of the use of faringalised voice is a commercial where a human has an ability to talk to animals https://www.youtube.com/watch?v=7UHKB6nQrzM.

Auditory symbols are frequently used in synchronizations of cartoons. For instance, the director or screenwriter may have a clear idea as to how a donkey in the film *Shrek* should sound, and what personality traits need to be conveyed with his prosodic features. Or in the cases of synchronizations of political debates for purposes of satire, certain specific prosodic qualities are used to make fun of someone, to picture him/her as a less bright, less competent person. Of course, in a specific context an auditory symbol may become an essential part of argumentative discourse.

Groarke & Tindale (2013, p. 143) explain visual arguments as those that "convey premises and conclusion with a nonverbal visual image which one finds in drawings, photographs, film, videos, sculpture, natural objects and so on." Is it possible to have an auditory correlate to visual argument, i.e., *auditory argument*? At least not entirely, for the auditory component, i.e., a given prosodic feature, is always connected to the verbal part of the message. While one can extract prosodic features (e.g., intonation couture) by means of acoustic tools, in the context of rhetoric and argumentation this would not be of immediate value.

One can nevertheless distinguish between prosodic features that form an *essential* part of an argument (i.e., without which the argument would substantially change) and prosodic features that *contribute* to the strength of an argument (i.e., that are replaceable but with the result of weakening the argument).

Prosodic features can contribute to the strength of argument in various ways. Good examples are commercials for Viagra (https://www.youtube.com/watch?v=y1ZqQ55T25c) or Porsche (https://www.youtube.com/watch?v=KRbzJ0L1Zn8), where prosodic features of a male voice-over contribute to the argument. In both commercials, the male voice shares similar qualities – deep, low pitched voices, with volume, being perceived as attractive male voices connected with an attractive appearance but also with masculinity, self-esteem, confidence and similar broadly positive character traits. Besides voice quality, features such as pauses, tempo and word emphasis are also important in these commercials.

Each of the commercials can be reconstructed. The Viagra commercial claims that an erectile dysfunction is much like any other difficulty that can occur in life, and that mature men can handle by using the right solution – Viagra.

> (4) Stated premise: Mature men learn how to deal with difficulties in life – this is the age of knowing what needs to be done
> (difficulties supported by visuals – broken sail; mature age supported by visual image of a man as well as prosodic features)
> Stated premise: Erectile dysfunction is one of the difficulties of mature age.
> Implied conclusion: Mature man should resolve this difficulty just as any other difficulty.
> Implied conclusion: Viagra is a solution for erectile dysfunction.
> Stated conclusion: Talk to your doctor about Viagra.
> Final (implied) claim: Use Viagra.

Prosodic features here support the premise of the mature man who can deal with difficulties. Not just the fact that he is mature by age but also his problem-solving ability. His voice quality supports confidence, strength, his tempo and intonation support a calm and determined approach to problems. Different prosodic features would result in the lack of such support. For example, a high pitched nasalized voice, which is generally perceived as negative, or a different intonation (high rise couture) and faster tempo, which are common for panic, fear, etc. Such prosodic changes would make the argument weaker and less convincing. Thus, the reconstruction of the argument would remain the same, regardless of its prosodic features, but its persuasive effect would change.

Similarly to Viagra, Porsche commercials target older, mature men, describing a car so special that desiring it lasts for decades, thus remaining desired from boyhood until manhood. And as a man becomes successful, he buys the car because a Porsche ignites passion in drivers like no other car. The argument can be reconstructed in following way:

> (5) Implied premise: When you want something passionately you will wait even for decades to have it.
> Stated premise: For some people decades pass between wanting and owning Porsche.
> Stated conclusion: Porsche ignites passion in drivers like no other car.
> Implied conclusion: Buy Porsche.

This is another example where prosodic features of a voice-over contribute to argument strength. Here, the voice (representative of all men who feel the same about the Porsche) is that of a mature man, a deep, low-pitched, attractive voice with a bit of harshness to it is (again stereotypically) connected to a confident, cultivated, educated, and successful male. This is a voice of someone who can afford a Porsche, someone usually connected to an upper social stratum. Similar to the Viagra commercial, we can imagine another type of voice, with other prosodic features, but it would not have the same effect. How can we assume that? Although the connection between prosodic features and perceived qualities of a speaker are mostly based on stereotypes, numerous empirical researches have suggested that such findings likely hold in real-world situations.

For instance, we can try to reconstruct the commercial for an energy drink (https://www.youtube.com/watch?v=iFEmtJRSrvw), which is an example of a multi-modal discourse and combination of verbal, visual and auditory mode. The implied conclusion stating that this specific drink can help you get through the most disturbing moments of life. No matter what happens, whichever worst-case scenario you have in mind, drink Irn-Bru and you will feel better. Everything that would drive a "normal" person crazy, you - the consumer of the Irn-Bru– will survive easily.

Those disturbing moments are presented visually through different situations. The main character in the commercial has a wife who changed the decoration in his house (everything is pink and fluffy), took his TV so that he is unable to watch football games, his mother-in-law moved in with them (she walks out of the bathroom straight from the shower, informing him that she borrowed his shaving tools). Between every scene he sips a bit of that drink and calmly receives all changes. Prosodic features of his wife play a crucial role in the disturbing scenario he has to deal with. She speaks with high pitch and her intonations have a high rising contour; her voice can be described as squeaky and very irritating. The role of voice in this example is more important than in the previous ones, because it is an equal mode to the visual (redecorating and moving of mother-in-law). The wife herself (as supported by prosodic features) is one of the disturbing moments being cured by drinking Irn-Bru.

Finally, in the commercial for Booking.com (https://www.youtube.com/watch?v=VG6Lt7_8uEw), prosodic features are essential in reconstructing the argument.

Booking.com is the best option for you when planning your holyday because it offers you excitement, thrill, offers you adventure, and offers something new and unique, something no one else can provide. The commercial is humorous because it lists all the usual things (like showers, ice cubes, slippers and beds), but all of the usual things are more exciting via booking.com. Usual things thus become a source of excitement, arousal. And this is supported by a voice-over which represents that excitement. Without it, the shower is just a shower and a hairdryer just a hairdryer. But with higher pitch, wider pitch-range, faster tempo and higher intensity (all specific for excitement) a consumer perceives the sprit level that she is about to experience by going on a booking.com.

The commercial can be reconstructed on the Toulmin model:

> (6) Ground: Booking.com gives you unique, exciting experience.
> Supported with pitch, pitch range, tempo, intensity, intonation
> Warrant: Something that is unique and special is better that something ordinary.
> Claim: Go to booking.com.

If prosodic features were removed or altered, in this example the argument would be altered also. It would not function as reconstructed. It is therefore obvious that in some examples prosodic features have an important argumentative function with respect to persuasive effect.

4. CONCLUSION

Prosodic features have thus clearly been shown to be of importance for the assessment of a speaker's personality, persuasiveness and emotional state. This paper has shown that because of its importance in nonverbal communication, prosodic features have their function in an argumentative discourse. It has been argued that they should not be neglected in the analysis of multi-modal discourse.

The term "prosodic features" covers all aspects of the manner of speech, including voice quality, accent and pronunciation (e.g., of vowel and consonants), tempo, rhythm, intensity, intonation, word emphasis, and (non-)fluencies, and can thus be regarded as the auditory part of argumentation. The starting point for an analysis can be the theoretical settings provided by visual argumentation scholars. However, it has to be taken into account that an appeal to the ear functions differently from an appeal to the eye, because information is processed in a different way and there are significant differences between visual and auditory memory (Hollien, 2002). Auditory elements in an argumentative discourse often rely on stereotypes and frequently work on a subconscious level (e.g., we hear someone on a radio and perceive him/her as deceptive, irritating, appealing etc.); not being aware why we feel that way, we cannot explain it. But often the reason is not *what* someone is saying but the *how* – the manner of his speech. Perhaps even because of their elusive character, auditory elements are frequently used in public discourse (politics, advertising, business). They should be taken into account in serious argumentative analyses of a multi-modal discourse. Without diminishing the importance of verbal argumentation, which is its visual correlate, the auditory part of arguing may sometimes be essential to persuasive effect.

REFERENCES

Banissy, M. J., Sauter, D. A., Ward, J., Warren, J. E., Walsh, V., & Scott, S. K. (2010). Suppressing sensorimotor activity modulates the discrimination of auditory emotions but not speaker identity. *Journal of Neuroscience, 30*(41), 13552–13557.

Bartsch, S. (2009). "What sounds beautiful is good?" How employee vocal attractiveness affects customer`s evaluation of the voice-to-voice service encounter? *Aktuelle Forschzngsfragen in Deinstleistungsmarketing*, 45-68.

Berry, D. S. (1991). Accuracy in social perception: Contributions on facial and vocal information. *Journal of Personality and Social Psychology, 62*, 298-307.

Berry, D. S. (1992). Vocal types and stereotypes: Joint effects of vocal attractiveness and vocal maturity on person perception. *Journal of Nonverbal Behavior, 18,* 187-197.

Birdsell, D., & Groarke, L. (1996). Toward a theory of visual argument. *Argumentation and Advocacy, 33*(1), 1-10.

Birdsell, D., & Groarke, L. (2007). Outlines of a theory of visual argument. *Argumentation and Advocacy, 43,* 103-113.

Blair, A. (1996). The possibility and actuality of visual arguments. *Argumentation and Advocacy, 33*(1), 23-39.

Burgoon, J. K., Birk, T., & Pfau, M. (1990). Nonverbal Behaviors, Persuasion and Credibility. *Human Communication Research, 17*(1), 140-169.

Coupland, N., & Bishop, H. (2007). Ideologised values for British accents. *Journal of Sociolinguistics, 11*(1), 74-93.

Davitz, J. R. (1964). *The communication of emotional meaning.* New York: McGraw- Hill.

Esling, J. H. (1978). The identification of features of voice quality in social groups. *Journal of the International Phonetic Association, 7,* 18-23.

Fleming, D. (1996). Can pictures be arguments? *Argumentation and Advocacy, 33,* 11–22.

Giles, H., Bourhis, R., & Davies, A. (1975). Prestige speech styles: the imposed norm and inherent value hypothesis. In W.C. MaCormack, S. Wurm (Eds.), *Language and Anthropology. IV Language in many ways.* Hague: Mouton.

Godden, D. (2015). Images as arguments: Progress and problems, a brief commentary. *Argumentation, 29,* 235-238.

Groarke, L., & Tindale, C. W. (2013). *Good Reasoning Matters! A constructive Approach to Critical Thinking.* Oxford: Oxford University Press.

Groarke, L. (2015). Going multimodal: What is a mode of arguing and why does it matter? *Argumentation, 29,* 133-155.

Hickson, M., Stacks, D., & Moore, N. (2004). *Nonverbal communication – Studies and Applications.* Los Angeles: Roxbury Publishing Company.

Honey, J. (1989). *Does Accent Matter?* London-Boston: The Pygmalion Factor, Faber&Faber.

Hughes, S. M., Farley, S. D., & Rhodes, B. C. (2010). Vocal and physiological changes in response to the physical attractiveness of conversational partners. *Journal of Nonverbal Behavior, 34,* 155–167.

Juslin, P. N., & Laukka, P. (2003). Communication of emotions in vocal expression and music performance: Different channels, same code? *Psychological Bulletin, 129,* 770–814

Kjeldsen, J. E. (2012). Pictorial argumentation in advertising: Visual tropes and figures as a way of creating visual argumentation. In F.H. van Eemeren & B. Garssen (Eds.), *Topical Themes in Argumentation Theory: Twenty Exploratory Studies* (pp. 239-255). Amsterdam: Springer.

Kimble, C. E., & Seidel, S. D. (1991). Vocal signs of confidence. *Journal of Nonverbal Behavior, 15,* 99–105.

Kišiček, G. (2012). Forensic Profiling and Speaker identification on urban varieties of Croatian language. *Doctoral Thesis,* Zagreb.

Knapp, M. L., Hall, J. A., & Horgan, T. G. (2013). *Nonverbal communication in Human Interaction*. Boston: Wadsworth Publishing Company.

Kramer, E. (1964). Personality stereotypes in voice: A reconsideration of the data. *The Journal of Social Psychology, 62*, 247–251.

Kramer, C. (1978). Female and male perceptions of female and male speech. *Language and Speech, 20*, 151-161.

Labov, W. (1966). *The social stratification of English in New York City*. Washington: Center for Applied Linguistics.

Lake, R. A., Pickering, B. (1998). Argumentation, the Visual, and the Possibility of Refutation: An Exploration. *Argumentation, 12*, 79–93.

Laukka, P. (2008). Research on vocal expression of emotion: State of the art and future directions. In K. Izdebski (Ed.), *Emotions in the human voice. Vol 1. Foundations* (pp. 153–169). San Diego, CA: Plural Publishing.

Lippi-Green, R. (1997). *English with an accent: Language, ideology, and discrimination in the United States*. London: Routledge.

Meyer, M., Baumann, S., Wildgruber, D., & Alter, K. (2007). How the brain laughs. Comparative evidence from behavioral, electrophysiological and neuroimaging studies in human and monkey. *Behavioural Brain Research, 182*(2), 245–260.

Neumann, R., & Strack, F. (2000). "Mood Contagion": The automatic transfer of mood between persons. *Journal of Personality and Social Psychology, 79*, 211-223.

Rezlescu, C. et al. (2015). Dominant voices and attractive faces: The contribution of visual and auditory information to integrated person impressions. *Journal of Nonverbal Behavior, 39*(4), 355-370.

Scherer, K. R. (1972). Judging personality from voice: A cross-cultural approach to an old issue in inter-personal perception. *Journal of Personality, 40*, 191–210.

Scott, S. K. (2008). Voice processing in monkey and human brains. *Trends in Cognitive Sciences, 12*(9), 323–325.

Smith, B., Brown, B., Strong, W., & Rencher, A. (1975). Effects of Speech Rate on Personality Perception. *Language and Speech, 18*(2), 145-152.

Surawski M. K., & Ossoff, E.P. (2006). The effects of physical and vocal attractiveness on impression formation of politicians. *Current Psychology, 25*(1), 15-27.

Vroomen, J., Collier, R., & Mozziconacci, S. (1993). Duration and intonation in emotional speech. In *Proceedings of Eurospeech 1993*, Berlin, Germany, vol. 1 (pp. 577-580). Baixas, France: International Speech Communication Association (ISCA).

Vukovic, J., Jones, B. C., Feinberg, D. R., DeBruine, L. M., Smith, F. G., Welling, L. L., & Little, A. C. (2011). Variation in perceptions of physical dominance and trustworthiness predicts individual differences in the effect of relationship context on women's preferences for masculine pitch in men's voices. *British Journal of Psychology, 102*(1), 37–48.

Willis, J., & Todorov, A. (2006). First impressions: Making up your mind after a 100-ms exposure to a face. *Psychological Science, 17*(7), 592–598.

Zuckerman, M., & Driver, R.E. (1989). What sounds beautiful is good: The Vocal attractiveness stereotype. *Journal of Nonverbal Behavior, 13*(2), 67–77.

Zuckerman, M., Hodgins, H., & Miyake, K. (1990). The vocal attractiveness stereotype: Replication and elaboration. *Journal of Nonverbal Behavior, 14*, 97-112.

Zuckerman, M., & Miyake, K. (1993). The attractive voice: What makes it so? *Journal of Nonverbal Behavior, 17*, 119-135.

Zuckerman, M., & Sinicropi, V. (2011). When Physical and Vocal Attractiveness Differ: Effects on Favorability of Interpersonal Impressions. *Journal of Nonverbal Behavior, 35*, 75-86.

38

Adjudication and Justification: To What Extent Should the Excluded Be Included in the Judge's Decision?

BART VAN KLINK
Vrije Universiteit Amsterdam, The Netherlands
B.van.Klink@vu.nl

As follows from the Rule of Law, the judge has to justify her decision. In contemporary legal and social theory, it is argued that she should somehow give recognition to arguments and viewpoints that have been excluded from the final decision. In my paper, I will address the question why, to what extent and in what way the judge has to give recognition to the arguments and viewpoints that she has excluded from her decision.

KEYWORDS: Arendt, authority, decisionism, decision-making, jurisprudence, legal argumentation, Luhmann, rule of law, Schmitt, system theory

1. THE BURDEN OF RATIONALITY

Adjudication is about deciding. Two (or more) parties quarrel over the right interpretation of the law and by appealing to a judge, they give her the authority to determine the law's meaning, at least temporarily, for the *case* at hand. However, adjudication is not just about deciding. Not every decision will do: the judge's decision, including the way in which it is reached and how it is justified, has to meet some standard of rationality. As Lon L. Fuller, 1978, p. 380, argues, "adjudication is a form of social ordering institutionally committed to 'rational' decision." He even believes that, compared to other forms of social ordering, adjudication carries the heaviest burden of rationality:

> Adjudication is, then, a device which gives formal and institutional expression to the influence of reasoned argument in human affairs. As such it assumes a burden of rationality not borne by any other form of social ordering. A decision

D. Mohammed & M. Lewiński (eds.) (2016). *Argumentation and Reasoned Action: Proceedings of the 1st European Conference on Argumentation, Lisbon, 2015. Vol. II, 643-655.* London: College Publications.

> which is the product of reasoned argument must be prepared itself to meet the test of reason.[1]

The question is why adjudication has to meet this severe test of reason and when it can be said to have passed this test. Interestingly, Fuller puts "rational" between inverted commas when he claims that adjudication is "institutionally committed to 'rational' decision". Could this perhaps indicate some reservation on his side towards the supposedly rational character of adjudication?

It may be questioned that a judge should justify her decision extensively. One could argue that by giving reasons, the judge does not necessarily strengthen her authority, but could also weaken or undermine it. According to Hannah Arendt, 2006, p. 93, authority excludes *potestas* or power ("where force is used, authority has failed") as well as persuasion: "Authority (...) is incompatible with persuasion, which presupposes equality and works through a process of argumentation. Where arguments are used, authority is left in abeyance." By giving reasons authority is deferred, because the acceptance of authority is made to depend on the always insecure outcome of persuasion: the parties involved may or may not accept the arguments offered by the judge.

In our Age of Reason, it is very difficult to accept that adjudication may not be fully "rational". For good reasons, we as citizens expect the judge to give good reasons for her decision. One reason follows from the indeterminate quality of the law. The law, necessarily phrased in general terms, can never determine fully its application on concrete cases. Inevitably, the judge has to make a choice among competing interpretations – a choice which cannot be made solely on the basis of the settled law. Because multiple applications are possible, the judge has to justify why she favours one application over the other. Other reasons have to do with the need for accountability, controllability and predictability: the parties involved want to know whether she has considered their arguments seriously, higher judges must be able to assess whether she has applied the law correctly, scholars have to construe a doctrine on the basis of the various decisions, lawyers need to calculate their chances in similar future cases, and so on. So some amount of justification seems to be justified.

To a growing extent, in contemporary social and legal theory the burden of rationality that rests on the judge is raised. Some scholars have argued that she should somehow give more recognition to arguments and viewpoints that have been excluded from the final

---

[1] Fuller, 1978, pp. 366-367.

decision. According to Luhmann, 2000 and 2005, the judge provides information not only on the decision itself but inevitably also on the solution rejected, why this constitutes an alternative and no alternative at the same time. Following Luhmann, Fischer-Lescano & Christensen, 2005, claim that in the judge's decision the possibility of a different solution should be kept open. By giving recognition to all arguments put forward by the parties involved, the decision would lose some of its violent character. In my paper I will address the question why, to what extent and in what way the judge has to give recognition to the arguments and points of view that she has to exclude from her decision. Or does this inclusion of the excluded undermine the authority of the judge?

I will defend here a moderate decisionist position, which acknowledges that in the final analysis the judge has to take a decision that aims at reaching legal closure, that is, at ending - for the time being and for the case at hand - the on-going discussion on the meaning of the law. At the same time, it provides some room for reasoning and argumentation, albeit within institutionally defined and defensible limits. Below I will address the following three questions: in what sense can the judge's ruling be understood as a decision? (section 2); why does adjudication need any justification? (section 3); and what can be reasonably be expected from a justification by the judge? To what extent should the excluded be included in the final decision? (section 4).[2] My aim is to show that Fuller was right in putting inverted commas around "rational": the judge's decision should meet some reasonable standard of rationality, but should not be overburdened by the insatiable demands of Reason.

2. DECISION AND DECISIONISM

The concept of decision has been brought into discredit due to its connection to decisionism (see Lübbe, 1971, p. 7). Decisionism is usually seen as a political theory that favours an authoritarian and arbitrary style of governance, more suited to dictatorship than to the Rule of Law. It is associated with Carl Schmitt, the German philosopher and jurist who, for a short period of time, worked for the Nazi regime.[3] Schmitt himself did not always defend a decisionist position. In one of his earliest works, *Gesetz und Urteil* (*Law and Decision*), originally published in 1912, he argues that legal practise is directed at achieving "legal

---

[2] This paper is partly based on earlier publications on the same topic in Dutch (see, for instance, Van Klink, 2012).

[3] A short biography can be found in Müller, 2003, pp. 15-48.

clarity" ("Rechtsbestimmtheit").[4,5] The judge's decision has to be "foreseeable and calculable" (Schmitt, 1969, p. 73). That does not mean that the judge is required to simply apply the law by way of subsumption – that is, according to Schmitt, no longer possible –, but she has to establish whether another judge in the same case would have arrived at the same result: "A judge's decision is nowadays correct, if it can be expected that another judge would have decided in the same way" (Schmitt, 1969, p. 71). Moreover, the decision has to be justified: "There is no decision without justification; the justification belongs to the decision" (Schmitt, 1969, p. 69). In the justification it has to be explained why the decision in the given circumstances is right.

In his later work Schmitt gives up this intersubjective criterion for the correctness of a decision. Moreover, he breaks the connection which he previously had forged between decision and justification. Within the context of his theory of sovereignty, it is the sovereign who decides on the state of exception. By declaring the state of exception, the sovereign suspends the existing legal order in its entirety. The sovereign's decision is a subjective act of volition that, from the perspective of the law, seems to come from nowhere: "Looked at normatively, the decision emanates from nothingness" (Schmitt, 2005, pp. 31-32). From a legal point of view, no grounds can be given for the decision, since there are no legal norms anymore that can be applied; it is exactly the decision which has to prepare the ground for a return to the normal situation where it is possible again to apply legal norms. However, the exception appears to be not that exceptional. Also in daily legal practice a decisionist element can be found in the judge's ruling:

> Every concrete juristic decision contains a moment of indifference from the perspective of content, because the juristic deduction is not traceable in the last detail to its premises and because the circumstance that requires a decision remains an independently determining moment.[6]

The general norms of the law have to be concretized or "transformed" in order to be applicable to a concrete case. This "transformation" presupposes an "*auctoritatis interpositio*", a determination of who has the authority to decide, which cannot be derived from the legal norms

[4] Some of the texts I refer to in this article are written in German and have not been translated into English. Whenever I quote from these texts (like in this case), I have made my own translation (as indicated).

[5] Literally, "Bestimmtheit" means determinateness.

[6] Schmitt, 2005, p. 30, see also Schmitt, 1996, p. 46.

themselves (Schmitt, 2005, p. 31); nor does the law determine how the authority has to apply the norms at hand. According to Schmitt, the decision is in a literal sense a "de-cision" ("Ent-Scheidung") in which norm and fact which are separated as much as possible in the "ordinary" legal order are re-united.[7] In his view, the point of a legal decision is not so much to offer an "overwhelming argumentation" but to take away the doubt in an authoritarian way.[8] In other words, to acquire force of law the decision does not have to rely upon its justification or "substantiation" (Schmitt, 2005, p. 32).

Hermann Lübbe, a critical student of Schmitt, has tried to rescue the concept of decision, despite its politically charged origin, for political theory. He develops a decisionist position that does not result in a defense of authoritarian or down-right dictatorial ruling. He accuses Schmitt and his followers - whom he, not without irony, calls the "Romanticists of the state of exception"[9] - that they have created a false opposition between "real life" and the daily legal practice based on routine. According to Lübbe, ordinary life also consists of extraordinary circumstances and decisive moments in which decisions have to be taken. In the course of history, due to the erosion of common traditions, human existence has increasingly become decisionist in character. That is, to a growing extent, modern man lives under the pressure to take decisions. A decision is a choice one feels oneself compelled to make between two incompatible options. Only one of the possible options can be realized, so the other option has to be excluded. There are no "decisive" grounds available to make a well-founded choice between both alternatives. That does not mean, however, that the decision is irrational:

---

[7] According to Schmitt, 1934, p. 18, in the normativist variant of legal positivism, facts and norms are separated in the most rigorous way.

[8] Schmitt, 1996, p. 46 (my translation).

[9] Schmitt never gets tired of blaming his liberal opponents to hold on to the romantic ideal of the "eternal conversation": instead of taking decisions, they rather prefer to start a debate (see, for instance, his polemic against political romanticism in Schmitt, 1925). Lübbe, 1971, p. 29, counters this criticism by pointing out matter-of-factly that politicians in liberal democracies do take decisions, albeit in another way than a dictator would do.

> The decision transcends a lack of rational determining grounds for action. It is not, for that reason, irrational. The rationality[10] of the situation of deciding resides exactly in the fact that one determines a course of action, although there are no sufficient reasons to act in one way and only in this way.[11]

The situation in which one has to decide always is a state of exception, because shared standards drawn from law, morality, tradition, public opinion and so on do no longer offer enough guidance and, therefore, one is left to one's own judgment (Lübbe, 1971, p. 21). The authority who has to take a decision, has to carry herself the "weight of the exception."[12]

Likewise, the judge's judgment can be understood as a decision in this sense. Building on Oakeshott's conception of the Rule of Law (Oakeshott, 1975 and 1999), I consider it to be her task to assess whether the actions in a specific case are in accordance with the general conditions set by the law. The judge does not evaluate these conditions – that is up to the legislature –, but primarily deals with their correct application on the case at hand. The general norms of the law are never directly applicable on the case; there are no easy cases in which the law is simply given. The relation between the judge's ruling and the law is contingent, that is, multiple applications are always possible. Every application of a higher legal norm necessarily gives the judge to a greater or lesser extents freedom. Inevitably, adjudication involves choosing. The settled law does limit the freedom to choose, but can never take it away entirely. The judge has to make a choice among the various options that the law offers and this choice cannot be made on the basis of the settled law itself. There is no established method that can produce "one right answer" for the case at hand.[13] From the perspective of the settled law, as Hans Kelsen, 1994, p. 96, argues, the various possible decisions are of equal value, because the choice among them is based on a subjective evaluation from another normative, for instance moral or political, point of view. In the moment of decision, the judge finds herself in the state of exception as described by Lübbe: with

---

[10] In German "Vernunft," which, among other things, also refers to reason, intellectual power or capacity.

[11] Lübbe, 1971, p. 21 (my translation).

[12] This expression is taken from Heidbrink, 2007, p. 178 (my translation).

[13] Also Dworkin, who defends the famous "one right answer thesis" (see, e.g., Dworkin, 1978, pp. 279-290) does not offer a method of that kind. His appeal to general legal principles makes it even harder, if not impossible, to arrive at a univocal determination of the law's content in a specific case.

no generally shared, unequivocal standards at her disposal, she is the one who has to take the decision and carry its weight on her own.

3. DECISION AND JUSTIFICATION

Inevitably, the judge has to take a decision. However, that does not imply by necessity that she does not have to offer reasons for her decision. On the contrary, one could argue, if the law allows for multiple applications, she has to justify why she prioritizes one possible application over the other. There is not much to explain when the law would dictate univocally one right answer. In our modern age, adjudication is no longer seen as a matter of sheer subsumption (as if it ever was) but it is commonly acknowledged that the judge has nowadays to a greater or lesser extent discretion when interpreting the law. In the decision, the subjectivity of the subject who decides comes to the fore more and more explicitly. According to the German sociologist Ludger Heidbrink, 2007, this is due to the erosion of our normative orders. Normative orders, such as law or morality, do not have a self-evident validity anymore and their norm content is increasingly uncertain and indeterminate. As a result, the subject who has to decide, is forced to find a justification on her own and within herself. As Heidbrink argues:

> The loss of traditional orientations, the decay of moral certainties and the erosion of social norms and rules have contributed to the situation in which the individual human being is forced to justify his actions and decisions from within himself ("aus sich heraus") and to implement them into practice.[14]

In the absence of shared norms prescribing clear duties, one needs a justification: "Justification does matter in particular in situations where the validity, the determinateness and the information content have changed, where perfect duties are transformed into imperfect obligations" (Heidbrink, 2007, pp. 77-78; my translation). The growing need for justification is, therefore, caused by the increased complexity and insecurity in modern society, both on a normative level (which norms are valid?) and a cognitive level (how do the valid norms have to be applied?).

Decision and justification are thus no contrary concepts, as decisionism in its most radical form seems to suggest, but are

---

[14] Heidbrink, 2007, p. 182.

inextricably linked: exactly *because* on has to decide, one has to justify oneself. At the same, an uneasy paradox occurs: due to the lack of generally valid standards, the subject increasingly has to justify herself and yet, again due to the lack of generally valid standards, she has less and less means at her disposal to do so in a persuasive way. According to Heidbrink, 2007, p. 114 (my translation) justification becomes to a growing extent an "Atopia" or a "non-place" which "can no longer be located, localized or captured."[15]

Niklas Luhmann, 1995, also assumes that the application of law is not or not only a cognitive operation. In most cases, the law applicable in a specific case cannot easily be deduced from the legal text. Since the legal text fails to provide unequivocal answers, often additional sources of information such as principles are invoked (like in Dworkin's theory of interpretation, see e.g. Dworkin, 1978). However, principles of law and morality neither produce one right answers. In a pluralist society, as Luhmann argues, conceptions of justice differ fundamentally so justice cannot decide upon a conflict between competing principles. Therefore, the settled law cannot be conceived of without the concept of decision: "The concept of positivity suggests that it can be understood through the concept of decision. Positive law is supposed to be validated through decisions" (Luhmann, 2004, p. 76). Although law's validity is based on decisions, Luhmann dismisses any suggestion of decisionism in law:

> This leads to the charge of "decisionism" in the sense of a possibility to decide in an arbitrary fashion, dependent only on the coercive force behind such decisions. Thus, this leads in fact to a dead-end; after all, everybody knows that in law decisions are never simply made arbitrarily.[16]

According to Luhmann, 2000, p. 134, the decision has to be accompanied by a justification in which it is explained that whoever takes the decision either has the right or the authority or good reason for deciding in the way she has decided. In the justification information has to be provided not only about the decision itself, but also about the non-chosen option – why it constitutes simultaneously an alternative and no alternative.[17] By doing so, the authority gives herself or another

---

[15] Original text: "sie lässt sich nicht mehr verorten, eingrenzen, dingfest machen."

[16] Luhmann, 2004, p. 76.

[17] The original text runs as follows: "Die Entscheidung muss über sich selbst, aber dann auch noch über die Alternative informieren, also über das Paradox,

authority the opportunity to reconsider the decision and to include the excluded alternative. So in Luhmann's view decision and justification are necessarily connected: because one could have decided *otherwise*[18] (and will do so in the future), one has to justify oneself.

Building on Luhmann's systems theory as well as Derrida's deconstructivism, Fischer-Lescano & Christensen, 2005, reject the decisionist approach as advocated by Schmitt. They accuse Schmitt of having misrepresented the decision as a purely subjective act which takes place in splendid isolation, as a creation from nowhere. In their view, the decision by the judge is part of a social communication process that does not end with the decision; on the contrary, it continues the communication process. The judge's decision is a suspension ("Aufschub") rather than a determination of law: "The law does not terminate the conflict between citizens by offering a stable decision, but [these disputes] delay the law continuously and force it into metamorphoses" (Fischer-Lescano & Christensen, 2005, p. 230; my translation). The decision taken has to be acceptable for society. For that purpose, the judge has to be open towards law's environment consisting of other social subsystems such as economy, morality, religion and science. However, she has to prevent that one or more of these systems will dominate the law.[19] The deconstructivist systems-theoretical concept of decision is both pluralist and "extremely formalistic" (Fischer-Lescano & Christensen, 2005, p. 236; my translation). It aims at involving the law as an "empty signifier" in a process of eternal semiosis which includes as much as possible the excluded alternative meanings. Because the judge has a duty to justify her decision, she has to offer reasons which make the decision transparant and controllable and which explicitly leave room for a different decision. By taking into account all arguments presented by the parties, she is able to diminish or "relativize" the violent character of the decision (Fischer-Lescano & Christensen, 2005, p. 232; my translation).

---

dass die Alternative eine ist (denn sonst wäre die Entscheidung keine Entscheidung) und zugleich keine ist (denn sonst wäre die Entscheidung keine Entscheidung)" (Luhmann, 2000, p. 134). This can be related to what Agamben, 1998, p. 18, calls the "relation of exception," that is, "the extreme form of relation by which something is included solely through its exclusion" which, in his view, characterizes the state of exception.

[18] This phrase is taken from Fischer-Lescano & Christensen, 2005, p. 241.

[19] Or, as Fischer-Lescano & Christensen, 2005, p. 237 (my translation), put it, "to instrumentalize one-sidedly" the law.

So it seems that in contemporary legal and social theory, the concept of decision has not disappeared but is stripped from its decisionist connotations. Increasingly, decisionism is transformed into a deliberative approach that requires from the judge to give good reasons for her decision and to include somehow the viewpoints that she has excluded in her decision.

4. THE FINALITY OF DECISION

However, at some point the discussion has to stop. Although decisions in law often come about after and through lengthy debate, they aim at establishing legal closure. The law offers an institutionalized method for putting an end to legal discussions in society and in academia that, in the absence of an ultimate authority that is entitled to speak the "last word", may continue endlessly. From an abstract academic perspective, *sub specie aeternitates* so to speak, this "end" may appear, a temporary halt in the process of eternal semiosis; for the parties involved in a case it is the final stop, if there are no legal possibilities left to re-open the debate. And this may be welcomed, since otherwise legal procedures would go on forever. In order to achieve legal closure, the law presupposes authority – authority that does not owe its validity to good reasons. According to Arendt (as quoted in section 1), "authority is incompatible with persuasion"[20] because, in her view, persuasion involves argumentation which equals a deferral or suspension of authority. Authority would become very volatile, when on every occasion it would depend on the willingness of the parties to accept the arguments given by the judge.[21]

The judge is not a participant in an ongoing discussion, as followers of the deliberative approach would have it, but an instance of authority that determines, by means of the law and in a legally binding way, what the law means here and now, for the case at hand. The judge's decision aims, as the early Schmitt indicated, at providing "legal clarity". Violence cannot be avoided, because alternative decisions are excluded, at least in this case. However, this does not mean that the judge as an instance of authority has to take way doubt in an *authoritarian* way, as the later Schmitt argued.[22] On the contrary, it is reasonable to expect that she gives reasons for her decision. The justification given by the

---

[20] On Arendt's view of authority, see Honig, 1993, chapter 4.

[21] As Oakeshott, 1999, p. 149, argues, the Rule of Law is based on "the recognition of the authenticity of the law," independent of its content.

[22] The two prior references can be found in section 2.

judge cannot deprive the decision of its violent character though (or reduce the violence, as Fischer-Lescano & Christensen suggest): notwithstanding all considerations and caveats, ultimately legal consequences will be attached to the decision. Robert Cover, 1995, p. 203, expresses the relation between legal interpretation and violence very clearly: "Legal interpretive acts signal and occasion the imposition of violence upon others: A judge articulates her understanding of a text, and as a result, somebody loses his freedom, his property, his children, even his life."

As indicated earlier, every application of a legal norm involves to a greater or lesser extent freedom. Therefore, it can be expected that the judge, when applying general norms to a specific case, demonstrates that the interpretative choices made follow from a *possible and defendable*, not necessarily the 'one right' application of the legal norms at hand. She needs to justify her decision, because she could have taken a different decision (that is exactly why it is a decision). Strictly speaking, the validity of law in a concrete case does not depend on the size and quality of the justification; also short and poorly reasoned rulings are legally binding for the parties involved (as long as the decision is not overruled by a higher court). However, for the social acceptance of the general legal order it is important that the law, both when it comes to its creation and its application, manifests itself as a "product of reflexive intelligence."[23] In the justification, the different opinions of the parties involved must be reflected, so they are and that their case is taken seriously. Moreover, other judges who have to take a decision in the same or a similar case must have access to the relevant considerations of the judge. In my view, the judge does not have to address every argument that has been put forward in the social or academic debate on the issue at hand or to balance extensively the pros and cons of every solution thinkable. Moreover, I would be very cautious with including the excluded, as some authors have suggested (see section 3). The judge should, of course, justify why a certain view point, argument or solution is excluded, though not in order to keep the possibility of a different decision open (in the line of Luhmann and Fischer-Lescano & Christensen), but to achieve legal closure, however temporarily or provisionally. The justification should demonstrate convincingly and forcefully why the judge has favoured this decision over other possible decisions.

I would therefore like to argue for a justification that is concise, consistent and modest in that it presents itself, not as *the* right answer (that may not exist), but as *a possible* right answer. The faculty of reason

[23] Oakeshott, 1999, p. 256, borrows this notion from Hegel.

will always find reasons to question the judge's decision. Undoubtedly, from some normative point of view, other decisions will appear to be better or better justified. For lack of generally shared values,[24] we continue to argue about the meaning of the norms involved, the values at stake and their internal relation. But at a certain point the case has to be closed. What can be required in any case, is that the judge shows that her ruling is based on a possible application of the relevant legal norms, that it connects to earlier applications of these norms by other judges and is supported by the legal system as whole, its underlying aims and values and its historical development (that is, in Gadamer's terms, its effective history[25]). As Odo Marquard, 2003, p. 78 (my translation) states, "the burden of proof is on the one who changes something." When deciding a case, the judge can appeal to the variety of *topoi* or common places, accepted by many people or the wise, that can be found in the various subsystems of society.[26] *Topoi*, such as equality or equity, can give some normative orientation when difficult choices have to be made, but they can never dictate a specific outcome. *Which* topoi are applicable in a concrete case and *how* they have to be applied, cannot be derived from the *topoi* themselves. When the cognitive operations have been carried out, the moment of choice has come. Ultimately, it is the judge who has to cut the knot.

REFERENCES

Agamben, G. (1998). *Homo sacer. Sovereign power and bare life* (translated by D. Heller-Roazen). Stanford: Stanford University Press.

Arendt, H. (2006). *Between past and future. Eight exercises in political thought.* New York: Penguin Books.

Cover, R. (1995). Violence and the word. In M. Minow, M. Ryan & A. Sarat (Eds.), *Narrative, violence, and the law. The essays of Robert Cover* (pp. 203-238). Ann Arbor: The University of Michigan Press.

Dworkin, R. (1978). *Taking rights seriously.* Cambridge, MA: Harvard University Press.

Fischer-Lescano, A., & Christensen, R. (2005). *Auctoritatis interpositio.* Die Dekonstruktion des Dezisionismus durch die Systemtheorie. *Der Staat, 44*(2), 213-241.

Fuller, L.L. (1978). The forms and limits of adjudication. *Harvard Law Review, 92*(2), 353-409.

---

[24] Caused by the "disenchantment of the world", as analysed by Weber, 1996.

[25] See Gadamer, 2013, pp. 312 ff (in German: *Wirkungsgeschichte*).

[26] On the role of *topoi* in political decision-making, see Lübbe, 1971, p. 61.

Gadamer, H.-G. (2013). *Truth and method* (translation revised by J. Weinsheimer & D.G. Marschall). London: Bloomsbury Academic.

Heidbrink, L. (2007). *Handeln in der Ungewissheit. Paradoxien der Verantwortung*. Berlin: Kulturverlag Kadmos.

Honig, B. (1993). *Political theory and the displacement of politics*. Ithaca: Cornell University Press.

Kelsen, H. (1994). *Reine Rechtslehre. Einleitung in die Rechtswissenschaftliche Problematik* (1. Auflage). Aalen: Scientia Verlag.

Lübbe, H. (1971). *Theorie und Entscheidung. Studien zum Primat der praktischen Vernunft.* Freiburg: Verlag Rombach.

Luhmann, N. (2000). *Organisation und Entscheidung*. Wiesbaden: VS Verlag für Sozialwissenschaften.

Luhmann, N. (2004). *Law as a social system* (translated by K.A. Ziegert). Oxford: Oxford University Press.

Luhmann, N. (2005). The paradox of decision making (translated by K. Finney-Kellerhof & D. Seidl). In D. Seidl & K.H. Becker (Eds.), *Niklas Luhmann and Organization Studies* (pp. 85-106). Kristianstad: Liber & Copenhagen Business School Press.

Marquard, O. (2003). *Zukunft braucht Herkunft. Philosophische Essays.* Stuttgart: Philip Reklam Jun.

Müller, J.-W. (2003). *A dangerous mind. Carl Schmitt in post-war European thought.* New Haven: Yale University Press.

Oakeshott, M. (1975). *On human conduct.* Oxford: Clarendon Press.

Oakeshott, M. (1999). The rule of law. In M. Oakeshott, *On history and other essays* (pp. 129-178). Indianapolis: Liberty Fund.

Schmitt, C. (1925). *Politische Romantik*. München: Verlag von Duncker & Humblot.

Schmitt, C. (1934). *Über die drei Arten des rechtswissenschaftlichen Denkens.* Hamburg: Hanseatische Verlagsanstalt.

Schmitt, C. (1969). *Gesetz und Urteil. Eine Untersuchung zum Problem der Rechtspraxis.* München: C.H. Beck'sche Verlagsbuchhandlung.

Schmitt, C. (1996). *Der Hüter der Verfassung*. Berlin: Duncker & Humblot.

Schmitt, C. (2005). *Political theology. Four chapters on the concept of sovereignty* (translated by G. Schwab). Chicago: The University of Chicago Press.

Van Klink, B. (2012). De finaliteit van het rechterlijk oordeel. Naar een motivering van het genoeg. In E.T. Feteris et al. (Eds.), *Gewogen oordelen. Essays over argumentatie en recht* (pp. 159-169). Den Haag: Boom Juridische uitgevers.

Weber, M. (1996). *Wissenschaft als Beruf.* Berlin: Duncker & Humblot.

39

"Doctor, I disagree!" Development and Initial Validation of a Scale to Measure Patients' Argumentativeness in Medical Consultation

NANON LABRIE
University of Lugano, Switzerland
nanon.labrie@usi.ch

ANNEGRET HANNAWA
University of Lugano, Switzerland
annegret.hannawa@usi.ch

PETER SCHULZ
University of Lugano, Switzerland
peter.schulz@usi.ch

This paper describes the development and validation of a theory-driven instrument to measure patients' trait argumentativeness, i.e., patients' propensity to engage in a critical discussion procedure with their physicians. Exploratory factor analysis (n = 183) confirmed a two-factor scale structure, representing the tendency to avoid (7 items) and approach argumentation (8 items). The instrument can be used in medical practice to critically assess patients' argumentative preferences and, thereby, facilitate the communicative interaction between doctors and their patients.

KEYWORDS: doctor-patient communication, medical consultation, patient-centeredness, scale development, trait argumentativeness

1. INTRODUCTION

1.1 Argumentation in general practice consultation

During medical consultations, differences of opinion may arise between doctors and patients concerning, e.g., the nature of patients' symptoms,

D. Mohammed & M. Lewiński (eds.) (2016). *Argumentation and Reasoned Action: Proceedings of the 1st European Conference on Argumentation, Lisbon, 2015. Vol. II, 657-670.* London: College Publications.

the interpretation of test results, or the applicability of certain treatment methods. Contemporary patient-centered approaches to medical interaction encourage patient involvement in decision-making discussions and stipulate that doctors and patients should cooperate in a rational deliberation process that aims to attain a shared treatment decision – given that the patient is willing and capable to do so (Charles, Gafni, & Whelan, 1997; 1999; Frosch & Kaplan, 1999; Sandman & Munthe, 2010). As a result, medical consultation can be viewed as a communicative activity type in which argumentative discourse is characteristically present (van Eemeren, 2010; Labrie, 2012).

Several argumentation scholars have demonstrated that argumentation indeed plays a crucial role in the context of medical consultations and, more specifically, within the context of general practice. Pilgram (2010; 2012) and Bigi (2010; 2012), for instance, qualitatively explore doctors' use of authority argumentation, while Snoeck Henkemans and Wagemans (2012), Labrie (2012), as well as Rubinelli and Schulz (2006; Schulz & Rubinelli, 2008) analyze doctors' strategic argumentative maneuvering in light of patient-centered communication and, in particular, the shared decision-making model. In a series of quantitative and experimental studies, Labrie and Schulz (2014b; 2014c; 2015a; 2015b) show that doctors' use of reasonable argumentation positively affects consultation outcomes, such as perceived patient involvement, perceived doctor credibility, patients' acceptance of, and adherence to the doctor's advice, and satisfaction. The emerging study of medical argumentation, thus, is grounded in a solid theoretical starting point and contains a strong empirical and practical component. Moreover, research conducted within the realm of argumentation theory has resonated in the field of health communication. To cite Salmon (2015, p. 544):

> Bringing the concept of shared decision-making together with argumentation theory could help develop a normative, as well as a theoretical framework for doctor-patient communication. That is, we will be able to expose the details of persuasion and influence and subject them to empirical and ethical scrutiny – which ones work or are counter-productive, and which ones are ethically acceptable or inadmissible.

Notably – and somewhat paradoxically given the patient-centered ideal – to date the majority of studies on medical argumentation focus on the doctor's contribution to the argumentative discussion in general practice consultation rather than on the patient. However, patients' communication preferences may differ depending on numerous factors, such as age, gender, socio-economic status, culture, disease condition, and psychological state. For example, young, healthy, female patients who have a higher educational degree typically prefer an active role in decision-making (Levinson et al., 2005). Similarly, it can be assumed that patients' argumentation preferences vary. Patients' competence and willingness to engage in an argumentative discussion with their doctor in order to resolve potential disagreements in respect to a diagnosis or treatment plan may differ depending on both the medical context and patient characteristics.

While measurement instruments to assess people's general argumentativeness exist, to date no tools have been developed to measure patients' specific argumentativeness when interacting with their doctor. Due to the specific communicative characteristics of medical consultations, such tools could provide relevant, contextualized information about communicators' argumentativeness. Therefore, starting from Infante and Rancer's (1982) general argumentativeness scale, this study aims to provide an overview of the development and preliminary validation of a patient argumentativeness scale (P-ARG scale). Doing so, first, the concept of patient argumentativeness will be defined, followed by a rationale for the development of a new measure. Then, the procedures for item development and testing will be described, followed by a discussion of the study findings. Finally, the study's practical implications will be outlined, together with recommendations for further research that could lead to refinement of the proposed measure.

1.2 Patient argumentativeness

Based on Infante and Rancer's (1982) conceptualization of general argumentativeness, patient argumentativeness is conceptualized as a generally stable trait that predisposes the patient in medical consultation to engage in a critical discussion process with the doctor when a difference of opinion arises concerning treatment. Entering such a critical discussion process entails that the patient aims to ultimately resolve the disagreement at hand by means of advancing arguments. These arguments may be intended to support the patient's own standpoint or to counter the doctor's point-of-view. The patient who scores high on argumentativeness perceives this activity as an intellectual challenge that offers a feeling of motivation and satisfaction. Moreover, this patient feels confident about his or her ability to engage in an argumentative discussion with the doctor. In contrast, the patient who is low in argumentativeness prefers to avoid argumentative situations in medical consultation. When faced with an argumentative discussion, this patient experiences feelings of nervousness, discomfort, and a lack of confidence regarding their ability to argue.

Patient argumentativeness is expressed as the difference between a patient's tendency to approach argumentation in the medical context (mean score) minus the tendency to avoid argumentation (mean score) when talking to a doctor, as such that: P-ARG = P-ARG*ap* - P-ARG*av*. Patient argumentativeness predisposes patients to defend their views and preferences concerning their medical care and to explicitly voice their doubts about, and disagreement with, their doctor's position when necessary. Patients are classified as high or low in argumentativeness when their score is one standard deviation above or below the mean for the sample. Patients with a score that falls within one standard deviation of the mean are considered as moderate in patient argumentativeness.

It should be emphasized that the conceptualization of patient argumentativeness as sketched above is not intended to refer to any process of bickering or quarreling between doctor and patient. In the English language, the term *argument* often bears the connotation of a verbal fight (van Eemeren, 2010, p. 26). However, in this paper, starting from the pragma-dialectical theory of argumentation developed by van Eemeren and Grootendorst (1982; 1992; 2004), the term argumentation is used to refer to a rational and reasonable discussion process in which the participants ideally strive to reach mutual agreement. When doctor and patient fail to reach mutual agreement, however, ethically the patient retains the right to settle the discussion by autonomously deciding between the relevant treatment options. As such, the use of argumentative discourse in the interaction between doctors and patients is in accordance with patient-centered approaches to medical consultation. Sandman & Munthe (2010, p. 78) refer to such an argumentation-based decision-making procedure as a *shared rational deliberative joint decision* (or, if settled by the patient, *shared rational deliberative patient choice*) model. In this model both doctor and patient should be given the opportunity to take part and should be allowed to express their relevant needs, preferences, and reasons. Moreover, each of the parties should be open to seriously and openly consider the arguments of the other party and allow their own views to be critically questioned - regardless of their role in the medical consultation (medical expert versus lay expert). As such, the notion of medical argumentativeness is void of negative connotations.

1.3 Rationale for a new measure

While an individual may be highly argumentative within, for instance, the context of communication with a partner or spouse, this does not necessarily imply that this person is argumentative when talking to a doctor as well. The particularities of the medical context - which is characterized by distinctive authority roles, imbalance of knowledge, undesirable and involuntary aspects of the health condition, as well as time constraints - may affect a person's capability and willingness to argue. The unique characteristics of medical consultations call for the development of a specific measure that assesses patient argumentativeness. Moreover, Infante and Rancer's (1982) original argumentativeness scale has received considerable criticism for its language ambiguities and lack of theoretical consistency in describing the process of argumentation (Dowling & Flint, 1990; DeWine, Nicotera, & Parry, 1991). The proposed P-ARG scale therefore additionally aims to resolve these issues by relying on the pragma-dialectical theory of argumentation (van Eemeren & Grootendorst, 1984; 1992; 2004) for item development. The particular procedures are described next.

2. METHODS

2.1 Scale development

Infante and Rancer's (1982) original argumentativeness scale included 20 items (10 items for each tendency, ARG*av* and ARG*ap*), and was measured on a 7-point Likert scale. In order to fit the original scale to the medical consultation context and to capture the resolution-oriented character of patient argumentativeness, a number of alterations were necessary. First, an effort was made to remove the ambiguity of the original scale's terminology, which has been criticized by several authors for compromising the measure's face validity (Dowling & Flint, 1990; DeWine, Nicotera, & Parry, 1991). Dowling and Flint (1990), for instance, note that the term argument, as it is used in the original scale, carries relational undertones. This connotation was considered unfitting within a resolution-oriented definition of patient argumentativeness. In reformulating the scale items, therefore, consistent use was made of the pragma-dialectical conceptualization of an argumentative discussion, as well as its associated terminology (van Eemeren & Grootendorst, 1984; 1992; 2004). Additionally, the commonly criticized, content-bound term *issues* was avoided in the new scale. In some cases, the original measurement instrument referred to having arguments (or, rather, having differences of opinion) about *controversial issues*. Because engaging in a medical discussion is certainly not restricted to controversial topics, this term was not included in the P-ARG scale. Finally, ten out of the twenty general argumentativeness items were adapted to reflect the context of treatment decision-making discussions in general practice to create a proper patient argumentativeness scale. In particular, five items measuring the tendency to avoid argumentation were matched to the medical context and, likewise, five items pertaining to the tendency to approach argumentation were reformulated to fit the particular context.

2.2 Participants

The new scale was administered to a sample of 183 English-speaking Master's and Ph.D. students from a variety of study programs at the Università della Svizzera italiana (USI) in Lugano, Switzerland. Respondents were recruited via email to participate in a survey on doctor-patient interaction. As a participatory incentive, students were entered into an opportunity drawing for three department store vouchers of 30 Swiss Francs. The sample consisted of 64.5% females and 35.5% males and ranged between 21 and 55 years of age (M = 26.4, SD = 3.9). The sample was ethnically diverse, comprising 29.5% Italian, 22.4% Swiss, and 48.1% students from other origins. Almost all students (96.2%, n = 176) judged their health as either good or excellent (M = 3.74, SD = .80) on a 5-point scale ranging from 1 (poor) to 5 (excellent).

2.3 Additional measurements

Besides standard demographic questions, participants responded to Degner and Sloan's (1992) scale, selecting their preferred decision-making style (including active, collaborative, and passive roles), and Levinson, Kao, Kuby, and Thisted's (2005) patient participation scale. The latter scale includes three items, with a 5-point Likert scale: "I prefer to rely on my doctor's knowledge without trying to find out about my condition on my own," "I prefer that my doctor offers me choices and asks my opinion," and "I prefer to leave decisions about my medical care up to my doctor". These questions concerning participants' personal communication preferences in the context of medical consultation were included to assess the new scale's concurrent validity.

3. RESULTS

3.1 Factor structure

Given the significant alterations that were performed on the original argumentativeness scale to adapt its items to the medical context and the orthogonal nature of its original two dimensions, principal components analyses with oblique rotation was conducted to identify the components of the measure. Examination of the Kaiser-Meyer Olkin measure of sampling adequacy suggested that the sample was indeed factorable (KMO = .77). The number of components extracted was based on eigenvalues greater than unity as well as careful scrutiny of the scree plot. While analysis of eigenvalues suggested a five-factor scale structure, the scree plot clearly indicated the presence of two components that mirrored the original scale's two-dimensional structure, with the first factor describing patients' tendency to avoid argumentation (P-ARG*av*) and the second factor indicating patients' tendency to approach argumentation (P-ARG*ap*) in the context of medical consultation.

The two components of the new scale, P-ARG*av* and P-ARG*ap*, showed acceptable internal consistency (both $\alpha = 0.77$) and were uncorrelated, $r(181) = .007$, $p > .05$, as in the original scale. Five items had to be removed from the initial item pool because they did not reach a primary factor loading of .50 and had a secondary loading smaller than .30. Most of these five items referred to some form of pleasure, enjoyment, or happiness either in engaging in an argumentative discussion or in keeping it at bay. As the argumentativeness scale was designed to assess patients' argumentativeness, it appears that the terminology used in these items – as it was used in Infante and Rancer's (1982) original scale – was unfit to the context of medical consultation. Because none of the five items added significantly to the internal consistency of the P-ARG*av* and P-ARG*ap* factors, they were removed from the scale, resulting in a final measure that consisted of seven and eight items, respectively (see table 1).

Item		Factor Loadings	
		P-ARGap (α = .77)	P-ARGav (α = .77)
1.	I feel refreshed and satisfied after engaging in an argumentative discussion.	**.71**	
2.	I feel excitement when I expect that a conversation I am in is leading to a difference of opinion.	**.67**	
3.	I consider having a difference of opinion with my doctor about medical treatment an exciting intellectual challenge.	**.65**	
4.	I feel energetic when I defend my standpoint regarding treatment during medical consultation.	**.63**	
5.	I enjoy engaging in a discussion.	**.62**	
6.	I have the ability to do well when I defend my standpoint regarding treatment during medical consultation.	**.57**	
7.	Engaging in an argumentative discussion to resolve a difference of opinion improves my intelligence.	**.50**	
8.	I do not like to miss the opportunity to discuss my standpoint regarding treatment with my doctor.	**.50**	
9.	Engaging in an argumentative discussion with my doctor about treatment creates more problems than it solves.		**.70**
10.	I get an unpleasant feeling when I realize that I am about to defend a standpoint that opposes someone else's standpoint.		**.66**
11.	I try to avoid getting into a treatment-related argumentative discussion with my doctor.		**.65**
12.	When I finish an argumentative discussion with my doctor about medical treatment, I feel nervous and upset.		**.61**
13.	Once I finish an argumentative discussion, I promise myself that I will not get into another.		**.59**
14.	When I explicitly voice a treatment preference that differs from my doctor's treatment advice, I worry that my doctor will form a negative impression of me.		**.59**
15.	I find myself unable to think of effective arguments during an argumentative discussion with my doctor about my treatment plan.		**.57**

Note. Loadings <.30 were masked to enhance interpretability.

Table 1. P-ARG item pool and factor loadings

3.2 Scale validity

The fact that the P-ARG scale mirrored Infante and Rancer's (1982) argumentativeness scale both in structure and content provides evidence for the new measure's convergent validity. Participants scored a 3.81 average on P-ARG*ap* (*SD* = .85) and a 3.57 average on P-ARG*av* (*SD* = .95). Males scored significantly higher on patient argumentativeness than females, $F(1, 181) = 4.37$, $p < 0.05$, $\eta^2 = .02$). This implies, despite a small effect size, that male patients ($M = .52$, $SD = 1.24$) are more prone to engage in an argumentative discussions concerning treatment with their doctors than female patients ($M = .09$, $SD = 1.35$). Because prior studies demonstrate similar patterns (Infante, 1982), these findings provide further proof for the scale's convergent validity.

The measure's concurrent validity was assessed by exploring the relationships between participants' participation (Levinson, Kao, Kuby, & Thisted, 2005) and shared decision-making preferences (Degner & Sloan, 1992) with their patient argumentativeness scores. As expected, respondents' tendency to avoid argumentation in the medical consultation context was directly associated with a preference to rely on the doctor's knowledge ($r = .16$, n = 183, $p < 0.05$) and to leave all decisions to the medical expert ($r = .24$, n = 183, $p < 0.01$). However, the relationship between the tendency to approach argumentation and the preference to be offered choices was not significant. Finally, one-way analysis of variance with Bonferroni correction indicated a significant, albeit small group difference ($F(4, 178) = 2.84$, $p < 0.05$, $\eta^2 = .06$) between participants identifying themselves with a paternalistic decision-making style ("I prefer to leave all decisions regarding treatment to my doctor", $M = .087$, $SD = 1.40$) and participants identifying with an informed choice decision-making style ("I prefer to make the final selection of my treatment after seriously considering my doctor's opinion", $M = .85$, $SD = 1.31$), implying that participants who preferred to make an informed decision after having discussed and weighed all possible options tended to be more argumentative than patients who preferred a paternalistic approach to medical consultation. These findings confirmed concurrent validity of the scale.

4. DISCUSSION AND CONCLUSION

4.1 Discussion

In light of contemporary prevalent patient-centered models of communication that increasingly encourage argumentative discussion between doctors and patients to enhance the quality of decision-making, this study developed a first, theory-based trait P-ARG to measure patients' willingness to argue with doctors in the context of medical consultations. Factor analysis confirmed that the new scale consists of two uncorrelated, reliable factors (P-ARG*ap* and P-ARG*av*) that mirror the structure and contents of the original argumentativeness scale developed by Infante and Rancer (1982). Similar to existing findings that have been associated with the original scale, men on average displayed higher patient argumentativeness than females. As predicted, the tendency to avoid argumentation was associated with a preference to leave all treatment-related decisions to medical professionals. Moreover, participants who identified themselves with a paternalistic approach to healthcare communication scored significantly lower on patient argumentativeness than those who indicated a preference for a shared rational deliberative patient choice Sandman & Munthe, 2010). This is particularly notable, because the latter model most closely resembles a resolution-oriented, argumentative ideal model of doctor-patient interaction – retaining patients' autonomy. Together, these findings provide evidence that P-ARG is a theoretically grounded scale with convergent and concurrent validity.

A validated measurement instrument for patient argumentativeness has considerable value for researchers of argumentation and health communication alike. Development of such measure fits within the trend to study argumentative discourse both contextually and empirically, using both qualitative and quantitative methods (Bigi, 2010; 2012; Labrie, 2012; Labrie & Schulz; 2014b; 2014c; 2015a; 2015b; Pilgram, 2010; 2012; Schulz & Rubinelli, 2008; Snoeck Henkemans & Wagemans, 2012; Rubinelli & Schulz, 2006). Providing quantitative information about patients' predisposition to engage in argumentative discussions, the scale can be used to further our understanding of the argumentative activity type as well as to provide explanatory power to argumentative analyses that focus on patients' (potential) argumentative moves. Whereas studies focusing on argumentation in medical consultation to date have predominantly focused on the doctor's contributions to the argumentative discussion, the P-ARG scale places an emphasis on the patient's discursive role in doctor-patient interactions. In light of patient-centered models of care, shifting the focus towards the patient not only seems to be a logical, but also a necessary step forward in the study of medical argumentation.

Health communication researchers can use the P-ARG scale to further explore the importance and impact of argumentative discourse in medical practice, such as for consultation outcomes. Several studies have advocated doctors' use of argumentation in support of their medical recommendations in order to enhance their credibility, facilitate participatory decision-making, secure patients' agreement and intended adherence, and to increase their satisfaction (Labrie & Schulz, 2014a; 2014b; 2015a; 2015b; Salmon, 2015). Intervention studies focusing on doctors' use of argumentation should be combined with an assessment of patients' argumentativeness in order to allow for tailored communication and decision-making, ensuring "argumentative fit" of doctors' messages with respective patient preferences. That is, future studies should investigate the extent to which argumentative patients achieve better consultation outcomes when paired with doctors who provide arguments to support their recommendations in a way that allows their patients to actively engage in a critical discussion concerning treatment. Moreover, the P-ARG scale should be used to explore and explain why some patients engage more actively in treatment decision-making discussions than others.

The present study evidently also has some limitations and leaves opportunities for further elaboration. First, despite the fact that Infante and Rancer's (1982) scale was carefully revised to account for earlier criticisms and to adapt the scale items to the context of medical consultations, five items had to be removed from the pool due to insufficient factor loadings. Possibly, these items should have been revised even more extensively to fit the particular setting. Focus group studies and further testing could shed light on the suitability and comprehensibility of these scale items with the objective to explore their potential roles in the P-ARG scale. Furthermore, the scale respondents were relatively young, highly educated, and healthy. This may have affected the study results. Future studies need to retest this scale on a more heterogeneous sample of actual patients.

While the present study investigated the scale's concurrent and convergent validity, future studies need to extend this initial assessment. Confirmatory studies should be employed to evaluate the concurrent and convergent validity of the P-ARG scale in association with measures of general argumentativeness, patient advocacy, patient autonomy, and patient empowerment. For divergent and predictive validity tests, the P-ARG scale could be associated with measures of verbal aggressiveness and intention to argue. Such investigations would be important to extend and strengthen the findings of this current study, and to identify concrete steps for further refinement of the scale.

Lastly, a question that inherently underlies the present study is whether patient argumentativeness should be conceptualized as a stable trait – a predisposition of character. It could be argued that such conceptualization – which was based on Infante and Rancer (1982) –

opposes a view of argumentativeness as a teachable skill and, thereby, disregards the possibility of improving one's capacity and willingness to argue over time through learning. Moreover, it seems to overlook situational influences, such as the patient's general health literacy levels, knowledge of a particular health condition, and mental and physical fitness at the time of the discussion. These factors can all impact the patient's capacity and willingness to enter a discussion with the doctor at a given moment. However, it should be emphasized that views of patient argumentativeness as a stable trait and as a learned skill are not mutually exclusive. While the definition of patient argumentativeness as a character trait implies that one's predisposition to argue is relatively stable, it does not completely rule out the effect of learning or contextual influences. Rather, one's patient argumentativeness should be seen as a baseline value that can increase or diminish under particular circumstances or over time.

4.2 Conclusion

The present study forms a first and important step towards a validated, theory-driven instrument that can be used to measure patients' argumentativeness. The development of such instrument fits within the increasing focus on the role of argumentative discourse within the context of patient-centred medical interactions (Labrie & Schulz, 2014a) and provides health communication researchers as well as argumentation scholars with a scale that has the potential to advance the joint field of medical argumentation.

4.3 Practice implications

Upon further needed validity assessments, the P-ARG scale can be applied by practitioners, too, to assess a patient's general tendency and preference to engage in treatment decision-making discussions. Such an implementation could provide useful information for doctors to adopt their communication to each patient's needs. At that point, the medical context no longer forms a mere context in which one can study the complex workings of argumentative discourse – but argumentation becomes a means through which better health outcomes can be achieved.

REFERENCES

Bigi, S. (2010, July). Institutional constraints on the (un) sound use of the argument from expert opinion in the medical context. In F. H. van Eemeren, B. Garssen, D. Godden, & G. Mitchell (Eds.), *Proceedings of the 7th Conference of the International Society for the Study of Argumentation*. Amsterdam: Rozenberg.

Bigi, S. (2012). Evaluating argumentative moves in medical consultations. *Journal of Argumentation in Context, 1*(1), 51-65.

Charles, C., Gafni, A., & Whelan, T. (1997). Shared decision-making in the medical encounter: What does it mean? (or it takes at least two to tango). *Social Science & Medicine, 44*(5), 681-692.

Charles, C., Gafni, A., & Whelan, T. (1999). Decision-making in the physician-patient encounter: Revisiting the shared treatment decision-making model. *Social Science & Medicine, 49*(5), 651-661.

Degner, L. F., & Sloan, J. A. (1992). Decision making during serious illness: What role do patients really want to play? *Journal of Clinical Epidemiology, 45*(9), 941-950.

DeWine, S., Nicotera, A. M., & Parry, D. (1991). Argumentativeness and aggressiveness: The flip side of gentle persuasion. *Management Communication Quarterly, 4*(3), 386-411.

Dowling, R. E., & Flint, L. J. (1990). The argumentativeness scale: Problems and promise. *Communication Studies, 41*(2), 183-198.

Eemeren, F. H. van (2010). *Strategic maneuvering in argumentative discourse: Extending the pragma-dialectical theory of argumentation*. Amsterdam: John Benjamins.

Eemeren, F. H. van, & Grootendorst, R. (1984). *Speech acts in argumentative discussions: A theoretical model for the analysis of discussions directed towards solving conflicts of opinion*. Dordrecht: Foris.

Eemeren, F. H. van, & Grootendorst, R. (1992). *Argumentation, communication, and fallacies: A pragma-dialectical perspective*. Mahwah, NJ: Lawrence Erlbaum.

Eemeren, F. H. van, & Grootendorst, R. (2004). *A systematic theory of argumentation: The pragma-dialectical approach*. Cambridge: Cambridge University Press.

Frosch, D. L., & Kaplan, R. M. (1999). Shared decision making in clinical medicine: past research and future directions. *American Journal of Preventive Medicine, 17*(4), 285-294.

Infante, D. A. (1982). The argumentative student in the speech communication classroom: An investigation and implications. *Communication Education, 31*(2), 141-148.

Infante, D. A., & Rancer, A. S. (1982). A conceptualization and measure of argumentativeness. *Journal of Personality Assessment, 46*(1), 72-80.

Labrie, N. H. M. (2012). Strategic maneuvering in treatment decision-making discussions: Two cases in point. *Argumentation, 26*(2), 171-199.

Labrie, N. H. M., & Schulz, P. J. (2014a). Does argumentation matter? A systematic literature review on the role of argumentation in doctor–patient communication. *Health Communication, 29*(10), 996-1008.

Labrie, N. H. M., & Schulz, P. J. (2014b). The effects of general practitioners' use of argumentation to support their treatment advice: Results of an experimental study using video-vignettes. *Health Communication* (Online First). DOI:10.1080/10410236.2014.909276

Labrie, N. H. M., & Schulz, P. J. (2014c). Quantifying doctors' argumentation in general practice consultation through content analysis: Measurement development and preliminary results. *Argumentation, 29*(1), 33-55.

Labrie, N. H. M., & Schulz, P. J. (2015a). Exploring the relationships between participatory decision-making, visit duration, and general practitioners' provision of argumentation to support their medical advice: Results from a content analysis. *Patient Education and Counseling, 98*(5), 572-577.

Labrie, N. H. M., & Schulz, P. J. (2015b). The effects of reasoned shared decision-making on consultation outcomes: Results of a randomized controlled experiment among a student population. *Studies in Communication Sciences 15*(2), *182-189.*

Levinson, W., Kao, A., Kuby, A., Thisted, R. A. (2005). Not all patients want to participate in decision making. *Journal of General Internal Medicine, 20*(6), 531-535.

Pilgram, R. (2010, July). A doctor's argumentation by authority as a strategic manoeuvre. In F. H. van Eemeren, B. Garssen, D. Godden, & G. Mitchell, *Proceedings of the 7th Conference of the International Society for the Study of Argumentation.* Amsterdam: Rozenberg.

Pilgram, R. (2012). Reasonableness of a doctor's argument by authority: A pragma-dialectical analysis of the specific soundness conditions. *Journal of Argumentation in Context, 1*(1), 33-50.

Rubinelli, S., & Schulz, P. J. (2006). "Let me tell you why!". When argumentation in doctor–patient interaction makes a difference. *Argumentation, 20*(3), 353-375.

Salmon, P. (2015). Argumentation and persuasion in patient-centred communication. *Patient Education and Counseling, 98*(5), 543-544.

Sandman, L., & Munthe, C. (2010). Shared decision-making, paternalism and patient choice. *Health Care Analysis, 18*(1), 60-84.

Schulz, P. J., & Rubinelli, S. (2008). Arguing 'for' the patient: Informed consent and strategic maneuvering in doctor–patient interaction. *Argumentation, 22*(3), 423-432.

Snoeck Henkemans, A. F., & Wagemans, J. (2012, June). The reasonableness of argumentation from expert opinion in medical discussions: institutional safeguards for the quality of shared decision-making. In J. Goodwin (Ed.), *Between Scientists & Citizens. Proceedings of a Conference at Iowa State University.* Ames, IA: Great Plains Society for the Study of Argumentation.

40

Temporality in Rhetorical Argumentation

ILON LAUER
Western Illinois University, USA
mi-lauer@wiu.edu

This paper advocates a robust conception of rhetorical argumentation that will more elegantly clarify both the content and context of rhetorical argumentation. I propose defining rhetorical argumentation as public argument presented during public time. The concept, public time, denotes arguments that are presented in time as opposed to outside of time. The notion of public argument denotes that the argumentation addresses issues of public importance.

KEYWORDS: demonstration, Elizabeth Warren, Perelman, Public Time, rhetorical argumentation, temporality

1. INTRODUCTION

Even though Aristotle clearly specified rhetoric as the counterpart to dialectic, his corpus contains a more elaborate definition of dialectical approaches to reasoning than rhetorical ones. Over the past decade, a flourishing in scholarship detailing the scope and content of rhetorical argumentation (Bermejo-Luque, 2010; Kock, 2013; 2008, Tindale, 2013, 2004) has coincided with a broader reappraisal of the value of rhetorical topics (Braet, 2005; Crosswhite, 2008; Gutierrez, 2012; Rigotti, 2009; Rubinelli, 2009; Zompetti, 2006) and study of the argumentative capacity of rhetorical figures (Oswald & Rihs, 2013; Fahnestock, 2011, 1999; Tindale, 2004). Collectively, this scholarship has clarified the scope, form and nature of rhetorical argument, drawing attention to the need to define rhetorical argumentation with enough clarity to differentiate it from formal and informal argumentation processes and to facilitate the teaching and study of the unique social dynamics that emerge when audiences, publics, rhetors, and arguments interact.

Such an endeavor is complicated by competing impulses. It seems possible to systematically detail some features of rhetorical

D. Mohammed & M. Lewiński (eds.) (2016). *Argumentation and Reasoned Action: Proceedings of the 1st European Conference on Argumentation, Lisbon, 2015. Vol. II, 671-681.* London: College Publications.

argumentation if we accept the broadly shared notion that rhetorical argumentation "as a process" (Tindale, 2004, p. 4), but attempts to codify guidelines and rules for understanding this process reduce it to a finite set generalizable and teachable principles, all too often casting it as a public variant of dialectic. Though not problematic, in and of itself, we should be mindful of Perelman and Olbrechts-Tyteca's caution that generalizing from particular circumstances "is, in some ways, a technique of avoiding time and a technique of rendering norms incompatible." (Perelman & Olbrechts-Tyteca, 1969, p. 329). As I hope to demonstrate, avoidance of time inevitably hampers efforts to understand the type of responses to specific controversies that typify rhetorical argumentation. Because the study of rhetorical argumentation demands an understanding of its temporal circumstances, argument cannot be reduced to a particular set of reasoning schemes—language distinguishing between probabilistic and absolute reasoning, for instance.

This paper advances a conception of rhetorical argumentation as public argument presented during public time. Two subsets of this definition deserve clarification: First, "public time" denotes the discourse that frames public arguments in relation to a temporal state. Such framing, whether implicit or explicit is a precondition to public discourse; second, "public argument" denotes reasoning pertaining to issues of public concern. This is a quasi-topical condition; determining whether argumentative discourse fulfills this condition requires consideration of how the discourse manifests its public importance *sui generis*. Including public time and public argument into a definition of rhetorical argument elegantly encapsulates the indissoluble link between content and context in rhetorical argumentation.

An appreciation for the temporal dimensions of public argument can redress some of the problems arising when limiting rhetorical argumentation's concern to a finite subject domain, namely civic argumentation. Kock's designation of rhetorical argumentation's concern with public, or at least social, questions shifts concern away from its propositional truths/falsities and towards consideration of its justification for better/worse choices. Kock advocates a "domain-based view of rhetorical argumentation, which sees it as centrally concerned with choice of action, rather than with any issue at all..." (Kock, 2009, p. 65). Emphasizing rhetoric's civic origins, he defines rhetorical argumentation as civic argumentation, a definition that emphasizes rhetorical argumentation's content area and locale to the exclusion of any particular method, nullifying in effect discussion of its "aims and means." But foreclosing means and aims analysis is highly problematic to the study of rhetorical argumentation, because there are so many distinctive features of the type of argumentation that arises in public deliberation, features that have less to do with the subject matter of rhetorical argumentation than with its spatial and temporal contingencies.

Among its many distinguishing features—its use of non-standard reasoning schemes, audience-oriented focus, distinct pedagogic tradition, and linkages between rhetor and audience—rhetorical argumentation is signaled by its aggregative and imperfect features. Its aggregative elements include seemingly repetitious and overlapping enunciations to create, reinforce, and modify appeals to multiple audiences with diverse and varied background beliefs, values, understandings, perspectives, and agreements. The same conditions that cultivate redundancy in rhetorical argumentation also ensure its enduring imperfect state. In a fluctuating context, individual and group adherence is constantly subject to revision and modification. The issues, positions, and arguments at any particular point in time are always subject to repetition, reconceptualization, and refutation—processes that foster a spectrum of responses ranging from stronger adherence to rejection of an earlier commitment.

2. ARGUMENTATION AND TEMPORALITY

Whether addressing the past, present or future, rhetorical argumentation emerges from specific contexts and is framed in regards to particular points in time. Before delving into the specific temporal conditions of rhetorical argumentation, it would be helpful to address the fundamental connections between argumentation and time. These conditions, though not sufficient, are necessary components of any definition of rhetorical argumentation and also supply some sense of its parameters. Chaim Perelman and Lucy Olbrechts-Tyteca have detailed some of the temporal dynamics attendant to the process of producing, conceptualizing, and responding to argumentation. Argumentation is constantly subject to restatement, revision, and reconsideration and as it proceeds, audience adherence is constantly modulating. Demands of time influence the arguer's determination of what positions can be reasonably addressed and offered and, accordingly, it influences argument prioritization. Quite often, argumentative discourse invokes time explicitly, expressing it in the form of a particular commonplace. Perelman and Olbrechts-Tyteca have classified several commonplaces with strongly temporal elements, for instance, the rule of justice as well as arguments of waste and unlimited development.

Perelman and Olbrechts-Tyteca contrast the atemporal conditions of demonstration with the temporal flux of argumentation.

They refer to arguments of demonstration as timeless arguments, because such arguments are unconstrained by an evaluative time-frame. But the type of argumentation that people use to engage their world emerges from and addresses specific time periods. The qualities they attribute to argumentation "hesitation, doubt, [and] freedom of choice" are necessitated by the contingency of time; time then is a critical defining feature of argumentation, the feature "that best allows us to distinguish argumentation from demonstration." (Bolduc & Frank, 2010, p. 316) So inescapable is the influence of time upon argumentation that Perelman and Olbrechts-Tyteca stress that: "...there is hardly an argument that does not receive its significance or its force from the place that it occupies, from the moment that it is initiated." (Bolduc & Frank, 2010, p. 326).

Temporality influences both the content and conduct of argumentation because premises and conclusions that occur in time are always subject to change. As time progresses, conditions change, new reasons emerge, assumptions are revised, and information is revealed, forgotten, and modified—actions all contributing to the impermanence and redundancies of argumentation. Perelman and Olbrechts-Tyteca describe this flux as a "perpetually changeable context" and they invoke the notion of an always already present, a condition in which the status of arguments are constantly revised and renewed:

> We know the effects of argumentation are not definitive, that adherence is changeable with time, that it has a tendency to weaken, although on occasion, it might receive unexpected reinforcement. In any case, while memory suffices to retain a demonstration, argumentation must be lived anew. At most, adherence gained through argument will be provisional and subject to revision. (Bolduc & Frank, 2010, p. 318)

The need for reinforcement, the challenge of revision, the contest between strengthening and weakening premises and assumptions are characteristic of argumentation.

Temporal conditions affect the argument process in two particular ways: first, temporal limitations influence the prioritization of certain arguments—at times, catalyzing tremendous acts of inventive genius even to the favor of the "weaker" position. Second, the temporal process both reinforces and destabilizes adherence to various positions. Both the selection and judgment of argument are prompted and influenced by time and time's constraints. As Perelman and Olbrechts-Tyteca elaborate, often "arguments of timeliness are considered, calling for a choice: there is an occasion to seize or to miss, circumstances that will never be reproduced, into which the sought-after decision must be introduced." (Bolduc & Frank, 2010, p. 320)

Because arguments are presented during particular times and in particular places they use limited premises to support contingent conclusions. Even when such premises reflect the best of what is known in a particular time, they are still subject to revision. Emphasizing the strength of this connection between the conditions of argumentation and its manifestation, Perelman and Olbrechts-Tyteca stress that "...there is hardly an argument that does not receive its significance or its force from the place that it occupies, from the moment that it is initiated." (Bolduc & Frank, 2010, p. 326). Temporal interruptions arising in the form of judgment become critical points for stabilizing argument forms. Perelman and Olbrechts-Tyteca note that "Even if they do not alter its changing character and its insertion into a full time perpetually bringing in novelty, the factors influencing the formation of argument—including temporal interruptions of decision making, a focus on the issues to be judged, a pausing on the object of judgment, and precedents that have become models—all structure the argument." (Bolduc & Frank, 2010, p. 327)

Perelman and Olbrechts-Tyteca resuscitated the study of dialectical and rhetorical approaches to argumentation. But, although their work is of particular utility to the study of dialectic, particular elements are more germane to the study of rhetoric. The broad overlap between the New Rhetoric Project's (NRP) approach to argumentation and Aristotle's dialectic has resulted in the NRP being studied more in depth by informal logicians endeavoring to establish a framework for argumentation that takes a different approach from the "traditional" approach to instruction in logic. Not sharing any particular commitment to rhetorical scholarship other than referencing relevant concepts as they are presented in rhetorical texts, this approach to argumentation scholarship has steadily erased many of the critical features of rhetoric, namely its connections to language, audiences, and contingent time-bound circumstances. While not exclusive of these features, contemporary approaches to dialectic have rendered these features less relevant to the study of argumentation.

3. PUBLIC TIME AND RHETORICAL ARGUMENTATION

The connection between the temporal and topical domains of rhetorical argumentation are so tightly bound that the argumentation regarding one usually implicates the other. Consequently, the temporal domain of rhetorical argumentation is often contested during the argumentation process. The concept of public time clarifies the connections between such contestation and temporality. Thomas Goodnight defines public time as a discursive product, an implicitly or explicitly invoked temporal state that cannot be reduced to "a mere taxonomic scheme." (Goodnight, 1987, p. 429) Instead, understanding of particular connections between public time and given controversy lies in the discourse and deliberative norms which create both possibilities and constraints. In creating possibilities and constraints, public time, along with public space, is a necessary antecedent to public discourse:

> In sum, the public may be understood as that domain of discursive practices open to those whose opinions count in contesting a decision of consequence to a community. The public sphere is made available by public space, the locations of common meetings and discussion where the discourse of the community is held open in principle to all who have a say in a matter of common urgency, and by public time, the temporal structures and processes that uphold traditions of collective decision making, alternatively disrupting and preserving private and social temporal patterns. (Goodnight, 1987, p. 431)

Public time enables public deliberation. Roger Stahl analogizes public time to a computer's operating system, underscoring time's profound influence upon the conduct of public deliberation. Supplying an underlying cognitive design for public deliberation, public time structures deliberative conduct by "...circumscribing boundaries and openings for discursive action." (Stahl, 2009, p. 74) In his own study of the distortions in public time that hampered deliberation in the lead-up to the Iraq conflict, Stahl directs critical attention to the way public time creates the crises, perceptions, senses and assumptions that motivate a host of discursive responses ranging from recalcitrance to accommodation.

So how does public time influence rhetorical argumentation? First, public time is an ever-changing and plastic concept; even the identified absence of it can be identified as a crisis, supplying the basis for new policy arguments and appeals. Second public time is the existential precondition to rhetorical argument. Just as an operating system needs to load before the computer can operate, the operation of time is a precondition for deliberation. Third, time and rhetorical argumentation coevolve and have a mutual dependence. Shifts in the way time works alter the way the argumentation operates and vice versa, argumentation can modify the workings of time. Temporal constraints and possibilities establish what may be labeled the "para-normative" elements that operate during deliberation (e.g. order, inclusivity, and exclusivity). Fourth, time strongly governs topical possibility; just as certain programs can only be run on certain operating systems, the limits of what can and cannot be argued—the types of potential proofs, are imposed by the temporal conditions of the argument. Finally, to understand the relationship between public argumentation and public time, we should look towards Perelman's discussion of hierarchical forms of argument; shifts in the hierarchy of issues can alter the conduct of public time. Time constrains choice and choice influences the conduct of rhetorical argumentation. Accordingly, we should be more attuned to the temporal constraints that influence deliberation.

4. OPERATING WITH AND IN TIME

In two speeches presented on the Senate floor by Massachusetts Senator Elizabeth Warren, invocations of different temporal conditions enable argumentation and strongly determine its form and conduct. Reviewing Warren's two speeches; one where no time remains for deliberation and one in which time for deliberation was never granted, we can see the ever-present influence of time.

Warren's remarks in December 2014, offer an interesting provocation; they begin from her rhetorical disadvantages to warrant her position, an argumentative move bound to her temporal framing. She infuses her floor speech with temporal deixis markers from the very beginning, drawing attention to her position "back on the floor" to oppose the "last minute" provision to the appropriations bill. Warren identifies her speech as a culminating activity, completing her advocacy "Wednesday" when she spoke with her fellow Democrats about the bill and "Thursday," when she engaged the Republicans on this issue. Complementing her temporal deixis markers are spatial references to her physical position on the Senate floor as a non-partisan opponent of Citibank: "Today I'm coming to the floor not to talk about Democrats or Republicans, but to talk about a third group that also wields tremendous power in Washington: Citigroup." Warren's references to her physical and temporal circumstances emphasize her standing as a vigorous and active opponent against Citibank.

Warren's speech falls outside any practical deliberative time-frame. She is out of time and identifies this as the justification for her topical choice to address Citi's power and its relation to democracy. Augmenting her standing as plutocracy's deadly foe, her main comments implicitly point to the futility of her advocacy, appearing to undercut her main position. The more she is able to demonstrate Citigroup's iron grip has on the political process, the more she casts doubt on her and Congress's ability to resist. As Warren inveighs against the inordinate level of power wielded by Citigroup, she concedes the immediate battle, but uses its loss to advocate a broader and more sustained set of measures against Citigroup. Warren declares that Citi's "grip over economic policymaking in the executive branch is unprecedented" and then recites a litany of current executive-branch appointments that secured Citi's ability to capture adverse regulations. She amplifies this discussion by accounting for the extensive lobbying Citigroup funds and the immediate benefits of these efforts. Resorting to an apostrophe, she addresses Citi directly. She argues: "You know, there's a lot of talk lately about how Dodd-Frank Act isn't perfect. There's a lot of talk coming from Citigroup about how Dodd-Frank Act isn't perfect. So let me say this to anyone who is listening at Citi: I agree with you: Dodd-Frank isn't perfect: It should have broken you into pieces." (Warren, 2014, np.) But it is doubtful that this direct address casts Citi as her argumentative adversary. Earlier Warren noted that her direct queries to Citigroup were ignored, so there is little reason to believe she is intending to elicit a response in the form of dialectical opposition from Citi.

Warren's address invokes time and attempts to initiate deliberation regarding the suitability of too big to fail as a policy governing the banks. Her conclusion contains several additional temporal deictic markers. Warren places herself in the position of the public who are "now" observing the passing of the provision designed to enable a future bail-out for Citi, should it be needed. Her mantra-like repetition of the phrase "enough is enough" to construct the present as a crisis moment to motivate a large-scale roll back of Citigroup's power: "Enough is enough. Enough is enough with Wall Street insiders getting key position after key position and the kind of cronyism that we have seen in the Executive Branch. Enough is enough with Citigroup passing eleventh-hour deregulatory provisions that nobody takes ownership over but everybody will come to regret. Enough is enough." (Warren, 2014, np.)

Warren's overt references to time depict Citibank as a threat to democracy. Her argumentation is both redundant and imperfect. It is imperfect in that it lacks a clear course of action to take against the Citi corporation and it is redundant in her amplification of Citigroup's control over the political process. If anything, she concedes that Citicorp has more power than she is able to deal with. In addition to her redundancy and her imperfection, her rhetoric demonstrates the move to invoke or summon a crisis moment, using deictic markers to amplify the sense of crisis and transform appropriations debate into a discussion of Citi's threat to America's democratic institutions.

As Warren advocates legislation enabling debate over granting Obama "fast-track" authority to negotiate the Trans-Pacific Partnership (TPP) she was at an even greater rhetorical disadvantage. Because the granting of formal "fast-track" authority to negotiate the TPP agreement in secret essentially authorized the executive to negotiate on behalf of congress, some members of congress were concerned about approving an agreement that they would not be able to see before it was ratified. Lacking a text of the TPP trade agreement—none was made public—Warren simply lacked anything to refute. She chose to argue for the necessity of time, and her remarks on May 21, 2015 dissociated the empty deliberation of second hand knowledge with a more genuine deliberation that could take place provided enough information regarding the TPP. Warren's office had released a document, "Broken Promises," in the lead up to the congressional deliberation, but this paper was limited to documenting the record of false and misleading claims associated with past trade agreements. Because it could only account for past agreements, the proponents of the TPP could always claim that this agreement was different.

We can see how temporality guided topical choice and argument position as Warren argued that the informational deficit warranted allocation of time to deliberate the merits of the TPP. She distinguished the more reliable understanding derived from seeing the agreement with the more limited knowledge of "hearing about" an agreement. She pointed out that "the public has heard a lot" about the TPP, reiterating the general arguments offered in favor of the agreement, but she then noted that no member of the public had actually seen the deal. In essence, she conceded that opponents of the deal were unable to offer meaningful refutation because they were at an informational disadvantage. Acknowledging this disadvantage did not prevent Warren from offering both a qualified rejection of granting fast-track authority, and finally and a proposal to enable longer deliberation over the TPP (60 days). In other words, Warren identified her weaker position as a justification for the allocation of more public time. Time becomes the crux of her argument and securing more of it the goal of her advocacy:

> That's why I have introduced a simple bill with my friend from West Virginia, Senator Manchin. This bill would require the President to publicly released the scrubbed bracketed text of a trade deal at least 60 days before Congress votes on any fast-track for that deal. That would give the public, the experts, the press an opportunity to review the deal. It would allow for some honest public debate, and it would give Congress a chance to actually step in and block any special deals and give-aways that are being proposed as part of this trade deal before

> Congress decides whether the grease the skids to make the deal the law. (Warren, 2015, np)

Operating at a rhetorical disadvantage, Warren's oratory pursues as its object the norms of public deliberation and the necessity of public time. In essence, she seeks to reboot a frozen deliberative operating system.

5. CONCLUSION

Warren's attempts to advance a critique of the administration's economic policies demonstrated some of the fundamental ways time intersects with rhetorical argumentation. Her advocacy invokes time as a flexible condition necessary for rhetorical argumentation. The allocation of public time for rhetorical argumentation is a normative expectation. The absence of this time signals the absence of deliberation and her arguments promoting deliberation begin with the crafting of time for debate. At the same time the absence of time has constrained her ability to offer a more robust refutation of the bills she opposes, this absence catalyzes a range of arguments advancing her position. In sum, her advocacy is highly suggestive of the ways argument scholars can begin to consider the workings of time in rhetorical argumentation.

This essay has warranted more consideration and understanding of temporality in our models and definitions of rhetorical argument and it has advanced the notion of public time as a useful tool in such an endeavor. Identifying temporality as a precondition to deliberation does not define and explain the essential features that constitute public time, but it does establish deliberation as a consequence of public time, a sign of its existence.

REFERENCES

Bermejo-Luque, L. (2010). Intrinsic versus instrumental values of argumentation: The rhetorical dimension of argumentation. *Argumentation, 24*, 453-474.

Bolduc, M. K., & Frank, D. A. (2010). Chaïm Perelman and Lucie Olbrechts-Tyteca's "On Temporality as a Characteristic of Argumentation": Commentary and Translation. *Philosophy and Rhetoric, 43*(4), 308-336.

Braet, A.C. (2005). The common topic in Aristotle's Rhetoric: Precursor of the argumentation scheme. *Argumentation, 19*, 65-83.

Crosswhite, J. (2008). Awakening the *Topoi*: Sources of invention in the New Rhetoric's argument model. *Argumentation and Advocacy, 44*, 169-184.

Fahnestock, J. (1999). *Rhetorical figures in science*. Oxford: UP.

Fahnestock, J. (2011). *Rhetorical style: The uses of language in persuasion.* Oxford: UP.

Gambra Gutiérrez, J. M. (2012). The topoi from the greater, the lesser and the same degree: An essay on the σύνκρισις in Aristotle's *Topics. Argumentation, 26*, 413-437.

Goodnight, G. T. (1987). Public Discourse. *Critical Studies in Mass Communication, 4*, 428-432.

Kock, C. (2009). Choice is not true or false: The domain of rhetorical argumentation. *Argumentation, 23*, 61-80.

Kock, C. (2013). Defining rhetorical argumentation. *Philosophy and Rhetoric, 46*, 437-64.

Oswald, S., & Rihs, A. (2014). Metaphor as argument: Rhetorical and epistemic advantages of extended metaphors. *Argumentation, 28*, 133-159.

Perelman, C., & Olbrechts-Tyteca L. (1969). *The New Rhetoric: A Treatise on Argumentation* Trans. John Wilkinson & Purcell Weaver, Notre Dame: University of Notre Dame Press.

Rigotti, E (2009). Whether and how classical topics can be revived within contemporary argumentation theory. In F. H. van Eemeren & B. Garssen (Eds.), *Pondering on Problems of Argumentation: Twenty Essays on Theoretical Issues* (pp. 157-180). Dordrecht: Springer.

Rubinelli, S. (2009). *Ars Topica: The classical technique of constructing arguments from Aristotle to Cicero.* Dordrecht: Springer.

Stahl, R. (2009). Why we 'Support the Troops': Rhetorical evolutions. *Rhetoric and Public Affairs, 12*, 533-570.

Tindale, C. W. (2004). *Rhetorical argumentation.* Thousand Oaks: Sage.

Tindale, C. W. (2013). Rhetorical argumentation and the nature of audience: Toward an understanding of audience-issues in argumentation. *Philosophy and Rhetoric, 46*, 508-532.

Warren, E. *Remarks.* Washington Post, December 12, 2014. http://www.washingtonpost.com/news/wonkblog/wp/2014/12/1enough-is-enough-elizabeth-warrens-fiery-attack-comes-after-congress-weakens-wall-streetregulations/ Downloaded 9/15/2015.

Warren, E. Remarks, May 21, 2015. *https://www.youtube.com/watch?v=SnwIeoHe9UE*

Zompetti, J. P. (2006). The value of topoi. *Argumentation, 20*, 15-28.

41

Is Reasoning Universal? Perspectives from India

KEITH LLOYD
Kent State University at Stark, USA
kslloyd@kent.edu

Ancient India developed its own method of reasoned argument, *Nyāya,* which though comparable in influence to Aristotle, differs in structure, emphasis, motivation, and goals from much of Western reasoning. *Nyāya* joins a claim and reason with a common analogy, while stressing *vada*, positive discussion, above *jalpa*, arguing to win, and *vitanda*, arguing to disprove. The presentation explores the implications of *Nyāya* for a cross-cultural understanding of human reasoning.

KEYWORDS: analogy, comparative rhetoric, India, *Nyāya*, *vāda*, rhetoric, reasoning

1. INTRODUCTION

Though the physical gestures and postures of verbal arguments may be readable across cultures, the careful study of culturally-influenced habits of reasoning associated with processes of persuasion and argumentation, as well as each culture's possible implied or articulated theories of reasoning, has only recently begun. Our understanding of the "rhetorical" practices around the world proceeds culture by culture, language group by language group. This essay enhances our understanding of human reasoning by exploring implications found in an ancient India(n) tradition of argumentation and debate called *Nyāya.* Such investigations provide engaging possibilities toward global perspectives on the ways human beings use reason to persuade.

In this context, the term "reasoning" refers mental processes used by individuals or communities to hypothesize, weigh, and establish contestable perspectives. These processes may include assumptions as well as concepts intentionally applied by members of a culture or subculture. "Rhetoric" is taken in the broadest sense as the practices and/or techniques each culture brings to private and public reasoning

D. Mohammed & M. Lewiński (eds.) (2016). *Argumentation and Reasoned Action: Proceedings of the 1st European Conference on Argumentation, Lisbon, 2015. Vol. II, 683-693.* London: College Publications.

in order to bring others to acknowledge, question, or adopt points of view established through the processes of reasoning.

2. *NYĀYA'S* BACKGROUND AND CONTEXT

Nyāya, though in India comparable in influence to Aristotle in the West, differs in structure, emphasis, motivation and goals from much of Western reasoning. It is recognized as one of six orthodox schools of Hindu thought, and its methods were adopted by Buddhists, Jains, and Muslims as well; its influence is still apparent in Indian culture today. *Nyāya* is both a method (*techne*) and philosophy of logical reasoning, and its principles were codified in the *Nyāyasūtra*, by *Akṣapāda Gotoma* about 150 CE. Its roots trace from about 550 BCE in the teachings of *Medhatithi Gotama*, traditionally ascribed at the author of the *Nyāyasūtra*.

Nyāya reasoning joins a claim and reason with a common analogy and stresses *vāda,* "fruit bearing" discussion, rather than *jalpa,* arguing to win, or *vitaṇḍā,* arguing only to disprove.

Term	Literal Translation	Nyāya Translation
Vāda	"kindly speak" or "please speak"	fruitful discussion
Jalpa	"mad talks," or "playful speech"	wrangling (arguing to win)
Vitaṇḍā	counterarguments	cavil (arguing against)

(Literal Translations Vedabase.net; English translations Vidyābhūṣaṇa)

Nyāya favors *vāda* because it leads interlocutors to seek motivation beyond individual fear and desire to focus on positive and creative perspectives. The *Sūtra* details that "Discussion (*vāda*) is the adoption of one of two opposing sides. What is adopted is analyzed in the form of the five members [the Nyāya method, outlined below], and decided by any of the means of right knowledge [*pramāṇa*] while its opposite is assailed by confutation…" (*NS* I. 2. 42). *Vāda* also implies liberation from doubt, misapprehension, and even the cycles of life and rebirth (NS I.1.2.). Clearly the motive for argument differs here from most Western assumptions, which would be more *jalpic*.

The word *Nyāya* translates as "justice" and means "word" or "rule" as in a "plumbline." "In classical Sanskrit, [Amartya Sen] writes, the word 'justice' has two forms: '*niti*', which refers to organisational propriety and behavioural correctness, and '*Nyāya*', which stands for a comprehensive concept of realised justice…" (Majoribanks). Sen uses the term to refer to an approach to reasoning focused on positive outcome for all involved rather than application of abstracted rules. As will be noted later, this perspective is rooted in *Nyāya*'s case-based origins. The term also relates analogically to the Ancient Greek idea of the *logos*, the divine rule of the cosmos, as well as the Indian concepts of *ṛta* (truth/reality) and *dharma* (divine order, destiny). In *Nyāya*, debate and dialogue are used to create a balanced and just society.

Though the terms are not specifically *Nyāya*, as a philosophy it reflects a Hindu Vedic view of language as consisting of four layers or expressions. The surface level of spoken or written language is *Baikharī*, "speech-in-thought." The second level is *Madhyama*, "the abstract-conceptual level where one might work out a math problem, plan a to-do list, design an essay, or repeat a poem." The third level is *Paṣyantī*, from the root "*Pashya,*" "to see," "the sense is more felt than articulated, as if in a glance," similar to what is often called "felt sense." In the fourth level, we "leave human speech behind for the infinite silence and infinite dynamism of *Parā*, the transcendental field, the source of speech" (20-21).

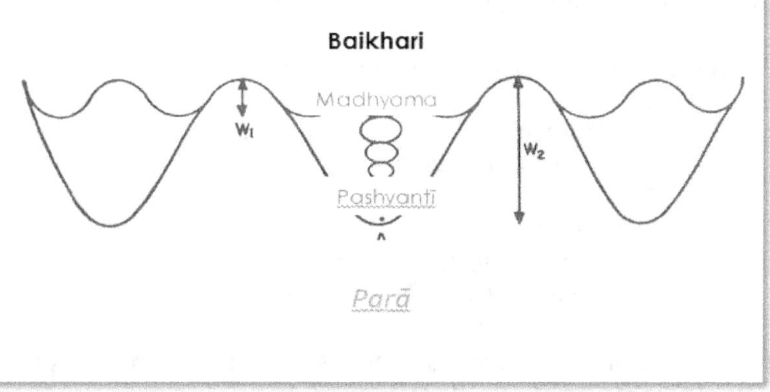

From Anne Melfi's Dissertation, "Understanding Indian Rhetoric on its own terms: Using a Vedic Key toUnlock the Vedic Paradigm" (Georgia State University, 2016)

The process of *Nyāya* reasoning similarly implies a dialogue in which speakers use the spoken *Baikarhari* level to stimulate thought connections in *Madhyama*. The combination of the claim, reason, and analogy lead to *Paṣyantī*, a felt sense. The process ideally ends in *Parā*, the transcendental field, the source of speech, when the interlocutors come to a shared realization that liberates them from delusion.

Before the rhetor speaks, however, she or he begins with *Parā*, with "meditation," and moves through the same levels to speech. For these reasons, *The Nyāyasūtra* identifies the processes of debate with *mokṣa*, liberation from reincarnation's cycles of death and rebirth. As the NS I.1.2 notes, "pain, rebirth, activity, faults, and misapprehension—on the successive annihilation of these in reverse order, there follows release." *Nyāya* reasoning reflects the *Vedic* notion of *mokṣa* adapted from Hindu *Vaiśeṣika* philosophy. The method is then used internally and dialogically to release us from misapprehension and faulty perceptions to open true understanding.

This release, we learn, stems from a combination of meditation, discussion, and study: "there should be a purifying of our soul by abstinence from evil and observance of certain duties as well as by following the spiritual injunctions gleaned from the Yoga institute" (NS IV, II, 456, p. 172). Rhetorical preparation thus encompasses a way of being: "To secure release, it is necessary to study and follow this treatise on knowledge as well as to hold discussions with those learned in that treatise" (NS IV, II, 457, p. 172).

3. WHAT IS *NYĀYA* REASONING?

The *Nyāya* method is taught using the following exemplar:

Hypothesis (*pratijñā*): *The hill (pakṣa) is on fire (sādhya)*
Reason (*hetu*): *Because there is smoke (hetu)*
Example (*dṛṣṭānta*)
Positive: *Like in the hearth*
Negative: *Unlike a lake*
[Discussion—Nirnaya]
Reaffirmation*:* (*upanaya*)*: This is the case*
Conclusion*:* (*nigamana*)*: The hill is indeed on fire.*

There is no general first premise because of a distrust in the provability of general statements—we all *experience* only *specifics*, never universals, which may be inferred, but not proven. *Nyāya* later develops the idea of *vyāpti* ("pervasion"), but it is used to qualify the analogy: "In the case of the hearth, 'Where there is smoke there is fire.'"

The claim, reason, and example are somewhat analogous to Aristotle's enthymeme and example (Lloyd, "Culture" 81-92). Aristotle notes that "rhetorical study, in its strict sense, is concerned with modes of persuasion. Persuasion is clearly a sort of demonstration, since we are most fully persuaded when we consider a thing demonstrated. The orator's demonstration is the *enthymeme*, and this is, in general, the most effective of the modes of persuasion" (*Rhetoric* I.2.1355a 3-9, p 22). He explains that an enthymeme is "a rhetorical syllogism, and example a rhetorical induction. Everyone who effects persuasion through proof does in fact use either enthymemes or examples" (*Rhetoric* I.2. 1356b 4-7 p 26).

Aristotle' s work is not clear as to what exactly enthymemes are, but if they are, as he says, analogous to the syllogism, they are likely three-part interlocking claims and reasons about probable and debatable issues—one or more of which may be omitted because of audience familiarity (*Rhetoric* I.2, 1357a 23, p. 28). By the term example (*paradigma*), Aristotle means arguing a point by applying a series of supportive instances, as when one lists three historical figures that asked for bodyguards and became despots to imply that this will be the case in a similar current situation (see *Rhetoric* I.2 1357b 28-1358a 1 p. 30).

In *Nyāya,* the two part claim and reason structure is similar to an enthymeme, but lacks a third premise, and *dṛṣṭānta* (example) is analogue rather than inductive support for a conclusion. It also includes both positive and negative examples. While to Aristotle an example is a supporting instance of general principle, the Nyāya example is analogical. For instance, the *Caraka Saṃhitā* notes that the example "describes the subject, e.g. hot as 'fire,' stable as 'earth,' etc., or just as the 'sun' is an illuminator so is the text of the Samkhyas." In analogy, two sets of terms are shown to be proportional due to a common property. The *Saṃhitā* similarly offers that just as the sun illuminates the world, the Samkhyas illuminate the reader.

Sun		the Samkhyas
	Illuminates	
The world		the reader

As Jonardon Ganeri puts it, "a perceived association between the symptoms of one case provides a reason for supposing there to be an analogous relation in other, resembling cases" (328; see also Lloyd "Re-Thinking", 372-376). Interestingly, Aristotle also traces rhetoric's origins to case-based reasoning: "Rhetoric's "function is not simply to succeed in persuading, but to discover the means of coming near such success as each particular case will allow" (Rhetoric I. 1. 1355b 10-12, 23; see Lloyd, "Re-Thinking" 376).

And, as in Aristotle's enthymeme, the Nyāya method is understood in India as a type of verbal demonstration. As Vidyabhusana notes, "The *Caraka Saṃhitā*, as far as we know, contains for the first time an exposition of the doctrine of syllogism [the Nyāya method] under the name of *sthāpanā* (demonstration)..." (Vid *History* 42).

The *Saṃhitā* not only details the five-part method, it offers examples, such as the following:

The soul is eternal.
Because it is a non-product.
Just as ether being a non-product is eternal.
The soul being similar to ether is a non-product.
Therefore the soul is eternal.
Caraka Samhita (Vidhabysana *History* 32).

As in the other instances, the *dṛṣṭānta* (ether), supports the claim and reason by way of analogy:

Soul		Ether
	Non-product (eternal)	
Humans		Elements

In sum, the claim, reason and example function analogously to Aristotle's concepts of the enthymeme as *demonstration* and paradigma as *support.*

Nyāyaikas (practitioners of Nyāya philosophy) considered the Nyāya method the perfect form of argument because it expressed the four basic modes of knowing (the *pramāṇa*)—ways by which we comprehend the world.

> *Pratyakṣa* (perception) NS I.1.4: "that knowledge which arises from the contact of the sense with the object"
> *Upamaya* (comparison) NS I.1.6: the knowledge of a thing through its similarity to a thing previously known"
> *Anumāna* (inference) NS I.1.5: "knowledge preceded by perception, and is of three kinds, 'from cause to effect,' (*śesa-vat*), effect to cause (*nānyto-dristam*), and 'commonly seen' (*cha and*)."
> *Śabda* (word) NS I.1.7: "the instructive assertion of a reliable person"

As Bimal Matilal notes, the *pratijñā* (claim) is "helped in some way by śabda *pramāṇa* or verbal testimony." The *hetu* (reason) is a "skeleton form of *anumāna* or inference," the *dṛṣṭānta* (example) is a kind of "perception" [*pratyakṣa*], and the *upanaya* (conclusion) "bears some distant similarity with the implication of *upamaya* or comparison" (23).

The *pramāṇa* are the first of sixteen "categories," concepts the Nyāyaika/rhetor meditates upon and comprehends in order to effectively debate. The second category is *prameya* (objects of valid knowledge), which include "the soul, body, senses, objects of sense, intellect, mind, activity, fault, transmigration, fruit, pain, and release." (*NS* I.1.9).

Once the ways and subjects of knowing are understood, the next categories outline the reasoning process:

saṁśaya (doubt)
prayojana (aim)
dṛṣṭānta (example)
siddhānta (conclusion)
avayava (the Nyāya method)
tarka (hypothetical reasoning)
nirṇaya (settlement),
vāda (discussion)[1]

For Nyāyaikas, this process represents the way we argue with ourselves by testing and eliminating various possible conclusions in order to bring them to the last step, discussion. However, in many recorded dialogues, discussions (*vāda*) begin when one interlocutor expresses *saṁśaya* "doubt," and together speakers begin to establish the *prayojana* (aim). One speaker usually functions as a teacher who offers various *pratijñā, hetu, and dṛṣṭānta* in response to the questions and concerns of the respondent. The conversations lead interlocutors intentionally toward *mokṣa*, liberation from doubt and misapprehension, and ultimately cycles of death and rebirth.

Indian rhetoric, at least in this context, is not persuasion in a traditional sense, i.e. winning someone over to one's point of view. It functions as a type of revelation, and respondents are not so much "convinced of" as "convinced to" look for and find the truth within themselves. Roy M. Perrett calls the end result of Nyāya reasoning as "knowing episodes," which consist of "awareness or experience" that is the "end-product of a perceptual or inferential episode" (320).

3.1 Examples of Nyāya Reasoning in Indian Texts

Both Ancient and modern Indian texts of all sorts apply Nyāya reasoning. Even the most ancient *Upaniṣad* (book of Hindu teachings), *the Brihadāraṇyaka Upaniṣad* (770 BCE), offers interesting examples. In the following excerpt, a speaker named Yajnavalkya responds to his wife Maitreyi's questions about immortality.

> *Na hāsya udgrahaṇāyeva syāt, yato yatas tv ādadīta lavaṇam eva evaṁ vā ara idam mahad bhūtam anantam apāraṁIdam mahad bhūtam anantam apāraṁ vijñāna-ghana eva* (BU 4: 12)

> As a lump of salt thrown in water dissolves and cannot be taken out again, though wherever we taste the water it is salty, even so beloved, the separate self dissolves in the sea of pure consciousness, infinite and immortal. Separateness comes from identifying the Self with the body, which is made up of the elements; when this physical identification dissolves, there can be no more separate self (Translation Easwaran 102).

We can outline the Nyāya patterns of his remarks:

> *Pratijñā: "the separate self dissolves in the sea of pure consciousness, infinite and immortal"*
> *Hetu:* [Because] *"when this physical identification dissolves, there can be no more separate self"*
> *Dṛṣṭānta: "As a lump of salt thrown in water dissolves and cannot be taken out again, though wherever we taste the water it is salty"*

Such early examples prove that Nyāya reasoning was a part of Hindu/Indian reasoning well before it was codified in the Nyāyasūtra.

Nyāya arguments also appear in the *Bhagavad Gita*, which traditionally dates as from the third or fourth millennium BCE, reaching its present from about 400 CE. It is part of a larger epic, the *Mahabarata*, where Krishna, speaks as, unbeknownst to his interlocutor Arjuna at the beginning of the book, an avatar (god in human form) of the God Vishnu. The impetus of the dialogue is a civil war that Arjuna must fight to regain his rightful kingship. He hesitates, however, because he must fight relatives, companions, and former teachers. Krishna points out that the war is just, and that we are all eternal beings and our lives here are temporary—no one will truly die. He encourages him to fight by explaining to Arjuna to act without expectation of either failure or glory:

> *aapuuryamaaNamachalapratishhTha. nsamudramaapaH pravishantiyadvat. htadvatkaamaa yaM pravishanti sarve sa shaantimaap noti na kaamakaamii* (BG 2:70-71)
>
> Even as all waters flow into the ocean, but the ocean never overflows, even so the sage feels desires, but is ever one in his infinite peace.
> For the man who forsakes all desires and abandons all pride of possession and of self reaches the goal of peace supreme.
> (Translation Juan Mascaró)

The Nyāya form is clear in this passage:

> Pratijñā: "the sage feels desires, but is ever one in his infinite peace"
> Hetu: "For the man who forsakes all desires and abandons all pride of possession and of self reaches the goal of peace supreme."
> Dṛṣṭānta: "Even as all waters flow into the ocean, but the ocean never overflows..."

In both examples above, the speaker focuses a series of declarative statements with a rhetorical turn, the application of the Nyāya method, which serves to encapsulate the meaning of the entire section of dialogue. The analogy creates a sensory palpability to what is being said, enabling the shared perceptual experience for both interlocutors.

Even in modern times, Indian speakers and writers employ Nyāya arguments. For instance, Eknath Eswaran writes in his book, *The End of Sorrow*,

> In Sanskrit, the language of the Gita, the underlying Reality of life is called by a simple but very powerful name: *advaita*, 'not'two.' ... There is no division, no fragmentation in life at all; no matter how much we may appear on the surface, the welfare of each one of us is inseparable from the welfare of all others. Even on the level of the body, we know that in cancer the whole organism is eventually destroyed when even a single cell begins to pursue its own course independently of rest. Similarly, the Bhagavad Gita tells us, you and I cannot fulfill ourselves by going our own way. We can find lasting fulfillment only by contributing to the joy and fulfillment of others, in which our own joy and fulfillment are included.

If we sketch out his argument, it easy fits a Nyāya structure:

> *Pratijñā*: "you and I cannot fulfill ourselves by going our own way"
> *Hetu:* Because "the welfare of each one of us is inseparable from the welfare of all others"
> *Dṛṣṭānta*: Like "the body" (Negative example: cancer)

Again we see the Nyāya claim, reason and analogy summarizing and establishing a common sensory and perceptual connection. These passages summarize earlier statements, invite the interlocutor to experience perceptively the argument, and to weigh what is said by her or his own experience. Together the interlocutor's seek shared truth and release.

4. CONCLUSIONS

Nyāya involves a full approach to reasoning embedded in the context of human liberation from doubt, fear, and misapprehension. Ideally, rhetors adopt a life of study, meditation, and inquiry leading them to use rhetoric not just to convince, but also to enlighten. Rhetorical reasoning is not something we simply do, but something we embody, something that enriches our experience of life.

The Nyāya method embodies the four ways of knowing, applies claim and reason without being fully enthymematic, and offers claims and reasons conjoined with an analogy rather than a first major premise. It began as type of case-based demonstration that is analogous to Aristotle's vision of the enthymeme: "The orator's demonstration is the *enthymeme*, and this is, in general, the most effective of the modes of persuasion" (Aristotle. *Rhetoric* I.2.1355a 3-9, p. 22). In the Indian context, claim and reason and analogy combined into one dominant model of reasoning making no distinction between de- and induction, metaphysical and physical matters, or rhetoric and dialectic. It promotes revelation over persuasion; demonstration/reason is used to point to something within the respondent that they have to see and experience themselves.

We can assert then that though they differ in many ways, Nyāya is analogous to Aristotle in terms of its rhetorical function:

Nyāya		Aristotelian Reasoning
	Method of (rhetorical) demonstration	
India		Greece

India developed its own unique methods of persuasion *parallel* to Greek practices, rhetorical methods in the sense that they are analogous in function as (ideals for) persuasive speech that both reflect and create cultural practice.

So, is reasoning universal? Yes and no. The example of Nyāya implies that various cultural practices may be analogous in purpose, but reflect and embody specific socio-cultural context, goals, and methods. Our Western terminologies cannot adequately express the nuances of those of the *Other*, so we can no longer assume the inherent pre-imminence of the terminologies of Western reasoning. Instead, we would do well to immerse ourselves in the ways unfamiliar cultures reason until we begin to see the world from different perspectives. Only then can we begin to understand human reason in its myriad expressions.

REFERENCES

The Bhagavad Gita. (2003). Trans. Juan Mascaró. New York: Penguin.

Easwaran, E. (1993). *The End of Sorrow: The Bhagavad Gita for Daily Living, Volume I [India's timeless and practical scripture presented as a manual for everyday use]*. Tomales, CA: Nilgiri Press.

Ganeri, J. (2004). Indian Logic. *Handbook of the History of Logic*. Vol. 1 Greek, Indian, and Arabic Logic. Eds. Dov M. Gambay and John Woods. Amsterdam: Elsevier North Holland.

Lloyd, K. (2004). Culture and rhetorical patterns: Mining the rich relations between Aristotle's enthymeme and example and India's Nyāya method. *Rhetorica, 29*(1), 76-105.

Lloyd, K. (2007). Rethinking rhetoric from an Indian perspective: Implications in the Nyāya Sūtra. *Rhetoric Review, 26*(4), 365–384.

Matilal, B. K. (1985). *Logic, Language and Reality. Indian Philosophy and Contemporary Issues*. Delhi: Motilal Benarsidass,

Marjoribanks, D. (14 June 2010). Review of The Idea of Justice. *Marx and Philosophy Review of Books*. Marx and Philosophy Society. http://marxandphilosophy.org.uk/reviewofbooks/reviews/2010/144

Melfi, A. Dissertation (in Progress Georgia State University): The Root of Vedic Rhetoric: *Ṛta* and the Deep Levels of Speech, A Terministic Key to the Ancient South Asian Paradigm and its Practices.

Perrett, R.W. (Oct. 1999). History, time and knowledge in ancient India. *History and Theory, 38*(3), 307-321.

The Upanisads. (2007). 2nd Edition. Trans. Eknath Easwaran. Canada: Nilgiri Press.

Vidyābhūsana, S. C. (1921). *A History of Indian Logic*. Delhi: Motilal Banarsidass.

Vidyābhūsana, M. M. S. C. (1930, 1990). *The Nyāya Sūtras of Gotama*. Nanda Lal Sinha, Ed. Delhi: Motilal Banarsidass Publishers.

42

The Argumentation of H.L.A. Hart on Legal Positivism: The Descriptivist Stance and its Categories

ANTÓNIO MARQUES
IFILNOVA, Universidade Nova de Lisboa, Portugal
marquesantoni@gmail.com

Our aim is to identify the main lines of Hart's defense of his conception of legal positivism against what in his opinion is a misrepresentation of Dworkin of his concept of positivism. We want also to explore in this context what can be understood as Dworkin's interpretitivist account of positivism as an alternative to the moderate conception of Hart´s legal positivism.

KEYWORDS: Dworkin, H.L.A Hart, legal positivism

The kind of legal positivism defended and developed by H. L. A. Hart is based on a methodological stance, the descriptive stance, whose application, not only to legal issues, but also to ethical problems, I'll try to evaluate. From an analysis of the descriptive stance (that is also defended by a philosopher like Wittgenstein) it is possible to design what I call an **elementary descriptive tool** with correspondent categories. In other words, my aim is to identify, yet in a very preliminary way, the utility and applicability of such descriptive tool, namely in areas such as ethics and legal reasoning.

In his answer to the objections of Ronald Dworkin regarding the definition of a legal system (of what a legal system is), H.L.A. Hart, in a Postscript of the 1991 edition of his influential *The Concept of Law*, presents a clear and rather complete perspective of two major methodological stances, the descriptive one and the interpretive/ justificatory one. With help of the mentioned text of Hart against Dworkin, let's go deeper in the structure of what can be designated the descriptive methodological stance contrasting it with the referred to justificatory one. I'll focus mainly on the descriptive side, since it corresponds to philosophical and methodological conceptions that are near of the moderate positivism of Hart and uses philosophical

D. Mohammed & M. Lewiński (eds.) (2016). *Argumentation and Reasoned Action: Proceedings of the 1st European Conference on Argumentation, Lisbon, 2015. Vol. II, 695-704.* London: College Publications.

categories rooted in a recent tradition inaugurated by the late Wittgenstein. Of course I'm not suggesting that there is a pure descriptive model that we could find among different versions of descriptive stances. What is at stake is a possible characterization of a descriptive stance, which is a common ground for the understanding and argumentation of different issues, either in morals, or in politics and law.

Hart's philosophical enterprise is well formulated in the following terms: "My account is *descriptive* in that it is morally neutral and has no justificatory aims: it does not seek to justify or commend on moral or other grounds the forms and structures which appear in my general account of law" (Hart, 1994, p. 240). The implicit presence of the late Wittgenstein and his new method in philosophy is obvious and it deserves to be mentioned. In fact, one finds in his *Philosophical Investigations* (2000. First published in 1953) the ground of the antinomy description/ explanation (explanation in the sense of the above mentioned justificatory stance). The following passage of the is quite clear.

> All *explanation* must disappear, and description alone must take place. And this description gets its light - that is to say, its purpose – from the philosophical problems. These are, of course, not empirical problems; but they are solved through an insight into the workings of our language, and that in such a way that these workings are recognized – *despite* an urge to misunderstand them. The problems are solved, not by coming up with new discoveries, but by assembling what we have been familiar with. Philosophy is a struggle against the bewitchment of our understanding by the resources of our language (Wittgenstein, 2000, sec. 109).

Perhaps in an even more radical tone, Wittgenstein adds that "Philosophy just puts everything before us, and neither explains nor deduces anything. - Since everything lies open to view, there is nothing to explain. For whatever may be hidden to us is of no interest to us" (2000, sec. 126). The superiority of the descriptive stance consists prima facie in its capacity to solve problems through "an insight into the workings of our language", which contrasts with the explanatory acts that focus essentially on hidden or inner processes of what we call *interpretation*. Furthermore, descriptions are tools, indeed *schematic tools* that are designed for particular uses (2000, sec. 291) in order to bring before us the event or a concept. "Think of a machine-drawing, a cross-section, an elevation with measurements, which an engineer has before him" (2000, sec. 291). Instead saying that one explains or interprets a form of life, a ritual, a custom, Wittgenstein prefers to say that one describes those things through an overview of the uses of language at work, which justifies his famous formula, "to know a language is to know a form of life". Important to our issue, that is the descriptive stance of Hartian positivism and the refutation he does of the evaluative legal philosophy, is the characterization of the interpretative element. Both authors, Hart and Wittgenstein, refuse the interpretive method to understand what means "to follow a rule". In fact, this point of view is to be found in Wittgenstein by occasion of discussing what is following a rule: "... in this chain of reasoning we place one interpretation behind another, as if each one contented us at least for a moment, until we thought of a way of grasping a rule which is *not* an interpretation behind another ..." (2000, sec. 201). The refutation of the interpretative stance stresses the fact that an interpretative approach corresponds to a chain of reasoning where one places one interpretation behind another and so *ad infinitum*. The delusion consists in confounding just one more interpretation with a last justification, which the moral or legal interpreter fixes as *the* last rule or norm. But what happens is, in the terms of Wittgenstein, that we just substitute an expression of a rule for another expression and follow the last one as *the ultimate rule.*

Now we have identified the main theoretical concepts that we need to design an **elementary descriptive tool** that can work either as an explanatory or as an argumentative task regarding any kind of practices or institutions in ethics or law. This can be seen contradictory with was characterized as the antinomy between a descriptive stance and a justificatory one. At this point I would only add that the descriptive stance *has also an explanatory feature* if by "explanatory" *is not meant* "interpretive" or justification through an ultimate or grounding rule. This is Hart's view: "My aim in this book was to provide a theory of what law is which is both general and descriptive. It is *general* in the sense that it is not tied to any particular legal system or legal culture, but seeks to give an explanatory and clarifying account of law as a complex social and political institution with a rule-governed (and in that sense 'normative') aspect" (Hart, 1994, p. 239).

There are some central concepts or categories that Hart identifies in the course of his work, categories that operate as the main concepts for an **elementary descriptive tool:**

Categories of an Elementary Descriptive Tool or Descriptive Stance:

duty-imposing rules	power-imposing rules	rules of recognition	internal-external rules

So we have the conceptual shape of a descriptive stance, where the following categories:

- duty-imposing rules
- power-imposing rules
- rules of recognition
- internal-external aspect of rules

are tools to understand a multiplicity of issues and to apply in various argumentative situations. For example, is this pattern of behavior the expression of a legal rule? Is it a habit? Individuals accept it or simply do they obey it under threat?

Further theoretical and methodological work is required but the design of an **elementary descriptive tool** with this conceptual structure, under the inspiration of the methodological thought of Hart and Wittgenstein, seems to offer promising approaches for various argumentation situations. In the perspective of Hart these are methodological concepts for the study of a legal system, but it can also be used in the description of a moral form of life.

For example, we can ask whether this or that pattern of behavior expresses a legal rule, how it is institutionalized, with which other rules does it form a legal system of a concrete form of life, and so on. To give an example, in legal system, which *doesn't reduce itself to coercion*, John shall accomplish his promise to pay his debt to Thomas since debts are to be paid and since he (John) recognizes that in the legal framework of his community values he must pay his debt. Then it is possible to find in this argument-pattern the categories of the descriptive stance: "Debts are to be paid" (power-imposing rules", "John shall accomplish the payment of his debt" (duty-imposing rule); "John recognizes (accepts) that this act is to be accomplished" (rule of recognition) and "the rule is obeyed by John from a first person perspective" (internal aspect of legal/ ethical rules).

So categories of the descriptive stance play a decisive role in this argumentative situation; some of them they play even the role similar of backing tools in the sense of Toulmin's baking warrants.

1) Debts must be paid,
2) John has a debt to Thomas,
3) so he must pay it to Thomas, since
4) he, John, *recognizes* that this is a valid legal rule (debts must be paid) and
5) people in general *recognizes* such a rule as valid in that particular legal system.

Note that 4) and 5) express the *category of recognition*, considered by Hart as essential in order to qualify a system of rules as a *legal* system of rules, which are not simply *coercion* rules. Of course one can extend this category to ethical issues.

The descriptive stance, according to Hart, joins all these propositions in a framework within which the mentioned categories are at work. Now the question is where we trace the boundaries between, on one side, the descriptive stance with the associated argumentation, in the sense above mentioned, and on the other side the interpretive stance. We have already seen that the former (descriptive stance) is also explanatory yet not seeking for the ultimate reason, rule or interpretation.

A legal positivism, which makes use of the Hartian categories, doesn't give up the role played by the rule of recognition carried out by members of a community with their internal point of views. As already mentioned a descriptivist view of a legal or ethical practice, using the above mentioned categories is an excellent resource, not only to understand a system of ethical and legal values. If we assume that the Hartian categories of a descriptivist stance are good instruments to identify the status of any norm in a legal or ethical system (system of values or axiology), so it will be also a relevant tool in a variety of argumentative situations. Also soft positivism with its descriptivist method allows us to look at the relationship between morals and law in a different way, far from the classical pure positivism view that is to be found in Hans Kelsen. So let's try to apply the descriptive tool of the Hartian positivism to the problem of the relationship between morals and law.

In our days there is a common perception in all domains of our life that our societies need more moral reflection, a deeper insight into the so called universe of values and a new understanding of the priority of the requirements of the ethical life vis-à-vis the constraints of life in all diverse areas, such as politics, economy, and so on. So one can consider the discussion about the relevance of ethical life in our contemporary experience, a chief point and it leads us to the clarification of the relationship between morals and law.

The relationship between morals and law is a classical and controversial issue among philosophers of law or law authorities, judges and lawyers. Often in the practice and the application of law, many problems come to light that involve discussions that can only develop towards a point where law seems to be determined by morals. Yet the common understanding of the nature and function of law in our societies tend to separate on one side normative realities or normative practices, which are specific of law and on the other side those norms of morals. In this sense we need to identify in the contemporary discussion some relevant points, starting from some representations of positivist stances of the most important positivist author, Kelsen, who set the agenda of the discussion.

In this sense let's see some Kelsen's declarations defending the complete separation of law from morals as it is argued by standard positivist views such as that of Hans Kelsen's well known "Pure Theory of Law". Then I'll refer to and comment some issues of the philosophy of law of H.L.A Hart, who in his influential work, *The* Concept *of Law* , preserves the autonomy of law but doesn't eliminate the moral stance from the best understanding of the former[1], finally it will be exposed some common features between morals and law, as a result of the descriptive stance of the kind of Hart's positivism.

An introductory reflection allows me to point out what seems to shape a general perception existing in our societies in relation to the prevailing and irreversible separation between morals and law. Let's start with the legal positivist comprehension of our issue. Despite the differences among philosophers or theories of law, that belong to this tradition, is it possible to identify a general thesis? How is it possible to characterize it? The answer is that law has acquired an autonomous status, and even a scientific status. Therefore, not only from the point of view of its foundations, but also from the point of view of its methodology law is a separate domain, which doesn't depend whatsoever from morals. And in fact this is a kind of common opinion well rooted in our contemporary western societies: a legal system must be independent of ideological representations, metaphysical ideas or communitarian values. Everyone knows how often laws do not coincide with a so called popular feeling and judiciary authorities decide against this or that predominant value of people, for the sake of the state of right. Also if we think particularly on the typical agency of judges and lawyers it seems to us obvious that it cannot be determined by pure moral judgments; decision making processes in law have certainly a specificity related to protocol procedures, then of rules of law cannot be submitted to moral values or moral judgments. But if it is obvious to our contemporary perception that law and the authorities that create or apply the law should not be determined by morality, it also obvious that

[1] In the words of Hart, "In all societies which have developed a legal system there are, among its non-legal rules, some to which supreme importance is attached, and which in spite of crucial differences have many similarities to its law. Very often the vocabulary of 'rights', 'obligations', and 'duties' used to express the requirements of legal rules is used with the addition of 'moral', to express the acts or forbearances required by these rules" (1994, p. 170).

the pressure of morals on law is a reality, what shapes an ambiguous and complex situation. From Kelsen's point of view, the separation of both domains is a clear and necessary condition for the scientific status of law. These are some statements expressing his pure doctrine of law:

> What is rejected is simply the view that the law as such is part of morality, and that therefore every law, as law, is in some sense and to some degree moral (Kelsen, 1996, p. 15)
>
> What makes certain human behaviour illegal – a delict (in the broader sense of the word)- is neither some sort of immanent quality nor some sort of connection to a metalegal norm, to a moral value, a value transcending the positive law. Rather, what makes certain behavior a delict is simply and solely that this behavior is set in the reconstructed legal norm as the condition of a specific consequence, it is simply and solely that the positive legal system responds to this behavior with a coercive act.
>
> ...
>
> The law is a coercive apparatus having in and of itself no political or ethical value...
>
> ...
>
> The Pure Theory of Law preserves its anti-ideological stance by seeking to isolate representations of the positive law from every natural law ideology of justice. It does not discuss the possibility of the validity of a system higher than the positive law (Kelsen, 1996, p. 26).

Now if we look at Hart's doctrine of law exposed in *The Concept of Law*, we found as already mentioned another version of legal positivism and although he doesn't place law on the dependence of morals (like any other positivist) he refuses to see the former as a science completely immune to the moral element.

For the descriptivist stance one of the most striking features that one finds in a legal rule is its "internal aspect" contrasting with the type of coercion of other rules, such as rules imposed by violence, social pressure, and so on. You cannot simply understand what a legal system is and how it works if this internal perspective of any norm is nor identified. Furthermore, the reality of this internal aspect of law is required by the rules of recognition at work in every legal system. In fact, in a legal system works always what Hart designates "rules of recognition", that is the way, in which particular rules are identified, either by courts, or by other officials, private persons and their advisers. In a descriptivist approach *a legal system is not or doesn't limit itself to the observable regularities of behavior.* If we reduce the legal system only to regularities of behavior that an observer can observe it *ab extra*, then it would be not possible to make sense of how rules function as rules of law in the lives of those who normally are the majority of society[2].

The relevant point in this version of soft legal positivism of Hart is the central role that is given to the interior aspect of rule of law, which means its recognition or acceptance by those sharing the same legal system. This element introduced by Hart in the characterization of a legal system, supposes a shared acceptance/ recognition of rules of law and represent an important difference in relation to the positivism of Kelsen. In first place the ground of the validity of a norm or rule of law doesn't depend of a *last* norm upon which all the system of rules is built. In fact, the *criteria of recognition* by the members of the community are indeed the ultimate foundation of the validity of norms.

I don't enter here in more technical and philosophical details regarding the status of a "Grundnorm" in Kelsen that I invoke only to stress the difference from the view of Hart about the recognition as the ultimate ground of a legal system[3]. What is relevant for our discussion is the fact that this element of a rule of recognition of law is a condition for its validity, shared by the community members. This is true regarding a legal system, but also regarding the ethical life. In fact, it is not conceivable that a legal system and a moral system of a community grounded on the criteria of recognition, although belonging to different domains, don't cross each other or influence reciprocally. In this sense the approach of Hart is quite interesting and allows us to contrast on one side and to differentiate on the other side the relation between both domains, the moral one and the legal one, what was possible in the framework of a radical positivist stance like that of Kelsen.

Contrasting moral rules of a moral system (the system of moral values) and the norms of a law system Hart meets four cardinal differences (1994, pp. 175-180):

a) *Importance* – the essential feature of any moral rule or standard is that it is regarded as something of great importance. This is a truism but if we compare the place of a moral rule with other rules (namely of law) it is remarkable the fact that moral standards are maintained against the

---

[2] This rule of recognition of a legal rule is so important from Hart's point of view that he claims that the juridical validity of the rule depends on its recognition or acceptance expressed in "internal statements". Hart makes explicit this essential link between validity and recognition in the following terms: "To say that a given rule is valid is to recognize it as passing all the tests provided by the rule of the recognition..." (Hart, 1994, p. 103).

[3] See Hart (1994, pp. 105-6).

drive of strong passions which they restrict, at the cost of sacrificing considerable personal interests. Hart remarks that it is not essential to the characterization of the status of legal rules,

b) *Immunity from deliberate change* – It is characteristic of a legal system that new norms can be introduced and old ones change by deliberate enactment. That is not the case of moral rules of which one cannot say that they start to be valid from such or such date on. But of course along the time morals are not completely immune to any impact of law: some values or practices considered virtues can cease or disappear under the pressure of juridical norms. Standards of honesty and humanity can and often are introduced in society through the enactment of juridical norms and more or less slowly are incorporated as real moral rules,

c) *Voluntary character of moral offences* – A relevant contrast between the two kinds of norms stresses the point that "moral blame" regarding an action is excluded as far as it is demonstrated that the individual did all he could do. The legal evaluation is not so open to the consideration of such excuses, and

d) *The form of moral pressure* – a distinguishing feature of morality is the singular form of moral pressure, which is exerted in its support. It doesn't consist mainly in threats, since such pressure is, say, incorporated by the members of the community. By contrast the typical form of legal pressure consists mainly in threats.

Now if we leave the contrasting point of view and seek some common conditions of human societies or individuals, it is possible to determine a "minimal content of natural law" that seems to give to both law and morals a kind of common ground. Without pretending to go deeper in the discussion of this ground I wish only to stress how this minimal content of natural law is accepted by a positivist like Hart. He describes the common ground as the main requirements of law and morality to answer to certain, say, anthropological conditions. He enumerates five requirements:

a) human vulnerability – if men were to lose their vulnerability to each other there would vanish one reason for the most characteristic provision of law and morals, that is, *Thou shall not kill* (Hart, 1994, p. 195),

b) approximate equality – it is a fact of major importance for the understanding of different forms of law and morality that no individual is so much more powerful than others, that he is able, without cooperation, to dominate or subdue them for more than a short period" (Hart, 1994, p. 195),

c) limited altruism – men are not devils but they are also not angels and the basic rules of law and morals set the limits of actions and a system of

mutual forbearances necessary to survive,
d) limited understanding and strength of will – this is a continuation of the late requirement and
e) limited resources – law and morals need to create institutions that set up criteria of distribution, property (individual or collective), and in consequence principles of justice.

What is relevant here is that the application of the **elementary descriptive tool** above described to law and morals discussions, carried out in the context of a set of differences but also a set of common conditions (a minimal content of natural law in the words of Hart). The principal claim of a moderate positivism regarding the complicated relationship between morals and law can be formulated in the following terms: rules of law are not dependent on moral rules but as Hart remarks, "moral and legal rules of obligation and duty have therefore certain striking similarities enough to show that their common vocabulary is no accident" (Hart, 1994, p. 172). Only research and discussions based on the descriptive stance can reveal the complex articulations and reciprocal influence between the both domains of law and morals.

REFERENCES

Hart, H.L.A. (1994). *The Concept of Law*, second ed. Oxford: Clarendon Press.
Kelsen, H. (1996). *Introduction to Problems of the Legal Theory*. Oxford: Clarendon Press.
Wittgenstein, L. (2000). *Philosophical Investigations*, second ed., Oxford: Blackwell.

43

Arguing in the Healthcare: On the Discourse of Web-Based Communication to Patients

DAVIDE MAZZI
University of Modena and Reggio Emilia, Italy
davide.mazzi@unimore.it

A growing body of research has recently been devoted to argumentative discourse in healthcare settings. Within this framework, this study carries out a corpus-based investigation on web-based resources employed in Ireland to communicate to the public about cancer. The qualitative and quantitative evidence of the investigation establishes a correlation between the deployment of argument forms, phraseological tools and the sections in which argumentative discourse is most likely to cluster.

KEYWORDS: argumentation, booklets, cancer, corpus, discourse, Ireland, lexical bundle, phraseology

1. INTRODUCTION

Over the last twenty years, healthcare discourse has generated a spate of interest in scholars from a wide range of disciplinary backgrounds. As a broad area of investigation, the use of language in healthcare settings has been analysed across genres, e.g. medical case presentations (Schryer, Lingard, Spafford, & Garwood, 2003), doctor-patient interaction (Rubinelli & Schulz, 2006) and informative materials to patients and families (Wizowski, Harper, & Hutchings, 2014).

Among more specific aspects covered by recent research into medical discourse, two are worth mentioning. First of all, the study of discourse strategies behind the expression of empathy as "a cognitive [...] attribute that involves an understanding of the inner experiences and perspectives of the patient, combined with a capability to communicate this understanding to the patient" (Hojat, Gonnella, Nasca, Mangione, Veloski, & Magee, 2002, p. 58). Secondly, a major area of concern to analysts of communication in healthcare settings has been

D. Mohammed & M. Lewiński (eds.) (2016). *Argumentation and Reasoned Action: Proceedings of the 1st European Conference on Argumentation, Lisbon, 2015. Vol. II, 705-718.* London: College Publications.

the study of argumentation, which has resulted in two deeply interconnected research directions.

First of all, studies have focused on widespread argument schemes, including analogical reasoning and argument from ignorance in scientific inquiries (Cummings, 2005 and 2009) and pragmatic argumentation in health brochures (Van Poppel, 2012). In second place, scholarly research has been devoted to sharpening existing knowledge of explicit and implicit arguments in the overall argumentative structure of doctor/patient interaction (Rubinelli & Schulz, 2006). In the analysis of argumentation in healthcare domains, the importance of cultural variables in decoding the specificity of communicative practices has been forcefully stressed in recent years (Bigi, 2014, p. 63). With reference to single countries, for instance, a growing body of research has appeared about socio-demographic perceptions of illness causation and broader attitudes to the health care in the Republic of Ireland (cf. MacFarlane & Kelleher, 2002).

In light of these multi-faceted research perspectives, the aim of this work is to substantiate the findings in the literature so far, by bringing a genuinely discourse-based perspective to them. With the aim of achieving this goal, a corpus investigation will be carried out of web-based resources employed by a leading nationwide organisation – the Irish Cancer Society – to communicate to the public. Rather than focusing on medical discourse as expert-to-expert communication, therefore, the study addresses such discursive aspects of expert-lay communication as: (a) are there any recurrent discourse patterns that tend to be reiterated across the sections of informative healthcare materials?; (b) how are patients conceptualised, or else their needs taken into account through the language of and the argumentation in such materials?.

2. MATERIALS AND METHODS

The study was based on a corpus of 11 healthcare "booklets" issued by the Irish Cancer Society (henceforward, the "ICS_Corpus"). These were all downloaded from the Society's official website at http://www.cancer.ie/publications#sthash.mRjZN6Aa.dpbs. Taken together, they cover a time span of four years (2011-2014), they amount to a total of 173,689 words, and they cover four widespread illnesses, i.e. bowel cancer, breast cancer, lung cancer and prostate cancer.

From a methodological point of view, two strands of analysis were combined and carried out in parallel. The first strand was a qualitative text-based study pursuing two objectives: to begin with, the identification of any *prima facie* recurrent "discourse pattern" in the data. By discourse pattern, reference is made here to "units of communication that are larger wholes than just words" (Scollon & Scollon, 2001, p. 60), i.e. "X, because of Y". These patterns are considered to be of great value to understand "the basic principles of communication between members of different groups" (Scollon & Scollon, 2001, p. 2), with due regard to communication within professional settings. Furthermore, the second objective of this stage of the analysis was the retrieval of recurrent argumentative schemes along with the underlying argumentative structure (Snoeck Henkemans, 2003; Van Eemeren, Houtlosser, & Snoeck Henkemans, 2007).

At the same time, the second strand of the investigation consisted in a quantitative corpus-driven study (Tognini Bonelli, 2001) of phraseology as a concomitant principle of discourse organisation, whereby words tend to go together and make meaning by virtue of their combination (Sinclair, 1996). In order to examine key instances of phraseology in context, emphasis was laid on "lexical bundles" as "multi-word sequences that occur[red] most frequently in particular genres, regardless of whether or not they constitute[d] idioms or structurally complete units" (Breeze, 2013, p. 230). For the purpose of the present study, lexical bundles were retrieved by using the linguistic software package *AntConc* (Anthony, 2006). An "n-gram list" was thus generated for the ICS_Corpus, in the attempt to extract the top-ten most frequent bundles. Once these were identified, they were studied in context with the aim was to uncover their main discourse function.

The two strands of the analysis mentioned above fruitfully integrated each other within a full-relief investigation of the discourse tools and the argumentation deployed in the ICS_Corpus. For instance, therefore, evidence was collected of the presence of lexical bundles (e.g., "if you have") within and at the heart of larger discourse patterns – e.g., "if a, then b" (cf. Section 3.1) – or otherwise, of bundles (e.g., "you may feel") embedded within empathic strategies that trigger the onset of variants of pragmatic argumentation (Section 3.2). The main findings of the research are presented in the upcoming section.

3. RESULTS

3.1 Discourse patterns: forms and functions

The n-gram list generated for the ICS_Corpus allowed for the retrieval of the bundles displayed in table 1 below as the most frequent ones at a wider corpus level.

Bundle	Frequency (raw)	Frequency (per 1,000 words)
if you have	270	1.55
if you are	250	1.44
your doctor or nurse	212	1.22
you would like	111	0.64
you may be	104	0.59
your doctor will	92	0.53
there is no	82	0.47
you do not	65	0.37
for you to	63	0.36
you may feel	63	0.36

Table 1 – Most frequent lexical bundles of the ICS_Corpus

Among the salient findings emphasised by the occurrence and collocational[1] trends of the selected lexical bundles, the widespread presence of discourse patterns deserves to be mentioned as a first, prominent feature. More specifically, data point to the even distribution of three main patterns across the various sections into which booklets are subdivided. The first pattern is one that can be schematised as "if a, then b", which amounts to 27.4% of the occurrences of the bundle "if you have" in the ICS_Corpus. When the bundle describes the pattern, there seem to be two main functions involved: the first is to document any cause-effect relationship about health risks correlated with the onset of cancer, as in (1) below; the second function is to inform patients about the treatment they are most likely to be prescribed under specific circumstances, as in (2):[2]

(1) **If you have** a brother or father with the disease, your risk is higher.
(2) **If you have** a high recurrence score, you will be advised to have chemotherapy as well as hormone therapy.

---

[1] The term "collocation" (and its derivative adjective "collocational") will be used in the paper to indicate the tendency of words to co-occur on a regular basis (Sinclair, 1996).

[2] In all numbered examples, the lexical bundle through which evidence of the highlighted discourse pattern was collected is underlined.

With "if a, then b", the organisation's expert voice informs patients that "risk/treatment b" will follow in the event that "situation a" is the case. The message is rendered fairly straightforward by the fact that the pattern is embedded within sentences with low lexical density, i.e. an average of 18.3 words per sentence.

The second discourse pattern disclosed by corpus data is "(Even) if a, you MOD", where "MOD" stands for any modal operator, whether a modal verb itself (typically "could") or any other comparable indicator of the same category (e.g., "it is possible to"). This pattern applies to 26.2% of the 65 entries of "you do not", and its primary function is to offer patients polite suggestions about any course of action through which they might find the relief they are seeking (see 3 and 4 below):

> (3) If **you do not** feel like eating during treatment, you could replace some meals with special high-calorie drinks.
> (4) Even if **you do not** consider yourself a religious or spiritual person, it is still possible to take comfort and support from these practices.

On a qualitative plane, an attractive aspect emphasised in Section 3.2 as well is the religious or spiritual dimension to the patient's experience of illness and therapy, which might indeed stand out as a peculiarity of Irish booklets and their underlying socio-cultural context.

The third major discourse pattern observed in the ICS_Corpus is "If a, do b". The pattern implies such a form of direct address to the patient as the imperative: interestingly, whether in the more direct (e.g., "do x") or the polite version (i.e., "let x know"), the imperative is an invitation to the patient to take up an active role in both discussing any relevant clinical or therapeutic aspect with doctors and nurses, and finding out more about available treatment options:

> (5) If **you would like** more information on brachytherapy, call the National Cancer Helpline on 1800 200 700 for a free copy of the booklet Understanding Radiotherapy.
> (6) If you want to refuse treatment, let **your doctor or nurse** know your concerns first.

The use of "you would like" in statements concerning an all too desirable quest for information on the patient's part concerns 38.7% of the corpus tokens of the bundle, whereas the collocational pattern of "your doctor or nurse" with imperative forms – with "ask", "tell", "talk to" and "discuss with" in prominent position – holds for 79.5% of the occurrences of it as the only 4-word bundle in the corpus.

3.2 The expression of empathy: strategies, signals and argumentative patterns

In addition to the general discourse patterns analysed in the previous section, a second major area disclosed by corpus data is the expression of empathy. In this respect, the use of bundles can be associated with discourse strategies designed to help patients cope with uncertainty of a mainly emotional kind.

From a discourse perspective, empathy was identified as being expressed through a wide array of tools. These include and are constructed around the lexical bundles listed in table 1, which fundamentally act as "signals" (Bres & Mellet, 2009, p. 6) of empathic engagement. As a matter of fact, the expression of empathy in the booklets by the Irish Cancer Society tends to take the shape of signals of interlocutive dialogism (cf. Bachtin, 1984): the expert voice therefore anticipates and addresses the emotional needs of patients acknowledged through answers to imaginary questions by patients themselves. In the rest of this section, the questions patients are expected to come up with, and the related empathic answers to reduce their perceived uncertainty, are presented with the aim of showing the relationship between the use of bundles in context and the voicing of empathy.

The first question could be easily schematised as "I feel x. Is anything wrong with me?", where "x" invariably denotes an unpleasant state of mind. In this respect, 11.1% of the occurrences of "for you to" collocate with two main adjectives – i.e. "natural' and "normal" – in the larger pattern "It is natural/normal for you to...", followed by a reference to the patients' putative feelings when diagnosed with cancer.

> (7) It is natural **for you to** be afraid or concerned about the future too. You may wonder if you will be cured. Living with this uncertainty can make you feel anxious and fearful. You may not wish to make any plans or decisions.

Also in the context of the emotional trauma faced by patients after diagnosis, the second question envisaged in empathic statements is arguably "I have been diagnosed with x. How am I going to feel about that?/What am I to expect?". Not surprisingly, the answer is overwhelmingly provided through "you may feel". The bundle is followed by an indication of the feeling patients are supposed to go through in 42.8% of its corpus entries: in descending order of frequency, examples include "sad", "devastated", "upset", "disappointed", "guilty", "lost" and "under pressure", although the top collocate of "you may feel" is "angry". It is a variety of subjects or entities patients are conceptualised as being angry with, from doctors and nurses to the diagnosis itself. Still, what might be hypothesised to be a peculiarity of Irish booklets compared to materials from other countries is anger against God, as in the last fragment of example (8):

> (8) Many aspects of your illness can result in anger and distress. Anger can often hide other feelings such as fear, sadness or frustration. **You may feel** angry that you got lung cancer through smoking. **You may feel** angry towards the doctors and nurses who are caring for you. Or if you have a religious belief, **you may feel** angry with God for allowing cancer to occur.

What is highly interesting about passages such as (8) above is the fact that the expression of empathy is often the preamble to the occurrence of pragmatic argumentation, as in (9) below:

> (9) Every family deals with cancer in a different way. **You may feel** that **you do not** want your illness to upset family life, or feel guilty that you cannot do activities with your children or grandchildren or that you're letting them down. These are all natural feelings to have at this time. Be honest. The main thing to remember is that being honest with your family really helps. Keeping your illness a secret may not be the best thing for your children. It can put added pressures on your family and lead to confusion. Young children are very sensitive to stress and tension and if you try to protect them by saying nothing, they may feel isolated. In fact, they may have greater fears if told nothing.

In (9), the expressed standpoint (S) is "be honest", namely the advice that patients should avoid bottling up their feelings, no matter how distressful that might be. The standpoint is supported by two coordinate "variant-II" pragmatic arguments (Van Poppel, 2012, pp. 99-100) schematised in figure 1, below.

Both arguments support the standpoint by pointing to the undesirable effects that the opposite course of action – i.e. telling no one about one's illness – is supposed to cause. Moreover, the two arguments are coordinated because taken together, they "constitute a single attempt at defending the standpoint", while the argument displayed on the right-hand side is added to the first "to overcome the doubt or answer the criticism that it is insufficient" (Snoeck Henkemans, 2003, p. 410). In this case, the strength of the first argument – notably, that keeping one's illness a secret is contra-indicated on the grounds that it would both put added pressures on one's family, and lead them to confusion – may be resisted by a challenging view. This is the notion that children, at least, should be dispensed with the news about their parent's condition, possibly because of their increased sensitivity: in turn, however, this potential criticism is reversed by the second argument, whose repairing function is to show that telling nothing to children also produces unpleasant results, i.e. leading them to a feeling of isolation and greater fears.

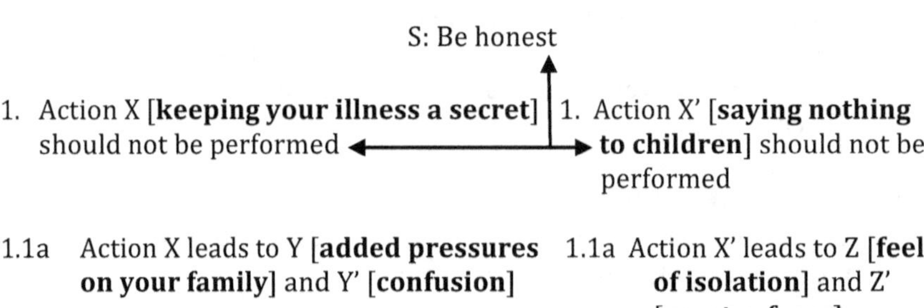

1.1b Y and Y' are undesirable

1.1b Z and Z' are undesirable

1.1a-1.1b' (If Action X leads to Y/Y', and Y/Y' are undesirable, then Action X should not be performed]

1.1a-1.1b' (If Action X' leads to Z /Z', and Z/Z' are undesirable, then Action X' should not be performed)

Figure 1 – Coordinative pragmatic argumentation (Variant II) in the ICS_Corpus

The third question explored by booklets is related to a vitally important dimension of the experience of illness, notably timing. An effective way to schematise it is "It is taking me a long time to achieve y. Is anything wrong with me?", where "y" indicates a desirable state of affairs patients are believed to be aspiring to. The issue raised here is crucial, because it focuses on the mind-set required by patients to adjust to new routines while attempting to restore old and altogether desirable habits. In response to patient doubts, the discourse of booklets mainly deals with timing through "for you to": in 15.8% of its tokens, the bundle collocates with items sharing a semantic preference of "time", such as "it may take a while" or "it may take a year". Whether it is about getting over the side effects of treatment or the yearned-for return of sexual appetite, the emphasis is invariably on the idea of taking as much time as each and every individual's condition may require, before achieving normality (cf. 10):

> (10) Even though you may want to stick to your old routines, sometimes this may not be possible. It may take a while **for you to** adjust to your new routine.

The expression of empathy is, finally, intertwined with the extensive use of a contrastive pattern, which could be rendered as "People say/feel x, but there is no evidence to prove it". Here, the writer's voice is set to reverse an undesirable element "x" that putatively worries patients, scares them, or else makes them feel guilty (11):

> (11) Some women may feel guilty for thinking about reconstruction and that it might even seem vain. This is not so - reconstruction can be an important part of treatment that helps emotional recovery and well-being.

However phrased in different ways - in other cases, the forms "there is no evidence" and "there is no truth" are attested - the pattern is used in contexts where the writer's primary concern is to disprove commonplaces that might have solidified over time in the patients' conscience. Examples such as (11) indeed leave one with pertinent questions: What is the status of research findings in the materials under investigation? Are references to them common and reported across the sections of materials, or rather more circumscribed and swiftly alluded to? At a more general level, the study of Irish booklets indicates that clinical research contributions, although common, mainly concern two sections, i.e. "Alternative therapies" and "New treatments". In the former, taken here as an illustrative case, the effectiveness of megavitamin therapy, herbalism and in particular diet therapy is discussed:

> (12) Most doctors do not believe that such treatments can cure or control cancer. Diet therapy can often be restrictive. This means it does not allow you to eat foods that could be nutritious for you. Some restrictive diets can harm your health and may even cause malnutrition.

What is noteworthy about (12) and other similar instances is the argument structure (cf. figure 2 below). The standpoint is that such alternative therapies are not reliable, and it is supported by two coordinate pragmatic arguments. The first is a "variant IV" argument

(Van Poppel, 2012, p. 101), the aim of which is to stress that choosing alternative therapies is wrong or at least potentially harmful, because it fails to lead to the desirable effect of curing or controlling cancer. The second pragmatic argument is a "variant II"one that more specifically addresses diet therapy: it suggests that diet therapy should be avoided in that it would lead to the adverse effects of harming the patient's health and causing malnutrition. This argument is coordinated with the first "variant-IV" one, to which it is intended to lend support in response to a predictable criticism, i.e. "Alternative therapies may well be contraindicated on the whole, but what is wrong with diet therapy? After all, it involves the use of no drugs and it only impacts eating habits".

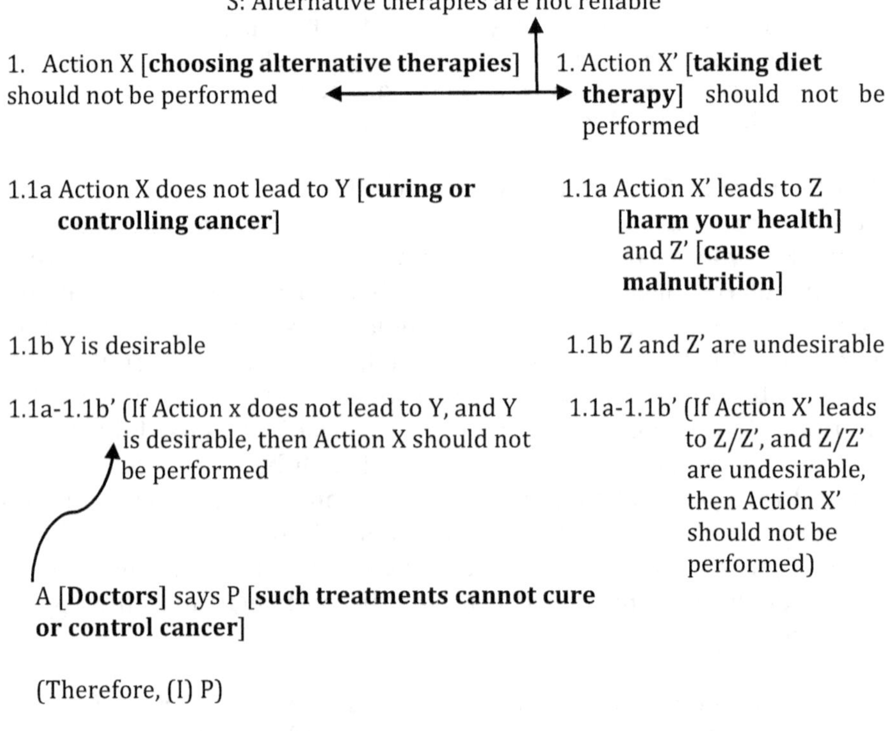

Figure 2 – Coordinative pragmatic argumentation (Variants IV and II) and subordinative argumentation from authority in the ICS_Corpus

Of equal interest is that the "variant-IV" pragmatic argument schematised below on the left-hand side is supported by a subordinate argumentation from authority (Schellens & De Jong, 2004): here, the authority A is represented by doctors, P is the view that alternative treatments are unable to cure or control cancer, and (I) stands for "interpretation of" accounting for conclusions that go back to the authority cited (Schellens & De Jong, 2004, p. 312). Following Snoeck Henkemans (2003, p. 408), therefore, the subordinate argumentation here does not support an unexpressed premise, but rather another argument: in so doing, it "lends additional support to the standpoint" in response to a potential rejection of the "variant-IV" argument – i.e., something along the lines of "What makes you think that alternative therapies are ineffective?".

4. CONCLUSIONS

In response to the first research question posed at the outset (cf. Section 1) – i.e., are there any recurrent discourse patterns that tend to be reiterated across the sections of informative healthcare materials? – the data presented in Section 3.1 suggest a certain drive towards standardisation in the discourse strategies underlying the creation of healthcare materials. In fact, the frequency of patterns and the presence of lexical bundles within them became quantitatively apparent from the first stage of the analysis.

With regard to the second research question – i.e. how are patients conceptualised, or else their needs taken into account through the language of and the argumentation in such materials? – corpus data pointed to the centrality of empathy (Section 3.2). In particular, the discursive management of empathy was observed to be a distinctive feature in devising healthcare materials that "reflect the way they [patients and families] think about their health conditions" (Wizowski, Harper, & Hutchings, 2014, p. 85) and engage them as constructively as possible.

In that regard, findings provide evidence of the strongly dialogic nature of the discourse of the selected materials, in the attempt to acknowledge and effectively address patients' hopes, fears, expectations and emotional needs. When it comes to entering text-internal dialogues with patients and/or else clearing up reported doubts or fears, the discourse of booklets tends to reflect careful consideration of the relevant content. Accordingly, the contexts where bundles were retrieved are indicative of the effort to mediate between the acknowledgment of what patients would like to know about their own health condition, and the careful review of clinical practice guidelines.

It is also in that light that the expression of empathy can be interpreted as an interface between the narrative and the more inherently argumentative sections of the analysed materials, so that for instance, the ideal path along which Irish booklets were observed to develop is a kind of triad: first of all, "we tell you the bare facts about the illness"; secondly, "we anticipate your likely (emotional) reactions"; and finally, "we provide warranted guidance through argumentation".

When argumentation was detected with specific reference to the practice of reporting research findings about alternative therapies and new treatments in Irish booklets, furthermore, corpus data point to argument structures that operate as instances of strategic manoeuvring. Consistent with the literature, these appear to be "directed at diminishing" the "tension between the claimed data-driven objectivity of the medical advice and the authority" of an expert voice that has "reasons for considering a certain treatment to be more appropriate than the others (Schulz & Rubinelli, 2008, p. 428).

On a conclusive note, the depth of the analysis performed here may allow for a few tentative generalisations concerning the Irish context. First of all, what tends to lie behind the discourse of the Irish Cancer Society's booklets is an attempt to produce materials that address a perceived need "for a health-literate health care system" as a major public health issue (Coughlan, Turner, & Trujillo, 2013, p. 167). By defining "health literacy" as the degree to which individuals are in a position to obtain, process and understand health information required to make appropriate health decisions, the authors point to the well-established correlation between low literacy and poorer health outcomes as well as poorer use of healthcare services. More surprisingly, though, they refer to the outcome of the 2011 EU Health Literacy Survey as showing that people with higher incomes and more education are still at risk of low health literacy in the Republic of Ireland. In this vein, the discourse and the structuration of sound healthcare materials might have been the driving concerns behind the publication of recent booklets such as those analysed here.

Furthermore, such an interpretation of data would be in keeping with the findings about lay perceptions of cardiovascular disease and associated risk factors in NicGabhainn, Kelleher, Naughton, Carter, Flanagan and McGrath (1999, p. 626), who aptly argue that

> the importance placed on advice from medical practitioners and the need for clear unequivocal messages, together with the frequency and intensity of anecdotes related by participants that appear to conflict with public health messages, fits well with the hypothesis that lack of behaviour change is related to scepticism about the scientific message.

Again, the quest for rhetorically more adequate materials that tell patients about the bare facts while at the same time successfully addressing key areas of concern, might have underlain the concept of the Society's booklets as an instance of best practice.

In order to evaluate whether such an instance is fully or partly replicable in other countries, it appears wise to consider the broad cultural background materials are intended to be perused in. Although the attention was primarily devoted to Irish booklets as the focus of this investigation, and despite the ensuing limitations, the present study was intended to develop and implement a number of useful methodological tools and analytical principles. From the combination of corpus and discourse views to the quantitative and qualitative examination of lexical bundles, discourse patterns and argument forms, these can fruitfully be set against the background of other widespread illnesses and national/cultural contexts.

REFERENCES

Anthony, L. (2006). *AntConc 3.2.1*. http://www.laurenceanthony.net/ (accessed 27 May 2015).

Bachtin, M. (1984). Les genres du discours. In M. Bachtin, *Esthétique de la création verbale* (pp. 265-308). Paris: Gallimard.

Bigi, S. (2014). Evaluating argumentative moves in medical consultations. In S. Rubinelli & A. F. Snoeck Henkemans (Eds.), *Argumentation and health* (pp. 51-66). Amsterdam: John Benjamins.

Breeze, R. (2013). Lexical bundles across four legal genres. *International Journal of Corpus Linguistics, 18*(2), 229-253.

Bres, J., & Mellet, S. (2009). Une approche dialogique des faits grammaticaux. *Langue française, 163*(3), 3-20.

Coughlan, D., Turner, B., & Trujillo, A. (2013). Motivation for a health-literate health care system – Does socioeconomic status play a substantial role? Implications for the Irish policy maker. *Journal of Health Communication, 18*, 158-171.

Cummings, L. (2005). Giving science a bad name: Politically and commercially motivated fallacies. *Argumentation, 19*, 123-143.

Cummings, L. (2009). Emerging infectious diseases: Coping with uncertainty. *Argumentation, 23*, 171-188.

Hojat, M., Gonnella, J. S., Nasca, T. J., Mangione, S, Veloksi, J. J., & Magee, M. (2002). The Jefferson scale of physician empathy: Further psychometric data and differences by gender and specialty at item level. *Academic Medicine, 77*(10), 58-60.

MacFarlane, A., & Kelleher, C. C. (2002). Concepts of illness causation and attitudes to health care among older people in the Republic of Ireland. *Social Science & Medicine, 54*, 1389-1400.

Nic Gabhainn, S., Kelleher, C., Naughton, A. M., Carter, F., Flanagan, M., & McGrath, M. J. (1999). Socio-demographic variations in perspectives on cardiovascular disease and associated risk factors. *Health Education Research, 14*(5), 619-628.

Rubinelli, S., & Schulz, P. J. (2006). "Let me tell you why!". When argumentation in doctor-patient interaction makes a difference. *Argumentation, 20,* 353-375.

Schellens, P. J., & De Jong, M. (2004). Argumentation schemes in persuasive brochures. *Argumentation, 18,* 295-323.

Schulz, P. J., & Rubinelli, S. (2008). Arguing 'for' the patient: Informed consent and strategic maneuvering in doctor-patient interaction. *Argumentation, 22,* 423-432.

Schryer, C. F., Lingard, L., Spafford, M. M., & Garwood, K. (2003). Structure and agency in medical case presentations. In C. Bazerman & D. Russell (Eds.), *Writing selves/Writing societies: Research from activity perspectives.* Fort Collins, CO: WAC Clearinghouse. http://wac.colostate.edu/books/selves_societies/ (accessed: 28 May 2015).

Scollon, R., & Scollon, S. W. (2001). *Intercultural communication: A discourse approach.* Oxford: Blackwell.

Sinclair, J. (1996). The search for units of meaning. *Textus, 9*(1), 75-106.

Snoeck Henkemans, A. F. (2003). Complex argumentation in a critical discussion. *Argumentation, 17,* 405-419.

Tognini Bonelli, E. (2001). *Corpus linguistics at work.* Amsterdam: Benjamins.

Van Eemeren, F. H., Houtlosser, P., & Snoeck Henkemans, A. F. (2007). *Argumentative indicators in discourse. A pragma-dialectical approach.* Dordrecht: Springer.

Van Poppel, L. (2012). Pragmatic argumentation in health brochures. *Journal of Argumentation in Context, 1,* 97-112.

Wizowski, L., Harper, T., & Hutchings, T. (2014). *Writing health information for patients and families.* Hamilton, ON: Hamilton Health Sciences.

44

Phronesis and Fallacies

TIMOTHY MOSTELLER
California Baptist University, USA
tmosteller@calbaptist.edu

In this paper I argue: 1) Informal fallacies are primarily practical and particular, as opposed to theoretical and general. 2) The technique that one uses to identify fallacies will be deeply intertwined with one's practical wisdom or *phronesis*. 3) The way in which students learning the skills of argumentation are taught to identify and avoid fallacies as well as restructure arguments containing such fallacies must take this into account.

KEYWORDS: argumentation, Aristotle, fallacies, phronesis, wisdom

1. INTRODUCTION

Standard logic textbooks (e.g. Patrick Hurley's *A Concise Introduction to Philosophy*) often distinguish between formal fallacies (e.g. denying the antecedent or undistributed middle) and informal fallacies (e.g. appeal to pity or straw man) in ordinary language argumentation. In evaluating an argument, formal fallacies can be identified by looking at the logical structure of an argument, while informal fallacies require a different type of insight in order to see that they are fallacious or unreasonable. This paper will explore the type of rational insight needed to identify informal fallacies. Logic texts such as Hurley's usually help students identify informal fallacies by providing a definition of a particular informal fallacy followed by clear cases which exemplify the fallacy. The mode of insight for students learning to identify informal fallacies seems to be one in which they pick out the fallacy in ordinary occurrences (e.g. television commercials, debates, interviews, letters to editors of newspapers, etc...). Students simply look back to the definitions and examples in a textbook context.

In this paper I will argue that such an approach to argumentation and fallacy identification is incomplete. I will argue for

D. Mohammed & M. Lewiński (eds.) (2016). *Argumentation and Reasoned Action: Proceedings of the 1st European Conference on Argumentation, Lisbon, 2015. Vol. II, 719-731.* London: College Publications.

the following three ideas. First, informal fallacies (as opposed to formal fallacies) fall within the category of reasoning and argumentation which is primarily practical and particular as opposed to theoretical and general. Second, since informal fallacies deal primarily with practical reasoning, the technique that one uses to identify them will be deeply intertwined with one's practical wisdom or *phronesis*. Third, since informal fallacies deal with practical reasoning and practical wisdom (*phronesis*), the way in which students learning the skills of argumentation are taught to identify and avoid fallacies as well as re-structure arguments containing such fallacies must take this into account. This will entail that philosophers (among others) who teach students the skills of argumentation must help their students to identify how to cultivate practical wisdom (*phronesis*) alongside the kind of theoretical wisdom associated with formal logic.

2. FORMAL AND INFORMAL FALLACIES

The teaching of argumentation is commonly divided between the formal (e.g. categorical logic, symbolic logic), and the informal (e.g. informal fallacies). In each division, one job of someone teaching is to help students identify and avoid logical errors or fallacies. Patrick Hurley defines a fallacy as, "a defect in an argument that arises from either a mistake in reasoning or the creation of an illusion that makes a bad argument appear good" (Hurley, 2014, p. 122). He continues by distinguishing between formal and informal fallacies. Formal fallacies, "may be identified by merely examining the form or structure of an argument." (p. 122).

> FF1:
> All executions are grotesque rituals. (All E are G)
> All bullfights are grotesque rituals. (All B are G)
> Therefore, all bullfights are executions. (All B are E)

This is an AAA-2 invalid syllogism which commits the fallacy of undistributed middle. One can know that is invalid by drawing a Venn diagram, or by reflection on the meaning of the logical relations which compose the argument's structure. Informal fallacies "are those that can be detected only by examining the content of the argument" (p. 123). Consider three examples:

> IF1:
> All plants are things that contain chlorophyll. (All P are C)
> All factories are plants. (All F are P)

> Therefore, all factories are things that contain chlorophyll. (All F are C)

This is an AAA figure 1 valid syllogism, but it is informally fallacious due to the ambiguity of the word "plant." Thus, this commits the fallacy of "equivocation." One can only know that this is informally fallacious if one understands the ambiguity in the word "plant" as it changes from the first to the second premise.

> IF2:
> Atoms are invisible. (All A are I)
> The Brooklyn Bridge is made of atoms. (All B are A)
> Therefore, the Brooklyn Bridge is invisible. (All B are I)

This is an AAA figure 1 valid syllogism, but it is informally fallacious due to the presumption that the properties of the parts of a physical object necessarily transfer to the whole object. Thus, it commits the fallacy of composition. Hurley indicates, that to see that this is fallacious "one must know something about bridges— namely, that they are large visible objects" (p. 123)

> IF3:
> A chess player is a person.
> Therefore, a bad chess player is a bad person.

While this is not a formally valid argument, Hurley indicates "to detect this fallacy one must know that the meaning of the word 'bad' depends on what it modifies, and that being a bad chess player is quite different from being a bad person" (p. 123). Thus, this commits the fallacy of equivocation on the meaning of the word good and its application to persons and chess players.

3. INFORMAL FALLACIES AND PHRONESIS

The question that arises here is this: How does the way in which one comes to know that an argument is fallacious differ between a formal and an informal fallacy? Given these examples, it seems fairly straightforward to say that formal fallacies can be known generally and theoretically. S can know that argument A is fallacious by a general test of a formal criteria, e.g. Venn diagrams, or truth tables. However, when it comes to informal fallacies, the means by which S comes to know that A is fallacious, is only by particular and practical means. For example, in IF1, S only knows that the argument is fallacious if S knows the particular word in this case is ambiguous in the English language. In IF2,

S must have experience with actual bridges (or at least have conceptual clarity about the nature of bridges). Finally in IF3, S must have some understanding of the relationship between the essential nature of chess players, human beings and good members of each kind. It seems especially the case that in IF3, the way in which we have knowledge that this argument commits the fallacy of equivocation (or perhaps even begging the question) is by means of a kind of wisdom that is not merely, theoretical, but is also a kind of practical wisdom, what Aristotle calls *phronesis*.

Second, since informal fallacies deal primarily with practical reasoning, the technique that one uses to identify them will be deeply intertwined with one's practical wisdom or *phronesis* (also translated as prudence). Let me indicate four things which Aristotle says about practical wisdom, and then say how these ideas connect with the skill of identifying informal fallacies.

> A1: Practical wisdom is "quality of mind concerned with things just and noble and good for man" (Aristotle, 1984, 1143b20).
> A2: "The function of man [viz. rational activity] is achieved only in accordance with practical wisdom as well as with moral excellence [virtue]" (1144a5).
> A3: "It is impossible to be practically wise without being good" (1144a31-35).
> A4: "It is not possible to be good in the strict sense without practical wisdom, nor practically wise without moral excellence" (1145a1).

A1-4 give us a picture of the connection between practical wisdom, the essential nature of being human and the particular inferences that are made in cases like informal fallacies. According to Aristotle, we are essentially rationally active beings (e.g. the notion of human function is discussed in Book 1 Chapter 7 (1097a15-1098a20) of the *Nichomachean Ethics*), whose essences can only be fully actualized with a combination of theoretical, practical and moral excellences. In fact, Aristotle seems to make a very strong bi-conditional claim: S only has practical wisdom IFF S is morally good. If Aristotle is right about this, then we have the following syllogism:

> P1: All things that require practical wisdom are things that require moral excellence.
> P2: All analysis of informal fallacies are things that require practical wisdom.
> C: All analysis of informal fallacies are things that require moral excellences.

4. APPEALS TO PITY VS. ARGUMENTS FROM COMPASSION

Let us turn to a particular fallacy: the appeal to pity. Consider Hurley's definition: "The appeal to pity fallacy occurs when an arguer attempts to support a conclusion by merely evoking pity from the reader or listener" (Hurley, 2014, p. 127). Here is a textbook example:

> Taxpayer to judge: Your Honor, I admit that I declared thirteen children as dependents on my tax return, even though I have only two. But if you find me guilty of tax evasion, my reputation will be ruined. I'll probably lose my job, my poor wife will not be able to have the operation that she desperately needs, and my kids will starve. [Therefore], surely I am not guilty. (Hurley, 2014, p. 127).

The point of this argument is to show that the external conditions of a loss of reputation or job, not being able to support a family is irrelevant to the question of whether the person is guilty or innocent of tax evasion due to overstating his dependents. We can tell whether this argument is fallacious by understanding the conditions of what it means to be guilty (or innocent) of tax evasion. In this case, following Aristotle's notions of practical wisdom, evaluating the argument would require *knowledge* of those things which are just and noble for the good of the person in question. In fact, some facility of the notion of juridical guilt and innocence which goes beyond the merely procedural would surely be required in this case.

The thing which makes this particular fallacy, the appeal to pity, so helpful in drawing out Aristotle's ideas of the role of practical wisdom in fallacy identification, is that according to Hurley, the appeal to pity has a clear cut exception with its own name. Such non-fallacious arguments are according to Hurley, "Arguments from Compassion" (AC). These arguments will:

> AC1 "evoke compassion on behalf of some person;"
> AC2 "supply information about why that person is genuinely deserving of help or special consideration;"
> AC3 "show that the person in question is a victim of circumstances" and is
> AC4 "not responsible for the dire straits he finds himself in;"
> AC5 show "that the recommended help or special consideration is not illegal or inappropriate,"
> AC6 and "will genuinely help the person in question" (Hurley, 2014, p. 127).

Let us consider these criteria AC1-6. First, there is information needed about why a person deserves help. But, to understand why a person needs assistance requires just the type of practical wisdom and moral background information which Aristotle suggests. Second, Hurley indicates that the person seeking help is a victim of circumstances and not responsible. This is a very easy thing to say, but how easy a thing is it to determine? How much background information must someone need in order to determine which circumstances are relevant or what kind of straits one is in, and how much responsibility falls on the person in question? Answers to these kinds of questions require the kind of prudence or *phronesis* which Aristotle indicates is a necessary fulfilment for human nature. Third, sometimes giving certain kinds of help to someone might be illegal but nevertheless good or moral (e.g. giving refuge in one's home to a dissident fugitive from a vicious political regime). Similarly, some help might be legal but possibly not morally appropriate (as we shall consider in the next case). Sorting this all out cannot come merely by examining textbook definitions, and I would suggest requires practical wisdom in order to distinguish between a fallacious appeal to pity and a non-fallacious argument from compassion.

> As you might know, I care deeply about stem cell research. In Missouri, you can elect Claire McCaskill, who shares my hope for cures. Unfortunately, Senator Jim Talent opposes expanding stem cell research. Senator Talent even wanted to criminalize the science that gives us a chance for hope. They say all politics is local, but that's not always the case. What you do in Missouri matters to millions of Americans — Americans like me (Fox, 2006).

Here is one plausible way to construct this political advertisement as a formal argument, which possibly commits the informal fallacy of appeal to pity:

> P1: If MJF (and many other) are suffering from horrible diseases, then if it is possible that stem cell research is able to help people like MJF then we should vote for McCaskill who supports stem cell research.
> P2: MJF (and many others) are suffering from horrible diseases.
> P3: It is possible that stem cell research is able to help people like MJF.
> C: We should vote for McCaskill.

Is this an appeal to pity or an argument from compassion?

If it were an appeal to pity, then the conclusion would be supported "merely evoking pity from the reader or listener." If it were a non-fallacious argument from compassion, then A1-A6 would have to be met. My claim is this: being able to tell whether this is a fallacious appeal to pity or a non-fallacious argument from compassion, will hinge on practical insight, practical wisdom into two moral questions implicit in this argument regarding opposition to stem cell research:

> MQ1: Does embryonic stem cell research kill human persons?
> MQ2: Is it morally permissible to kill some human persons in order to help other persons who are suffering from debilitating diseases?

If the answer to the first question (MQ1) is, "No" or if the answer to the first question (MQ1) is, "Yes" and the answer to the second question (MQ2) is "Yes" then, this may simply be an Argument from Compassion (assuming that such research will actually help people like MJF, see A6 above). However, if the answer to the first question is "Yes", and the answer to the second question is "No," then while this argument appears to meet AC1-4, and possibly AC6, it would certainly violate AC5, even if such research were legal, it would be (to put it mildly) "inappropriate" to kill some people in order to help others who are suffering, in the same way that it would be (to put it mildly) "inappropriate" to experiment on homeless people in order to find a cure for spinal cord injuries (see the film *Extreme Measures*).

5. ARGUMENTATION AND WISDOM

So which is it: Appeal to Pity or Argument from Compassion? It is not my purpose in this paper to debate the answers to MQ1 and MQ2 here in this paper. It is my purpose to propose this point: determining how to tell (in specific cases) whether an argument is informally fallacious requires much more than simple textbook definitions, even when those definitions are followed with textbook cases. Deciding in this case whether or not the MJF argument is fallacious requires the kind of practical wisdom which is able to answer specific moral questions such as: What is a human person? Can some persons be destroyed in order to help others?

To return to Aristotle, finding the answers to such questions will require hard work for any student of argumentation and logic. It will require as Aristotle says, a "quality of mind concerned with things just and noble and good for man." How this happens is of great concern, not only for students of argumentation, but also for those who teach it. Yet, it must be done if Aristotle is right that we can have practical wisdom if and only if we are morally good.

6. SOME CHALLENGES TO TEACHING FALLACIES EVEN WITH PHRONESIS

Trudy Govier in her article, "Who Says There Are No Fallacies?" (Govier, 1983) argues persuasively against the idea that there are no informal fallacies, and thus they simply can't be taught at all, even if practical wisdom is had by someone analyzing a particular fallacy. Govier identifies at least three problems of teaching fallacies. These problems, if left unaddressed could lead one to the conclusion that fallacies just aren't real or important when teaching the basics of logic.

6.1 The Standards Problem

The first problem of teaching fallacies is what I want to call the "standards" problem. Govier identifies this problem in this way. She writes,

> In order to say that the reasoning, as characterized, embodies a mistake, it may be necessary to invoke a standard of good reasoning. Thus it is no simple matter to demonstrate that someone has argued in such a way as to have committed a fallacy. And this point, which should be remembered, is of considerable pedagogic significance. It may be an important reason against trying to teach critical skill by teaching fallacies (Govier, 1983, p. 3).

Govier does not address this problem as thoroughly as she does the others (listed below) in her paper. I believe that this problem may be a specific case of relativism about knowledge or about logic itself. It has been my experience that relativism is a problem for many students not only in teaching ethics (where one would expect relativism to show up), but also in logic, especially critical thinking. When I ask students whether there are "right ways to think" they generally say, "No."

Govier is correct that if there are *mistakes* (i.e. fallacies) in reasoning, then there must be *correct* ways to reason, which does require "a standard of good reasoning." In order to ameliorate this problem, I suggest a two-pronged strategy with students. First, demonstrate the problems of relativistic thinking (e.g. its self-refuting nature, etc.). Second, show that there are in fact reasonable standards for good reasoning, such as those required by the traditional square of opposition in deduction (i.e. non-contradiction), or coherence or probability (i.e. the general concept of inductive strength and weakness). These concepts, while they can be debated among philosophers, do have a prima facie and common sense reasonableness to them, which can be validated in countless individual experiences. Presenting these to the students can move the ball forward in learning that 1) it is *possible* that there are right and wrong ways to think (by refuting relativism), and 2) fallacies do not meet reasonable standards apparent in common sense and traditional views of deduction and induction. This may help them on their way to obtaining the third and crucial thing (discussed above) which is *practical wisdom*.

6.2 The Limitation Problem

The limitation problem is a problem which arises from the mistaken view that logical analysis is limited to formal deductive analysis. I believe that this problem is a vestige of logical positivism and any *programme* seeking to reduce thought to language, language to symbols, and symbols to matter. Logic seems to me broader than the deductive and not on the surface reducible to the linguistic (see Willard, 1973).

Govier analyzes Lambert and Ulrich's claim in *The Nature of Argument* that the only form of logic that only deductive analysis is worthwhile. Those authors argue that studying fallacies does not help students to learn reasoning skills, because reasoning is solely or primarily deductive in nature. However, according to Govier, this seems unfairly to limit reasoning to deduction, specifically deductive logical structure rather than inductive, analogical or logical content. Govier argues that good reasoning does include non-deductive reasoning and much of the study of informal fallacies can fall into these other areas. In addition, she argues that even in deductive analysis fallacies can be informal. For example, what is commonly called a false dichotomy or false dilemma can occur in valid instances of a disjunctive syllogism.

6.3 The Interpretive Problem

Govier cites Finocchiaro who asks, "The problem I wish to raise here is, do people actually commit fallacies as usually understood? That is, do fallacies exist in practice? Or do they exist only in the mind of the interpreter who is claiming that a fallacy is being committed?" (1983, p. 5). Let us call this the Interpretive Problem. This problem centers on the issue faced by many after being familiarized with quite a few textbook examples of common fallacies. They begin to see fallacies everywhere they look. However, perhaps there is something one could call an *argumentum ad fallacium*. (I'm sure this idea is not original to me, but I'm not sure to whom to attribute it). This fallacy would be committed when an arguer attempts to refute another argument by mistakenly/wrongly accusing it of committing a fallacy, when it in fact does no such thing. Of course, this could occur by a simple interpretive mistake. When an arguer does not carefully consider both the context of an argument, an arguer's meaning and the structure of an argument itself, the interpretation can be botched and the *argumentum ad fallacium* committed.

Regarding the interpretive problem, Govier also points out that, "In practice it is often difficult to tell whether people are offering arguments or not... It is entirely possible that some alleged appeals to force and pity are most plausibly interpreted in a way which does not make them out to be arguments" (1983, p. 6).

Govier responds to the interpretive problem raised by Finocchiaro by stating,

> We should see that there is a fair interpretive basis for regarding her [i.e. the arguer] as having offered an argument and we should try to get some sense for the certainty with which she wishes to assert her conclusion. Sensitive interpretation is a basic component of good argument assessment-and it is a skill under-emphasized in much standard pedagogy (1983, p. 7).

Govier suggests that something like the principle of charity in interpreting arguments can be helpful in solving the interpretive problem.

7. HOW TO TEACH FALLACIES

Philosophers have been teaching their students to avoid mistakes in reasoning since the beginning of western philosophy. The teaching of logical fallacies can be traced back to Aristotle's *Sophistical Refutations*. It seems that the ability to reason poorly has been with us for a long time. Aristotle was dealing with the same issues that arise in 21st century textbooks.

Aristotle claims that "for some people it is better worthwhile to seem to be wise, than to be wise without seeming to be" but for those who would be reasonable, "it is the business of one who knows a thing, himself to avoid falsities in the subjects which he knows and to be able to show up the man who makes them" (Aristotle, 1984, p. 279). He cites fallacies which "depend on the assumption of the point at issue" (p. 283), e.g. begging the question. He argues against fallacies which depend "upon treating as a cause what is not a cause" (p. 283), i.e. false cause fallacy. He argues against fallacies which depend on "homonymy and ambiguity" (p. 301), i.e. equivocation and amphiboly. He argues against those "refutations which make several questions into one" (p. 309), i.e. complex question.

From Aristotle to the most recent editions of logic and critical thinking textbooks, fallacies have been taught as mistakes to be avoided. There remain philosophical questions about the place of teaching fallacies should have in full courses in critical thinking. First, should they be taught at all? Second, if so, how?

7.1 The Minimalist Approach

David Hitchcock argues that teaching fallacies should (at best) be taught very minimally. He argues that there are areas of fallacious deception in ordinary discourse (1995, p. 324) but the best way to teach students to avoid fallacies is not simply through an enumeration and exemplification of mistaken modes of reasoning. He makes an analogy with sports. Teaching fallacies "is like saying that the best way to teach somebody to play tennis without making the common mistakes... is to demonstrate these faults in action and get him to label and respond to them" (p. 324). Hitchcock argues that "it makes more sense to teach the analytical apparatus for correct reasoning... than to begin with fallacies" (Hitchcock, 1995, p. 325).

Hitchcock's sports analogy is a good one, but it is limited solely to those courses or texts which only teach critical thinking or reasoning skills by means of fallacy identification and analysis. Hitchcock indicates that a teacher ought to be one who should focus on "how to do it correctly" (p. 324). I agree with Hitchcock here. In my courses, antecedently to teaching fallacies, I expect students to have some facility in argument identification and analysis of deductive validity using traditional Aristotelian deduction.

An additional concern raised by Hitchcock is that focusing on fallacies can foster in students "an attitude of looking for the mistake, and of stopping once one has found something one can pin a fallacy label on, rather than coming to grips with the substance of what one is discussing" (p. 326). I can attest to this in my own teaching of fallacies, especially when using video clips, which limit the range of discussion to rather brief (1-2 minute) sound-bites. This problem can be solved in several ways. First, having students watch an entire debate in context and identifying fallacies which may occur within that context. Second, having students write out an argument with clear premises and a conclusion in order to show how the arguer is arguing so that a charitable interpretation of the argument can be assessed.

7.2 The Non-Minimalist Approach

Anthony Blair suggests that one way to approach teaching fallacies consists in "supplying paradigm examples and by identifying the typical conditions constitutive of the fallacies in question" (Blair, 1995, p. 332). However, Blair indicates that alongside teaching fallacies, one must simultaneously introduce "the cogent arguments that the fallacious ones approximate or counterfeit, and the principles of their cogency" (p. 332). This will allow students to identify fallacies *and* know the difference between those that are not (p. 332).

Blair's suggestion then does some work in overcoming Hitchcock's concern about teaching fallacies in the analogy of sports. One does *both*. One shows common mistakes, how to avoid them in the context of the right moves of the game. In addition, one can avoid the interpretive problem above by having a facility for good argument construction which allows one to have the ability to interpret other arguments for their strength or weakness.

Blair concludes that there are three options for teaching fallacies in critical thinking type courses. First, he advocates that the teaching of fallacies in courses that are designed around them is a good thing (1995, p. 338). Second, he advocates teaching a few fallacies in a course that is not designed around them. Third, he does not advocate teaching fallacies "as a separate short unit in a course in argumentation or in reasoning" (p. 338). Blair believes that due to the complexity, the need for qualifications, and the need for "practice and coaching" (p. 335), fallacies cannot be taught well in a small portion of a course on reasoning or argumentation. Blair claims,

> A two- or three-lecture packet risks educational irresponsibility. No two- or three-week unit can do them all justice. Almost certainly more harm than good is done in briefly sketching a distortingly oversimplified conception of fallacy, in ignoring or underemphasizing the difficulties of interpreting working arguments, and in rushing over the carful discriminations and qualifications that are part of appropriate argument evaluation using fallacies (1995, p. 335).

I agree with Blair on this general point. Rushing through teaching fallacies, as does rushing through teaching anything is harmful for students. A one or two day lecture on ten or more fallacies would not be particularly helpful to students. In my courses, I usually spend 4-5 weeks on five fallacy types with about twenty five fallacies covered. This is meant as an introduction to fallacious forms of reasoning, and does require careful coaching and analysis of fallacious arguments especially with respect to the issue of argument interpretation.

8. CONCLUSION

I believe that with careful interpretative training, by using concrete examples from ordinary discourse, including videos of debates and commercials, students can receive an initial training that allows them to have the beginnings of discriminatory abilities to distinguish between good arguments which might appear at first glance to be fallacious and those which really are fallacious. In some cases, the question of the fallaciousness of an argument *is* a matter of debate, as we have seen above. This analysis and debate itself can be a useful heuristic device challenging the students to give good (i.e. non-fallacious) arguments for their positions on why a particular argument should be understood as fallacious or not. This is the type of habituation and practice required for the kind of phronesis and practical wisdom that our world so desperately needs.

REFERENCES

Aristotle (1984). *Nichomachean Ethics*. Translated by W.D. Ross, in *The Complete Works of Aristotle*, edited by Jonathan Barnes. Princeton: Princeton University Press.

Blair, J. A. (1995). The place of teaching informal fallacies in teaching reasoning skills or critical thinking. In H. V. Hansen & R. C. Pinto (Eds.), *Fallacies* (pp. 328-338). University Park, PA: Pennsylvania State University Press.

Fox, M. J. (2006). Michael J. Fox. www.youtube.com/watch?v=a9WB_PXjTBo.

Govier, T. (1983). Who says there are no fallacies? *Informal Logic, 5*, 2-10.

Hitchcock, D. (1995). Do fallacies have a place in the teaching of reasoning skills or critical thinking? In H. V. Hansen & R. C. Pinto (Eds.), *Fallacies* (pp. 319-327). University Park, PA: Pennsylvania State University Press.

Hurley, P. (2014). *A Concise Introduction to Logic, 12th Ed.* Stamford, CT: Cengage.

Willard, D. (1973). The absurdity of thinking in language. *The Southwestern Journal of Philosophy, IV*, 125-132.

45

An Agentive Response to the Incompleteness Problem for the Virtue Argumentation Theory

DOUGLAS NIÑO
Universidad Jorge Tadeo Lozano, Colombia
edison.nino@utadeo.edu.co

DANNY MARRERO
Universidad Jorge Tadeo Lozano, Colombia
danny.marrero@cij.edu.co

This paper outlines an agent-centered theory of argumentation. Our working hypothesis is that the aim of argumentation depends upon the agenda agents are disposed to close or advance. The novelty of this idea is that our theory, unlike the main accounts of argumentation, does not establish a fixed function that agents have to achieve when arguing. Instead, we believe that the aims of argumentation depend upon the purposes agents are disposed to achieve (agendas).

KEYWORDS: agenda, agent, agent-centered theory of argumentation, function of argumentation

1. INTRODUCTION

Virtue argumentation theorists claim to adopt an *agent-centered* approach to argumentation. From our perspective, an agent-centered approach to argumentation should provide an explanation of what an agent is and the role arguments have in agents' cognitive economies. Therefore, such approach should provide an account of "what agents are like, what their interests are, and what they are capable of" (Gabbay & Woods, 2009, pp. 70-71). Virtue argumentation theorists seem not to satisfy this requirement, because their focus is the virtues an arguer might or might not have, but not the arguer as such. This is our argument:

D. Mohammed & M. Lewiński (eds.) (2016). *Argumentation and Reasoned Action: Proceedings of the 1st European Conference on Argumentation, Lisbon, 2015. Vol. II, 733-741.* London: College Publications.

(1) The traditional virtue argumentation approach claims it is an agent-centered approach.
(2) The traditional agent-centered approach studies the virtues an agent has.
(3) The virtue argumentation approach is a virtue-centered approach, and not an agent-centered approach.

For instance, Andrew Aberdain claims: "We have seen that a virtue theoretic approach to argument must focus on agents rather than actions. This entails distinguishing good from bad arguers rather than good from bad arguments ... This raises the question of what the virtues of the ideal arguer are expected to track" (2010, pp. 171-173). This quote confirms the allegedly methodological option of the virtue argumentation approach, as it is stated by proposition (1). Proposition (2) shows that the virtue argumentation approach does not study agents, but their virtues. Therefore, this is a virtue-centered approach, and not an agent-centred approach, as it is stated in proposition (3).

From our perspective, the focus on virtues and not on agents, leads the virtue argumentation approach to the Incompleteness Problem (IP). First, they cannot deal with conflicts of argumentative virtues without using extra-virtue theoretic considerations. Second, they do not justify why it is good to be a virtuous arguer without using utilitarian considerations (MacPherson, 2013, p. 1). The aim of this paper is to provide an agentive response to IP. First, we will outline Daniel H. Cohen's virtue argumentation approach as it was presented in "Virtue Epistemology and Argumentation Theory" (2009). Second, we will show how Cohen's virtue argumentation theory is vulnerable to IP, as it was suggested by Brian MacPherson in "The Incompleteness Problem for a Virtue-based Theory of Argumentation" (2013). Third, we will propose an agentive solution for IP. Our working hypothesis is that a theory of argumentation clarifying the cognitive agendas of the agents arguing not only provides a response for IP, but also shows the virtues of argumentation in agents' cognitive enterprises.

2. COHEN'S VIRTUE ARGUMENTATION APPROACH

Cohen's main motivation to adopt a virtue approach to argumentation takes the form of an analogy: As the virtue approach has been theoretically fruitful for ethics and epistemology, it could be fruitful for argumentation theory. Let's explore the first term of the analogy. On one hand, virtue ethics is "better situated than [its] consequentialist and deontological counterparts to recognize, accommodate, and appreciate ethical but non-moral values without flattening them into moral values."

On the other, virtue epistemology is "perfectly situated to recognize, accommodate, and appreciate *cognitive* but *non-epistemic* values without having to flatten them into the standard epistemological categories" (Cohen, 2009, p. 2). For Cohen, virtue ethics takes into account "important goods" that alternative theories such as deontology and consequentialism do not. For example, virtue ethics is able to explain why family and friendship are *ethically good* but not *morally obligatory*: family and friendship are goods that contribute to the enjoyment of life, but someone lacking family or friendship networks is not morally blameworthy. Virtue epistemology seems to have a similar explanatory power. While the received epistemological view exclusively focuses on the attainment of propositional knowledge, virtue epistemology recognizes both that "[t]here are many different cognitive achievements in addition to knowledge and justified belief" and that "[t]here are different cognitive *abilities* leading to those achievements that are not reducible to propositional knowledge" (Cohen 2009, p. 3). For instance, traditional epistemologists cannot see the virtue of closed-mindedness and the vice of open-mindedness because they do not seem to contribute to the justification of our beliefs. However, virtue epistemology would claim that close-mindedness is virtuous and the opposite is vicious in cases in which a justified belief has been achieved, and it is counter-productive to re-open unnecessary questions (Cohen, 2009, p. 4).

If Cohen is right, virtue ethics and virtue epistemology have a distinctive element giving them their allegedly explanatory power. For him, that is the agent-centered approach adopted by both views. Firstly, "[v]irtue ethics focuses broadly on agents and their lives, rather than narrowly on just their actions, just their motives, or just governing principles". Secondly, virtue ethics had "a change in focus from beliefs to believers" (Cohen, 2009, p. 2), that is, "from *what to think* to *what kind of thinker to be*" (Cohen, 2009, p. 4). Consequently, Cohen's contention is that if argumentation theory wants to be as fecund as virtue ethics and virtue epistemology seem to be, it has to adopt an agent-centered approach, too. From his view, this means that theories of argumentation should agree on the claim that "a good argument is one that has been conducted virtuously" (Cohen, 2009, p. 1). Spelling out Cohen's concept of a good argument calls for the clarification of the concept of "virtue." In his words, virtues are "the conditions that are conducive to the desired ends" (Cohen, 2009, p. 4). Therefore, a good argument is the one satisfying the conditions for the achievement of the desired end. This second formulation of Cohen's concept of a good argument requires some precision on the concept of "desired end." Because one of the things arguments can do is to transform agent's doxastic states from disbelief to belief via persuasion, arguments have to do with cognitive ends (Cohen, 2009, p. 5). Good, or virtuous, arguments, as a result, are catalysts of cognitive achievements. For instance, along with persuasion, other argumentative virtues allowing cognitive advancements are:

> • a deepened *understanding* of one's own position;
> • the *improvement* of one's position;
> • the *abandonment* of a standpoint for a better one - *other than the opponent's*;
> • a deepened understanding of the *opponent's* position;
> • a deepened *appreciation* of the opponent's position;
> • *acknowledgement* of (the reasonableness of) another's position;
> • greater *attention* to previously over-looked or under-valued details;
> • better *grasp* of connections and how things might be fit together in a big picture;
> • *entitlement* to one's own position. (Cohen, 2009, p. 7)

3. THE INCOMPLETENESS PROBLEM FOR THE VIRTUE ARGUMENTATION THEORY

Virtue argumentation theories, such as Cohen's, have been recently attacked by the Incompleteness Problem (*IP*). According to MacPherson, the incompleteness of virtue argumentation theories are two sides of the same coin. On one side, the virtue approach to argumentation is incomplete because it has to recur to elements external to the virtue approach to resolve conflicts of argumentative values. On the other side, if virtue argumentation theorists confine themselves to the limits of their approach, they are not able to "provide us with a reason for why it is good to be a virtuous arguer" (MacPherson, 2013, p. 1). This being so, the virtue-centered approach to argumentation is vulnerable to *IP* because its theoretical tools are not enough for dealing with some of the problems arising from its own theoretical framework. What is more, when virtue argumentation theorists deal with problems such as the mentioned above, they are forced to use elements belonging to their antagonistic theories such as the utilitarian or the dialectical-obligation approach to argumentation.

MacPherson illustrates the problem of conflicts of argumentative virtues with the following example:

> Suppose two evolutionary biologists [Deborah and Ibrahim] are deeply divided on the issue of whether evolution is a result of mutation, natural selection, migration, and genetic drift

> (traditional theory) or whether evolution is explainable simply in terms of genetic drift (the "neutral" theory). ... [S]uppose that Ibrahim like Deborah values both tenacity with respect to one's position along with keeping an open mind. Ibrahim is committed to the neutral theory of evolution on the grounds that it can offer new insights into the evolution of sexual reproduction in eukaryotic organisms, which remains somewhat of a mystery in the context of a more traditional account of evolution. However, because he also values being open-minded, he does not carry his tenacity to a vicious extreme. He is willing to concede to Deborah's arguments for the traditional view, provided that they are cogent and closer to the truth than the neutral theory. He is even willing to completely abandon his own account of evolution if it does not stand up to the traditional account of evolution, but not to the point of simply abandoning his position without putting up a serious defense of his own views. (MacPherson, 2013, pp. 3-4)

According to MacPherson, this case shows the main elements of *IP*. First, in this case, there is a clash of argumentative values. On one hand, Ibrahim values the tenacity with respect to his own position given increasing evidence in the scientific literature supporting the neutral theory. On the other hand, he considers it important to keep an open mind so that he recognizes arguments that could defeat his position. None of these argumentative values are endorsed to an extreme position making them a vice. That is, Ibrahim neither defends his position to the point of becoming dogmatic, nor is he as open-minded as to accept any kind of argument against his point of view. Secondly, given that the virtue argumentation theory does not provide criteria for valuing an argumentative value over another, Ibrahim cannot, with the virtue argumentation approach, decide whether to prefer to be tenacious defending his position or to be open-minded to recognize possible defeaters against his theoretical weaknesses. Finally, if one of the values under conflict had to be preferred over the other, extra virtue theoretical elements should be used, such as if Ibrahim would have preferred to stay open-minded because of the utilitarian criterion of the progress of science, or if he would have chosen to be tenacious defending his view because he believes there is a moral responsibility to defend a neutral theory of evolution.

To clarify *IP*'s second aspect, MacPherson recalls that Cohen grounds argumentative virtues on cognitive achievements as the understanding and improvement of one's own position, the abandonment of a standpoint for a better one, the understanding and appreciation of the opponent's position, and so on. In this sense, "open-mindedness is an argumentative virtue grounded in the cognitive achievement of more deeply appreciating other points of view. On the other hand, tenacity is an argumentative virtue grounded in the cognitive achievement of entitlement to one's own position" (MacPherson, 2013, p. 5). Consequently, MacPherson interprets that Cohen claims that we should be virtuous agents because it leads to the attainment of important cognitive ends "that are universal and regarded as important by all human beings". The problem for Cohen is that his cognitive achievements are not necessarily universal. For instance, the deepened appreciation of another's position might be different for different individuals. Additionally, the cognitive achievements suggested by Cohen can conflict with each other. This takes us back to the lack of criteria for resolving conflicts of argumentative values.

4. AN AGENTIVE RESPONSE TO THE INCOMPLETENESS PROBLEM

Our working hypothesis is that a theory of argumentation clarifying the cognitive agendas of the agents arguing not only provides a response for *IP*, but also shows the virtues of argumentation in agents' cognitive enterprises. Our inspiration comes from the theory of fallacies as cognitive virtues of Dov Gabbay and John Woods (2009), more precisely, from their multi-volume of *Practical Logic of Cognitive Systems* (2003; 2005; Woods, 2013). Shortly, from this view, some fallacies are cognitive virtues when they meet the conditions of resolution of the agenda, or agendas, an agent is disposed to close or advance. To clarify, an *agent* is an entity with capacity for acting and the objective he/she is trying to obtain is his/her *agenda* (Gabbay & Woods, 2003, pp. 183-185; pp. 195-219; Niño, 2015, p. 39).[1] According to Dov Gabbay and John Woods, there are two factors that determine the different types of cognitive agents (2005, p. 11). Firstly, there is the degree of command of resources (time, information and computational capability) an agent needs to advance or close his/her agendas. Secondly, is the height of the cognitive bar that the agent has set for him/herself. With this in mind, Gabbay and Woods incorporate a *hierarchical approach to agency*. It postulates a hierarchy in which agents are placed in light of their interests and their capacities. In this model, individuals would be placed

---

[1] There is a remarkable difference between Gabbay and Woods's and Niño's approaches. While in Gabbay and Woods's proposal, the agenda is understood as the aim plus the plan for its realization, in Niño's account, the agenda is just the objective, and it is distinguished from its conditions of resolution. As a consequence, while in the former, if something is not planned, it is not an agenda, in the latter, everything that appears as an aim for an agent, either deliberately planned or not, is recognized as an agenda.

towards the bottom of the hierarchy and theoretical agents would be higher up. While individuals "perform their cognitive tasks on the basis of less information and less time than they might otherwise like to have", theoretical agents "can wait long enough to make a try for total information, and they can run the calculations that close their agendas both powerfully and precisely" (2005, pp. 11-12).

The standard of exigency for determining the precision of a response changes in each case: a major precision involves more specific information or more strict reasoning methods. This is why it is exaggerated to call for inductive standards (sampling, probabilities calculus, etc.) for practical agendas of practical agents such as us. For instance, whereas in the traditional view the arrival to a generic proposition such as tigers are striped, from the fact that one has seen a striped tiger would be considered a fallacy, in Gabbay and Woods approach, this reasoning is considered as a real cognitive virtue because without a reasoning of this kind, our species would have been not survived (2009, p. 86). On the contrary, inductive standards are exigible to theoretical agents such as NASA who have a great amount of time, information and computational capacity.

Let us examine MacPherson's response to *IP*. In his words:

> Suppose Ibrahim finds out that his research proposal for the [Natural Sciences and Engineering Research Council (NSERC) of Canada] defending the neutral theory of evolution has a good chance of being short-listed since there is increased interest in the scientific community regarding the neutral theory. Then it may be in his best interest to pursue tenacity with respect to his own theory and to focus less on being open-minded with respect to defenses of traditional evolutionary theory. If Ibrahim's project is funded, he can employ graduate students to help out with the research along with bringing prestige to his institution, and it may advance significantly the field of evolutionary theory. (...)

For MacPherson, the conflict among virtues (open-mindedness vs. tenacity) is solved when one keeps in mind Ibrahim's agendas. However, MacPherson takes for granted what an agent is, particularly, an agent who is a biologist researcher. Certainly, his discussion is about the conflict among virtues, but it is important to establish how an agent adopts certain goals or entertain certain interests and not others. The social role of a scientist in his/her community implies that in order to advance a hypothesis he/she should take into account theoretical factors such as simplicity, consilience and coherence (cf. Thagard, 1978; Lipton, 2004), compacity, falsiability, and consequences on other hypotheses. Considering *all* those factors is part of the standard of exigency and precision a good scientist should take into account. In MacPherson's example, Ibrahim does not contemplate all those factors, but he can choose tenacity because, among other things, he can fund his research and bring prestige to his university. Thus, some relevant theoretical considerations are not taken into account. Ibrahim's consideration, instead, seems to be advantageous for the institutional or administrative community, but it does not seem evident that it is so for the scientific community. On the contrary, this would imply giving priority to some agendas related to other roles, rather than the one he is supposed to fulfill. The point here is not only to keep in mind the agendas an agent has at any moment, but also if this agent adopts the agendas (and their standards) he/she should have adopted at each moment and how he/ she behaves up to it.

Notice that in certain social roles (i.e., being a lawyer, a medical doctor, a biological researcher, etc.) the exigible standard for closing the agendas is established socio-historically. To illustrate, the determination of a certain body temperature as fever requires two exigency and precision standards if the agent is Jimmy's father or Jimmy's pediatrician. In this sense, an agent performance *evaluation* relates the efficiency and effectiveness of the deployment of his/her cognitive resources in reference to the standard that his/her agendas demand for their proper closure. When an agent accepts or adopts a social role (or at least, when it is imputable to him/her), it is exigible to him/her that his/her performance meets those standards. This shows that in order to behave properly, the agent needs to align his/her own agendas to the agendas specified by the role he/she is supposed to embody. When this is the case and the agent systematically succeeds in achieving those role-related agendas, with their socially standardized levels of exigency and precision, the agent is not only skillful, but also virtuous.

5. CONCLUSION

To conclude, from our view, an agent is an entity with capacity for acting, which means, to have agendas and to try to carry them out according to the resources at his/her disposal. Someone is a virtuous agent -in the sense of being doxastically reliable - when, under his/her actual circumstance, systematically and successfully uses the resources at his disposal as a means of adjustment and compensation in order to obtain his/her agendas in the best way. The evaluation of "the best way" occurs as a relationship to the exigency and precision the standards of the agendas that are intended (or should be intended) to close according to the roles he/she embodies, and this involves another type of virtuosity: doxastic responsibility. This implies, in itself, also to choose which agendas to give priority, and accordingly, this involves that he/she should align her personal agendas with those imposed and expected by the role.

REFERENCES

Aberdein, A. (2010). Virtue in Argument. *Argumentation, 24,* 165–179.

Cohen, D., (2007). Virtue epistemology and argumentation theory. In H.Hansen & C. Tindale (Eds.), *Dissensus & The Search for Common Ground. Proceedings of the 7th International Conference of the Ontario Society for the Study of Argumentation* (OSSA), Windsor, 6-9 June 2007, (pp. 1-9). http://scholar.uwindsor.ca/ossaarchive/OSSA7/papersandcommentaries/29/

Gabbay D., & Woods, J. (2003). *Agenda relevance: A study of formal pragmatics. A practical logic of cognitive systems,* Vol. 1. Amsterdam: Elsevier.

Gabbay D., & Woods, J. (2005). *The reach of abduction: Insight and trial. A practical logic of cognitive systems,* Vol. 2. Amsterdam: Elsevier.

Gabbay, D., & Woods, J. (2009) Fallacies as cognitive virtues. In O. Majer, A. Pietarinen & T. Tulenheimo (Eds.), *Games: Unifying logic, language and philosophy* (pp. 57-98). Dordrecht: Springer.

Lipton, P. (2004). *Inference to the best explanation.* London/New York: Routledge.

MacPherson, B. (2013). The incompleteness problem for a virtue-based theory of argumentation. In D. Mohammed & M. Lewiński (Eds.), *Virtues of Argumentation. Proceedings of the 10th International Conference of the Ontario Society for the Study of Argumentation* (OSSA), Windsor, 22-26 May 2013, (pp. 1-8). http://scholar.uwindsor.ca/ossaarchive/OSSA10/papersandcommentaries/111/

Niño, D. (2015). *Elementos de semiótica agentiva.* Bogotá: Universidad Jorge Tadeo Lozano.

Thagard, P. (1978). The best explanation: Criteria for theory choice. *The Journal of Philosophy, 75,* 76-92.

Woods, J. (2013) *Errors of reasoning: Naturalizing the logic inference.* Milton Keynes: College Publications.

46

Narrativity, Narrative Arguments and Practical Argumentation

PAULA OLMOS
Universidad Autónoma de Madrid, Spain
paula.olmos@uam.es

I explore the relationship between narrative arguments and practical argumentation. Going beyond the intrinsic narrativity of envisaging future scenarios related to argued-for actions, I focus in arguments involving explicit narratives. The continuum between argument and meta-argument is probably more evident in practical argumentation than in any other field, due to the complexity and loosely regulated character of the realm of human action. Narrations may be part of our societies' long-standing solutions to such an intricacy.

KEYWORDS: Aristotle, classical rhetoric, deliberative genre, narrative argument, narrative rationality, philosophy of literature, *paradeigmata*, practical argumentation, proposal, Wittgenstein

1. INTRODUCTION

In this paper I want to explore the apparently privileged relationship between narrative arguments and practical argumentation that, in my opinion, has been rather intuitively assumed and considered.

I've been lately working on different aspects of the relationship between narratives and argument: a) on the argumentative and persuasive possibilities of narrative discourse (Olmos, 2012, 2014); b) on the characteristics, structure and assessment possibilities of certain narrative arguments in context, as fables (Olmos, 2014b); or on the factors/reasons supporting the credibility of narratives and their evidential role (Olmos, 2015). Here I would like to explore, even if in a rather tentative and speculative way, the relationship between narrative discursive modes and one particular argumentative field, the realm of practical argument in contexts such as politics and ethics, which is the focus of this conference.

D. Mohammed & M. Lewiński (eds.) (2016). *Argumentation and Reasoned Action: Proceedings of the 1st European Conference on Argumentation, Lisbon, 2015. Vol. II, 743-753.* London: College Publications.

As in other contributions, I would like to make clear that my ultimate reference, my focus is on arguments involving explicit narratives as a manifest linguistic strategy, i.e. "arguments in which the particular linguistic features and genre-specific qualities of narration play a significant role" (Olmos, 2015, p. 156). In this sense, an argument involving a one-step outcome for an action (i.e. the typical example of a "practical syllogism"), supported or supportable by a simple rule as a warrant (typically the normative, evaluative, or means/ends premise of a traditional "practical syllogism"), would not count for me as a "narrative argument"; although in some sense we could say that there is an underlying *narrativity* or *narrative rationality*, according to Walter Fisher's account (Fisher, 1989 [1987]), in such characterizations of our purposeful actions.

The labels "narrative discourse", "narrative argument" or "argumentative/persuasive narrative discourse" will be saved for cases that explicitly involve a more complex (typically non-linear) succession or compound of events, typically with several agents/factors involved presented with the intention and expectation to be understood and assessed somehow *as a whole*. I claim that the less obvious bond between such modes of persuasive discourse and practical argumentation needs to be clarified.

Of course there is a long tradition in the use of apologues, fables and even novels as part of moral discourse, starting with Aristotle's suggestion about the use of *paradeigmatic* arguments in the deliberative genre, the genre whose avowed objective is "to exhort <people to> or dissuade <people from>" *sc.* engaging in actions[1]; the genre of practical argument. Aristotle claims that *paradeigmata* —including both fictive and historical "examples", generally, although not necessarily, with a narrative structure, according to my previous characterization— are *appropriate* for public and political argumentation. He says: "Fables are suitable for public speaking" (1394a1), "lessons conveyed by fables are easier to provide, and those <*paradeigmata*> derived from real facts are <*even*> more useful for deliberative oratory" (1394a7-8). I would like to explore why this is so.

Aristotle takes it somewhat for granted. That is, he states why he thinks examples taken from real facts are *even better* than fables in such argumentative field (fables being nevertheless fairly suitable)[2] but not

---

[1] *"hai te protropai kai hai apotropai" Rhet.* 1360b10: literally "activating motion/action and countering motion/action", that is, "mobilizing and demobilizing".

[2] "Because, for the most part, what's coming will be similar to what's already happened" (*"hómoia gar hōs epi to polu ta méllonta tois gegonósin"*, 1394a7).

why narrative arguments are appropriate at all for deliberative (practical) oratory. Moreover, Aristotle makes these remarks when explicitly speaking about such narrative *paradeigmata* (Book II, Chap. 20) while nothing at all about this explicit kind of persuasive discourse appears when he is supposedly describing the characteristics of deliberative argumentation (Book I, Chap. 4-8).

2. PRACTICAL ARGUMENTATION

Let's say something about "practical argumentation". Regarding human actions, human affairs and persuasive discourse about them: leaving aside, for the moment, "prediction or foreseeing" which is not our aim here (in general, we don't have much confidence in predictive arguments regarding human actions, although, for example, psychiatrists may be asked to reasonably engage in it in certain contexts), we might be interested in *a posteriori* "explaining" human action or in planning/preparing/advising/proposing human acts through what is generally called practical reasoning (which is of a teleological character according to von Wright, 1971) or its communicative counterpart, i.e. "practical argument".

As has been assumed in the recent literature, practical arguments present certain complications as opposed to theoretical arguments. Practical arguments argue for or support either individual "purposes" (plans, designs) or more collectively understood "proposals" (Kock, 2007; Vega, 2013). According to Luis Vega's account (2013, pp. 2-3):

> - An expressed "purpose": "I intend (plan, set out) to do A" involves
>   (i) The description of an action or course of action (A) *plus*
>   (ii) A pro-active attitude towards it,
>
> - For a "purpose" to become a "proposal": "I propose we do A" we need an additional condition:
>   (iii) An invitation to the interlocutor(s) to share the commitment regarding the proposed action.

It's the kind of commitment and issuing rights and duties thereof that make all the difference from "purpose" to "proposal". On the other hand, while "purposes" may remain the object of "practical reasoning", "proposals" are always the object of "practical argument" as they involve communication (Marraud, 2013, pp. 11-12). Political and deliberative argumentative discourse (as the one envisaged by Aristotle) would be typically about "proposals".

In any case, we are probably more interested in the contrast between proposals (as well as purposes) and propositions also framed by Vega (2013, p. 3) in Table 1 (in which I assume that a rather "conservative" account of propositions is presented in order to enhance the "obvious" differences with proposals):

Propositions	**Proposals**
a) Might be characterized as true/false. These alethic values belong to the propositions as *objective semantic attributes*	a) Might be evaluated as convenient, appropriate, expedient. Such values are: (i) gradual, comparative (ii) acquired in the course of deliberation
b) Are investigated/examined in order to determine their T/F value. Belong within the Cognitive/Theoretical field	b) Are considered, confronted, compared or weighed in order to establish their viability and acceptability. Belong within the Normative field
c) Present a word-to-world direction of fit	c) Present a world-to-word direction of fit
d) Are supported by theoretical argument (different schemas)	d) Are supported by practical argument (different schemas)

Table 1: Comparison between Propositions and Proposals (Vega, 2013, p. 3)

The argumentative complexities of "proposals" and of practical arguments have to do with:

a) the obviously gradual (neither plausibly nor even ideally bivalent) and comparative assessment of their claims: we have to assume we will have better and worse arguments (not valid/invalid, good/bad) about the greater or lesser convenience of our claims (not about the truth/falsity of our claims);
b) the usually value-laden character of the grounds supporting them which can make the comparison/weighing rather difficult. C. Kock (2007, 2012) talks about a "multidimensionality" of values that may cause, in certain cases, at least apparent or initial "incommensurability";
c) together with (in many cases) the "institutional" requirement to reach a decision (the necessity to set a policy, to engage in some action or other);

d) restricted, again in most cases, by the "material incompatibility" of taking at the same time different actions that allegedly would jointly reach an optimum of value-satisfaction (Kock, 2007).

Consideration c) forces us to look for some solution to the assessment and decision conflicts mentioned by Kock, among others. And there are several possibilities: some of a more argumentative character than others. Voting, for example, would be the most obvious non-argumentative or past-argumentative practice.

In order to illustrate and discuss the problem of "multidimensionality", Kock (2006, p. 255) has referred to the *Rhet. to Alexander* (1421b) mentioning the different (and not easily comparable) criteria listed by this ancient rhetorical treatise for supporting the eligibility of an action: i.e. that the action is either "just", "lawful", "expedient", "noble", "pleasurable", "easy to accomplish", or, if difficult, "practicable", or *in some way* "necessary".

Again, in a more recent article (2012, p. 282), Kock has also made use of Classical Rhetoric (in this case the analysis provided by the *Status theory*) to point up similar questions:

> Ancient status theorist, notably Hermogenes, also included lists of so-called practical status, basically naming the main warrants that could be invoked in political reasoning, such as legality, justice, advantage, feasibility, honor, consequence; the earliest such list is to be found in the anonymous *Rhetorica ad Alexandrum*. But the items on the list are the same multidimensional norms that clash in political debates today. For example, one debater recommends a policy as advantageous, while another opposes it as dishonorable.

I claim that Aristotle's account in his *Rhetoric* is as illustrative and, in my opinion, much subtler. Aristotle does *not only* take into account the first stage of deliberation in which the arguers may chose possibly incompatible claims and defend them with reasons based on possibly incommensurable values. He *also* tries to advance (at least a little bit) into the means to break the impasse and *continue* the discussion. He begins with a very simple and general schema (almost akin to a "practical syllogism") which is just applicable whenever there is an ample consensus on certain starting points and then he starts complicating the matter (and it gets rather complicated) with a sensibility to certain aspects and problems of arguments about "proposals" which I find lacking in a more schematic and less self-reflective or philosophical text as that of the *RaA*.

In the relevant passages in Book I of the *Rhetoric* (1359a30-1366a22) Aristotle characterizes arguments for actions to achieve agreed-upon ends, then arguments to support the eligibility of disputable ends and even meta-arguments to support the relative priority of different ends based on the weighing of the force of the arguments supporting their eligibility. Thus, Aristotle reveals not just the well-known value-laden multidimensionality of arguments aiming at supporting "what's advisable" (correctly emphasized by Kock, 2006, 2012), but an almost inescapable multi-level-multidimensionality, as it is here, more than in any other part of the *Rhetoric*, that he feels compelled to engage in listing topics and criteria for explicitly comparative meta-argumentation (1363b5-1365b20).

The "continuum between argument and argument criticism", mentioned by Pinto (2001) is probably more evident (and this was already evident for Aristotle) in such kind of practical argumentation than in any other field, mainly because of the debatable, exceptionable and multidimensional character of almost every possible allegation.

3. NARRATION AND PRACTICAL ARGUMENT

In this situation, narration and storytelling may come to our help. Narrations may be part of our societies' long-standing solutions to such an intricacy as a means to "shortcut" (in a somewhat less argumentative but not necessarily illegitimate way) a detailed and cumbersome multi-level argumentation by means of the persuasive use of our (value-laden) imagination. Our narrative abilities (so engraved in our human nature as Fisher's expression *homo narrans* suggests) might let us both:

> - *learn* to handle multidimensional situations, acquiring the necessary experience and experiential prudence (*phronesis*) to take sensible decisions in them;
> - and *argue* in a relatively economical way for policies and actions, forcing those who want to oppose our argumentation to handle an elaborate compound whose assessibility is not based on any well-ordered system of rules.

As Wittgenstein suggests (*Philosophical Investigations* 355-356), due to the complexity and just loosely regulated character of the realm of free human action, we cannot expect such a field to provide us with a well-ordered backing system for simple and easily applicable rules (and, thus, argumentative warrants):

> "Can one learn "knowledge of human nature" <*Menschenkenntnis*>? Yes, some can. Not, however, by taking a course in it, but through 'experience'. - Can someone else be a man's teacher in this? Certainly. From time to time he gives him the right tip.- This is what 'learning' and 'teaching' are like here.- What one acquires here is not a technique; one learns correct judgement. There are also rules, but they do not form a system, and only experienced people can apply them right. Unlike calculating rules. What is most difficult here is to put this indefiniteness, correctly and unfalsified, into words".[3]

Several things should be highlighted in this paragraph. First the lack of a "system of rules" (a well-ordered *backing* theory) that makes claims and grounds in the realm of human action "endlessly debatable" (in principle). We have already discussed that. But, second (and tending somehow in the opposite direction), the conviction that:

1. There *are* rules (argumentation is thus possible, as we have seen, although it might be endless, multi-level and branched-off). There are in fact too many rules (warrants) and even opposing rules (positive/negative variants related to the same criteria)[4].
2. There *is* "something to be learnt" (not easily, though: mark the expressions "some can", "from time to time"): what is learnt is "correct judgement" (*richtige Urteile*): being able to make sensible decisions in such a field.

---

[3] *Philosophical Investigations* 355-356: „Kann man Menschenkenntnis lernen? Ja; Mancher kann sie lernen. Aber nicht durch einen Lehrkurs, sondern durch "Erfahrung". - Kann ein Andrer dabei sein Lehrer sein? Gewiss. Er gibt ihm von Zeit zu Zeit den richtigen Wink. - So schaut hier das „Lernen" und das „Lehren" aus. - Was man erlernt, ist keine Technik; man lernt richtige Urteile. Es gibt auch Regeln, aber sie bilden kein System, und nur der Erfahrene kann sie richtig anwenden. Unähnlich den Rechenregeln. Das Schwerste ist hier, die Unbestimmtheit richtig und unverfälscht zum Ausdruck zu bringen".

[4] Aristotle (1362b 30ff, 1363b5ff) remarkably emphasizes the multiple possibilities (both positive and negative) of the use of certain allegations: e.g. that something is extraordinary and difficult can be used as a reason for its advisability or, on the contrary, become a reason for avoiding it (the same with "something that is easy and accessible for everybody": "that's the easy solution" as a pro/con allegation). The ambivalent character of certain argument schemes, originally mentioned here by Aristotle, has not been, to my knowledge, properly explored in argumentation theory. An exception to this neglect is Marraud (2012).

3. This is, however, done through "experience". We can paraphrase Wittgenstein thus: "unlike calculating rules, only experienced people can apply *Menchenkenntnis* rules right" (it's not only a question of knowing the rules, taking in account that they are not well-ordered).
4. Nevertheless, such experienced people may (*von Zeit zu Zeit*) "teach" others.

Experience and experiential good sense (*phronesis*) is of and is acquired by reflecting on "particular cases", without necessarily extracting "universally applicable generalizations" or "laws". One has to learn to be able to deal in usual (but exceptionable) patterns and mechanisms, without losing the sense of uniqueness (what is applicable in one case might be inapplicable in another, which is nevertheless apparently similar). Some recent works on "the cognitive value of literature" (Bouveresse, 2007; Green, 2010; Mikkonen, 2013), stress the peculiar way and the peculiar things we may learn from literature emphasizing, precisely, its role in the schooling of *phronesis*.

Green and Mikkonen explicitly discuss *literary cognitivism*: the thesis that "literary fiction can be a source of knowledge in a way that depends crucially on its being fictional". Such "knowledge", though, can belong to different types (Green, 2010, p. 352; Mikkonen, 2013, p. 9). It may be:

> (a) propositional knowledge –knowledge that such and such is the case (know-that);
> (b) phenomenal knowledge –knowledge of what an experience is like, or how an emotion or mood feels (know-what-it-is like);
> (c) knowledge of how to do something, where the *doing* in question may include (as Green emphasizes) not only bodily actions, but those involving use of the imagination (a broadly understood know-how).

Mikkonen's book disregards the discussion of b) and c) and becomes a defence of what he calls "moderate propositional literary cognitivism". Green, instead, without discounting the possibility that literature may provide "propositional knowledge", focuses on the possibilities of "phenomenal knowledge" as the characteristic produce of literature:

> an author can show how an experience, emotion, or mood feels by inviting the reader into a *de se* imagining. In this case, however, it is crucial that the author knows her subject matter, either by personal experience or careful study. The qualified author can then provide her reader knowledge of what an experience is like by serving, in effect, as a source of testimony. (Green, 2010, p. 364)

Finally, Bouveresse basing most of his reflections on a Wittgensteinian approach (he stresses once and again the significance of Wittgenstein's relevant *PI* paragraph as quoted above) tries to attribute to literature the provision of something more akin to c) "practical or operational knowledge": a kind of know-how (knowledge of how to do something, including, as Green suggest, with the use of our imagination):

> Precisely because literature is probably the most appropriate medium to express without falsifying them, the indeterminacy and complexity of moral life, it can possibly teach us to grasp something essential in that realm. Recalling another of Wittgenstein's expressions, it can teach us to look and to see - to look and see many more things than real life would allow us- when precisely we would tend, too early, too quickly to think. (Bouveresse, 2013 [2007], p. 81).

For Bouveresse the knowledge provided by literature is not only "practical" in the sense of being a kind of "broadly understood know-how" (knowing how to do something). It is, in fact, twice practical because it is the kind of wisdom we need for acting (for taking practical decisions) in real life:

> Hence, the link between content and form <*sc.* in literature> cannot be apprehended just by taking in account its relation with the problem of knowledge, one needs to refer as well to volition and action. (Bouveresse, 2013[2007], p. 95).

For Bouveresse, thus, the main cognitive role of literature (there can be others of course) is "enriching and clarifying life" for the sake of volition and action. The use of narrations becomes a way to face human practical reasoning and argument without "falsifying the indeterminacy and complexity of moral life" in Bouveresse's words or in Wittgenstein's putting "indefiniteness, correctly and unfalsified, into words".

4. CONCLUSION

Trying to sum up and relate some of the things I have tried to mention:

i. In human practical decision making we have a very complex realm where we cannot expect to deal with "universal" or "widely applicable generalizations" or "laws" (in scientific terms). Paraphrasing Wittgenstein: we have warrants (lots of them), but they do not form a system. It is a realm of particulars, uniqueness reigns but this doesn't need to be considered a methodological weakness of our *Menscherkenntnis.*[5]

ii. In order to learn to handle such a realm we need *experience* providing *phronesis,* we need to learn to deal in usual (but exceptionable) patterns and mechanisms without losing the sense of uniqueness.

iii. Most of our learning is acquired through real/personal experience of human life/action but we can also count on "surrogate experience", that is, learning from the "experienced": (remember Green's words: "it is crucial that the author knows her subject matter, either by personal experience or careful study").

iv. Narration (imagining possible stories and framing and making manageable the factual ones) is a means to acquire experience and training judgement "without falsifying the indeterminacy and complexity of moral life". A means to acquire imaginative (not-so-obvious) alternatives, understand embodied values, and explore the workings of practical judgment.

v. It is also (and accordingly) a possible means to discursively handle such a realm and try to argue in it in a sensible and expedient way, as an alternative to engaging in a complex multidimensional multilevel argumentation (which is, of course always possible).

ACKNOWLEDGEMENTS: This contribution has been made possible by funds provided by the Spanish Ministry of Economy and Competitiveness: Research Project FFI2011-23125.

REFERENCES

Bouveresse, J. (2007). *La connaissance de l'écrivain. Sur la littérature, la vérité et la vie.* Marseille: Agone. (Spanish translation (2013) *El conocimiento del escritor. Sobre la literatura, la verdad y la vida.* Barcelona: Ediciones del Subsuelo).

[5] Discussions about the use of "narrative explanations" in Evolutionary Biology, for example, have led to some authors to defend their scientific relevance as a means to explain phenomena in a realm with "no laws", a field that has acquired its maturity precisely by assuming its own complexity. According, for example, to Robert J. Richards: "When the barriers are down, we will see, not that historical narrative fails as a scientific explanation, but that much of science succeeds only as historical narrative" (1992, p. 40).

Fisher, W.R. (1989 [1987]). *Human communication as narration: Toward a philosophy of reason, value, and action.* Columbia, SC: University of South Carolina Press.

Green, M. (2010). How and what we can learn from fiction. In G.L. Hagberg & W. Jost (Eds.), *A Companion to the Philosophy of Literature* (pp. 350-366). Wiley-Blackwell.

Kock, C. (2006). Multiple warrants in practical reasoning. In D. Hitchcock & B. Verheij (Eds.), *Arguing on the Toulmin Model. New Essays in Argumentation Analysis and Evaluation* (pp. 247-260). Dordrecht: Springer.

Kock, C. (2007). Dialectical obligations in political debate. *Informal Logic, 27*(3), 233-247.

Kock, C. (2012). A tool for rhetorical citizenship: Generalizing the *status* system. In C. Kock & L. Villadsen (Eds.), *Rhetorical Citizenship and Public Deliberation* (pp. 279-295). University Park, PA: The Pennsylvania State University Press.

Marraud, H. (2012). Las razones del necio, *Bajo palabra. Revista de filosofía*, Época 2, No. 7, 533-541.

Marraud, H. (2013). *¿Es logic@? Análisis y evaluación de argumentos.* Cátedra: Madrid.

Mikkonen, J. (2013). *The cognitive value of philosophical fiction.* London: Bloomsbury Academic.

Olmos, P. (2012). La preceptiva sobre la *narratio* en los rétores latinos. *Revista de Estudios Sociales, 44*, 62-74.
http://res.uniandes.edu.co/view.php/803/index.php?id=803

Olmos, P. (2014). Narration as argument. In D. *Mohammed & M.* Lewiński *(Eds.), Virtues of Argumentation. Proceedings of the 10th International Conference of the Ontario Society for the Study of Argumentation (OSSA), 22-26 May 2013, CD edition. Windsor: University of Windsor.*

Olmos, P. (2014b). Classical fables as arguments: Narration and analogy. In H. Jales Ribeiro (Ed.), *Systematic Approaches to Argument by Analogy* (pp. 189-208). Amsterdam: Springer.

Olmos, P. (2015). Story credibility in narrative arguments. In F.H. van Eemeren & B. Garssen, (Eds.), *Reflections on Theoretical Issues in Argumentation Theory* (pp. 155-167). Amsterdam: Springer.

Richards, R. J. (1992). The structure of narrative explanation in history and biology. In M.H. Nitecki & D.V. Nitecki, (Eds.), *History and Evolution* (pp. 19-54. Albany: State University of New York Press.

Vega, L. (2013). Argumentando una innovación. *Revista Iberoamericana de Argumentación, 7*, 1-17.

Wittgenstein, L. (1999 [1953]). *Philosophical investigations.* Upper Saddle River, NJ: Prentice Hall.

von Wright, G. H. (1971). *Explanation and understanding*, Ithaca, NY: Cornell University Press.

47

Maneuvering Strategically by Means of an Allegorical Beast Fable in Political Communication

AHMED OMAR
University of Amsterdam, The Netherlands
A.A.A.M.H.Omar@uva.nl

With the help of the extended pragma-dialectical theory and narrative speech act analysis, this paper aims to analyze and evaluate how arguers may maneuver strategically by means of an allegorical beast fable in the domain of political communication. As an illustration, the political column "A Story for Adults and Children" of Alaa Al Aswany (Al Shorouk, May 25, 2010) is analyzed in the light of the specific predicament Al Aswany attempts to overcome.

KEYWORDS: conceptual metaphors, fiction-making, narrative speech act analysis, pragma-dialectics, strategic maneuvering

1. INTRODUCTION

This paper is based on my research interest in the Egyptian political columns that paved the way to the revolutionary uprising of 2011 in Egypt, especially within the call for democracy made by Alaa Al Aswany, an Egyptian novelist and columnist. In one of his columns entitled "A Story for Children and Adults"[1], Al Aswany predicts that massive demonstrations against Mubarak's regime will erupt led by El Baradei, the prominent opposition figure, and will succeed in ousting Mubarak's regime. Al Aswany's political message is conveyed by means of a story in which an old elephant, a fox, a wolf, a pig, a donkey, and a giraffe correspond to Mubarak, the intelligence, the police commanders, the stakeholders, the media hypocrites and El Baradei respectively.

I aim to investigate how an allegorical beast fable (henceforth ABF) can function as an argumentative discourse and to identify the main choices available for arguers to strategically maneuver with using

[1] The English translation of this column is in *On the State of Egypt* (2011, pp. 47-51).

D. Mohammed & M. Lewiński (eds.) (2016). *Argumentation and Reasoned Action: Proceedings of the 1st European Conference on Argumentation, Lisbon, 2015. Vol. II, 755-768.* London: College Publications.

this form in their pursuit to maintain the balance between reasonableness and effectiveness (van Eemeren, 2010). To this end, I shall give a definition of an ABF in section 2. In section 3, I will give a pragmatic description of an ABF from a speech act perspective as a series of fiction-making acts in order to systematically explain how this fictional form is politically interpreted and how it can function as an argumentative discourse. Using the lens of pragma-dialectics, I will explain in section 4 how an ABF can be aimed at resolving a difference of opinion. In section 5, I will highlight the particular selections resulting from the use of ABFs in terms of topical choices and presentational devices that can be effectively adapted to audience demand.

2. BEAST FABLE, ALLEGORY, AND ALLEGORICAL BEAST RFABLE

In *Oxford Dictionary of Literary Terms*, "beast fable" is defined as follows:

> The commonest type of fable, in which animals and birds speak and behave like human beings in a short tale *usually* [my italics] illustrating some moral point. The fables attributed to Aesop in the sixth century B.C. and those written in verse by La Fontaine are the best known, along with the fables of Bier Rabbit adapted by the American journalist Joel Chandler Harris from black folklore.

The word "usually" indicates that the goal intended to be achieved by means of a beast fable is not limited to moral points, albeit this is the case in the average examples. The only indispensable requirement for calling a narrative text a "beast fable" is the fact that its characters are animals behaving like humans.

Blackham emphasizes that the main goal of telling a beast fable is to convey a moral point, calling this point a "truth" and a "conceptual meaning": "Fable generates conceptual meanings, does not merely furnish an illustration in a particular instance... In so far as it is a fable, it is so because it can stand on its own with more general application... a narrative device, to provoke and to aid concrete thinking, focused on some general matter of concern." The degree of generalization is according to Blackham, however, changing. It varies from representing a 'truth' of the human nature in general to less generalized representations. This point is apparent when modern instances are discussed. (Blackham, 1985, pp. xv-xvii)

According to Blackham, a less generalized story with speaking animals and birds is still a "fable". Although what *Animal Farm* is believed to say (in its most interpretations at least) is not a universal truth that is applicable to the human nature in general, *Animal Farm* is still viewed as a fable, even if it is a "more ambitious" one. So, any narrative form with uttering animals and birds can be called a fable, regardless of the degree of generalization its point has.

Oxford Dictionary of Literary Terms gives the following definition of an "allegory": "a story or a visual image with a second distinct meaning partially hidden behind its literal or visible meaning. An allegory may be regarded as a metaphor that is extended into a structured system." Putting several and coherent metaphors together in a narrative form which constitutes a meaningful story is how the process of allegorical structuring is carried out.

Crisp gives a distinctive illustration of what is meant by allegory as a radically extended metaphor. Conceived as a conceptual metaphor (Lakoff & Johnson, 1980), an allegory is based upon mapping between a source domain (animals for example) and a target domain (political actors for example). An allegory is a result of extending the source domain through narrative means and with no overt reference to the target domain. The target domain's conceptualization is of course enriched by the source domain's activation, as with any form of conceptual metaphor. This is the ultimate point of allegory. (Crisp, 2008, p. 333)

Although Blackham (1985) differentiates between "fable" and "allegory" in a quite clear manner, in his book (and in other sources), *Animal Farm,* for instance, is described as a beast fable and an allegory in an exchangeable manner. It is sometimes called a "fable" and an "allegory" in other times.

In this paper, I use the term "allegorical beast fable" to denote the combination of these two features in a specific way that may convey any kind of messages (e.g. political). A politically ABF refers to a fictional, narrative form which is intended by its author to be interpreted (and actually read) at two levels: a direct (literal/fictitious/source domain) level at which animal characters, settings, happenings and actions are recognized as reflecting themselves as fictional values, and an indirect (figurative/real/target domain) level at which the same elements are recognized as paralleled real elements of some political situation.

3. A SPEECH-ACT ORIENTED APPROACH TO ABF

A speech-act-based analysis of the fictional form at issue is useful for justifying the allegorical interpretation of beast fables, and reconstructing it as an argumentative discourse aimed at resolving a difference of opinion. This kind of analysis is carried out by identifying the specific type of the speech acts performed by uttering the text. There are two main levels at which speech acts are performed. The first is a fictional level at which fictional characters do things with words. The second level is that of the author. Every segment of the text indicates that the author performs a specific speech act with a propositional content to which he can be held as committed to. Three types of deriving speech acts at this level shall be discerned in section 4.

I here adopt some pragmatic insights advanced by other scholars on the nature of fictive communication and the differences between fiction and non-fiction. In *The Nature of Fiction,* Currie (1990) advances fiction-making as a speech act performed by the speaker aiming at inviting the listener to involve imaginatively in some propositional content, or to make believe that the story as uttered is true[2]. Currie states a number of conditions required for an utterance *U* uttered by *S* with the propositional content *P* to be taken as fictive:

> S's utterance of U is fictive if and only if S utters U intending that the audience will
> (1) recognize that U means P;
> (2) recognize that U is intended by S to mean P;
> (3) recognize that S intends them (the audience) to make believe that p;
> (4) make believe that P. (Currie, 1990, pp. 30-35)

Friend (2011) makes his own differentiation between fiction and non-fiction as corresponding to the difference between imagining and belief:

> The guiding intuition is that belief, rather than imagining, is appropriate for non-accidentally true content ... the kind of imagining prescribed by fiction must be imagining without belief. Call this attitude mere-make-believe. (p. 165)

García-Carpintero (2013) makes a contribution on the correctness conditions of the speech act of fiction-making. He discusses a variation of ideas presented by Walton (1990) and Currie (1990), elaborating the

---

[2] Currie himself admits that talk about make-believe tends to be "loose and unsystematic". Yet, he gives some commonly acknowledged generalizations which are discernible: "[M]ake-believe allows us to achieve in imagination what we are denied in reality." (Currie, 1990, pp. 20-21)

view that fictive utterances are characterized by a specific form of illocutionary force in the family of directives – a proposal or invitation to imagine. According to García-Carpintero, "Fictions are proposals addressed to those with a general mindset of interests, abilities and dispositions, and fictions of specific kinds are proposals addressed to those with a correspondingly specific mindset." Accordingly, he puts the main correctness condition of the speech act of fiction-making in the following way: "For one to fiction-make *p* is correct if, and only if, *p* is worth imagining for one's audience, on the assumption that they have *relevant desires and dispositions*." (García-Carpintero, 2013, pp. 350-351)

Based on the contributions of these three scholars, the speech act of fiction-making can be formulated in a Searlean form. A Searlean characterization (as developed by van Eemeren and Grootendorst) of this speech act amounts to viewing fiction-making as an illocution which is recognized by the listener when its identity conditions are fulfilled and which is successful when its correctness conditions are happily fulfilled. The identity conditions in turn consist of the propositional content condition and the essential condition. The correctness conditions are divided into two categories: the preparatory conditions and the sincerity conditions. (van Eemeren & Grootendorst, 1984, pp. 42-44)

For a situation in which a speaker (or a writer) *S* utters a fictive utterance *U* in a communicative situation *C*, the felicity conditions of the speech act of fiction-making are formulated as follows:

The identity condition:
The propositional content condition:
The utterance *U* consists of the propositional content *P*
The essential condition:
By uttering *U*, *S* invites the listener (or reader) to make believe *p*
The correctness conditions:
The preparatory conditions:
(1) *S* believes that the listener (or reader) believes that *p* is not a fact.
(2) *S* believes that the listener (or reader) considers *p* as worth imagining within *C*(i.e. the listener has relevant desires and dispositions to *p*).
The sincerity conditions:
(1) *S* believes that *P* is not a fact.
(2) *S* believes that *P* is worth imagining for his listener within *C*.

For a fictional work to be accepted as such, it must consist of a number (at least two) of fiction-making speech acts of which the felicity conditions are happily fulfilled. By applying this insight to the case at issue, the speech acts of fiction-making that Al Aswany performs by uttering his text as a whole seems to be incorrect. Inviting the readers to imagine the propositional content of animals speaking and acting seems inappropriate. That is, "adult" readers can obtain no value from imaginatively attending to the content of the speech acts, if not in general, at least within the communicative situation of reading a political column in a daily elite newspaper. Thereby, the second preparatory condition is inappropriately fulfilled. Al Aswany, even though he is stating in the title that he addresses adults and children, knows well that his readers, in this specific situation where political non-fiction is dominantly addressed, are not interested in such content. So, and similar to the preparatory conditions, the second sincerity condition is inadequately met.

However, the fact that readers consider these speech acts at first glance as incorrect does not disrupt communication. It just induces the readers to reconsider each speech act as another indirect speech act. Van Eemeren and Grootendorst integrated the Searlean conditions for the performance of speech acts with general conversational rules such as Grice's maxims. This integration was an adaptation of standard speech act theory. The Gricean Co-operative Principle was replaced by a general Principle of Communication that applies to speech acts. As van Eemeren and Grootendorst explain, in accordance with Grice's Co-operative Principle (Be clear, honest, efficient and to the point), people who are communicating with each other generally try to ensure that their communication goes as smoothly as possible. A listener or a reader acting cooperatively must treat any apparent violation of any of these rules in a way that makes the speech act at issue relevant to communication. (van Eemeren & Grootendorst, 1992, pp. 49-51; 2004, pp. 75-80)

Performing the given fiction-making speech acts which are intended to invite the readers to imagine animals talking and behaving in a human way in a political column is considered as insincere. Therefore, a violation of the second rule of the Principle of Communication occurs. Suggesting an allegorical interpretation is the only way in which the violation can be removed: the allegorical interpretation enables the reader to derive from the fictional utterance a series of connected indirect assertives whose propositional contents are related to the political reality.

4. THE ARGUMENTATIVENESS OF AN ABF

Whether a fictional form is argumentative or not is identified by examining the fulfillment of the identity conditions of the speech act complex of argumentation performed by uttering the text at issue. Van Eemeren and Grootendorst formulate these conditions as follows:

> The propositional content condition: The constellation of statements S1, S2 (,...., Sn) consist of assertives in which propositions are expressed.
> The essential condition: Advancing the constellation of statements S1, S2 (,...., Sn) counts as an attempt by S [the speaker] to justify O [expressed opinion or a standpoint] to L's satisfaction, i.e. to convince L [the listener] of the acceptability of O. (van Eemeren & Grootendorst, 1984, p. 43)

In order to show that the propositional content condition is fulfilled, it should be indicated how, systematically, a constellation of assertives can be claimed to be performed by the author of a fictional form. There are three specific types of derivation can be discerned in the case of ABF (without any claim of exhaustiveness):

> (1) *Derivation from identification of characters:* To find out which political actors the animal characters parallel implies a series of propositions.[3] In putting an appropriate allegorical interpretation of a beast fable, the reader derives this series from the organizing conceptual metaphors and the related conceptual scenarios. Each of these propositions has the formula (political actor X has some trait T as animal Y). To take the pig as an example, as a consequence of accepting an allegorical interpretation, a reader will think of the pig in terms of, firstly, his dominant characteristics according to the shared cultural views of his community, and, secondly, the narrated description of it:
>
> > *The pig,* ***whose body gave off a foul smell****, squealed to object.*
>
> Thinking of this phrase in the light of the cultural set of values connected to pigs results in highlighting the pig's filthiness. Consequently, a reader will search for a political group which shares the pig's generic characteristic of "being greedy for satisfying desires in any possible way even by using impure means." This shared characteristic is translated at a political level as being corrupt. Hence, a

[3] For more on the integration of speech act theory into narratology, see Bernaerts (2010, pp. 278-285).

conceptual metaphor is established, in which the pig functions as source domain, and the corrupt net of interests benefiting from the regime as target domain. The mapping is between eating feces (or any other example of glutton and filth) and taking advantage of corruption. (2) *Derivation based on a slight change of references:* Some other assertives are uttered in an (almost) fully explicit way within the narrative dialogues. In the final scene, for instance, the giraffe addresses the King elephant, saying:

> *Old elephant, your reign has come to an end today. I still remember how the animals had high hopes at the beginning of your reign, but you surrounded yourself with the worst and dirtiest animals, and now you can see the result for yourself*

In this fragment, the giraffe performs a fictional assertive, among others, with the propositional content "you surround yourself with the worst and dirtiest animals." At the non-fictional level of the text, a parallel assertive can be derived with the following propositional content: "Mubarak has been surrounding himself with the worst and dirtiest assistants." The latter proposition is a support for the sub-argument "The influent people in Mubarak's regime seize all the country's bounties."

(3) *Derivation based on inferring from longer segments:* The actions of some character (nonverbal physical act, speeches, thoughts, or feelings, perceptions and sensations) may imply some approximate proposition pertaining to his or her traits. The way in which the giraffe, for instance, talks shows its courage and leadership. These traits cannot be justified by referring to the conceptual metaphors and scenarios as in other cases, since the giraffe (in the shared cultural views of the Egyptians) is an animal that has no distinctive stereotypical traits. The following fragment illustrates this point:

> *At this point, the wolf snarled and said: "Since when did we have to take those wretched animals into account? We decide what we want and they just obey our orders"*
> *Suddenly, the wolf shouted out, "Who are you and what do you want?"*
> *The giraffe shouted back, "We are the inhabitants of this jungle and we have grievances we want to submit to the elephant king,"*
> *"This isn't the time for grievances. The king is tired and busy. Go away."*
> *The giraffe swung his long neck right and left, "We won't go away until we've submitted our grievances."*
> *"How dare you be so bold!"*

The wolf knows well who they are and what they want. This can be easily inferred from the initial dialogue between the four assistants of the King Elephant (the fox, the wolf, the pig and the donkey) in which the fox says:

> *...I have heard that all the animals in the forest are coming this way in a protest march led by the giraffe... The problem is not with the giraffe. All the animals are disgruntled and we have to negotiate with them.*
> *The wolf's reaction to what the fox says is related only to opposing the former's preference of negotiating, and therefore it can be concluded that the wolf agrees on the other propositional contents the fox has uttered. The wolf replies:*
> *I am sorry, fox. We won't negotiate with anyone... Now we need to be tougher than ever. We own everything. We have a trained army of dogs fierce enough to subdue any animal that lifts its head against us.*

Accordingly, one of the felicity conditions for performing these two directives ("who are you and what do you want?") is not fulfilled as the wolf can be considered as insincere in raising his questions. Since asking about what they want is a sequel of asking about who they are, it can be inferred that the speaker may make some allusion. The wolf may make an indirect evaluative statement asserting that such demands related to political power and distribution of benefits(what they want) should not be made by such ordinary animals (who they are). So, the wolf can be held by the reader to performing an indirect assertive with the propositional content "You should not do that."

However, the giraffe's reaction to this indirect assertive reflects that it chooses to neglect this fact. Its answer is an assertive with a propositional content stating who they are and what they want, in a manner that implies its ability to run the conversation for its own sake.

By performing the directive "Go away" (fragment B), the wolf presupposes that he has an institutional position over inhabitants of the jungle that allows him to give orders to them. The giraffe's reaction, however, reflects that this order of social positions is an arena of conflict between the two camps: the elephant's camp insists on maintaining this kind of hierarchy, whereas the furious animals led by the giraffe challenge it and are willing to change it. Through the utterance of the giraffe "We won't go until we've submitted our grievances", a commisive is performed. One of its correctness conditions is that the giraffe is allowed to speak on behalf of other animals. The giraffe's speech manner, as explained by these two examples, reflects the character of a leader.

Examining whether the felicity conditions of some specific fictive speech acts performed by the narrative characters are fulfilled or not can be instrumental in deriving certain speech acts performed by the columnist. Through this kind of speech act analysis applied to fragments B, C and D, the columnist can be held committed to perform an indirect assertive. The propositional content of this assertive is on El Baradei, the political counterpart of the giraffe. It can be formulated as such: "El Baradei will be an effective leader of the protesters."

With regard to the *essential condition*, these derived assertives must be explained as justifying some opinion: the standpoint put forward. There are no explicit verbal indicators that indicate justification. However, presenting the argumentative discourse in a narrative form provides some indicators of justification; of supporting an opinion by advancing an argumentation. Viewing the narrative structure as causal chains leads to showing how narratively connecting events entails justification.

A narrative plot requires causality, not only chronology. The resolution event of a story does not manifest itself in an event which succeeds the complication events and settles the conflict involved, but it is also an event that is caused by the complication and other inserting factors which can be categorized as "happenings". "The working out of a plot (or at least some plots) is a process of declining or narrowing possibility. The choices become more and more limited, and the final choice seems not a choice at all, but an inevitability." (Chatman, 1978, pp. 46–51)

In cases in which argumentation is presented in a narrative form, the standpoint is derived from the resolution event. This standpoint is justified by arguments derived from some (if not all) preceding events and the characters' motives and traits that lead up to the resolution. An allegorical narrative constitutes the presentation of an argumentative discourse in which a standpoint is derived from the story resolution, and this standpoint is justified by a constellation of arguments which is derived from the events and motivations that simultaneously lead to this resolution.

In the case of "A story for Children and Adults", for instance, the victory of masses of animals over the King elephant and his clique, which is depicted in the resolution event, is allegorically interpreted as an assertive with the propositional content "the Egyptian people will succeed in ousting Mubarak's regime."

5. IDENTIFYING THE RESULTS ACHIEVED BY STRATEGICALLY USING AN ALLEGORICAL BEAST FABLE

In order to analyze and evaluate the argumentative moves made by Al Aswany as strategic choices, I will explain the specific situational predicament which Al Aswany attempts to overcome by means of using an ABF. In "A Story for Children and Adults", Al Aswany aims to defend the acceptability of a prediction that has no equivalents in the modern Egyptian history witnessed by the vast majority of his readers: middle-class, educated, and trans-ideologically Egyptian youth.[4]

In order to identify the results achieved by arguing by means of an ABF in the domain of political communication, an analytic overview of the text must be given in order to bring together systematically everything that is relevant to the resolution of the difference of opinion. The overview states exactly the parties involved in the dispute, their procedural and material starting points, the arguments advanced, and how they are connected to each other. (van Eemeren & Grootendorst, 2004, p. 118)

I will concentrate on reconstructing the standpoint put forward and the arguments supporting it because some of these propositions are idiosyncratically formulated as a consequence of selecting an ABF as a form of communication. The overview of the argumentation adduced in support of the standpoint at issue can be schematically described as in Figure 1 below.

Using the narrative form of an ABF allows for using opportune topical choices and can as a whole function as an effective presentational device. With regard to topical choices, it is evident from the argumentation structure that the sub-standpoints (1.1a.1c.1a.1a-1.1a.1c.1a.1b-1.1a.1c.1a.1c-1.1a.1c.1b-1.1a.1c.1c-1.1a.1c.1d-1.1c.1a-1.1c.1b) represent prospective events supported by propositions based on conceptual metaphors (written in bold). Making use of shared conceptual metaphors enhances the acceptability of the notion that pro-Mubarak political institutions will act in an exactly predictable way, and that each institution is a one-sided entity motivated by a single trait: severity, gluttony, hypocrisy, etc. Formulating some arguments in the

---

[4] The most recent instance of a revolutionary action were the 1977's so called "bread riots" which forced President Sadat (1918-1981) to withdraw his economic decisions regarding termination of state subsidies on basic foodstuffs. In fact, the demonstrations of 1977 reflected an expression of anger at some specific political decision, and did not involve any ambition for a wider political change.

form of widely-shared conceptual metaphors creates and enhances a simplified vision of a complicated political reality. Al Aswany's readers are unlikely to cast doubt on his predictions regarding how pro-Mubarak institutions will act in case of massive protests because these predictions are justified by shared conceptual metaphors.

1 The Egyptian people will succeed in ousting Mubarak's regime
1.1a Many, diverse Egyptians will participate in anti-regime demonstrations
1.1a.1a The Egyptian people are no longer afraid from oppression
1.1a.1b The regime's clique deprived them of all bounties
1.1a.1b.1 The regime's clique is the worst and the most unscrupulous
1.1a.1c The regime will make no compromise with opposition representatives
1.1a.1c.1a Most of the regime's components prefer to use violence
1.1a.1c.1a.1a The security apparatuses prefer using violent solutions
1.1a.1c.1a.1a.1 **The security apparatuses are grim as wolves**
1.1a.1c.1a.1b The stakeholders prefer using violence
1.1a.1c.1a.1b.1 **The stakeholders are gluttonous as pigs**
1.1a.1c.1a.1c The media supporters prefer using violence
1.1a.1c.1a.1c.1 **The media supporters are thoughtless as donkeys**
1.1a.1c.1b Only the intelligence agency prefers peaceful solutions
1.1a.1c.1b.1 **The intelligence agency is cunning as a fox**
1.1a.1c.1c Mubarak has no considerable role in taking decisions
1.1a.1c.1c.1 **Mubarak is feeble as an aged elephant**
1.1a.1c.1d Most of the pro-regime actors over-trust the capacity of security personnel
1.1a.1c.1d.1a The number of security personnel is big
1.1a.1c.1d.1b **Security personnel are fierce and loyal as guard dogs**
1.1b El Baradei will be an effective leader of the protesters
1.1c None of pro-regime actors will be able to support Mubarak to the end
1.1c.1aThe media supporters and stakeholders will be too confused to act
1.1c.1a.1**The media supporters and stakeholders are stupid as donkeys and pigs**
1.1c.1b The intelligence will sidestep from the scene
1.1c.1b.1 **The intelligence sidestep losing battles as a fox does**
1.1c.1c The security forces will not be able to suppress the masses
1.1c.1c.1 Some of the security personnel will retreat from attacking the protesters
1.1c.1c.1.1 The security personnel suffer from poverty and injustice as well

Figure 1

The genre of fable is rooted in a long history of stories that claim presenting the "truth": what is general and universal in human nature.[5]

[5] For more on the historically established connotations of fables, see: Stewart (1991, p. 16).

As a presentational device, using the fable in presenting an argumentative discourse bestows the impression of generality and absolute wisdom-possession on the argumentation advanced. Consequently, the prediction of the people´s rebellion against Mubarak´s regime is viewed as incontestable truth: an acceptable proposition in the most effective manner.

6. CONCLUSION

I have argued that an allegorical beast fable can function as an argumentative discourse aimed at resolving a difference of opinion in the domain of political communication. With the help of a developed Searlean formulation of felicity conditions and narrative speech act analysis, I have explained how readers can systematically derive a series of indirect assertive from the fiction-making speech acts performed within the fictional form at issue. Arguing using this fictional form results in advancing propositions based on conceptual metaphors as arguments presented in a form that is strongly related to wisdom and absolute truth. Consequently, making use of this fictional form helps in effectively convincing the readers of the acceptability of a simplified version of the political reality in an incontestable manner. Further research is needed to address the differences between using an allegorical beast fable within different communicative activity types: columns, novels, films and cartoons in terms of the reconstruction of argumentative discourse and the choices available to maneuver with.

REFERENCES

Al Aswany, A. (2011). *On the state of Egypt* (J. Wright, Trans.). Cairo: The American University in Cairo Press.

Baldick, C. (2008). *Oxford dictionary of literary terms.* New York: Oxford University Press.

Bernearts, L. (2010). "Cuckoo's Nest": Elements of a narrative speech-act analysis. *Narrative, 18*(3), 276-299.

Blackham, H. J. (1985). *The fable as literature.* London: The Atholne Press.

Chatman, S. (1978). *Story and discourse: Narrative structure in fiction and film.* Ithaca: Cornell University Press.

Crisp, P. (2008). The pilgrim's progress: Allegory or novel? *Language and Literature, 21*(4), 328–344.

Currie, G. (1990). *The nature of fiction.* Cambridge: Cambridge University Press.

Eemeren, F.H. van, & R. Grootendorst (1984). *Speech acts in argumentative discussions.* Dordrecht: Foris Publications.

Eemeren, F.H. van, & R. Grootendorst (1992). *Argumentation, communication, and fallacies: A pragma-dialectical perspective.* Hillsdale, NJ: Lawrence Erlbaum Associates.

Eemeren, F.H. van, & R. Grootendorst (2004). *A Systematic Theory of argumentation: The pragma-dialectical approach.* Cambridge: Cambridge University Press.

Eemeren, F.H. van (2010). *Strategic maneuvering in argumentative discourse: Extending the pragma-dialectical theory of argumentation.* Amsterdam: John Benjamins.

Friend, S. (2011). Fictive utterance and imagining. *Proceedings of the Aristotelian Society Supplementary, lxxxv*, 163-180.

García-Carpintero, M. (2013). Norms of fiction-making. *British Journal of Aesthetics, 53*(3), 339–357.

Lakoff, G., & Johnson, M. (1980) *Metaphors we live by.* Chicago: Chicago University Press.

48

Algorithms in Argumentation: Implications for Reasoned Decision Making

MARCUS PAROSKE
University of Michigan-Flint, USA
paroske@umflint.edu

RON VON BURG
Department of Communication, Wake Forest University, USA
vonburrl@wfu.edu

Recent trends toward utilizing algorithms to analyze data sets challenge our assumptions about how argumentation works. While the power of algorithmic analysis is clear, the moment of human judgment and choosing between competing arguments remains elusive. A fully developed theory of reasoned decision making would need to account for these final judgements. This essay discusses how the cultural movement embracing algorithms can inform our understanding of the role of reason when judgements must be made.

KEYWORDS: algorithmic argumentation; big data; decision making; human judgment; reasonability and argumentation

1. INTRODUCTION

The desire to make our public deliberations and policy decisions more analytical, rigorous and data driven is increasingly prevalent. The trend is strengthened by our exponentially greater access to the computational power necessary to process vast amounts of data. The influence of "big data" on decision making is evident in numerous institutions: government, commerce, journalism, and medicine, to name a few. Public policy analysts employ data-driven decision making to recommend government action that many assume to be fair and free of bias. Health care professionals practicing evidence-based medicine privilege extensive data networks of research and clinical trials when

D. Mohammed & M. Lewiński (eds.) (2016). *Argumentation and Reasoned Action: Proceedings of the 1st European Conference on Argumentation, Lisbon, 2015. Vol. II, 769-780.* London: College Publications.

offering prognoses. Climatologists utilize powerful General Circulation Models to predict climate change based on numerous anthropogenic and natural variables. However, these "big data" approaches only become manageable for decision making through the use of sophisticated algorithms—computational codes that promise to quickly and objectively (within the parameters of their coding) discern patterns in that data and offer predictions. In *Automate This*, Steiner (2012) explains how these tools work by noting that, "at its core, an algorithm is a set of instructions to be carried out perfunctorily to achieve an ideal result. Information goes into a given algorithm, answers come out" (p. 54). In this sense, with regard to argumentation, algorithms should be understood as complicated decision trees, processes that rely on a series of if/then or other conditional statements to yield binary answers.

The virtue of algorithms resides in a hope that their amalgamation and evaluation of data—offering predictions, evaluating trends, and informing decision making—avoids human prejudice and error. While there are robust disputes over the technical definition of the term "algorithm" that unfold within fields such as computer science, we are most concerned with algorithms as a conceptual category used in decision making based on computational processes. For example, MacCormick points out that "an algorithm is a precise recipe that specifies the exact sequence of steps to solve a problem" (2012, p. 3). Likewise, Berlinski argues that "an algorithm is . . . a way of getting something done in a finite number of discrete steps" (2000, p. *xvi*). Despite differing understandings of algorithms as a technical computational device, our analysis underscores the role algorithmic processes play in rendering decisions and inflecting changes on human judgment. This attention to the growing force of algorithms as decision rules, as ways to guide public argumentation, animates our critical focus. Since this is the dominant popular conception of the term, it has particular weight for our analysis of the impact of algorithms on understanding public argumentation and policy making.

This proliferation of algorithms in public life and policy deliberations invites the argumentation community to better understand how we use them in articulating premises and rendering decisions. As we cede decision-making authority to algorithmic processes in a variety of contexts, scholars must find ways to come to terms with the guiding assumptions and practices of algorithmic argumentation. The conference theme "Argumentation and Reasoned Action" identifies the importance of reasoning and argumentation at moments of decision. To that end, this essay argues that applying the expectation that the premises of our arguments be data driven and algorithmically analyzed creates challenges for our subsequent assessments of those arguments when deciding a course of action. This essay does not seek to provide a systemic treatment of the relationship between algorithms and argumentation. Rather, we draw lessons from other attempts to "rationalize" human judgments to discern what problems we may encounter when we try to bridge the gulf between premise generation and arriving at conclusions. To do this, we first critically examine the increasing prevalence of algorithms in guiding reasoned decision making in public policy. Next, we explore how two other disciplines incorporate the role of rationality and decision rules in human judgment. Finally, we offer a series of questions that a fully developed theory of reasoned decision making needs to deal with in an algorithmic age.

2. ALGORITHMS IN SOCIETY

The role of algorithms in influencing human judgment has a long, storied history. From the Sumerians who used a basic algorithm to evenly divide wheat across a variable population, to contemporary doctors who rely on algorithms to propose treatment protocols, humans have long used rules to guide "objective" and "data driven" decision making. The increasing reliance on algorithms to inform decision making highlights an effort to minimize the influence of ideology or personal experience on human judgment. However, calling decisions generated by algorithms "objective" ignores the inherent bias of any computational tool. After all, humans program algorithms, and the operational procedures of any algorithm incorporate normative assumptions on how to assess data. For example, the Sumerians supplied the premise that a wheat surplus should be divided into equal shares. While a reasonable premise, that animating assumption reflects a particular social value in rendering a decision. Likewise, the idea that less invasive medical procedures are preferable to more invasive ones is a subjective, normative parameter set by the programmer for a diagnostic algorithm to inform medical practice. Hence, coding by humans inevitably influences how to organize, process, and evaluate large amounts of data.

Nonetheless, the allure of algorithms is potent. In terms of their appeal for decision making, algorithms offer at least three advantages over humans: speed, objectivity, and the ability to incorporate vast quantities of data. Although the speed of decision making is not a variable in argumentation theory used to assess validity, the utility of rapid judgment is appealing. For example, the first use of algorithms in the stock market resulted in faster trades and consequently greater profits (Steiner, 2012, pp. 11-52). Likewise, self-driving cars or autopilot functions can assess variables quickly enough to take actions that improve the safety of passengers. In each case, the algorithms incorporate warrants for action based on the presented data. The warrants for action are already coded into the algorithm, allowing the decision to act to be made nearly instantaneously. This is not to suggest that operating at such speed is unquestionably good. Indeed, the algorithmic code guarantees a static relationship between the contextual data and the pre-programmed warrant. Speed is only possible through the parameters already programmed into the algorithm. To borrow an example from Carr, a self-driving car can quickly react to a deer in the road. The direction, speed, and size of the deer enables the self-driving car to respond in a way that yields the least harm to the car and its passengers. However, if the deer is pursued by the family dog, the algorithm cannot discern between the deer and the beloved pet as a meaningful creature which would affect how the human driver responds. Whereas the algorithm might judge it safer to hit the dog as opposed to the larger and more destructive deer, the human driver would likely render a different decision. In this case, speed is not the issue, so much as the contextual moral calculation between the two animals. More important in that moment is the moral warrant for action, not the rapidity of the decision. This dovetails well with the second comparative advantage: objectivity.

As the norms of rational argument privilege objectivity and neutrality over subjectivity and bias, algorithms offer the promise of putatively fair and rapid assessments of data. The fear of human bias and fallibility informs many uses of algorithms in our decision-making processes. For example, in the late 1990s, after an outcry over perceived voting bias, the college football conferences in the United States turned to computer polling to determine who ought to play in the national title game. Turning over this decision to computers drew significant public backlash a few years later when, ironically, the computers failed to select championship contenders that mirrored public perception of the most worthy teams. A similar dynamic informs many public policy decisions. The expectation to let data (and by extension to let the algorithms that process such data) determine economic policy, for example, is driven by our desire for objective assessments of market forces. Truly ceding judgment to algorithmic analyses, however, overlooks the very real role humans play in setting those decision protocols. Algorithmic argumentation requires us to be even more vigilant in evaluating public arguments, as the subjective elements that drive algorithm creation are often hidden behind a patina of objectivity. This challenge increases as algorithms are frequently proprietary, keeping the programming used in the decision-making process opaque to preserve profits.

Finally, there is the question of data quantity. The ability to process more data points seems to strengthen the validity of arguments produced by algorithmic analysis. Data have always been an instrumental feature in argumentation theory. Data, in Toulmin's sense, provide the grounds on which the claim and warrant rest. We might assume that more data is better, but quantity does nothing to ensure the efficacy of that data to the decision at hand. The savvy humanist would rightly contend that "data do not speak for themselves." Processing data incorporates programming that presumes a variety of value propositions. The selection of types of data, and how they weigh against one another, has far reaching implications for the subjects of any public policy decision. Indeed, the norms of argumentation are changing as a result; to many, the more data that drives your position, the more reasonable your advocacy appears. Hence, we often sacrifice qualitative assessments of data under a "more is better" default paradigm.

The Chicago Police Department's tactic of utilizing an algorithm created by engineer Miles Wernick to compose a "heat list"—a collection of 400 or so names of residents who fit a profile that predicts future criminal behavior—demonstrates these complications. Although a number of American police departments employ similar predictive models, the Chicago Police Department uses the algorithm to quickly develop a list people who Wernick says "clearly have a high likelihood of being involved in violence" (qtd. in Stroud, 2014). The algorithm employs social network variables, suggesting that the location where one lives or the company one keeps, in addition to past behavior, is predictive of future criminal behavior, hence informing how police interact with such individuals. Even though the model offers predictive direction, Wernick notes "the recommendations of the mapping system will not replace the expertise of police officers, but instead highlight potential concerns so that police officers can take them into account" (qtd. in Stroud, 2014). Beyond an unsettling similarity to the pre-crime unit of the film *Minority Report* and the possibility of employing prejudiced (e.g., racist) assumptions under the cover of algorithmic objectivity, there is the increasing likelihood that the individual officers could conflate probable criminal behavior with actual criminal behavior and act accordingly.

Social welfare and education policy making echoes similar stories, where predictive algorithms instantiate various assumptions and relationships that become problematic when the algorithmic results become the driver of public policy action. As Eubanks (2015) argues,

> the algorithms that dominate policymaking—particularly in public services such as law enforcement, welfare, and child protection—act less like data sifters and more like gatekeepers, mediating access to public resources, assessing risks, and sorting groups of people into "deserving" and "undeserving" and "suspicious" and "unsuspicious" categories.

For that reason, we join those (Berry, 2014) who call for the critical examination of how algorithms and computation more generally impacts democratic society.

For the purposes of this essay, we are most interested in the relationship between the algorithmic results (functionally, the generation of premises for argument invention) and the dimensions of human judgment. The role of algorithms in shaping judgment presents a series of challenges to our understanding of argumentative practice. Even though algorithms can process large stocks of data, their limitations of producing only binary answers undermines the ability to accommodate nuance and consider multiple perspectives simultaneously. Even the most sophisticated algorithms that "learn" from anomalies or the presence of previously unaccounted for data rely on existing coding operations to know what to look for in that data, creating a closed system that only accounts for prescribed variables in rendering a result. It is up to the human aspects of argumentation as a communicative act to supply parts of practical reasoning that no algorithm can fully calculate. For example, algorithms are not very effective at predicting or understanding irrational human behavior (Steiner, 2012, p. 132) or providing ethical judgment (Carr, 2014, pp. 183-199). Likewise, an algorithm combing through census data may find a relationship between poverty and education level down to the city block, but it cannot tell us why that is the case, or what to do about it. Mirroring actual human behavior and answering qualitative questions about what data actually means may still require messy and subjective judgment of arguments. To see why, we turn to some other disciplines for guidance.

3. LESSONS FROM OTHE DISCIPLINES

The attempt to automate judgment is not new. Indeed, two key 20th century intellectual movements have toyed with the idea, albeit from rather different disciplinary perspectives. The first, experimental psychology, sought in part to make rationality a model for human cognition, pushing into policy formation as a space to consider how to make reasonable decisions. Inspired by positivist assumptions of thinking and rationality, this project challenged the role of intuition in decision making. The second, digital humanities, recognized the power of computation in analyzing data, and wondered what role, if any, it might play in the types of subjective, critical judgments that have defined humanistic understandings of culture and art. Both projects give some clues to argumentation scholars as to what challenges might attach to a fully developed theory of reasoned decision making.

For the sake of brevity, we let one author stand in for each of these approaches. Kenneth Hammond, former director of the Center for Judgment and Policy at the University of Colorado-Boulder, was a proponent of empirical psychology and strongly argued for rationality as a preferred model for human cognition and guide for decision making. While his early work pre-dates our broad access to algorithms, Hammond did see ways that computational power could sift through large data sets, greatly expanding the potential for data-driven psychology to further understand rationality.

Hammond argues that the application of this rational psychology to public policy formation would eliminate bad or arbitrary decisions. This project is not just about determining the public policy choice that, for example, optimizes economic productivity; it is also about encouraging policymakers to utilize a rationalist approach to decision making that encourages them to adopt that optimized policy. He critiques the discipline of policy analysis for lacking a theory of human cognition to bridge the gap between isolating optimal decisions with actually inducing humans to adopt them and act upon those decisions. A marriage of computationally powered argument analysis with a theory of human cognition that could convince policy makers to adopt the most rational course for decision making seems, for Hammond, the best route for solving social problems.

However, Hammond has a problem. Since he chooses policy adoption as his artifact of analysis, Hammond enters into a space where contextual elements stymie the kinds of perfect knowledge necessary to guide his rules-based approach to decisions. He notes that policy is controlled by "irreducible uncertainty." We must always make actual decisions with imperfect knowledge. No amount of rationality can fill in the information gaps necessary to theorize moments of decision. There is still some intuition that must come into the picture. He notes:

> If one is faced with a set of circumstances that demand action, it may be possible to turn to a rule. Rules are of various types—some are better than others—and generally constitute the core of a professional person's activity. But circumstances that require a social policy—developing a transportation system, a health care system, a method for reviving a

> neighborhood or a city—are not dealt with so easily; there are no proven if-then rules for these circumstances. It is these conditions that involve irreducible uncertainty; it is these conditions that make judgment necessary (Hammond, 1996, p. 20).

Hammond's if/then, decision rule language shows his predilection for an algorithmic approach to argumentation. But the irreducible uncertainty of the moment of decision threatens his bid for a fully rational approach to public policy. Before seeing Hammond's solution, we can see another challenge to algorithmic decision making from a very different theoretical perspective.

The second approach to algorithmic arguments is digital humanities. Here, Professor of English Steve Ramsay (2011) and his book *Reading Machines: Toward an Algorithmic Criticism,* represent the digital humanities approach. Ramsay wrestles with the impact of software and technology on provinces traditionally seen as the entire purview of human judgment and reflection such as literary criticism. If Hammond's foundational premise is that rationality trumps all other considerations, then Ramsay begins with an opposite viewpoint, that subjective human judgment is the representative paradigm for the types of decisions that he is looking to make. Because critical judgment is subjective, humanistic criticism has embraced a view of decisions where arriving at one final, determinative answer is not the point of the exercise. Instead, it is to proliferate a number of possible answers to the question and to spark discussions about the meaning or relevance of those answers. This argumentative exchange is how we learn about rich artifacts.

Nonetheless, Ramsay is faced with the unquestionable power of algorithms to sift through text, to crunch data at a rate far greater than any human. In particular, algorithms excel at spotting patterns in large bodies of text. This leads to a non-controversial form of algorithmic criticism, namely activities such as author attribution or word association. Ramsay calls using algorithms to test human judgment "verifying intuition." Ramsay, though, is faced with a deeper question of how to align those techniques with different sorts of questions, judgments beyond empirical questions and toward deeper levels of the meaning of texts, of their historical, political and ideological force. Can an algorithm tell us whether a work is feminist? He notes the pressure toward algorithms, the importance of using their utility to help us understand texts, but still is faced with the fundamental subjectivity of the types of judgments he is looking to render from a humanist perspective. He observes the dilemma in considering whether a particular work by Virginia Woolf is feminist or not. Ramsay contends:

> There is no control group that can contain "current feminist reconfigurations." And surely there is no metric by which we may quantify "pertinence" either for Woolf or for the author's own judgment. The hermeneutical implications of these absences invoke ancient suspicions toward rhetoric, and in particular, toward the rhetorical office of *inventio*: the sophistic process of seeking truth through the dialectical interplay of trust, emotion, logic, and tradition, which has, since the seventeenth century, contended with the promises of empiricism (2011, p. 7).

This is a different but related challenge to Hammond. Ramsay must make his predilection toward humanism fit with the mistrust of rhetoric that is amplified as we become increasingly reliant on technology to identify premises. Hammond must make his desire for rational decision making harmonize with irreducible uncertainty, for the inability of data to capture everything. Both Hammond and Ramsay find algorithms useful tools to meet their goals (social policy, humanistic criticism), but are still left short. In other words, algorithms could not address everything, and they both return to human judgment to bridge the gap.

Hammond's solution is to accept that, at the point of decision, irreducible uncertainty requires us to commit to a final, intuitive leap where we act based on our human assessment of arguments. Common sense kicks in, allowing us to consider data inferred from context and moment. In a space of reasoned decision making, reason never gets us all the way to the final decision. Hammond described this as

> cognition that is as analytical as it can be, and as intuitive as it must be, or the converse, depending on the inducement for task conditions . . . When the limits of one's rationality is encountered, one begins to draw upon intuitive cognition, and vice versa (1996, p. 150).

Hammond refers to this step as quasi-rationality.

Ramsay is happy to cede over premise generation to computers to some extent. In digital humanities, even this is controversial as it may betray the fundamental humanness of judgments. Ramsay will embrace a version of *inventio* that harnesses machines to expand the amount of premises that can be generated. This "let a thousand flowers bloom" approach, enabled by the generative power of algorithmic textual analysis, seems grist for the humanist mill. But choosing between these now proliferated premises is still a subjective, human activity. Both Hammond and Ramsay, surprisingly, appreciate the use of algorithms to generate argumentative premises, but not to make decisions. Reasonability at that particular moment where we choose between A and B still requires some level of subjective human intervention, and invites the expertise of the argumentation community to understand such a dynamic.

4. CONCLUSION

We argue, as a result, that reasoned decision making is a process of premise generation, more and more significantly enabled by contemporary algorithmic technologies, leading to a quasi-rational moment of decision where humanistic judgment is the best and only tool in rendering a choice. Just as Hammond and Ramsay acknowledge, rationality at those final moments must take into account context, including the amount of available information, and taste in order to make the most reasonable decision.

This approach presents a series of questions for the argumentation community. First, if the thrust of argumentation studies is not merely to refine abstract rules for logic, but also to make those rules applicable to real world decisions and solving actual problems in a democratic society, then the particular moment of decision requires a deeper theorization. Hammond finds that social policy is always marked by irreducible uncertainty at the moment of decision, and that there was no purely rational way to guide decisions that must be made at moments of imperfect knowledge. If this is the case for reasoned decision making, then a view of cognition as applied to argumentation analysis must contain humanistic elements.

Second, if it is the case that many, if not all, decisions need to be made with the acceptance of some uncertainty, then should that be factored into our rules for reasoning? Doing so requires developing a companion set of argumentative practices that do more than provide rationalist guidance for interactions, but aid the intuitive selection of rival premises when decisions are nigh. This intuitive theory may be much more akin to humanistic criticism than it is to rationalist psychology, since it involves the kinds of reflexive human judgment that must be made without the assistance of hard and fast decision rules or guides. Computation can help us analyze data and come to empirical descriptions of the world (including its social problems) but irreducible uncertainty demands some form of humanistic criticism at the end, where decisions are made. Ultimately, this invites questions of timing, context, and other fluid data points that inflect new dimensions on standards of judgment.

Finally, as we are increasingly asked to consider the theoretical implications of algorithmic tools in digesting public argumentation, we should consider the end to which those tools are used. Ramsay compromises with algorithms by utilizing them to expand the possible interpretations of a text, as opposed to constraining decision making. If we think of decision making not as a process of narrowing down choices until only the most rational remains, but of expanding the number of choices so that human judgment has more options to choose from, then we may need to consider different directions for argumentation studies in the age of algorithms. This puts pressure on deliberation and compromise to sift through the expanded set of premises. We cannot lean on rationality to do our thinking for us. Better theorizing decision making may help avoid the potential problems of an overly algorithmic society. To wit, the Chicago Police Department would need more than a computer program to identify suspects.

All told, we need to continue to employ deliberation to hone our judgment skills, as opposed to ceding those decisions to an elusive rationalist paradigm likely never possible. Not only does this approach provide a more honest assessment of how reasoned decision making occurs, but it resists the temptation to overinvest in the supposedly purer, more objective impulses that are driving the increasing popularity of algorithmic argumentation. Carr (2014) argues that in our automated society, where we cede decision making to computers and algorithms, we sacrifice cognitive abilities that are characteristically human. Like a muscle that atrophies due to disuse, outsourcing decision making dulls our ability to interpret data and render judgment. As algorithms become increasingly sophisticated and can appear to replicate a reasoned decision a human would make, the means by which those decisions are reached become decidedly less important and assessments of intelligence and good argumentation fall by the wayside. There is something similar afoot in our argument theorization as well, where we will need to continue to exercise our critical judgment in order to arm ourselves for making better decisions.

ACKNOWLEDGEMENTS: Funding for this essay was provided by the Office of Research at the University of Michigan-Flint and the Dingledine Grant at Wake Forest University.

REFERENCES

Berlinski, D. (2000). *The advent of the algorithm*. New York: Harcourt.

Berry, D. M. (2014). *Critical theory and the digital*. New York: Bloomsbury Academic.

Carr, N. (2011). *The shallows*. New York: W.W. Norton.

Carr, N. (2014). *The glass cage: Automation and us*. New York: W.W. Norton.

Edwards, P. N. (2013). *A vast machine*. Cambridge: MIT Press.

Eubanks, V. (2015, April 30). The policy machine: The dangers of letting algorithms make decisions in law enforcement, welfare, and child protection. *Slate.com*. Accessed on May 24, 2015. http://www.slate.com/articles/technology/future_tense/2015/04/the_dangers_of_letting_algorithms_enforce_policy.single.html.

Hammond, K. R. (1996). *Human judgment and social policy: Irreducible uncertainty, inevitable error, unavoidable justice*. New York: Oxford University Press.

MacCormick, J. (2012). *Nine algorithms that changed the future: The ingenious ideas that drive today's computers*. Princeton: Princeton University Press.

Miller, C. R. (2004). Expertise and agency: Transformations of ethos in human-computer interaction. In M. J. Hyde (Ed.), *The ethos of rhetoric* (pp. 197-218). Columbia: University of South Carolina Press.

Ramsay, S. (2011). *Reading machines: Toward an algorithmic criticism*. Urbana: University of Illinois Press.

Silver, N. (2012). *The signal and the noise*. New York: Penguin Books.

Steiner, C. (2012). *Automate this*. New York: Penguin Books.

Stroud, M. (2014, February 19). The minority report: Chicago's new police computer predicts crime, but is it racist? *The verge*. Accessed on May 15, 2015. http://www.theverge.com/2014/2/19/5419854/theminority-report-this-computer-predicts-crime-but-is-it-racist.

Zeide, E. (2015, May 4). Algorithms can be lousy fortunetellers: But potential employers could take them seriously anyway. *Slate.com*. Accessed on May 24, 2015. http://www.slate.com/articles/technology/future_tense/2015/05/crystal_app_algorithmic_fortunetelling_for_employers_and_potential_customers.html.

49

Whose function? Which normativity?

SUNE H. PEDERSEN
University of Copenhagen, Denmark
rnf436@hum.ku.dk

Many argumentation theorists agree that argument has a function, but they do not agree on what this function is. Here I ask how we could come to know this function. I explain how it is possible for different argumentation theorists to arrive at different evaluative norms due to their different approaches to function ascription. The analysis thus reveals a new perspective from which to appreciate the differences between argumentation theorists with regard to their evaluation of argumentation.

KEYWORDS: argumentation, evaluation, function, normativity

1. INTRODUCTION

In her thought-provoking paper *Argument Has No Function* (2007), Jean Goodwin takes up a central problem in argumentation theory, namely whether argument (or argumentation) has a function.[1] While Goodwin claims (with like-minded theorists like Jackson, 1998, 1999; Jacobs, 1999, 2000; Aakhus, 2002, 2003) that argument has no function (at least not in the sense that several mainstream argumentation theorists claim), she cites a host of other scholars who claim that it does. Goodwin thus identifies an important divide in argumentation theory between two camps that we might label the *functionalists* and the *non-functionalists.*

Instead of focusing on the point of *agreement* between the functionalists (namely that argument has a function), I wish to draw attention to a point of *disagreement* between the functionalists. This is

[1] As is well known, some theorists distinguish sharply between argument (product) and argumentation (process and procedure). Due to space considerations, I will not go into the implications of this distinction for my agenda here.

D. Mohammed & M. Lewiński (eds.) (2016). *Argumentation and Reasoned Action: Proceedings of the 1st European Conference on Argumentation, Lisbon, 2015. Vol. II, 781-799.* London: College Publications.

the issue of what the function of argumentation actually is. As it turns out, functionalists hold a range of different positions with respect to this issue. Importantly, as I will show, these positions are sometimes contradictory. This raises the question of how the functionalists could know what the function of argument is.

Why is this important? Why does it matter whether argument has a function, what it is and how we could know this? I will argue that these issues are of fundamental importance for argumentation theory. For the functionalists not only assume that argument has a function, they also use this assumption as an important steppingstone for deriving norms for argument evaluation. In doing this, they follow the thinking of Aristotle (and others) in saying that "for all things that have a function or activity, the good and the 'well' is thought to reside in the function" (NE1.71097b26–27). An implication of this way of thinking is that one needs to know the function of the thing one is evaluating. Get the function wrong, and the norms derived from the function are also likely to be off target. In other words: It matters whether argument has a function, it matters what this function is, and it matters how we could know. The very possibility of formulating well-founded norms for the evaluation of argumentative discourse hinges on it.

To shed light on these questions, I will proceed in the following way. In *Section 2*, I illustrate that argumentation theorists commonly commit themselves to the claim that argument has a function. Further, I draw attention to the fact that argumentation theorists *disagree* about the function of argumentation. The disagreement motivates the following question: How is it possible to know the function of something (anything)? Before addressing this question, I attend to some possible objections to function theory in *Section 3*. In *Section 4*, *5* and *6* I then consider how insights from the literature on function theory could apply to argumentation theory. The point of these sections is to show that the literature on function theory contains not one, but several approaches for ascribing functions to objects. I construct three simplified approaches to function ascription and use these to show that it is possible to reconstruct the way argumentation theorists go about ascribing functions to argument and how they use these functions as the basis for their evaluation of argumentation. An important finding to emerge from this discussion is that there are several ways of understanding the idea that argument "has a function" and these different ways influence our understanding of what its function might be. I discuss these findings in *Section 7* before concluding the paper in *Section 8*. I submit here that argumentation theorists can and should use insights from the extensive literature on functions as a valuable resource to improve our discussions about argument evaluation.

2. FUNCTION CLAIMS IN ARGUMENTATION THEORY

The defining feature of functionalists as I use the term here is their commitment to the claim that argument has a function. While this is a common view among argumentation theorists, there is no agreement on exactly what this function is. Importantly, the disagreement seems not merely to be superficial. In at least some of the cases to be reviewed below, the function claim put forth by one theorist *contradicts* the function claims of others. In order to see this, consider the following function claims from selected argumentation theorists.

Christoph Lumer is explicitly committed to the idea that argument has a function. In his article on "Pragma-Dialectics and the function of argumentation" (2010), Christoph Lumer criticizes pragma-dialectics for what he calls "its consensualistic view of the function of argumentation." The thrust of the criticism is that the norms of pragma-dialectics are not appropriate for the evaluation of argumentative discourse, because they are derived from a mistaken view of the function of argument. As an alternative to the pragma-dialectical view of the function of argumentation, Lumer claims that "the standard function of arguments shall be [...] to lead to justified belief" (Lumer, 2005).

Lumer's position as explained above can be reconstructed as containing two distinct function claims. First of all, Lumer holds that pragma-dialectics (see, e.g., van Eemeren & Grootendorst, 1984, 1992, 1993, 2004) is committed to a mistaken "consensualistic" function claim—let us call this *Function PD*.[2] We may rephrase this function claim (FC) in the following way:

(FC1) The function of argument is not *Function PD*.

Secondly, Lumer also puts forward his own view of the function of argument. This is the view that the function of argument is to lead to justified belief. Let us call this *Function JB*. We may rephrase this function claim in the following way:

(FC2) The function of argument is *Function JB*.

---

[2] "Consensualistic" is Lumer's term. Pragma-dialectics is more frequently understood to claim that the function of argumentation is to "resolve a difference of opinion". I shall not go into the possible differences of these formulations here, since this is not crucial to the general point of my argument in this paper.

Consider now another functionalist, namely Douglas Walton. In several of his papers and books (e.g., Walton, 1998, p. 259), Walton explains his central idea that arguments should be understood and assessed by considering the variety of functions (or goals or aims) they perform in specific dialogic contexts. We may rephrase this function claim in the following way:

(FC3) The function of argument is *Function* $W, c_1 \ldots c_n$.

Consider yet another position on the question of the function of argument. As is obvious from the title and text of her paper (*Argument Has No Function*, 2007), Goodwin disagrees with all of the above function claims. Her position is therefore not exactly a function claim, but rather an assertion to the effect that the set of functions for argument is empty. Goodwin's position can be rephrased as follows:

(FC4) The function of argument is *Ø*.

Lumer, pragma-dialectics, Walton and Goodwin thus exemplify four different views on whether argument has a function and what this function is. Moreover, these positions do not seem compatible: The claim that argument has no function contradicts the claims that argument has a function. The claim that argument has one function (whatever this function might be) contradicts the claim that argument has no function *and* the claim that argument has multiple functions. And the claim that argument has one single *Function PD* contradicts the claim that argument has another single *Function JB*.

Whether the positions as I have laid them out here are entirely accurate is beside the point at this moment. The only thing I am trying to make clear is that there seems to be genuine disagreement among argumentation theorists as to the right or proper view of whether argument has a function and what this function is. This plurality of positions on the matter indicates that there is room for clarification with respect to the concept of function in argumentation theory. What could we mean when we say that argument "has a function"? And if we belong to those who think that argument has a function, how do we know that argument has a function?

I think these are foundational questions that need attention from argumentation theorists. And we may assume that it is possible to find answers to them. Argumentation theorists who claim that argument has a function must be using some kind of approach for arriving at their view that argument has a function. What approaches could these be? I propose some answers to this question in *Sections 4, 5* and *6*.

3. FUNCTIONALISM DEFENDED

But at this point I must address an outstanding issue. As we have seen, Goodwin (2007) holds that argument has no function. If this were true, then it would be pointless to investigate how we could know what the function of argument is. Before I move on, it is therefore necessary to take a closer look at the reasons Goodwin put forth for dismissing the idea that argument has a function.

As a way of substantiating her position, Goodwin explains with admirable clarity that functionalism has been tried and shown to be a failure. This leads her to conclude that argumentation theorists are being lead astray by all their thinking about functions. While Goodwin's objection to functionalism deserves more attention than I am able to devote to it here, it seems to me on first approximation that Goodwin is a little too quick to dismiss the possible usefulness of function theory. The problem with Goodwin's argument is, briefly put, that she dismisses function theory *en bloc*, because a particular, dubious sociological perspective associated with Parsons, Merton and others happened to be very fond of making function claims. But the idea that function claims *as such* are problematic would only follow from this if *all* function claims could be said to be relevantly similar to the function claims made by the dubious sociological functionalists. But this is not the case. As Crilly (2010), Perlman (2009) and others have pointed out, there is a plethora of different approaches to function theory. Function theory can no longer be considered to be primarily a variety of sociological theory. It has become an umbrella term encompassing a large variety of different theoretical perspectives with roots going back to the *ergon* arguments of Plato and Aristotle. Sociology is responsible for *some* of the contributions to function theory, but philosophers of mind, science, biology and technology have also provided a lively context for the discussion of functions in recent years. Scholars such as Wright (1973), Cummins (1975), Millikan (1984), Dennett (1990), Godfrey-Smith (1993, 1994), Searle (1995, 1998), Preston (1998), Franssen (2006), Vermaas & Houkes (2006), Kroes (2012), Thomasson (2014) and many others have made contributions to this discussion that move far beyond the sociological functionalism rejected by Goodwin.

I think there is a lot to be learnt from this literature, and argumentation theorists with an interest in pragmatics and interpretation could certainly also offer something in return. In sum, it seems to me to be premature to rule out the possibility that something valuable could come of a dialogue between argumentation theory and function theory, not least in connection with ongoing work on the potentially *multi-functional* nature of argument (see e.g., Mohammed, 2013; Lewinski, 2012; Lewinski & Aakhus, 2014). And the case for this seems to me to be even clearer given the ubiquity of function talk in argumentation theory that Goodwin rightly responds to.

I therefore move forward on the assumption that some insights for argument may be gained from function theory. And let me repeat why I think this is so: Evaluation of argumentative discourse frequently proceeds on the assumption that argument has a function, because the norms necessary for argument evaluation are assumed to be in part derived from this function. So unless we are quite clear about what it means for argument to have a function and how we could know what this function is, the theoretical basis of the norms used for argument evaluation becomes slippery indeed. For this reason, it seems to me worthwhile to attempt to get clearer on what function theorists from the host of fields mentioned above have to offer.

In *Sections 4, 5* and *6* below, I therefore explicate three different, stylized approaches to function ascription. (Function ascription is, roughly, the activity of somehow homing in on the function of an object). All approaches make use of basic analytic categories similar to those found in argumentation theory such as speaker ("designer"), message ("artifact") and audience ("user").

The first approach outlined below—the "designer intention approach"—views the function of an artifact as flowing from the intentions of the designer. The second approach—the "user intention approach"—views the function of an artifact as flowing from the intentions of the user. The third approach—the "optimality approach"—views function as flowing from considerations about which capacity the artifact would be able to optimally fulfill. All three positions have their bases in function theory, although I have simplified them considerably here in order not to confuse the points I wish to make. And these points are: (1) that function ascription is going on in argumentation theory; (2) that function ascription is going on in different ways that we can understand and reconstruct by way of insights from function theory; and importantly (3) that our approach to function ascription influences the evaluative norms we view as appropriate.

4. FUNCTION FROM DESIGNER INTENTION

4.1 The designer intention approach illustrated

According to the "designer intention approach" of function ascription, the function of an artifact flows from the intentions of the designer or maker. To illustrate this idea, consider the following short scenario. If I make you a small clay bowl and *intend* it to be an ashtray, then it simply *has* the function of an ashtray. If I make an identical clay bowl and *intend* it to be for holding keys, then it simply *has* the function of a key holder. On this approach, the function of the artifact flows from the intentions of the designer. If an inquisitive guest visiting your home asks you what the function of the clay object on your table is, presumably you would reply that the function of the clay object is to contain ashes (or hold keys). Further, if your guest asked you how you know, you would tell him that you know this because the designer of the object has told you that this is what he intended the function of the object to be.

4.2 The theoretical basis of the designer intention approach: Neander

The above story is meant to be illustrative rather than accurate. Perhaps no existing function theorist would subscribe to a simple intentionalist approach to function ascription such as the one I propose here. Nevertheless, this kind of theory is not completely made up. Consider this quotation from Karen Neander (1991, p. 462):

> [T]he function of an artifact is the purpose or end for which it was designed [or] made [...] by an agent.

The quotation, even if taken out of its context, illustrates that traces of the "designer intention approach" can be found in the literature on functions. Since my ashtray example above is a case of an artifact made and designed with a purpose by an agent, it follows from the reasoning quoted above that my clay structure has, due to my intentions in making it, the function of an ashtray.

4.3 Neo-Aristotelian Criticism as a designer intention approach

The point of illustrating the above approach to function ascription is to see whether the approach, or something similar to it, can be found in argumentation theory. To my knowledge, no current argumentation theorist subscribes to exactly this kind of function ascription. Even so, the history of rhetorical criticism includes an important epoch during which the standard type of discourse evaluation began from the same kind of designer-intentionalist assumptions I have just sketched. The kind of criticism I am thinking of is the kind typically referred to as 'Neo-Aristotelian criticism'. The guiding assumption for a critic engaged in this kind of criticism is that the speaker intends his discourse to persuade. The function of the discourse (to persuade) follows from the intention of designer, which for discourse usually means the speaker. And subsequently, this function is then used to derive the norms by which the critic must evaluate the discourse. These norms, as is well known, evolve around questions of persuasive effect. As Kuypers (2001) says in one passage, Neo-Aristotelian critics "stressed the importance of speaker intention, especially when critical judgments on effectiveness are made."

Neo-Aristotelian criticism has of course been largely abandoned as a critical enterprise after Edwin Black (1965) synthesized and rearticulated the then growing problems of this kind of speaker-centered approach to criticism. But this does nothing to detract from my main point: If one follows the "designer intention approach" to function ascription, the function of (and thus the norms of) argumentative discourse can be derived from the intentions of the speaker or "designer". This is the central move in Neo-Aristotelian criticism, and as such this type of criticism is based on a version of the *designer intention approach to* function ascription.

5. FUNCTION FROM USER INTENTIONS

5.1 The user intention approach illustrated

I turn now to another approach to function ascription, this time based on *user intentions* rather than *designer intentions*. To illustrate this approach, consider the following story. In the late 1970s, the Japanese company Yamaha designed a particular pair of loudspeakers called the NS10s. The speakers were of modest size, and they were black with an aesthetically appealing white speaker cone as the centerpiece. These design choices were made because the designers intended the speaker to be for home use. The Yamaha NS10s, however, never became a success amongst home users. Some users cited problems known as "listener fatigue" when listening to music from the speaker; home users simply did not like the way it sounded. But this did not lead to a negative evaluation of the speakers on the whole. On the contrary, the NS10s have been dubbed "the most important speaker you have never heard of", and they were sold in hundreds of thousands of copies for almost 25 years after their introduction. The reason for this is that sound engineers in recording studios quickly discovered that the particular frequency response of the speaker (the way in which the speaker conveys bass, middle and treble frequencies) made it very useful for so-called nearfield monitoring in recording studios. The rule of thumb for sound engineers was that if they could get the recordings to sound good on the NS10s, then the recordings would sound good on most other playback systems. The speakers thus never fulfilled the function intended by its designers. But they ended up being appropriated for a different purpose by their (unintended) users—professional sound engineers. Eventually, the speakers were rebranded as the NS10Ms with the letter M added to the name to signify its "new" function as a (near-field) monitor.

So, what is the function of the Yamaha NS10s? On the user intention approach to function ascription, their function is clearly to "near-field monitor". The point of this story is to illustrate that *designer intentions* sometimes do not seem to matter for the way we ascribe functions to objects. User-intentionalists would use a story like this to substantiate that function ascription really starts from the intentions of the *users*, not the designers. Consequently, if we want to evaluate the speakers, we should not begin from the function that flows from the intentions of the designers (this would presumably lead to a negative evaluation), but instead from the function that flows from the intentions of the users (this would presumably lead to a positive evaluation).

5.2 The theoretical basis of the user intention approach: McLaughlin

Just like in the case of the designer intention approach, it is difficult to find a clear-cut explication of the user intention approach in the function literature. Like the designer intention approach, the user intention approach is construct obtained by simplifying some considerably more complex lines of reasoning found in the function literature. Nonetheless, this quotation from McLaughlin (2001, p. 47) illustrates fragments of the user intention approach:

> The function or purpose of an artifact is the end to which it is a means–whether successful or unsuccessful–for whoever [...] acquired it [or] used it [...]

On this line of reasoning, the fact that the Yamaha NS10 speakers were successfully "acquired" and "used" for near-field monitoring by predominantly sound engineers simply makes it the case that near-field monitoring is their function. The intentions of the designers, though well documented, are of no importance. Accordingly, from the perspective of the user intention approach, we should evaluate the speakers from norms flowing from the question of how well the speakers perform the function of near-field monitoring.

5.3 *Christian Kock as a user intentionalist*

Is it possible to find an example of a "user intentionalist" in current argumentation theory? I believe it is possible to at least come close. In one of his books on public political debate (intended for a general audience), Christian Kock (2011) argues that argumentation in a democratic context should be evaluated by taking the "user" (i.e. the citizen) into account. Leading up the passage I quote below, Kock has considered some assumptions prevalent among social scientists associated with the Chicago School of economics. Simplifying the Chicago School's line of thought, the School holds that politicians in some cases might *appear* to be interested in governing the polity out of concern for the common good, but *really*—underneath the surface—the motives and driving forces of politicians across the bar are in fact selfish: Politicians intend only to maximize their own power, reputation, wealth, etc. (Kock, 2011, pp. 261-262, my translation):

> But regardless of whether politicians are really driven solely by strategic power considerations (which seems implausible), this would not change a single thing about the message of this book. [...] It is in fact utterly unimportant what the facts are about the inner motives of the politicians—the needs of the citizens concerning the political debate would still be that arguments facilitating deliberation are put forth. The politicians may be driven by any intention whatsoever; this does not change what the citizens need from the public debate: that politicians argue [in the manner laid out in the book] and respond to arguments [in the manner laid out in the book].

The important thing to note here is this: The politicians' intentions are not relevant inputs for the analyst. Assuming one thing or the other about the intentions of the politicians (the makers/designer/inventors of the arguments to be evaluated) does not change the approach to argument evaluation. Rather, what matters is how the (idealized) users intend to use the arguments. These (idealized) intentions are to collect deliberative input from politicians about the possible courses of action available to the polity. On this model, the *idealized user intentions* determine the function of argument. The *designer intentions* do not matter. Argumentation should thus be evaluated using norms flowing from this user-intended function, and *not* the function flowing from the possibly strategic self-interested intentions of the politicians ("designers"). (Incidentally, many political commentators—at least in some countries—start from the assumption that strategic intention is the only thing that counts when determining the function of political argument. See e.g., Bengtsson, 2014).

6. FUNCTION FROM OPTIMALITY CONSIDERATIONS

6.1 The optimality approach illustrated

Above, I have introduced the "designer intention approach" and the "user intention approach" to function ascription. I turn now to the "optimality approach". To understand this approach to function ascription, consider the following short story: The "Antikythera Mechanism" has been dubbed the world's first analogue computer. It was recovered in 1900-01 from the Antikythera shipwreck off the Greek island of Antikythera. It is believed to be more than 2500 years old. For long, researcher did not have any clear idea as to the function of the artifact. After its recovery in 1902, archeologist Valerios Stais proposed that it was a kind of astronomical clock, but this function ascription did not catch on. Investigations into the function of the artifact were soon dropped and only began anew when British science historian and Yale University professor Derek J. de Solla Price became interested in the device again more than 50 years later. After thorough investigations, a consensus emerged that the Antikythera Mechanism must have had the function of an orrery (a kind of planetarium). How did the researchers manage to ascribe this function to the device without access to designer or user intentions? They did so by demonstrating that the device has the optimal capacity to serve this function. Researchers have calculated the periods of the rotation of the wheels on the Antikythera mechanism and found out that the artifact would have made an accurate representation of what was then known about planets and their motion paths.

On the "optimality approach" to function ascription, we determine the function of an object through so-called optimality considerations rather than from designer or user intentions. Simply put, the function of an artifact is found by asking what the artifact would be good at. Various candidate tasks or capacities are considered and the task or capacity that the device would be able to "best" perform is taken to be its function. We know that the Antikythera Mechanism was an orrery, simply put, because it would be good orrery.

6.2 The theoretical basis of the optimality approach: Dennett

Contrary to my stylized "designer intention approach" and "user intention approach", the "optimality approach" actually exists in function theory. Daniel Dennett is perhaps the most noted "optimalist" in function theory. According to Dennett, we ascribe functions to artifacts by considering using the "optimality approach" (my term). To the best of my knowledge, Dennett has never formulated a positive account of function ascription, but his position can be inferred from the following quotation: (Dennett, 1990, p. 184):

> It counts against the hypothesis that something is a cherry-pitter, for instance, if it would have been a demonstrably inferior cherry-pitter.

Dennett is not the first "optimalist". Plato provides us with another example. In *The Republic,* towards the end of Book One, Socrates debates with Thrasymachus:

> I do not understand, [Thrasymachus] said.
>
> Socrates: Let me explain: Can you see, except with the eye?
> Thrasymachus: Certainly not.
> S: Or hear, except with the ear?
> T: No.
> S: These then may be truly said to be the ends of these organs?
> T: They may.
> S: But you can cut off a vine-branch with a dagger or with a chisel, and in many other ways?
> T: Of course.
> S: And yet not so well as with a pruning-hook made for the purpose?
> T: True.
> S: May we not say that this is the end of a pruning-hook?
> T: We may.
> S: Then now I think you will have no difficulty in understanding my meaning when I asked the question whether *the end of anything would be that which could not be accomplished, or not so well accomplished, by any other thing*? [my italics]
>
> I understand your meaning, he said, and assent.

And a few dialogue turns later on:

> S: Well; and has not the soul an end which nothing else can fulfill? For example, to superintend and command and deliberate and the like. *Are not these functions proper to the soul, and can they rightly be assigned to any other?*

Plato's reasoning here seems to be this: If a given artifact is able to accomplish something which no other type of artifact could accomplish or accomplish as well, then what the artifact can accomplish can be assigned to the artifact as its proper function. To repeat the story of the Antikythera Mechanism: We know it is an orrery, because it would be a good orrery.

6.3 Christoph Lumer as an optimalist

The optimality approach is perhaps the most difficult of my three approaches to exemplify with a case from argumentation theory.[3] But, at least in some passages, Christoph Lumer's work seems to be reconstructable as employing a (partially) optimalist approach to function ascription. I should say from the outset of this analysis that Lumer is the most explicitly functionalist argumentation theorist I know of, and that in several of his papers (1991, 2005, 2010) he devotes quite a lot of effort to the issue of how argument can be understood to have a function. His views on the topic are subtle, and I do not pretend to do justice to them all in my brief illustration here. Nonetheless, I believe to be able to show at least *traces* of the optimalist approach in Lumer's thinking.

As we saw in Section 2 above, Lumer explicitly claims that argument has a function (namely to lead to justified belief). Further, he claims that this is the *standard function* of argument. He illustrates the concept of *standard function* through an analogy:

> [T]he standard-function [of a drill] is: to drill holes. Another function of a drill—however not a standard-function—is to whip cream; the input, in this case, is to supply the drill with a whisk, to put this into the cream and to supply the machine with electric current; the output is whipped cream or butter.

The point of the analogy is to explain how to distinguish the *standard function* of something from its mere *accidental functions*. Accidental functions are all those functions that an artifact would be capable of performing, but that cannot be said to be the function of the artifact: I can use a screwdriver to open a can of paint, but this does not make the screwdriver into a can opener. For more on this distinction, see e.g. Houkes & Vermaas (2004). Lumer makes the distinction in order to be

---

[3] Steven W. Patterson (2011) provides another and more clear-cut example of an "optimalist" approach to function ascription. I came across Patterson's work too late to include it in this paper, but for the actual ECA presentation I substituted the analysis of Lumer's work with an analysis of Patterson's work.

able to home in on the specific function that should be used as the basis for developing norms for argument evaluation. The question to be pondered here is this: How does Lumer distinguish the standard function of argument from its accidental functions? There must be some kind of basis for this distinction. What could it be?

One possible way to make this distinction between standard and accidental functions is to use *optimality considerations* as proposed by Dennett and Plato above. Instead of beginning from designer or user intentions, this kind of function ascription prompts the function analyst to ask what argumentation would be "best at." Using Plato's words, in order to ascribe a proper function to argument, we need to search for the task that nothing could do better than argument. There is some evidence in Lumer's papers indicating that this is the strategy he is following. See, for instance, how he singles out (epistemologically designed) argumentation as the kind of argument "best at" reaching true/truth-like/justified belief (Lumer, 2005, pp. 238-239):

> [E]pistemologically conceived arguments do and should prevail in most domains of our lives: of course in science, but also in political decisions, courts of law and daily life, because here we are interested in truth and in finding out best solutions. *And such arguments provide this much better than mere rhetoric.*

In other words, there is something (namely *to lead to justified belief*) that (good) argumentation can do, which other kinds of (not so good) arguments—like rhetoric or other types of argumentation—cannot do. Seen from the optimality approach of Dennett and Plato, this something must then be the function of argument. Accordingly, the norms we use to assess argumentation should flow from this function (Lumer, 2005, pp. 213-214):

> An epistemological theory of argument is characterized by two features. 1. It takes the *standard function of arguments to be*: to lead the argument's addressee to (rationally) justified belief [...] [and] 2. It develops criteria for good arguments and argumentation on this basis, i.e., it designs them in such a way as to fulfil their epistemic function.

This discussion has concerned the issue of how Lumer can arrive at a "standard function of arguments". I have tried to show that Lumer can be shown to use at least traces of an optimalist approach to answer this question. From an optimalist approach, the function of an argument is what the argument would be "best at" achieving. Whatever this turns out to be, the norms appropriate for evaluation should flow from this function.

7. DISCUSSION

The illustration and exemplification of the three approaches to function ascription serve to show the following points: (1) The questions of whether argument has a function and how we know can be answered in *several* ways. The way we answer the question can be considered an approach to function ascription. (2) Argumentation theorist are rarely explicit about their approaches to function ascription, but their approaches can be reconstructed using insights from function theory. (3) Argumentation theorists frequently evaluate argumentation by considering how well the argumentation in question *fulfils its function*. Ultimately, then, the norms that argumentation theorists use to evaluate argumentation are therefore at least partly a result of their specific approach to function ascription.

A few remarks on the analysis of function ascription are in order. The three approaches to function ascription are *analytical categories*. They have been constructed for illustrative purposes only and lay no claim to be exact reconstructions of how function ascription actually takes place. I use them here to show that it is possible to analyze how argumentation theorists go about ascribing functions to arguments.

And I think a lot can be learned from conducting such analyses. Much more work on this topic could readily be undertaken with promising prospects. For instance, it is possible to extend the understanding of function ascription by considering the context of the artifact to be evaluated. As we saw in the case of Christian Kock above (our representative of the "user intentions approach"), certain idealizing assumptions were necessary in order to make his actual approach to function ascription align with my own construction. These idealizing assumptions stem from the fact that Kock considers argumentation in the specific context of a democratic system. This context gives rise to agentive roles such as "politicians" and "citizens" and these roles in turn are associated with certain rights and responsibilities. One such responsibility is the responsibility of a citizen to help govern the polity through the usual paths of democratic agency such as voting. In order for this responsibility to make sense, the citizen has a corresponding right to receive the necessary inputs to make it possible for the citizen to carry out his responsibility meaningfully. This right in turn creates a responsibility for politicians and other powerful agents (all those capable of actually steering the polity in one direction or another) to provide these inputs for the citizen in such a way that the citizen is able to carry out his or her responsibility.

Ideas like these follow from more subtle considerations of how the context of an artifact affects its function—a topic that I am unable to address here, but which function theorists also provide valuable insights into (see e.g., Preston, 2006). When considerations about how the local context of arguments may affect which functions we ascribe to arguments and how we ascribe them, the division we started out with in this paper between functionalists and non-functionalists begins to crumble. If we are open to a wide range of approaches to function ascription—which I think we should be—then even Goodwin begins to look like a functionalist (a point that Goodwin (2007, p. 70) herself acknowledges when she says that her objections to the idea of function only holds for the specific sense of function she employs in her paper).

In sum, the discussion of approaches to function ascription explains how it is possible for argumentation theorists to reach incompatible views about the function of argument. Whether argument has a function and what it is depends on the approach we use to ascribe functions to arguments.

8. CONCLUSION

With this insight in mind, it is time to review the points made in this paper. I began by drawing attention to a situation of both agreement and disagreement in argumentation theory: Several theorists agree that argumentation has a function, but they do not agree on what this function is. This situation prompted a search for an answer to the questions of whether and how argumentation can be said to have a function—and how we could know. On the basis of insights from the literature on function theory, I constructed three basic approaches to function ascription: The "designer intentions approach", the "user intentions approach", and the "optimality approach." By reviewing some ways in which different rhetoricians and argumentation theorists go about constructing norms for argument evaluation, I then showed how the reasoning from these theorists can be reconstructed as exemplifying the three approaches to function ascription. From the ensuing discussion, it becomes possible to see that functions can be ascribed to artifacts in several ways, and that this insight also holds for argumentation.

The implication of this finding is that the disagreement between different argumentation theorists with regard to evaluative norms can be appreciated from a new perspective. Since evaluation of argumentation often begins from the assumption that there is a function to be fulfilled, it makes a difference to be able to actually ascertain how things such as arguments have functions and why it can make sense to disagree about what this function is. In other words, in order to understand argument evaluation, it is crucial to ask *whose function* we are talking about: The function flowing from the designer, the user or the artifact? When we know this, we can answer the question: *Which normativity*?

The concern in this paper has been with norms for evaluation. But the manner in which this issue has been addressed has been descriptive only. A crucial outstanding issue to be addressed, therefore, is this: What makes a given function ascription *more justified* or *more appropriate* than others in a given context? Until this question has been addressed, some important insights with regard to argument evaluation seem to be beyond our grasp.

ACKNOWLEDGEMENTS: First of all, I would like to thank Christian Kock, Charlotte Jørgensen and William Keith for their constructive criticisms and comments on this paper prior to the ECA conference. Secondly, the audience at the ECA including David Godden, Jean Goodwin, David Hitchcock, Constanza Ihnen Jory, Jens Kjeldsen, Amnon Knoll, Dima Mohammed, and Frank Zenker gracefully pointed to many points in need of further argument and research, and were most helpful. Also, a big thanks to Harvey Siegel for comments and additional insights after the ECA conference. This version of the paper has not been altered substantially after the conference, but I hope to be able to include many of the exacting comments and criticisms I have received in a future version.

REFERENCES

Aakhus, M. (2002). Modeling reconstruction in groupware technology. In F.H. van Eemeren (Ed.), *Advances in pragma-dialectics* (pp. 121-136). Amsterdam: Sic Sat.

Aakhus, M. (2003). Neither naïve nor critical reconstruction: Dispute mediators, impasse, and the design of argumentation. *Argumentation, 17*, 265-290.

Aristotle. *Nicomachean Ethics.*

Black, E. (1965/1978). *Rhetorical criticism: A study in method.* Wisconsin: The University of Wisconsin Press.

Bengtsson, M. (2014). *For borgeren, tilskueren eller den indviede?: En praksisorienteret retorisk kritik af avisens politiske kommentarer.* Københavns Universitet, Det Humanistiske Fakultet.

Crilly, N. (2010). The roles that artefacts play: Technical, social and aesthetic functions. *Design Studies, 31*(4), 311–344.

Cummins, R. (1975). Functional analysis. *The Journal of Philosophy, 72*(20), 741–765.

Dennett, D. C. (1990). The Interpretation of texts, people and other artifacts. *Philosophy and Phenomenological Research, 50*, 177–194.

Dipert, R. R. (1993). *Artifacts, art works, and agency*. Temple University Press.

Eemeren, F.H. van, & Grootendorst, R. (1984). *Speech acts in argumentative discussions: A theoretical model for the analysis of discussions directed towards solving conflicts of opinion*. Dordrecht: Floris Publications.

Eemeren, F.H. van, & Grootendorst, R. (1992). *Argumentation, communication, and fallacies: A pragma-dialectical perspective*. Hillsdale, NJ: Lawrence Erlbaum Associates.

Eemeren, F. V., Grootendorst, R., Jacobs, S., & Jackson, S. (1993). *Reconstructing Argumentative Discourse*. Tuscaloosa, AL: University Alabama Press.

Eemeren, F.H. van, & Grootendorst, R. (2004). *A systematic theory of argumentation: The pragma-dialectical approach*. Cambridge: Cambridge University Press.

Franssen, M. (2006). The normativity of artefacts. *Studies in History and Philosophy of Science, Part A, 37*(1), 42–57.

Jackson, S. (1998). Disputation by design. *Argumentation, 12*, 183-198.

Jackson, S. (1999). The importance of being argumentative: Designing disagreement in to teaching/learning dialogues. In F. H. van Eemeren, R. Grootendorst, J. A. Blair, & C. A. Willard (Eds.), *Proceedings of the Fourth International Conference of the International Society for the Study of Argumentation* (pp. 392-396). Amsterdam: Sic Sat.

Jacobs, S. (1999). Argumentation as normative pragmatics. In F. H. van Eemeren, R. Grootendorst, J. A. Blair, & C. A. Willard (Eds.), *Proceedings of the Fourth International Conference of the International Society for the Study of Argumentation* (pp. 397-403). Amsterdam: Sic Sat.

Jacobs, S. (2000). Rhetoric and dialectic from the standpoint of normative pragmatics. *Argumentation, 14*, 261–286.

Godfrey-Smith, P. (1993). Functions: Consensus without unity. *Pacific Philosophical Quarterly, 74*, 196-208.

Godfrey-Smith, P. (1994). A modern history theory of functions. *Noûs, 28*, 344-362.

Goodwin, J. (2007). Argument has no function. *Informal Logic, 27*(1), 69–90.

Houkes, W., & Vermaas, P. E. (2004). Actions versus functions: A plea for an alternative metaphysics of artifacts. *The Monist, 87*, 52-71.

Kock, C. (2011). *De svarer ikke [They Are Not Answering]*. 2nd Ed. Gyldendal.

Kroes, P. (2012). *Technical artefacts: Creations of mind and matter* (Vol. 6). Dordrecht: Springer Netherlands.

Kuypers, J. A., & King, A. (2001). *Twentieth-century roots of rhetorical studies*. Westport, Conn.: Praeger.

Lewiński, M. (2012). Public deliberation as a polylogue: Challenges of argumentation analysis and evaluation. In H. J. Ribeiro (Ed.), *Inside arguments: Logic and the study of argumentation* (pp. 223-245). Newcastle upon Tyne: Cambridge Scholars Publishing.

Lewiński, M., & Aakhus, M. (2014). Argumentative polylogues in a dialectical framework: A methodological inquiry. *Argumentation, 28*(2), 161–185.

Lumer, C. (1991). Structure and function of argumentations. An epistemological approach to determining criteria for the validity and adequacy of argumentations. In *Proceedings of the Second International Conference on Argumentation* (Vol. 1, pp. 98–107).

Lumer, C. (2005). The epistemological theory of argument–how and why? *Informal Logic, 25*(3), 213-242.

Lumer, C. (2010). Pragma-Dialectics and the function of argumentation. *Argumentation, 24(1),* 41-69.

McLaughlin, P. (2001). *What functions explain: Functional explanation and self-reproducing systems.* Cambridge: Cambridge University Press.

Millikan, R. G. (1984). *Language, thought, and other biological categories: New foundations for realism.* Cambridge, MA: MIT Press.

Mohammed, D. (2013). Pursuing multiple goals in European Parliamentary Debates: EU immigration policies as a case in point. *Journal of Argumentation in Context, 2*(1), 47–74.

Neander, K. (1991). The teleological notion of 'function'. *Australasian Journal of Philosophy, 69*(4), 454–468.

Patterson, S. W. (2011). Functionalism, normativity and the concept of argumentation. *Informal Logic, 31*(1), 1-26.

Perlman, M. (2009). Changing the mission of theories of teleology: DOs and DON'Ts for thinking about function. In U. Krohs & P. Kroes (Eds.), *Functions in biological and artificial worlds: Comparative philosophical perspectives* (pp. 17-25). Cambridge, MA: MIT Press.

Plato. *The Republic.*

Preston, B. (1998). Why is a wing like a spoon? A pluralist theory of function. *The Journal of Philosophy, 95*(5), 215–254.

Preston, B. (2006). Social context and artefact function. *Studies in the History and Philosophy of Science, Part A, 37*(1), 37-41.

Walton, D. (1998). *The New Dialectic. Conversational contexts of argument.* Toronto: University of Toronto Press.

Wright, L. (1973). Functions. *The Philosophical Review, 82*(2), 139–168.

Thomasson, A. L. (2014). Public artifacts, intentions, and norms. In M. Franssen, P. Kroes, T. A. C. Reydon, & P. E. Vermaas (Eds.), *Artefact Kinds* (pp. 45–62). Springer International Publishing.

Searle, J. (1995). *The construction of social reality.* New York: The Free Press.

Searle, J. (1998). *Mind, language and society, philosophy in the real world.* New York: Basic Books.

Vermaas, P. E., & Houkes, W. (2006). Technical functions: A drawbridge between the intentional and structural natures of technical artefacts. *Studies in History and Philosophy of Science, Part A, 37*(1), 5–18.

50

An Annotated Corpus of Argumentative Microtexts

ANDREAS PELDSZUS
Applied Computational Linguistics, University of Potsdam, Germany
peldszus@uni-potsdam.de

MANFRED STEDE
Applied Computational Linguistics, University of Potsdam, Germany
stede@uni-potsdam.de

We present a freely available corpus of argumentative "microtexts", featuring short and dense authentic arguments, annotated according to a scheme for representing text-level argumentation structure. The corpus consists of 112 German texts plus professional English translations that preserve linearization and argumentative structure. We provide statistics of the variety and the linguistic realization of argumentation structure in the corpus. We hope the data release serves the needs of data-driven approaches to argument mining and qualitative analysis alike.

KEYWORDS: argument mining, argumentation structure, dialectical argument, informal logic, text corpus

1. INTRODUCTION

Argumentation can, for theoretical purposes, be studied on the basis of carefully constructed examples that illustrate specific phenomena, but for many researchers, the link to authentic, human-authored text is highly desirable. This is obviously the case for the Computational Linguistic discipline of "argumentation mining", which in recent years has attracted a lot of attention, but also for research aiming to uncover the linguistic features of argumentative text and the specific mechanisms of various argumentative moves.

For these reasons, the interest in argumentation-oriented corpora of monologue text as well as spoken dialog is rising. In the work reported here, we address this need by making a resource publicly available that is designed to fill a particular gap. So far, there exist only a

D. Mohamı ... :d Action: Proceedings of the 1st
European (... ndon: College Publications.

few resources with annotated argumentation structures over monologue texts, as e.g. the AIFdb, the former Araucaria corpus (Reed et al., 2008) with in large parts newswire articles, furthermore a small set of commentaries analysed in (Stede & Sauermann, 2008), and a corpus of student essays (Stab & Gurevych, 2014). While authentic text from social media or newspapers is ultimately the target for automatic argumentation mining, these sources are often not ideal for more qualitatively oriented research. In newswire text, the language can be quite complex, while in social media and language learners text, it is often ill-formed. This also has an impact on the underlying argumentation structure, in some cases it is quite trivial, and in other cases quite intransparent.

Our contribution is a collection of 112 "microtexts" that have been written in response to trigger questions, mostly in the form of "Should one do X". The texts are short but at the same time "complete" in that they provide a standpoint and a justification, by necessity in a fairly dense form. Hence, the underlying argumentation structure is relatively clear. We collected the texts in German and then had them translated to English; both versions are available to interested researchers.

In addition to the raw texts, we provide manually-created annotations of the argumentation structure, following a scheme that is inspired by the informal-logic tradition. Thus, argumentation researchers will find a resource of simple, authentic natural language texts together with suggestions of structural representations of the underlying argument. At the same time, the data can also be used for building models in automatic argumentation mining.

The paper is structured as follows. In Section 2, we describe the process of gathering the data, and Section 3 provides a brief summary of the annotation scheme. The process of creating the annotations is described in Section 4. Some statistics on the corpus and the argument structures are presented in Section 5, and Section 6 gives information on the form and availability of the corpus. Finally, some conclusions are presented in Section 7.

2. DATA COLLECTION AND CLEANING

2.1 Collection

The microtext corpus consists of two parts. On the one hand, 23 texts were written by the authors as a "proof of concept" for the idea. These texts also have been used as examples in teaching and testing argumentation analysis with students. An example text is given in (1):

(1) Energy saving light bulbs contain a significant amount of toxins. A commercially available bulb may contain for example up to five milligrams of mercury. That's why they should be taken off the market, unless they're unbreakable. But precisely this is unfortunately not the case.

On the other hand, 90 texts have been collected in a controlled text generation experiment, where normal competent language users wrote short texts of controlled linguistic and rhetoric complexity.

To this end, 23 probands were instructed to write a text on a topic that was to be chosen from a given set of trigger questions. All probands were native speakers of German, of varying age, education and profession. They received a short written instruction (about one page long) with a description of the task and three sample texts. The probands were asked to first gather a list with the pros and cons of the trigger question, then take stance for one side and argue for it on the basis of their reflection in a short argumentative text. Each text was to fulfil three requirements: It should be about five segments long; all segments should be argumentatively relevant, either formulating the main claim of the text, supporting the main claim or another segment, or attacking the main claim or another segment. Also, the probands were asked that at least one possible objection to the claim should be considered in the text. Finally, the text should be written in such a way that it would be understandable without having its trigger question as a headline. Regarding these triggers, we offered a number of questions to the probands to choose from, and the five most frequently selected issues were:

- Should the fine for leaving dog excrements on sideways be increased?
- Should shopping malls generally be allowed to open on Sundays?
- Should Germany introduce the death penalty?
- Should public health insurance cover treatments in complementary and alternative medicine?
- Should only those viewers pay a TV licence fee who actually want to watch programs offered by public broadcasters?

2.2 Cleaning

Since we aim for a corpus of texts featuring authentic argumentation but also regular language, all texts have been corrected for spelling and grammar errors. As a next step, the texts were segmented into elementary units of argumentation. Most probands already marked up in some way what they regarded as a segment. Their segmentation was corrected when necessary, e.g. when only complex noun phrase conjuncts or restrictive relative clauses had been marked, or when subordinate clauses had not been split off. All remaining texts were segmented from scratch. Due to this step of (re-) segmentation, not all of the final texts conform to the length restriction of five segments; they can be one segment longer or shorter.

Unfortunately, some probands wrote relatively long texts. We decided to shorten these texts if possible by removing segments that appeared less relevant. This removal also required some modifications in the remaining segments to maintain text coherence, which we made as minimal as possible.

Another source of problems were segments that did not meet our requirement of argumentative relevance. When writers did not concentrate on discussing the thesis, but moved on to a different issue, we removed those segments, again with minimal changes in the remaining segments. Some texts containing several of such segments remained too short after the removal and thus have been discarded from the dataset. After the cleanup steps, 90 of the original 100 written texts remained for annotation of argumentation structure.

2.3 Translation

To supplement the original German version of the collected texts, the whole corpus has been professionally translated to English, in order to reach a wider audience of potential users. Our aim was to have a parallel corpus, where annotated argumentation structures could represent both the German and the English version of a text. We thus constrained the translation to preserve the segmentation of the text on the one hand (effectively ruling out phrasal translations of clause-type segments) and to preserve its linearization on the other hand (disallowing changes to the order of appearance of arguments). Besides these constraints, the translation was free in any other respect. Note that the translator had only access to the segmented source text, but not to an argumentative analysis of the text.

3. ANNOTATION SCHEME

For all 112 (23+90) texts, the argumentation structure has been annotated manually. Our representation of it is based on Freeman's theory of the macro-structure of argumentation (Freeman, 1991, 2011), which aims to integrate the ideas of Toulmin (1958) into the argument diagraming techniques of the informal logic tradition (Beardsley, 1950;

Thomas, 1974) in a systematic and compositional way. Its central idea is to model argumentation as a hypothetical dialectical exchange between the proponent, who presents and defends his claims, and the opponent, who critically questions them in a regimented fashion. Every move in such an exchange corresponds to a structural element in the argument graph. In Figure 1, we show the representation for one of our microtexts. The nodes of this graph represent the propositions expressed in text segments (grey boxes), and their shape indicates the role in the dialectical exchange: Round nodes are proponent's nodes, square ones are opponent's nodes. The arcs connecting the nodes represent different supporting (arrow-head links) and attacking moves (circle/square-head links). By means of recursive application of relations, representations of relatively complex texts can be created.

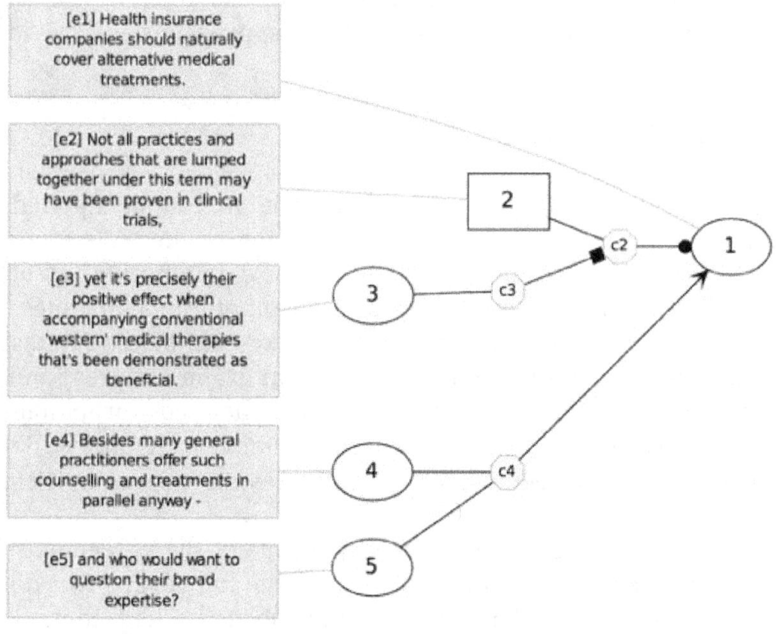

Figure 1: sample text and argumentation graph

The scheme distinguishes several different supporting and attacking moves, or argumentative functions of a segment. Besides the "standard" case of a premise supporting a claim, there can be support by example. For the attack moves, the scheme distinguishes rebuttals (challenging the acceptability of a proposition) from undercutters (challenging the acceptability of an inference between two propositions). In our example text shown in Figure 1, the second segment rebuts the first segment, and this rebutting move is then undercut by the third segment. Furthermore, the scheme allows for combining multiple premises in one move. In the example, segment four and five jointly support the main claim, which corresponds to linked premises in Freeman's theory. Note, that this combination of premises is not only possible for supporting moves but also for attacking moves.

Our move inventory could be specified further with a more fine-grained set, as provided for example by the theory of argumentation schemes (Walton et al., 2008). Still, we focus on the coarse grained set, since we see it as providing a reasonable "backbone" of the argumentation, and since it reduces the complexity for the task of automatic argument identification and classification, which is one central target application of the corpus.

Our adaption of Freeman's theory and the resulting annotation scheme is described in more detail and with examples in (Peldszus & Stede, 2013), where comparisons to related approaches are provided as well.

4. ANNOTATION PROCESS

In order to show that the annotation scheme can be applied in a reproducible fashion, we conducted annotation studies. We found that trained annotators can determine the argumentation structures reliably: On the basis of written 8-page long annotation guidelines, three annotators achieved an agreement of Fleiss k=0.83 for the full task (i.e. the segment-wise annotation of full argument graph features) and even higher agreement for the basic distinctions between proponent and opponent, or supporting and attacking moves. A more detailed explanation of this agreement study and its results is given in (Peldszus, 2014).

After verifying our approach by means of the agreement study, the markup of argumentation structures in the full corpus was done by one expert annotator. All annotations have been checked, controversial instances have been discussed in a reconciliation phase by two or more expert annotators. The annotation of the corpus was originally done manually on paper. In follow-up annotations, we used GraPAT (Sonntag & Stede, 2014), a web-based annotation tool specifically dedicated to constructing graph structures.

All annotation studies and the annotation of all texts have been done on the original German version of the corpus. Since the professional translation preserves linearization and argumentation structures, all annotated graphs represent both the German original and the English translation of the argument.

The bilingual texts and the annotations are publicly available in a suitable XML format; see Section 6.

5. CORPUS STATISTICS

The corpus features a wide range of different argumentation patterns. In the following, we will present detailed statistics on these, including distribution of roles and argumentative moves, positioning of the central claim in the text, as well as forward (from premise to conclusion) and backward linearizations of arguments.

5.1 General statistics

In the corpus, there are 112 texts, with in total 576 segments. Table 1 shows the length of texts in the corpus measured in segments: The great majority of texts are four, five or six segments long (the average being 5.1), with only a few exceptions.

text length	number of texts
3	3
4	11
5	71
6	26
7	2
8	0
9	0
10	1

Table 1: Length of the texts in segments

5.2 Central claim

In the English-speaking school of essay writing and debating, there is a tendency to state the central claim of a text or a paragraph in the very first sentence, followed by supporting arguments. To some extent, we can expect to find this pattern also in other languages. To investigate whether the tendency also holds in our corpus, we divide each text into five equal parts and count the occurrence of the central claim in this position. As Table 2 shows, the dominant position is indeed the beginning of a text, directly followed by the end of text. Note however, that the overall majority of central claims (57%) is at positions other than the beginning.

position	number of central claims
1/5	48
2/5	18
3/5	16
4/5	3
5/5	27

Table 2: Position of the central claim

5.3 Argumentative role

As we indicated earlier, the scheme distinguishes two argumentative roles: the proponent and the opponent. Of the 576 segments, 451 are proponent ones and 125 are opponent ones. While there are 15 texts where no opponent segment has been marked (either because the author did not conform to the requirement to consider at least one objection or because he phrased it indirectly in a non-clausal construction), the majority of texts (74) have exactly one opponent segment. Two opponent segments can be found in 18 texts, and three of them in five of the texts. Furthermore, Table 3 shows the position of opponent segments:

Position	number of objections
1/5	20
2/5	29
3/5	22
4/5	36
5/5	18

Table 3: Position of opponent segments (objections)

It turns out that the dominant place to mention a potential objection is right before the end of the text, thus giving the author the possibility to conclude his text with a counter of the potential objection.

For a comparison of the distribution of argumentative roles between this corpus and a corpus of longer newspaper commentaries, see (Peldszus & Stede, 2015a).

5.4 Argumentative function

The frequency of argumentative functions annotated in our corpus is shown in Table 4: Most segments are normal support moves. Examples are used only rarely. About a third of the segments have an attacking function (either the opponent challenging the central claim or the proponent countering these objections), with overall more rebutters than undercutters. Linked premises are usually found in supporting arguments, and only rarely in attacks.

Type	Number	sub-type	number
Support	272	normal	263
		example	9
Attack	171	rebut	108
		undercut	63
Linked	21		
central claim	112		

Table 4: Frequency of argumentative function

It is noteworthy that rebutters and undercutters are not equally distributed over both argumentative roles. This is shown in Figure 2: The opponent typically rebuts, and the great majority of these rebuttals is directed against the central claim, while only a few work against supporting arguments. In contrast to that, the proponent usually undercuts. We attribute this to the common strategy of the authors to first concede a possible objection, thereby demonstrating that their presentation is not fully biased, and then render it irrelevant.

Also notice that a possible objection (an attack of the opponent) does not necessarily need to be counter-attacked by the proponent: The total number of attacks by the proponent is significantly smaller than the total number of attacks by the opponent (63 vs. 108). This is not too surprising – an author might rather choose to present just another good reason in favour of the central claim, and thereby outweigh the objection, or he might pose the possible objection in an unalluring manner signalling that counter-attacking or outweighing is not even necessary.

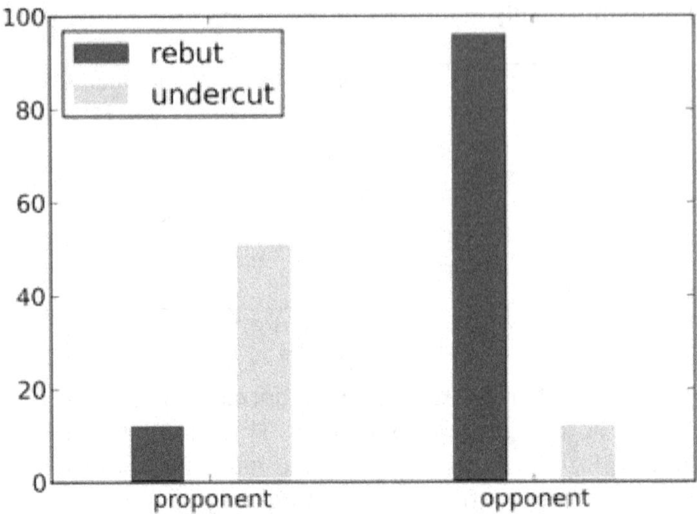

Figure 2: Attack moves against argumentative role

5.5 Attachment distance

One aspect of argumentation structure that makes its automatic recognition especially challenging is the possibility of long distance dependencies. Although segments are often connected locally, i.e. they are supporting or attacking adjacent segments, there may very well be direct argumentative relations across the whole texts, even between the very first and the very last segment of a text. It is thus worthwhile to investigate the degree to which we find these non-local relations in our corpus.

To this end, we calculate the distance and the direction of attachment for every relation annotated in the corpus (464 in total, the remaining segments functioning as central claims). An attachment distance of -1 means that the target of the argumentative relation directly precedes the source, a distance of +1 corresponds to a target immediately following the source. For segments targeting a relation instead of another segment, as it is the case for undercutters and linked premises, we considered the position of the source of the targeted relation. For example, the undercutting segment 3 in the graph in Figure 1 has an attachment distance of -1, as it undercuts the relation of the previous segment 2.

The distribution of distances and directions of attachment found in the corpus is shown in Figure 3. The great majority (45%) of argumentative relations attach to the immediately preceding segment.

Another 11% attach to the following segment. In total, 56% of the relations hold between adjacent segments, so conversely nearly half of the segments do not attach locally. Considering that our texts are relatively short, it is to be expected to find even more non-adjacent relations in longer texts. E.g., Stab & Gurevych (2014) report a rate of 63% of non-adjacent relations in their corpus of student essays.

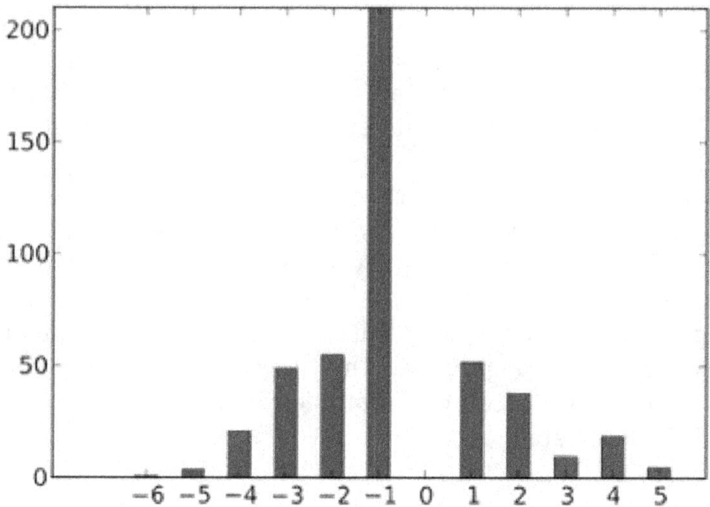

Figure 3: Attachment distance and direction (negative distances directed backwards and positive distances directed forwards)

5.6 Linearization strategies

The final feature of the argumentation graphs we want to investigate is how authors linearize their arguments in the text. This has already been covered to some degree in the sections 5.3 and 5.4 when we studied at which positions in the text the central claim and objections are typically expressed. In the following, we combine this with the direction of attachment and distinguish four different simple linearization strategies, which are summarized in Table 5.

The first strategy involves only backward relations, where the author open his text with the central claim (c) and then presents a series of reasons, possible objections, and counters, all of them directed backwards (b), targeting propositions made in prior segments. The second linearization strategy unfolds the argumentation the other way around, with only forward relations. The author first starts with premises and successively draws conclusions from them (f) until he finally reaches the central claim of the text. The third strategy combines these two patterns, presenting the central claim in the middle of the text. It naturally involves a switch of attachment direction after the central claim. All other texts not matching one of these three strategies involve a change in the direction of argumentation independent of the presentation of the central claim.

linearity strategy	pattern	frequency
Backward	c b^+	50%
Forward	f^+ c	5%
forward-backward	f^+ c b^+	13%
Other	other	31%

Table 5: Ratio of texts matching different linearization strategies

As shown in Table 5, the first strategy which opens with the central claim and argues for it with only backward relations, is the dominant one found in half of the texts. The reverse strategy is used only rarely, while the mixed strategy appears at least in 13% of the texts. Most interestingly, about 31% of the texts do not follow these strict patterns. As an example, see Figure 4: This text's linearization pattern corresponds to "fbfbcb", featuring multiple changes in direction before the central claim is stated.

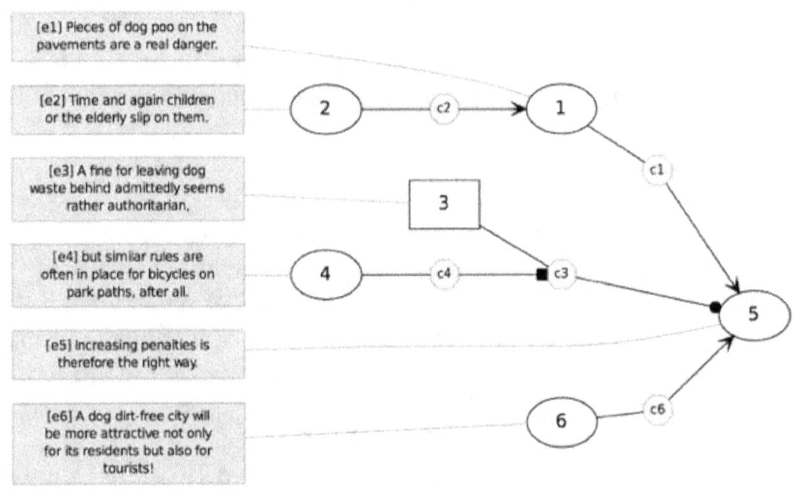

Figure 4: An example text with multiple direction changes.

6. CORPUS DELIVERY

The corpus is published online[1] and freely distributed under a Creative Commons BY-NC-SA 4.0 International License[2]. The annotated graph structures are stored in the Potsdam Argumentation XML format (PAX), a both human- and machine-readable format, similar to GraphML (Brandes et al, 2002). The corpus repository contains a specification of the format in form of a document type definition.

For both versions of the corpus, the German and the English one, we provide the raw source text, the annotated argumentation graph in PAX (primarily for machine reading), as well as a graphical argument diagram such as the one in Fig. 1, in order to facilitate human inspection of the structures. An importer for the PAX format has also recently been added to the Carneades Tools[3], allowing to map and evaluate the graphs of our corpus.

Finally, notice that first results on automatic recognition of the argumentation structures annotated in the corpus are presented in (Peldszus & Stede, 2015b).

7. CONCLUSION

We presented a freely available parallel corpus of short argumentative texts. Our microtexts are "authentic" in the sense that the vast majority was written by probands not involved in the research, and the trigger questions concern issues of daily life and public interest. At the same time, they are "constrained" because we provided the probands with some instructions on target length and form. This was done in order to obtain a relatively homogeneous data set that allows for studying properties of the argumentation. For the same reason we decided to do a moderate "cleaning" of the texts, which on the one hand reduces "authenticity" but on the other hand contributes to uniformity and – for many purposes – usability.

Research in automatic argument mining typically targets social media contributions in their original form and often focusses on the task of argument identification and local relation identification. While the design of our corpus differs from this orientation, we still think that the data can be useful for purposes of feature engineering and as supplemental training data. Finally, we consider our data set to be a reasonable starting point for the task of global relation identification, i.e. for the automatic prediction of text-level argumentation structure, before tackling more complex text genres.

---

[1] *https://github.com/peldszus/arg-microtexts*

[2] *https://creativecommons.org/licenses/by-nc-sa/4.0/*

[3] *https://carneades.github.io/*

ACKNOWLEDGEMENTS: We would like to thank the anonymous reviewers and the conference attendees for their helpful feedback. Thanks to our probands and annotators for participating in our studies, as well as to Regina Peldszus and Kirsten Brock for the translations. The first author was supported by a grant from Cusanuswerk.

REFERENCES

Beardsley, M. C. (1950). *Practical Logic*. Prentice-Hall, New York.

Brandes, U., Eiglsperger, M., Herman, I., Himsolt, M., & Marshall, M. S. (2002). GraphML progress report structural layer proposal. In: *Graph Drawing* (pp. 501-512). Berlin / Heidelberg: Springer.

Freeman, J. B. (1991). *Dialectics and the macrostructure of argument*. Berlin: Foris.

Freeman, J. B. (2011). *Argument structure: Representation and theory*. Dordrecht: Springer.

Peldszus, A. (2014). Towards segment-based recognition of argumentation structure in short texts. In: *Proc. of the First Workshop on Argumentation Mining* (pp. 88-97). Baltimore, Maryland: ACL.

Peldszus, A., & Stede, M. (2013). From argument diagrams to argumentation mining in texts: a survey. *International Journal of Cognitive Informatics and Natural Intelligence*, *7*(1), 1-31.

Peldszus, A., & Stede, M. (2015a). Towards detecting counter-considerations in text. In: *Proc. of the 2nd Workshop on Argumentation Mining* (pp. 104-109). Denver, CO: NAACL-HLT.

Peldszus, A., & Stede, M. (2015b). Joint prediction in MST-style discourse parsing for argumentation mining. In: *Proc. of the Conference on Empirical Methods in Natural Language Processing* (EMNLP) (pp. 938-948). Lisbon, Portugal.

Reed, C., Mochales Palau, R., Rowe, G., & Moens, M. F. (2008). Language resources for studying argument. In *Proc. of the 6th conference on language resources and evaluation* (LREC 2008) (pp. 91-100).

Sonntag, J. & Stede, M. (2014). GraPAT: A tool for graph annotations. In: *Proceedings of the Language Resources and Evaluation Conference* (LREC), Reykjavik.

Stab, C., & Gurevych, I. (2014). Annotating argument components and relations in persuasive essays. In: *Proc. of the 25th International Conference on Computational Linguistics* (COLING 2014) (pp. 1501-1510).

Stede, M., & Sauermann, A. (2008). Linearization of arguments in commentary text. In W. Ramm & C. Fabricius-Hansen (Eds.), *Linearisation and Segmentation in Discourse. Multidisciplinary Approaches to Discourse 2008* (MAD 08), February 20-23, Lysebu, Oslo, Norway.

Thomas, S. N. (1974). *Practical reasoning in natural language*. Prentice-Hall, New York.

Toulmin, S. (1958). *The uses of argument.* Cambridge University Press, Cambridge.

Walton, D., Reed, C., & Macagno, F. (2008). *Argumentation schemes.* Cambridge University Press.

51

Approximate Syllogism as Argumentative Expression for Knowledge Representation and Reasoning with Generalized Bayes' Theorem

M. PEREIRA-FARIÑA
CiTIUS, Universidade de Santiago de Compostela, Spain
martin.pereira@usc.es

A. BUGARÍN
CiTIUS, Universidade de Santiago de Compostela, Spain
alberto.bugarin.diz@usc.es

In this paper, we propose an argumentative model, based on approximate syllogisms, for dealing with Generalized Bayes' Theorem (GBT). We show how probabilistic notation can be expressed by means of quantified statements with the form "Q A are B", where conditional probabilities are included in the premises of the argument and the variable to be estimated in its conclusion. Our main aim is to facilitate the understanding of GBT to non-specialized users.

KEYWORDS: argument, generalized Bayes' theorem, common-sense reasoning, fuzzy quantifiers, approximate syllogism

1. INTRODUCTION

In daily life, humans deal with taking decisions which involve vagueness and uncertainty. Although classical logic is defined as the science of correct reasoning, decision-making problems cannot always be addressed by it, since for instance logical concepts of soundness and completeness are non-compatible with vagueness and uncertainty.

However, most of these situations can be approached from the perspective of common-sense reasoning, but applying other types of rules and mechanisms, which are effectively compatible both with vagueness and with uncertainty. For instance, let us consider the following statement of a weather forecasting: "today is a very hot day". This involves the concept *very hot*, which is vague but it provides the

D. Mohammed & M. Lewiński (eds.) (2016). *Argumentation and Reasoned Action: Proceedings of the 1st European Conference on Argumentation, Lisbon, 2015. Vol. II, 817-830.* London: College Publications.

sufficient information to perform a subjective representation of the temperature for today. In addition, it is context-dependent; i.e., for a person that lives in the Sahara Desert, the prediction lead him to infer that the temperature will be more than 50ºC. If this person lives in Siberia, the same statement leads him to infer a temperature around 25ºC.

The study of common-sense reasoning has been present in the branch of Artificial Intelligence from the very beginning, involving different fields such as philosophy, mathematics, argumentation theory or psychology, among others. One of the most fruitful frameworks for addressing this task is Bayesian epistemology (Oaksford & Chater, 2007) or Bayesian rationality. It is founded on Bayes' theorem (Bayes, 1764), a well-known simple mathematical formula used for calculating conditional probabilities. In computer science, Bayesian Networks (BNs) (Korb, 2010) are a very useful method for representing and reasoning about complex problems involving probability or uncertainty.

Notwithstanding, modelling problems through Bayesian notation in order to be solved is a non-trivial task which demands specialized knowledge. In the literature, we can find several different fallacies that come from mistaken interpretation or misapplications of Bayes' theorem (Cosmides & Tooby, 1996). Some examples are the Prosecutor's fallacy (Thompson & Schumann, 1987) in the area of legal reasoning which has appeared in several real cases[1]; affirming the consequent, probabilistic relevance (Korb, 2004), base-rate fallacy (Tversky & Kahneman, 1982), etc.

From our point of view, one of the roots of this problem is the gap between natural language, the usual way of people to represent and express any type of matter, and Bayesian mathematical notation, which may be difficult to understood and obscure for non-specialized users. For instance, let us consider the following example (adapted from Casscell et al., 1978):

> If a test to detect a disease whose prevalence is 1/1000 has a false positive rate of 5%, what is the chance that a person found to have a positive result actually has the disease, assuming that you know nothing about the person's symptoms or signs?

---

[1] One of the most studied cases of misapplication of Bayesian probability is the so-called Sally vs. Clark case. A document from the Royal Statistical Society about it can be consulted in http://www.rss.org.uk/uploadedfiles/documentlibrary/744.pdf.

The answer of the question entails to apply Bayes' theorem. This must be mathematically formulated using the standard Bayesian notation, being carefully with the correct identification of the probabilities corresponding to each variable. Thus, its right formulation is as follows:

> **Variables:** Prevalence of the disease (D), person with a positive test (PT).
> **Assumptions:** There are not false negative results.
> **Evidence:** Person with a positive test
> **Question:** What is the chance of having cancer of a person with a positive test?
> Bayesian expression:

$$\mathrm{P(PT)} = \mathrm{P}(PT|D) \cdot \mathrm{P(D)} + \mathrm{P}(PT|\overline{D}) \cdot \mathrm{P}(\overline{D}) = 1 \cdot 0.001 + 0.05 \cdot 0.999 = 0.5095$$

$$\mathrm{P}(D|PT) = \frac{\mathrm{P}(PT|D) \cdot \mathrm{P(D)}}{\mathrm{P(PT)}} = \frac{1 \cdot 0.001}{0.05095} = 0.0196$$

The right answer for this problem is "1.96% of probability of having the disease for a patient with a positive test". It is worth noting that the *a priori* probability of having cancer is 0.1%, while, when evidence is incorporated to the inference, this probability is increased by a factor of 20, obtaining a probability around 2%. However, as can be observed, the gap between the linguistic expression of the problem and its mathematical notation is not trivial. In addition, more that 50% of people asked about this question answer erroneously, stating higher results (Cosmides & Tooby, 1996, p. 27). Therefore, from our point of view, methods that make easier the understanding and representation of Bayesian problems is a task that must be addressed.

There are several different proposals in the literature that specifically deal with this question. Some of these approaches assume the frequentist interpretation of probability, such as *frequency formats* (Hoffrage & Gigerenzer, 1998). In essence, this model is a graphical representation of the problem in order to show the dependences between the variables (i.e., conditional probabilities) and how the modification in one of them affects to the remaining ones.

Figure shows the frequency formats tree corresponding to Casscell et al.'s example. To calculate the chance that a person with a positive test actually has the disease, we have to obtain the proportion of the people with a positive test and the disease between all the people with positive test (with and without the disease); i.e., 1/51 = 0.0196.

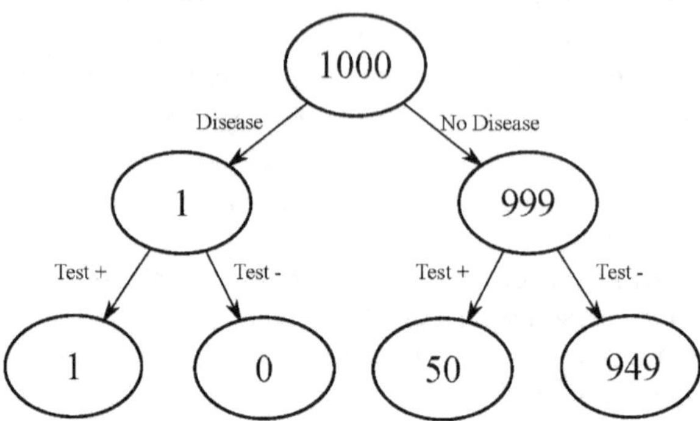

Figure 1. Frequency formats for Casscell et al.'s example

Other approaches, such as Bayesian Networks (BNs) (Korb, 2004), are still supported on Bayesian rationality. BNs codify the information in a graphical structure (in particular, a direct acyclic graph) and allow us to perform different types of inferences (e.g., predictive, diagnostic, etc.). They are composed of nodes, that represent the variables of the problem domain, and arcs, which link pairs of nodes showing the direct dependencies between them. In addition, many computational tools that implements BNs. Figure shows the BN corresponding to the Casscell et al.'s modelled and solved with Netica software (Netica, 2013). In this case, an evidence has been added to the BN (i.e., a patient with a positive test; P(Positive test)=1), and we have to compute the probability of cancer given it (this example is a case of diagnostic inference); as can be observed, the obtained result is the same.

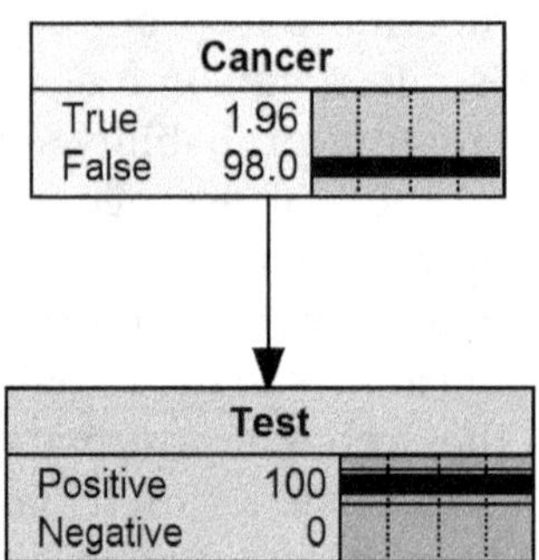

Figure 2. BN for Casscell et al.'s example modelled and solved using Netica software (Netica, 2013)

Notwithstanding, from our point of view, the main contributions of these approaches are in the simplification of the calculations rather than overcome the gap between the linguistic expression and the Bayesian formulation.

Thus, we propose to use a general approximate syllogism model (Pereira et al., 2014) in order to represent and solve problems that involve Bayesian probability. As we will describe, approximate syllogism allows us to preserve the linguistic expression of the problems, expressing probabilities as binary quantified statements; and to use the standard argumentative structure, premises and conclusion, and, consequently, overcome the aforementioned gap.

In this paper, we focus on a particular type of Bayesian problems: those where only conditional probabilities are available. The information of *a priori* probabilities, mandatory to apply Bayes' theorem in its basic version, is not available. This type of problems has been previously addressed using fuzzy approaches (Dubois et al., 1993), applying the so-called Generalized Bayes' Theorem (GBT), a version of Bayes' theorem which only needs conditional probabilities in order to calculate updated conditional probabilities. Therefore, there already exists a model to address this type of probability problems that can be computationally implemented. However, since it involves an iterative process combining the GBT with a particular pattern of syllogistic reasoning (Dubois et al., 1990), its application may become difficult.

For the sake of simplicity, we illustrate our approach by using a simple GBT example. The focus of this paper is on defining a procedure for modelling a GBT problem by means of an approximate syllogism and, for that reason, we consider that a complex case could obscure this objective. In addition, for the same reason, we only use precise probability values represented by precise percentages. Although the approximate syllogism frame is able to deal with vague quantifiers, such as *most, few, almost all*, etc., here we only consider crisp values in order to facilitate the understanding of the procedure and the comparison with the approach developed in (Dubois et al., 1993). Considering fuzzy quantifiers is a straightforward extension (Pereira et al., 2014) that will be addressed in future works.

This paper is organized as follows: in section 2, the preliminary concepts about GBT and approximate syllogism are introduced as well as the ideas that support the use of syllogisms to deal with GBT problems; section 3 describes the procedure to perform a syllogism through an example; finally, section 4 summarizes the main conclusions of the paper.

2. PRELIMINARY CONCEPTS

2.1 Generalized Bayes' Theorem

As stated before, GBT is a version of Bayes' theorem that only uses conditional probabilities instead of *a priori* and conditional ones, as the basic version of Bayes' theorem does. Thus, let us consider a problem with *k* chained variables, where each variable corresponds to a node. Figure illustrates the basic structure of the graph corresponding to the GBT, where $A_1, A_2, \ldots, A_k$ are the nodes corresponding with the variables of the problem.

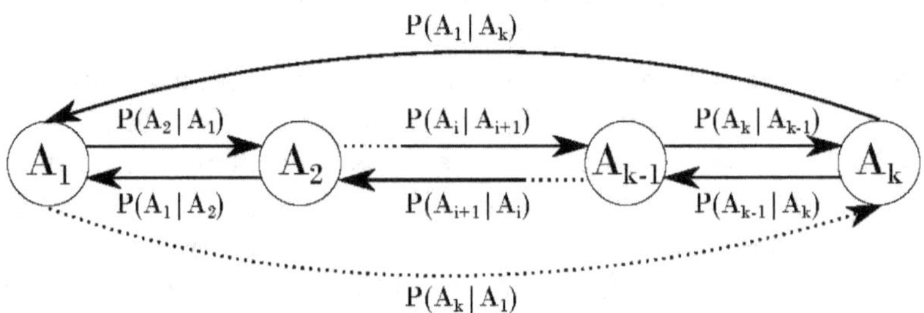

Figure 3. GBT represented in graph theory

For instance, for k = 4, for calculating P(A4|A1), we need to know P(A1|A4), P(A1|A2), P(A2|A1), P(A2|A3), P(A3|A2), P(A3|A4) and P(A4|A3). The mathematical expression of GBT is as follows:

$$\forall A_1, \ldots, A_k, P(A_1|A_k) = P(A_k|A_1) \prod_{i=1,k-1} \frac{P(A_i|A_{i+1})}{P(A_{i+1}|A_i)}$$

2.2 Approximate Syllogisms

Syllogisms, in their classical definition, are arguments composed by three terms, two premises and a conclusion, where the conclusion is inferred by necessity from the premises (Aristotle, 1949). Of the three terms, two of them appear in the premises and in the conclusion and the third one (the middle term) only appears in the premises. The type of statements involved in a syllogism are the so-called "categorical statements"; i.e., sentences headed by *all, none, some* or *some... not*, that is, quantified statements.

In Aristotelian syllogistics, the first model of syllogistic reasoning in the history of logic, moreover to these four categorical statements, four different *Figures* (according the position of the middle term in the premises) are defined, which generate the twenty-four Aristotelian *moods*. A well-known example of Aristotelian syllogism is shown in Table , which involves the terms *humans*, *Greeks* and *mortal*. *Greeks* and *mortal* appear both in the premises and in the conclusion while *humans* only appears in the premises (i.e, the middle term) and it chains the other two ones in order to obtain a valid conclusion.

PR1:	All humans are mortal
PR2:	All Greeks are humans
C:	All Greeks are mortal

Table 1. Aristotelian mood: Barbara

In the twentieth century, several proposals that expanded Aristotelian syllogistic appeared. In this paper, we shall use the model named approximate syllogistic reasoning (Pereira et al., 2014), which introduces notions from fuzzy logic inside the syllogistic framework. The main novelties introduced by this model with respect to the Aristotelian one are three: i) the four Aristotelian categorical statements involving crisp quantifiers (*all*, *no*, *some* and *some... not*) are substituted by the more general notion of "quantified statements"; this means to introduce all the binary types of quantifiers defined by the Theory of Generalized Quantifiers (TGQ) (Barwise & Cooper, 1981), such as proportional ones (e.g., *most*, *few*, *many*, etc.), exception ones (e.g., *all but three*, *all but around five*, etc.), etc. inside a specific inference schema; ii) vagueness implicit in most of quantifiers (e.g., *few*, *most*, *almost all*, etc.) can be addressed contextually, tailoring in each case the cardinality associated to the corresponding quantifier; iii) syllogistic inference can be addressed computationally, which allows us to deal with arguments involving any number of premises and terms.

In order to model linguistic quantifiers, the general fuzzy framework for quantification (Glöckner, 2006) is adopted. In order to make the syllogism treatable computationally, the argument is transformed into an inequations system, where the quantifier of the conclusion is computed through an optimization process over the space of solutions defined by the premises.

The general form of an approximate syllogism follows the pattern shown in Table . It comprises an argument with N premises, $PRn; n = 1,\ldots, N$; where Q_n, $n = 1,\ldots,N$ are the linguistic quantifiers in the N premises; $L_{n,j}$; $n = 1,\ldots,N$; $j = 1,2$ denotes an arbitrary boolean combination among the terms considered in the syllogism, Q_C stands for the quantifier of the conclusion (which is the value to be calculated) and $L_{C,1}$ and $L_{C,2}$ stand for the subject-term and the predicate-term in the conclusion fixed by the user.

Once the premises are transformed into the corresponding inequations system, the quantifier of the conclusion is obtained by means of a mathematical optimization procedure. In particular, in this model Simplex and linear fractional programming are applied.

PR1:	Q_1 $L_{1,1}$ are $L_{1,2}$
PR2:	Q_2 $L_{2,1}$ are $L_{2,2}$
PRn:	...
PRN:	Q_N $L_{N,1}$ are $L_{N,2}$
C:	Q_C $L_{C,1}$ are $L_{C,2}$

Table 2. General form of approximate syllogism

In (Pereira et al., 2014), several examples with the application of approximation syllogism explained in detail can be found. For the sake of simplicity and brevity, in this paper we only introduce the main idea underlying it.

2.3. Ideas that support the use of approximate syllogism for dealing with GBT problems

The approach to GBT problems proposed here uses approximate syllogisms for knowledge representation and reasoning. This idea is supported on the fact that quantified statements constitute a linguistic equivalent way to express conditional probabilities (Schwartz, 1997). For instance; let us consider the probability of having lung cancer given a smoker patient. In Table , the both alternative notations are described.

Probabilistic notation	Linguistic notation
P(Cancer|Smoker) =0.25	Twenty-five percent of smokers have cancer

Table 3. Probabilistic and Linguistic notations

Natural language is the main human tool for explaining a problem that involves probability. The procedure for its representation using the canonical Bayesian notation is not easy and it is a source of fallacies (Cosmides & Tooby, 1996). Therefore, we assume that if the linguistic notation can be preserved, the transformation process will be easier. Approximate syllogisms, based on quantified statements, allow us to preserve the linguistic notation organizing the information needed to solve the problem into the two parts of any argument: 1) premises, which is the known information provided by the problem; 2) conclusion, which terms are the variables to obtain the conditional probability values and the quantifier is the probability to be calculated. Thus, only very basic logic concepts, known by anybody, are demanded in order to represent a GBT problem preserving its natural language expression.

The third idea that supports the formalization of GBT problems with approximate syllogisms is our model for syllogistic reasoning (Pereira et al., 2014). This can deal with quantified statements headed both linguistic quantifiers and percentages; these also can be precise, imprecise or fuzzy and it is implemented into a software tool (SEREA, 2014) which allows us to deal with complex problems, something non-possible with Aristotelian framework.

3. PROCEDURE FOR PERFORMING AN APPROXIMATE SYLLOGISM TO ADDRESS GBT PROBLEMS.

We will illustrate the procedure for performing an approximate syllogism by an example. In particular, we will use an adapted version of a GBT problem which is solved applying an alternative method (Amarger, 1991). The problem is formulated as follows:

> Let us consider the people in a university campus. A sociological study is performed by two students. They want to discover the different links among those who are students, single, young, practice sports and have children. To ensure the most reliable values, different students ask the same questions and obtain slightly different results. To preserve these differences, they are represented by means of intervals. Table 4 summarizes the data collected by the students, where empty brackets ("[]") stand for unknown values.

	Student	Sport	Single	Young	Children
Student	[1,1]	[0.7,0.9]	[]	[0.85,0.95]	[]
Sport	[0.4,0.6]	[1,1]	[0.8,0.85]	[0.9,1]	[]
Single	[]	[0.7,0.9]	[1,1]	[0.6,0.8]	[0.05,0.1]
Young	[0.25,0.35]	[0.8,0.9]	[0.9,1]	[1,1]	[0,0.05]
Children	[]	[]	[0,0.5]	[0,0.5]	[1,1]

Table 4. Data collected by the students

Figure shows a representation of the problem by means of a graph.

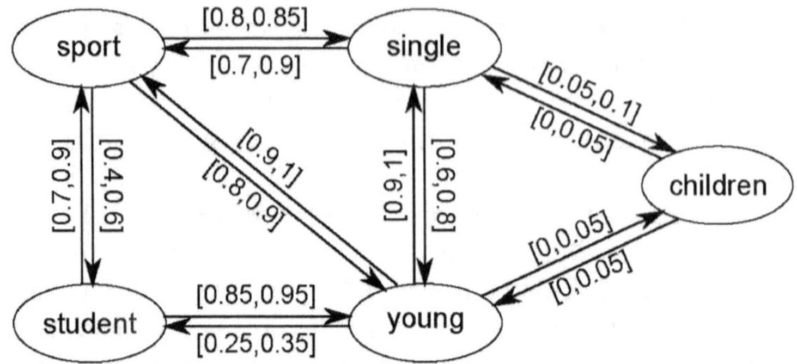

Figure 4. Graph corresponding to the data collected by the students

This is a typical example of GBT problem since there is not *a priori* probability value associated to any variable of the problem; the only data that we have are the relationships between the variables.

The procedure for performing the syllogism corresponding to this problem is divided into three steps:

- Step 1. Defining the set of premises of the argument.
- Step 2. Defining the conclusion of the argument.
- Step 3. Solving the inequation system.

3.1 Step 1. Defining the set of premises of the argument

Given the equivalence between probability notation and binary quantified statements described in section 2.2, it is possible to generate a sentence for each cell of the table, following this scheme:

Q S are R,

where **S** stands for the variables of the row, **R** stands for the variables of the column and **Q** stands for the probability value expressed as an interval (Q= [*a*,*b*]), where *a* stands for the lower bound and *b* stands for the upper one). For instance, P(young|student)=[0.85,0.95] can be expressed as "Between 85% and 95% of students are young".

For each non-empty cell of the table a premise is generated (with exception of diagonal values and from upper left to bottom right), obtained the set of premises shown in Table .

In general terms, given the conditional probabilities table of a GBT (or the corresponding graph), for each non-empty cell (or node) a premise is generated (**Q S are R**), where the variable of the rows (or the node origin of the arc) is the subject-term of the statement (**S**), the variable of the columns (or the node destination of the arc) is the predicate-term (**R**), and the quantifier (**Q**) is the associated probability value, which can be precise, an interval (as in this example) or fuzzy; i.e., P(R|S)=[*a*,*b*] is equivalent to say "Q S are R", where Q=[*a*,*b*].

PR1	Between 70% and 90% of students play sports
PR2	Between 85% and 95% students are young
PR3	Between 40% and 60% sport players are students
PR4	Between 80% and 85% sport players are single
PR5	Over 90% of the sport players are young
PR6	Between 70% and 90% single people are sport players
PR7	Between 60% and 80% single people are young
PR8	Between 5% and 10% single people have children
PR9	Between 25% and 35% young people are student
PR10	Between 80% and 90% young people are sport players
PR11	Over 90% of the young people are single
PR12	Less than 5% of the young people have children
PR13	Less than 5% of the people with children are young
PR14	Less than 5% of the people with children are single
C	Q S are R

Table 5. Set of premises of the syllogistic argument

3.2 Step 2. Defining the conclusion of the argument.

Once the set of premises is defined, the step 2 is to define the conclusion of the argument. In this case, the values to be calculated correspond to empty cells and the conclusion has the standard form, where **S** stands for the row value (origin node in a graph), **R** stands for the column one (destination node in a graph) and **Q** is the quantifier to be calculated. Table shows all the possible conclusions that can be calculated.

C1	Q students are single
C2	Q students have children
C3	Q sport players have children
C4	Q single people are students
C5	Q people with children are students
C6	Q people with children play sports

Table 6. Conclusions to be computed

In general terms, for each empty cell of a CPT, a quantified statement is generated (**Q S are R**), where the variable of the row (or the origin node) is the subject-term of the statement (**S**), the variable of the column (or the destination node) is the predicate-term (**R**), and the quantifier (**Q**) is the probability value to be computed; which can be precise, an interval or fuzzy. Conclusions have the same form of the premises except that the interval associated to Q is unknown.

3.3 Step 3. Solving the inequations system.

Lastly, the step 3 is the procedure to compute the value of Q in each premise. Using a software tool (SEREA, 2014), which implements our model of approximate syllogism (Pereira et al. 2014), the correct results can be easily achieved. Table shows the corresponding values for each conclusion.

C1	Q= At least 60.7%
C2	Q= At most 27.1%
C3	Q= At most 15.3%
C4	Q= Between 22.2% and 36.6%
C5	Q= At most 9.9%
C6	Q= At most 12.6%

Table 7. Results for each conclusion

The new values obtained after the resolution of the syllogism can be incorporated to Table 5. In this case, the previous values included in it also can be updated and the network saturated. This can be obtained applying and iterative process.

In general terms, each conclusion is computed independently applying the model defined in (Pereira et al., 2014).

4. CONCLUSIONS

We have shown a proposal in order to overcome the gap between the linguistic expression of a problem involving probability and its Bayesian notation. This consists of the use of approximate syllogism, an inference scheme that allows us to deal with quantified statements and arguments composed by any number of premises and terms.

Our proposal is supported on the equivalence between Bayesian notation and quantified statements to express probability, on the use of the standard structure of arguments and on the computational model that implements approximate syllogism framework. We have defined a procedure with the steps for modelling correctly an approximate obtained from and is equivalent to GBT. In order to illustrate how it works, an example is described in detail. We showed how it can be addressed in a linguistic way, closer to common-sense knowledge representation and reasoning than canonical Bayesian notation.

As future work, we propose to introduce linguistic quantifiers in order to manage vague probabilities. In addition, other types of probabilistic reasoning, such as BNs, also will be considered for its linguistic expression.

ACKNOWLEDGEMENTS: This work was supported by the European

Regional Development Fund (ERDF/FEDER) and the Spanish Ministry for Economy and Competitiveness under grant TIN2014-56633-C3-1-R and by the European Regional Development Fund (ERDF/FEDER) under the projects CN2012/151 and GRC2014/030 of the Galician Ministry of Education.

REFERENCES

Amarger, S., Dubois, D., & Prade, H. (1991). Constraint propagation with imprecise conditional probabilities. In B. D'Ambrosio, & P. Smets & P. Bonissone (Eds.), *Proceedings of the 7th Conference on Uncertainty in Artificial Intelligence* (pp. 26-34). San Mateo: Morgan Kaufmann.

Aristotle (1949). *Prior and posterior analytics.* Oxford: Clarendon Press.

Barwise, J., & Cooper, R. (1981). Generalized quantifiers and natural language. *Linguistics and Philosophy, 4*, 159-219.

Bayes, T. (1764). An essay toward solving a problem in the doctrine of chances. *Philosophical Transactions of the Royal Society of London, 53*, 370-418.

Casscells, W., Shoenberger, A., & Graboys, T. B. (1978). Interpretation by physicians of clinical laboratory results. *New England Journal of Medicine, 299*(18), 999-1001.

Cosmides, L., & Tooby, J. (1996) Are humans good intuitive statisticians after all? Rethinking some conclusions from the literature on judgment under uncertainty. *Cognition, 58*(1), 1-73.

Dubois, D., Godo, L., López de Mántaras, R., & Prade, H. (1993). Qualitative Reasoning with Imprecise Probabilities. *Journal of Intelligent Information Systems, 2*, 319-363.

Dubois, D., Prade, H., & Toucas, J. M. (1990). Inference with imprecise numerical quantifiers. In Z.W. Ras & M. Zemankova (Eds.), *Intelligent Systems. State of the Art and Future Directions.* London: Ellis Horwood.

Glöckner, I. (2006). *Fuzzy quantifiers. A computational theory.* Berlin: Springer.

Hoffrage, U., & Gigerenzer, G. (1998). Using natural frequencies to improve diagnostic inferences. *Academic Medicine, 73*, 538-540.

Korb, K., & Nicholson, A. E. (2010). *Bayesian Artificial Intelligence, Second Edition (2nd ed.).* Boca Raton, FL: CRC Press, Inc.

Koons, R. (2014). Defeasible reasoning. In E. N. Zalta (ed.), *The Stanford Encyclopedia of Philosophy*, URL = <http://plato.stanford.edu/archives/spr2014/entries/reasoning-defeasible/>.

Netica (Version 5) [Software]. (2013). Norsys Software Corporation, obtained from https://www.norsys.com/download.html.

Oaksford, M., & Chater, N. (2007). *Bayesian rationality: The probabilistic approach to human reasoning.* Oxford: Oxford University Press.

Pearl, J. (1988). *Probabilistic reasoning in intelligent systems: Networks of plausible inference.* Morgan Kaufmann, California.

Pereira-Fariña, M., Vidal, J.C., Díaz-Hermida, F., & Bugarín, A. (2014). A fuzzy syllogistic reasoning schema for generalized quantifiers. *Fuzzy Sets and*

Systems, 234(1), 79-96.

Perira-Fariña, M. & al. (2014). SEREA (Version 2) [Software]. Obtained from http://proxectos.citius.usc.es/serea.

Thompson, W. C., & Shumann, E. L. (1987). Interpretation of statistical evidence in criminal tries: The prosecutor's fallacy and the defense attorney's fallacy. *Law and Human Behavior, 11*, 167-187.

Tversky, A., & Kahneman, D. (1982). Evidential impact of base rates. In D. Kahneman, P. Slovic & A. Tversky (Eds.), *Judgement under Uncertainty. Heuristics and Biases* (pp. 153-160). Cambridge: Cambridge University Press.

Schwartz, D. G. (1997). Dynamic reasoning with qualified syllogisms. *Artificial Intelligence, 93*, 103-167.

Zadeh, L. A. (1965). Fuzzy sets. *Information and Control, 8*(3), 338-353.

Zadeh, L. A. (1975). Fuzzy logic and approximate reasoning. *Synthese, 30*(3-4), 407-428.

52

Comparing Words to Debate about Drinking Water: Textometrics for Argumentation Studies

CLAIRE POLO
ICAR Laboratory, Lyon, France
claire.polo@ens-lyon.fr

CHRISTIAN PLANTIN
ICAR Laboratory, Lyon, France
christian.plantin@univ-lyon2.fr

KRISTINE LUND
ICAR Laboratory, Lyon, France
kristine.lund@univ-lyon2.fr

GERALD PETER NICCOLAI
ICAR Laboratory, Lyon, France
gerald.niccolai@ens-lyon.fr

In ten videotaped socio-scientific debates related to water, students from Mexico, the USA and France tend to focus on a few alternative positions. On the basis of Grize's definition of schematization, we followed their reasoning by studying how they cast light on specific aspects of the discursive object "water". Through textometrical analysis of debate transcripts, we specified 6 characteristics of "water" that are more or less emphasized depending on the prevailing national argumentative scenario.

KEYWORDS: argumentative scenario, comparative analysis, education, discourse object, framing, schematization, socio-scientific issues, textometrics

1. INTRODUCTION

Argumentation is a great object of interest in education (e.g. Andriessen, Baker, & Suthers, 2003; Driver et al., 2000; Erduran & Jiménez-

D. Mohammed & M. Lewiński (eds.) (2016). *Argumentation and Reasoned Action: Proceedings of the 1st European Conference on Argumentation, Lisbon, 2015. Vol. II, 831-848.* London: College Publications.

Aleixandre, 2007; Von Aufschnaiter et al., 2008; Muller Mirza, 2008). At the crossing of science, environmental and citizenship education, some topics challenge usual didactical perspective on argumentation: the socio-scientific issues (SSI). During a scientific *café* activity, students in Mexico, the USA and France proved capable of building complex arguments in response to SSI related to drinking water management, using knowledge, norms, values and emotions (Polo, 2014). In this paper, we follow how they elaborate their reasoning, by analyzing the *schematization processes* affecting the *discourse object* "water". We base our work on Grize's perspective of *schematization*, as a discursive construct giving clues about the underlying cognitive process (1997, p. 65). A comparison among the 3 national corpuses allows us to identify the prevailing orientation focuses of the *schematization* of "water" and characterize a typical *argumentative framing* for each country. While intercultural work in argumentation generally emphasizes matters of rhetorical style (e. g. Disson, 2002; Oetzel et al., 2001; Taft et al., 2011), our approach engages into a comparative study based on the substantial objects under discussion.

After detailing our theoretical orientations and research questions (2), we specify the context and data of our study (3). Then we detail our methodology (4), and present our main results (5), which significance is discussed in a final section (6).

2. THEORETICAL ORIENTATIONS AND CHALLENGES

2.1 "Schematization" (Grize, 1997)

Grize's "natural logic" relies on the (re)construction of discourse objects through operations of *schematization* in dialogic communication. "Schematization" refers both to the process and the result, a specific discursive representation of what the discussion is about (Grize, 1997, p. 29). Then, a *schematization* is inherently subjective and individual, even among argumentatively aligned participants. Nevertheless, people defending the same view tend to build similar schematization. This approach somehow extends the argumentative orientation of language described by Anscombre and Ducrot (1997). Grize studies both explicit and explicit argumentative moves in discourse:

> As I understand it, in argumentation considers the interlocutor is not considered as an object to be manipulated, but rather as an alter ego who must be brought to share one's vision. Acting on him, it's trying to modify the diverse representations that he might have, by emphasizing some aspects of things, hiding

> others, suggesting new ones, and all this thanks to an appropriate schematization.[1]

An appropriate *schematization* cats *light* on specific aspects of a *discourse object* to serve an argumentative purpose, by creating axiological inferences (Grize, 1997, p. 48). Last but not least, a *schematization* does not consist of an isolated utterance but corresponds to a whole "system" (*Id.*, p. 73).

Grize considers the *discourse objects* from an extensive perspective, including their "faisceau" (object's "ray cluster"), which is to say all the usual things generally associated to the object (intrinsic characteristics, typical relations to other objects, action schemes). A participant's contribution to the "ray cluster" (*faisceau*) of a *discourse object* is argumentatively orientated to the defense of a preferred conclusion. Tracking and comparing such *schematizations* is key to do substantial argumentation analysis[2].

2.2 Research questions: comparing schematizations

Still, how can such tracking be operated? Generally, comparing *schematization* is done through qualitative discourse analysis, at a microscopic level. For instance, we proposed a monography of two rival *schematizations* in a ten-minute debate, with a special focus on the diverging emotional framing (Polo, Plantin, Lund, & Niccolai, 2013). But when it comes to comparing *schematizations* at a larger scale, especially for intercultural analysis, other tools are needed. The use of textometrics can give us interesting clues, with a first automatic

---

[1] Own translation from the French "Telle que je l'entends, l'argumentation considère l'interlocuteur, non comme un objet à manipuler, mais comme un alter ego auquel il s'agira de faire partager sa vision. Agir sur lui, c'est chercher à modifier les diverses représentations qu'on lui prête, en mettant en évidence certains aspects des choses, en en occultant d'autres, en en proposant de nouvelles et tout cela à l'aide d'une schématisation appropriée." (Grize, 1997, p. 40).

[2] This approach is similar to the Anglo-Saxon frame analysis. As a discursive construct, "frames", as *schematizations*, "induce us to filter our perceptions of the world in particular ways, essentially making some aspects of our multidimensional reality more noticeable than other aspects." (Kuypers, 2009, p. 181). As a process, both "framing" and "schematization" "encourages the facts of a given situation to be interpreted by others in a particular manner." (Kuypers, 2006, p. 8). Nevertheless, frame analysis is related to the critical tradition in discourse analysis, which does not fit with our radically descriptive epistemological approach.

quantitative analysis. In this perspective, we defined three precise research questions:

> 1. Is the specific *light* cast on a *discourse object* embodied in the words used to argue?
> 2. If so, can we describe *argumentative frames* using textometrics?
> 3. How do such *argumentative frames* vary in different cultural and linguistic contexts?

To better understand the corresponding methodology (4), we now specify the context and the empirical data characterizing this study (3).

3. EMPIRICAL STUDY: CONTEXT AND DATA

Our data consists of video-taped debates run during a scientific-*café*-type activity in three countries, in 2011 and 2012. The two-hour-long pedagogical activity is extracurricular but organized at school, in usual class groups of students aged 12-14. The session is led by a duo of trained elder students, aged 15-18. It is based on a multiple-choice questionnaire, alternating quiz questions giving basic information and "opinion questions". The activity is oriented toward a final debate, on the "main question" (MQ, figure 1), which is also presented at the beginning to introduce the topic. After following three thematic steps, the students discuss it in group, choose a common answer, and then debate it with the whole-class. Finally, each student expresses an individual answer through an anonymous vote. Our corpus consists of 10 final debates, at the class level, on the MQ, 2 from each of the two Mexican schools, 3 from the US school, and 3 from the French school.

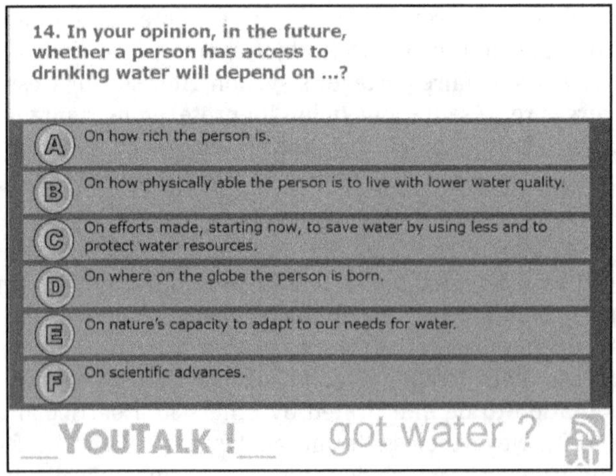

Figure 1 – Main question of the café (English version).

It is essential to take into account the nature of the topic. In the literature, SSI are characterized upon 4 main properties (e.g. Albe, 2009; Driver, Newton, & Osborne, 2000; Gayford, 2002; Kolsto, 2001; Legardez & Simonneaux, 2006; Simonneaux & Simonneaux, 2009; Zeidler et al., 2005): interdisciplinarity, knowledge hybridizing, subjectivity (Oulton, Dillon, Grace, 2004) and controversy (Albe, 2009; Legardez, 2006, pp. 19-20). As a result, the debates studied here are very open compared to usual classroom discussions. This semi-formal activity lets the students use a great diversity of argumentative resources, and intends to have them feel comfortable with not all sharing the same view. The final debates on the MQ varied a lot among our data, with variations of the *schematization* of "water". Section 4 details our methodological approach to track these differences.

4. METHODOLOGY

Our research questions (see 2.2) can be operationalized into the three following problems:

> 1. Which words is the *object* "water" associated with, in students' argumentative discourse?
> 2. Do such co-occurrences group into characteristic orientation focus or "lights" ("éclairages")?
> 3. Can we observe regularities among national subcorpora?

To answer these questions, we analyzed all the ten transcripts of the debates with TXM software[3], which automatically lists the co-occurrences between words. In order to get a complete picture of the use of the *discourse object* "water", and not only the word itself, we proceeded into 3 analytical steps.

4.1 Researching contexts for the lemma "agua", "water", "eau"

The first analytical step consists in researching co-occurrences for the lemmas corresponding to "water", in each of the three subcorpora. We did not find any occurrence of "water" using the plural. In the Mexican corpus, we found 117 occurrences of "agua", which was the 5th more frequent word. In the US corpus, "water" was only the 9th most frequent word, with a total of 91 occurrences. In the French transcripts, only 43 occurrences were found, "eau" ranging as the 32th most frequent word.

---

[3] This open-source tool is freely available and fully documented here: http://textometrie.ens-lyon.fr/.

Such global differences can be partly explained by the fact that the debates were not of equal duration: they lasted between 4 minutes and 04 seconds for the shortest, to 9 minutes and 47 seconds for the longest. Moreover, French debates were characterized by many topical digressions.

LOC- time - café	Left context	Pivotal term
ROS - 1:29:23, 2	advances in order to: save er:	water
MAR - 1:26:24, 2	move somewhere else to have access to:	water
STA - 1:33:40, 2	x \ I thought one about D /	water
CAT - 1:26:39, 2	move from where they were to get a	water
ERI - 1:32:54, 4	only does it advance er water consumption and	water
CAT - 1:33:53, 2	would need the more purified water \ bad	water
CAT - 1:26:39, 2	to like move toward an area with better	water
ROS - 1:28:13, 2	and it would xx more: available **drinking**	water
ROS - 1:28:13, 2	you know who has access to water **drinking**	water
ERI - 1:32:54, 4	in science not only does it advance er	water
JIM - 1:30:36, 2	give them \ then they can interchange for	water
STE - 1:23:58, 3	the other parts of the world can **get**	water
MAR - 1:30:05, 2	up finding ways like you could probably **get**	water
SYD - 1:32:34, 4	so: it would be easier to **get**	water
THE - 1:27:09, 3	possibility that that it can happen to **get**	water
JAD - 1:35:33, 4	they would not really be willing to give	water
MIC - 1:32:00, 4	like Africa that they do not have **good**	water
ABI - 1:33:49, 2	was drinking like em: not very **good**	water
CAT - 1:31:32, 2	go to an area that does not have	water

Table 1 – First occurrences of 'water' in left-context order, US corpus.

These lemma-based requests in TXM gave us the terms associated to each occurrence of the pivotal term "water" (*agua, eau*), which can be alphabetically organized following the left context or the right context. Such visualization enables the analyst to start identifying, at a glance, repeated co-occurrences. Table 1 illustrates this type of data directly exported from the software, here twenty occurrences of "water" in the US corpus, the first ones following the left context alphabetical order.

4.2 Researching pronouns referring to water

To get a more complete idea of the way the *object* "water" is constructed in students' discourse, we also took into account the occurrences of pronouns standing for water. We identified only a few ones actually used: "la", "lo", "se", "esa", "le", "que", "qué" in the Spanish data; "it", "that" and "ours" in the US corpus; "ça", "la", "l'", "en", and "elle" for the French transcripts. This second analytical step enabled us to catch other occurrences of the *discourse object* water. Below are examples taken from the three corpuses, with the locutor name, the time when it is uttered in the video, the number of the debate, and own translation into English when necessary:

(1) LAU 1:48:09, 4	**elle** va devenir rare (**it**'s gonna become scarce)
(2) LIS 1:30:05, 1	a la vez **la** estamos contaminando (at the same time we're polluting **it**)
(3) ABI 1:25:52, 2	they might not be able to get **it** and **ours** is very good\

4.3 Researching contextual terms

In order to make sure that no occurrences of the *object* "water" were missing, we followed a 3rd analytical step, in TXM, making requests based on the results of steps 1 and 2. We researched any contextual term that has been identified as co-occurring with the lemma "water" (3.1), or a pronoun standing for it (3.2). These requests were generally made using a radical form, since the automatic grammatical tagging tool used by TXM was not equally developed for the three languages. This last step, of course, mostly repeated results previously found, but it also enabled us to find new occurences, such as:

(4) OCE 1:49:28, 1	il y aura du **manque** (there will be a **shortage**)
(5) MON 1:32:09, 1	y va a **haber** muy poquita (**there** will **be** very little)
(6) ROS 1:29:23, 2	maybe for the future we will **have** enough

This three-step methodology enabled us to get a quite complete picture of the discourse construction of the object "water" among students' transcripts. The last stage of our methodology, interpretation, is based on two sets of results: a) identifying and specifying the different orientation focuses or *lights* ("éclairages") conferred to "water" by grouping co-oriented terms, and b) comparing their relative weight in each subcorpus. It is presented together with our main results, in section 5.

5. MAIN RESULTS

Our first result was to characterize precisely the alternative discursive constructions of the object "water" in relation to the argumentative *light* being emphasized (5.1). Then, comparing the relative weight of each of these perspectives, we could identify coherent systems of predominant *argumentative framing* in each linguistico-cultural field (5.2).

5.1 Specification of 6 orientation focuses or lights *("éclairages")*

On the basis of TXM results, we qualitatively grouped the contextual terms co-occurring with the *discourse object* "water" into 6 main *lights* characterizing alternative schematizations.

1. A first group corresponds to rewording the general issue of **accessing water**.
2. A second group of associations made with water is characterized by a **focus on natural water resources**.
3. The third group of terms corresponds to apprehending water as related to the **satisfaction of human needs**, either from a qualitative or a qualitative perspective. In addition to the contextual terms listed in table 2, two global expressions (defined as including both a left context and a right context) are also part of this group: "as much water as we want, lower the water quality".
4. Another set of contextual terms casts light on the more or less environmentally friendly **human uses of water**. Two global expressions fall into group 4: "reduce the water that we search, leave the water on".
5. The fifth group of associated words considers water from the **viewpoint of the water producer**, concerned by technical processing problems or/and the matter of providing water to the consumers. This group, in addition to the terms listed in table 2, includes 3 global expressions: "change salt water into soft (*convertir el agua salina en dulce*), make water available, give water to other countries".
6. A sixth set of contextual words put emphasis on the monetary and **commercial exchange related to access to water**. It consists of the contextual terms listed in table 2, more the three global expressions: "use water as an economic thing (*ocupar el agua como algo económico*), see water as an economic thing (*ver el agua como algo económico*), income of the water to pay (*ingreso del agua por pagar*)".

The full list of contextual terms associated to each group is reproduced in table 2 (except the above mentioned global expressions).

Focus	Left context	Right context
1. Accessing water	access to, get, how much, availability of, there is, there is less, half of the, about, have, have a little, accessibility of, get, get a lot more of, get that much, wrong with their	be, little, is out to bigger countries, issue, problem, depends on where we are born, become, 'stay available' (*rester accessible*), and the expression 'the issue of having water equally spread' (*le problème que l'eau soit repartie équitablement*)
2. Natural water resources	there is scarcity of (*haber escasez*), without, populated with, no, salt, ocean, lot of, ton of, available, vapor (*vapeur d'*)	end (*acabarse*), exhaust (*agotarse*), be a resource, be scarce, sources, supplies, supply, supplied, resources, scarce, has been taken, from the polar icecaps, goes, become scarce, from the air
3. Satisfaction of human water needs	have enough, need for, needs for, the right to, lack, the lack of, run low on, run out of, thin, purified, needed amount of, good, better, bad, drinking, the worst of the, need, adapt to the, drink	of lower quality, to drink, to live, potable, be a vital resource, of good quality, of bad quality
4. Human uses of water	preserve, save, saving(s) of (*economies d', ahorro del*), waste, use, usage of, use less, pollute, protect, shampoo on, economize, consume, conserve, abuse, keep, take care of (*cuidar*), share, take advantage of (*aprovechar*)	to, to take care of (*de cuidar*), consumption
5. Viewpoint of the water producer or provider	distribute, make, cool down (*refroidir*), reuse, renew, provide, produce (and the French 'fabriquer'), out of, give, give them, desalinize, take the salt out of, desalinization of, doing, channel the, get from (*sacar*), stay with (*quedarse con*), gifts (*donaciones de*), offer, share	purification, plumbing, lines
6. Water trade	sell, give expensively (*dar caro*), steal, price of, money of, pay for, pay with, buy, interchange for, pay to get, the riches get, rich people have, bill of (*facture d'*)	become expensive, be as money, depends on the richness, expensive, cost, cheap, affordable, be valued (*valer*), rise, of a high price (*a un precio elevado*)

Table 2 – Grouping of contextual terms into 6 specific orientation focuses, or 'lights' (*éclairages*).

Whether the students' discourse casts light on one or the other of these aspects has different consequences in terms of *argumentative framing* of the debate. At first sight, group 1 associations do not seem to provide strong argumentative orientation toward a specific answer or argumentative conclusion. Still, the way the general issue is called gives a first insight about which orientation the speaker is trying to bring the debate toward. In combination with the analysis of the relative place of the 5 other groups of contextual terms, the analyst can specify the alternative argumentative *framing* being displayed. Since *schematization* doesn't rely on isolated utterances, but works as a meaning and orientating discursive network, a global picture is needed to actually describe this construct. We developed a synthetic word-cloud type visualization, for each subcorpus, of the textometrical analysis of co-occurrences, which makes it easier to combine these parameters and draw conclusions about the typical *argumentative framing* that emerges from the data (5.2).

5.2 Alternative schematizations *of "water": results for each country*

Figures 2, 3 and 4 synthesize our results, respectively for the Mexican, the US and the French corpora. We designed this word-cloud visualization in order to present a global picture that articulates the analysis of co-occurrences made in TXM with the first step analysis consisting of qualitative grouping of occurrences into the 6 specific focuses described in 5.1. In the center of the visualization appears the pivotal lemma "agua", "water" or "eau". Then, left-context occurrences of the *discourse object* "water" are reproduced on the left part of the figure, and the right-context occurrences in the right section of the plane, both in alphabetical order. When a contextual term co-occurred several times with the *discourse object* "water", we reproduce the lemma (infinitive of verbs, adjectives and nouns in singular masculine), except if there was no variation. For instance, 3 occurrences of "les riches" are reproduced as "les riches", but the thirteen occurrences of the verb "to get" used at different persons appear with the lemma "get". In order to get a meaningful representation of our results, contextual terms are bigger when they are more frequent, proportionally to the total number of co-occurrences. More frequent terms are also slightly more centered, appearing closer to the pivotal lemma. For the French corpus, it corresponds to terms used at least five times to characterize the discourse object "water", while the threshold is three for the US corpus and four for the Mexican corpus. A number is associated to each contextual term, corresponding to its frequency in the corpus. Terms that share the same radical (ex: "access", "accessibility") or are semantically synonymous (ex: "desalinize" and "take the salt out") are reproduced close to each other, and in the police and place corresponding to the sum of their frequencies. Some occurrences cannot be classified as specifically left or right context terms but rather consist of a global expression including left and right contexts for "water". These expressions were placed directly next to the central pivotal word. Last but not least, a color code was used to identify the different orientation focuses (*éclairages*) characterizing the *schematizations* of the *discourse object* "water":

1) in black appears the terms corresponding to the general problem of access to water;
2) blue is used for terms referring to water as a natural resource;
3) red is used for terms characterizing water in reference to human qualitative and quantitative needs;
4) in green are reproduced the terms dealing with human more or less environmentally friendly uses of water;
5) brown is the color corresponding to words describing water from the viewpoint of the water producer or provider;
6) purple is used for terms referring to water trade and the commercial transaction determining someone's access to water.

These visualizations reveal a specific *argumentative framing* of the issue in each subcorpus.

Five key features characterize the Mexican corpus in terms of prevailing orientation focuses of *schematization* of "water" (see figure 2). First, when water is considered as a natural resource (in blue), what is emphasize is how scarce (4 occurrences of "escasez" or "escasa" - *scarcity* or *scarce*), absent (2 occurrences of "sin agua" -*without water*), or endangered (18 occurrences of "acabarse" or "agotarse" –*extinguish* or *dry up*) it is.

Secondly, among human use of water (green), the idea that water must be saved (37 occurrences of "ahorrar", "guardar" or "usar/utilizar menos" - *save, keep*, or *use less*) is very frequent. This focus is characterized by the extensive use of the verb "cuidar" (*take care*), with 21 occurrences. Even if the light cast on human use of water here mostly consists of prescribing good practices, some behaviors that waste water are also mentioned (9 occurrences of "desperdiciar", "gastar" or "desgastar").

Third characteristic of the Mexican corpus: when water is considered for its capacity to satisfy human needs (in red), two aspects are emphasized, the fact that water is vital (two occurrences of "necesitar" –*need*- and one of "recurso vital" - *vital resource*), and the capacity of people to adapt to the evolution of water resources (3 occurences of "adaptar"). The expression "recurso vital" is surrounded by a blue frame because the substantive also refers to water as a natural resource. The emotional tonality associated to the mentioned human needs is quite high, as they correspond to matters of life or death: 3 occurrences of "tomar" or "beber" (*drink*), and one occurrence of "vivir" (*live*).

On the contrary, this corpus presents very few contextual terms corresponding to the viewpoint of the water producer (in brown). A unique occurrence among the three national corpuses appears in Mexican data: the use of the verb "compartir" (*share*), which is reproduced in green and surrounded by a brown line: here water provider is also presented as a co-user.

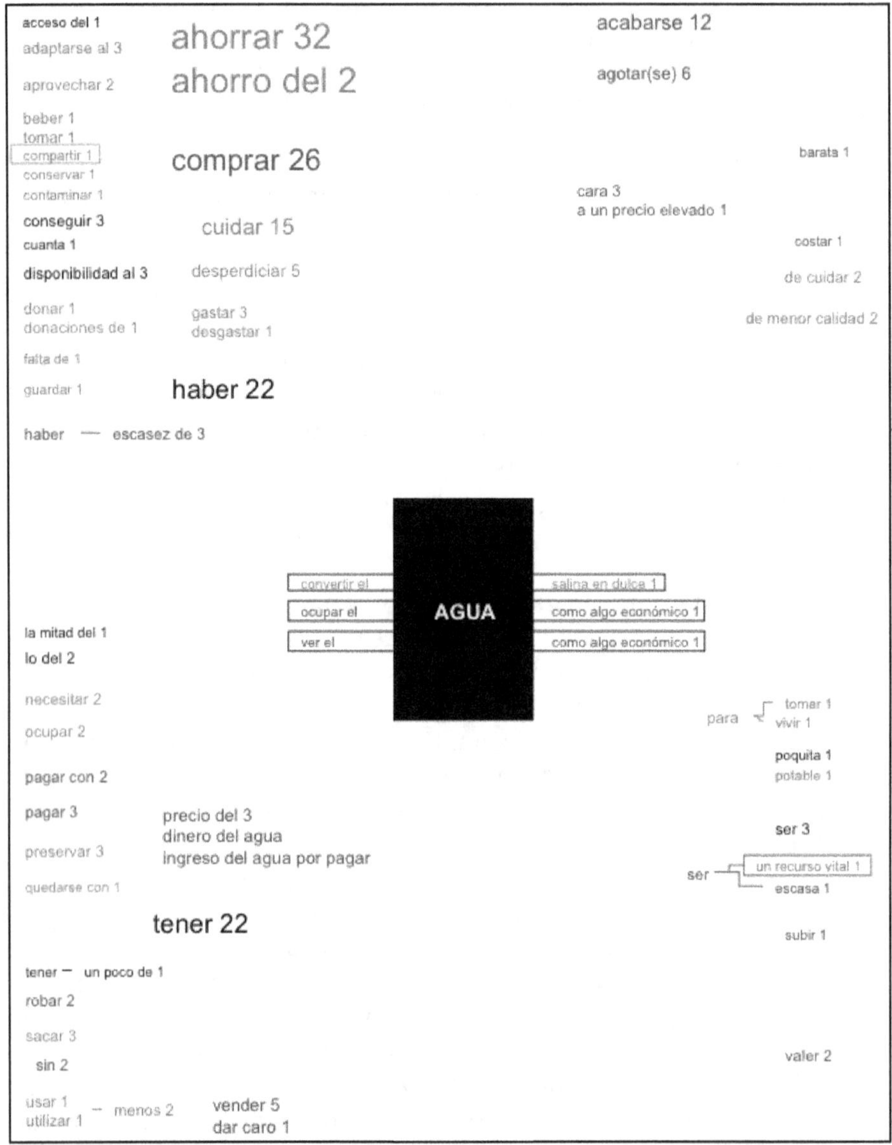

Figure 2 – *Schematization* of 'water' in the Mexican data: visualization of co-occurring terms, their focus and frequency.

Fifth characteristic of the Mexican data: a large number of occurrences referring to water trade (in purple). Both selling and buying water are mentioned, but the students mostly present themselves from the viewpoint of the consumer (30 occurrences of "comprar", "pagar" or "costar" –*buy, pay,* or *cost,* versus 6 occurrences of "vender" or "dar caro" -*sell* or *give at a high price*). Moreover, 13 terms deal with water price, 5 of them emphasizing how expensive it is, or mentioning that it is going up. Water is even considered as a potential currency twice ("pagar con" - *pay with*). Last but not least, the verb "robar" (*steal*) appears twice. Drinking water, thought as a commodity, can be monopolized, traded, or stolen.

Figure 3 synthesizes the results obtained for the US corpus. Here, the focus on water as a natural resource (in blue) is much more frequent than it is in the two other subcorpora. This perspective is embodied both in a general lexicon (10 occurrences of "sources", "resources" or "supplies") and by reference to specific water sources: polar icecaps (1 occurrence), seawater (5 occurrences). If 3 occurrences mention that water is scarce or absent (2 occurrences of "no water"; one of "scarce"), there is also 2 expressions emphasizing its abundance ("a lot of", "a ton of"). Two contextual expressions are ambivalent in this aspect: "has been taken" and "out to bigger countries", because they establish scarcity in a place corresponding to more water in other places, the USA falling into the second category (the "bigger countries").

Such dichotomy is consistent with what appears in red, corresponding to water characteristics for the satisfaction of human needs. In quantitative terms, 5 occurrences mention the lack of water ("lack of" or "run out of/low on"), 2 rather refer to an appropriate amount of water ("enough"; "the needed amount"), and one occurrence describes a situation of abundance ("as much water as we want"). Qualitatively, water is also framed as more or less good (6 occurrences of "good", "better", "purified" or "thin" versus 4 occurrences of "bad", "worst", or "lower water quality").

When it comes to the uses of water (in green), 7 occurrences are quite neutral ("consume", "consumption", "usage" or "use"), while 20 describe good practices to follow and 5 bad practices to be avoided. For good practices, the most frequent verb is "save", referring to both preserving water and economizing it.

The US corpus is also characterized by a very frequent *schematization* of the discourse object "water" from the viewpoint of the water producer or provider (in brown). Water production is extensively discussed, with 4 general occurrences ("produce"; "make water available"; "make"; "do"), and the mention of a large number of production techniques: the desalinization of seawater (4), the recycling of water ("reuse"; "renew"), water purification (1), and even channeling a river (1). A lexicon referring to water distribution also embodies this orientation focus: "lines"; "plumbing"; "give" (4); "provide" (1).

Water trade appears in the US corpus, as much from the viewpoint of the seller (2 occurrences of "sell") as from the viewpoint of the buyer ("pay to get", "pay for"). One occurrence of the neutral term "interchange for" was also inventoried. About the water price, the emphasis is more on how cheap it is or might be (3 occurrences of "cheap", one of "affordable") than on how expensive it is or might be (only one occurrence of "expensive").

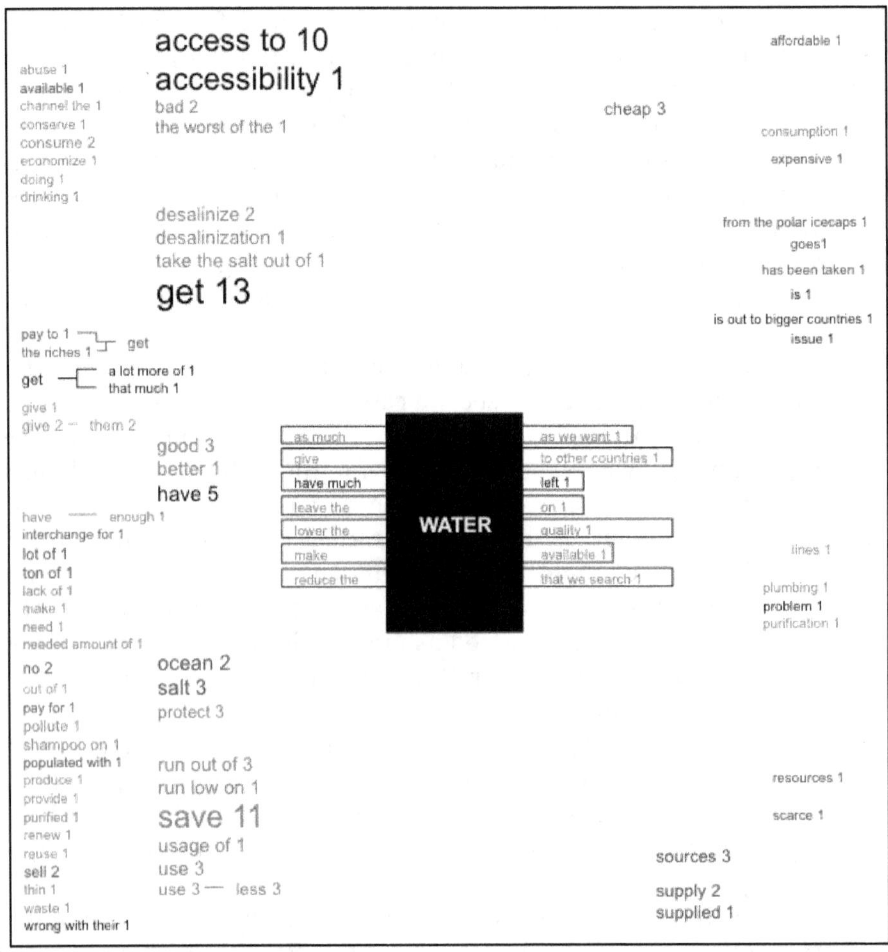

Figure 3 – *Schematization* of "water" in the US data: visualization of co-occurring terms, their focus and frequency.

Besides, in the US corpus, two words employed for the general problem of access to water deserve a specific attention. The first one is the massive use of the verb "get", which embeds an idea of voluntary action that is absent in the verb "have", for instance. Rather than considering access to water as determined by fatality, this framing is emotionally less intense than the French or Mexican ones. This tendency is confirmed by another specificity of the way the global problem is referred to, in the US corpus: the use of abstract words and a bigger emotional distance to the issue ("water problem", "water issue", "it is wrong with their water").

In figure 4 are visualized the main results for the French corpus. Here, water is hardly considered as a natural resource (only 3 occurrences in blue). When such framing occurs, it is mainly to put emphasis on the fact that water is rarifying.

"Economiser l'eau" is THE motto that stands out of the French data, when it comes to water uses (in green). In total, 16 occurrences refer to good practices to use less water (11 occurrences) or preserve it (5 occurrences). But, contrary to the other corpora, here no concrete situation is presented.

In terms of *schematization* of water for its capacity to meet human needs (in red), both qualitative and quantitative aspects are mentioned in the French corpus. A unique, interesting occurrence frames the need for access to water as a right to water.

This is consistent with the fact that, among the 12 occurrences corresponding to the viewpoint of the water producer or provider (in brown), one specifically focuses on water distribution considered as a matter of public service. Some contextual terms also refer to the general action of producing water ("faire", "fabriquer" - *make*), while the specific word "refroidir" (*cool down*) refers to getting water from the air.

The orientation focus on water trade (in purple) is quite well represented among the French data, with 9 occurrences referring to the price to be paid by the consumer to access water. Moreover, 10 occurrences embody a concern for the socioeconomic inequalities among people for access to water.

At the end of the day, the French corpus is strongly characterized by addressing the problem of access to water at a macro-social level, with a global motto ("économiser l'eau") rather than concrete, micro-level examples of good practices, the definition of a universal right to water, and the subsequent concerns of dealing with social inequalities and the matter of providing water to everyone.

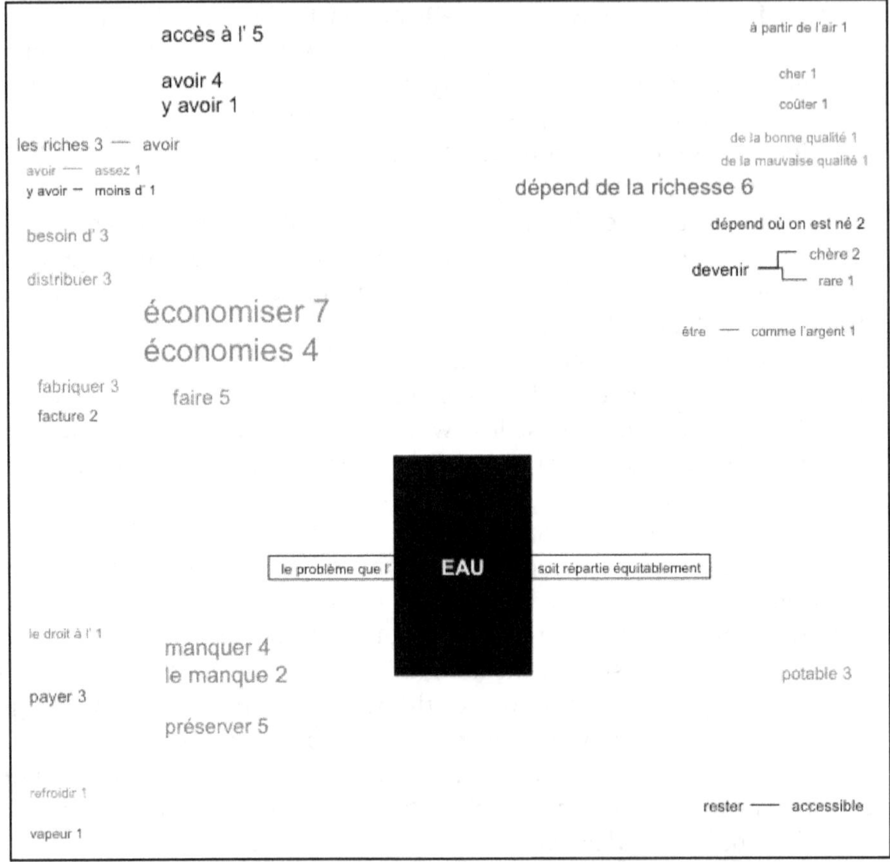

Figure 4 – *Schematization* of "water" in the French data: visualization of co-occurring terms, their focus and frequency.

6. CONCLUSION

The analysis of co-occurrences in TXM proved useful to specify and compare students' *schematization* of "water" as a *discourse object*, between our three national corpora (Mexico, USA and France). Beyond effects of cultural rhetorical styles, substantial differences in terms of orientation focuses (*éclairages*) distinguish the debates hold in the three countries. In the Mexican debates, water is mostly apprehended from the viewpoint of the consumer, worrying about daily practices for getting water at an affordable price and saving the vital endangered resource. On the contrary, in the US corpus, the viewpoint of the water producer prevails. Here what counts is finding technical solutions to provide water at a low price to the consumers, making the best out of the natural water resources of each territory. In France, social inequalities are mentioned from a macro-social viewpoint, and the addressed solutions mostly rely on a better global organization of the distribution of water, considered as a universal right and a public service. These results about the specific *argumentative framing* converge with results obtained by two other methods, about the cognitive models of "water" used by the students, the prevailing knowledge area in the debates in the three countries, and the analysis of students' votes during the activity (Polo, 2014, pp. 359-424). They contribute to the comparative characterization of the prevailing "*argumentative scenario*" (Plantin, in press, pp. 424-425) in each setting. Innovative word-cloud visualizations gives a global picture of the automatic textometrical results and the complementary qualitative analysis. They were very helpful for the comparative dimension of our analysis.

More generally, we believe that the operationalization of Grize's concept of *schematization* (1997), as a descriptive framing approach, can be applied in other argumentative situations and enrich the methodological toolbox of the community of argumentation studies. The textometrics provide promising ways of doing comparative analysis in argumentation that really focuses on the *objects* under discussion. Moreover, the word-cloud-type visualization of mixt-method results that we developed in this study can be useful even in monographies, out of any comparative aim. It might be especially relevant to study the role of emotions in a specific argumentative corpus, enabling the analyst to see at a glance the prevailing orientations given to the debate, and the co-substantial emotional framing (Polo et al., 2013).

ACKNOWLEDGEMENTS: This research is supported by LABEX ASLAN (ANR-10-LABX-0081) of *Université de Lyon*, within the program "Investissements d'Avenir" (ANR-11-IDEX-0007) of the French National Research Agency (ANR).

REFERENCES

Albe, V. (2009). *Enseigner des controverses*. Presses universitaires de Rennes.

Andriessen, J., Baker, M., & Suthers, D. (2003). *Arguing to learn: Confronting cognitions in computer-supported collaborative learning environments*. Springer Netherlands.

Anscombre, J. C., & Ducrot, O. (1997 [1981]). *L'argumentation dans la langue*. Editions Mardaga.

Disson A. (2002). D'une culture l'autre : argumentation et stratégies discursives au Japon, *Études de Linguistique Appliquée, 126*(2), 181-188.

Driver, R., Newton, P., & Osborne, J. (2000). Establishing the norms of scientific argumentation in classrooms. *Science Education, 84*(3), 287–312.

Erduran, S., & Jiménez-Aleixandre, M. P. (2007). *Argumentation in science education: Recent developments and future directions.* Dordrecht: Springer.

Gayford, C. (2002). Controversial environmental issues: A case study for the professional development of science teachers. *International Journal of Science Education, 24*(11), 1191–1200.

Grize, J. B. (1997 [1990]). *Logique et langage.* Ophrys.

Kolstø, S. D. (2001). Scientific literacy for citizenship: tools for dealing with controversial socio-scientific issues. *Science Education, 85*(3), 291-310.

Kuypers, J. A. (2009). *Rhetorical criticism: Perspectives in action.* Lexington Press.

Kuypers, J. A. (2006). *Bush's war: Media bias and justifications for war in a terrorist age.* Rowman & Littlefield Publishers, Inc.

Legardez, A., & Simonneaux, L. (2006). *L'école à l'épreuve de l'actualité. Enseigner les questions vives.* Paris: ESF.

Muller Mirza, N. (2008). Préface. In C. Buty & C. Plantin (Eds.), *Argumenter en classe de sciences* (pp. 7-16). INRP.

Oetzel, J., Ting-Toomey, S., Masumoto, T., Yokochi, Y., Pan, X., Takai, J., & Wilcox, R. (2001). Face and facework in conflict: A cross-cultural comparison of China, Germany, Japan, and the United States. *Communication Monographs, 68(3),* 235–258.

Oulton, C., Dillon, J., & Grace, M. M. (2004). Reconceptualizing the teaching of controversial issues. *International Journal of Science Education, 26*(4), 411–423.

Plantin, C. (in press). *Dictionnaire de l'argumentation - Une introduction notionnelle aux études d'argumentation, ENS Editions,* Lyon.

Polo, C. (2014). *L'eau à la bouche: ressources et travail argumentatifs des élèves lors de débats socio-scientifiques sur l'eau potable. Etude comparée de 10 cafés scientifiques menés au Mexique, aux USA et en France, en 2011-2012.* Doctoral dissertation, Lyon 2 University, France.

Polo, C., Plantin, C., Lund, K., & Niccolai, G. (2013). Quand construire une position émotionnelle, c'est choisir une conclusion argumentative. *Semen, 35,* 41-63.

Simonneaux, L., & Simonneaux, J. (2009). Students' socio-scientific reasoning on controversies from the viewpoint of education for sustainable development. *Cultural Studies of Science Education, 4*(3), 657–687.

Taft, M., Kacanas, D., Huen, W., & Chan, R. (2011). An empirical demonstration of contrastive rhetoric: Preference for rhetorical structure depends on one's first language. *Intercultural Pragmatics, 8*(4), 503–516.

Von Aufschnaiter, C., Erduran, S., Osborne, J., & Simon, S. (2008). Arguing to learn and learning to argue: Case studies of how students' argumentation relates to their scientific knowledge. *Journal of Research in Science Teaching, 45*(1), 101–131.

Zeidler, D. L., Sadler, T. D., Simmons, M. L., & Howes, E. V. (2005). Beyond STS: A research-based framework for socioscientific issues education. *Science Education, 89*(3), 357–377.

53

The Philosophical and Literary Argumentation Methods in the Ancient Egyptian Rhetorical Systems

HANY RASHWAN
SOAS, University of London, UK
540887@soas.ac.uk

The ancient Egyptian and the Greco-Roman cultures have two distinct perspectives for viewing language and its rhetorical system, as a result, differing in structuring their persuasive messages. The paper investigates the possibility of offering closer analytical readings of ancient Egyptian Argumentation methods, by shedding new insights into their persuasive aesthetic richness, and confirms the persuasive structural integrity of the ancient Egyptian language.

KEYWORDS: ancient Egyptian argumentation, comparative rhetoric, eurocentric, literary rhetoric, non-Western logic

1. INTRODUCTION

The connections between argumentation and rhetoric have often been extremely close in the Western and non-Western traditions. Modern scholars have to first understand the more dominant rhetorical system in order to better understand the nature of the argumentation. The best approach to achieve this objective can be provided under the umbrella of an emerging discipline called "Comparative Rhetoric"[1]. It is a new discipline that deals with the study of rhetoric across different cultural traditions, and it is a "potentially rich, extremely challenging, and thus,

---

[1] The real beginning of this young discipline goes back to 1966 when the English professor Robert Kaplan offered an article that examines how the non-Western students in American universities write their arguments in English and how that may reflect the rhetorical characteristics of their own native languages. His essay thus pioneered an area of study now called "Contrastive Rhetoric". Afterwards, it has been developed to establish the discipline of "Comparative Rhetoric" (LuMing, 2003).

D. Mohammed & M. Lewiński (eds.) (2016). *Argumentation and Reasoned Action: Proceedings of the 1st European Conference on Argumentation, Lisbon, 2015. Vol. II, 849-863.* London: College Publications.

largely untouched area of study." (Garrett, 1998, p. 431) There are few studies about reconstructing the ancient Egyptian Rhetoric (henceforth AER) but they were hampered at the start by number of preconceptions that have long been embedded in the discourse as scientific or empirical facts. Most such preconceptions centered around a primary definition of AER as part of a public oral persuasive practice; behind this concept lies the hegemonic tradition of speeches in the assemblies and Senates of the ancient Greek and Roman world. This Eurocentric methodology still surrounds all the writings about non-Western rhetorical systems. The sinologist Marry Garrett criticized this hegemonic approach reviewing one of the fundamental books of the field (Kennedy, 1998) saying:

> Kennedy gives pride of place to the terminology and theories of Western rhetoric, not just as a heuristically convenient starting point, but also as the limit of his inquiry. From Kennedy's perspective, the project is one of "test[ing] the applicability of Western rhetorical concepts outside the West (p. 5). Specifically, to what extent can the rhetorical terminology of the Greco-Roman tradition describe the practices of other traditions?" (Garrett, 1998, p. 431)

The main expected result of such Eurocentric methodology is to turn all the "other" rhetorical systems into an ugly replica of the "perfect" Greco–Roman system. An example of such Eurocentric views is James Murphy's statement: "There is no evidence of an interest in rhetoric in the ancient civilization of Babylon or Egypt, for instance neither Africa nor Asia to this day produced a rhetoric." (Murphy, 1981, p. 3) Another is Michael Fox's speculation about the whole of the non-Western societies and their illogical communication systems: "Non-Western rhetoric doesn't teach how to formulate arguments because it is not argumentation but rather the ethical stance of the speakers that will maintain harmony in the social order, and that is the ultimate goal of Egyptian rhetoric." (Fox, 1983, p. 21) George Kennedy supported Fox's claim saying that he did not find in the ancient Egyptian literature "any good examples of argument from probability. Neither in Egypt nor elsewhere outside classical Greece are full syllogisms stated, but enthymemes... are ubiquitous." (Kennedy, 1998, p. 183)

Apparently, there is little interest from Western "Comparative Rhetoric" specialists in studying the "other's" rhetorical system by using that system's own concepts, without any reliance on the Eurocentric application used in their own methodology. If we begin our discussion by adopting the Greek concepts and definitions, we lose a genuine ability to understand the other's systems, as they have been situated and embedded according to their own culture and language by their own intellectual figures. Using Eurocentric classic typology with its terms and concepts, as a methodology to treat non-Western rhetoric, is something the "dead" ancient Egyptian and even the alive Arabic languages suffer from. The classic traditions were developed within the Greco-Roman world to express linguistic and literary minutiae that related only to the Greek and Latin world. The European linguistic schools have the full right to use them for studying all the minutiae of their kindred languages. However, the situation should be different when we deal with a non-Western language. Imposing the Western terms and concepts obscures the character of the studied language, provides problematic answers, implies that there is nothing more to be said and gets in the way of developing a new closer approach to better understand these ancient non-Western cultures.

> Rhetoric, derived from Greek philosophy, is traditionally conceptualized as a method of persuasion through logical/rational arguments. This Eurocentric perspective has privileged, at the expense of other traditions of orality and orator, a narrow understanding of the ways by which cultures communicate meaning and traditions through the spoken word. (Young, 2004, p. 81)

A recent study offered insightful thoughts about the two rhetorical systems that always existed alongside each other in every culture: Philosophical and Literary Rhetoric. Philosophical Rhetoric deals with argumentation and regulation of public oral speech and is strongly represented in the Greek rhetorical system that is based on Plato and Aristotle's theories. Literary Rhetoric deals with the conveying of meaning in the best of literary verbal forms or the study of aesthetic effectiveness and is strongly represented in the Arabic system (*Balāgha*) which mainly arose from the text of the Quran, whose literary inimitability pushed the Grammarians very early on (9th-10th century) to enumerate, define, exemplify and classify the literary and grammatical peculiarities of the Revelation[2]. "The two rhetorical systems can be distinguished by their goals, methods, programs and

[2] Fortunately, the two systems existed in Arabic traditions. The argumentation one is more related to the science of Arabic *khataba* which literally means public oral speech. This Arabic discipline has been heavily influenced by Greek-Roman rhetoric through Aristotle's translations and has been developed later by the Arabic speech practitioners, under what has been called علم الكلام - science of Speech. This discipline is still studied too little in the West. (Halldén, 2005, p. 20)

sources." (Woerther, 2009, p. 10) There is no comprehensive study yet that tries to understand how the two rhetorical systems influence each other in order to form the persuasive arguments for their native receivers, especially in the non-Western cultures.

Furthermore, the two offered rhetorical systems in this monograph reflect also two mental faculties of producing arguments: one is of oral nature and the other is of written nature. In writing the author has time to give full concentration to moulding a succession of related ideas into a more complex, coherent and integrated unity, while the oral conversation could be more fragmented in nature and the way that it establishes its cohesion is different, since it can rely more on non-verbal communication tools, such as modulations of the voice, body and facial gestures and the direct reactions of the receiver, in addition to depending on a shared situation between the speaker and the addressee. Cohesion in writing is generally attributed to the effective way that the author uses his linguistic-literary background, to express the intended message to the reader:

> This greater complexity is generally attributed to two distinctive characteristics of writing: the lack of strict timing constraints during the production and the need to establish the cohesion strictly through the lexical-syntactic channel. (Al-Ansary, 2001, p. 150)

In general writing, as a human activity in comparison with speech, is claimed to be more structurally complex and elaborate since it is more deliberately organized and planned than speech. These two factors can be related to the ancient Egyptian materials of writing, since writing on expensive materials such as Papyrus or a hard stone was not an easy process and requires more detailed attention to the way that the author carefully constructs his intended message. Therefore, the literary-rhetorical critics should recognize the different linguistic characterizations of oral and written arguments, as the two modes may require different analytic natures.

Therefore, the scholars of non-Western argumentation should be aware that: in the strong literary rhetoric societies, such as ancient Near Eastern cultures, the interrelationships between argumentation and the literary devices used to express the intended persuasive message are strong enough to bury the main logical argument. All the arguments produced in such cultures are heavily influenced by the literary perfection of the way in which they are expressed. The author uses the highest literary mechanism to deliver his argument, and without this literary cover it would not be accepted or considered effective for the native readers. This well-knit relation can be at once obvious but could be unclear to the modern scholars in the translation realm. The original poetic form of the argument's language is crucial to better understand the argument nature and its own impact. Scholars are thus encouraged to avoid the divorce between form and content, which can result from just depending on a translation.

This paper offers two examples to illustrate each rhetorical system[3]. The first is a song that is narrated by a harp singer, using a voice of dead king, in order to advise the live receiver to enjoy earthly life and stop sacrificing it for afterlife expectations. The author carefully chooses rational arguments that ask the mind's eye to see their truth or to challenge their claimed accuracy logically. In this song the logical-rational rhetorical system will overcome the literary metaphorical one, but still the author uses high literary language.

2. SONG OF THE HARPER

This song was found written on a papyrus that belongs to the Ramsesside period (1295-1070 B.C)[4]. The writer or the copier of the song stated in the introduction, which is unusual, that the song originally existed on the walls of a royal tomb-complex belonging to one of the unidentified dead kings named Intwf, who could be one of the kings of that name from the 11th dynasty (ca2134-2061 B.C.) or 17th dynasty (ca1585-1560 B.C.).

ḥsw nty m ḥwt (intwf) m3ʿ ḫrw nty m b3ḥ p3 ḥsyw m bnt

The song that is in the tomb of (King Intwf) the justified which is located in front of the singer with harp.

1.

w3ḏ pw wr pn nfr š3w nfr ḥḏy

A lucky one is this great person, with good fate and good ending.

---

[3] For ease of reference, I have divided the texts into semantically separated full sentences. The offered reading here based on my translation and without heavy philological treatments, given the nature of the focus on Argumentation.

[4] The paper follows the Hieroglyphic transcription of Michael Fox, 1985, pp. 378-80.

The writer here stresses two happy encounters that this great person could achieve by being wr-rich: the happy earthly life and wealthy ending which suppose to pave the way for a joyful afterlife too.

2.

ẖt z ḥr| sbit ktyw ḥr mn ḏr rk imyw-ḥ3t

The human body is always perished but others remain from the time of the early ancestors.

The writer here forces the receivers to compare the status of people whose bodies are already gone and vanished, because they could not afford to have excellent equipment to preserve their bodies for the afterlife, with that of people who have access to super mummification process and preservation instruments which enable their bodies to fight time and survive for thousands of years.

3.

nṯrw ḫprw ẖr-ḥ3t.i ḥtpw m mrw.sn

The Deifieds became in front of me resting peacefully in their pyramid-tombs.

4.

sꜥḥw 3ḫw m mitt ḳrsw m mrwt.sn

The glorified nobles are buried as well in their pyramid-tombs.

5.

ḳd ḥwwt nn wn swt.sn ptr irw im.sn

The tomb builders, their burial-places do not exist anymore and look what happened in them.

The writer here compares two opposite social levels, mainly regarding the survival of their graves and how these religious teachings were mainly designed to serve those lucky nTrw-kings and saHw Axw the glorious nobles - who could dedicate many poor workers qd Hwwt to build their huge tombs. In a sarcastic tone, the writer forces the imagination of his receivers to see by their own eyes the two examples, to demonstrate the irony that these religious teachings created, wondering about the destiny of those poor workers after their simple unplanned graves (lit. places) have been completely destroyed. The big irony here is that those poor people had devoted all their earthly life to secure the afterlife paradise for other "glorified" rich people. The question raised here is: Are they still eligible to meet the condition of afterlife paradise, in comparison with those rich who used them to secure their own happy destiny after their death, or because of their predestined poverty will they have to face another hard situation during their afterlife? The writer challenges the accuracy of these religious teachings that ask all the people to prepare strong decorated tombs to secure the afterlife paradise, and thus he questions the reality of the afterlife existence.

6.

iw sḏm.n.i mdt ii-m-ḥtp ḥnˁ ḥr-ddf sḏd.ti m sḏdwt.sn r-sy

I have heard the speeches of Imhotep and Hordedef that have been narrated from all their narrations.

The author shows his full familiarity with these famous Old Kingdom teachings, thereby inducing the receivers to think well of his personal knowledge and thus to accept more readily what he has to say afterwards. In other words, the speaker thus extols his own virtue and at the same time casts doubt on his opponents.

7.

ptr swt iry inbw.sn fḫw nn wn swt.sn mi nty nn ḫpr.sn

What of their own burial-places? Their walls are destroyed, their burial-places do not even exist anymore like those who have never come to existence before.

The writer extends here the attack on the influential teaching that strong tombs should be built that could survive and fight time to secure the afterlife, by challenging the long-established fame of the authors of these teachings. It seems that those specific authors were famous by instructing the people about the importance of building and furnishing the graves, as a temporary bridge for their bodies, in order to be transferred to the promised paradise. The writer asks the receiver to see by his eye what happened to their own burial-places, as they could not survive against time. The writer held here a comparison between the tomb builders, whose burial-places could not survive because of their limited resources, and religious-minded teachers who devoted their wealth to build great tombs with huge walls. The irony here is: both of them became equal as time made them suffer from a similar condition: the lack of graves that secure the afterlife.

8.

bw ii im sḏd.f ḳdw.sn sḏd.f ḫrtw.sn

No one came back from there to narrate their after-death-nature or to narrate their after-death-concerns.

9.

stmt.f ib.n r ḥnt.tn r bw šmt.sn im

or to comfort our hearts regarding your eagerness for the place they went to there.

The writer extends his argument against those famous religious figures, asking the receiver if they can prove that those religious-minded teachers themselves have found the paradise which they have always promised to the people. He asks the receiver to use his logical thinking to assess the misleading content of those long-established teachings.

10.

wḏꜣ.k ib.k r.s mht ib z ḥr sꜣḫ n.k

May you heal your heart about it and may the heart of every one neglect being turned to a glorified-deceased by you.

11.

šms ib.k wnn.kwi imy ꜥntyw ḥr tp.kwi

Follow your heart while you exist. Put the myrrh on your head.

12.

wnḫ t[w] n.k m pꜣḳt gs.ti m biꜣw mꜣꜥw n ḫtw-nṯr

Put on dress for you from fine linen that is anointed with true luxury items that belong to God's things.

The writer recalls the traditional religious practice of people offering many luxury items to the temple's priests, mistakenly thinking that they are offering to the Gods, and he asks the receiver to wear them instead of offering them.

13.

imy ḥꜣw nfrw.kwi [m] bꜣgi ib.k

Increase your happiness and [do not] let your heart be weary.

14.

šms ib.k ḥnꜥ nfrw.k

Follow your heart with your happiness.

15.

ir ḫtw.k tp tꜣ m ḥḏ ib.k iw n.k hrw pfꜣ n sbḥwt

Do your things on earth (be active) but do not upset your heart until that day of wailing comes to you.

16.

bw sḏm.n wrd-ib sbḥꜣt.sn

The deceased (lit. the one with weary heart) does not even hear their wailing.

17.

bw šd nꜣy.sn biꜣk ib.i z im m ḥꜣt.i

Their mourning did not rescue my human heart from being there in my tomb.

mꜣwt

Refrain

1.

ir hrw nfr m wrd.n z im.f

Make a happy day that a man does not tire from it.

2.

mk nn rdi.n z iṯt ḫttw.f ḥnꜥ.f

Behold no one is allowed to take his things with him.

3.

mk nn rdi.n z iṯt ḫttw.f ḥnꜥ.f

Behold myself I could not go and come back again.

3. THE DEBATE BETWEEN A MAN AND HIS SOUL

To illustrate a literary argument this paper uses a short story that has been used in a poetic debate between a man and his soul, in which the man complains about his miserable life, which belongs to the Middle

Kingdom (2000 B.C.-1700 B.C.).[5] The man tries to convince his soul to commit suicide, to end his misery, thinking that his afterlife life might be better. In the third speech, the soul used a sad melodramatic story to persuade this man that his miserable situation is nothing if he is compared with another person who faced a harder catastrophic situation.

67.

mk nn rdi.n z iṯt ḫttw.f ḥnʿ.f

May you listen to me. Behold it is good to listen to people

68.

šms hrw nfr smḫ mḥ

Follow the happy time and forget the sadness (lit. the state of being worried and nervous about something in the future)

68-69.

iw nḏs sk3.f šdw.f

The lad cultivates his small land by himself,

69-70.

iw.f 3[t]p.f šmw.f r ẖnw dpt

He carries himself and his harvest to the cubbyhole of the boat

70-71.

st3s.f sḳdwt ḥb.f tkn

He hauls the sail, as his festival is approaching soon

71-72.

m3.n.f prt wḫt nt mḥyt

when he sees the disappearance of the darkness of the northern rainy wind

72.

rs m dpt

He stayed awake in the boat

---

[5] The paper follows the Hieroglyphic transcription of James Allen (2011, pp. 282-287).

73-74.
rꜥ ḥr ꜥḳ pr ḥnꜥ ḥmt.f msw.f
and when the sun began to enter he went out with his wife and children

74-75.
3ḳ tp š šn m grḥ ẖr mryt
Annihilation is atop the lake. The crocodile encircled (them) during the night, under the riverbank of the crocodiles.

75-76.
ḏr.in.f ḥms psš.f m ḫrw ḥr ḏd
Then he sat down (like a small child) ululating in a loud voice saying:

76-78.
n rm.i n tf3 mst nn n.s prt m imnt r kt ḥr t3
I am not crying for this young mother who will not emerge from the west in comparison with the other (woman who died) on earth

78.
mhy.i ḥr msw.s
but I am sad about her young children

79-80.
sdw m swḥt m3w ḥr n ḫnty n ꜥnḫt.sn
They have just broken the egg and they will see the crocodile's face of god Khenty. They had not yet lived.

To fully appreciate the emotional persuasive power of this short story, I will rephrase it in my words, reconstructing faithfully what the soul narrated:

There was a humble-young man who was fighting the hard circumstances of having no permanent job or income. This man got married to a younger girl who accepted to share with him this harsh life, surviving day by day. They used to go to the temple together, asking the god to help them and ensuring a better afterlife for them. He was happy that his young wife brought some children to his life. To feed them, he used a small piece of fertile land, by a Nile channel, to plant some crops in order to sell them and earn a little money that help him to survive with his small family another rotation of crops. He used to play with his young children after his hard work to make them happy. He was doing all that was possible to take care of his small family. He was by himself working this small piece of land. It was a difficult job for him, since his wife was occupied with taking care of their very young children, but he managed to achieve success and he carried all the harvest by himself to the boat. This boat had a protective cubbyhole that was designed to protect him and his family from the night wind and rain. In the evening, he erected the mast of the boat in order to use the wind for moving the boat forward to the market, and with every move of the boat the man thought that he was going forward towards his feast, the deserved reward of working hard. It was a windy and rainy night but he stayed all the night awake, dreaming about how to celebrate with his small family, after selling the crops and feeling the money in his hand. He said to himself: this dream will soon become true when the darkness of this windy, rainy night finishes and the boat reaches the market place. With the first light of the day, they all went out of the boat's cubbyhole, thinking that the boat had safely reached the market place during the night, but they were all shocked and scared to death when they saw that the boat was astray on the Nile, surrounded by hungry crocodiles and that the nearest riverbank was full of crocodiles as well. The man could hardly speak but he just said this short sentence "Annihilation is atop the lake". The man realized that they were trapped beside the dwelling place of crocodiles. He could not even move the boat or ask help from any one. He just sat down like a small child feeling helpless, looking at his scared young wife with her children. They were looking at him as well, asking him to do anything or may be to blame him that he should not make use of the mast during the night since he was staying with them in the cubbyhole of the boat. The man did not care about what might happen to him as much as he was sad about the destiny of his beloved young wife in the afterlife. He was wondering if she would be able to have a happy afterlife destiny like her peers who died on the earth. But his most heartbreaking concern was related to his young children who had just been born (lit. broke their eggs); and they had not seen much happiness in their short lives, but instead they would see the face of the crocodiles that would devour them without any mercy.

4. ANALYSIS

These persuasive stories of suffering persons are not only designed to be an entertaining element in the discourse but mainly as a script for performance, in which the reader or the listener become its speaking voice. The receiver adopts the main story voice, by standing behind the suffering situations that are expressed in emotional poetic words and speaking them as his own. These persuasive sad stories can be related to the correlation between sadness and aesthetic enjoyment or pleasure. "The emotion of sadness often plays a crucial role in aesthetic experience: from stage tragedies and painterly representations of sad motifs to 'tear-jerking' literature and filmic genres such as the melodrama. These various art forms not only represent sadness on the content level, but their audiences often feel the emotion as well." (Hanich, 2014, p. 130) This emotional union with the suffering character of the story forces the mental mechanism of the receiver to compare his actual suffering situation with those who suffered more than him. Consequently he feels psychologically satisfied that there are many people in much worse situations and thus considers himself fortunate because he had not gone through the same horrible situations. This way of designing the satisfactory argument can be really effective for broken people, i.e. who suffer from many hard situations and just need to find an encouraging-strong model to follow, in order to face and hence overcome their own hardship. It is a successful poetic argumentation practice, which has a long history in the ancient Near Eastern cultures, especially the ones created by a sacred voice. The Bible and Quran used many similar suffering stories, where the God shows how His chosen prophets suffered, in order to carry the holy message. The God thus uses them not only as a leading model for people who needs such encouraging power but also to feel more relaxed and satisfied that they did not have equal hardship. Although the receivers feel sad about those suffering figures, this sadness is always mixed with a mysterious enjoyment that forces the mental mechanism of the receiver to appreciate his current miserable status without complaint. The main poetic argument in these persuasive stories is: compare your claimed miserable situation with those figures in order to see how lucky you still are. The persuasive emotional effect of such suffering stories is equal to the modern sad genres:

> Sadness plays an important role in the enjoyment of melodramas and other sad genres precisely because it contributes its intensifying or "energizing" share to the inherently pleasurable feeling of being moved. (Hanich, 2014, p. 140)

The two rhetorical systems illustrated in these two texts reflect the conflict between the two systems, which exist in every nation, especially within the religious discourse: one uses logic to answer questions and the other uses emotional, metaphorical, poetic statement, (which are considered from this point of view "facts" as well) to support his answers. It can explain why ancient Egypt was described by the Greek historian Herodotus as conservative "to excess, far beyond any other race of men" when he visited Egypt in the fifth century BC (*History* II, ch. 37). Although, he acknowledged also another group of Egyptians who were using a visual-logical argument to convince the receivers to enjoy the earthly life as much as they can, because their bodies after death will be trapped in tight wooden coffins:

> In social meetings among the rich, when the banquet is ended, a servant carries round to the several guests a coffin, in which there is a wooden image of a corpse, carved and painted to resemble nature as nearly as possible, about a cubit or two cubits in length. As he shows it to each guest in turn, the servant says, "Gaze here, and drink and be merry; for when you die, such will you be". (*History* II, ch. 78)

This rhetorical paradox can be well observed in the case of searching for logical authority of religious metaphysical "statements" or even questioning God's existence. The religious-minded figures will always try to answer logically but at a certain point, they would declare their abhor of such logical approaches that ask them to develop logical answers for such religious-poetic statements. This dislike can be well illustrated in one of the sayings claimed to be of Imam al-Shāfiʿī (767-820 A.D./ 150-204 A.H.): "People did not become ignorant nor differed except after their abandonment of the Arabic language and their inclination to the language of Aristotle" (Al-Suyuti, n.d., p. 48) Critical Education is the key to a flourishing logical-rational system; the Education system which asks the student to memorize without any critical understanding of the studied content is the factor behind the flourishing of the literary emotional system.

5. CONCLUSION

It is misleading to assess any text belonging to the literary rhetorical system according to the rules of the philosophical-rational system, or vice versa. For example we cannot assess any religious text using the Greek-Roman labels, as the text will seem logically abortive. The Western studies will not be able to evaluate or genuinely appreciate the literary arguments and especially the ones offered by the sacred voices of ancient Near Eastern cultures by using the modern Western or the Greek-Roman rhetorical system, as these non-Western texts mainly belong to the literary rhetorical system, where there is no divorce between the creativity of employing all the available literary forms and producing an eloquent coherent content which matches their receiver's own taste and thus can easily penetrate his heart in order to deliver the required effect. The original literary form of the content, in the ancient Near Eastern cultures, is one effective tool in the persuasion process and without recognizing its effective role semantically, we will not be able to understand how once these texts have guided and influenced many ancient generations, leaving impressive impact on their religious-social-political cultures and understandings.

ACKNOWLEDGEMENTS: I wish to acknowledge my gratitude to Stephen Quirke-UCL, Elizabeth Thornton-UCLA, and Peter Philips-SOAS for their helpful comments.

REFERENCES

Al-Ansary, S. (2003). NP Structure Types in Spoken and Written Modern Standard Arabic (MSA) Corpora. In D. Parkinson & S. Farwaneh (Eds.), *Perspectives on Arabic Linguistics XV: Papers from the Fifteenth Annual Symposium on Arabic Linguistics, Salt Lake City 2001* (pp. 149–180). Amsterdam Studies in Theory and History of Linguistic Science. Series IV, Current Issues in Linguistic Theory. Amsterdam: Johm Benjamins.

Allen, J. (2011). *The Debate between a Man and His Soul: A Masterpiece of Ancient Egyptian Literature. Culture and History of the Ancient Near East, v. 44.* Leiden: Brill.

Al-Suyuti, J. A. (n.d.). *Sawn al-mantiq wa al-kalam an fann al-mantiq wa al-kalam, taHqiq: Ali al-nashar wa Souad abd al-Raziq.* Cairo: salsalah aHyaa al-turath al-islamy.

Fox, M. (1983). Ancient Egyptian rhetoric. *Rhetorica: A Journal of the History of Rhetoric, 1*(1), 9-22.

Fox, M. (1985). The Song of Songs *and the ancient Egyptian love songs.* Madison: University of Wisconsin Press.

Garrett, M. (1998). Review of *Comparative rhetoric: An historical and cross cultural introduction. Rhetorica: A Journal of the History of Rhetoric, 16*(4), 431-433.

Halldén, P. (2005). What is Arab Islamic rhetoric? Rethinking the history of Muslim oratory art and homiletics. *International Journal of Middle East Studies, 37*(1), 19-38.

Hanich, J., Wagner, V., Shah, M., Jacobsen, T., & Menninghaus, W. (2014). Why we like to watch sad films. The pleasure of being moved in aesthetic experiences. *Psychology of Aesthetics, Creativity, and the Arts, 8*(2), 130-143.

The History of Herodotus. (George Rawlinson, ed. and tr., 1885, vol. 2 New York: D. Appleton and Company.) Herodotus' Description of Egypt and the Egyptians: http://www.shsu.edu/~his_ncp/Heroegy.html (Checked on 25/09/2015)

Kennedy, G. (1998). *Comparative rhetoric: An historical and cross-cultural introduction.* New York: Oxford University Press.

Mao, LuMing. (2003). Reflective encounters: Illustrating comparative rhetoric. *Style, 37*(4), 401-425.

Murphy, J. (1981). *Rhetoric in the Middle Ages: A history of rhetorical theory from Saint Augustine to the Renaissance.* Berkeley: University of California Press.

Woerther, F., (Ed.). (2009). *Literary and philosophical rhetoric in the Greek, Roman, Syriac, and Arabic Worlds. Europaea Memoria. Reihe 1, Studien,* Bd. 66. Hildesheim, NY: Olms.

Young, C. (2004). Book Reviews. *Southern Communication Journal, 70*(1), 81-89.

54

Evaluation of Pro and Contra Argumentation

MAGNE REITAN
Norwegian University of Technology and Science, Norway
magne.reitan@ntnu.no

The aim of this paper is to present some basic principles for a theory of evaluation of pro and contra argumentation with respect to plausible argumentation. We take some fundamental notions of evaluation from Næss: *tenability* and *relevance*. Next we propose a reconciliation of these with Walton's rules for evaluation of linked and convergent arguments. Finally we propose how to give an overall evaluation of both sides of an argument with respect to a thesis.

KEYWORDS: argument evaluation, plausible argumentation, pro and contra argumentation, relevance, strongest line, tenability, weakest link

1. INTRODUCTION

We shall make the common distinction between three different areas of argument evaluation; classical formal logic, probative argumentation and plausible argumentation. In classical formal logic, arguments are analysed deductively as either valid or not valid. If an argument is deductively valid, then, if the premises are all true, the conclusion is also bound to be true. Secondly, one defines a notion of soundness as a valid argument with true premises. Since these notions are well understood in formal logic, one is tacitly led to think that formal logic gives the paradigm of how to evaluate arguments.

In the second area, probative argument evaluation, one exploits the resources of probability calculus. This might be done by using Bayes' theorem. But using probability theory presupposes that we can assign a probability to each proposition, and that quantitative assignments follow the probability theory.

The area of plausible argumentation turns out to be more problematic to define. Walton (2001, p. 166) defines a plausible

D. Mohammed & M. Lewiński (eds.) (2016). *Argumentation and Reasoned Action: Proceedings of the 1st European Conference on Argumentation, Lisbon, 2015. Vol. II, 865-877.* London: College Publications.

proposition as "something that seems to be true". This notion of plausible proposition is pretty vague, but it captures the non-clearcutness of plausibility. In our argumentation, we commonly use propositions that we neither know if they are true nor can estimate any probability of their being true. A plausible argument is an argument from a plausible proposition. Although plausible arguments have been with us from antiquity, they have been given little attention. Reasons for this may be that there has been a strong focus on formal logic, and that probability calculus has taken over in the study of weaker arguments.

In plausible argumentation the weight of the premises makes the conclusion likely without there being any ability to give quantitative probabilities, neither to the premises nor to the conclusion. Further, it may be that there are good arguments that make both the *pros* and the *contras* to appear likely to be true, as in the case of the well-known example of the weak and the strong man from the Antique (See for example Aristotle, *Rhetoric*, 1402a17-1402a28).

But it might turn out that as we get more evidence, one side becomes more strongly supported than the other. Nevertheless, it might be possible that we are left with both sides being equally supported.

When we try to develop principles for evaluating plausible reasoning, it is easy to be misled by the clear and precise understanding in the former two models of reasoning and assume that the notion of plausibility behaves in a similar way.

2. RESCHER AND WALTON ON PLAUSIBLE REASONING

Let us consider two different structures of arguments, linked and convergent structure (for simplicity arguments with two reasons), respectively Figure 1 and 2 below:

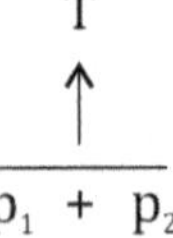

Figure 1 – Linked argument structure

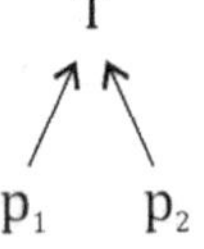

Figure 2 – Convergent argument structure

In a linked structure as in Figure 1, the premises are dependent on each other in the support for the thesis. In a convergent structure as in Figure 2, the premises are independent support for the thesis.

As an example of a convergent structure, suppose person **a** has been murdered. Let T be: "**b** committed the murder." Reasons to believe this are the following two: p_1: "there are forensic evidences that connect **b** to the murder scene", and p_2: "a passerby **c** says that he saw **b** leaving the murder scene shortly after the time of the murder." None of p_1 and p_2 depends on the other for its support of that **b** committed the murder.

2.1 Linked arguments

Let us start with the linked structure. In a valid argument, if the conjunction $p_1 \wedge p_2$ is true, T is bound to be true also, and if it is sound, T is simply true. In a probative argument the value of the thesis is given by the product $(Pr(p_1) \times Pr(p_2)) \times Pr(T|(p_1 \wedge p_2))$, given that p_1 and p_2 are independent. We see from this that the probability value of the premises taken together must be less than or equal to the weakest premise. And consequently the thesis must also be evaluated to less than or equal to the weakest premise. How much less, depends both on the value of the other premises and on the value of the conditional probability.

In his discussion of plausible reasoning Rescher (1976) assumed that we have entailment from the premises to the conclusion, and that the problem is how to evaluate the argument when the premises are to be considered as plausible. Rescher demonstrated that plausible reasoning does not follow probability calculus. For instance, the principle $Pr(\neg p)=1-Pr(p)$ of probability calculus does not hold for plausible reasoning, —there might be good reasons for both p and ¬p that make them both plausible. Thus the principles for evaluating plausible arguments cannot follow those of probative argumentation. Rescher formulated a system for plausible reasoning where "[t]he modality of the conclusion [...] must follow that of the weakest premiss." (Rescher, 1976, p. 24, the Theophrastos' Rule). He modifies this as "cannot be weaker than that of the weakest premiss (but could possibly be stronger)". Walton (1992) accepts this weakest link principle.

But what are the motivations for the weakest link? The Theophrastos' Rule refers to the modalities in an argument. The modalities belong to the language of arguments, not something that comes into account in the evaluation of arguments. The Theophrastos' Rule is a sort of general inference rule. Rescher's move to plausibility grads transforms the Theophrastos' rule to a principle of evaluation. He says that when making deduction from plausible premises, "the epistemic status of the conclusion must be taken as lying on the plane of the most vulnerable premise needed for its establishment" (p. 23). Thus he grounded the evaluation of plausible reasoning on epistemic considerations.

Walton's motivation for the weakest link lays in the pragmatics of argumentation. When a linked argument is put forward, the proponent cannot expect to convince to a stronger degree than the weakest point of the argument (Walton, 1992, p. 41). From the antagonist point of view it will normally be the best strategy to attack the weakest premise. If the weakest premise falls, the whole argument falls. The weakest link is simply the awareness of the weakest point of a linked argument —a linked argument will not be stronger than this.

But what is the case if we do not have entailment, but some weaker relation of support? Suppose that a thesis is supported by a set of premises. Such a relationship might be of the sort that the thesis is only likely to be true if all premises in the set are true. Or given that all the premises are strongly plausible, the thesis although having some plausibility will be less than strongly plausible (Walton, 1992, pp. 35, 39, 44, calls such relations "might-conditionals").

Walton (1992, p. 44) formulates a special principle for evaluating rules (might-conditionals), what he calls the Reduction rule:

> Rules of inference are to be assigned numerical plausibility values in arguments and counted in at the last stage of plausibility adjustment by being treated as a premise linked to the argument.

That is, in the evaluation process of an argument, we first evaluate the premises according to the weakest link; then we evaluate the result of this together with the plausibility value of the rule, again according to the weakest link. Thus the value of the rule comes in as a second step in the process of evaluation.

In the citation above Walton is clearly aware of the difference between rule of inference and premise, but in the evaluation he treats them as similar. In our view they have to be sharply distinguished, not only in structure, but also in the evaluation process. We think that the type of plausibility of a proposition and that of an inference rule are fundamentally different.

2.2 Convergent arguments

Let us now turn to convergent argument structures. Since the argument lines in a convergent structure are independent reasons for the thesis, one line of argument in such a structure does not influence the other argument lines. Neither can it make any other argument line stronger or weaker. And since one argument line cannot be strengthened by another argument line, this means that argument strengths do not add up: Many weak arguments for a conclusion cannot add up to become one single strong argument. In this case we have merged the independent argument lines to be a linked structure and the weakest link applies.

There is of course a psychological matter of the issue at hand. Comparing a case with five weak arguments against a case with just one argument which is also weak, one will tend to take the case with the five arguments to give stronger support for the thesis than the one with just one. This is not so —it gives more support when considering the numbers of arguments, but not stronger support. A huge number of arguments may have psychological effect, but does not amount to argumentative strength.

On the condition that there is entailment within each line of argument from the premises to the thesis, Walton (1992, p. 42) claims that in a convergent argument structure one should take the plausibility value of the thesis to be at least as strong as the most plausible premise. From the foregoing considerations this seems to be reasonable if we look at the premises alone.

But the relationship of support between a reason and the thesis can be weaker than entailment. In that case one should also to look at the value of the rule of the argument line. Consequently we should for each argument line consider the plausibility value of the premise(s) and the value of the rule of the argument in that line. The considered value of an argument line cannot be stronger than the value of the rule of the argument line. Neither can it be stronger than the value of the premise, —in case of several premises, nor stronger than the result of the weakest link applied to the premises.

This means that for each argument line we first apply the weakest link for that argument line, then we use the Reduction Rule. Consequently we end up with a considered value for each argument line.

The next step in the evaluation of a convergent structure is to compare the considered values of all argument lines. We take as the value of the thesis to be equally to the strongest considered value of the argument lines within the whole structure of argument lines. The reason for this is the same as above: since an argument line is independent of other argument lines, a weaker considered value of some other argument line in the structure cannot influence the strength of the best argument line.

3. TOULMIN: DATA AND WARRANT

Toulmin (1964) proposed that arguments for a thesis (claim) fall into one of two groups: First there is the data, reasons that are descriptions of some sort of facts. Second there are warrants, which make it explicit as to why the data supports the claim. The warrant can be seen as a kind of rule that legitimates the movement from the data to the thesis. This is usually put up schematically as follows:[1]

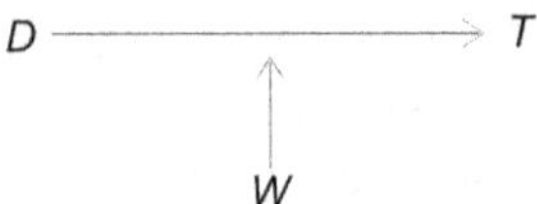

Figure 3 – Toulmin structure

Let us see how a concrete argument, an argument from likeness, fits into this scheme.

> p_1: The drug A curers sickness S in rats.
> p_2: There is likeness in the bodies of rats and human beings, and sickness S and sickness S' in human beings is similar.
> T: Therefore, drug A curers sickness S' in human beings.

This reasoning is an example of plausible reasoning. It is not logically valid, neither is it probable. To use Toulmin's terms, p_1 is to be considered as data, p_2 is the warrant, and T is the claim (the thesis). The two reasons p_1 and p_2 work together in the support for the thesis. Neither of them can stand alone in support of the thesis. Thus this type of structure is an example of a linked argument structure.

4. TENABILITY AND RELEVANCE

Suppose we have a simple case of one reason in support for a thesis. In his informal logic, Næss (1966) characterized two notions in the evaluation of how good an argument is; tenability and relevance. Tenability has to do with the acceptability of a reason, whether the reason is true, probable or plausible in itself, independent of its role as a reason for something.

---

[1] A similar distinction is made in the tradition of Næss (1966). Here data is called "argument of tenability" with respect to a thesis, and the warrant is called "argument of relevance", since it makes it explicit as to why the argument of tenability is relevant to the thesis.

The relevance has to do with the relation of support between a reason and the thesis, or more general, a relation of support from one proposition to another. This relation is independent of whether the first proposition is true or probable. The relevance can strengthen or weaken the second proposition, given that the first proposition is true, probable or plausible.

If a reason is acceptable at all, it has to have both tenability and relevance. If either is lacking, it will not be acceptable as a reason at all for the thesis. If it is not tenable, it will not support anything. If it is not relevant with respect to the thesis, it will not support the thesis, regardless of being tenable in itself.

In this paper we ignore the grounds for relevance and tenability; we treat them as general notions that apply to all types of arguments. We are here concerned with the overall impacts of these notions in the evaluation of arguments, not the grounds for them.

Rescher used numbers between 0 and 1.0 to represent the plausibility of a proposition. We think that this is too fine-meshed to suit the nature of plausibility. Since plausibility is tentative in its nature, it may for a certain proposition be too arbitrary to choose between e.g. 0.6 and 0.7 to represent its plausibility. For this reason we choose to differentiate the strength of plausibility in three classes, from the strongest to the weakest, with the names of the classes "s", "g" and "w", with the obvious interpretations, respectively, strong, good and weak.

The strengths of tenability and relevance combine equally since they are property and relation connected to one and the same proposition. Taken together they will make up the strength of an argument. Thus the result of this combination has impact on the plausibility of the thesis.

We define an evaluation function V with respect to a reason A for a thesis T as in Figure 4.

$$V(A \rightarrow T) = \begin{cases} \text{s if the tenability of A is s and the relevance of A} \\ \quad \text{with respect to T is s,} \\ \text{w if the tenability of A is w or the relevance of A} \\ \quad \text{with respect to T is w,} \\ \text{g otherwise.} \end{cases}$$

Figure 4 – The evaluation function V.

Given one single reason A in support of a thesis T, the function V gives the plausibility of T given the tenability and relevance of A. If case of a linked structure, let A represents the conjunction of the premises in the argument, and its plausibility value the result of the weakest link on it.

Let us now turn to the evaluation of a Toulmin-structure. First tenability. The tenability of the data (D) is simply its quality. In general this is a question of whether the description is true, probable or plausible.

The tenability of the warrant (W) is the soundness of the rule that it expresses. This rule goes from this type of data to such a thesis. Thus the tenability of the warrant is the truth, probability or plausibility of the rule in itself.

The relevance of the warrant is the application of the rule to the present case. The point can be seen more clearly when the warrant is a certain statistical method. The statistical method can be sound in itself, that is tenable, but it is another matter whether the method can be applied to the case at hand. We have to separate the question of soundness of the method and its applicability, and likewise to separate the tenability and the relevance of the warrant.

The combination of the tenability and the relevance of the warrant, gives the relevance of the data with respect to T. This is given by the function V on the tenability and relevance of the warrant. When we have worked out the relevance of the data, given both the tenability and relevance of the warrant, using the function V, we can use the function V a second time on the tenability and relevance of the data. The result of this will be the plausibility of the thesis, given the data.

In this way, to evaluate a Toulmin structure, we see that the plausibility of the thesis depends both on the tenability of the data and the warrant, and also on the relevance of the warrant. When all these factors are given, we can calculate the plausibility of the thesis with respect of these arguments. From the function V, we see that if both the data and the warrant have strong tenability, but the relevance of the warrant is weak, this only supports a weak plausibility of the thesis.

Fundamentally Walton (1992) has one single basic rule for the evaluation of linked arguments. His Reduction Rule is a kind of heuristic rule that makes it possible to also apply the weakest link to the inference rule of an argument.

Although the function V may look like Walton's rules, it is actually more sophisticated. His notion of plausibility can be considered similar to that of tenability alone. He uses a notion of a rule to establish the connection between the premises and the conclusion. He uses the terms "rule" and "conditional" equally, and thus he does not seem to make any principal distinction between being a premise, like the conditional of the modus ponens, and being an inference rule. He often uses the term "rule", but he never states clearly what a rule is, whether it is to be considered a special sort of premise, or that it has an analogical role to that of a rule of inference in formal logic, and that there is an analogy to the proof theory of formal logic. This unclear status of the rule of an argument confuses the matters. If it is a premise, then what is the rule (the warrant) of the argument?

The idea of a common set of rules in plausible reasoning is rather problematic. Plausible reasoning is fundamentally different from formal logic in that it does not exploit formalized rules, but normally relies on content, e.g. when considering arguments based on likeness: Not only does likeness vary from one argument to another, but the very notion of likeness also change from area to area, from arguments dealing with theoretical knowledge, to arguments about humour, about aesthetics, about ethics, and so on. One can say: "you will like this film, because you liked the other one; it is your type of humour". Or in casuistic ethics where one compares an ethical problematic case to an ethical well-understood case, based on likeness in ethical relevant features. It is problematic to claim that the notion of likeness is of one and the same sort in all such different areas. The legitimation of rules in plausible reasoning is based on the understanding of nature, society and human beings. This gives a huge variation in possible rules in this type of reasoning.

In plausible reasoning rule-like connections between premises and the thesis should be spelled out as a warrant. The point of the warrant is to make explicit the connection between the data and the thesis. The warrant would then both have tenability and relevance for the case at hand. Making a distinction between the notions of tenability and relevance of the warrant, makes a distinction between the warrant being plausible in itself, and being relevant for the concrete argument from the data to the thesis. The distinction comes clear in cases where the warrant is a general statement, and the data is of particulate nature. Thus, to equate the plausibility simply with the tenability of the rule, and ignore its relevance, might give us too high estimate to the overall plausibility of the argument. The rule might have high plausibility in itself, but not having high relevance for the concrete argument. Therefore something crucial is missing in Walton's analysis.

The function V breaks with weakest link in that the overall plausibility never can become stronger than the weakest link, but that it may become weaker than the weakest premise, depending on the relevance of the warrant.

In section 2.2 above we have discussed the basic structure of convergent arguments and how Walton evaluate such arguments. This has now to be modified in the light of tenability and relevance: Each convergent argument will first be evaluated with respect to both its tenability and its relevance. The outcome of this evaluation gives a certain plausibility value to each argument. Then we apply the rule of the strongest argument line, which gives the plausibility value for the thesis.

We formulate the following rules for evaluating convergent arguments:

C1: Evaluate each argument with respect to its tenability and relevance.
C2: The overall strength of the arguments is independent of each other and do not interact with each other.
C3: The weights do not add up.
C4: Always choose the strongest value given by the independent argument lines.

5. PRO- AND CONTRA-ARGUMENTS: T VS ¬T

A *contra*-argument c with respect to a thesis T, is simply a direct argument for ¬T. But if c is acceptable as an argument for ¬T, it has to have a certain tenability in itself and a certain relevance with respect to ¬T. If the function $V(c \rightarrow \neg T)$ gives a value within [w, .., s], ¬T comes out with this value as its plausibility given c.

Even though we find that ¬T has a certain plausibility, which does not mean that we have to reject T. As we have seen above, two propositions where one is a negation of the other, might both be considered as plausible. In order to compare the plausibility of a *pro*-argument p and a *contra*-argument c, both with respect to a thesis T, we will give the *pro*-argument a positive plausibility with respect to T, and the *contra*-argument a negative plausibility with respect to T. But the strength of the plausibility of c comes from considering it as an argument with respect to ¬T. We turn this to a negative plausibility with respect to T.

Let us now consider a case with just one *pro*-argument p for a thesis T, and one *contra*-argument c with respect to T. In the evaluation we start with the *pro*-argument. We first have to consider the warrant for p to T. We take into account both the strength of its tenability and its relevance, and end up with the relevance of p to T. Based on the tenability and the relevance of p, we can work out the plausibility for T given p. Next we have to do the same for the argument c. We have to consider its warrant with respect to ¬T. After working out the results of the warrant's tenability and relevance, we can work out the plausibility from c with respect to ¬T.

The final task is to give an evaluation of the overall plausibility of T, given the results from the arguments p and c. This is given by Figure 5:

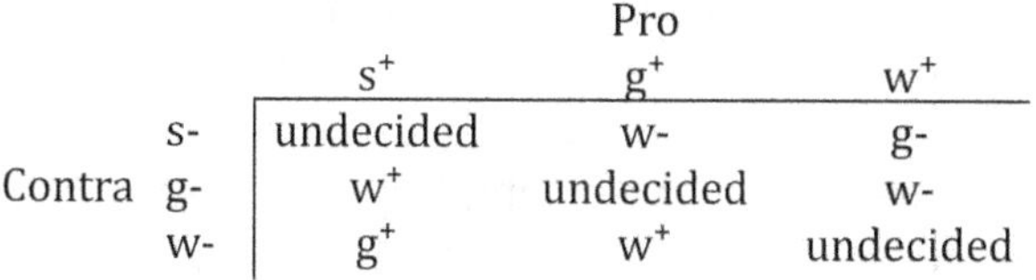

		Pro		
		s^+	g^+	w^+
	s-	undecided	w-	g-
Contra	g-	w^+	undecided	w-
	w-	g^+	w^+	undecided

Figure 5 – Pro and contra evaluation.

If p and c are evaluated as unequal in their support for T and ¬T, respectively, we reduce the strongest one with the strength of the other. E.g. if p is strong and c is weak, the overall plausibility for T will be g. Thus we reject ¬T on the basis of p and c. If c is strong and p is weak, the overall plausibility for ¬T will be g, and we reject T on the basis of p and c. ¬T will be the strongest one.

If p and c are evaluated as equal in their plausibilities with respect to T or ¬T, we simply have to accept that there are equal plausibilities of the opposites. This means that as reasons are considered, it is not possible to decide which side is the strongest one. So for argumentative reasons, we should leave the case for T or ¬T undecided. There might of course be other considerations that make us prefer one rather than the other.

In case we have several linked premises that function as one *contra*-argument, we can treat this case as similar to several premises that function as one *pro*-argument. We use the weakest link and then combine with the evaluation from the warrant.

6. SEVERAL CONVERGENT *PRO*- AND *CONTRA*-ARGUMENTS

Suppose we have a situation with several convergent *pro*-arguments and also several convergent *contra*-arguments with respect to the same proposition p. We will represent this as the structure in Figure 6. Here we treat the *pro*-arguments in the normal way. But we will treat the *contra*-arguments as *pro*-arguments with respect to ¬p.

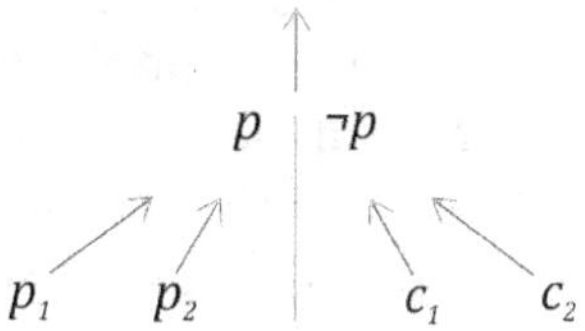

Figure 6 – Pro- and contra-structure

The process of evaluation such a structure falls into two steps: First, we evaluate each of the two convergent sub-structures: the one with respect to p, and the one with respect to ¬p. In each of these we use the strongest link. The value toward p gives us a positive strength, and the value toward ¬p gives us a negative strength. The next step is to give an overall evaluation of these two results to find the strongest side, p or ¬p, according to Figure 5. The strongest side is the winner with that considered strength. And that value is passed further up in the structure as the tenability of p (or ¬p). In case of ¬p we are dealing with a *contra-argument.*

If the two sides of p and ¬p are equally strong, positive and negative, the argumentation is undecided. This means that we neither can give any tenability to p nor to ¬p, and this node in the argument structure has to be left out in the evaluation.

7. BOTH WARRANT AND UNDERCUTTING DEFEATER

While the warrant is an argument which legitimate the step from the data to the thesis, an undercutting defeater is an argument directed against the support relation from the data to the thesis (see Pollock, 1987).

To evaluate a case with both warrant and undercutting defeater we simply apply the normal procedure: First, to work out the evaluation of the warrant and the undercutting defeater, both with respect to their tenability and relevance, using the function V. Then we compare the strength of the warrant and the undercutting defeater according to Figure 5, where the warrant is positive and the undercutting defeater is negative. If the overall result is positive, we have a case for the thesis. If the overall result is negative, the data fail to support the thesis. If equal in strength, this simply means that we cannot tell anything about whether the data support the thesis or not.

If the undercutting defeater wins over the warrant, this does not mean that the data is an argument for ¬T. Of course, it may be part of a strategy for arguing for ¬T, in the way that it weakens T.

Suppose that we not only have one warrant, but several, and not only have one undercutting defeater, but several. The procedure is to evaluate all the warrants, find the overall strength, and so evaluate all the undercutting defeaters and find the overall strength of them. Then we evaluate the warrant side and the undercutting defeater side according to Figure 5.

8. CONCLUDING REMARKS

In this paper we have merged insights from Walton and Næss into a unified theory of evaluation of plausible reasoning. We have assumed the common distinction between linked and convergent argument structures together with Toulmin's distinction between data and warrant as supports for a thesis. In the evaluation we have adjusted Walton's principles for linked and convergent arguments with Næss' notions of tenability and relevance. Thus we have the resources to evaluate both sides in a pro and contra argumentation. Finally we have proposed how to weight the pro- and contra sides against each other.

REFERENCES

Næss, A. (1966). *Communication and argument. Elements of applied semantics.* London: Allen and Unwin.

Pollock, J.L. (1987). Defeasible reasoning. *Cognitive Science, 11*, 481-518.

Rescher, N. (1976). *Plausible reasoning.* Allsen: Van Gorcum.

Toulmin, S. (1964). *The uses of argument.* Cambridge: Cambridge University Press.

Walton, D. (1992). Rules for plausible reasoning. *Informal Logic, 14*, 33-51.

Walton, D. (2001). Abductive, presumptive and plausible arguments. *Informal Logic, 21*, 141-169.

55

An Exploration of the Relatedness Problem between Arguments: Combining the Generative Lexicon with Lexical Inference

PATRICK SAINT-DIZIER
IRIT-CNRS Toulouse, France
stdizier@irit.fr

Given a controversial issue, argument mining from natural language texts is extremely challenging: domain knowledge is often required together with appropriate forms of inferences. This contribution explores the use of the Generative Lexicon viewed as both a lexicon and a domain knowledge representation.

KEYWORDS: argument mining, generative lexicon, knowledge representation, natural language processing

1. AIMS AND CHALLENGES OF ARGUMENT MINING

One of the main goals of argumentation is, given a controversial issue, to identify in various texts the arguments for or against that issue. These related arguments act as supports or attacks, depending on their orientation. Arguments may also attack or support the other arguments which support or attack that controversial issue in order to cancel out their impact. An argumentation is represented by a graph where supports and attacks are linked.

Arguments are difficult to identify and to characterize from a linguistic point of view, in particular when they are not adjacent to the controversial issue, possibly not in the same text. They are often standard natural language statements which get the status of arguments because of the specific relations they have with a controversial issue. For example, the statement from the Nepali Times: *we now see long lines of your girls with school bags along the roads* could be just factual, but if the controversial issue at stake is: *the situation of women has improved in Nepal*, then, the first statement becomes an argument supporting this issue, which is then interpreted as a controversial issue which can then

D. Mohammed & M. Lewiński (eds.) (2016). *Argumentation and Reasoned Action: Proceedings of the 1st European Conference on Argumentation, Lisbon, 2015. Vol. II, 879-893.* London: College Publications.

be supported or attacked. Except in specific contexts, and for certain forms of arguments (e.g. warnings, threats, advice, requirements), most arguments do not have any specific linguistic mark that would allow their identification. Furthermore, it is difficult to identify whether a statement is a support or an attack of a controversial issue, and what it precisely attacks or supports. In the above example, "school bags" means education: it is a means to improve women's condition because it leads to jobs and more independence. It does not talk about anything else about women's conditions. Relating a controversial issue with arguments requires knowledge, lexical semantics data, and appropriate inferential patterns. It is therefore much more complex in terms of semantics and reasoning than e.g. standard opinion analysis based on evaluative expressions analysis.

Argument mining is an emerging research area with new challenges that requires the combination of linguistic analysis and language processing with artificial intelligence. Argument mining is at the moment applied mainly to written texts, e.g. (Mochales Palau, & Moens, 2009), (Kirschner, Eckle-Kohler, & Gurevych, 2015), opinion analysis, e.g. (Villalba & Saint-Dizier, 2012) dialogue analysis, e.g. (Budzynska, Janier, Reed, Saint-Dizier, Stede, & Yakorska, 2014), (Swanson, Ecker, & Walker, 2015). Annotated corpora are being made available, such as the AIFDB dialogue corpora at Dundee University and (Walker, Anand, Fox Tree & Abbott, & King, 2012). These corpora are very useful to identify argumentative discourse units (ADUs), linguistic cues, as in e.g. (Nguyen & Litman, 2015), and argumentation strategies, in a more concrete way than abstract argumentation schemes, as shown in e.g. (Feng & Hirst, 2011).

The goal of argument mining is in our work, given a controversial issue, to identify in a set of texts, statements which can be interpreted as supports or attacks. In opinion analysis, the benefits are not only to identify the customer satisfaction level, but also to characterize why customers are happy or unhappy. Abstracting over arguments allows to construct summaries and to define customer value systems (e.g. low fares are preferred to localization or quality of welcome for some categories of hotel customers).

Argument mining from full natural language texts is extremely challenging: given a controversial issue, the identification of the relations between arguments and that issue is often more complex than just the bipolar support or attack view. For example, given the issue *Vaccination against Ebola is necessary*, and the argument: *There are almost no cases of Ebola in Europe*: is this argument a support, an attack or something else, such as e.g. a concession? Identifying the "conceptual link" between a controversial issue and an argument is of much importance to have a clear analysis of the role of an argument.

The main questions addressed in this paper are:

- The identification of relatedness factors between a controversial issue and a statement, potential argument: what types of inferences and knowledge are necessary, how much are they related (for example the importance of the argument, its persuasion effect), and which aspects or facets of the issue are concerned ?
- The investyigation of the exact nature of the relations between an issue and an argument. Besides supports and attacks, other types of relations such as various types of causality, specializations, concessions and supports must be taken into account.
- The identification of the relations that hold between arguments: neutral, attack, support, concede, etc.
- The identification of the linguistic cues that may be relevant to identify arguments in texts, when they exist, via a linguistic, statistical or hybrid approach,
- The identification of the types of knowledge and inferences which may be needed to identify arguments. This raises the problem of re-usability or scalability of an argument mining system.

In this paper, via two case studies, we show that the Generative Lexicon (GL) (Pustejovsky, 1995) paired with a few generic inferential patterns is a useful approach to argument mining. The GL provides both language resources and the knowledge that is required in a large number of cases to deal with the above questions. This contribution is essentially exploratory at this stage.

In the next section, we briefly present the GL. The next two sections develop the two use cases and the inferences that are involved. We conclude this article by an analysis of the scalability of this approach in order to develop large-scale argument mining systems.

2. AN INTRODUCTION TO THE GENERATIVE LEXICON

The Generative Lexicon (GL) (Pustejovsky, 1995) is an attempt to structure lexical semantics knowledge from several perspectives. It allows to explain a number of language phenomena such as various types of metonymies via a *decompositional* view of lexical meaning. Various forms of so-called "generative aspects of lexical combinations" have been characterized via the operation of type shifting, where the original type that is expected in a proposition has been coerced to another type, allowing metaphors such as "to devour books".

The GL develops some original forms of semantic typing, such as dotted types, that allow to account for the different facets of an entity (e.g. the physical and contents facets of a book), the development of a specific argument structure, an event structure and the Qualia structure, which is the structure that is considered in our investigations. In our approach to argument mining, the Qualia structure is used as a lexical knowledge repository for a given domain.

Very briefly, the Qualia structure of an entity is composed of four fields called roles:

> - the *constitutive* role describes the various parts of the entity and its physical properties, it may be decomposed into subroles such as shape and dimensions, color, etc.
> - the *formal* role describes what distinguishes the entity from other objects, i.e. the entity in its environment, in particular the entities which are more generic,
> - the *telic* role describes the entity functions, uses, roles and purposes. It is the most crucial role in our approach.
> - the *agentive* role describes the origin of the entity, how it was created or produced.

The elements in each of these roles are constants, predicates or formula. The GL has remained a relatively theoretical view of language. Relatively few resources have been produced to validate the approach, except for the EEC SIMPLE project, carried out about two decades ago.

However, as will be seen below, the GL is a very relevant means to structure lexical knowledge to characterize some forms of language production. The predicates used in the different roles are a priori defined from a domain ontology or from a general purpose ontology. From the examples are given in the two next sections, it is clear that analysis and description methods are necessary to further develop Qualia structures of the GL on a wide range.

3. CASE STUDY A: *SHOULD I GET VACCINATED AGAINST EBOLA*?

3.1 Analysis of the controversial issue

Let us consider the controversial issue:

> *The vaccine against Ebola is necessary.*

This statement means (1) that the vaccine must be developed and (2) that populations must be vaccinated.

The Qualia structure of the head term of this statement, vaccine, is represented as follows:

Vaccine(X):

$$\begin{bmatrix} \text{CONSTITUTIVE: } \begin{bmatrix}\text{ACTIVE\_PRINCIPLE, ADJUVANT}\end{bmatrix}, \\ \text{TELIC: } \begin{bmatrix}\text{INJECT(Z,X,Y), PROTECT\_FROM(X,Y,D), AVOID(X,DISSEMINATION(D))}\end{bmatrix}, \\ \text{FORMAL: } \begin{bmatrix}\text{MEDICINE}\end{bmatrix}, \\ \text{AGENTIVE : } \begin{bmatrix}\text{DEVELOP(T,X), TEST(T,X), SELL(T,X)}\end{bmatrix} \end{bmatrix}$$

Where X is the variable that represents the vaccine, Y is the person that is vaccinated, T is the biologist or company that develops the vaccine, Z is the doctor that makes the injection, and D is the disease associated with the vaccine, D = disease of(X) = Ebola in our case.

The agentive role develops the way the vaccine is created while the telic role develops its functions and roles. This GL representation can be further organized, in particular to develop causal and temporal chains. For example, the predicates in the agentive and telic roles could be structured as follows, using event-denoting variables *Ei*:

$$develop(E1,T,X), test(E2,T,X), sell(E3,T,X) \wedge E1 \leqslant E2 \leqslant E3.$$

Similarly, a causal chain can be developed in the telic role:

$$inject(E1,Z,X,Y) \Rightarrow protect\_from(E2,X,Y,D) \Rightarrow avoid(X,dissemination(E3,D)).$$

with a kind of partial overlap between E2 and E3.

Next, the Qualia structure of Ebola (and more generally, of a virus) can be defined as follows:

EBOLA:

$$\begin{bmatrix} \text{FORMAL: } \begin{bmatrix}\text{VIRUS, DISEASE}\end{bmatrix}, \\ \text{TELIC: } \begin{bmatrix}\text{INFECT(E1,EBOLA, P)} \Rightarrow \text{GET\_SICK(E2,P)} \Rightarrow \\ \Diamond\ \text{DIE(E3,P)} \wedge \text{E1} \leqslant \text{E2} \leqslant \text{E3.}\end{bmatrix} \end{bmatrix}$$

Where P represents here the patient that gets the disease. The purpose of Ebola is to infect people (P) who get sick and may die. There is no agentive role since there is no volition in the Ebola virus.

The meaning of the controversial issue (*people must be vaccinated*) has the following semantic representation, based on the domain ontology predicates:

$\Box(\forall Y, patient(Y), \exists Z, X, doctor(Z), vaccine\_against(X,Ebola) \wedge inject(Z,X,Y)).$

where 'vaccine against Ebola' in the controversial issue is analysed as a metonymy for 'inject vaccine against Ebola'. The metonymy is reconstructed from the predicates in the telic role of 'vaccine'. This controversial issue has a positive orientation, since vaccination is a positive action, inferred from the lexical properties of the predicate 'protect' in the telic which is reinforced by the modal 'necessary'.

3.2 Argument A1: ***The Ebola adjuvant is toxic for humans***

The constitutive role of *vaccine(X)* tells us that the adjuvant is part of the vaccine. The role of the adjuvant is specified in its GL entry:

Adjuvant(Y, X1):

$$\begin{bmatrix} \text{FORMAL}: \begin{bmatrix}\text{VACCINE, MEDICINE, CHEMICALS}\end{bmatrix}, \\ \text{TELIC}: \begin{bmatrix}\text{DILUTE}(Y,X1) \ldots\end{bmatrix} \end{bmatrix}$$

where Y is the adjuvant of X1, which is the active principle of the vaccine X. Let us now investigate whether this argument attacks or supports the controversial issue and how:

> (1) The informal lexical semantics of dilute(Y,X1) is that Y and X1 are mixed together and form a single entity: the vaccine X.
> (2) *upwards inheritance of a property in a part-of relation*: this inference rule says that:
> if a constitutive part K1 of an object K has a property P (i.e. toxic), then (probably) the entire object K has P (is toxic):
> $has\_property(K1,P) \wedge part\_of(K1,K) \Rightarrow has\_property(K,P).$
> (3) since Y and X1 are parts of X, given that Y is toxic, therefore X is toxic, then the injection of X is toxic for humans, which has a negative orientation, due to the lexical properties of *toxic*. More formally:
> $\forall X, P, human(X), product(P) \wedge has\_property(P,toxic) \Rightarrow \neg (inject(X,P))$

The global statement A1: *The Ebola adjuvant is toxic for humans* **attacks the controversial issue**. This statement may also be interpreted as a **contrast to the controversial issue**: 'the vaccine is necessary BUT it is toxic'. According to the semantics of a contrast (e.g. (Winterstein, 2012), the second part of the contrast, Q, in: 'P but Q', wins without contradicting or fully attacking the controversial issue P.

3.3 *Argument A2:* ***Seven persons died during the Ebola vaccine tests***

This statement reports die events that occurred during tests. 'Die' has obviously a negative polarity. If we consider again the GL structure of vaccine(X), the 'test' activity is related to the agentive role. An axiomatization of the GL structure says:

> *by definition, the agentive role is pre-telic: the events it describes occur before the functions or events given in the telic role. There are a priori no causal relations between these two sets of events.*

In more formal terms, where P and Q are predicates, and E event variables:

$$\forall P(E) \in agentive\text{-}role,\ \forall Q(E1) \in telic\text{-}role,\ E \leqslant E1 \land \neg(P \Rightarrow Q).$$

From that point of view, the statement A2 about tests does not say anything about the vaccine roles, functions and consequences once it has been fully tested and approved.

From an argumentation point of view, strictly speaking, Argument A2 is **irrelevant or neutral w.r.t. the controversial issue** since tests fully precede injection. However, due to its very negative orientation it may be possible to say that it very weakly attacks (concession) the first interpretation of the issue: the vaccine must be developed, BUT tests are dangerous.

3.4 *Argument A3:* ***Vaccine is important to avoid Ebola epidemic***

The main concept in this statement that motivates the vaccine is 'epidemic'. Its GL Qualia structure can be roughly defined as follows, where X is the disease (Ebola in our example):

Epidemic(X):

$$\begin{bmatrix} \text{AGENTIVE: } \left[\text{LOW\_PREVENTION\_AGAINST}(X)\right], \\ \text{TELIC: } \left[\textit{several}\ Z,\ \text{HUMAN}(Z) \land \text{INFECT}(X, Z)\right] \end{bmatrix}$$

To have an accurate representation, an informal quantification, 'several', has been added to the usual GL representation. The predicate in the agentive role is a kind of 'prerequisite' or preliminary step before the infection occurs. Let us now analyse argument A3 from the Epidemic and Ebola GL representations:

(1) From epidemic: *several Z, human(Z)* $\wedge$ *infect(Ebola, Z)*
(2) from Ebola, *infect(Ebola, Z)* can be expanded using the causal chain given in the Ebola telic role, binding the variables. The telic role defines the purpose of epidemic(X):

several Z, human(Z) $\wedge$ *infect(E1,Ebola, Z)* $\Rightarrow$ *several Z, human(Z)* $\wedge$ *get_sick(E2,Z)* $\Rightarrow$ *several Z, human(Z)* $\wedge$ $\Diamond$ *die(E3,Z)* $\wedge$ *E1* $\leqslant$ *E2* $\leqslant$ *E3.*

(3) from the statement 'Ebola epidemic', *human(Z)* $\wedge$ $\Diamond$ *die(E3,Z)* is inferred. The conclusion is negative, it is a lexical property of the verb 'die'.
(4) *avoid(Event)* reverses the orientation of the statement: *avoid(die(E3,Z))* is positively oriented,
(5) finally, the intensifier adverb 'important', applied to 'avoid' reinforces the strength of the orientation.

To conclude, from the inference sequence and the polarity of its different terms, it turns out that **Argument A3 is a strong support** to the event of vaccination, and therefore to the controversial issue.

3.5 Argument A4: ***No one is infected by Ebola in Europe***

The representation of this statement is the following:

$$\forall Z, human(Z), \wedge in(Z,Europe) \Rightarrow \neg\ infect(E,Ebola,Z).$$

(1) This representation is in contradiction with the telic role of epidemic(Ebola). It therefore does not meet the purpose or goal of epidemic(Ebola): the semantic representation of argument A4 is not compatible with the telic role of epidemic(X).
(2) Therefore, since there is no one infected by Ebola in Europe, it follows that there is no risk of dissemination at the moment.
(3) Therefore, Argument A4 is an attack of the controversial issue, but with several restrictions: it is valid only in Europe and it relates a fact that occurs at the present time, which may be different in the future. Note that argument A4 does not contradict the other elements of the telic role of vaccine(X).
(4) Therefore, argument A4 only partially attacks the controversial issue, it is analysed as a **contrast:** *vaccine is indeed necessary as a general principle, BUT since there are no cases in Europe at the moment it may not be necessary in Europe.*

3.6 Argument A5: ***The vaccine is not always efficient: 3 vaccinated people died in Monrovia***

The higher-order adjective 'efficient' applies to the following predicates in the telic role of vaccine:

protect_from(X,Y,Ebola), avoid(X, dissemination(X,Ebola)).

Then:

(1) Argument A5 means that among the patients Y that got vaccinated, a few of them got sick and died:

$\exists$ *Y, patient(Y), (inject(E1,X,Ebola_vaccine,Y)* $\wedge$ *get_sick(E2,Y)* $\wedge$ *die(E3,Y)).*

(2) Then, the following proposition is infered:

$\neg$ *(protect_from(X,Y,Ebola).*

which contradicts the purpose of vaccine(X) (telic role).

(3) Since the protection of the vaccine is not always guaranteed, Argument A5 weakly attacks the controversial issue, which says that vaccines protects everyone that is vaccinated. However, since the number of non-protection cases is very limited, this statement is preferably **interpreted as a concession**:

Vaccine protects the population HOWEVER there are a few cases where it does not work.

The concession basically supports the controversial issue, but adds some restrictions.

3.7 Argument A6: ***The vaccine is too expensive for some countries***

This last argument refers to a property of a vaccine: its cost. This property is defined in its constitutive role: cost is a property of artefacts. A6 is informally represented as follows, assuming the predicates has_property and has_value:

vaccine(X) $\wedge$ *has_property(X, cost)* $\wedge$ *has_value(cost, high).*

Let us now investigate the orientation of A6 w.r.t. the controversial issue:

(1) A general purpose rule says that 'an object X that is expensive is not accessible to everyone':

has_property(X, cost) $\wedge$ *has_value(cost, high)* $\Rightarrow \neg$ ($\forall$ *Y, human(Y)* $\wedge$ *accessible(X,Y))*.

(2) the controversial issue says that the vaccine is necessary. This is represented in 3.1. The predicate *inject(Z,X,Y)* combined with the modal 'necessary' presupposes or suggests that the vaccine is or must accessible to everyone:

$\forall$ *X, vaccine(X)* $\Box$ *(inject(Z,X,Y))* $\Rightarrow$ $\forall$ *Y, human(Y)* $\wedge$ *accessible(X,Y)*.

(3) the representation in (1) is in contradiction with the presupposition in (2), therefore, argument A6 attacks the controversial issue. However, A6 does not fully attack the controversial issue, but it says that the necessity that is evoked is difficult to realize: it can be analysed as a contrast.

3.8 Summing up

The above cases have revealed various types of relations between the argument and the controversial issue:

- attack: arguments A1, A6
- support: argument A3,
- neutral or irrelevant: argument A2 (even if the topic is addressed)
- contrast: arguments A1, A4, A6
- concession: argument A5

The analyses that are provided show that the boundaries between these six cases are not very clear-cut.

The other main interest of this analysis is that, given a set of independent statements, coming from different origins, it is possible to precisely relate them (or not) to the main issue and to identify, for each of them, how they are related, their argumentative orientation and strength. This is realized by means of three main knowledge and reasoning sources, which are, informally:

- domain knowledge: encded via the formalism of the Generative Lexicon, via the Qualia structure, including event structures and causal chains that make the set of predicates in the roles more structured.
- lexical semantics: some basic semantic features associated with lexical items are necessary, in particular polarity, e.g. for verbs (avoid), intensifiers (for adverbs),
- reasoning: several types of inferences have been informally identified:

1. inferences related to the semantics of the qualia roles in the GL structure, e.g. agentive events occur before telic events and do not form a priori causal chains,

2. inferences related to lexical semantics structures, e.g. the upward inheritance of properties for the part of relation, which is not as systematic than the downward one for the isa relation,
3. inferences related to general purpose domain knowledge and to presuppositions,
4. inferences dedicated to argumentation, that allow to compute relations and their strength between the controversial issue and the argument at stake.

The cases presented above are essentially descriptive. They obviously need to be further elaborated. Finally, another challenge is to identify those strategy principles that indicate which rule or data must be triggered to analyse the relation between the potential argument at stake and the controversial issue.

4 CASE STUDY B: *HAS THE POLITICAL SITUATION IMPROVED*?

4.1 The controversial issue: ***The political situation has improved.***

Let us now consider another type of situation that shows (1) the use of higher-order representations and (2) how the GL is used when relatively complex and abstract concepts are considered instead of concrete ones, as in Case Study A. The GL is clearly less efficient to represent abstract than concrete concepts, but this is a general trend for most knowledge representation systems. Nevertheless, several forms of inferences can be developed on a relatively superficial level of description.

The controversial issue is the following (from the Nepali Times): *The political situation has recently improved.*

The term 'situation' that heads the subject NP is a kind of support term, the main conceptual term of the NP is the adjective 'political'. 'Political situation' refers to the features of the generic and abstract term 'politics', which, in this statement, are said to have improved recently. The verb 'improve' is higher-order: it roughly means that the 'level of realization or satisfaction' of the properties of the telic of 'politics' have increased. This is obviously extremely difficult to represent in general. At a coarse-grain level, political aspects or features can be summarized by the following Qualia structure, using a simplified logical form:

Politics:

$$\begin{bmatrix} \text{CONSTITUTIVE: } \begin{bmatrix} \text{POLITICAL\_PARTIES(P), EXECUTIVE\_FORCE(X),} \\ \text{LEGISLATIVE\_FORCE(Y), ...} \end{bmatrix}, \\ \text{TELIC: } \begin{bmatrix} \text{GUARANTEE(X, SAFETY\_OF(Z)),} \\ \text{MANAGE(X,ENFORCEMENT\_OF(LAW)),} \\ \text{MANAGE(X,DEVELOPMENT\_OF(ECONOMY)), ...} \end{bmatrix}, \\ \text{AGENTIVE: } \begin{bmatrix} \text{ELECTION\_OF(Y), VOTE(Z,Y), CONFRONTATION\_BETWEEN(P)} \end{bmatrix} \end{bmatrix}.$$

In this Qualia, the agentive role describes, roughly, how a political party comes to power, the constitutive role describes the structure of power in a government while the telic role enumerates the different responsibilities a political party in power has, in other words, its role or purpose. The variable Z represents the citizens.

4.2 Argument B1: ***less people get murdered every day***

The main predicate in this statement is the verb 'murder'. Let us see how verbs are represented by a Qualia:

murder(X,Y):

$$\begin{bmatrix} \text{AGENTIVE: } \begin{bmatrix} \text{LACK\_OF(SAFETY\_OF(Y)), ...} \end{bmatrix}, \\ \text{TELIC: } \begin{bmatrix} \text{GET\_RID\_OF(X,Y), USE(X,WEAPON), DIE(Y), ...} \end{bmatrix} \end{bmatrix}.$$

In this Qualia, the agentive role does not describe the agent of the murder but the reasons of the murder. This role may also contain e.g. presuppositions, facilitators, sources, etc. One of the reasons of 'murder' is a 'lack of safety', whatever it may cover. The Telic role describes the way the action is carried out, its consequences, and any form of manner or tool. Argument B1 is processed as follows:

(1) Let us first analyze 'people get murdered'. The causes of the murders is represented by the predicate: *lack_of(safety_of(Y))*.
(2) This predicate is in contradiction with the telic of 'politics' via the predicate:
guarantee(X,safety_of(Y)). This can be shown as follows:

If *safety_of(Y)* is a property then *lack_of(safety_of(Y)* entails: *for_most E, ¬ (safety_of(E,Y))*.
On the other hand, *guarantee(Property)* means, roughly, 'Property satisfied to a high level: *for_most E, P(E) = for_most E, safety_of(E,Y)*, or, equivalently, very few instances of *¬ (safety_of(E, Y))*.

(3) So far, the statement is in opposition with the controversial issue because of the conflict between the predicates in the telic role,

(4) However, argument B2 says that 'less people get murdered', the quantifier 'less' applied to the proposition 'people get murdered' reverses the polarity of the expression since the number of situations of lack of safety decreases. B1 is then a **support of the controversial issue**.

4.3 Argument B2: ***shops remain closed****

The telic role of 'politics' contains the predicate:

manage(X, development_of(economy)).
Roughly, the telic of 'shop' is, for its 'activity' facet:

Shop(T):

$$\left[\text{TELIC: } \left[\text{SELL}(T,P,Y),\ \text{BUY}(T,P,Y),\ \text{MAKE\_PROFIT}(T), \ldots\right]\right]$$

The different predicates in the Telic of 'shop' are instances or a contribution to *development_of(economy)*. The verb 'closed', in its aspectual dimension, means that these economic activities do not occur. Therefore, B2 contradicts *development_of(economy)*. **Then B2 attacks the controversial issue** since there is a lack of management of economy.

The argument **B3**: *corruption and favouritism have not decreased* is analysed in a similar way, it concerns the predicate *manage(enforcement_of(law))* since corruption and favouritism are related to legal aspects, via their agentive role (what brings them about, i.e. lack of law enforcement). **B3 also attacks the controversial issue**.

4.4 Argument B4: ***the police is less present in the streets***

This argument is more subtle. Its interpretation is ambiguous. The Qualia structure of Police is, e.g.:

Police(Z):

$$\begin{bmatrix} \text{TELIC: } \begin{bmatrix} \text{CONTROL(Z,SAFETY\_OF(Y)), PREVENT(Z,CORRUPTION), ...} \end{bmatrix} \\ \text{AGENTIVE: } \begin{bmatrix} \text{'ROLE AND BEHAVIOR OF POLICE DEFINED BY POLITICAL FORCE'} \end{bmatrix} \end{bmatrix}$$

The statement 'less present in the streets' means that the police has a lower degree of activity concerning its telic aspects, in particular safety controls. This statement can be interpreted in two opposite ways, which can be paraphrased as follows:

> (1) there is less need of control, therefore, the safety situation has improved. **This interpretation supports the controversial issue**.
> (2) the police is less committed and safety is less controlled, therefore the safety situation has probably worsened. **This second interpretation attacks the controversial issue**.

5 PERSPECTIVES: MOVING TOWARDS A REAL APPLICATION

The experiments presented in this paper are preliminary, but they show some challenges and possible tracks to reach an efficient and expressive argument mining system. To go from the previous use-cases to a real argument mining application, several points are challenging and need to be considered in depth, among which: the definition of GL Qualia structures on a relatively large scale (scalability of knowledge and lexical data), the specification of the various types of inferences which have been presented, and, finally, strategy issues: how the Qualia structure data and the inferences are triggered. What are the principles that govern such a system?

Our preliminary investigations tend to show that the number of Qualia structures for a given domain such as vaccination is not very large, around 50 to 70 structures, which is feasible by hand given a domain ontology of predicates, functions and constants. Lexical semantics data need to be specialized for argumentation such as polarity. The most complex task is to identify, categorize and structure the different types of inferences which may occur in the argument mining process. Besides the categorization of inferences, another challenge is the definition of an adequate and generic processing strategy, as developed in TextCoop (Saint-Dizier, 2012), that is able to trigger inferences in an efficient and relevant way.

REFERENCES

Boltuzic, F., & Snajder. J. (2015). Identifying prominent arguments in online debates using semantic textual similarity. In *Proc. of the Second Workshop on Argumentation Mining.*

Budzynska, K., Janier, M., Reed, C., Saint-Dizier, P., Stede, M., & Yakorska, O. (2014). A model for processing illocutionary structures and argumentation in debates. In *Proc. of the 9th edition of the Language Resources and Evaluation Conference (LREC).*

Feng V. W., & Hirst, G. (2011). Classifying arguments by scheme. In *Proceedings of the 49th Annual Meeting of the Association for Computational Linguistics: Human Language Technologies* (pp. 987–996). Portland, OR.

Kirschner, C., Eckle-Kohler, J., & Gurevych, I. (2015). Linking the thoughts: Analysis of argumentation structures in scientific publications. In *Proceedings of the 2nd Workshop on Argumentation Mining* (pp. 1-11). Denver, CO.

Mochales Palau, R., & Moens, M.F. (2009). Argumentation mining: The detection, classification and structure of arguments in text. In *Twelfth international conference on artificial intelligence and law (ICAIL 2009)* (pp. 98-109). Barcelona.

Nguyen, H., & Litman, D. (2015). Extracting argument and domain words for identifying argument components in texts. In *Proceedings of the 2nd Workshop on Argumentation Mining* (pp. 22-28). Denver, CO.

Pustejovsky, J. (1995). *The generative lexicon.* Cambridge, MA: MIT Press.

Saint-Dizier, P. (2012). Processing natural language arguments with the TextCoop platform. *Argument & Computation, 3*(1), 49-82.

Swanson, R., Ecker, B., & Walker, M. (2015), Argument Mining: Extracting Arguments from Online Dialogue, In *proc. SIGDIAL 2015, USA.*

Villalba M. G., & Saint-Dizier, P. (2012). Some facets of argument mining for opinion analysis, *COMMA.* Vienna: IOS Publishing,

Walker, M., Anand, P., Fox Tree, J.E. &, Abbott, R. & King, J. (2012). A corpus for research on deliberation and debate. *Proc. of the Language Resources and Evaluation Conference (LREC),* Istanbul.

Winterstein, G. (2012). What but-sentences argue for: An argumentative analysis of but. *Lingua, 122,* 1864-1885.

56

Argument Compound Mining in Technical Texts: Linguistic Structures, Implementation and Annotation Schemas

PATRICK SAINT-DIZIER,
IRIT-CNRS Toulouse, France
stdizier@irit.fr

JUYEON KANG
IRIT-CNRS Toulouse, France
stdizier@irit.fr

In this paper, we motivate and develop the linguistic characteristics of argument compounds. The discourse structures that refine or elaborate arguments are analysed and their cognitive impact in argumentation is developed.
An implementation is then presented. It is carried out in Dislog on the TextCoop platform. Dislog allows high level specifications in logic for fast and easy prototyping at a high level of linguistic adequacy. Elements of an indicative evaluation are provided.

KEYWORDS: discourse structure, language processing, linguistic analysis, logic programming

1. INTRODUCTION AND AIMS

Language expressions of arguments are often very diverse and complex, making their automatic identification in texts a very challenging task. Besides language complexity, a large number of arguments are not clearly marked by specific linguistic cues, therefore, it is often necessary to have recourse to semantics and pragmatics to identify, delimit and understand them and then identify the relations within and between compounds. Indeed, an argument for or against a given controversial statement can be just a fact if the relation with that controversial issue is not established. If it is established, knowledge may be necessary to identify whether it is an attack or a support, and then its strength.

D. Mohammed & M. Lewiński (eds.) (2016). *Argumentation and Reasoned Action: Proceedings of the 1st European Conference on Argumentation, Lisbon, 2015. Vol. II, 895-906.* London: College Publications.

Technical documents (e.g. procedures, product manuals, specifications, business rules) form a linguistic genre with restricted linguistic constraints in terms of lexical realizations, including business aspects, grammar, style and overall organization. These documents are designed to be as efficient and unambiguous as possible. For that purpose, they tend to follow relatively precise authoring principles concerning both their form and contents. Technical documents abound in various classes of arguments, in particular recommendations, warnings, advice, requirements and regulations.

Each argument can be associated with several supports, possibly contradictory, and various forms of explanation. We call this kind of clustering and ***argument compound***. Automatically identifying argument compounds in technical texts and producing a conceptual representation adequate for subsequent treatments is the major concern of this paper. For that purpose, we develop a discourse grammar from a corpus of technical texts that accounts for the conceptual structure of argument compounds. The modelling is based on logic, logic programming and constraint satisfaction, as implemented in the TextCoop platform via the Dislog language.

This paper further elaborates on results presented in (1) (Saint-Dizier, 2012) where processing isolated warnings and advice are presented together with their implementation in Dislog, (2) (Villalba & Saint-Dizier, 2012) where we show that discourse structures, for which a detailed semantic analysis is developed, can be interpreted as argument supports in opinion analysis, and (3) (Kang & Saint-Dizier, 2013) dedicated to requirement mining.

2. CONCEPTUAL AND LINGUISTIC ANALYSIS

2.1 Conceptual analysis

The linguistic structure of arguments as isolated utterances or as networks of arguments has been investigated in a number of works in linguistics and cognitive semantics (e.g. van Eemeren & Grootendorst, 1992; Walton, Reed, & Macagno, 2008; Walton, 2011). Much less has been developed from a technical perspective in computational linguistics, but there are now several works in this direction. Difficulties come from the large diversity arguments may have in language, the need of contextual information to identify them and the difficulty to relate arguments with their supports or with other arguments, in particular when they are not adjacent in a text or in a dialogue.

In terms of discourse, the RST (Mann & Thompson, 1988), (Taboada & Mann, 2006) has been very influential over the last two decades.

However, identifying discourse structures in general is a challenge since linguistic cues are relatively limited or ambiguous between relations (see e.g. http://www.sfu.ca/rst/).

Several approaches, based on corpus analysis with a strong linguistic basis, are of much interest for our approach. Besides the Penn Discourse Treebank, relations have been investigated together with their linguistic markers in e.g. Delin, Hartley, Paris, Scott, & Vander Linden (1994), Marcu (1997), Miltasaki, Prasad, Joshi, & Webber (2004). Saito, Yamamoto, & Sekine (2006) among others developed an extensive study on how markers can be quite systematically acquired. Finally, Stede (2012), developed a useful typology of markers.

Our approach to structure argument compounds merges argument and discourse structure analysis. In this context, the typical configuration of an argument compound can be summarized as follows:

FRAME(S)
CIRCUMSTANCE(S) / CONDITION(S), PURPOSE(S)
  [ARGUMENT CONCLUSION + SUPPORT(S)]*
    PURPOSE(S), CONCESSION(S) / CONTRAST(S), ELABORATION(S)

The kernel of this structure is the organized set of arguments and their supports. The main argument occurs in general first, it is then followed by secondary arguments; their functions are developed below. A number of sections or paragraphs in technical documents start by a frame that describes the scope or the domain of the section (e.g. *for pumps X45....).* Frames are often not adjacent to argument compounds, they are comparable to focus and will not be investigated here.

The compound starts with circumstances and conditions, possibly purposes, when they have a wide scope over the arguments. Then follows the set of arguments and their supports. The compound ends by purposes, concessions or contrasts and elaborations.

At the language realization level, this conceptual organization may not be realized straightforwardly. In particular, we observed that:

- the initial group, that should logically precede the set of arguments, may be inserted between arguments,
- the last group, that should also logically follow the set of arguments, may be inserted between these arguments,
- purposes may be realized as supports,
- an argument may have several supports, possibly with different orientations, supports may not be adjacent to their related conclusion,
- supports may be inserted within their conclusion, instead of following or preceding it.

Let us illustrate argument compounds, where a few tags have been inserted to facilitate the analysis:

> (1)<ArgCompound> <purpose> Cleaning your leathers. </purpose> <advice> <conclusion > Prefer natural products. </conclusion>
> <support polarity="-"> they are more expensive </support> but <support polarity ="+"> they will have a longer effect and make minor repairs. </support> </advice> </ArgCompound>>.

> (2)<ArgCompound> <definition> Inventory of qualifications refers to norm YY. </definition>
> <mainArg> Periodically, an inventory of supplier's qualifications shall be produced. </mainArg>
> <secondaryArg> In addition, the supplier's quality department shall periodically conduct a monitoring audit program. </secondaryArg>
> <elaboration> At any time, the supplier should be able to provide evidences that EC qualification is maintained. </elaboration> </ArgCompound>

> (3)<ArgCompound> <warning> <conclusion> Products X and Y, <support> because of their toxicity, </support> are not allowed in this building. </conclusion> </warning>
> <concession> In case of emergency, a special permission is needed to use them in buildings. </concession> </ArgCompound>

Example (1) illustrates the case where an argument of type advice has several supports with different orientations, positive or negative, but these are not contradictory, they just reflect the various facets of the concept at stake. The contrastive connector "but" introduces the inversion of the polarity in the discourse. The first support is not really an attack, but a kind of contrast, which is a weak form of attack.

Example (2) is a requirement compound (or business rule compound). It shows how a definition makes the requirements more accurate. A secondary requirement complements the main one, which is further elaborated in the last sentence. This latter sentence is not a requirement because of the modal "should be able to" which is not injunctive.

Example (3) illustrates the case where a support is inserted into the middle of a conclusion. The second sentence is a concession that allows exceptional situations.

2.2 Linguistic characterization

Let us first develop an illustrated analysis of a few types of arguments, usually found in technical texts. This analysis is illustrated by (*i*) typical patterns that identify arguments and (*ii*) related lexical resources for which we have developed specific linguistic categorizations.

Requirements and regulations requirements (Hull, Jackson, & Dick, 2011) and regulations form a special class or arguments, with specific linguistic forms and a very injunctive orientation. Their support(s) must not be confused with purpose clauses: their role is to justify the requirement, its importance, and the potential risks and difficulties that may be encountered. Their identification in English is quite simple since requirements must follow very precise authoring guidelines. A requirement is injunctive, it is based on precise patterns in a sentence (Kang & Saint-Dizier, 2013) such as:

[modal(shall, must, have to) + infinitive verb].

Supports are introduced by a purpose connector, e.g. *to, for, in order to*.

A comprehensive requirement is e.g. *an inspection shall be carried out monthly for a correct cleaning of the universal joint shafts*.

Prevention arguments or warnings basically explain and justify a fact, an information, an instruction or a group of instructions. These are very frequent in most types of technical documents. Formulations with a negative polarity are frequent since the main goal is e.g. to warn users against misuses of products, their structure is given in (Saint-Dizier, 2012) and summarized here. The structure of a conclusion is:

(1) prevention verbs like avoid' NP / to VP (*avoid hot water*)
(2) do not / never / ... VP(infinitive) ... (*never expose this product to the sun*)
(3) it is essential, vital, ... (to never) VP(infinitive). (it is essential that you switch off electricity before starting any operation).

Supports are realized by one of the following syntactic schemas:

(1) negative causal connector + infinitive risk verb,
(2) negative causal mark + risk verb,
(3) positive causal connector + VP(negative form),
(4) positive causal connector + prevention verb.

The grammatical and lexical elements in these constructions are in particular:

- negative connectors: otherwise, under the risk of, (e.g. *otherwise you may damage the connectors*),
- risk verb class: risk, damage, etc. (e.g. *in order not to risk to hurt your fingers*) or verbs of a "conservative" type : *preserve, maintain*, etc. (e.g. *so that the axis is maintained vertical*),
- prevention verbs: *avoid, prevent*, etc. (e.g. in order to prevent the card from skipping off its rack),
- positive causal mark and negative verb form: *in order not to*, (e.g. *in order not to make it too bright*),
- modal SV: may, could, (e.g. because it may be prematurely stop due to the failure of another component).

These are stored in the system lexicon with their semantic characteristics.

Threatening arguments are less frequent than warnings. The reader and the author of the threat are directly involved in the consequences of the action or the incorrectness of the information that is given, whereas warnings are more neutral and only concern the action being carried out. These arguments have a strong impact on the user's attention when he realizes the instruction. These arguments follow one of the following syntactic schemas:

(1) otherwise connectors + consequence proposition,
(2) otherwise negative expression + consequence proposition, with, e.g.:
– otherwise connectors: e.g. *otherwise*,
– otherwise negative expression: if ... do not ...} (e.g. if you do not pay your registration fees within the next two days, we will cancel your application).

2.3 Discourse Relations in a compound

In an argument compound, as shown in section 2.1 above, the different utterances are linked by means of discourse relations. This defines a kind of network of relations. The relations between arguments are essentially contrasts, concessions and specializations. The other relations structure the compound with non-argumentative utterances, the aim is to give more details about e.g. the compound facets.
The structure and the markers and connectors typical of discourse relations found in technical texts are developed in e.g. (Stede, 2012) and

(Saint-Dizier, 2014). These have been enhanced and adapted to the compound context via several sequences of tests on our corpus. The main relations found are the following:

- **contrast**, (Wolf & Gibson, 2005) and (Spenader & Lobanova, 2007), is a relation between two arguments that introduces one or more equivalent but alternative views, but which refer to a unique situation. Formally, the apparent contradiction that results motivates the use of a defeasible inference logic and semantics to preserve the coherence of the whole structure. Contrast is introduced by *however, although, but* combined, in the utterance, with e.g. adverbs such as *also*, modals or specific verbs expressing choice.
- **concession** states a general requirement followed by an apparently contradictory argument that could be admitted as an exception (e.g. Ex. 3.). The contradiction with the implicit conclusion which can be drawn from the first argument is partial (e.g. (Couper-Kuhlen & Kortmann, 2000)). Concessions are often categorized as denied phenomenal causes or motivational causes. Typical marks are, e.g.: *however, although, even though, despite*, or modal constructions such as: *may be, could be*. We observe a kind of continuum between contrast and concession. The ambiguity is represented in our approach by the *polymorphic relation* "contrast-concession". Ambiguities may then be resolved via knowledge and inferences.
- **specializations**, and subsequent constraints develop the concepts or rules that are presented. These often involve domain knowledge to be identified as such, the kind of specialization or constraints that is invoked and how it affects the main statements,
- **information and definitions** mainly occur before the main argument. They anticipate and develop notions given in the main argument which may be complex or insufficiently clear to the reader or may contradict his beliefs. Definition identification has been largely developed in various information retrieval systems (e.g. in TREC), its identification is often based on marks or specific syntactic forms.
- **elaborations** follow an argument, they develop some of its facets to facilitate its understanding. Elaborations may play the role of supports. Since this relation is very generic and under-specified, we consider it as the by-default relation in the compound. A categorization of the main functions covered by elaboration are in particular: *localization, precision, focus, future actions, application domains, constraints, prerequisites*. An automatic identification of these functions is ongoing and beyond the scope of this paper.
- **illustration** provides related examples. It is characterized by simple marks such as: *this includes, for example, an example, examples* or punctuation associated with an enumeration. Illustration can also be analysed as a form of support.

- **result** specifies the outcome of an action. Its linguistic structure is basically the active-inchoative alternation that describes the expected result, implemented via the use of the theme combined with the main verb past participle or with an aspectual verb denoting completion or quasi-completion.
- **circumstance** introduces a kind of local frame under which the argument compound is valid or relevant. Circumstances often appear before the argument(s) they apply to. Circumstances introduce temporal, spatial or factual contexts or particular events or occasions.
- **purpose** expresses the underlying motivations of the argument compound. It must not be confused with argument supports. Purpose clauses are introduced by purpose connectors, causal verbs, purpose verbs (e.g. *demonstrate*) or by various types of expressions such as: *the objective is.*

3. IMPLEMENTATION IN DISLOG

Let us now briefly show how these linguistic elements are implemented in a running system and what the performances are. So far evaluation is essentially indicative since the system is in an early development stage.

3.1 TextCoop: a platform for discourse analysis

The TextCoop platform and the Dislog language (standing for Discourse in Logic) have been primarily designed for argumentation and discourse processing (Saint-Dizier, 2012).
TextCoop is based on Logic Programming, it is a platform that includes:

(1) **Dislog**, a logic-based language designed to describe in a declarative way discourse structures and the way they can be bound via selective binding rules,
(2) **an engine** associated with a set of processing strategies. Dislog rules are processed according to a cascade that specifies their execution order. This engine offers several mechanisms to deal with ambiguity and **concurrency** when different discourse structures can be recognized on a given text fragment,
(3) **a set of active constraints**, in the sense of Constraint Logic Programming, that state well-formedness constraints typical of discourse structures (e.g. precedence, dominance, bounding nodes); these can be parameterized by the grammar writer,
(4) **input-output facilities** (XML, MS Word), and interfaces with other environments, but so far in a relatively limited way,
(5) a set of **lexical resources** which are frequently used in discourse analysis (e.g. connectors),

(6) a set of about 180 **generic rules** that describe 12 frequently encountered discourse structures such as reformulation, illustration, cause, contrast, concession, etc.

The system designed for argument compound analysis is very declarative. It is composed of a set of rule clusters, associated lexical entries, and constraints.

To deal with "scrambling" situations as illustrated in Example (3), rules are non-deterministically decomposed under constraints by the TextCoop engine. Therefore, these strategy elements are transparent to the user or grammar writer.

In TextCoop, rule clusters are activated one after the other with an order specified in a cascade. This cascade allows, among other things, to specify priorities (a cluster must be fully processed before another one is activated) and to avoid ambiguities.

3.2 Indicative evaluation

The following indicative evaluation is designed to identify improvement directions. The evaluation has been realized on our test corpus on a total of 255 argument compounds, which have been first manually annotated by human annotators.

Since this is a difficult task, the result has been realized via discussion among annotators, to guarantee a certain quality. Compound identification produces the results given in Table 1:

Criteria	Precision	Recall
Identification of compound	83%	77%
Opening boundary	96%	90%
Closing boundary	88%	78%

Table 1. Result evaluation

The closing boundary is more difficult to identify because some terms out of the compound can be interpreted as theme variants. The accuracy of a compound identification could be improved by adding more theme variants, but there is a trade-off to elaborate in order to avoid noise. Our strategy is so far to favour precision.

The identification of discourse structures in a compound produces the results given in Table 2:

Relation	Number of rules	Number of annotated structures	Precision	Recall
Contrast	14	29	84	88
Concession	11	62	83	85
Specialization	6	39	74	71
Information	6	29	84	76
Definition	9	87	85	74
Elaboration	14	118	84	80
Illustration	20	53	91	84
Result	16	99	84	80
Circumstance	15	112	88	80
Purpose	17	112	89	81

Table 2. Evaluation of discourse analysis structure

Some relations have more elaborated sets of rules because they have been reused and improved from previous experiments. This explains the differences in number of rules. Some sets of rules may need further expansion to produce more accurate results, this is the case for "specialization" which remains somewhat vague. Information and definition are not necessarily identified on the basis of marks but on their position in the compound, which is also a vague criterion. In general, however, results are good for discourse analysis.

4. PERSPECTIVES

In this paper, we have developed a linguistic model for the analysis and the representation of argument compounds. This contribution illustrates and investigates the complexity of argument constructions and the development of a conceptual model.

Our results form a kind a ***discourse grammar*** dedicated to argument compounds. The specific discourse relations we have identified are conceptually characterized, with the functions they play, so that inferences can be drawn within and between argument compounds. We feel this work can be further refined but also extended, gradually, to other textual genres and other types of arguments. This is not an easy task, but we propose in this paper a simple method which could be reused, with adaptations.

Besides going on improving the recognition of argument compounds, we aim at investigating other forms of arguments in texts which have a relatively controlled language forms (e.g. didactic texts, contracts). Another important direction is the development of a conceptual model that allows various forms of inferences so that sets of argument compounds can be analysed for example w.r.t. their coherence or overlap in a text. Identifying arguments based on knowledge and inference is a bottleneck in argument mining. In this volume, we present a simple and preliminary investigation on this topic based on the Generative Lexicon that seems promising since it merges lexical knowledge with domain knowledge.

The implementation of the work presented here is carried out in Dislog on the TextCoop platform. Dislog allows high level specifications in logic that allow fast and easy prototyping. Elements of an indicative evaluation are developed: results are good for a discourse processing task. Most of the code of this project is freely available under a Creative Commons BY License and can be obtained from the author.

ACKNOWLEDGEMENTS: I am grateful to several reviewers that helped improve this work.

REFERENCES

Blakemore, D. (2002) Relevance and linguistic meaning: The semantics and pragmatics of discourse markers. Cambridge: Cambridge University Press.

Couper-Kuhlen, E., & Kortmann, B. (2000). Cause, condition, concession, contrast: Cognitive and discourse perspectives, Topics in English Linguistics, 33, Mouton de Gruyter.

Delin, J., Hartley, A., & Paris, C., & Scott, D., Vander Linden, K., (1994). Expressing procedural relationships in multilingual instructions, Proceedings of the Seventh IWNLG, 61-70.

Eemeren, F.H., & van Grootendorst, R. (1992). Argumentation, communication, and fallacies: A pragma-dialectical perspective. Hillsdale, NJ: Lawrence Erlbaum Associates.

Grosz, B., & Sidner, C. (1986). Attention, intention and the structure of discourse. Computational Linguistics, 12(3), 175–204.

Hull, E., Jackson, K., & Dick, J. (2011). Requirements engineering, Springer Verlag.

Kang, J., & Saint-Dizier, P. (2013). Discourse structure analysis for requirement mining, International Journal of Knowledge Content Development and Technology, 3(2), 67–91.

Mann, W., & Thompson, S. (1988). Rhetorical Structure Theory: Towards a functional theory of text organisation, TEXT, 8(3), 243-281.

Marcu, D. (2000). The theory and practice of discourse parsing and summarization. MIT Press.

Miltasaki, E., Prasad, R., Joshi, A., & Webber, B. (2004). Annotating discourse connectives and their arguments, Proceedings of new frontiers in NLP.

Saint-Dizier, P. (2012). Processing natural language arguments with the TextCoop platform. *Argument & Computation, 3*(1), 49-82.

Saint-Dizier, P. (2014). Challenges of discourse processing: The case of technical documents. Cambridge Scholars Publishing.

Saito, M., Yamamoto, K., & Sekine, S. (2006). Using phrasal patterns to identify discourse relations. In Proceedings ACL06.

Spenader, J., & Lobanova, A. (2007). Reliable discourse markers for contrast. In Eighth International Workshop on Computational Semantics, Tilburg.

Stede, M. (2012). Discourse processing. Morgan and Claypool Publishers.

Taboada, M., & Mann, W. C. (2006). Rhetorical Structure Theory: Looking back and moving ahead. Discourse Studies, 8(3), 423-459.

Villalba M. G., & Saint-Dizier, P. (2012). Some facets of argument mining for opinion analysis. COMMA, Vienna: IOS Publishing.

Walton, D., Reed, & C., Macagno, F. (2008). Argumentation Schemes. Cambridge: Cambridge University Press.

Walton, D. (2011) Argument mining by applying argumentation schemes. Studies in Logic, 4(1), 38-64.

Wolf, F., & Gibson, E. (2005). Representing discourse coherence: A corpus-based study. Computational Linguistics, 31(2), 249-288.

57

When Subjectivity Arises in a Swiss Criminal Court: How Intensifiers Can Work as Pragmatic Markers in Argumentative Discourse

CAMILLIA SALAS
University of Neuchâtel, Switzerland
Camillia.salas@unine.ch

THIERRY RAEBER
University of Neuchâtel, Switzerland
Thierry.raeber@unine.ch

This paper is interested in examining how 'loss' of objectivity can reveal itself in the context of criminal courts (Switzerland). Our goal is to identify the role of intensification in argumentative discourse, and how it may strengthen argumentative aims. Our present task is to show how speakers in this context can (1) make use of intensifiers and (2) use them as markers of subjectivity when it comes to disputed facts.

KEYWORDS: argumentative discourse, evidence, intensifiers, pragmatics, subjectivity

1. INTRODUCTION

Argumentation plays an important role in law. One of the main concern in legal argumentation is often related to decision-making (Aarnio, 1987; Alexy, 1989; Feteris & Klooterhuis, 2009). Legal decision-making is highly associated with an ideal of the Rule of Law – that calls for legal certainty, predictability and reasonableness. Somehow, legal decision-making requires an ideal of objectivity. However, this paper aims to examine how a "loss" of objectivity may reveal itself during the hearings of a Swiss criminal trial. Indeed, specific institutional constraints allow legitimacy of the decision-making, but it seems that interactional and discursive constraints highlight also a kind of subjectivity.

D. Mohammed & M. Lewiński (eds.) (2016). *Argumentation and Reasoned Action: Proceedings of the 1st European Conference on Argumentation, Lisbon, 2015. Vol. II, 907-920.* London: College Publications.

The markers of subjectivity we are talking about are related to the theory of evidence in law (Wigmore, 1940; McCormick, 1978; Walton, 2002). In Switzerland, the judge is actively involved in the legal decision-making process – since he conducts the investigation for matters of law as well as matters of facts. Thus, the way the judge looks for information, in order to collect evidence, can reveal a certain type of subjectivity. Indeed, when tracking a lead, it is sometimes part of the game to insist on some elements of presumed facts, in order to confirm/infirm the lead. This is how we define subjectivity in a legal context. Moreover, this subjectivity may be experienced by all the participant at a trial (judge, lawyers, prosecutor, the accused, character witness, *etc.*). Indeed, trial interactions rely on power dynamics where participants – disregarding the different goals they are pursuing – can influence each other. And in the case in which facts are disputed, each participant might make use of markers of subjectivity.

This paper is focusing on a specific kind of markers of subjectivity, i.e. intensifiers. Intensifiers can be defined as linguistic elements which are used to indicate intensity, such as degree modifiers, metaphor, hyperbole, *etc.* The phenomenon of intensification has been studied through different perspectives, from a sociolinguistic approach (Labov, 1984; Athanasiadou, 2007; Cacchiani, 2007) to semantics (Kleiber, 2007, 2013; Lenepveu, 2013) and speech acts theory (Romero, 2001). Yet, linguistic studies of intensification are very heterogeneous, since (1) intensity can be associated to different linguistic phenomena, and (2) its definition is a bit vague (Anscombre & Tamba, 2013). Indeed, linguistic phenomena that are related to intensification can be identified on different levels (*Idem*. p.3):

- Lexical level: adverbial quantifiers ('perfectly', 'completely', 'totally', 'really', *etc.*), some verbs ('to scream', ' to shout'), gradation of a qualitative appreciation ('good', 'best', 'better than', 'excellent'), some prefix ('hyper', 'extra', 'super');
- Semantic and syntactic level: hyperbole ('extraordinary luck'), metaphor ('it's raining cats and dogs', 'to be cold as a stone'), antiphrasis ('you look happy with your grumpy face');
- Prosodic level: intonation, emphasis, *etc.*;
- Uttered and argumentative level: illocutionary act, negation, relevance weight, *etc.*

In this paper, and following initial considerations presented in Raeber (forthcoming), we consider that (1) intensification is a pragmatic process through which the lexical expression targeted by the intensifier is contextually rebuilt (*ad hoc* concept, Wilson & Carston, 2007) with additional properties/consequences relevant in context; and (2) intensifiers can work as procedural expressions with a metalinguistic scope, leading the addressee towards a more specific interpretation of the targeted concept. Intensifiers do not describe the concept itself, but rather the way the lexical expression has to be interpreted. Our main hypothesis is that these additional effects might direct toward an argumentative standpoint. In other words, making use of intensifiers during the hearing works as one of the linguistic means that expresses the inquiry of information considered the most relevant. The inquiry of such information would *in fine* help the judge to collect final evidence. At the end of the hearing, once evidence has been collected, the judge will use it in his final decision and give arguments to support it – in the light of the Rule of Law. Consequently, the final decision and the hearing are strongly associated: the hearing – thanks to linguistic means – orient toward argumentative standpoints; the final decision put those arguments down.

For that reason, and in order to account for argumentative effects, we suggest to (1) use a semantic/pragmatic framework to describe how information is built up and structured in context and (2) complete it by a discourse-analytical framework to identify interactional and argumentative effects triggered by linguistic markers. Since markers of subjectivity are expressed under interactional and discursive constraints, we suggest approaching argumentation from a descriptive linguistic angle. This approach takes into account linguistic moves as well as interactional ones (Plantin, 1996; Jaquin & Micheli, 2012; Doury, 2003).

In such a context, we assume that intensifiers may serve argumentative goals, such as:

(1) Leading toward an evaluative (generally negative) conclusion
(2) Forcing the addressee to defend himself
(3) Testing the addressee's stance (related to the reliability of the delivered information)

In order to understand how intensifiers can be performed in such a way, let us first say how the inquiry and collecting evidence stages can reveal themselves during a Swiss criminal trial.

2. COLLECTING EVIDENCE AND THE EXAMINATION DIALOGUE IN A SWISS CRIMINAL CASE

Courtroom is a prototypical context for argumentation. The way law is applied in American or Continental countries influence the way facts are going to be interpreted and how evidences are going to be questioned and collected.

From an interactional viewpoint, in both types of countries, trials are institutionalized interactions that are typically asymmetrical, since power and control are located in the institutional participant, rather than being equally distributed. (Coulthard & Johnson, 2007, p. 15). However, in Swiss criminal trials, most of interactions are based on a question-answer relationship that forces the answerer to answer, but "where the answerer is pressured to answer in the way the questioner wishes by means of a range of linguistic means" (Gibbons, 2003, p. 128).

Since question of facts and collecting evidence are a judge's concern, question-answer relationship works as an examination dialogue[1] (Salas, forthcoming.). According to Walton (2006), this examination dialogue process works *simultaneously* as a two steps structure: (1) information seeking, and (2) testing out the reliability of the delivered information. In such a context, argumentative discourse can occur in different ways. The way the questioner extracts information and tests its reliability is somehow a way for requesting information (*Idem.*), and test out presumed facts. Since, the questioner has also the responsibility to test out the reliability of the information he is trying to get, he cannot explicitly ask for it: that's why some information can be expressed and other unexpressed. Consequently, the questioner is going to draw inferences, based on common form of inference. Consequently, the examination dialogue is strongly related to how the delivered information by non-institutional members is interpreted by institutional one. At the end of the hearing, those inferences are going to deliver intuitions about arguments (Mercier & Sperber, 2011).

In the following parts of this paper, we are going to analyse how the prosecutor will track a specific lead regarding the nature of intent in a stabbing case. On one hand, by tracking this lead, institutional members will make use of subjectivity, and thus use certain kind of intensifiers. On the other hand, the accused is also going to react to the multiple requests of the prosecutor who is trying to collect evidence.

---

[1] In Walton's paper, examination dialogue occurs mostly in the cross-examination stage. However, in Switzerland, such stage - among others (opening statements, examination stage, etc.) are not so well identified since we have to deal with an inquisitorial model in which an inquiry is conducted by the judge. For that reason, legal-decision making process in the light of Swiss criminal procedure can be viewed as an examination dialogue (Cf. Salas, forthcoming).

3. THE NATURE OF INTENT AS DISPUTED EVIDENCE

Let us say that Samuel Moraz[2] is a suspect of common assault among other offenses in the stabbing of Benjamin Friché, the present victim. Because of a drug concern, Samuel Moraz and Benjamin Friché started arguing, and at some point, the accused punched the victim in the face and then stabbed him in the abdomen with his 8 cm long knife. The known facts that can be taken for granted are the following:

- Samuel Moraz did punch the victim in the face
- He did stab him in the abdomen

When it comes to the written statement, the prosecutor decides to charge the accused with three kinds of felonies: (1) common assault, (2) endangering life[3] and (3) acts of aggression[4]. But, when the case is presented to the fact finder (2) and (3) is not questioned by the defendant, but (1) is contested because the nature of the intention regarding the commission of the act was not clear. In such a context, some elements need to be discussed and questioned by the judge so that he can make his mind regarding the legal qualification of facts and thus formulate an applicable rule[5]. At some point, during the examination dialogue, elements of information are going to be questioned such as:

- Did the accused intentionally cause injury to the victim in order to kill him?
- Did the accused intentionally cause injury to the victim in order to inflict life-threatening injuries?
- Did the accused intentionally cause the injury to the victim in order to inflict non-fatal injuries?

For instance, while the judge and the prosecutor start interrogating the accused, they are especially evaluating his intent of hurting Benjamin Friché. In other words, if the accused intentionally inflicts a life-threatening injury to the victim he will not be charged with common

---

[2] All names are fictional.

[3] (Art.129 SCC) Endangering life: any person who unscrupulously places another in immediate life-threatening danger is liable to a custodial sentence not exceeding five years or to a monetary penalty.

[4] (Art 126. SCC) Acts of aggression: any person who commits acts of aggression against another that do not cause any injury to the person or health is liable on complaint to a fine.

[5] The way a judge, in a Swiss hard case, interpret facts so as to find the best way to apply the rule has been developed in Salas (forthcoming).

assault anymore (art.123 CP)[6] but with serious assault (art.122 CP)[7]. On the same level, if the accused intentionally caused injury to the victim in order to kill him, the judge can charge him with attempted murder. So, the main goal of the institutional members, during the examination dialogue, is to test the accused's "effective" intent and make a decision about the appropriate felony the accused would be charge with. Moreover, as we can expect it, the way the prosecutor and the judge are interrogating the accused draws the way they expect him to respond. That means, if they are not satisfied with the answer of Samuel Moraz, the judge as well as the prosecutor, are going to do their best to clarify, to rephrase and so forth their questions. The main goal is to test out the reliability of the accused's responses.

We are going to see now how facts are going to be interpreted and how intensifiers can work as pragmatic markers since they (1) express the reliability-testing step that is performed by the judge and/or the prosecutor and (2) can also reveal the way the person interrogated reacts to such an investigation's lead. The analysis focuses on 44'91 seconds hearing and is a part of the interaction between the prosecutor and the accused himself. We present here the transcription of the extract (Figure 2 below), and we will focus on two main elements: (1) the use of the adverb "completely" (line 7) as an intensifier that serves to strengthen the guilt stance of the accused, and (2) the adverb "seriously", used by the accused to increase his reliability.

In the following section, we examine how intensifiers may be used as a discursive strategy in order to strengthen the guilty stance of the accused. In section 5, we will examine how, in return, intensifiers can be used by the accused to strengthen his reliability, and improve his credibility.

---

[6] Any person who willfully causes injury to the person or the health of another in any other way is liable on complaint to a custodial sentence not exceeding three years or to a monetary penalty.

[7] Any person who intentionally inflicts a life-threatening injury on another, any person who intentionally inflicts serious injury on the person, or on an important organ or limb of another, makes an important organ or limb unusable, makes another permanently unfit for work, infirm or mentally ill, or who disfigures the face of another badly and permanently, any person who intentionally causes any other serious damage to the person or to the physical or mental health of another, is liable to a custodial sentence not exceeding ten years or to a monetary penalty of not less than 180 daily penalty units.

1	Pg	avec le recul maintenant/ vous avez eu le temps/ d'y réfléchir euh
	Pr	looking backward now you've had enough time to think about it er
2		vous avez pris conscience un peu de ce que vous avez fait//
		you became a little bit aware of what you did
3		de la gravité de vos actes/
		of the seriousness of your acts
4	P	ben oui// ça aurait pu déraper (2.0) gravement quoi
	A_{cc}	well yes things could have seriously got out of control you know
5 6	Pg	ouais (2.0) vous auriez pu le tuer/ (3.0) en plus <u>lui</u> il l'a pas vu le couteau//
		yeah you could have killed him on top of that he didn't see the knife
7		donc c'est complètement par surprise (si vous voulez)
		so it was completely unexpected
8	P	oui ((P hoche la tête))
	A_{cc}	yes ((Acc is nodding in agreement))

Pr : Prosecutor
A_{cc} : the accused

Figure 2: Extract of an interaction between the prosecutor and the accused

4. HOW INTENSIFIERS CAN REINFORCE THE GUILT POSITION OF THE ACCUSED: THE EXAMPLE OF "COMPLETELY"

We begin by describing the effect of the intensifier from a purely linguistic viewpoint (i.e. meaning effects), and we will show that these particular effects serve as an argumentative tool to diminish the credibility of the accused. The element we want to focus on here is the use of the adverb "completely" used in line 7 of the extract, retyped here:

> "so [the assault] was *completely* unexpected" (l.7)

Let's start by explaining the effect of the adverb from a purely linguistic and conceptual viewpoint. An adverb like "completely" is classified by Molinier & Levrier (2000) in the category of "adverbs of completeness" (*adverbes de complétude*), with "totally", "fully", "entirely", *etc.* As Lenepveu (2013) showed, these adverbs can have at least two different uses, depending on the nature of their argument (in a syntactic sense). The first use, called "quantitative" (illustrated by example (1) below), occurs when the argument of the adverb may form a whole, and when the degree of the property is quantifiable. In this case, the adverb indicates that the property is fully reached, and can thus be replaced by the expression "in its totality".

(1) John has completely repainted his house.
John has repainted his house in its totality.
* John has repainted his house in every aspect.

The second use, called "qualitative" does not mean that the property is fully applied to its target, since it cannot be seen as a whole. Instead, the group "completely different" (example (2) below) expresses that the property applies to its target in every possible way. Whatever the definition, the description "the neighborhood is different" still remains adequate.

(2) Compared to when I was a child, the neighborhood is completely different.
* Compared to when I was a child, the neighborhood is different in its totality.
Compared to when I was a child, the neighborhood is different in every aspect.

As presented in more details in Raeber (forthcoming), the qualitative use of these adverbs requires a pragmatic adaptation in order to retrieve the intended representation of the speaker. Saying that the neighborhood is completely different, i.e. different in every possible regard, presupposes that there are various aspects that characterize the difference. And in order to fulfill the expectation of relevance of the utterance (Sperber & Wilson, 1995), the hearer has to identify what may be these different aspects in order to properly recover the speaker's intention. These properties are context sensitive, and the hearer has to build an *ad hoc* concept that contains the properties contextually selected.

In the extract above, the use of the adverb "completely" is qualitative. With the utterance "[the assault] was completely unexpected", the prosecutor communicates that the assault was unexpected in every conceivable regard, and that all the clues that might have led the victim to anticipate this assault were hidden. The knife was dissimulated, the accused was not ostentatiously aggressive at the time of the assault, their relationship was not particularly contentious, *etc.* By reinforcing the degree of surprise of the assault, the prosecutor points out the fact that the victim had no chance to anticipate the threat, and thus no chance to defend himself. It also implies that the action was not spontaneous, but on the contrary prepared, measured, and performed in cold blood, and that it was not an act of self-defense. All these additional elements can be derived from this marker of intensity, and serve the argumentative aim of the prosecutor and/or the judge to present a depreciative evaluation of the accused, and reinforce his guilt stance.

5. HOW INTENSIFIERS CAN REDUCE THE GUILT POSITION OF THE ACCUSED: THE EXAMPLE OF "SERIOUSLY"

We turn now to the case where the accused uses intensifiers to reinforce its own credibility. We focus on the following utterance:

> "things could have *seriously* got out of control" (l.4)

5.1 *Meaning effects of the intensifier "seriously"*

The question we might address here is "How does the situation change if we add this adverb?" What is the difference between a situation that gets out of control and a situation that *seriously* gets out of control? Of course, what follows are general considerations about the meaning effect of the adverb that will be specified in a particular context. We have no intention to speculate about the speaker's intimate intentions. We simply want to sketch out the general effects such an adverb will have when associated with its complement.

So first, let us give an abstract description of what an expression such as "the situation got out of control" expresses, in order to see more clearly how the intensifier can modify this description. A good start could be to identify the essential features a situation should gather in order to be considered as "gone out of control". The main feature is that in such situation, you initially have a good ability to predict what will happen, and for some reason, you lose this ability to anticipate the future of events. This definition is very wide and different reasons can lead to this loss of control. As a consequence, many different properties can be targeted by an adverb such as "seriously". It is then important to see how the different ways of representing "a situation [that] gets out of control" can be built in context. To see that, let us consider two different situations:

> (1) You are driving your car. Hopefully, you efficiently anticipate the trajectory of your car. If you "lose control of your vehicle" (and by extend, of the situation), you actually lose the ability to predict where the car is going (because the road was slipper, the engine was faulty, *etc*).

> (2) You have to lead a school outing and take care of twenty pupils. Everyone obeys you submissively until two of them see a family of swans nearby, and decide to have a quick look. You head toward them to bring them back but in the meantime, another group takes advantage of your distraction and walk away as well, whatever the reason. In no time, the entire group split, and you cannot manage to gather them. Again, you lost control of the situation.

These two different scenarios present two different ways of losing control. In both cases, you have a task to deal with: driving your car properly, and keeping the pupils under guard. In the first case though, this single task becomes impossible to carry out because the necessary conditions required (for example, a functional engine) are not fulfilled anymore. So a single element can lead to the loss of control. In the other scenario, mastering the global task of keeping the kids in the ranks requires dealing with many different sub-tasks. One needs to keep an eye on every kid, deal with their needs, look at the map, etc. The loss of control corresponds here to too many elements that require attention at the same time. The amount of separate sub-tasks becomes overwhelming, and the organization collapses.

With these distinctions in mind (and there might be many other aspects not mentioned here), we can see more efficiently how much contextual the effect of an intensifier is. As we said earlier, we consider that intensifiers serve to guide the process during which a hearer creates a representation of the informative intention of the speaker. In all cases, the intensifier is used to "strengthen" part of the expression. But it can target different components. In the car scenario, by saying "the situation got *seriously* out of control" the adverb might target different aspects. It can target the lack of response of the car, making your efforts to bring it back on the tracks useless. So you lost control in such a way that you crossed the entire road, and ended in the nearby field, in spite of your efforts to impose a direction. It might also point out the consequences of the car spinning out of control. You didn't get out of the road, but you hit a car that hit another one, leading to a chain reaction, and a disastrous accident involving many people. But it could also mean that the trajectory issue, although reasonable in itself, made you panic in such a way that you froze and were unable to make a single move to take back the control of your car. Similarly, in the school outing case, you could target the number of sub-tasks involved in the process, and thus communicate that every single kid went away in different directions. You could as well communicate that even though only a few of them went away, the consequences were *serious*. Or, that this situation made you lose your nerves, and you started screaming and harming them.

This apparent digression from our initial case serves to show that there are many different accessible representations from a single expression, and in every case, the intensifier is used to help the hearer(s) access the proper representation the speaker has in mind when he says that the situation got out of control. More precisely, it serves to specify that at least one of the elements that form the loss of control was *serious*, in the sense that it was not anodyne. In the given case, it might once again target different elements (the number of people involved, the accused committing actions that he would deeply regret, the consequences leading to death, *etc.*). Predicting which properties will be selected in context is difficult. Nevertheless, some of the effects on the accused's credibility can be identified.

5.2 *Argumentative effects of the intensifier "seriously" on the accused's credibility*

The fact that the accused intensifies the degree of seriousness his actions might have led to is an efficient way to increase his reliability. By doing that, he shows himself as an honest and lucid man. He is not denying what happened. He is able to see clear. He acknowledges his responsibility. The use of the intensifier can be an efficient signal to indicate that the accused does not minimize the gravity of the situation, and takes the case *seriously*. This kind of strategy is a good way to show himself as being in good faith.

In addition to that, we notice that the use of the adverb is, to some extent, a rewording of the previous utterance of the prosecutor. We can see that the expression "It could have *seriously* got out of control" directly refers to the initial description "you became aware of the *seriousness* of your acts". By using the same expression, and then expressing the same level of intensity ("This is a *serious* matter"), the accused validates (part of) the evaluation made by the prosecutor.

But of course, the intensifier is not the only element that serves to reduce the guilt position of the accused in this extract. Saying that the situation got out of control communicates that the negative consequences were not planned nor desired. As previously said, losing control corresponds to losing the ability to predict what will happen, and to prevent it. In that sense, the accused has less responsibility in what happened, since he was not able to maintain control of the events. Moreover, the fact that he uses the conditional tense (it *could* have got seriously out of control) is a convenient way to express that it *did not* get seriously out of control.

6. CONCLUSION

In this paper, we wanted to explore how intensifiers can work as pragmatic markers in a legal context. Under interactional and discursive constraints, intensifiers function as subjective markers of high degree. Indeed, with these intensifiers institutional members can point out some part of known facts or leading track on which they want to insist. In such a context, intensifiers can (1) suggest the orientation of argumentative standpoints such as dropping the self-defence hypothesis. It can also (2) announce which lexical expression the institutional member wants to focus on and (3) force the addressee to "take a stand in such a way as to hold out against contention" (Doury, 2009, p. 143). In the latter case, it has to deal with face-threatening acts (Goffman, 1974).

At the very end, it seems that intensifiers might work as discursive clues that show which part of the uttered sentence the hearer has to pay attention to. In the given hard case, in which a certain number of facts and inferences are tested, becoming aware of the role of intensifiers would help institutional members as well as non-institutional ones to foresee in a way the forthcoming elements that would be refuted, accepted or just dropped because of lack of reliable information.

REFERENCES

Aarnio, A. (1987). *The rational as reasonable. A treatise of legal justification.* Dordrecht: Reidel.

Alexy, R. (1989). *A theory of legal argumentation. The theory of rational discourse as theory of legal justification.* Oxford: Clarendon Press. (Translation of: *Theorie derJuristischen Argumentation. Die Theorie des Rationalen Diskurses als Theorie der Juristischen Begründung,* Suhrkamp, Frankfurt a.M., 1978, Second edition 1991 with a reaction to critics).

Anscombre, J.-C., & Tamba I. (2013). Autour du concept d'intensification. *Langue française, 177*(1), 1-3.

Athanasiadou, A. (2007). On the subjectivity of intensifiers. *Language Science, 29*, 554-565.

Cacchiani, S. (2007, July). From narratives to intensification and hyperbole: Promotional uses of book blurbs. In *Proceedings of the Corpus Linguistics Conference, University of Birmingham*, 1-15.

Coulthard M., & Johnson A., (2007). *An introduction to forensic linguistics: Language in Evidence.* Routledge.

Doury, M. (2003). L'évaluation des arguments dans les discours ordinaires. Le cas de l'accusation d'amalgame. *Langage et société, 3*(105), 9-37.

Doury, M. (2009). Argument SShemes typologies in practice. The case of comparative argumentative norms. In F.H. van Eemeren & B. Garssen (Eds.), *Pondering on problems of argumentation* (pp. 141-155). New York: Springer.

Feteris, E. T. (2011). The study of legal argumentation in argumentation theory and legal theory: Approaches and developments. *Cogency, 3*(2), 11-32.

Feteris, E. T., & Kloosterhuis, H. (2009). The analysis and evaluation of legal argumentation: Approaches from legal theory and argumentation theory. *Studies in Logic, Grammar and Rhetoric, 16*(29), 307-331.

Goffman, E. (1974). *Les rites d'interaction*. Paris: éditions de Minuit.

Gibbons, J. (2003). *Forensic Linguistics: An Introduction to Language in the Justice System*. Australia: Blackwell Publishing,

Jaquin, J., & Micheli, R. (2012). Entre texte et interaction: Propositions méthodologiques pour une approche discursive de l'argumentation en sciences du langage, *Congrès Mondial de Linguistique Française, EPD Sciences*, 599-611.

Kleiber, G. (2007). Sur la sémantique de l'intensité. In J. Cuartero Otal & M. Emsel (Eds.), *Vernetzungen. Bedeutung in Wort, Satz und Text. Festschrift für Gerd Wotjak zum 65. Geburtstag* (pp. 249-262). Berlin: Peter Lang.

Kleiber, G. (2013). À la recherche de l'intensité. *Langue française, 177*(1), 63-76.

Labov, W. (1984). Intensity. In D. Schiffrin (Eds.), *Meaning, Form, and Use in Context: Linguistic Applications*, GURT, (pp. 43-71) Washington, D.C: Georges University Press.

Lenepveu, V. (2013). De la complétude à l'intensité: totalement, entièrement et complètement. *Langue française, 177*(1), 95-109.

MacCormick, N. (1978). *Legal reasoning and legal theory*. Oxford University Press.

Mercier, H., & Sperber, D. (2011). Why do humans reason? Arguments for an argumentative theory. *Behavioral and Brain Sciences, 34*(2), 57-74.

Molinier, C., & Levrier, F. (2000). *Grammaire des adverbes: Description des formes en –ment*. Genève-Paris : Librairie Droz.

Plantin, C. (1996). *L'argumentation*. Paris: Editions du Seuil.

Raeber, T. (forthcoming). Intensification et concepts *ad hoc*, le cas des adverbes de complétude. *Syntaxe et sémantique*, 17.

Romero, C. (2001). *L'intensité en français contemporain: Analyse sémantique et pragmatique*, Thèse de doctorat de l'Université Paris VII.

Salas, C. (forthcoming.). Oral argument based on inferences in a Swiss criminal case: How the judge accounts for the way he collects evidence in the light of the justification of legal reasons. *Proceedings of the International Conference on Legal Argumentation and the Rule of Law*. Eleven International publishing.

Sperber, D., & Wilson, D. (1995). *Relevance: Communication and cognition*, Second Edition, Oxford/Cambridge: Blackwell Publishers.

Walton, D. N. (2002). *Legal argumentation and evidence*. University Park, PA: Pennsylvania State University Press.

Walton, D. N. (2006). Examination dialogue: An argumentation framework for critically questioning an expert opinion. *Journal of Pragmatics, 38*, 745-777.

Wilson, D., & Carston, R. (2007). A unitary approach to lexical pragmatics: Relevance, inference and ad hoc concepts. In N. Burton-Roberts (Ed.), *Pragmatics*. Palgrave Macmillan.

Wigmore, J.H. (1940). *A Treatise on the Anglo-American System of Evidence, vol. I* (of 10 volumes), 3Med. Boston: Little, Brown, and Company.

58

Rustic Scepticism as Argumentation

VITOR HIRSCHBRUCH SCHVARTZ
University of São Paulo, FAPESP, Brazil
vschvartz@hotmail.com

Since Antiquity, sceptic philosophers were known to be great debaters, and figures such as Carneades were known to argue for many and conflicting positions on various subjects, as it is known from the testimony of his visit to Rome. Sextus Empiricus, a great ancient sceptic and our main source for greek Pyrrhonism, defines scepticism as an "ability to oppose (arguments)". The aim of this paper is to reflect, on the one hand, on the meaning of this account of the Sceptic School via a "rustic" interpretation of Ancient scepticism, and, on the other hand, to consider the significance of such a philosophical position in contemporary thought.

KEYWORDS: argumentation, arguments, scepticism, Sextus Empiricus, suspension of judgement

1. INTRODUCTION

The challenge to overcome scepticism, or whatever each philosopher has understood by "scepticism", characterized the works of several authors of the Modern Age such as, among others, Locke, Kant, Berkeley and Hume. Also in contemporary times, the so-called "sceptical challenge" is at the centre of some of the most important debates of Epistemology, about, for instance, the possibility of knowledge (or the possession of so-called "true and justified beliefs") or knowledge of the "outside world", among other issues. Scepticism made an imprint in the philosophy of late Wittgenstein, and has also been used by many teachers to train their students, as evidenced by the success of the book decades of Barry Stroud, "The Significance of Philosophical Skepticism" as an introduction to epistemology in numerous philosophy courses in English-speaking countries. The first translations of Sextus Empiricus into Latin in the early modern era, following the rediscovery of the work of philosopher manuscripts, encouraged the formation of rows of

D. Mohammed & M. Lewiński (eds.) (2016). *Argumentation and Reasoned Action: Proceedings of the 1st European Conference on Argumentation, Lisbon, 2015. Vol. II, 921-931.* London: College Publications.

philosophers who, regardless of the different ways in which reconciled faith and reason, developed works partially or fully sceptical; through Montaigne and reaching Descartes, whom he calls *sceptique malgré lui*, Popkin describes how the Cartesian hyperbolic doubt had an impact beyond its solution inside the work of Descartes.

If it is true, on the one hand, that since the rediscovery of the work of Sextus until present day, the object of debate surrounding scepticism are arguments, doubts, assumptions and sceptical objections - and rarely the much-discussed "scepticism" takes the form a sceptical philosophical system - it is also true that the ancient sources have bequeathed us with only one complete sceptical work, depriving us of the direct reading of the great philosophers known to have assumed the reins of the Platonic Academy in later centuries to the death of Plato. These philosophers, especially Arcesilaus, Carneades, and Clitômachus, formulated a theoretical basis without which there could have been no the foundation of Pyrrhonian scepticism by Aenesidemus in the first century BC, philosopher who left the Academy, thus founding the school known as "*Skepsis*", the Pyrrhonian scepticism.

The philosophy of the Middle and New Academies (names that was assigned to the Academy to refer to Arcesilaus leadership on) would anachronistically later be called "academic scepticism", as only Aenesidemus and his followers would have called themselves *skeptikoí*, that is, sceptics, word that literally means "those who observe", "the examinators." The etymological origin of the term "sceptic" has nothing to do with the idea of "doubt" that so frequently defines modern varieties of scepticism. These philosophers, from Aenesidemus, also titled "Pyrrhonians" because they saw in the figure of Pyrrho, a philosopher who accompanied Alexander to India, a precursor of scepticism.

Sextus Empiricus however, the one of whom we have a few complete books, a physician and philosopher who lived in the second half of the II century AD in Rome, Alexandria or Athens, was an important sceptic philosopher. His work today is our main source for the study of Pyrrhonian scepticism. It consists of the Pyrrhonian Outlines and a series of other texts, best known by its Latin title *Adversus Mathematicos* (Against the Men of Science). The Outlines are a general introduction to Pyrrhonism, and the rest a more detailed questioning of dogmatic philosophies in different fields of knowledge. The differences and similarities between the philosophy of the New Academy and the Pyrrhonism have been the subject of several studies. This talk does not address these interesting and profound historical questions, by the contrary: being the work of Sextus the only set of books a sceptic author to which we have access, for "scepticism", here,

I`m taking the philosophy of that author, regardless of what is original or not in his work when we compare it to previous sceptics or rival philosophers. My opinion is that we find, in Sextus, a defensible and fully articulated sceptical philosophy, whose study could be proved useful to the contemporary philosophical debate. The objectives here are two: to defend a specific interpretation of Sextus and to raise some arguments in favour of its strength as a philosophical position. The impact of Sextus works went beyond its role in the genesis and development of modern philosophy and in the establishment of many contemporary epistemological problems. More recently, when the contemporary exegetical debate about the Greek scepticism began, starring scholars as Michael Frede, Myles Burnyeat and Johnathan Barnes, the philosophical community witnessed an interpretive discussion of Sextus' work that embarked on philosophical paths and was a good example of a discussion in which it is almost impossible to distinguish the practice of the history of philosophy from philosophical reflection. Issues such as the "insulation", introduced by Myles Burnyeat, which pointed to the divorce between philosophy and life, characteristic of the contemporary way of doing philosophy, made of the interpretation of the Greek sources a pretext for wider judgements about all the history of philosophy as well as the introduction of new philosophical problems. That exegetical and philosophical debate over the old scepticism is often of great impact to scholars.

When met with a complete sceptical philosophy such as Sextus, the contemporary philosopher is faced with the possibility of adhering to scepticism. This is reflected in the habit, common to many experts in the subject, to argue in favour or against of Pyrrhonian scepticism, which culminates in a growing number of scholarly and creative reconstructions of Pyrrhonism. The contemporary debate offers us a huge formulation and reformulation of objections, going much further than just arguing against a caricature of scepticism never defended by any actual philosopher, but against a very interesting Greek philosophy rebuilt and sometimes updated. And the philosophical citizenship of the sceptic school, obscure for centuries among other reasons for the shortage of systematically sceptical philosophers, has so been rescued by both studies of Sextus as the work of contemporary philosophers such as Michael Williams and Robert Fogelin, among others.

2. RUSTIC SCEPTICISM AND THE ACCOUNT OF THE SCEPTICAL SCHOOL

By "rustic scepticism" scholars have taken a radical form of scepticism in which the suspension of judgement is directed against every sort of belief, even the simplest everyday belief that, for instance, the wall is white. Urbane scepticism, on the other hand, is the sort of scepticism in which the beliefs rejected by the sceptic are only of a specific sort – complex beliefs, theoretical or scientific put forward by dogmatic thinkers.

The Pyrrhonist self-image as someone free from *doxa*, free from opinions as the result of a certain dialectical practice, his self-image as someone who lives a life "without beliefs," in Sextus own words, formed a major obstacle for scholars of Pyrrhonism. Some of them have judged some of Sextus passages as ambiguous. The paradigmatic example to illustrate this apparent ambiguity is HP I, 19- 20. The texts seem ambiguous because often Sextus presents two restrictions: it restricts the extent to which the sceptic gives assent, on the one hand, stating that the sceptic adheres solely to apearences; and, on the other hand, presents a restriction Pyrrhonian question (in this case to the *logos*, which has been translated as "philosophical theory", and in other cases the restriction is to the "non-evident objects investigated by sciences", to give two examples).

Whether the ancient sceptic of Sextus allowed for any sort of belief or, in other words, the question of the scope of the suspension of Pyrrhonian judgment, is the most controversial point of the work of Sextus. Following Jonathan Barnes, in his article "The Beliefs of the Phyrrhonist" we can divide the different interpretations of the scope of Pyrrhonian *epoche* into two main groups: those who understand the sceptic Sextus as a "rustic sceptic" and those that understand as "urban sceptic". Nuances apart, all rustic interpreter supports the idea that the sceptic does not have common beliefs and the Pyrrhonian questioning is not restricted to the philosophical theory and science, but also affects the beliefs and everyday assertions of ordinary life. The urban interpretation, in its turn, distinguishes clearly between types of belief, considering some as dogmatic therefore targeted by Pyrrhonian combat, and others as non-dogmatic, shared by the sceptic with most men.

In recent papers I have argued for a rustic interpretation of the scepticism of Sextus Empiricus. The core of my argument has always been that, in light of the expressed suspension of judgement that we see in Sextus in so far as the opinions of the plain man, and also in light of the sceptic path towards *epoche* (suspension of judgement) the rustic interpretation was more convincing than its rival, the urbane interpretation.

Let us remember a few passages of Sextus that, from the very beginning of the Outlines of Pyrronhism, cause some perplexity by invoking, without further explanation, the concept of *phainomenon*. Sextus says that the septic does not pretend to affirm that things are just as he says they are, but, in PH I, 4: "we report (*apangellomen*), like a chronicler, that which appears to us at the time" (*tò nûn phainómenon hemîn*)". And the sceptic will say that the scope of 'that which appears' is immune to suspension of judgement. The *phainomenon* is the criterion of action of scepticism (PH I, 21), and the so called "positive" side of pyrrhonism is permeated by what I'll call here "phenomenism".

The "life without beliefs" of the sceptic has traditionally been understood as an "adherence to appearances". Many translators and commentators of Sextus have opted to translate *phainomenon* as "appearance". I do not intent here to evaluate the merit of such an option, but that translation bares the risk of attributing to Sextus a few anachronistic thesis, that might originate from the mind of a reader influenced by the millenary history of the concept of "appearance" in the history of philosophy, and not so much in the spirit of ancient scepticism – and this is often pointed out by people who chose that translation. Such a translation does not reflect the reach of the sceptic notion of *phainomenon*, a scope that has Sextus work and life as witnesses (since he was a physician and, I assume, saw no contradiction between his practice and his ordinary life, in one hand, and his philosophy, in the other). Even a superficial reading of the Outlines and of Sextus' other books would reveal a very large amount of descriptions of customs, of places, of people and of arguments, and a very strange philosophical maneuver would be required to define it all as a mere "expression of appearances".

Any reader of the Outlines that wishes to take "*phainomenon*" for "appearance" will find discomfort in the large scope of things Sextus is happy to write under the umbrella of a phenomenic language. Thus it would be interesting for us to explore what sort of things Sextus is talking about when he speaks of the apparent things or *ta phainomena*.

Which brings us to the passage I wish to interpret here today. In the very beginning of The Outlines Sextus defines scepticism as an ability or disposition (dynamis) to oppose. Since Antiquity, sceptic were known to be great debaters, as is known for instance of Arceliaus visit to Rome, when senator Cato tried to expel the philosopher for persuasively defending different and contradictory positions thus corrupting roman youth. The passage where Sextus first defines scepticism goes as follows:

> What scepticism is: The Sceptic Way is a disposition to oppose phenomena and noumena to one another in any way whatever, with the result that, owing to the equipollence among the things and statements thus opposed, we are brought first to epoché and then to ataraxia. We do not apply the term "disposition" in any subtle sense, but simply as

> cognate with "to be disposed." At this point we are taking as phenomena the objects of sense perception, thus contrasting them with the noumena. (PH I 8-9)

And Sextus goes furthermore:

> Does the sceptic deny appearences? And even when we do present arguments in opposition to the appearances, we do not put these forward with the intention of denying the appearances but by way of pointing out the precipitancy of the Dogmatists; for if the theory is so deceptive as to all but snatch away the appearances from under our very eyes, should we not distrust it in regard to the non-evident, and thus avoid being led by it into precipitate judgements? (PH I, 19-20)

A lot of things could be said about those passages. That ability to oppose arguments that, according to Sextus, defines scepticism, opposes that witch appears to thing said. Of course this opposition is made within language. Say, some eleatic philosopher raises an argument against the reality of motion: the sceptic will use such an argument to confront the ordinary belief in motion and also the more complex Aristotelian account of motion, thus suspending judgement. That is what Sextus means by opposing phainomena to noumena, things that appear to things that are thought. The important thing here is that, although the fact that motion appears is a good point in favour of its existence, it is not sufficient to establish its existence and the arguments against its existence leads the sceptic to complete suspension of judgement about everything he sees. The sceptic, therefore, has no opinion on those things, but acts without opinion according to the way things appear.

When Sextus says he does not abolish appearences, he means that he will act, in the above example, according to the existence of movement, even though he knows it might be an illusion. The existence of motion, in this example, can be thought of as a *phainomenon* in two senses of the word. In a way it is a perceptual *phainomenon*, sense impression leads us to believe in the reality of motion. But it is also a common sense belief, and in this sense it is also a *phainomenon*, something that appears to everyone. Insofar as his criterion to act, the sceptic will act accordingly, but insofar as it is a matter of truth, the sceptic will suspend his judgement. Dogmatists, sceptics and ordinary people share the *phainomenon* of motion, but the sceptic is the one who refrains from hypostatizing it.

The rustic interpretation of the ancient scepticism of Sextus Empiricus accounts for the enormous distance between the pyrrhonists and the ordinary people: the sceptic knows that things can always be different from the way which they appear, while the plain man usually trust his experience to formulate his everyday beliefs about reality.

3. RADICAL SEXTAN SCEPTICISM

One of the main controversies surrounding the interpretation of ancient scepticism is, nevertheless, the question of what is considered "dogmatic" by the sceptics, and thus targeted by the sceptic arsenal, and what would be considered "phenomenic" and therefore immune to sceptical arguments. Although all recent scholarship has shown us that the sceptic philosophy of the past is profoundly different from the caricatures made of it in ancient and modern times, scholars still seek to assess to what extent the alleged life without beliefs of the Sextus sceptic actually differs from ordinary life.

The interpretation of Sextus I`ve defended in previous papers was one in which even everyday propositions such as "the wall is white" or "the honey is sweet" are rejected by the sceptics. Passages in the Outlines of Pyrrhonism explicitly include the opinions of the plain men in the controversies subject to the sceptic suspension of judgment. When introducing the famous Aggripean five modes leading to the suspension of judgment, in PH I, 165, Sextus states that: "And the (mode) of controversy is the one according to which, to every subject proposed, we find an undecidable controversy both among philosophers and ordinary people (...)"

Fredian interpreters would say that the fact that common people`s opinions are included in this controversy poses no problem for their interpretation, since Sextus could be referring to moral and religious opinions in which everyday proposition are considered dogmatic even when innocently stated by ordinary men, in any interpretation. But in PH, 169 Sextus writes that anything that can be investigated is subjected to the five modes, and we know that philosophical investigation transcends moral and religious opinions. In other passages, the plain man's opinions are included in the controversy about the existence of motion (PH III, 65, 81), and the ordinary opinion is paraphrased by Sextus as "introducing the hypostasis of movement".

In common life, the existence and reality of movement are assumed, but this assumption does not take into account philosophical arguments against the existence of movement, such as we find in Eleatic philosophers, for example. Sextus is thus attributing a realistic stance to both realist philosophy and the plain man.

Passages like the aforementioned - where we find Sextus suspending judgment on the plain man's opinions - represent a insuperable obstacle to Frede, Porchat and Fogelin`s interpretation of ancient scepticism. I should also quote the text AM VIII, 362, in which Sextus makes reference to a general disagreement about things that appear between philosophers and ordinary men: "But we argued earlier that things that appear, whether sensible or intelligible, are the subject of great controversy both among philosophers and among ordinary people. (AM VIII, 362)"

But the text HP III, 65, certainly does not refer to moral or religious matters. In it, Sextus describes the conflicting positions and who suspends judgment then about the existence or non-existence of motion, of its reality or unreality. And the first of these positions in friction, the Sextus attributes to both specific philosophers and ordinary men.

In ordinary life, it is assumed that the movement exists, that it is real, but this way of thinking obviously does not take into account philosophical arguments against the existence of the motion. If this way of thinking is considered to be dogmatic by the sceptic, then we have here a good example of a common belief on which the sceptic suspends judgment. Sextus goes on by stating that ordinary people and some philosophers "introduce the hypostasis of the motion", which means in this context to substantiate it, say it existent and real - and the whole context, there is including the passages that precede the text shows that Sextus is assigning for philosophers and the common people this dogmatic stance. The use of an ontological term as "hypostasis", understood here as "reality" shows, in my view, a very incisive characterization of Sextus opinion in so far as common beliefs. If the sceptic claims to reject all dogmatism and at the same time understand as dogmatic the speech of ordinary people, Sextus is obviously rejecting the beliefs of ordinary life, daily expressed in his everyday speech.

The texts in which we find Sextus clearly suspending judgment about everyday life proposition are, in my view, one of the main aspects that favour a rustic interpretation of the work. The other major aspect is the idea, found in Sextus, of a sceptic path towards *epoche*. To explain it, the rustic interpretation is also more convincing than its rival, the urbane interpretation. But to what extent does a rustic approach to scepticism is a viable philosophical option? Is it possible to follow that which appears without having any belief?

4. REALLY SUSPENDING JUDGEMENT ABOUT EVERYTHING

Paulo Faria, in his paper "A encenação", explains the apparently impossible to solve problem put forward by the sceptic, that cleverly triumphs in accepting the description of objects, stating the unquestionable (*azetetos*) character of the phenomena, questioning only the claims of knowledge that go beyond what appears. The sceptic would actually be confessing his ignorance about objects which, in fact, allow for no discovery, since "everything we can know about objects is within their descriptions". In Faria's metaphor, the sceptic creates a scenario without solution: "(...) is as if the visitor of the Louvre wished to see the back of the Gioconda".

Nevertheless, as I see it, the sceptic phenomenic language has no claim of truth or knowledge whatsoever, so that the pyrrhonian strike also hits the simple belief of the Louvre visitor that really thinks he is before a famous painting. In light of the passage of M VII, 366, the sceptic can't know Gioconda's face any better that her back. The Louvre metaphor might give the impression that the critic of the *lógos* put foward by the sceptics is restricted to knowledge of essences, as in fact the sextian reflection shows that, ultimately, Gioconda's face is as unknowable as her back. And, in this respect, the sceptic totally differ from ordinary folks! Even though ordinary people don't have a complex epistemology to explain their beliefs in things, they are willing to maintain that they know at least some of them. Thus, a tower that seems round in a distance, but is verified to be square with closer observation, is judged by both the stoics (for example) and by ordinary people to be so.

Many passages in Sextus help us to see the radicalism of scepticism and helps us to understand why the sceptic needs a peculiar vocabulary to use the language: the phenomenic discourse, that has nothing to do with expressions like "is seems to me that it's going to rain".

Some contemporary philosophers call themselves neopyrrhosists, such as Porchat and Fogelin. They both argue that their philosophies don't depend on a historically correct interpretation of the neophyrronism of Sextus Empiricus, since the mere historic imprecision wouldn't affect the strength or coherence of their philosophies (See Fogelin, 2008, pp. 163-4; Porchat, 2006, preface). But one could ask if the old scepticism, rustically understood, can also be a source of inspiration for contemporary philosophical reflexions, since it offers a philosophical position aligned with the current tendency of praising the ancient schools of thought for not insulating their philosophies from the ordinary life.

The contemporary neopyrrhonist complains, apparently rightfully, of the caricature imposed on the sceptics thorough centuries, by ancient and modern philosophers. But the caricature that the neopyrrhonist makes of methaphisics is not smaller. Maybe a way to conceive a contemporary rustic neopyrrhonist would be to imagine someone much more susceptible to "metaphysical" reflexions, not so much unwilling to believe in the power of *lógos* to unveil truths as the urbane sceptic. Within his dialectic ability of opposing things that appear and thoughts, the rustic pyrrhonist in fact suspends judgement about the proposition "Gioconda's face is white", as well as about any other dogmatic proposition, because he recognizes no privilege of the senses over opinions, or vice-versa.

Such a pyrrhonism will have to deal, more than its counterpart, with the famous objection of *apraxia*, according to which it's impossible to act without having beliefs. I think that the sceptic phenomenic language, correctly understood, doesn't threaten the coherence of scepticism, but there is still much to say about this point. Anyhow, I invite the reader to an analogy: a fictitious eleatic philosopher that, convinced by his reading of Parmenides, "knows" that all those statements about things we make in our ordinary life are false, wouldn't be so much confronted with accusations of incoherence as the sceptic. Such a hypothetical eleatic philosopher leads a life within a world he "knows" to be an illusion, he believes in the falseness of most of the propositions he uses in his ordinary life. I ask why is it that we find more resistance against a more humble philosophy, that of a pyrrhonist that, having taken seriously many different philosophies and having found equipollence and consequently suspension of judgement, says he follows that which appears without believing or hypostasizing them. Not many people rise de accusation of incoherence against one who knows, at every second, that the world is totally different from the way in which it appears, but many seem to have a problem with the sceptic that only thinks that the world might be different from the way in which it appears.

ACKNOWLEDGEMENTS: This ongoing research on scepticism wouldn`t be possible without the support of a FAPESP grant.

REFERENCES

Burnyeat, M. (1980/1983). Can the skeptic live his skepticism? In M. Schofield, M. Burnyeat & J. Barnes (Eds.), *Doubt and dogmatism: Studies in Hellenistic epistemology* (pp. 20-53). Oxford. (Reprinted in M. Burnyeat (Ed.) (1983), *The Skeptical Tradition* (pp. 117-148). Berkeley.)

Faria, P. (2007). A encenação. *Revista Sképsis, 2*(1), 99-130.

Fogelin, R. J. (2004). The sceptics are coming! The sceptics are coming! In W. Sinnott Armstrong (Ed.), *Pyrrhonian Scepticism* (pp.161-173) Oxford: Oxford University Press.

Mates, B. (1996). *The sceptic way – Sextus Empiricus's outlines of Pyrrhonism.* New York / Oxford: Oxford University Press.

Porchat Pereira, O. (2006). *Rumo ao ceticismo.* São Paulo: Editora Unesp.

Schvartz, V. H. (2012). Epokhé e lógos no pirronismo grego. In: W. J. Silva Filho & P. J. Smith (Eds.), *As consequências do ceticismo* (pp. 75-94), 1ed. São Paulo: Alameda Editorial.

Sextus Empiricus (1976). In four volumes (v.1: Outlines of pyrrhonism; v. 2-4: Adversus Mathematicos VI-XI). Cambridge, MA & London: Loeb Classical Library, Harvard University Press and William Heinemann.

59

Multimodal Argumentation in a Climate Protection Initiative on Austrian Television

ANDREA SABINE SEDLACZEK
University of Vienna, Austria
a.sedlaczek@tele2.at

This paper examines multimodal argumentation in factual television programmes that were broadcast as part of a climate protection initiative in Austria. With the methodological approach of a multimodal critical discourse analysis the analysis investigates the television programmes as complex multimodal texts that include argumentation aimed at convincing the audience of the need for active engagement with climate change mitigation on the macro as well as the micro level.

KEYWORDS: climate change, critical discourse analysis, documentary film, multimodal argumentation, multimodality, normative argumentation, television

1. INTRODUCTION

This paper will contribute to the growing research and discussions about visual and multimodal argumentation in argumentation theory (Kjeldsen, 2015b), by providing a view of multimodal argumentation in the context of documentary television. The context of the television programmes, which form part of my research, is a climate protection initiative of the Austrian public service broadcaster ORF.[1] The aim of this initiative is to raise public awareness on the issue of climate change and to promote active engagement with climate change mitigation measures. This last goal is the most relevant to the focus of this paper, as it can be connected to the main theme of this conference – argumentation and reasoned action. I assume that in order to achieve

[1] For more information see the – since discontinued – web page of the initiative as captured on the Internet Archive: http://web.archive.org/web/20120214045528/http://klima.orf.at/.

D. Mohammed & M. Lewiński (eds.) (2016). *Argumentation and Reasoned Action: Proceedings of the 1st European Conference on Argumentation, Lisbon, 2015. Vol. II, 933-946.* London: College Publications.

the goal proposed by ORF's initiative, the television programmes broadcasted will advance normative claims aimed at convincing the audience of the need for active engagement with climate protection.

The paper will focus on normative argumentation in a television documentary about climate change and will take account of the complex interplay between different semiotic modes at work in this process. Adopting a view on multimodal argumentation that is informed by the theoretical and methodological framework of multimodal critical discourse analysis, I will argue that documentary television programmes are complex multimodal texts that may involve argumentation on the micro as well as the macro level.

2. MULTIMODAL ARGUMENTATION FROM THE PERSPECTIVE OF A MULTIMODAL CRITICAL DISCOURSE ANALYSIS

My view on multimodal argumentation is based on the theoretical and methodological framework of multimodal critical discourse analysis, which connects critical discourse analysis with approaches to semiotics and multimodality as well as argumentation theory.

My approach starts from the Discourse-Historical Approach (Reisigl & Wodak, 2009), which considers argumentation to be an intrinsic part of discourse. Discourse is seen as a set of context-dependent semiotic practices, which mainly relate to the argumentation of validity claims of truth and normative rightness between social actors with different perspectives. The Discourse-Historical Approach has incorporated different concepts and approaches to argumentation theory in its framework (Reisigl, 2014), both from the German as well as the English speaking scientific community (Kienpointner, 1992; Toulmin, 1958; van Eemeren & Grootendorst, 1992). In this paper, I will build on a basic functional model of argumentation, adapted from Toulmin's famous model, which distinguishes between three central elements of argumentation: a claim or conclusion, one or several grounds or arguments and a warrant or conclusion rule that connects the argument(s) to the claim (Reisigl, 2014, p. 75).

This basic view on argumentation is combined with a semiotic view on multimodal communication – informed by social semiotics (Kress, 2010; van Leeuwen, 2005) and the semiotic theory of Charles S. Peirce – as well as insights from the ongoing discussions about visual and multimodal argumentation in argumentation theory. These previous contributions to the study of multimodal argumentation have likewise built upon different approaches to argumentation theory, including Toulmin's model or Pragma-Dialectics (Groarke, 2009; Kjeldsen, 2011, 2015b). They have demonstrated that visual and other modes of expression can have an argumentative value similar to verbal argumentation.

Kjeldsen (2015b, p. 119) points out how scholars dealing with multimodal argumentation have moved between stressing two seemingly contradictory facts. On the one hand, argumentation is perceived as primarily a cognitive phenomenon, which takes place in the mind of the participants and can therefore be triggered by verbal as well as other forms of expression (van den Hoven, 2015). On the other hand, scholars have focused on the ways the different qualities of verbal and non-verbal forms of expression influence the nature of the argumentation (Kjeldsen, 2015a). I agree with both of these assumptions and conceptualize argumentation as a cognitive as well as a semiotic phenomenon (Reisigl, 2014, p. 70). Argumentation is embedded in a dialogic process in which one participant wants to give reasons for a contested claim and so produces a text, using whatever modes available. They then structure the text in a - more or less transparent - way to make the claims and arguments manifest and to guide the reconstruction of the argumentation by the recipients. The text, however, has to be interpreted by a recipient in order for it to function as argumentation.[2] This reconstruction by the recipient involves a cognitive process, which is influenced by contextual factors as well as previous knowledge. The importance of the recipient's interpretative performance is especially relevant for enthymematic argumentation, where not all the elements of argumentation are expressed explicitly in the text, but have to be completed in the mind of the recipient by inference. Such an enthymematic quality is essential for most visual and multimodal argumentation, as previous research has emphasised (Kjeldsen, 2015b).

While much of the previous research has approached non-verbal argumentation by reconstructing it in a verbal form (Blair, 2015; Groarke, 2009; Kjeldsen, 2011), the problems and shortcomings inherent in this "translation" are widely acknowledged (Kjeldsen, 2015b, p. 119). As work on visual and multimodal communication has shown, different modes of communication have different "affordances" and "constraints" (Kress, 2010; van Leeuwen, 2005) - therefore a verbal "translation" of a non-verbal text will never be equivalent to this text. In addition, the specific semiotic qualities of non-verbal forms of expression can make it difficult to clearly separate or identify the different elements of argumentation or their relationship to one another (Kjeldsen, 2011; 2015b, p. 119). Thus, interpretations and re-

---

[2] This mirrors the semiotic principle formulated by Peirce that a sign only functions as a sign if it is interpreted (Rellstab, 2007, p. 154).

constructions of argumentation in multimodal texts may vary considerably between different recipients.

As Blair (2015, p. 220) has argued, reconstructing multimodal argumentation in a verbal form could be more adequately seen as a "meta-argumentation", where the analyser provides reasoning for possible interpretations of the argumentation in the multimodal text. In order to arrive at such a reconstruction, the analysis of multimodal argumentation should closely investigate how the different modes of expression are used in constructing the text and what their argumentative value could be.

This analytical perspective applies to both the microstructure as well as the macrostructure of multimodal texts. This distinction is especially relevant for audio-visual filmic texts, which consist of a purposefully edited and structured succession of shots and which exhibit complex relationships between their visual, verbal and sound tracks on various levels. This structure of the film text can be investigated by using the different filmic levels that Iedema (2001) suggests in his social semiotic approach to the analysis of film and television. Apart from the basic levels of shots, scenes and sequences, which are established categories in film analysis, Iedema especially introduces the level of generic stage, which realises larger structures in the filmic text. Iedema relates the concept of generic stages to the genre that the film as a whole belongs to. He distinguishes between narrative and expository film genres. Narrative films display a predominantly narrative structure, which can be prototypically reconstructed as the generic stages of orientation, complication, resolution and coda. Expository genres, on the other hand, may have an argumentative or descriptive structure, which can consist of an introductory stage, several stages in between, which express arguments or facts, and a concluding stage[3] (Iedema, 2001, pp. 188-191).

While Iedema sees these different stage structures as mutually exclusive, dependent on the genre of the film, I argue that it is more useful to speak of the different structuring principles that may intersect in one film, regardless of its genre. To this end, I combine Iedema's concept of stages with a concept from text linguistics, where several textualisation patterns or patterns of text formation are distinguished, which in turn constitute and establish texts. These include narration, argumentation, description, explication and instruction (Heinemann, 2000, pp. 356-359; Reisigl, 2014, p. 72). In a given text, these basic

[3] These filmic stages should not be confused with the stages of argumentation (confrontation stage, opening stage, argumentation stage and concluding stage) identified in Pragma-Dialectics (van Eemeren & Grootendorst, 2004).

structuring principles may combine on different levels. A documentary film will usually display a stage structure that manifests one dominant pattern of text formation, e.g. argumentation or narration, but the structure of the stages themselves will combine argumentative, narrative, descriptive and explicatory parts.

To demonstrate this view of multimodal argumentation I will now analyse a documentary film about climate change, first focusing on a short segment of the film on the micro level and then investigating the argumentative structure of the film on the macro level.

3. MULTIMODAL ARGUMENTATION ON THE MICRO LEVEL

The analysis of multimodal argumentation on the micro level will concentrate on a short scene from the beginning of a documentary film on climate change (which can be viewed online in the context of its documentary film: https://www.youtube.com/watch?v=4H7OLpcO8PU, from 0:29 to 0:59). I will analyse the argumentation of this scene in two steps: first I will focus only on the visuals, music and sound and next I will incorporate the verbal content.

3.1 Visuals, music and sound

In order to approach the argumentation of this film segment, a description of its visual and auditory contents, excluding verbal information, follows: The scene begins with a satellite view of houses in a city, set to low, slow and monotonic violin music. At the same time the slow ticking of a clock begins, which accompanies the whole scene. With a quick zoom-out, which includes a camera pan to the right, the satellite view expands. It soon encompasses the whole city and then the whole west coast of North America, covered with white clouds. This quick zoom-out is enhanced by a rushing sound. As the zoom-out continues, the satellite view seamlessly transforms into a graphic representation of a rotating globe in a black and starry space. The surface of this globe is occupied by a number of toy-like three-dimensional and partly animated graphic objects: On the continents are cattle, farm houses, smoking power plants, skyscrapers and cranes; on the oceans are cargo vessels and oil platforms; bridges with moving cars and power lines span over the continents; and planes are flying across the Earth, trailing vapour trails behind them. A long cylindrical body in white bandages swirls around the globe. As it pauses to hover in front of the globe, the bandages fall off with a high tinkling sound, revealing a ghost-like figure. Its head, body and arms are made of grey smoke and it has two black dots as eyes. This smoke figure resumes swirling around the globe –

reinforced by a deep rushing sound – and trails a cloud of greyish-white smoke behind it. This smoke increasingly engulfs and enshrouds the Earth. The music swells with an electronically enhanced crescendo as the smoke figure rises from behind the globe. The scene ends with the smoke figure hovering above the smoke-covered Earth, with its arms extended horizontally on both sides. This is accompanied by loud dramatic music with hard drumbeats.

As can be noted from this description, the film segment relies heavily on tropes such as visual metaphor, metonymy and synecdoche in its construction of meaning. The highly stylized and toy-like three-dimensional objects on the globe constitute a simplified synecdochical representation of our modern industrial societies that release greenhouse gases. The greenhouse gas emissions themselves, which are in reality mostly invisible (with the slight exception of the vapour trails from the planes and the industrial smoke and vapour clouds of the power plants), are metaphorically personified as a ghost-like smoke figure, as well as metonymically represented as a cloud of smoke emanating from this ghostly figure and enshrouding the Earth.

Apart from its reliance on visual tropes, the film segment also sets a clear chronology, which will guide the possible reconstruction of argumentation. The smoke figure is at first bound in white bandages (in the manner of a mummy), but the bandages soon fall off (apparently of their own accord), releasing the smoke figure, which only then is able to enshroud the Earth in its cloud of smoke. This chronology is mirrored by the music. At first, the music is foreboding, insinuating an impending danger. In the end, the music becomes more dramatic with the use of a crescendo and hard drumbeats, emphasising the threatening nature of the smoke figure itself and of the Earth being "smothered" by greenhouse gases.

Previous work has already demonstrated how tropes such as metaphor and metonymy are particularly useful in constructing visual argumentation. They direct the attention towards certain aspects of their represented objects and thus guide the reconstruction of the intended argumentation (Kjeldsen, 2011). In the case of this film segment, the metonymic representation of greenhouse gases as thick fumes of smoke that fill the Earth's atmosphere, the metaphoric representation of greenhouse gases as a menacing spectre as well as the use of dramatic music all invite the reconstruction of a claim that greenhouse gases threaten our Earth. The visual narrative of the scene with its metaphoric representation of greenhouse gases being "released" and increasingly engulfing the Earth in turn offers a causal argument for this claim.

While the tropes and other multimodal resources employed in this film segment offer compelling arguments, they also represent the issue of climate change in a simplified way that obscures important causal relationships. The graphic representation of our industrial societies may allow the inference that these industrial activities are responsible for greenhouse gas emissions. However, the metaphoric and metonymic representation of greenhouse gases as a ghost-like figure that seemingly frees itself from its bonds and that engulfs the Earth in its smoke at the same time disguises this cause-effect-relationship. Through the process of personification, the agency and blame for the causes of climate change is externalised to this greenhouse gas figure that threatens us and our Earth.

One aspect of the audio-visual content of the film segment is still left to be mentioned: The whole segment is accompanied by the slow ticking of a clock. While this might indicate the passing of time, it can also be interpreted metaphorically as "the clock is ticking" or "time is running out". The sound effect of the ticking clock thus adds another argument to the previously established claim, which together can lead to a normative conclusion: The Earth is threatened by greenhouse gases and time is running out – therefore, something has to be done. Whether such a normative claim is likely to be inferred from the scene will, however, depend on its verbal content, to which I now turn.

3.2 Verbal voice-over

Before I come to the verbal part of the film segment, I have to clarify the context of the documentary that the analysed scene comes from. The scene is a segment from the beginning of a documentary film about climate change, which was originally produced in France by the French film maker Yves Billy. The French film was called *La post-carbone attitude* (The post-carbon attitude) and it was the second of two documentaries by the same film maker (the first one being called *Mister Carbone*) that were both broadcast on the Franco-German television channel Arte in November 2010 as part of a thematic evening on climate change (Billy, 2010).

The documentary was later adopted by the Austrian public-service broadcaster ORF and broadcast by ORF in February 2012 as part of its own climate protection initiative. In this context, the documentary was embedded in a programme of ORF called *Weltjournal*. It was, however, aired in a shortened and edited version under the title *Der gestresste Planet* (The stressed planet) (Billy & Neuhauser, 2012). This edited version of ORF has a new verbal voice-over commentary that differs from the original voice-over in some parts. As these different voice-overs may change the multimodal argumentation, I will briefly compare the original Arte version of the analysed segment with the version by ORF.

In both versions the voice-over begins while the bandaged figure is seen swirling around the globe and ends shortly before the drumbeats come in, which accompany the climax of the scene. In the original French version of Arte, the male voice-over reads as follows: [4]

> (1) *he remains invisible, -- but is on the rise -- and disrupts the climate. --- he saturates the oceans and the atmosphere. --- he quickly pervades - all sectors of human activity. --- the saga of mister carbon continues. --- the race against time has begun.*

This verbal voice-over takes up the metaphoric personification of greenhouse gases visualized as the smoke figure and specifies it with the name "Mister Carbon". Thus, carbon dioxide or CO2, which constitutes the most important greenhouse gas, is used to stand synecdochically for all greenhouse gases.

The argumentative content of this voice-over mostly mirrors or specifies the previously established audio-visual argumentation. The personified "Mister Carbon" is claimed to be a threat, as he "disrupts the climate, saturates the oceans and the atmosphere and pervades all sectors of human activity". As in the visual representation, the human causes of greenhouse gas emissions remain obscure. Instead the blame is shifted to a metaphoric character that conquers our human societies.

The statements that Mister Carbon is on the rise, that he is quick and that his "saga" continues, all mirror the sound effect of the ticking clock and lead to the final verbal claim "the race against time has begun". Thus, the multimodal argumentation of this version of the film segment clearly advances a normative claim: CO2 disrupts the climate, thus posing a threat. CO2 is increasing its influence rapidly, thus time is running out. Therefore, we have to race against time (and do something).

In contrast to the original Arte version of the documentary, the film segment in ORF's version of the documentary has a different, female, voice-over:

---

[4] All examples given are rough translations from the original French or German by the author. The transcriptions of the spoken voice-over are set in lower case throughout. Capital letters are used to indicate stressed syllables. A comma and a full stop both signal a falling intonation. One, two or three hyphens between words mark pauses of different length.

> (2) *the japanese satellite ibuki - measures the concentration of the most important greenhouse gases from space. -- without water vapour and ce-o-two the earth would be cold and uninhabitable. - too much of these greenhouse gases, - and the planet overheats.*

In this version, the visually metaphorical personification of the CO2 or greenhouse gas figure is not verbalised. The verbal voice-over instead keeps to the "real" greenhouse gas emissions, which are scientifically measured in the atmosphere by a satellite (thus linking this statement to the satellite view of the Earth in the beginning of the scene), and establishes a cause-effect-relationship: "too few greenhouse gases and the Earth would be cold and uninhabitable – too much greenhouse gases and the planet overheats". This cause-effect-relationship can be criticised for being overly simplistic, as it does not differentiate between the natural and the anthropogenic greenhouse effect. The former mostly relies on water vapour and prevents the Earth from being cold and uninhabitable. The latter is caused by man-made greenhouse gases, mostly CO2, and results in global warming. Thus, the precise contribution of human industrial activities to climate change is again left unclear.

In the argumentative context of the film segment the second part of the verbally established cause-effect-relationship – "Too much greenhouse gases and the planet overheats" – can be interpreted as a verbalized warrant. The visual depiction of the Earth being increasingly enshrouded in a thick layer of smoke provides the visual argument that there are too many greenhouse gases. Thus, the conclusion can be inferred that the planet will overheat – a conclusion that in turn can be connected to the visual representation of the Earth being "smothered" by thick fumes of smoke.

In this version of the segment, there is no explicit verbal normative claim. However, as this scene is only a small segment from a larger documentary, such a normative claim for action might still follow. To examine this assumption, I will now turn to an analysis of the macro-structure of the whole film and its argumentative value.

3. MULTIMODAL ARGUMENTATION ON THE MACRO LEVEL

The analysis of multimodal argumentation on the macro level will look at the argumentative macrostructure of the documentary that was broadcast by ORF as part of its climate protection initiative. In this context, the documentary was embedded in ORF's programme *Weltjournal* and introduced by a presenter in a television studio. This preliminary introduction by the presenter is important to consider, as it provides an interpretative framework that will guide the reconstruction of the argumentative structure of the film by the recipients. The introduction advances two main claims: First, our planet is stressed, because our lifestyle is overheating the Earth. And second, renewable energies are increasingly advanced as possible solutions, but they have little effect, because the US and China as the biggest environmental polluters exploit our Earth without any consideration of the impacts on the climate. Linking this introduction to the aim of ORF's initiative, one might ask, what normative claims for climate protection action could be advanced in this documentary.

The macrostructure of the documentary itself can be dissected into eight stages. The first introductory stage begins with the scene that I analysed on the micro level, which established the fact that too much greenhouse gases would lead to global warming. This scene is followed by two sequences, which are set to melancholic piano music. The first sequence consists of shots of a red sun and a dry branch, while the voice-over designates the causes of the rising CO2 emissions: "the fossil energy sources that have fuelled our progress of the last decades produce CO2". The second sequence includes a talking head shot of a meteorologist, who advances an argument, leading to a normative claim: "As long as we continue to emit greenhouse gases, the climate will change more and more", therefore "we don't have time to wait ten more years before we seriously do something about it."

The introductory stage ends with a shot, in which the smoke figure that was introduced in the beginning is seen spiralling in front of a black sky, accompanied by thunder and dissonant piano and violin music. Meanwhile, the voice-over states that signs of a warming climate are already noticeable, but CO2 emissions are still rising. The smoke figure acts as a recurring visual motif in the whole film, mostly appearing at the beginning or end of stages and serving as a reminder that the dominance and threat of CO2 still persist.

After introducing the problem and calling for a solution in stage one, six stages follow that explore the causes of CO2 emissions in our current fossil-based societies as well as the potentials and drawbacks in the developments of alternative energy sources. The stages are mostly composed of sequences that are structured in an argumentative or descriptive way, but also include a few narrative scenes and episodes as illustrating examples. The argumentative line advanced by these stages can – in a very broad way – be reconstructed as follows: (1) We have to reduce our dependency on fossil energy sources, but there are no quick solutions with alternative energy sources. (2) Even though investment in renewable energies is growing, CO2 emissions are still increasing. (3)

Renewable energies suffer from their current lack of energy efficiency. (4) China is significantly advocating renewable energies, but is still the biggest emitter of CO2. (5) The Western economic growth philosophy advanced by the US is the obstacle towards an energy efficient future. (6) Local initiatives, which try to set an environmentally friendly example, do not have a lasting effect on the global economy, but can act as blueprints for future social orders.

Following this argumentative exploration of the issue, the concluding stage eight in the end contemplates the likelihood of solutions to the problem. It includes the following remarks of the voice-over:

> (3) *optimists believe, that we will technically gain control of even climate change, and bank on technological solutions for humanity's energy thirst. idealists rely on self-denial, for example by leaving the car behind. the realists - calculate our future energy demand. oil plus thirty per cent. gas plus fifty per cent, coal plus forty per cent. the ce-o-two emissions will therefore surely rise too.*

While this verbal conclusion seems to advocate a more pessimistic view, in line with the initial introduction by the presenter, it is interesting to note that the music accompanying this sequence is not equally gloomy, but a lively and energetic guitar strumming. Visually the film ends with another shot of the smoke figure, which flies across the sky above a city, rising and falling, while its black smoke turns white. In the end, the smoke figure drops below the screen and the sound of a thud is heard, implying that the figure has crashed into the ground. Therefore, the verbal statement that CO2 emissions will surely rise is contradicted by the audio-visual scene, which seems to advocate a more optimistic view that CO2 levels may fall.

This ambivalent outlook is also taken up by the presenter in the studio, who adds his own concluding statement to the documentary and reinforces a normative claim for action:

> (4) *optimist, - realist, - idealist. it is in YOUR hands, even every single one of us can make one's contribution, so that FUTURE generations will find a yet - reasonably liveable environment.*

This normative claim for action – it is in your hands – is directly linked to the climate protection initiative of ORF, as this initiative was promoted under the title "Our climate – It lies in our hands".

4. CONCLUSION

In this presentation I wanted to show that documentary television programmes are complex, multimodal texts that involve argumentation on the micro as well as the macro level. Looking at argumentation at the micro level of documentary film texts, it is important to take account of all semiotic modes employed and to investigate the complexities of the multimodal interplay in the representation and argumentation process. As the analysis of two versions of a short scene from a documentary about climate change has shown, the visuals as well as the music and sounds of documentary film may provide claims and arguments, but the interpretation of argumentation will also depend on how these are contextualised in verbal statements.

Apart from the multimodal microstructures, documentary film texts also have complex multimodal macrostructures. These can involve different structuring principles and can be reconstructed with the concepts of stages and patterns of text formation. In the case of a predominantly argumentative film, the stage structure can be used to reconstruct the complex chain of argumentation of the film as a whole. A more detailed investigation of the argumentative macrostructure of the chosen documentary was beyond the scope of this paper.

My analysis focused on a television documentary that was broadcast as part of a climate protection initiative by the Austrian public service broadcaster ORF. By applying the aim of this initiative to the conference theme of argumentation and reasoned action, I was particularly interested in normative argumentation aimed at convincing the audience of the need for active engagement with climate protection in this documentary. The documentary advanced the normative claim that greenhouse gas emissions have to be reduced, and investigated the potentials and shortcomings of renewable energy sources as possible solutions. The conclusion of the documentary displayed an ambivalent relationship between a more pessimistic view expressed verbally and a more optimistic view implied by the visuals and music. The ways in which this particular documentary might help to motivate the audience to engage with climate protection therefore remains unclear.

REFERENCES

Billy, Y. (Writer/Director). (2010). La post-carbone attitude. In Auteurs Associés (Executive Producer), *Soirée Thema: Ce carbone qui nous enfume!* France: Arte.

Billy, Y. (Writer/Director) & Neuhauser, C. (Editor). (2012). Der gestresste Planet. In W. Erdelitsch (Executive Producer), *Weltjournal*. Austria: ORF.

Blair, J. A. (2015). Probative norms for multimodal visual arguments. *Argumentation, 29*(2), 217-233.

Groarke, L. (2009). Five theses on Toulmin and visual argument. In F. H. van Eemeren & B. Garssen (Eds.), *Pondering on problems of argumentation. Twenty essays on theoretical issues* (pp. 229-239). Amsterdam: Springer.

Heinemann, W. (2000). Vertextungsmuster Deskription. In K. Brinker, G. Antos, W. Heinemann, & S. Sager (Eds.), *Text- und Gesprächslinguistik: Ein internationales Handbuch zeitgenössischer Forschung. 1. Halbband* (pp. 356-369). Berlin: de Gruyter.

Iedema, R. (2001). Analysing film and television: A social semiotic account of Hospital: an Unhealthy Business. In T. van Leeuwen & C. Jewitt (Eds.), *Handbook of visual analysis* (pp. 183-204). London: Sage.

Kienpointner, M. (1992). *Alltagslogik: Struktur und Funktion von Argumentationsmustern*. Stuttgart: Frommann-Holzboog.

Kjeldsen, J. E. (2011). Visual tropes and figures as visual argumentation. In F. H. van Eemeren, B. Garssen, D. Godden, & G. Mitchell (Eds.), *Proceedings of the 7th Conference of the International Society for the Study of Argumentation: ISSA 2010* (pp. 949-960). Amsterdam: Sic Sat.

Kjeldsen, J. E. (2015a). The rhetoric of thick repesentation: How pictures render the importance and strength of an argument salient. *Argumentation, 29*(2), 197-215.

Kjeldsen, J. E. (2015b). The study of visual and multimodal argumentation. *Argumentation, 29*(2), 115-132.

Kress, G. (2010). *Multimodality: A social semiotic approach to contemporary communication*. London: Routledge.

Reisigl, M. (2014). Argumentation analysis and the Discourse-Historical Approach. A methodological framework. In C. Hart & P. Cap (Eds.), *Contemporary Critical Discourse Studies* (pp. 67-96). London: Bloomsbury.

Reisigl, M., & Wodak, R. (2009). The Discourse-Historical Approach (DHA). In R. Wodak & M. Meyer (Eds.), *Methods of critical discourse analysis* (2nd ed., pp. 86-121). London: Sage.

Rellstab, D. H. (2007). *Charles S. Peirce' Theorie natürlicher Sprache und ihre Relevanz für die Linguistik: Logik, Semantik, Pragmatik*. Tübingen: Narr.

Toulmin, S. (1958). *The uses of argument*. Cambridge: Cambridge University Press.

van den Hoven, P. (2015). Cognitive semiotics in argumentation. *Argumentation, 29*(2), 157-176.

van Eemeren, F. H., & Grootendorst, R. (1992). *Argumentation, communication, and fallacies: A pragma-dialectical perspective*. Hillsdale: Lawrence Erlbaum.

van Eemeren, F. H., & Grootendorst, R. (2004). *A systematic theory of argumentation: The pragma-dialectical approach.* Cambridge: Cambridge University Press.

van Leeuwen, T. (2005). *Introducing social semiotics.* London: Routledge.

60

Reasoning Types and Diagramming Method

MARCIN SELINGER
University of Wrocław, Poland
marcisel@uni.wroc.pl

By associating and combining Ajdukiewicz's classification of reasoning with argumentation diagrams and schemes we show how to represent a rich variety of reasoning types with different degrees of complexity as inference, derivation, justification, verification (i.e. confirmation and falsification) or explanation. We also indicate some meta-schemes concerning the process of reasoning itself, and we discuss diagrams and meta-schemes assigned to abduction.

KEYWORDS: abduction, argumentation diagrams, argumentation schemes, derivation, explanation, inference, justification, metalanguage, reasoning, verification

1. INTRODUCTION

One of the original achievements of the Lvov-Warsaw School in methodology, which can be an object of interest of the argumentation theory, is the development of the classification of reasoning (*cf.* Koszowy & Araszkiewicz, 2014). It was primarily introduced by Łukasiewicz (1912), who followed Twardowski's ideas. Then it was improved by Czeżowski (1946). Finally, the classification of reasoning was essentially modified and elaborated by Ajdukiewicz in (1955), which seems to be the most advanced approach. The publications and the ongoing discussion were carried out in Polish, but the main ideas and their development are exhaustively surveyed by Kwiatkowski (1993) in English.

The aim of this paper is to use Ajdukiewicz's classification in the argumentation theory in order to associate reasoning types with argumentation schemes and diagrams, which are usual tools for representing and analyzing natural language arguments in informal logic (*cf.* Reed, Walton, & Macagno, 2007; Walton, Reed, & Macagno, 2008).

D. Mohammed & M. Lewiński (eds.) (2016). *Argumentation and Reasoned Action: Proceedings of the 1st European Conference on Argumentation, Lisbon, 2015. Vol. II, 947-960.* London: College Publications.

First we recall the definitions of reasoning and some basic reasoning types as they were distinguished by Ajdukiewicz, namely the definitions of inference, derivation, justification, verification and explanation. In the next part of the paper we discuss the relationship between argumentation and reasoning, and in order to distinguish representations of the reasoning types considered by Ajdukiewicz we develop the notation of argument structure offered by 'the standard diagramming method' (*cf.* Jacquette, 2011). By determining epistemic and heuristic status of sentences we point out diagrams assigned to various reasoning types. We also show that some of these diagrams need to have schematic components. Finally, we indicate some meta-schemes concerning the process of reasoning itself, and we discuss diagrams assigned to abduction.

2. AJDUKIEWICZ'S CLASSIFICATION OF REASONING

Łukasiewicz and Czeżowski distinguished exactly four reasoning types, which have the same level of complexity: inference, justification, confirmation (positive verification) and explanation. All of them are simple in the sense that they do not consist of any proper part which is a reasoning too.

Ajdukiewicz (1955) presented a criticism of Łukasiewicz and Czeżowski's approach and developed his own classification, which is essentially different than the criticized one. In Ajdukiewicz's classification inference and derivation are basic reasoning types, which are components of all other ones. Thus, in order to grasp the whole variety of reasoning types we must begin with the definitions of inference and derivation.

2.1 Definition of reasoning

Ajdukiewicz defined inference as a kind of thinking (i.e. mental) act in which acceptance is transferred between sentences.

> To infer means to come, on the basis of some accepted sentences (propositions), to the acceptance of a new, not yet accepted sentence (proposition), i.e. conclusion, or to increase, on a basis of some accepted sentences, the certainty, with which another sentence is accepted (Ajdukiewicz, 1955, p. 282).

In contrast to inference from actually accepted premises, derivation is a suppositional form of reasoning, which involves sentences merely assumed, that is potentially, hypothetically accepted, i.e. suppositions. Ajdukiewicz called it expressively a "pseudo-inference". Thus, "to derive" means to come, on the basis of some potentially accepted sentences, to the potential acceptance of a new sentence.

Inference and derivation can be purely spontaneous, but they can be also goal-driven. Namely, they can be controlled by a kind of thinking task in this way that they are components of reasoning, which is used to complete such a task. Thus, Ajdukiewicz's definition of reasoning takes the following, non-classical, disjunctive form:

> we propose to classify as reasoning: 1) any processes of inference; 2) processes of derivation, that is of "pseudo"-inference, 3) processes of solving thinking tasks and problems, carried out by using inference or derivation (Ajdukiewicz, 1955, p. 294).

Reasoning is a thinking, i.e. mental act, which can be wrong, incorrect, erroneous, faulty or mistaken in any other sense, thus it does not need to be deductive or satisfy any conditions of logical or extra-logical correctness at all. Empirical course of reasoning processes, however, is the matter of psychological research. Since in this paper we consider the logical structure of reasoning, we do not focus on thinking acts themselves, but on their representations in a language, namely on sentences and their possible inferential orderings.

2.2 Thinking tasks and goal-driven reasoning

Thinking tasks can be expressed by a question or by an imperative sentence. First let us take into account justification, i.e.

> a thinking process of solving a task, which requests a sentence totally given in this task to be inferred from other already accepted sentences (Ajdukiewicz, 1955, p. 282).

The task of justification is given by the imperative 'prove (demonstrate) *Q*!'. In order to complete this task, a sentence (or sentences) *P* is to be found such that *Q* can be inferred from *P*, and moreover the process of inferring *Q* from *P* must be realized.

A type of reasoning controlled by a question is verification. It consists in

> the acceptance of a sentence [...] or of its negation [...] based on determining the truth or falsity of some consequences derived from this sentence (Ajdukiewicz, 1955, p. 282).

The question for verification has the form "is *P* true?". The given sentence *P* is not yet accepted, so that it cannot be a premise of any inference. Therefore, it is used as a supposition, and a sentence *Q* is derived from it, such that *Q* happens to be an either accepted or refused sentence (we assume that sentences are refused iff their negations are accepted). Finally, if *Q* is accepted then *P* is inferred, and if *Q* is refused then $\neg P$ is inferred (and thereby *P* is refused). Thus, verification can take any of two forms. It can be either confirmation or falsification, according to whether the sentence given in the task is eventually accepted or refused.

The task for explanation is given by the question "why *P* (why is *P* true)?". The thinking process leading to answer this question can be divided into three stages: 1) finding out such a sentence (or sentences) *Q*, that another given and already accepted sentence *P* can be derived from *Q*, and next, after deriving it, 2) inferring the obtained new sentence *Q* from the given sentence *P*, unless *Q* happens to be already accepted, and finally 3) inferring the answer of the form "*P* because *Q*" (*cf.* Ajdukiewicz, 1955, p. 283). Since the second step is not always necessary, we can distinguish two types of explanation. If, say, a theft explains to someone his car disappearance, then the sentence about the theft was not yet accepted by him, and it must be inferred. This is an example of the first type. But if the explanation is that his wife took the car, and he can even recall when she did it, then the explanatory sentence is already accepted, and it need not to be inferred. This is an example of the second type of explanation. The case when a well-known, accepted theory serves us to explain some phenomenon belongs to the second type as well.

3. DIAGRAMMATIC REPRESENTATION OF REASONING

3.1 Reasoning and argumentation. Classical diagramming method

In opposition to reasoning, which is a thinking act (or process), we understand argumentation as an act (or process) of communication, i.e. a speech act, in which some agent's thinking acts of reasoning are presumed and submitted to an audience for acceptance. The tasks assigned to reasoning types are explicitly present in argumentation dialogue, and they are also expressed in suitable speech acts. Following Hitchcock (2007) who claims that Pinto's "happy" phrase "Arguments are invitations to inference (Pinto, 2001)" applies also to suppositional reasoning (i.e. derivation), we are inclined to extend this suggestive characteristics and understand arguments as invitations to reasoning, which can take any of three forms specified by Ajdukiewicz's definition.[1]

If we disregard the persuasive and performative function of argumentation, we obtain propositional structures, which are the same in arguments and in the presumed acts of reasoning. Therefore the schemes and diagrams assigned to reasoning types and to the corresponding arguments are the same too. We will show how to adapt the classical diagramming method to represent the whole variety of reasoning types, which is potentially included in the definition of reasoning given by Ajdukiewicz. Trzęsicki (2011) proposed an alternative diagramming notation of reasoning, however, it corresponds to the classification formulated by Czeżowski. Firstly because derivation is absent in Trzęsicki's notation. Furthermore, in accordance with his proposal, inference is represented by one of four basic types of diagrams. Justification, explanation and confirmation (but not falsification, which is neglected) correspond to the remaining three. Thus inference in Trzęsicki's notation, as well as in Czeżowski's (and in Łukasiewicz's) classification, is one of four simple reasoning types, instead of being the component of the remaining three as it appears in Ajdukiewicz's approach.

We refer to the classical diagramming method as it was described by Jacquette (2011). Possible complexity of the classical diagrams is illustrated by Figure 1.

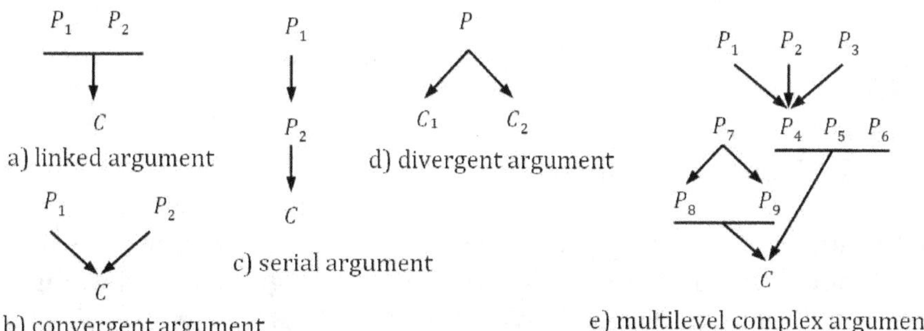

Figure 1 – Classical argumentation diagrams

[1] Separate derivations do not form arguments in a narrower sense of the term "argument". Since they do not have any explicitly accepted conclusion, in this sense they can be at most a kind of components of complete, proper arguments.

Complex argumentation structures, as the one presented in Figure 1e, can be obtained using the elementary operations corresponding to the constructions shown in Figures 1a-d.

3.2 Epistemic status of sentences. Inference and derivation

In order to adapt classical diagrams to Ajdukiewicz's classification first of all we must be able to distinguish the representations of inferences from the representations of derivations. Since sentences involved in these two fundamental reasoning types have different *epistemic status*, all the sentences on diagrams must be indicated in some way as being either actually accepted as in inference, or merely potentially accepted as in derivation. Freeman (1991) used a dashed line circles to represent suppositions. Here, in order to indicate the epistemic status of sentences, suppositions will be enclosed in square brackets (angle brackets, although not consequently, have been introduced by Jacquette (2011) for the aim of representing *ad absurdum* arguments).

Simple inference and derivation are represented by Figures 2a and 2b respectively. More complex examples of derivation are shown in Figures 2c-e. The dashed arrow in 2c means that there is a series of derivations between P_2 and P_n.

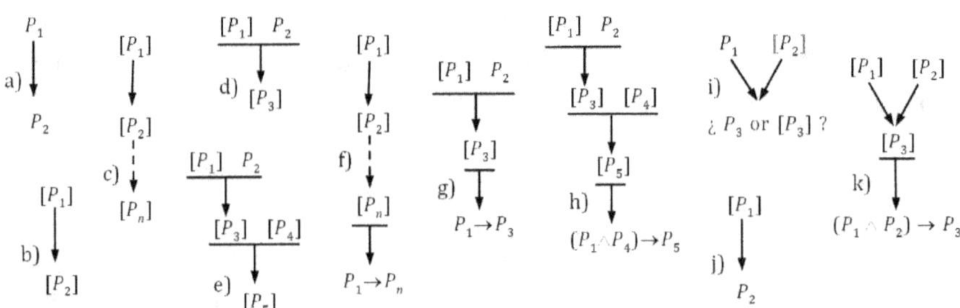

Figure 2 – Inference and derivation

Suppositions and actually accepted sentences can be combined (Figures 2d-k), however not every combination can be regarded as meaningful. Contrary to the example shown in Figure 2j, if a premise of some reasoning is a supposition, so should be the conclusion (in Jacquette's diagrams only first premises of derivations are in brackets, what is the reason for the claim that he did not develop his notation consequently). Also convergent combinations of inference and derivation (Figure 2i) seem to be meaningless, since the epistemic status of their conclusion is unclear. However, the following reasoning shows that perhaps it is a more problematic case: *God probably exists, because people believe in*

God, but if he would give us some sign now, his existence would be more certain. On the other hand, one can say that there are two separate arguments in this example, and the latter is actually a meta-argument about the first one. We will not discuss this issue here. Let us only note that a possible evaluation of such reasoning seems to require a kind of hybrid semantics, in which the acceptance of each sentence is a pair of values representing its actual and its potential component. Convergent derivations (Figure 2k) do not cause similar problems.

Since derivation itself only leads to suppositions, in order to produce an actually accepted conclusion, it must be involved in some more complex reasoning process as *ad absurdum* or practical reasoning for instance. Furthermore, derivation is often summarized by means of natural deduction. Thus, a sequence of derivations can be concluded by an implication of a special form. This implication has the conjunction of all the assumed suppositions as its predecessor, and the sentence eventually derived as its antecessor (*cf.* Freeman, 1991, p. 214). Derivations represented by Figures 2c–e have been completed by means of natural deduction and shown in Figures 2f–h, respectively. The horizontal line at the arrow preceding the final implication indicates that not merely the last supposition, but the whole underlined derivation plays the role of a premise in the final inference.

It should be emphasized that by using the term "natural deduction" we do not mean that all the compound, individual derivations must be deductive. They can be defeasible as well. Here, the word "deduction" is used by analogy to the deduction theorem[2], well-known from handbooks of formal logic. It is a meta-theoretic theorem which characterizes the relationship between formal derivation and implication. Thus, by "natural deduction" we mean an analogue of the deduction theorem for natural language. Obviously, the implication obtained as the conclusion of some derivation can be a premise of some further inference. It can be a premise of *ad absurdum* or practical reasoning, or of any other suitable reasoning as confirmation (see Sect. 3.4). So to speak, natural deduction is a "de-assumizer" – an uniform method to change the epistemic status of that what follows from suppositions.

[2] *If* $A \cup \{\alpha\} \vdash_S \beta$ *then* $A \vdash_S \alpha \rightarrow \beta$, where α and β are any sentences and A is any set of sentences formulated in the language of a fixed system S ($\vdash_S$ is the consequence relation of S). Obviously, it depends on the particular properties of a formal system if the deduction theorem holds for it.

3.3 Heuristic status of sentences. Justification and spontaneous inference

Standard diagrams do not allow us to distinguish inference from derivation. They do not allow us to distinguish spontaneous inference from justification as well. The cause is that they do not represent the *heuristic status* of the sentences involved in reasoning processes, namely they do not indicate which of these sentences are given at the starting point, and which occur at the subsequent stages of reasoning. In case of spontaneous inference the given sentences are premises, but in justification they are conclusions. In order to indicate this difference we use asterisks, so that the asterisked sentences are the starting points of reasoning. Thus Figure 3a provides an example of a simple spontaneous inference, while Figure 3b represents a simple justification. Figures 3c and 3d show a complex spontaneous inference and a complex justification, respectively.

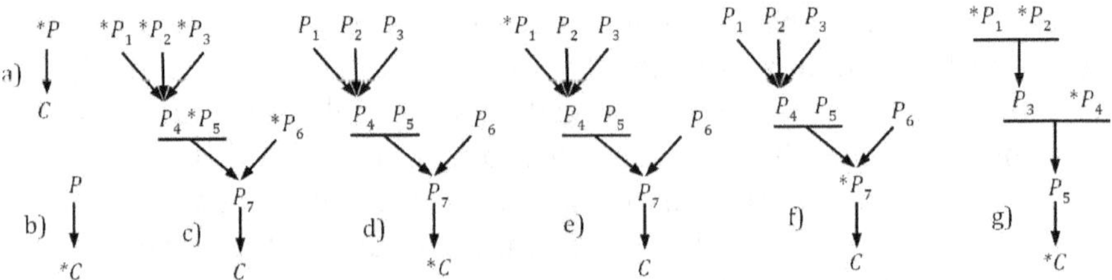

Figure 3 – Justification and spontaneous inference

Next two diagrams in Figure 3 represent some combinations of spontaneous inference and goal-driven justification. The course of reasoning can be read from Figure 3e as follows: P_1 is given, P_4 is inferred from it, then P_4 is additionally justified by P_2 and P_3, then a new premise P_5 is found, and P_7 is inferred from it and from P_4, then P_7 is additionally justified by P_6, and eventually the final conclusion C is inferred from P_7. In Figure 3f the sentence P_7 is justified firstly, and then the conclusion C is inferred from P_7. Figure 3g is the representation of some reasoning driven by a (e.g. mathematical) task "prove C from P_1, P_2 and P_4!".

3.4 Verification and practical reasoning

Since a supposition is the starting point of verification, diagrams assigned to this reasoning have to use both brackets and asterisks. Figures 4a and 4b show the diagramming of confirmation (without and with regard of natural deduction step, respectively). The remaining four diagrams in Figure 4 represent falsification, which can take any of two forms: *ad absurdum* as in Figures 4c and 4d, or *ad falsum* as in Figures 4e and 4f. The explicit use of natural deduction in Figures 4 b, 4d and 4f shows clearly that the last step of verification is an inference. This inference is deductive in falsification, while in confirmation it is not, however, since in real-life reasoning the derivative part of verification (represented by the dashed arrows) is mostly defeasible, neither confirmation, nor falsification can be absolutely conclusive in this case.

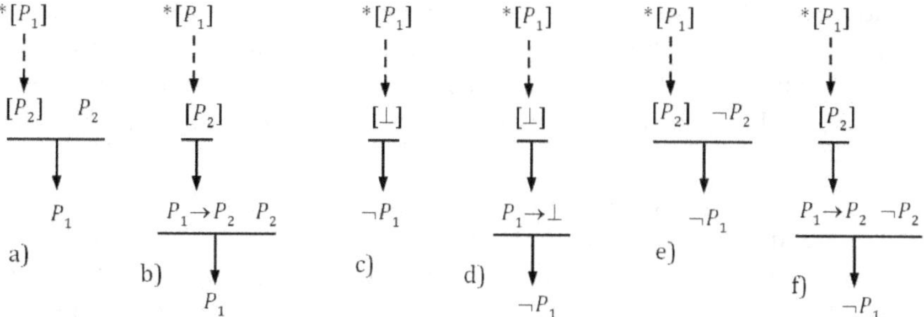

Figure 4 – Verification

Yet, let us consider representations of practical reasoning, which are shown by Figure 5. Practical reasoning is similar to verification, but the last inference has a different scheme. Also the suppositions that occur must have some special form, since they refer to some action and its possible effect.

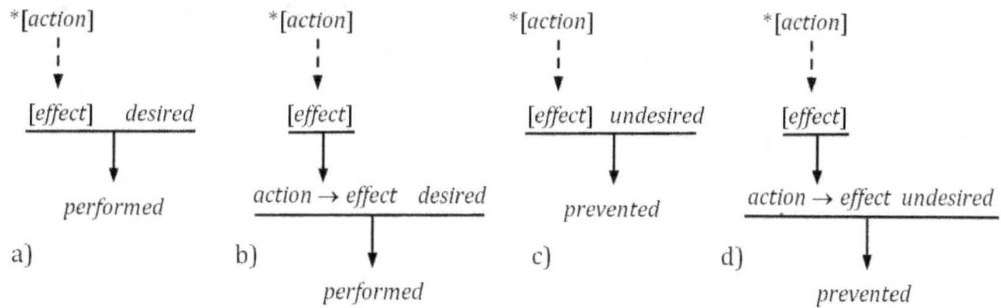

Figure 5 – Practical reasoning

Practical reasoning begins with the asterisked supposition assuming that we take up some action. Then we derive what will be an effect of this action. If this effect is desired, we conclude that the action should be performed (Figure 5a). If it is undesired, we conclude that the action should be prevented (Figure 5c). In Figures 5b and 5d natural deduction is used to explicitly reveal the scheme of the last, crucial inference. At the same time, the derivative part is still represented on the diagrams. Thus, by combining diagrams and schemes, the whole course of practical reasoning can be represented more accurately then by using schemes only.

3.5 Explanation and abduction

Diagrams of explanation are similar to those of confirmation, but they involve a special scheme of conclusion. Also the starting points of both reasoning types are different. While confirmation begins with a spontaneous derivation of consequences from a given sentence, the starting point of explanation is the sentence, which is to be supposedly justified. This type of explanation is shown in Figure 6a. The second type, in which the explanatory sentence is already accepted, is shown in Figure 6b.

The final conclusion in both cases is the sentence "P_2 because P_1" which means that P_1 is true, P_2 is true, and that P_2 follows from P_1. Therefore the sentence P_1 should be regarded as an implicit premise of the final inference. This premise has been added in Figure 6c specifying the second type of explanation (Figure 6d shows a version of this type, in which the explanatory sentence is used as a supposition in the initial derivation, but then it is not being inferred, because its veracity has been determined in another way, e.g. it was recalled). A detailed diagram for the first type of explanation is more complex. Actually, we have two steps of reasoning here. At the beginning we conclude with P_1 by showing that some given, acceptable sentence P_2 is derivable from it, and next on the same basis enriched with just accepted sentence P_1 (and with derivation replaced by inference), we draw the final conclusion 'P_2 because P_1'. This structure is shown in Figure 6e.

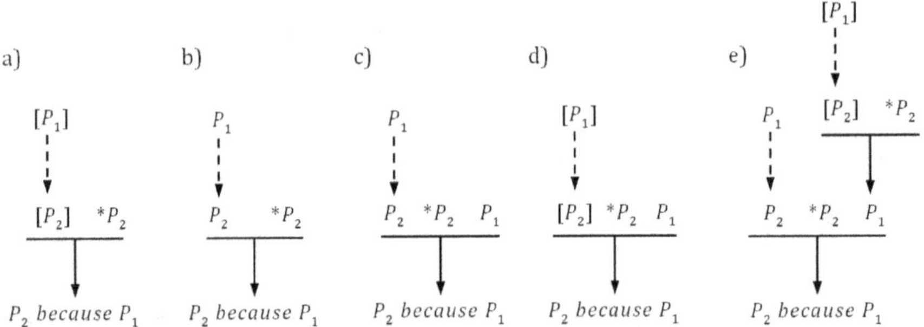

Figure 6 – Explanation

The diagrams in Figure 6 can be further specified. Namely, the derivations and inferences leading from P_1 to P_2 can be concluded explicitly by the sentence "P_2 follows from P_1" (the implication "$P_1 \rightarrow P_2$", offered by natural deduction, is too weak to entail the conclusion "P_2

because P_1"). Let us note, however, that "P_2 follows from P_1" belongs to semantics, so (unlike "$P_1 \rightarrow P_2$") it is a metalanguage sentence.

Since one and the same fact can have many different explanations, the inference of the explanatory sentence P_1 (Figure 6e) is rather unreliable, so that the whole reasoning can be regarded at most as an *ad hoc* explanation. Obviously, this unreliable inference can be supported by some other reasoning as e.g. falsification (Figure 7a). But it can be also enhanced by adding a premise which states that this particular explanation is the most successful one. Reasoning enriched in this way is called "abduction". In argumentation schemes of abduction this evaluative premise usually refers explicitly to some specified alternative explanations (*cf.* Walton, Reed & Macagno, 2008, p. 329). Figure 7b shows the diagram based on the backward scheme of abduction referring to *n* different explanations of the same fact.

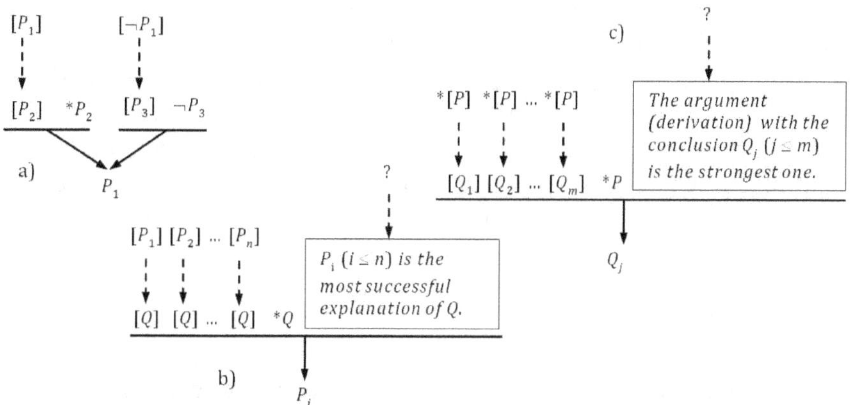

Figure 7 – Abduction

Walton, Reed, & Macagno also consider the forward scheme of abduction. Its diagrammatical representation is shown in Figure 7c. This reasoning aims at selecting the conclusion of "the most plausible (strongest) argument" having premises which belong to some given basis of plausible statements (*cf.* Walton, Reed, & Macagno, 2008, pp. 329-330). The term "argument" is interpreted in Figure 7c as denoting derivation, because the conclusions Q_1, Q_2, ..., Q_n can (or even should) be mutually exclusive, so that they must not be accepted at the same time, which would be the case if they were inferred from the basis *P*. So only one of them can be selected as the conclusion of the final inference, and thereby accepted. Let us note that these sentences often refer to some mutually exclusive possibilities, which are known in advance. The task for such reasoning might be formulated as the imperative "choose (i.e.

select the most plausible statement) among $Q_1, Q_2, ..., Q_n$ with respect to the basis *P*!". The representation of this reasoning is the diagram shown in Figure 7c, but with $Q_1, Q_2, ..., Q_n$ asterisked above the horizontal line (if Q_j was asterisked below the horizontal line too, the diagram would denote a sort of justification).

The question marks in Figures 7b and 7c mean that here we intend to discuss neither the issue of "the most successful explanation", nor that of "the strongest argument (derivation)". But it has to be underlined that evaluative premises referring to some reasoning cannot be avoided in abduction, what makes it being a meta-reasoning in fact. Thus meta-schematic components are indispensable in diagrams of abduction.

Finally, let us note that arguments based on meta-schemes, i.e. on the schemes referring to some reasoning, are frequently encountered in everyday argumentation. In addition to abduction and explanation, which need components of this kind, we have various meta-schemes referring to the fact that some act of reasoning was (or was not) performed, or that it was performed in some sense correctly (or incorrectly). The conclusion in these meta-schemes can be identical with the conclusion of the reasoning under consideration (or with its negation), but it can also contain an assessment of this statement (in particular, various forms of argument from ignorance fit this description) or of the person who reasons.

4. CONCLUSION

Classical diagramming method offers a certain concept of argument structure. Introducing indicators of epistemic as well as heuristic status of sentences allowed us to expand this concept in order to grasp more of important features of everyday argumentation. In one of my previous papers I proposed a formal definition of argument, which corresponds to the classical diagrams (*cf.* Definition 3: Selinger, 2014, p. 382). Thus, a question arises about an extension of this definition, which would enable us to cover the entire realm of arguments revealed by Ajdukiewicz's analysis of reasoning.

Obviously, a formal representation of epistemic and heuristic status of sentences has to be introduced in some way (e.g. by assigning pairs of Boolean values to the sentences). However, the crucial question is that of the character of argument premises. In the arguments representable by the classical diagrams only sentences can be premises. But this seems to be undermined due to the occurrence of arguments containing derivations as their counterparts. Since suppositions can lead only to suppositions, it is not the last derivative (e.g. P_2 in Figure

4a) that is the premise of the final inference. Hitchcock (2007) claims that also reasoning can be a premise of argument, and the whole derivation should be regarded as just such a premise in this case. This seems to be the most natural solution, however, it leads to a meaningful increase of the formal complexity of the definition in question. For instance, derivations may have premises which are again derivations, so that such a definition, unlike the one previously proposed in (Selinger, 2014), needs to be recursive in order to cover all possible levels of derivation. Thus, there is a question whether it is in some way possible to maintain the concept of argument, which employs only sentences operating as premises.

Another question addressed mostly to the future work is to estimate the extent to which metalanguage has to be introduced to the formal concept of argument structure. We have shown that meta-theoretic and meta-schematic elements are not dispensable in the diagrams of abduction. Introducing meta-theoretic elements to the diagrams of explanation is desirable, although not necessary. These are, however, only some partial results, therefore this issue needs a more elaborate, systematic and in-depth study.

ACKNOWLEDGEMENTS: I gratefully acknowledge the support of the Polish National Science Centre under grant 2011/03/B/HS1/04559.

REFERENCES

Ajdukiewicz, K. (1955). Klasyfikacja rozumowań (Classification of reasoning). *Studia Logica, 2*, 278-299.

Czeżowski, T. (1946). *Główne zasady nauk filozoficznych* (*Principles of philosophy*). Toruń: Księgarnia Naukowa T. Szczęsny i S-ka.

Freeman, J. B. (1991). *Dialectics and the macrostructure of arguments*. Berlin: Foris.

Hitchcock, D. (2007). Informal logic and the concept of argument. In D. Jacquette (Ed.), *Philosophy of logic* (pp. 101-130). Amsterdam: North Holland (Elsevier).

Jacquette, D. (2011). Enhancing the diagramming method in informal logic. *Argument, 1*(2), 327-360.

Koszowy, M., & Araszkiewicz, M. (2014). The Lvov-Warsaw School as a source of inspiration for argumentation theory. In K. Budzynska & M. Koszowy (Eds.), *The Polish School of Argumentation*, special issue of the journal *Argumentation, 28*(3), 283-300.

Kwiatkowski, T. (1993). Classification of reasonings in contemporary Polish philosophy. In F. Coniglione, R. Poli & J. Woleński (Eds.), *Polish scientific philosophy: The Lvov-Warsaw School* (pp. 117-167). Amsterdam - Atlanta: Rodopi.

Łukasiewicz, J. (1912). O twórczości w nauce (Creative elements in science). In *Księga Pamiątkowa ku uczczeniu 250 rocznicy założenia Uniwersytetu Lwowskiego przez króla Jana Kazimierza r. 1661 (Memorial Book to celebrate the 250th anniversary of the foundation of the Lvov University by King John Casimir in 1661)*, vol.1 (pp. 1-15). Lvov: Lvov University.

Pinto, R. (2001). The relation of argument to inference. In R. Pinto *Argument, inference and dialectic: Collected papers on informal logic with an introduction by Hans V. Hansen* (pp. 32-38). Dordrecht: Kluwer Academic Publishers.

Reed, C., Walton, D., & Macagno, F. (2007). Argument diagramming in logic, law and artificial intelligence. *The Knowledge Engineering Review, 22*(1), 87-109.

Selinger, M. (2014). Towards formal representation and evaluation of arguments. In K. Budzynska & M. Koszowy (Eds.), *The Polish School of Argumentation*, special issue of the journal *Argumentation, 28*(3), 379-393.

Trzęsicki, K. (2011). Arguments and their classification. *Studies in Logic, Grammar and Rhetoric, 23*(36), 59-67.

Walton, D., Reed, C., & Macagno, F. (2008). *Argumentation schemes.* New York: Cambridge University Press.

61

On the Ends of Argumentation

PAUL L. SIMARD SMITH
University of Connecticut, USA
paul.simard_smith@uconn.edu

Many argumentation theorists have endorsed the notion that a difference of opinion can only be successfully resolved in an argumentation when an agreement is reached between the participants of the argumentation about the rational status of the contested standpoint. I present and discuss some counterexamples to this view. I contend that the counterexamples provide cases of successful resolutions to differences of opinion even though disagreement about the contested standpoint remains.

KEYWORDS: agreement, logical pluralism, reasonable disagreement, resolution to a difference of opinion

1. INTRODUCTION

Several argumentation theorists have asserted that the purpose of argumentation is to resolve a difference of opinion. Moreover, it is also commonly claimed that a resolution to a difference of opinion is achieved only when there is agreement over the claim at issue in the argumentation (Godden, 2010; Hamblin, 1970; Walton & Krabbe, 1995; van Eemeren & Grootendorst, 2004). The purpose of this paper is to offer some support for the claim that a resolution to a difference of opinion can occur without agreement among the participants in an argumentation about the argumentation's contested claim.

First, I present a case of reasonable disagreement that may never, even in principle, be rationally bridged. I continue by contending that a resolution to an argumentation that revolved around such a reasonable disagreement is possible. Third, I sketch an argument that supports this contention. If such argumentations can resolve it follows that argumentations can resolve in which there is no agreement over the argumentation's contested proposition.

D. Mohammed & M. Lewiński (eds.) (2016). *Argumentation and Reasoned Action: Proceedings of the 1st European Conference on Argumentation, Lisbon, 2015. Vol. II, 961-972.* London: College Publications.

2. THE LOGICAL BASIS OF SOME REASONABLE DISAGREEMENTS

Following Goldman (2010, pp. 189-190) we adopt the following notion of disagreement,

> **(DIS).** Agents *S* and *Q* disagree if, and only if, *S* and *Q* hold *incompatible* doxastic attitudes toward a proposition *p*. (*Disagreement*)

A reasonable disagreement is as follows,

> **(R-DIS). A disagreement is reasonable if, and only if, *S* and *Q* disagree about a proposition *p* and the doxastic attitudes *S* and *Q* adopt toward *p* are reasonable. (Reasonable Disagreement)**

Now consider the following dialogue that could occur in the course of an argumentation over the claim that a prisoner, Ms. P, has good behavior. One agent in the dialogue holds the attitude of belief toward the contested claim and the other agent withholds from believing the contested claim.

(DIA).

Observer: Ms. P does not have good behaviour.
Ms. A: That isn't right! It's not the case that Ms. P doesn't have good behaviour.
Ms. B: I agree with you.
Ms. A: Great! Then logic tells us that it follows from what I said that she has good behaviour. You must then agree with me that Ms. P has good behaviour.
Ms. B: No, I don't agree with that. While it's not the case that Ms. P doesn't have good behaviour, one *can't claim* on that basis that she does have good behaviour.

The inference made by Ms. A can be represented as follows,

(INF).

1. It's not the case that Ms. P does not have good behavior.
2. Therefore, Ms. P does have good behavior.

(INF) is an instance of double negation-elimination (DNE).

In what follows I endorse the following principles of rational doxastic attitude formation,

> **(BEL).** If S reasonably believes $p_1 \ldots p_n$and S *legitimately endorses* the claim that $p_1 \ldots p_n \vDash_x q$ (while maintaining reasonable belief in $p_1 \ldots p_n$), then it is reasonable for S to believe q. (*Belief*)
>
> **(WIT).** If S reasonably believes $p_1 \ldots p_n$ and *legitimately endorses* the claim that $p_1 \ldots p_n \nvDash_x q$, then, *ceteris paribus*, it is reasonable for S to withhold from q (*Withholding*)[1]
>
> **(INC).** If it is reasonable for S to withhold from a proposition p, then it is not reasonable for S to believe p and if it is reasonable for S to believe p, then it is not reasonable for S to withhold from p. (*Incompatibility of Belief and Withholding*)

One can read '$p_1 \ldots p_n \vDash_x q$' as the argument with $p_1 \ldots p_n$ as premises and q as conclusion is valid$_x$ (and '$\nvDash$' as not generally valid$_x$ respectively).

There are certainly arguments that epistemologists have made against principles similar to BEL, WIT and INC. I do not claim that these are uncontroversial principles that are clearly true. Rather I explore possibilities that are available given the acceptance of these principles along with logical pluralism. However, for further reason to regard these principles as plausible consider the following observations. In regards to BEL, if one holds a reasonable belief in some premises and also believes that these premises logically entail a conclusion, then it is quite intuitive to hold that it would be reasonable for one to believe the conclusion. Of course, if the conclusion were absurd, known to be false or highly implausible, then one would not be able to maintain a reasonable belief in the premises while also reasonably believing the conclusion. Thus, it is important that in believing that the premises logically follow from the conclusion it is also warranted to maintain a reasonable belief in the premises. A notable exception to this norm may be cases of paradoxical reasoning where the premises are reasonable to believe and it is reasonable to believe the inference is valid but it is not reasonable to believe the conclusion. Thus, BEL may a defeasible norm in the sense that it is not operative over all cases (in particular it might not be operative in paradoxical cases). However it does seem to be operative in cases such as INF. In regards to WIT, if it is reasonable to

[1] Clearly this norm only applies to arguments or inferences that are deductive like INF. It can be reasonable to believe the conclusion of an invalid argument if the argument is inductively strong for example. However, if the argument is properly evaluated according to the standard of deductive validity, then (WIT) would apply.

believe some premises of a deductive argument and reasonable to believe that the conclusion does not follow from the premises, then presuming there is no additional evidence bearing on the conclusion or its negation, then it is reasonable to withhold from believing the conclusion. This epistemic principle is plausible since, in such a circumstance, for all one knows either the negation of the conclusion or the conclusion could be true. Given the lack of evidence supporting the conclusion or its negation it is reasonable for one to withhold from the conclusion. Finally INC is plausible given that it is odd to imagine a scenario in which the evidence for a conclusion rendered it reasonable to either believe a claim or withhold from believing a claim. Rather for a given agent one attitude or the other is reasonable given the evidence available to the agent.

The epistemic principles (BEL), (WIT) and (INC) have an interesting interaction with a view known as *logical pluralism*. In particular logical pluralism, together with these principles, can explain why some disagreements are reasonable.

What is logical pluralism? In its broadest formulation logical pluralism is the view that there are at least two correct systems of formal logic. In recent literature a variety of different attempts have been made to flesh out a pluralist thesis about logic (Beall & Restall, 2006; Bueno & Shalkowski, 2009; Cook, 2010; DeVidi, 2011; Field, 2009; Hjortland, 2012; Keefe, 2014; Lynch, 2008, 2009; Pedersen, 2014; Restall, 2002, 2014; Russell, 2008; Shapiro, 2014). The version of logical pluralism I adopt here is Beall-Restall pluralism.[2] Beall-Restall pluralism contends that the concept LOGICAL CONSEQUENCE has several legitimate specifications. According to Beall-Restall pluralism the core of the concept LOGICAL CONSEQUENCE is captured by the *Generalized Tarski Thesis*,

(GTT). $p_1 \dots p_n \vDash_x q$ if, and only if, in any case$_x$ in which $p_1 \dots p_n$ are true, q is true.[3]

---

[2] It is worthwhile noting that Beall-Restall pluralism has been subjected to several criticisms. The points I make here about the relationship between logical pluralism and reasonable disagreements can be made, in different ways, with most of these different variants of logical pluralism mentioned above. However, Beall-Restall pluralism has the virtue of being relatively well-known among current variants of logical pluralism and possessing a relatively straightforward characterization. Thus, I adopt this version of logical pluralism here for pedagogical purposes.

[3] This is a slight modification of GTT as it appears in Beall and Restall (2006). Their exact formulation of GTT is as follows,

By specifying $case_x$, and the relation of *being true in a case* in different ways (GTT) produces different *admissible* consequence relations. A specification of GTT is admissible if the specification is *necessary*, *normative* and *formal*.[4] Beall and Restall discuss several different admissible specifications of $case_x$. Of particular relevance to our discussion $case_x$ can be admissibly specified as Tarskian Models and is true in a case can be specified as the conditions that are need for sentences of the language in question to be true in a model. Alternatively $case_x$ could be specified as Stages Of Inquiry and is true in a case could be specified as the conditions that are required for sentences in the language in question to be true at a stage. Specifying $case_x$ as Tarskian Models renders classical logic correct and specifying $case_x$ as Stages of Inquiry renders intuitionistic logic correct (Beall & Restall, 2006). The details are not essential to understanding the primary point being made here. However, what is important to recall for present purposes is that in classical logic (DNE) is valid while in intuitionistic logic it is not, in general, valid.

Note that the epistemic principles (BEL) and (WIT) involve the use of the notion of *legitimately endorsing* an inference claim. One view would be that it is legitimate to endorse the claim that some inference is $valid_x$ (or not generally valid) if, and only if, the inference scheme is valid (or not generally valid) on *some* admissible specification of $case_x$. However, I submit that the validity of an inference scheme on some specification of $case_x$ is not sufficient for legitimately endorsing the corresponding inference claim. In order for an inference to be legitimately endorsed the inference must be valid on a specification of $case_x$ that is *appropriate for relevant contexts.*

In order to explain the notion of appropriateness of a specification of $case_x$ for relevant contexts I present an example of how varying obligations and practical goals can render different specifications of $case_x$ appropriate for different contexts.

(GTT). An argument is $valid_x$ if and only if in every $case_x$ in which the premises are true, so is the conclusion. (Beall & Restall, 2006, p. 29)

[4] Necessity, normativity and formality are constraints that specifications of $case_x$ must satisfy in order for the specification to be admissible (Beall & Restall, 2006, pp. 14-23). A specification of $case_x$ must determine a necessary connection between premises and conclusion of valid inferences, it must explain how rejecting the conclusion of a valid argument when one has accepted the premises is a *mistake*, and the notion of validity must be *formal* in one of several senses discussed by Beall and Restall. In order to develop the point at issue here it will not be necessary to explain these constraints in further detail.

Reflecting back on (DIA) consider if Ms. A is a family member of Ms. P. Ms. P is up for parole after having been in prison for some time. Assume that if Ms. P is deemed as having good behavior in prison, then it is much more likely for her to receive parole. Given Ms. A's close relationship to Ms. P she strongly supports Ms. P's parole. Now suppose that Ms. B is a lawyer representing the community. The community is not convinced that Ms. P has good behavior and is concerned that Ms. P will reoffend should she be released. Further suppose that it is clear, given the evidence pertaining to Ms. P's behavior, that it is not the case that Ms. P does not have good behavior. Thus, we are supposing that both Ms. A and Ms. B agree to the premise of (INF). In such a circumstance it would be appropriate for Ms. A to specify $case_x$ as Tarskian Models and, given (BEL), believe the conclusion of (INF) while it would be appropriate for Ms. B to specify $case_x$ as Stages Of Inquiry and, given (WIT), withhold from believe the conclusion of (INF). If either Ms. A or Ms. B adopted the alternative specification of $case_x$ it could impede achievement of their practical goals and obligations in their respective contexts. Given that both sets of goals are legitimate it is appropriate for Ms. A and Ms. B to adopt the logic that would best advance their goals.

If an argumentation between Ms. A and Ms. B revolved around the issues in DIA it could very well be the case, assuming the evidence pertaining to Ms. P's behavior is held constant, that no agreement should be reached between Ms. A and Ms. B. A natural response to this scenario would be to think that the argumentation between Ms. A and Ms. B could not be resolved.

There is an affinity with the view being articulated here and a broadly Carnapian approach to logical pluralism. In making this claim I do not mean to imply that Carnap adopted a version of pluralism that is identical to the one I am articulating here. Rather I simply point out some broad similarities between the view expressed here and Carnap's view that pragmatic balance of consideration reasoning should be used to determine which formal languages we adopt for various purposes (Carnap, 1950). Given one set of purposes one logic might be better than another. Given different purposes the other logic might be better. The basic thought is that given Ms. A's purposes a specification of $case_x$ as Tarskian models better serves the achievement of those purposes, but given Ms. B's purposes a specification of $case_x$ as stages of inquiry better serves her purposes.

Given that different specifications of casex are appropriate in the different contexts we can see how BEL, WIT, INC and logical pluralism can serve to explain how DIA can express a logically based reasonable peer disagreement. Given BEL, and given the contextual appropriateness of a specification of $case_x$ as Tarskian models in Ms. A's context, it is reasonable for Ms. A to believe that Ms. P has good behavior. However, given WIT, and given that stages of inquiry are a contextually appropriate specification of $case_x$ in Ms. B's context, it is reasonable for Ms. B to withhold from believing that Ms. P has good behavior. By INC, belief and withholding are incompatible doxastic attitudes. Thus, by DIS we have a disagreement between Ms. A and Ms. B. Given that Ms. A's Ms. B's attitudes are reasonable by R- DIS the disagreement is reasonable.

3. RESOLUTION

As mentioned several argumentation theorists claim that agreement is required for a resolution to a difference of opinion (Godden, 2010; Hamblin, 1970; Walton & Krabbe, 1995; van Eemeren & Grootendorst, 2004). Godden expresses this thesis as follows,

> **(RES).** A difference of opinion is resolved when the commitments of the disputants have reached a state of agreement with respect to the claim at issue. (Godden, 2010, p. 404)

Godden contends that Hamblin (1970), Walton and Krabbe (1995) as well as van Eemeren and Grootendorst (1984, 2004) endorse some variation of (RES). Godden's own point is to show that changes in beliefs and not just commitments are necessary for a successful resolution. I am not so sure that agreement is required for a successful resolution in all cases. I float the idea that an argumentation *can be resolved* if the following is true of argumentation participants *S* and *Q*,

> **(APP).** *S* disagrees with *Q* over the argumentation's contestedclaim *p* and *S appreciates* that *Q*'s doxastic attitude toward *p* isreasonable and *Q appreciates* that *S*'s doxastic attitude toward *p* is reasonable.

Of course it is possible for (APP) to be true in an argumentation that is not resolved. In certain cases various argumentative moves may be available to one or more participants in the argumentation that can produce a stronger reason-based agreement between the participants. However, in order to establish a case in which there is a resolution to an argumentation without agreement all that is required that APP is true of the participants in some resolved argumentations.

Note that it is (APP) is true of participants in an argumentation when those participants hold incompatible doxastic attitudes. When (APP) is true of participants in an argumentation those participants appreciate that a doxastic attitude the other participant in the argumentation has is reasonable without changing their own doxastic attitude toward the contested proposition. A participant in an argumentation *S* appreciates the doxastic attitude of another participant *Q* towards a proposition *p* without sharing the same attitude toward *p* if *S* recognizes that features of the contexts—such as different but reasonable goals or obligations— of *Q* render an attitude toward *p* reasonable for *Q* that is not reasonable for *S* given different features of *S's* context. The basic thought is that it is possible to appreciate the reasonability of an argumentation partner's doxastic attitude while maintaining an incompatible attitude through a process of counterfactual reasoning whereby one recognizes that if their obligations, goals, or other relevant features of their context differed, then a different attitude would be reasonable.

By way of illustration consider (DIA). The different doxastic attitudes that Ms. A and Ms. B take towards the proposition that Ms. P has good behavior are both reasonable. The different attitudes are justified by different but admissible and appropriate logics. Since they share all the same evidence there is no clear path to a reason-based agreement about Ms. P's behavior. Perhaps Ms. A or Ms. B could procure agreement through other strategic moves. However, it is not clear to me that this would be required for an argumentation about Ms. P's behavior between Ms. A and Ms. B to be resolved. If the argumentation expanded beyond (DIA) and Ms. A and Ms. B explained their differing objectives and relationships to Ms. P, then it seems plausible that Ms. A and Ms. B could appreciate the reasonability of the position that the other takes without changing their attitude or commitments towards the contested proposition itself. In such a circumstance (APP) tells us that their argumentation could be resolved.

4. WHY CAN (APP) BE TRUE OF A RESOLVED ARGUMENTATION?

What reason is there to think that when (APP) is true that there could be a resolution to a difference of opinion? Suppose the following three claims were true of some argumentation,

- **(i)** The participants have come to appreciate the reasonability of their argumentation partner's doxastic attitude toward the contested proposition.
- **(ii)** The participants have meaningfully changed their behavior towards the interlocutor's position in such a fashion that acknowledges the interlocutor's reasonability on the contested standpoint.

(iii) The participants have no reason-based agreement available to them.

I submit that it is intuitively more plausible to regard an argumentation in which (i)-(iii) are true as resolved than to regard it as unresolved. My sense is that the burden of proof is on someone who claimed that argumentations in which (i)-(iii) obtain are unresolved.

Note that if (APP) obtains it is possible that (i)-(iii) obtain. If the participants appreciate the reasonability of their interlocutor's attitude they could have; come to learn of this reasonability through the argumentation, meaningfully changed their behavior towards the interlocutor's position based on this acquired knowledge, and have no reason-based agreement available to them. If it is possible for (i)-(iii) to obtain in an argumentation when (APP) obtains in that argumentation, then it is possible for an argumentation in which (APP) obtains to be resolved. Therefore, in the absence of a compelling argument that a resolution cannot occur when (i)-(iii) obtain I contend that when (APP) obtains the argumentation could be resolved. Given that the participants in the argumentation disagree over the argumentations contested claim when (APP) obtains it is possible for an argumentation to be resolved in the presence of a disagreement.

One of the purposes of argumentation discussed in Ralph Johnson (2000) is to increase the overall level of rationality in the world. I submit that when (i) is true of an argumentation there has been an increase in the level of rationality of the participants in the argumentation with respect to issue under discussion. When (iii) is true of the participants in an argumentation it is not possible to increase their rationality with respect to the issue under discussion in the argumentation.

Why would one think that there has been an overall expansion of the level of rationality between the participants in an argumentation with respect to the issue under discussion when (i) is true of the participants in the argumentation? The reason there has been an increase in the level of rationality is that the participants have come to learn that there are rational positions that can be taken on the issue under discussion that differ from the position they take toward this issue. This knowledge plausibly can be regarded as increasing the level of rationality of the participant especially if they were previously unaware that there were rational position that can be taken to the issue under discussion that differed from their own position. Furthermore, if (iii) is also true of the participants in the argumentation and no stronger reason-based agreement is possible, then coming to learn about the rationality of different points of view on the issue under discussion is the greatest increase in rationality toward the issue under discussion that the argumentation can produce. Thus, If (i) and (iii) are true I contend that the argumentation would have achieved the purpose of increasing the overall level of rationality of the participants in the argumentation. I submit that achieving the goal of increasing the overall level of rationality of the participants in the argumentation is a reason to regard the argumentation as successfully resolved.

However, Godden contends that an argumentation is successfully resolved only if "commitments undertaken in [the] argumentation survive beyond its conclusion and go on to govern an arguer's actions in everyday life, e.g. by serving as premises in her practical reasoning" (Godden, 2010, p. 397). I contend that this condition can be satisfied when (ii) is true of the participants in an argumentation. Even though (ii) is compatible with the presence of disagreement between the participants in an argumentation coming to the view that there are rational positions, different from one's own, toward the issue under question in an argumentation can produce meaningful changes in how one reasons about the issue in question in the argumentation in ones everyday life.

Thus, I contend that when (i), (ii) and (iii) holds of the participants in an argumentation that (a) there is an increase in the overall level of rationality of the participants in the argumentation with respect to the issue under discussion (b) the participants change their everyday reasoning behavior with respect to their argumentation partner's position on the issue under discussion and (c) no greater increase in the overall level of rationality with respect to the issue under contention is possible. It strikes me that when features (a)-(b) hold of an argumentation the argumentation ought to be regarded as successfully resolved. The argumentation should be regarded as resolved since the argumentation achieved one of its important goals of increasing the level of rationality and it has also produced a meaningful change in the reasoning behavior of the participants in the argumentation. Since, as argued, (i), (ii) and (iii) can hold of the participants in an argumentation when (APP) holds of the participants in the argumentation, and since when (APP) holds there is lasting disagreement, it follows that argumentations can successfully resolve when there is lasting disagreement.

5. CONCLUSION

I have advanced an argument that it is possible for an argumentation to resolve in the presence of a disagreement over the argumentation's contested claim. Even in argumentations such as an appropriately extended DIA in which no reason-based agreement over the contested claim is possible the participants in an argumentation can come to learn that their interlocutor's attitude toward the proposition is reasonable given their interlocutor's context. In turn the participants can modify their behavior toward their interlocutor's standpoint. When such an appreciation of their interlocutor's standpoint occurs it is intuitively plausible to regard the argumentation as successfully resolved even though there is lasting disagreement.

ACKNOWLEDGEMENTS: I would like to thank the Social Sciences and Humanities Research Council of Canada for generous financial support through a postdoctoral fellowship (Grant number 756-2014-0556) during the period in which research for, and writing of, this paper took place. I would also like to than the University of Connecticut department of philosophy for hosting me during my postdoctoral studies. I have had several discussions with professors and graduate students at the University of Connecticut that impacted this paper. These include Michael Lynch, JC Beall, Nathan Kellen and Andrew Parisi. This paper stems from work I did while a PhD student at the University of Waterloo. Comments from professors at the University of Waterloo on earlier drafts of this paper also significantly impacted the shape of it currently takes. These professors include David DeVidi, Doreen Fraser, and Tim Kenyon. I am also grateful for comments and discussions on this paper with Andrei Moldovan of the University of Salamanca.

REFERENCES

Beall, J.C., & Restall, G. (2006). *Logical pluralism*. New York: Oxford University Press.

Bueno, O., & Shalkowski, S. (2009). Modalism and logical pluralism. *Mind, 118*(470), 295-321.

Cook, R. (2010). Let a thousand flowers bloom: A tour of logical pluralism. *Philosophy Compass, 5*(1), 492-504.

DeVidi, D. (2011). The municipal by-laws of thought. In D. DeVidi, M. Hallet, & P. Clark (Eds.), *Logic Mathematics, Philosophy: Vintage Enthusiasms* (pp. 97-112.). Dordrecht: Springer.

Field, H. (2009). Pluralism in Logic. *The Review of Symbolic Logic, 2*(2), 342-359.

Godden, D. (2010). The important of belief in argumentation: Belief, commitment and the effective resolution of a difference of opinion. *Sythese, 172*, 397-414.

Goldman, A. (2010). Epistemic relativism and reasonable disagreement. In R. Feldman & T. Warfield, (Eds.), *Disagreement* (pp. 187-215). New York: Oxford University Press.

Hamblin, C. (1970). *Fallacies.* London, U.K.: Methuen.

Hjortland, O. T. (2013). Logical pluralism, meaning variance, and verbal disputes. *Australasian Journal of Philosophy, 91*(2), 355-373.

Johnson, R. (2000). *Manifest Rationality.* Mahwah, N.J.: Lawrence Erlbaum Associates.

Keefe, R. (2014). What logical pluralism cannot be. *Synthese, 191*(7), 1375-1390.

Lynch, M. (2008). Alethic pluralism, logical consequence, and the universality of reason. *Midwest Studies in Philosophy, 32*(1), 122-140.

Lynch, M. (2009). *Truth as one and many.* Oxford: Oxford University Press.

Pedersen, N. J.L.L. (2014). Pluralism × 3 Truth, Logic, Metaphysics. *Erkenntnis, 79*(2), 259-277

Restall, G. (2002). Carnap's tolerance, meaning, and logical pluralism. *Journal of Philosophy, 99*(8), 426-443.

Restall, G. (2014). Pluralism and proofs. *Erkenntnis, 79*(2), 279-291.

Russell, G. (2008). One True logic? *Journal of Philosophical Logic, 37*(6), 593-611.

Shapiro, S. (2014). *Varieties of logic.* Oxford: Oxford University Press.

van Eemeren, F.H., & Grootendorst, R. (2004). *A systematic theory of argumentation: The pragma-dialectic approach.* New York: Cambridge University Press.

Walton, D., & Krabbe, E. (1995) *Commitment in dialogue: Basic concepts of interpersonal reasoning.* Albany, NY: State University of New York Press.

62

A Formal Model of Erotetic Reasoning in Solving Somewhat Ill-Defined Problems

MARIUSZ URBAŃSKI
Adam Mickiewicz University, Poznań, Poland
mariusz.urbanski@amu.edu.pl

NATALIA ŻYLUK
Adam Mickiewicz University, Poznań, Poland
nzyluk@gmail.com

KATARZYNA PALUSZKIEWICZ
Adam Mickiewicz University, Poznań, Poland
k.paluszkiewicz@amu.edu.pl

JOANNA URBAŃSKA
Adam Mickiewicz University, Poznań, Poland
joanna.urbanska@amu.edu.pl

In this paper we present some preliminary results on formal modelling of solving somewhat ill-defined problems in terms of logic of questions. Our evidence consisted of logs of gameplays of "Mind Maze" by Igrology. In analyses of episodes of erotetic reasoning we found that weak version of erotetic implication offers an adequate model for such processes.

KEYWORDS: erotetic reasoning, goal-directed reasoning, logic of questions, problem-solving, weak erotetic implication

1. INTRODUCTION

Our aim in this paper is to present some preliminary results on formal modelling of solving somewhat ill-defined problems in terms of logic of questions, which allows for analyses of erotetic reasoning, involving questions as both premises and conclusions.

D. Mohammed & M. Lewiński (eds.) (2016). *Argumentation and Reasoned Action: Proceedings of the 1st European Conference on Argumentation, Lisbon, 2015. Vol. II, 973-983.* London: College Publications.

2. EVIDENCE

"Mind Maze" by Igrology is a game in which, according to the manual, a gamemaster "describes a strange story and the players must determine why and how the story happened". Solution of each of the tasks is dependent on discovering key pieces of information (which are known to the gamemaster only) by asking auxiliary questions. In the original version only yes-no questions are allowed, with "not important" being a plausible answer as well. The players may collaborate in order to reach the solution.

Here is a simple toy-problem and a short gameplay ("P" stands for the player, and "G" for the gamemaster).

> A pilot jumped out of a plane and fell on the ground and did not die. How did this happened?
> P Did the pilot jumped with a parachute?
> G No.
> P Was the speed of the plane high?
> G No.
> P Did it take long to fell on the ground?
> G No.
> P Was the plane grounded at the airport?
> G Yes, that's the solution.

Thus the players face an abductive problem of making sense of a puzzling fact (see Thagard & Shelley, 1997; Urbański, 2016) given in a story and expressed in the initial question. The task is to find pieces of information accounting for this fact – the explanation of it (as there is the correct solution to each problem). The players are supposed to rely on their general knowledge as well as on their abilities to reason with questions as premises and conclusions.

One peculiarity of this kind of tasks is that typically while solving abductive problems reasoners aim at finding the best solution, or hypothesis, as evaluated against the class of epistemically ponderable criteria (including for example consilience, simplicity, coherence etc.; see Urbański, 2016). In the "Mind Maze" problems a solution is evaluated against criterion of being close enough to the prescribed one.

2.1 The data and the setup

We slightly modified the original rules of "Mind Maze", in order to allow for more cooperative behaviour of a gamemaster as well as to smoothen the process of data gathering:

1. The players are encouraged to think aloud and to take notes.
2. The players are asked to provide justifications for questions posed.
3. The gamemaster is allowed to ask questions which are clarification requests (see Purver, 2004).
4. The gamemaster is allowed to deliver additional information to the players, and they can opt-out.
5. The gamemaster is allowed to decide that a solution offered by a player is "close enough" to the correct one.
6. The gamemaster is allowed to use "critical" suggestions, or hints, pointing toward correct solution (because of ethical issues).
7. The gamemaster is allowed to ask for a summary of information available at a certain stage of the game.

After an initial pilot study 12 out of 65 original "Mind Maze" stories were chosen for this research. We decided not to use those which were heavily dependent on specific knowledge as well as those which were popularized on the Internet; also, some of the stories were evaluated as not engaging enough. Some of these 12 stories were also modified, either with respect to unequivocality of information conveyed or with respect to unequivocality of the questions posed to the players.

Each subject (or a pair of subjects) solved one problem in a single session, in the presence of an experimenter, who served as the gamemaster. Each session was audio recorded, with the players' permission. The sessions lasted from 5 to 38 minutes. Currently our data consist of 39 gamelogs, obtained from 42 players.

2.2 "Mind Maze" problems as somewhat ill-defined problems

It should be noted, that although "Mind Maze" problems are expressed in the form of open questions they are not strictly speaking ill-defined problems. Ill-defined (or ill-structured) problems are commonly characterized by the following three properties (see Sinnott, 1989):

1. There are no clear goals to be achieved by obtaining a solution to a problem.
2. There is no clear solution path.
3. There is no expected (or correct) solution.

In the case of "Mind Maze" problems there are no clear solution paths. However, a player knows in advance that there exists the correct solution and the goal is to identify it. Thus we may claim that, although not well-defined, the "Mind Maze" problems are only of somewhat ill-defined character.

Sinnott (1989) identifies five components of solving ill-defined problems: processes to construct problem space, processes to choose and generate solutions, monitors, memories, noncognitive elements. Each of these was present in solutions to the "Mind Maze" problems we obtained from the subjects. Shin Hong (1998) indicates that what correlates with fluency in solving ill-defined problems is structural knowledge, regulation of cognition, and justification skills. In our research we found evidence supporting these findings.

3. EXEMPLARY PROBLEM: `THE TRAVELLER'

We shall consider and describe a formal model of solutions to one of the "Mind Maze" problems, "The Traveller". The players were presented with the problem in the following form:

> A man, without a single visa, in one day visited eight different countries. Authorities of none of these countries tried to throw him out. What was his profession and how did he manage to do this?

For this particular problem we gathered evidence based on three single-player sessions (S1 – S3) and two two-players sessions (T1 and T2).

The solution to the considered problem is that the man in question was a courier delivering post to the embassies. Thus there are two key pieces of information which the players needed to identify in order to reach the solution. The first one concerns the profession of the protagonist. The second one concerns the fact that the protagonist visited embassies and not the countries themselves; this draws somewhat on a popular belief that all the embassies are sovereign territories of the represented state. Although in fact most of them do not enjoy full extraterritorial status, this did not raise any issues in this research.

What we found was that all the subjects in their solutions followed the very same pattern of first identifying what key pieces of information refer to and then establishing their actual content by means of going through a series of topics, introduced by asking questions. The notion of topic we use here is essentially van Kuppevelt's one, "which concerns the `aboutness' of (sets of) utterances" (van Kuppevelt, 1995, p. 111).

All the topics in our dialogues fall into one of the following five categories (we list them with exemplary topic-constituting questions):

1. Legality: "So, it was legal? No smuggling etc.?"
2. Geography: "The countries were on one continent? They were small?"
3. Means of transport: "Did he travel by plane? By ship? On foot?"
4. Profession: "He was someone of importance? A politician? A pope? Of common profession? A driver? A plumber?"
5. Concept of territory: "Did he physically enter the countries?"

In order to account for relative difficulty of the problem to the subjects we introduced two measures (see table 1). The first one is the length of a solution measured by the number of questions asked by the players to the experimenter. The second one is the distance between questions by answers to which key pieces of information were obtained.

	S1	S2	S3	T1	T2
Length of a solution (no. of questions)	11	24	13	12	37
K1	4th	22nd	6th	3rd	31st
K2	11th	24th	13th	12th	37th
Distance K2 - K1	7	2	7	9	6

Table 1 – The measures of relative difficulty of the problem

In the table 1 K1 and K2 stand for the first and the second key piece of information arrived at in respective dialogues. They represent only the order of the information, not their content; in three dialogues K1 was "courier" and K2 "embassies", in two the other way around. There seems to be no correlation between the length of a solution and the distance between arriving at key pieces of information. Also, there seems to be no correlation between the number of players (one vs two) and the number of questions asked. Thus we conjecture that in a collaborative setting the players do not solve the problems any faster or slower (that is, asking less or more questions, respectively) than in a solitaire setting. However, inspection of both measures does not allow for definite conclusions, as the considered sample is quite small.

All the dialogues exhibited similar structure. The subjects started with legality issue, followed by questioning if the protagonist was someone of importance. After that some subjects employed what we call "reconnaissance in force", trying to discover his profession. Only after unsuccessful guessing they started more systematic information processing, to which other subjects resorted earlier on. There consecutive questions were introduced on the basis of the previous ones and answers to them. What we claim is that these inferential processes of questioning (and thus the dynamic of topic constitution) can be modelled by means of logic of questions.

4. MODELLING THE SOLUTIONS

Solutions to all the "Mind Maze" problems consisted of two general phases. They correspond to Stenning and van Lambalgen's (2008) distinction of reasoning to vs reasoning from an interpretation. In the first one the subjects established interpretation of a problem; this phase can be also thought of as consisting of processes to construct problem space (Sinnott, 1989; see section 2.2). As most of the subjects were not very explicit about underlying reasoning processes, first part of our model is based on limited amount of data and offers a rational reconstruction of this phase rather than its full-fledged descriptive model. We employ here elements of Gabbay and Woods' (2005) formal schema of abductive reasoning and Kubiński's (1980) logical theory of numerical questions. In the second phase, which consisted of the actual dialogues with the experimenter, the subjects' information processing can be adequately modelled by means of weak erotetic implication of Inferential Erotetic Logic (see Urbański et al., 2015); all the remaining components of solving problems identified by Sinnott can be found in this phase. The particular solution we are going to reconstruct comes from the subject S1; it is typical with respect to the structure and has the advantage of being comparatively short.

2.1 The first phase: interpretation of the problem

During the first phase of the solution the subjects established their epistemic aims, fulfillment of which amounts to solving the problem. Formal reconstruction of this phase is given in the schema in figure 1. Following Gabbay and Woods (2005) we use the exclamation mark as an operator forming epistemic aims of the subject (we omit indices referring to the subjects, as they are not of crucial importance for our present purposes). "T" stands for solution to be obtained, "H" for hypotheses accounting for it. The arrow $\Rightarrow$ represents the attainment relation that holds between H and T; the subscripts p and c indicate whether it is partial or complete, respectively. Let us go through this schema step by step (which we numbered for the sake of readability).

The solution starts with setting the goal to be obtained (T) as the epistemic aim of the subject (step 1). In the steps 2 and 3 the key pieces of information are identified as partially accounting for T: H^a stands for profession of the protagonist, H^b – for the issue of visiting eight countries. In the step 4 it is settled that identifying H^a and H^b together completely account for *T*. Their conjunction is then set as the new epistemic aim of the subject and, consecutively, are the conjuncts (steps 5 and 6). Steps 8 - 11 are questions (in most cases not posed explicitly by the subjects), represented here in the formalism of Kubiński (1980). Question in the step 8 can be interpreted as follows: what is exactly one hypothesis H^a_i such that it is close enough to H^a; question 9 can be interpreted in a similar way. Question in the step 10, in turn, reads: what H^a_i's are such that they partially account for *T*? (and question in the step 11 reads analogously). Posing this questions concludes the first phase of the solution: interpretation of the problem.

$$
\begin{array}{cr}
T! & 1 \\
H^a \Rightarrow_p T & 2 \\
H^b \Rightarrow_p T & 3 \\
H^a \wedge H^b \Rightarrow_c T & 4 \\
(H^a \wedge H^b)! & 5 \\
H^a! & 6 \\
H^b! & 7 \\
(1)H^a_i(H^a_i \approx H^a) & 8 \\
(1)H^b_j(H^b_j \approx H^b) & 9 \\
0 < H^a_i(H^a_i \Rightarrow_p T) & 10 \\
0 < H^b_j(H^b_j \Rightarrow_p T) & 11 \\
\vdots &
\end{array}
$$

Figure 1 – Formal reconstruction of the first phase of the solution

2.2 The second phase: erotetic reasoning towards solution

The second phase of the solution consists of erotetic reasoning by which the subject S1 tries to identify the key pieces of information and, consequently, to find the correct answer to the initial question. This search, depicted in figure 2, is based on the subject's general knowledge and guided by obtaining answers to consecutive questions. The formulas of the form $H^n{}_m[A] \Rightarrow_p T$ represent activated elements of the subject's knowledge (or beliefs set), where *n* stands for the key piece of information being accounted for by the hypothesis, *m* being its consecutive number and *A* representing its content. Thus the formula $H^b{}_1[legality] \Rightarrow_p T$ represents subject's belief that the problem of whether the protagonist entered the countries legally partially accounts for the initial problem *T* in view of the key piece of information H^b to be identified – the issue of visiting eight countries. On this basis the subject asks the question represented by *?legality* ("So this person has visited these 8 countries fully legally, right?"). After the question represented by *?embassies* ("I wonder if, for example, this man just visited 8 embassies?") is asked and answered to the affirmative, the first key piece of information is obtained and the topics accounting for H^a are no longer relevant. Using van Kuppevelt's terminology we may consider this as an application of Dynamic Principle of Topic Termination: If a question is answered satisfactorily, the questioning process associated with it comes to an end and the relevant topic(s) loses its actuality in discourse (see van Kuppevelt, 1989, p. 131).

After the question represented by *?bodyguard* ("Or maybe he is a bodyguard for somebody?") the experimenter decided to deliver additional information, that the protagonist function is useful for the embassy staff (represented by *Useful*). Another additional information was delivered later on, that his profession is a common one. After that the subject is able to ask a question *?postman*, affirmative answer to which correctly identifies the second key piece of information that he is a postman; this is accepted by the experimenter as being close enough to the intended solution of the protagonist being a courier.

One important property of erotetic reasoning involved in the solution is that posing of consecutive auxiliary questions is not random but semantically justified. All these questions exhibits two properties:

1. if the question by which the initial problem is expressed is sound (i.e., there exists a true direct answer to this question) and all the declarative premises are true, then an auxiliary question is sound as well;
2. at least some answer to an auxiliary question is useful in answering the by which the initial problem is expressed (at least some answer to an auxiliary question narrows down the class of possible answers to it), provided that all the declarative premises are true.

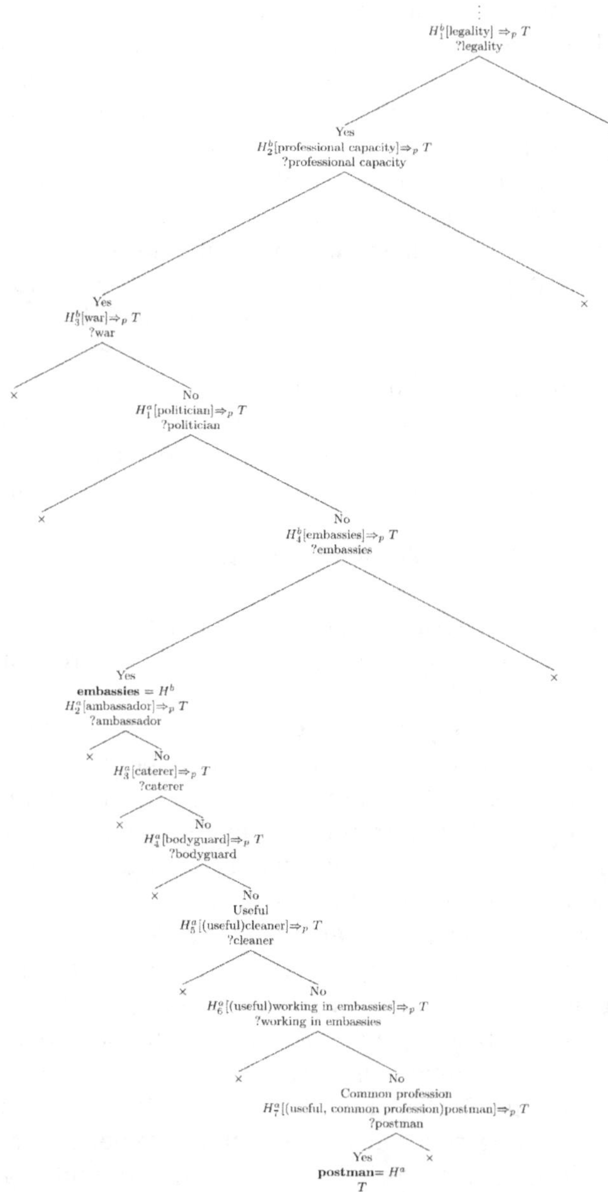

Figure 2 – Formal reconstruction of the second phase of the solution

The first property is called transmission of truth/soundness into soundness; the second one – partial cognitive usefulness. Together they define the relation of weak erotetic implication (see Urbański et al., 2015), which is a ternary relation between a question (question-premise), a set of declaratives (declarative premises) and a question

(question-conclusion). Partial cognitive usefulness warrants that at least some answers to an auxiliary question is useful in solving the initial problem; it may happen that some answers lead to nowhere (with respect to the goal of the reasoning), but from a prescriptive point of view (see Stanovich, 1999) there is some decent rationality involved in choosing partially useful solutions. Weak erotetic implication is a weakened version of canonical erotetic implication, which is an engine of formal models of reasoning involving questions as premises and conclusions in one of the most influential current paradigms in the logic of questions – Inferential Erotetic Logic (see Wiśniewski, 1995, 2013).

The search for solution can be represented in the form of a scenario, very much similar to the scenarios representing formal structure of abductive reasoning (see Urbański, 2016). Such a scenario describes the space of solutions to the initial problem. What is depicted in figure 1 is only one path of such a scenario, the one which is actualized in view of answers obtained to auxiliary questions.

5. CONCLUSION AND FURTHER RESEARCH

In this paper we presented some preliminary results on modelling solutions to a specific class of somewhat ill-defined problems by means of logic of questions. What we found is that these processes can be analysed in terms of topic management, for which a useful model is offered by weak erotetic implication. This supports the results of Shin Hong (see section 2.2). Structural knowledge determines the ability to identify the topics relevant to the solution. Regulation of cognition and justification skills manifest themselves in navigating through solution scenario, created on the go, with weak erotetic implication serving as a model for underlying prescriptive rationality. Thus our results support the claim that effective solving of such problems relies on applications of strategic rather than definitory rules (Hintikka, 1999).

Our further research will focus on development of a proper logic for this kind of reasoning processes as well as on their abductive interpretation. What is also of interest is the role of collaboration of the players in obtaining solution to the initial problem. Our preliminary results suggest that such a collaboration is not substantially helpful, but this issue requires more thorough investigation.

ACKNOWLEDGEMENTS: Research reported in this paper were supported by the National Science Centre, Poland (DEC-2013/10/E/HS1/00172)

REFERENCES

Gabbay, D., & Woods, J. (2005). *The reach of abduction*. London: Elsevier.

Hintikka, J. (1999). What is abduction? The fundamental problem of contemporary epistemology. In *Inquiry as Inquiry: A Logic of Scientific Discovery* (pp. 91–113). Dordrecht: Kluwer AP.

Kubiński, T. (1980). *An outline of the logical theory of questions*. Berlin: Akademie-Verlag.

van Kuppevelt, J. (1995). Discourse structure, topicality and questioning. *Journal of Linguistics, 31*, 109–147.

Purver, M. (2004). *The theory and use of clarification requests in dialogue*. PhD thesis. London: King's College.

Shin Hong, N. (1998). *The relationship between well-structured and ill-structured problem solving in multimedia simulation*. PhD thesis. The Pennsylvania State University.

Sinnott, J. D. (1989). A model for solution of ill-structured problems: Implications for everyday and abstract problem solving. In J. D. Sinnott (Ed.), *Everyday problem solving: Theory and applications* (pp. 72–99). New York: Praeger.

Stanovich, K. E. (1999). *Who is rational? Studies of individual differences in reasoning*. Mahwah: Lawrence Erlbaum.

Stenning, K., & van Lambalgen, M. (2008). *Human reasoning and cognitive science*. Cambridge: The MIT Press.

Thagard, P., & Shelley, C. P. (1997). Abductive reasoning: Logic, visual thinking and coherence. In M.-L. Dalla Chiara, K. Doets, D. Mundici & J. van Benthem (eds.), *Logic and scientific methods*, (pp. 413–427). Dordrecht: Kluwer AP.

Urbański, M. (2016). *Models of abductive reasoning*. Berlin: LiT Verlag. To appear.

Urbański, M., Paluszkiewicz, K., & Urbańska, J. (2013). *Deductive reasoning and learning: A cross-curricular study*. Research report. Poznań: Adam Mickiewicz University

Urbański, M., Paluszkiewicz, K., & Urbańska, J. (2016). *Erotetic problem solving: From real data to formal models. An analysis of solutions to Erotetic Reasoning Test task* (to appear).

Wiśniewski, A. (1995). *The posing of questions. Logical foundations of erotetic inferences*. Dordrecht: Kluwer AP.

Wiśniewski, A. (2003). Erotetic search scenarios. *Synthese, 134*(3), 389–427.

Wiśniewski, A. (2013). *Questions, inferences, and scenarios*. London: College Publications.

63

Dissociating between 'Is' and 'Ought': Recognizing and Interpreting Positions in Climate Change Controversies

MEHMET ALI UZELGUN
ArgLab, Universidade Nova de Lisboa, Portugal
uzelgun@fcsh.unl.pt

PAULA CASTRO
CIS-IUL, Instituto Universitário de Lisboa (ISCTE-IUL), Portugal
paula.castro@iscte.pt

This presentation focuses on the uses of dissociation in controversial debates. We report findings from an argumentative analysis of (N=22) interviews, in which participants were presented with contentious assertions concerning climate change action. We show how the interview responses were characterized by contrastive and concessive uses of the connective but, and explore the - temporal and spatial - patterns through which dissociation was used in enhancing the dialectical reasonableness together with the rhetorical effectiveness of the arguments.

KEYWORDS: appearance/reality pair, carbon offsetting, controversy, definition, dissociation, environmental discourse, temporality and spatiality

1. INTRODUCTION

In this paper we explore whether and how temporality and spatiality are used as organizing principles of dissociation in the discourse of climate change campaigners. We do this by examining argumentative *but*-constructions as linguistic vehicles of dissociation, and by considering the role of the *Appearance/Reality* pair (Perelman & Olbrechts-Tyteca, 1969) in sustaining the dissociations.

To achieve our goals, we focus on the argumentation carried out by climate campaigners in a controversy over the utility of carbon offsets. In order to instigate conflict and argumentation, an interview

D. Mohammed & M. Lewiński (eds.) (2016). *Argumentation and Reasoned Action: Proceedings of the 1st European Conference on Argumentation, Lisbon, 2015. Vol. II, 985-997.* London: College Publications.

study we use video-elicitation, presenting the participants with contentious arguments about the utility of carbon offsets.

A carbon offset is a financial instrument devised to compensate for carbon emissions created in one location through reductions in another (Lovell & Liverman, 2010). Voluntarily paying a small amount to compensate the emissions created by one's consumption is portrayed – in the current global climate change governance framework – as an efficient and economically viable way to mitigate climate change. This instrument has however also been contested as sustaining old habits and impeding engagement with the (unequal) political and production relations that lie at the heart of the climate problem (Bumpus & Liverman, 2008; Rathzel & Uzzell, 2009). The controversy over the utility of carbon offsets consequently gives us the chance to examine the ways through which different definitions and meanings are contrasted and negotiated, and political and ideological incompatibilities are dealt with.

Below, we first explain why paying attention to the temporal and spatial characteristics of climate change discourse is relevant, constructing these as two important characteristics of environmental discourse at large. Then we summarize the relevant literature on dissociation, drawing mainly on Perelman and Olbrechts-Tyteca's (1969) rhetorical, and van Rees's (2009) pragma-dialectical accounts. After the analysis of exemplary excerpts from our interviews, we discuss the relevance of the *Is/Ought* pair, through which environmental campaigners – in dissociating between certain notions – organize their criticisms of the status quo, while upholding some basic features of it.

1.1 Spatial and temporal characteristics of environmental discourse

Many authors have pointed out that environmental discourse has particular spatial and temporal characteristics, and climate change is a case example these accounts (e.g. Beck, Blok, Tyfield, & Zhang, 2013; Morton, 2010; Harré, Brockmeier, & Mühlhäusler, 1999; Adam, 1998). Two basic findings that specifically concern climate change communication may be summarized as follows.

First, regarding the spatial characteristics, the global, national and local scales of the threat and contexts of action are often distinguished or contrasted. For instance, it is known that in media portrayals of climate change in the industrialized countries, the threat is often represented as distant from the viewer (Nerlich & Jaspal, 2014; Smith & Joffe, 2009). In the press of industrializing countries, on the other hand, the representation of solutions and responsibility as situated elsewhere, "outside the national borders", is a repeated finding (e.g. Uzelgun & Castro, 2015; Billett, 2010). Representation of the threat and solutions as "external to the collective" is also regarded as part of a de-politicizing discourse, and part of "environmental post-politics" (e.g. Williams & Booth, 2013, p. 25; Swyngedouw, 2010).

Second, regarding temporality, it is possible to broadly say that environmental discourse is typically oriented towards the future (Morton, 2010; Beck, Blok, Tyfield, & Zhang, 2013), mainly to act upon the present, and to cover - and in some cases constrain - the potentialities emerging from use of miscellaneous metaphors (Harré, Brockmeier, & Mühlhäusler, 1999). This focus on the future has been criticized as evasion of problems and putting back of solutions (Uzelgun & Castro, 2014; Ramos & Carvalho, 2009), and also as part of a de-politicizing discourse on climate change (Demeritt, 2001). Furthermore, since problems like climate change are massively "distributed" in time (Morton, 2010), and our knowledge of them also extends to vastly different timescales (e.g. geological, cultural), the superimposition of different times and timescales in this type of discourse is almost inevitable (Harré et al., 1999).

Hence, looking at whether the imposition of value hierarchies in dissociating arguments is carried out through temporal or spatial categories may help to better understand the use of this technique in the given context.

1.2 Dissociation and its underlying pairs

Dissociation is an argumentative technique that consists of separating a (previously unified) idea into two elements, and imposing a value hierarchy on them (Perelman, 1982). The use of this technique is "always prompted by the desire to remove an incompatibility", and in dealing with a given incompatibility through dissociation, the speaker can "sacrifice one or even both of the conflicting values" (Perelman & Olbrechts-Tyteca, 1969, p. 413).

In their seminal work, Perelman and Olbrechts-Tyteca (1969) focus on the patterns through which the sacrifice of "conflicting values" is carried out in epistemic discourse. Their main concern is the *organising principle(s)* of dissociations and for this they identify a lengthy list of "philosophical pairs" (e.g. means/end, subjective /objective, particular/general). Among these, one is emphasised as the prototypical pair: *Appearance/Reality*. According to the authors, all other pairs boil down to this fundamental opposition (Jasinski, 2001), and dissociation is essentially founded upon the contrast created between the *apparent* and *real* interpretations of an idea, notion, or concept.

Researchers who pursued more elaborate accounts of dissociation (e.g. van Rees, 2006, 2009; Gata, 2007; Jasinski, 2001; Schiappa, 1993) maintain the proposition that the *Appearance/Reality* pair is central. That this fundamental pair manifests a differential valorisation of the dissociated notions is widely taken up in these accounts, however, it does not seem to have inspired research that focuses on its various formulations. Our goal in this paper is to pursue this task in a specific discursive context, and for this we mainly use van Rees' (2009) analytical treatment of this argumentative technique, introduced in the next section.

1.3 Basic characteristics and indicators of dissociation

So far we introduced that for Perelman and Olbrechts-Tyteca, dissociation involves, above all, a change in the organization of knowledge and values "used as the basis of argument" (1969, p. 413), and that this change is steered through certain conventional patterns or pairs. Compared with their explorations of different patterns of dissociation, van Rees' (2009) interest in the use of this technique is quite methodical. In an effort to devise a more systematic account, van Rees outlines three groups of clues for identifying dissociation. She does this by drawing on the three central characteristics of its use:

> (1) A distinction between two notions is made: "at the basis of dissociation there are two speech acts, distinction and definition" (van Rees, 2006, p. 474). These speech acts can be direct (e.g. I define...), indirect (e.g. We need to distinguish...), or implicit (e.g. The difference between...), and they are the essential features of dissociation.
> (2) The hierarchy of values associated with a given notion is altered: In this regard, the use of expressions such as "real", "true", "central", "technical", "pseudo", "quasi" etc. serve as the indicators of dissociation (van Rees, 2009, p. 39). It is through these adjectives that a value hierarchy is imposed on the previously undifferentiated meaning categories.
> (3) A contradiction is resolved or dealt with: In this regard, the indicator specifically expounded on by van Rees (2009, p. 42) is the presence of a contrastive or concessive *but* (see Snoeck Henkemans, 1995; Uzelgun, Mohammed, Lewiński, & Castro, 2015). While the *contrastive but* indicates a contrast about which the speaker has a clear preference such as negation (not-but), the *concessive but* indicates a limited preference about two contextually differentiated interpretations (yes-but). Overall, the mere presence of *but* signals that some contradiction or incompatibility is being dealt with.

Regarding the first set of clues, it can be said that these help the researcher to determine the degree of explicitness in externalizing this argumentative move; i.e., to decide whether one is dealing with performed vs. presupposed dissociations (van Rees, 2009; see also Jasinski, 2001; Gata, 2007). Furthermore, they also constitute linguistic means through which the most important – clarifying – function of dissociation is carried out. In this regard, from a rhetorical point of view, Schiappa (1993) argues that definitions indispensably involve questions such as:

> "How *ought* we use the word X?" or "What should be described by the word X?" Normative questions of this sort cannot be answered acontextually; they virtually compel interlocutors to address the *pragmatic* needs of a given community of language-users located in a particular historical moment. (Schiappa, 1993, p. 413)

This means that, especially in definitional disputes, the clarifying function of dissociation (van Rees, 2009) is brought to bear by normative considerations, through which value hierarchies are imposed (by the use of the second set of clues) in accordance with a certain pragmatic project, and in a certain point in time in the course of that collective project.

Our main interest in this study is at the intersection of the use of the second and third set of clues, that is, the differential valorization indicated or implied by the use of the connective *but*. To put it in one sentence, the literature on argumentative but-constructions suggests that the statements – and the underlying values – constituting the post-but segment of an argument is typically preferred against – or over – the statements constituting the pre-but segment (e.g. Billig, 1999). Building on this, we investigate the ways the underlying values are valorised or disqualified (stated before or after the *but*); or in other words, we will examine the organising principles of the dissociations: whether they are carried out with pertinence to certain characteristics of environmental discourse, namely temporality and spatiality.

2. METHOD

In-depth interviews (N=22) were conducted with climate change campaigners in Portugal and Turkey. The participants were referred members of environmental NGOs that are active in climate information and policy in the two countries, as well as internationally. In the last part of the interviews, they were presented with a series of short video-excerpts, and our assumption was that the people featured in the video-excerpts would be the main argumentative opponents of the interviewees. The video-excerpts were selected so as to contain minimal visual information – other than the speakers' faces – and to elicit conflict and contradiction.

The analysis presented below focuses exclusively on the video-excerpt that features a climate activist who contests the usefulness of carbon-offset mechanisms. The climate activist argues:

> *Carbon offsets are a fictitious commodity that have been created to exploit the rising levels of climate consciousness.* [...] *I think the more emphasis we put on individuals, we're moving away from what really needs to happen, in terms of people to come together in communities, to start organizing, to create political pressure for the bigger systemic changes that need to happen, in moving away from the growth based model, reigning in... eh, at the corporate self-interest...*[1]

In arguing this way, he contests (1) an existing practice adopted to abate climate change, (2) a hegemonic discourse that prioritizes privatized (individual) solutions, and (3) broader legitimizing norms – e.g. a growth based model, corporate self-interest. Hence, his discourse represents a minority perspective in the framework of global climate governance and action.

We expected this intervention to incite conflict and contradiction, and following the scholarly calls to pay attention to "small words" in the analyses of language and argumentation (Billig, 1999; Castro, 2006; van Rees, 2009), attended specifically to the argumentative uses of the connective *but.* The analysis of contrastive and concessive but-constructions was structured along the lines proposed by van Rees (2009): Although the use of a *but* itself indicates or implies a value hierarchy, we looked for further markers (e.g. real, true), as well as indirect and implicit speech acts. Finally, we also looked at whether and how the temporality of the *but*-constructions was organized in particular patterns, and whether it had any relevance in terms of the differential valorization of the dissociated notions.

3. ANALYSIS

---

1 The full video-excerpt (duration 2 minutes 20 seconds) can be found at http://www.youtube.com/watch?v=uk9Ev91jjQ8, starting from the beginning to 02'20".

We identified N=196 contextual uses of the connective *but* in our corpus. Almost half of these were concessive constructions, and the other half consisted of contrastive uses. In analyzing dissociations made through these, we did not focus on the argumentatively relevant differences between the uses of contrastive and concessive *but*-constructions. The attention paid to the temporal and spatial arrangements made by the interviewees in these constructions revealed some salient temporal patterns, and no spatial ones. Hence, below we focus only on the former. Due to space limitations, we here focus on only one conflict – between profit-making and costly climate action or solutions. We start with a response given to the interviewer's question concerning people's tendency to change their lives "only if they see a personal profit".

> *Example 1. Interview 3*
> I think we have to be very clear with definitions, what do we mean by profit? Because, when we're talking about profits, **today**, eh, like it or not, most of the profit goes to big huge corporations, and most of the people have... it's irrelevant for them, **that kind of profit**. We're talking about that... yeah, we don't need this, it's not beneficial. **But** if we're **looking for benefits**, to force people to act, to convince people to act, then it would be arranged in a way. [...] We are talking about two different things here. You know, profits of **corporations**, that's different from, from what it means for **individuals**.

In this example, a distinction between (corporate) profit and (personal) benefit is made, using a contrastive *but* located at the middle of the excerpt. The interviewee condemns "that kind of profit" which is "irrelevant" for people, "not beneficial", and not what we need. On the other hand, "benefits" are valorized – after the *but* – as they help to "convince people" to act upon climate change. The clarifying function of dissociation (van Rees, 2009), as well as indirect and implicit speech acts (e.g. "be very clear with definitions", "that's different from"), may be traced through the entire excerpt.

Regarding the underlying philosophical pairs, the individual/corporate pair seems obvious at first. However, this spatial organization of the dissociated terms – made in terms of scales of human economic activity – can be further pursued to identify a temporal organization as well: Against "today's" huge corporate profits, the interviewee seems to "looks for" (tomorrow's) benefits – in a future that is not pervaded by corporate interests but by action of convinced individuals. Arguably, it is this way that the interviewee resolves the incompatibility of asking people to take costly action without offering them any "profit" in return – by depicting the present *appearance* of benefits (i.e. profits) as "not beneficial", and reserving the proper meaning or the *reality* of benefits exclusively for individuals. Importantly, whether we "like it or not" the appearance *is* the present states of affairs, and the new definition is offered in an implicitly ("*if* we are looking") normative manner (Schiappa, 1993). The trouble with "profit", or its role in climate-relevant action, characterizes also our next example.

> *Example 2. Interview 8.*
> ...this growth, focusing too much on growth, growth-based mentality, it's too much you know it's, eh, that's too much. Profit, **I don't wanna use the word profit, but** I think it... that's why I said you know **well-being**, you know we, have to develop, not, not focusing on growth doesn't mean we have to stay, stand still. You know? **But there is a way**, for development to happen when responding, when taking eh, hm..., I mean I'm **trying to have a sustainable development**. Sustainable development is possible.

In this example, we find an old contradiction of the environmental movement, which was perhaps most at the fore in the 1970s: the incompatibility between development and environmental protection. From the 1980s on, the term "sustainable development" was instrumental in dissolving and overthrowing this incompatibility (Sachs, 1999; Lemke, 2002). Likewise, our interviewee criticizes the excessive focus on "growth, growth based mentality". As "profit" is associated with or seen within this mentality, "well-being" is offered as its alternative, which allows the interviewee to reconcile development and environmental protection. This reconciliation effort is quickly captured by the umbrella term sustainable development, which allows pursuing both goals. In short, a dissociation between *profit* and *well-being* is carried out and justified by organizing these concepts respectively in the framework of *growth* vs. (sustainable) *development*.

Concerning the temporality of this normative construction, we can say that sustainable development is not yet here and now, as it merely "is possible", constituting only "a way", and the interviewee is "trying to have" or achieve it. On the other hand, that there is "too much growth" is emphasized three times. Hence, a simplified reconstruction can be that there *is* too much profit or growth mentality *at present*, but what we ought "to have", *as a target*, is not a standstill but a mentality sustaining both well-being and development. Underlying this apparently concessive move is the notion that profit and growth at all costs is only the *present day appearance* of development, and the *reality* of development – to transpire in the future – is well being.

> *Example 3. Interview 4.*
> The thing is, people **think short term**, yeah it's not profitable, that's for sure. I mean you can't give them any profit in the short term, **but in the medium term or the long term**, the solutions are **more profitable actually**.

In this example, again the conflict is between climate solutions and profit. To manage this conflict, a dissociation is made between two interpretations of "profit". Here, the main difference from the previous examples is that the connective *but* is used as a *concessive* (yeah, it's not profitable... but...) and not as a *contrastive.* In the initial part of this *but*-construction, the interviewee admits that climate solutions are in conflict with "short term profit", which is negatively valorized or even presented as the problem ("the thing is..."). After the *but,* she asserts that the profitability of (real) climate solutions becomes evident when conceived in broader timescales; i.e., in the medium or long term. This way, by valorizing "thinking big" in time (Morton, 2010), the solutions are separated into *apparently* profitable (as understood in present time) and "*actually*" profitable (once understood properly). Our final example makes clear what may have remained cursory or obscure so far in this regard:

> *Example 4. Interview 15.*
> ...the way the profit is measured **nowadays**, you can't, you don't... climate and profit are not going on the same direction. I mean you can, **nowadays** you can **still** reduce... the effects on climate change, and still make profit, because of energy efficiency measures that you can **still** implement. And that's helping a company to improve their business case, eeh... **but you can only do it until a certain point, until** you reach a maximum level of efficiency. And then **after** that, the only way you can... eh, go in, avoid climate change is to actually reduce production. So energy efficiency is a way of fighting climate, but impro- and improving profit of the company, **but that's limited.** You can only do that **until a certain point** and **after** that then you need that systemic change, in the system, you need to change the way the companies do profit, or the way we see profit or measure profit. So that... a company can still try to do profit, **but profit is a different concept and therefore it's not conflicting with climate change**.

That the incompatibility dealt with in this example is between "climate and profit" is evident already from the first sentence. The interviewee argues that "nowadays", the companies can "still" make profit while fighting climate change; that is, by adopting no-regret energy efficiency measures. "But" one can do it "until a certain point", or for a relatively short period. In the long term, or "after" this certain point, the companies would have to (change the way they) see or make profit while "actually reducing production". In other words, it is through the dissociation between *profit as it is today* and *profit as it ought to be seen* in the future, the interviewee resolves the conflict he sets forth at the outset.

As in the previous example, the dissociation is carried out through a concessive but-construction. After conceding that "nowadays" it may *appear* as if one can make profit while fighting climate change, the interviewee argues that this appearance – the current state of affairs – is ephemeral. Is it possible to conclude that the interviewee places this present-time meaning of profit "outside" its proper meaning, disqualifying it "into a mockery of the real thing" (van Rees, 2009, p. 7)? An important constituent of the answer is that the interviewee does not define or distinguish explicitly the proper or "real" meaning of profit. As in example 2 above, there is rather a call, a "need" or a quest for a "different concept" that does not – more precisely *will not* – conflict with climate change. This temporal pattern can be described through the model of practical reasoning in political discourse suggested by Fairclough and Fairclough (2012), which we discuss in the following section.

4. DISCUSSION

In this study we explored how temporality is used as organizing principles of dissociation in the discourse of climate change campaigners. We did this by examining argumentative *but*-constructions as linguistic vehicles of dissociation, and by examining the role of the *Appearance/Reality* pair in sustaining the dissociations.

The pairs identified above, in the specific controversial context created via video-elicitation, were Profit/Benefit, Profit/Well-being, Short-term/Long-term conceptions of profit, and the definition of profit *At present/At a certain point in future*. The prototypical pair *Appearance/Reality* does not straightforwardly characterize these dissociations, rather it underlies such salient patterns (Perelman & Olbrechts-Tyteca, 1969). This "fundamental pair" is difficult to apply *directly*, when, for instance, profit "as it is defined today", "nowadays" or "in today's society" is disqualified for the search of a better notion; i.e., when the disqualified term reflects precisely the existing reality that – according to the speaker – has to be transformed.

The dissociations in our corpus seem to be characterized, in numerous occasions, by a linear arrow of time: from the (disqualified) interpretation of the notion in the *present state of affairs* to its new (valorized) interpretation in a *necessary future state of affairs*. In other words, the examples we analyzed suggest what may be called the *Is/Ought* pair, through which a unifying notion that structures the existing reality is disqualified for the search of a better notion, which *must* render the former notion somewhat irrelevant or trivial. Yet, it *can* only render the former notion in some future time. This seems in line with Fairclough & Fairclough's (2012) account of practical reasoning in political discourse. In this model, practical reasoning departs from *circumstances*, in order to attain *goals*, through the discursive and non-discursive *actions* of the speaker (and others). The temporal organization of this model – the arrow from present to future – mirrors the organizations in our corpus, however there may be some differences that require further consideration. Critically, in the model, such an arrow of time is established or obtained through the active efforts of the speaker and the collective. However, in many of the argumentative constructions in our corpus, the future is – represented as – rather certain, or determined by structural forces, or physical limits (e.g. you can do profit "until you reach a maximum level of efficiency").

More importantly, the suggestion of an *Is/Ought* pair as one of the possible organizing principles of dissociations needs to be clarified: can this pair be conceived as one of the main organizing principles of dissociations made in environmental discourse, or in controversial debates such as the one on climate action? Is its relevance limited only to definitional disputes, or can it apply more broadly to practical reasoning and argumentation? Surely the limitations of the present study – that it is based on a very limited corpus and a specific context – do not permit such conclusions. Further research that takes into account both the situated, interactional linguistic contexts (level of speech acts), and the underlying meaning systems that subsist in broader political-cultural contexts (level of implicit meanings, values, norms) is necessary to answer these questions.

It may be crucial to emphasize again that the existing reality or the present states of affairs cannot be simply rendered "*irrelevant*... to take into account" (Perelman & Olbrechts-Tyteca, 1969, p. 411, emphasis added). It can neither be changed right away, nor even in a future; and the success of the efforts to transform it depends largely on the relations and coalitions of these efforts with others. Arguably, this is among the reasons why our interviewees saliently resort to concessive but-constructions, and even when using *but* as a contrastive – as in example two – carry out dissociations and associations simultaneously.

5. CONCLUSION

In this paper we showed evidence for the use of the dissociative pair *Is/Ought*, through which the prototypical *Appearance/Reality* pair may allow speakers to disqualify the status quo, and some of the no-regret solutions used for mitigating climate change, as temporary or ephemeral (apparent). In our examples, this was done to support and emphasize the necessity (reality) of climate change mitigation, which, for our interviewees reside mostly in a future time rather than the present.

REFERENCES

Adam, B. (1998). *Timescapes of modernity: The Environment and invisible hazards.* London: Routledge.

Beck, U., Blok, A., Tyfield, D., & Zhang, J. (2013). Cosmopolitan communities of climate risk: Conceptual and empirical suggestions for a new research agenda. *Global Networks, 13,* 1-21.

Billett, S. (2010). Dividing climate change: Global warming in the Indian mass media. *Climatic Change, 99,* 1-16.

Billig, M. (1999). *Freudian repression: Conversation creating the unconscious.* Cambridge: Cambridge University Press.

Bumpus, A., & Liverman, D. (2008). Accumulation by decarbonisation and the governance of carbon offsets. *Economic Geography, 84*(2), 127–156.

Castro, P. (2006). Applying social psychology to the study of environmental concern and environmental worldviews: Contributions from the social representations approach. *Journal of Community & Applied Social Psychology, 16,* 247-266.

Demeritt, D. (2001). The construction of global warming and the politics of science. *Annals of the Association of American Geographers, 91,* 307-337.

Fairclough, I., & Fairclough, N. (2012). *Political discourse analysis: A method for advanced students.* London: Routledge.

Gata, A. (2007). Dissociation as a way of strategic manoeuvring. In F.H. van Eemeren, J.A. Blair, C.A. Willard & B. Garssen (Eds.), *Proceedings of the Sixth Conference of the International Society for the Study of Argumentation* (pp. 441–448). Amsterdam: Sic Sat.

Harré, R., Brockmeier, J., & Mühlhäusler, P. (1999). *Greenspeak: A study of environmental discourse.* London: Sage.

Jasinski, J. (2001). *Sourcebook in rhetoric.* London: Sage.

Lemke, T. (2002). Foucault, governmentality, and critique. *Rethinking Marxism, 14*(3), 49-64.

Lovell, H., & Liverman, D. (2010) Understanding carbon offset technologies. *New Political Economy, 15*(2), 255-273.

Morton, T. (2010). *The ecological thought.* Cambridge: Harvard University Press.

Nerlich, B., & Jaspal, R. (2014). Images of extreme weather: Symbolising human responses to climate change. *Science as Culture, 23*(2), 253-276.

Perelman, C. (1982). *The realm of rhetoric.* Notre Dame: University of Notre Dame.

Perelman, C., & Olbrechts-Tyteca, L. (1969). *The New Rhetoric,* J. Wilkinson & P. Weaver (transl.). London: University of Notre Dame.

Ramos, R., & Carvalho, A. (2008). Science as rhetoric in media discourses on climate change. In J. Strunck, L. Holmgreen & L. Dam (Eds.), *Rhetorical aspects of discourses in present-day society* (pp. 223-247). Cambridge: Cambridge Scholars.

Rathzel, N., & Uzzell, D. (2009). Changing relations in global environmental change. *Global Environmental Change, 19,* 326-335.

Sachs, W. (1999). *Planetary dialectics: Explorations in environment and development.* London: Zed Books.

Smith, N., & Joffe, H. (2009). Climate change in the British press: The role of the visual. *Journal of Risk Research, 12,* 647-663.

Schiappa, E. (1993). Arguing about definitions. *Argumentation*, 7, 403–418.

Snoeck Henkemans, A.F. (1995). 'But' as an indicator of counter-arguments and concessions. *Leuvense Bijdragen, 84*(3), 281–294.

Swyngedouw, E. (2010). Apocalypse forever?: Post-political populism and the spectre of climate change. *Theory Culture Society,* 27, 213-232.

Uzelgun, M.A., & Castro, P. (2015). Climate change in the mainstream Turkish press: Coverage trends and meaning dimensions in the first attention cycle. Pre-published in *Mass Communication and Society,* DOI: 10.1080/15205436.2015.1027407.

Uzelgun, M.A., Mohammed, D., Lewiński, M., & Castro, P. (2015). Managing disagreement through *yes, but...* constructions: An argumentative analysis. *Discourse Studies, 17*(4), 467-484.

van Rees, A. (2006). Strategic manoeuvring with dissociation. *Argumentation, 20,* 473-487.

van Rees, A. (2009). *Dissociation in argumentative discussions: A pragma-dialectical perspective.* Dordrecht: Springer.

Williams, S., & Booth, K. (2013). Time and the spatial post-politics of climate change: Insights from Australia. *Political Geography, 36,* 21-30.

64

"Vaccines Don't Make Your Baby Autistic": Arguing in Favour of Vaccines in Institutional Healthcare Communication[1]

ALESSANDRA VICENTINI
University of Insubria at Varese, Italy
alessandra.vicentini@uninsubria.it

KIM GREGO
University of Milan, Italy
kim.grego@unimi.it

This paper intends to explore argumentation as employed in institutional healthcare communication supranationally and in different developed countries to respond to distrust in vaccines as supported and spread by non-institutional sources like anti-vaccine movements. A corpus of institutional publications belonging to different genres are analysed from a critical discourse analysis perspective for: a) specific argumentative strategies employed to promote child immunisation; b) their linguistic realisation; c) their rhetorical relationship with anti-vaccination sources; d) their ethical aspects.

KEYWORDS: Anti-vaxxer campaigns, argumentation, critical discourse analysis, healthcare discourse, institutional discourse, medical discourse, vaccines, web-mediated communication

1. BACKGROUND

The most recent (but certainly not the first) movements against child vaccination can be traced back to the late 1990s, precisely to the effects on the public, as filtered and spread by the popular press, of the medical

---

[1] Research for this paper was conducted jointly by the two authors. In particular, Alessandra Vicentini is responsible for pars. 1, 3, 4.2; Kim Grego for pars. 2, 4, 4.1, 5.

D. Mohammed & M. Lewiński (eds.) (2016). *Argumentation and Reasoned Action: Proceedings of the 1st European Conference on Argumentation, Lisbon, 2015. Vol. II, 999-1020.* London: College Publications.

article Wakefield et al. (1998) (Poland & Jacobson, 2011). As is now widely known, this suggested a possible link between the measles–mumps–rubella (MMR) vaccine and the onset of a "developmental disorder", namely, forms of autism. The prestigious publication in *The Lancet*, together with the publicity the news received in the press and the historical timing (huge numbers of affordable personal computers connected for the first time by the advent of the Internet era) created a one-of-a-kind case of panic in developed countries, and led to the start of several popularly called "anti vaxxers" campaigns. As surprising as this may seem, these bottom-up movements achieved some considerable success, given that their "claim led to decreased use of MMR vaccine in Britain, Ireland, the United States, and other countries" (Poland & Jacobson, 2011). Not only, the logical though, again, unexpected consequence was an increase in measles outbreaks, as reported, for example, by Public Health England (2014) about England and Wales about measles cases between 1995 and 2013. So serious was the medical and social impact of antivaccinationism that an official investigation into the assumption was begun, and the original 1998 study was found to be flawed. As a result, the paper was retracted by *The Lancet* (2010) and Andrew Wakefield, principal investigator, was struck off the UK General Medical Register. The legacy of thr latest 17-year long anti-vaccination scare remains, though, and it is against this background that this study into institutional pro-vaccines argumentation is set.

2. AIMS AND METHOD

This paper aims to explore linguistic issues related to how healthcare institutions in developed countries respond to distrust in vaccines as supported and spread by non-institutional sources. In particular, it will explore what argumentative and discursive strategies they employ and how these are realised linguistically: do they engage rhetorically, directly or indirectly, with anti-vaccination sources? How do they deal with the ethical aspects of a sensitive debate that involves individual freedom of choice and children?

The analysis was of a qualitative nature, and combined methods in critical discourse analysis (Fairclough, 1989, 1995, 2003; van Dijk, 2008; Wodak 2013), genre analysis (Swales, 1990, 2004; Bhatia, 1993, 2004), text grammar (Werlich, 1983) and argumentation theory (van Emeren *et al.*, 1996; van Eemeren, 2010; van Eemeren & Garssen, 2012). How discourse analysis and argumentation theory can be seen to share methodological tools has well been shown, for example, by Degano (2012). As regards the aims of these perspectives, Degano (2012, p. 19) clarifies that, while "the scope of discourse analysis [...] is mainly descriptive. Argumentation theory, on its part, is primarily normative". However, Degano goes on to specify that "such a difference is not irreconcilable, in line of principle" (*ibid.*), and that, furthermore, one of the threads of discourse analysis, critical discourse analysis, "has an approach that can be considered ethical" (*ibid.*), because it is interested in relations of power in society and how these are expressed and influenced linguistically. It is precisely a critical discursive approach that is adopted here, supported by applied ethics in healthcare communication (e.g. Jonas, 1985) and with the specification that the level of interest and of competence is primarily the linguistic one.

3. MATERIAL

This small qualitative study focused on English-language supranational, British and US institutional websites, namely those of the World Health Organisation (WHO), the European Centre for Disease Prevention and Control (ECDC), the Centers for Disease Control and Prevention (CDC), NHS Choices (NHSC), MedlinePlus (Health A-Z + Medical Encyclopaedia), the European Commission. Four texts coming from the listed websites were selected as particularly representative of web-based institutional communication aimed at defending child immunisation from accusations of being linked to autism:

1. The World Health Organisation (WHO) > Programmes > Global Vaccine Safety > The Global Advisory Committee on Vaccine Safety > Topics > **MMR and autism** (Extract from report of GACVS meeting of 16-17 December 2002, published in the *WHO Weekly Epidemiological Record* on 24 January 2003), http://www.who.int/vaccine_safety/committee/topics/mmr/mmr_autism/en/ (353 words).
2. The European Centre for Disease Prevention and Control (ECDC), Health Topics > Immunisation > Let's talk about protection > Questions and answers > **Vaccines and autism** (latest reference in text from 2011), http://ecdc.europa.eu/en/healthtopics/immunisation/comms-aid/q-and-a/pages/vaccines-and-autism.aspx (745 words).
3. The Centers for Disease Control and Prevention (CDC) > Vaccine Safety > Common Concerns > **Vaccines Do Not Cause Autism** (latest reference in text from 2013; last reviewed and updated on 28 August 2015), http://www.cdc.gov/vaccinesafety/concerns/autism.html (463 words).
4. NHS Choices (NHSC) > Health News > **MMR vaccine 'does not cause autism'** (5 February 2008), http://www.nhs.uk/news/2007/January08/Pages/MMRvaccinedoesnotcauseautism.aspx (1318 words).

It is essential to provide a description of the broad communicative context in which the four texts analysed occur.

These are made of the main articles contained in the referred webpages, including titles, subtitles and references, but excluding any other additional text, as in side menus and top and bottom bars. The volatility of web-based documents is a well-known issue with researchers using this kind of material, so the four texts have been collected for reference in the Appendix. Volatility, though, tends to be less of a problem with institutional websites such as those considered here.

It ought to be specified that the adjective "institutional" is meant to refer to public institutions, specifically those concerned with health and healthcare, and not to mean "conventional" as in van Eemeren (2010). The distinction is necessary for contextualisation purposes, i.e. to clarify that "institutional communication" is understood here as that done by (healthcare) institutions, which of course also takes place subject to specific discursive conventions as per van Eemeren (2010)'s pragma-dialectical theory. The four texts on vaccines and autism are therefore defined as samples of "web-based communication by public health and healthcare institutions", which has specific communicative aims, linguistic features and ethical implications as identified for example in Grego and Vicentini (2011).

Formally, the length varies from 353 to 1318 words, and dates can be placed between 2002 and 2013, in line with the chronological background exposed in par. 1. Except for text 1 (WHO), which is also the shortest, the others include paragraphs or forms of subdivision into smaller units. WHO and ECDC do not contain hyperlinks to other hypertexts, while CDC and NHSC do.

The communicative purpose of the texts is, as in all institutional communication, informative. Nonetheless, given the ample debate surrounding the topic, there is also a clear persuasive function, as occurs in instances of communication having public health implications. The point supported by all four sources is the same, i.e. refuting any link between vaccines and autism. Because "institutional" here refers to public institutions, the authoritativeness of the addressor is evident, as is the fact that in WHO, ECDC and CDC no room for counterarguments is given, either discursively or physically (the communication is one-way, no comments or questions by the public are provided for). The NHS page is the only one allowing for comments (there were 7 of them at the time of writing, the latest from 2013). This may be due to the text being in the News section of the portal which, although still under the institutional umbrella and responsibility of the authority hosting it, is considered more informative and less prescriptive, and leaves room for public[2] debate.

Still on the addressor, a relevant note is that both the WHO and the NHSC choose to report texts originally coming from external sources. Actually, the WHO uses an extract of a report by the Global Advisory Committee on Vaccine Safety (GACVS) published in 2003 on its own *WHO Weekly Epidemiological Record*. NHSC, interestingly, edits and publishes a report by Bazian, a healthcare research consulting company, part of *The Economist* Intelligence Unit and one of NHS Choices' external partners[3]. This does not seem to affect the fact that the ultimate responsibility for and endorsement of the information given and the arguments expressed still lie with WHO and NHSC Choices, with the latter even specifying that it edited the text and thus claiming authorship at the re-writing (for popularising purposes) level.

Finally, as regards the addressee, this is strictly linked to the medium (the Internet) and, as such, presents with definitory issues typical of all web-based communication. Specifically, while it is easy to say that any text that is online and freely accessible is directed at the largest audience possible – and such a definition implies an unidentified and hardly identifiable mass of potential readers with heterogeneous backgrounds, needs and expectations – in fact, most online information will be reached by much smaller audiences with much narrower profiles. This, however, can only be certified by means of surveys on website userships, and anyway mostly applies to non-institutional websites. When it comes to these, the case is often that the presence of specific information is mostly author-centred, rather than reader-centred. In other words, institutional actors need to include certain information simply for the fact that they are public interlocutors in specific fields (health/healthcare, in this case), and cannot omit to cover certain topics within a showcase (their website) that may seem aimed mostly at the public but actually also dialogues with other (national and supranational) institutions, as well as with specialised communities of practice and discourse and, finally, the media – the point being that institutional communication is made conspicuous by what it says but also, and sometimes even more, by what it does not say.

---

[2] Only registered users may post comments, though.

[3] NHS Choices > About NHS Choices > NHS Choices partners > Bazian, http://www.nhs.uk/aboutNHSChoices/aboutnhschoices/partners/Bazian/Pages/Introduction.aspx.

4. LINGUISTIC STRATEGIES USED FOR ARGUMENTATION

The four texts reviewed all present with a comparable text structure that includes the following argumentative moves:

- reference to *the Lancet* article of 1998 (Wakefield *et* al., 1998);
- a definition of autism;
- the possible MMR-autism link;
- hints at anti-vaccination supporters;
- appeals to science/research (reference to scientific literature);
- the no MMR-autism link conclusion;
- a pro-vaccine appeal.

The detailed structure of the texts may be summarised as follows.

WHO	**ECDC**	**CDC**	**NHSC**
reference to *The Lancet* article of 1998	hints at anti-vaccination supporters	definition of autism	no MMR-autism link conclusion
possible MMR-autism link	possible MMR-autism link	possible MMR-autism link	appeals to science/research
definition of autism	no MMR-autism link conclusion	no MMR-autism link conclusion	possible MMR-autism link
appeals to science/research	definition of autism	appeals to science/research	reference to the *Lancet* article of 1998
no MMR-autism link conclusion	appeals to science/research	no MMR-autism link conclusion	appeals to science/research
pro-vaccine appeal	possible MMR-autism link	appeals to science/research	no MMR-autism link conclusion
	reference to the *Lancet* article of 1998	no MMR-autism link conclusion	no MMR-autism link conclusion
	appeals to science/research	appeals to science/research	no MMR-autism link conclusion
	MMR-autism link conclusion	no MMR-autism link conclusion	

Table 1 – Argumentative moves of the texts selected for analysis

4.1 Discursive strategies

Text 1 (WHO) contains a series of linguistic choices that may be grouped under several discursive strategies.

Depersonalisation, especially by means of the use of the passive voice, which is typical of specialised discourse and is aimed at creating a sense of objectivity by focusing on what is being discussed and not on the agent.

(1) Concerns [...] were raised
(2) the outcome of the review was presented
(3) Eleven epidemiological studies [...] were reviewed.

Nominalisation, or the preference and prevalence of noun phrases over verb phrases, is also a characteristic of specialised texts, which draws attention to the data rather than the processes leading to their retrieval.

(4) **publication** of studies
(5) on the **recommendation** of GAVS
(6) a better **understanding** of the **causes** of autism.

Hedging strategies are also widely employed, with the aim of reducing the illocutionary force of what is being said or, as in typical scholarly scientific writing, putting some distance between the sender and the message that, again, can also contribute to objectifying the discourse, starting from the premise that what is objective must be scientific, and what is scientific (according to the experimental method) is also true, as truth is understood in modern science. Hedging may be realised at different linguistic levels:

(7) Concerns about a **possible** link (lexicon)
(8) a strategy which **would** put children at risk (modality)
(9) **It** was concluded **that** (a cleft sentence it-clause).

Text 2 (ECDC) shows a high degree of subordination, related to the mix of specialised and non-specialised language adopted. This is the result of the Q&A format chosen to provide the information and argue against a MMR-autism link. The parent vs healthcare provider fictitious setting also means the power relationship where the source or writer is in charge is made explicit. Argumentatively, the authors formulate what they think the receiver is going to ask, in a dialogical exchange that only on the surface is such. The pretence of a dialogical interaction is of course also agreed on between the author and the reader, who agree on identifying themselves with the former or the latter party. In critical discursive terms, though, this also means a full overlapping and substitution of the receiver with the sender, who results in a dumb puppet impersonating not a stand-alone figure but a support for the main and only actor who is the healthcare provider. Indeed, the first question asked by the fictitious parent

(10) all those people who say that the MMR vaccine causes autism must be on to something

is made to sound colloquial (people, something), non-specialised (causes), generic, vague (**all** those people, something) and uttered by a suspicious (**must** be on to something) and misinformed (say) mind. The replies by the specialist are instead made to sound, at times, specialised and professional, adopting the same strategies as in Text 1 (WHO), which comes from a formal WHO report and therefore by all means pertains to a specialised genre:

> (11) Multiple studies have shown that (depersonalisation)
> (12) Thimerosal is a mercury-containing preservative (nominalisation and long pre-modified noun phrases)
> (13) Children with autism may not be affected in quite the same way (modality for hedging purposes).

At other times, however, the answers show colloquialisms, personalisation and evaluative language that possibly serve the function of maintaining the specialist vs non-specialist interaction:

> (14) Clinicians were also **confused** by the recommendation (choice of lexicon)
> (15) **Hundreds of thousands** of children (vagueness)
> (16) Autism is a **strong** and emotive issue and something **we all** care about (evaluative lexicon and inclusiveness)
> (17) **I** can show **you** study after study (personalisation)
> (18) **Unfortunately,** once a **seed of doubt** has been **planted** it tends to **grow** (evaluative lexicon and use of figurative speech).

Text 3 (CDC) shows prevailing parataxis, with many short sentences comprising only one main clause. Sentences are juxtaposed rather than joined by conjunctions, which is typical of specialised language and requires an inferencing effort on the reader's part. Depersonalisation is common throughout the text, with the use of several declarative clauses that allow the authors to avoid naming an agent as a subject while retaining a grammatical one:

> (19) Recent estimates [...] **found that**
> (20) research **showing that**
> (21) The results **showed that**
> (22) Research **shows that**
> (23) a 2004 scientific [...] **concluded that.**

On the other hand, relative clauses introduce denominations and definitions, which are a typical feature of the discourse of popularisation:

> (24) Autism spectrum disorder (ASD) is a developmental disability **that is caused by**
> (25) One vaccine ingredient **that has been studied specifically** is thimerosal.

The CDC text therefore shows linguistic aspects that make it closer to the popular article as may be found in a good-level scientific dissemination publication such as *Science*.

Text 4 (NHSC) is different, as anticipated in par. 3, because it is the edited version of a Bazier report. The tone is that of a broadsheet or quality weekly article (the link between Bazier and *The Economist* is apparent) and, after a short introduction, it also adopts the Q&A format:

> (26) There is no evidence for a link between the MMR jab and autism, say *The Guardian* and other news sources (free quote and citation of a popular source, *The Guardian*)
> (27) "No association between measles vaccination and ASD was shown.,", Gillian Baird, lead author (quotation box, typical of journalistic genres).

This text is also the one most characterised diatopically, as British:

> (28) MMR **jab** (UK, informal vs 'shot', US, informal)
> (29) **Sir Muir Gray** adds (a popular British doctor and scholar involved in public health)
> (30) say **the *Guardian*** / latest **UK** study / What does the **NHS** Knowledge Service make of this study? (culture- and country-specific references).

Syntactically, there is a high degree of coordination (especially through "and"), but both coordinative and subordinative conjunctions are used (however, but), which suggests signposting argumentative passages. Several disjunctive conjunctions also contribute to opening up rhetorically to the possibility of a dialogue:

> (31) Some parents wonder **whether**
> (32) To determine **whether** Wakefield's suspicion was correct
> (33) studies have been performed to determine **whether.**

Relative clauses for definition/explication are also present, with clear disseminating purposes:

> (34) specific genes that cause autism (a defining relative clause for definition)
> (35) Symptoms of autism, which typically appear during the first few years of life (a non-defining relative clause for explication, adding information deemed necessary and useful to help the reader).

4.2 Lexical strategies

The lexicon of web-based health institutions' communication offers more than a few interesting features, combining specialised and non-specialised terminology, and making use of popularising strategies such as re-writing and repetition that make it a valuable source for linguistic studies. The lexical aspect on which attention is drawn here is a specific one, chosen for its relevance for the argumentative nature of these texts, and to show how argumentation is also conveyed by lexical means. In detail, what follows reports on how "the other side", or the opponent, is represented linguistically in the four texts.

Firstly, a distinction must be made between those in doubt and those supporting the MMR-autism link. The texts all address, in different ways, the former, while reference to the latter remains in the background.

The extract published by WHO, coming from a specialised report, can be said to not formally refer to the dialogic opponent. In fact, this appears in the very first sentence, saying that

> **Concerns** about a possible link between vaccination with MMR and autism **were raised** in the late 1990s, following publication of studies claiming an association between natural and vaccine strains of measles virus and inflammatory bowel diseases, and separately, MMR vaccine, bowel disease and autism.

The claim is obviously that made by Wakefield *et al.* in their 1998 article (the "scientific", specialised opponent); the concerns, those had by parents and any actor who may have felt affected by the claim (the "popular", non-specialised opponent). This is the text presenting with the maximum detachment from the "other side", shown not only in the lexical (non)choices, but also in the depersonalisation strategies mentioned in par. 4.1 (passive voice + plural + absence of quantifiers).

ECDC opts for a personal syntactic construction, underlying a personal approach (in which people seem to be at the centre). Parents are introduced as actors, thus making them active agents with a right to choose:

> (36) Some parents of children with autism are concerned that vaccines are the cause.

In the second sentence, "concerns" is both the grammatical and the logical subject, in a personal syntactic structure having an impersonal logical subject:

> (37) **Their concerns** centre on three areas.

In the third sentence, an epistemic "must" is used to express the model parent's conviction; in this way, however, beliefs are attributed to the weaker actor by the stronger actor:

> (38) Parent: All those people who say that the MMR vaccine causes autism **must** be on to something.

In the fourth sentence, the sender's subjective position is clearly expressed, even using openly evaluative lexicon:

> (39) **sensational** media and internet coverage that isn't concerned with the facts.

In the fifth sentence, the healthcare operator's ideal wish is clearly expressed in a "wish + simple past" syntactic structure; and the lexical choice of the word 'victims' carries as clear connotations:

> (40) **I wish** the voices of those who have been **victims** of not getting vaccinated could be heard more loudly and clearly.

CDC also opts for a personal syntactic construction and approach:

> (41) Some people have had concerns.

Nonetheless, the indefinite quantifier "some" conveys a quantitative evaluation, i.e. not a large number, with the implication of little or no significance of such concerns. The use of the present perfect tense is also noticeable, and also in terms of relevance: it is the tense of indefinite information relevant *per se* and, thus, not placed in space and time. The implication could be that the concerns, although existing, are remote and detached from the readers' contexts. Incidentally, the CDC's is also the only text not referring, explicitly or implicitly, to the *Lancet* article of 1998.

Finally, Text 4 (NHSC) which, as indicated, was chosen, edited and endorsed by NHS Choices to inform its audience about vaccines and autism, but shows features of the journalistic genres due to its original authors, engages with "those in doubt" only indirectly

> (42) The jab **has been linked** to autism since 1998 ('those in doubt' have linked),

although it does openly engage with "those supporting the link", but indicating in figures the small number of children studied, the discredit it has drawn, and using absolute terms (any) to isolate it scientifically:

> (43) a **study** of **12** children published in *The Lancet* **link**ed the measles, mumps and rubella (MMR) jab to the development of autism. That research has since been **discredited** and two major studies have been published subsequently which also failed to show **any** link.

Not only, the lexical and semantic choice of term "link", typical of scientific intra-specialistic texts, here instead seems to be a form of hedging (i.e. meaning no certain relationship), thus suggesting a weak power position of 'those supporting' such a connection.

5. DISCUSSION AND CONCLUSIONS

In the pragma-dialectical model of argumentation, a protagonist and an antagonist try to solve a difference of opinion by entertaining a critical discussion that may help them solve their difference, producing speech acts chosen from their topical potential, according to their audience's demand and exploiting strategic presentational devices; they win the confrontation over a standpoint if their arguments prove reasonable and effective (van Eemeren, 2010). The reality is very different, as, in fact,

> Due to a variety of factors, argumentative reality seldom resembles the ideal of a critical discussion [...] in practice the disputants usually have to operate circumspectly [...] in argumentative discourse much remains unsaid. Not only is there seldom any mention of discussion rules and remain the material starting points often unstated, but other vital aspects of the resolution process, including even the perceived

> outcome, are often not clearly indicated either (van Eemeren, 2010, p. 13).

Web-based written communication definitely suffers from only partially resembling any traditional model of communication, heavily influenced as it is by a medium that can be as exclusive as it is inclusive. Another problematic aspect is that, when applying argumentative models such as pragma-dialectics to presenting arguments via online written texts (not short interactive ones but article-like ones), there is no antagonist or real-time debate, unless of course the instant participatory capacity of the Internet is exploited, for example by letting the public comment. If not, a text arguing in favour or against a certain point at issue runs the risk of remaining a one-off manifestation, a "quasi-discussion" which, just like in the case of TV debates, "is in fact a monologue calculated only to win the audience's consent to one's own views" (van Eemeren, 2010, p. 3), and certainly not made to persuade the other party: "this third party audience is in fact their primary addressee – or even their only 'real' addressee" (van Eemeren, 2010, p. 148). This is unless the online text is seen as one communicative event within a much larger (online) debate, spread out in space and time and even being carried on globally through different media. This can be the case with the issue of the MMR vaccine-autism link, one of the first global "scares" to be disseminated thanks to or because of the Internet. A comprehensive and diachronic account of the no-MMR vaccine question could well be from an argumentation theory perspective. A comprehensive collection of all that has been and is being written on the issue on the Internet would hardly be possible, due to the inherent vague limits of the medium itself. Taken alone, therefore, texts such as the ones examined can hardly be placed within an ideal argumentative framework, since they do not follow critical discussion stages consistently, for example leaving little space for the other party's doubts (see Table 1). Nonetheless, their argumentative nature is evident. A descriptive approach such as critical discourse analysis can contribute to an analysis of such material by incorporating aspects of argumentation theory within a mainly linguistic perspective, while providing insights as to responsibility for and power relations in the exchanges, of great relevance in institutional settings.

This particular kind of analysis has highlighted how the same topic, presented by comparable national and supranational health and healthcare institutions in online settings, is conveyed by means of different genres, at different levels of specialisation, making different strategic manoeuvring choices, and using different sources to appeal to authority. The discursive and particularly the lexical strategies employed – irrespective of any 'right vs wrong' considerations – point to an exploitation by the authors of their dominant social position as those in charge, representing and defending the status quo. Arguments are based on scientific evidence, where citing academic (but NHSC even quotes *The Guardian*) sources is used as an appeal to authority. This may appear strange, coming from public institutions, which should need no further authority to base their prescriptive stances upon. Institutional discourse is by definition endowed with both power over and responsibility toward the citizens, who are in democracies the ones granting them to the institutions. It is not so surprising if, however, the field at issue is health/healthcare, a particularly delicate domain, where the right of choice of the individual is increasingly being expanded, valued and guaranteed, meeting the needs of a better-educated and informed population. More freedom of choice, though, implies less authority on part of the institutions, who are now, in cases such as the MMR vaccine-autism debate, even obliged to defend the mainstream position by persuading the public.

To conclude, if one should go back to Aristotle's notion of persuasive appeals, to logos or pathos or ethos, in the analysed texts there is by definition an ethical appeal because it is institutional communication. The appeal to ethos is also in the scientific references provided (a higher authority than the authority itself). It would not be incorrect to say that a certain appeal to pathos also applies, especially in those texts (ECDC and NHSC) which adopt popularising strategies such as the Q&A (sub)genre. After all, though formally directed at anyone, the ideal and desired target audience of these communications are parents and recent parents at that. To be specific, the addressees may even be restricted to more mothers than fathers, in a society where women are increasingly educated, have growing power of choice about their parenting status and rights, become parents at an older age when their social status is better established, and who therefore tend to have much-sought only children, with a higher social and economic value.

The MMR vaccine vs autism debate is certainly a developed-world problem and not a first-rate medical emergency; nonetheless, a linguistic analysis from a critical discursive viewpoint of the argumentative aspects involved when it is communicated by institutions via the Internet can highlight interesting trends in the spread of (mis)information having ethical relevance for the individual and for society alike.

REFERENCES

Bhatia, V. K. (1993). *Analyzing genre: Language use in professional settings.* London: Longman.

Bhatia, V. K. (2004). *Worlds of written discourse: Genre-analytical view.* London: Continuum.

Degano, C. (2012). *Discourse analysis, argumentation theory and corpora: An integrated approach.* Milano: Arcipelago.

Fairclough, N. (1989). *Language and power.* London: Longman

Faiclough, N. (1995). *Critical discourse analysis.* Harlow: Longman.

Fairclough, N. (2003). *Analysing discourse: Textual analysis for social research.* London, New York: Routledge.

Garzone, G. (2007). Genres, multimodality and the World Wide Web: Theoretical issues. In G. Garzone, P. Catenaccio & G. Poncini (Eds.), *Multimodality in Corporate Communication. Web Genres and Discursive Identity* (pp. 15-30). Milan: Franco Angeli,

Gotti, M., & Salager-Meyer, F. (Eds.) (2006). *Advances in medical discourse analysis: Oral and written contexts.* Bern: Peter Lang.

Grego, K., & Vicentini, A. (2011). Holiday dialysis in Italy on the web: Multidimensional hybridization in institutional healthcare communication. In S. Sarangi, V. Polese, G. Caliendo (Eds.), *Genre(s) on the Move: Hybridization and Discourse Change in Specialized Communication* (pp. 393-406). Napoli: Edizioni Scientifiche Italiane.

Jonas, H. (1985). *Technik, medizin und ethik. zur praxis des prinzips verantwortung.* Frankfurt/M.: Insel.

The Lancet (2010). Retraction. Ileal-lymphoid-nodular hyperplasia, non-specific colitis, and pervasive developmental disorder in children, *The Lancet, 375*(9713), p. 445, 6 February 2010. http://www.thelancet.com/journals/lancet/article/PIIS0140-6736%2810%2960175-4/abstract.

Nature Neuroscience (2007). Silencing debate over autism, Editorial, *Nature Neuroscience, 10*(531). http://www.nature.com/neuro/journal/v10/n5/full/nn0507-531.html, 2014.

NHS (2014). Measles outbreak: What to do. http://www.nhs.uk/conditions/vaccinations/pages/measles-outbreak-advice.aspx, 2014.

Poland, G. A., & Jacobson, R. M. (2011). The age-old struggle against the antivaccinationists. *The New England Journal of Medicine, 364.* 97-99.

Public Health England (2014). Measles notifications (confirmed cases) England and Wales 1995 - 2013* by quarter, http://webarchive.nationalarchives.gov.uk/20140505192926/http://www.hpa.org.uk/web/HPAweb&HPAwebStandard/HPAweb_C/1195733811358.

Swales, J. (1990). *Genre analysis: English in academic and research settings.* Cambridge: Cambridge University Press.

Swales, J. (2004). *Research genres: Explorations and applications.* Cambridge: Cambridge University Press.

van Dijk, T. (2008). *Discourse and power.* Houndsmills: Palgrave.

van Eemeren, F. H. (2010). *Strategic maneuvering in argumentative discourse.* Amsterdam: John Benjamins.

van Eemeren, F. H., & Garssen, B. (Eds.) (2012). *Exploring Argumentative Contexts.* Amsterdam: John Benjamins.

van Eemeren, F. H., Grootendorst, R. & Snoeck Henkemans, F. (1996). *Fundamentals of argumentation theory. A handbook of historical backgrounds and contemporary developments.* Mahwah, NJ: Lawrence Erlbaum.

Wakefield et al. (1998). Ileal-lymphoid-nodular hyperplasia, non-specific colitis, and pervasive developmental disorder in children. *The Lancet,* 351(9103), 637-641, 28 February 1998.

Werlich, E. (1983). *A text grammar of English.* Hamburg: Helmut Buske Verlag.

Wodak, R. (Ed.) (2013). *Critical discourse analysis.* London: Sage Publications.

APPENDIX

1. WHO
MMR and autism

Extract from report of GACVS meeting of 16-17 December 2002, published in the WHO Weekly Epidemiological Record on 24 January 2003

Concerns about a possible link between vaccination with MMR and autism were raised in the late 1990s, following publication of studies claiming an association between natural and vaccine strains of measles virus and inflammatory bowel diseases, and separately, MMR vaccine, bowel disease and autism. WHO, on the recommendation of GACVS, commissioned a literature review by an independent researcher of the risk of autism associated with MMR vaccine; the outcome of the review was presented to GACVS for its consideration.

Autistic spectrum disorder represents a continuum of cognitive and neurobehavioral disorders including autism. The prevalence of autism varies considerably with case ascertainment, ranging from 0.7 - 21.1 per 10 000 children (median 5.2 per 10 000) while the prevalence of autistic spectrum disorder is estimated to be 1 - 6 per 1000. Eleven epidemiological studies (representing the most recent studies, mostly in the last 4 years) were reviewed in detail, taking into consideration study design (including ecologic, case control, case-crossover and cohort studies) and limitations. The review concluded that existing studies do not show evidence of an association between the risk of autism or autistic disorders and MMR vaccine. Three laboratory studies were also reviewed. It was concluded that the alleged persistence of measles vaccine virus in the gastrointestinal tract of children with autism and inflammatory bowel disease requires further investigation through independent studies before the laboratory findings of the published studies, which have serious limitations, can be considered confirmed.

Based on the extensive review presented, GACVS concluded that no evidence exists of a causal association between MMR vaccine and autism or autistic disorders. The Committee believes the matter is likely to be clarified by a better understanding of the causes of autism. GACVS also concluded that there is no evidence to support the routine use of monovalent measles, mumps and rubella vaccines over the combined vaccine, a strategy which would put children at increased risk of incomplete immunization. Thus, GACVS recommends that there should be no change in current vaccination practices with MMR.

2. ECDC
Vaccines and autism

Some parents of children with autism are concerned that vaccines are the cause. Their concerns centre on three areas:

1. the combination measles-mumps-rubella (MMR) vaccine;

2. thimerosal, a mercury-containing preservative previously contained in several vaccines;

3. the notion that babies receive too many vaccines too soon.

Parent: All those people who say that the MMR vaccine causes autism must be on to something.

Healthcare provider: Autism is a strong and emotive issue and something we allcare about. However, the link made by one doctor to autism has been firmly discredited, and I can show you study after study that demonstrates that there is no link between the MMR vaccine and autism. Unfortunately, once a seed of doubt has been planted it tends to grow, and is fuelled by sensational media and internet coverage that isn't concerned with the facts. The real issue here is the very real risks from not being protected. I wish the voices of those who have been victims of not getting vaccinated could be heard more loudly and clearly.

Q. What are the symptoms of autism?

A. Symptoms of autism, which typically appear during the first few years of life, include difficulties with behaviour, social skills and communication. Specifically, children with autism may have difficulty interacting socially with parents, siblings and other people; have difficulty with transitions and need routine; engage in repetitive behaviours such as hand flapping or rocking; display a preoccupation with activities or toys; and suffer a heightened sensitivity to noise and sounds. Autism spectrum disorders vary in the type and severity of the symptoms they cause, so two children with autism may not be affected in quite the same way (Bauman, 1999).

Q. What causes autism?

A. The specific cause or causes of autism in all children are not known. But one thing is clear: autism spectrum disorders are highly genetic. Researchers figured this out by studying twins. They found that when one identical twin had autism, the chance that the second twin had autism was greater than 90 per cent. But when one fraternal twin had autism, the chance that the second twin had autism was less than 10 per cent. Because identical twins have identical genes and fraternal twins don't, these studies proved the genetic basis of autism. More recently, researchers have successfully identified some of the specific genes that cause autism. Some parents wonder whether environmental factors - defined as anything other than genetic factors - can cause autism. It's possible. For example, researchers found that thalidomide, a sedative, can cause autism if used during early pregnancy. Also, if pregnant women are infected with rubella virus (German measles) during early pregnancy, their babies are more likely to have autism (Bailey, et al, 1995).

Adapted from information sheets developed by the Children's Hospital of Philadelphia www.vaccine.chop.edu and Wellington-Dufferin-Guelph Public Health, Canada, 2007.

Q. Does the MMR vaccine cause autism?

A. No. In 1998, a British researcher named Andrew Wakefield raised the notion that the MMR vaccine might cause autism. In the medical journal The Lancet, he reported the stories of eight children who developed autism and intestinal problems soon after receiving the MMR vaccine. To determine whether Wakefield's suspicion was correct, researchers performed a series of studies comparing hundreds of thousands of children who had received the MMR vaccine with hundreds of thousands who had never received the vaccine. They found that the risk of autism was the same in both groups. The MMR vaccine didn't cause autism. Furthermore, children with autism were not more likely than other children to have bowel problems (Deer, 2011; IOM, 2011).

Q. Does thimerosal cause autism?

A. No. Multiple studies have shown that thimerosal in vaccines does not cause autism. Thimerosal is a mercury-containing preservative that was used in vaccines to prevent contamination. In 1999, professional groups called for thimerosal to be removed from vaccines as a precaution. Unfortunately, the precipitous removal of thimerosal from all but some multi-dose preparations of influenza vaccine scared some parents. Clinicians were also confused by the recommendation. Since the removal of thimerosal, studies have been performed to determine whether thimerosal causes autism. Hundreds of thousands of children who received thimerosal-containing vaccines were compared to hundreds of thousands of children who received the same vaccines free of thimerosal. The results were clear: the risk of autism was the same in both groups (Gerber and Offit, 2009; Andrews, et al, 2004; Heron and Golding, 2004; Madsen, et al, 2003).

See more at: http://ecdc.europa.eu/en/healthtopics/immunisation/comms-aid/q-and-a/pages/vaccines-and-autism.aspx#sthash.KNTl5r2r.dpuf

3. CDC

Autism spectrum disorder (ASD) is a developmental disability that is caused by differences in how the brain functions. People with ASD may communicate, interact, behave, and learn in different ways. Recent estimates from CDC's Autism and Developmental Disabilities Monitoring Network found that about 1 in 68 children have been identified with ASD in communities across the United States. CDC is committed to providing essential data on ASD, searching for causes of and factors that increase the risk for ASD, and developing resources that help identify children with ASD as early as possible.

There is no link between vaccines and autism.

Some people have had concerns that ASD might be linked to the vaccines children receive, but studies have shown that there is no link between receiving vaccines and developing ASD. In 2011, an Institute of Medicine (IOM) report on eight vaccines given to children and adults found that with rare exceptions, these vaccines are very safe.

A 2013 CDC study added to the research showing that vaccines do not cause ASD. The study looked at the number of antigens (substances in vaccines that cause the body's immune system to produce disease-fighting antibodies) from vaccines during the first two years of life. The results showed that the total amount of antigen from vaccines received was the same between children with ASD and those that did not have ASD.

Vaccine ingredients do not cause autism.

One vaccine ingredient that has been studied specifically is thimerosal, a mercury-based preservative used to prevent contamination of multidose vials of vaccines. Research shows that thimerosal does not cause ASD. In fact, a 2004 scientific review by the IOM concluded that "the evidence favors rejection of a causal relationship between thimerosal–containing vaccines and autism." Since 2003, there have been nine CDC-funded or conducted studies that have found no link between thimerosal-containing vaccines and ASD, as well as no link between the measles, mumps, and rubella (MMR) vaccine and ASD in children.

Between 1999 and 2001, thimerosal was removed or reduced to trace amounts in all childhood vaccines except for some flu vaccines. This was done as part of a broader national effort to reduce all types of mercury exposure in children before studies were conducted that determined that thimerosal was not harmful. It was done as a precaution. Currently, the only childhood vaccines that contain thimerosal are flu vaccines packaged in multidose vials. Thimerosal-free alternatives are also available for flu vaccine. For more information, see the Timeline for Thimerosal in Vaccines.

Besides thimerosal, some people have had concerns about other vaccine ingredients in relation to ASD as well. However, no links have been found between any vaccine ingredients and ASD.

Featured Resource: Vaccines and Autism: A Summary of CDC Conducted or Sponsored Studies

4. NHS

MMR vaccine "does not cause autism"
Tuesday February 5 2008
There is no evidence for a link between the MMR jab and autism, say The Guardian and other news sources. The reports are based on a study that is the

"biggest review conducted to date", analysing "the blood from 250 children and concluded that the vaccine could not be responsible".

The jab has been linked to autism since 1998, when a study of 12 children published in The Lancet linked the measles, mumps and rubella (MMR) jab to the development of autism. That research has since been discredited and two major studies have been published subsequently which also failed to show any link.

In this latest UK study, researchers investigated whether the MMR vaccine contributes to the development of autistic spectrum. This study investigates some of the specific suggestions that have been put forward about the relationship between the MMR vaccine and autistic spectrum disorders in the past. These include the idea that the MMR vaccine was specifically associated with autism where children experienced a loss of developed skills (regression) and inflammation of the small intestine (enterocolitis); that autism is associated with an increased level of measles antibodies in the bloodstream; and that it's associated with an increased presence of genetic material from the measles virus in cells from the gut.

The researchers looked at three groups of children, one with autistic spectrum disorders, one with special educational needs but no autism and another who were developing normally. When blood samples were compared, there was no difference in any long-lasting signs of measles virus or increased levels of antibodies to the measles virus between the groups. They also found that enterocolitis was not commonly associated with autism. This study adds to the pool of evidence that suggests that there is no causal link between the MMR vaccine and autism.

Where did the story come from?

Professor Gillian Baird and colleagues from Guy's & St Thomas' NHS Foundation Trust, several Universities in the UK and Australia, the National Institute for Biological Standards and Control and the Health Protection Agency in the UK carried out this research. The study was funded by the Department of Health, the Wellcome Trust, the National Alliance for Autism Research, and Remedi. The sponsors did not play a role in study design, data collection, analysis or interpretation, or in writing the paper. It was published in the peer-reviewed medical journal: Archives of Disease in Childhood.

What kind of scientific study was this?

This was a case-control study which tested the possibility that the MMR vaccine may contribute to the development of autistic spectrum disorders (ASD). The researchers did this by comparing long-lasting signs of measles infection or immune response in children with ASD (cases) and children without ASD (controls). The researchers were particularly interested in looking at children who had lost some of their developmental skills (called regression) and children with specific digestive system problems (enterocolitis), as these are both phenomena that have been claimed to be linked to the MMR vaccine. This study was part of the Special Needs and Autism Project (SNAP), which enrolled 56,946 children from the South Thames region born between July 1 1990 and December 31 1991.

No association between measles vaccination and ASD was shown. Gillian Baird, lead author

There were 1,770 children from SNAP aged nine to 10-years-old, who had been classified as having special educational needs or had been diagnosed with ASD. A representative sample of 255 of these children were selected to have a standard in-depth diagnostic test for ASD. For this study, the researchers included only children who provided blood samples, and those who had received the MMR vaccine at least once. Information about whether a child had had the MMR vaccine was taken from parental report, GP and district records. This included 98 children (cases) with ASD and 52 children with special educational needs but not ASD (controls). They also selected another control group of 90 children from mainstream local schools who were developing normally, had received the MMR vaccine, and had agreed to have blood taken. All the children were aged between 10 and 12. The people who tested the blood samples did not know which were from cases and which from controls.

The researchers looked to see whether there were antibodies against the measles virus in the blood and investigated whether the levels of anti-measles antibodies a child had was related to the severity of their autistic symptoms. The children's blood samples were also tested for the presence of the measles virus by looking for genetic material from the virus. Previous studies have looked for the measles virus in cells from the stomach, however, as this is an invasive procedure it was considered unethical to carry out this procedure on the children, so instead the researchers looked at a particular type of white blood cell where viruses are known to replicate.

The researchers also asked the children's parents or guardian to complete a questionnaire about whether the children had symptoms of digestive system problems either in the past three months (current symptoms) or before (past symptoms). Children with persistent diarrhoea in the past, who did not have current constipation, and who had two or more of the following current symptoms were defined as having "possible enterocolitis": persistent vomiting, persistent diarrhoea, weight loss, persistent abdominal pain, or blood in stool.

The analyses were repeated to see whether their results differed in children who had received one compared with two doses of the MMR vaccine, or in children who had ASD with regression (defined as loss of five or more words in a three month period) compared with those with ASD but without regression.

What were the results of the study?

There was also no difference in the level of antibodies to measles in the bloodstream between children with ASD (cases) and children without ASD (controls). Also, there was no relationship between the level of measles antibodies a child had and how severe their autistic symptoms were. For the 23 children with ASD and regression, there was also no difference in the levels of antibodies between them and the pooled control group.

Genetic material from the measles virus was only found in one child with autism and two children who were developing normally. However, when they repeated the tests, the researchers could not find any measles virus genetic material in these samples.

Only one child had symptoms which could indicate enterocolitis, and this child was in the control group.

What interpretations did the researchers draw from these results?

The researchers concluded that there was no link between the MMR vaccine and autistic spectrum disorders.

What does the NHS Knowledge Service make of this study?

This study selected cases and controls from a large community based group, and the researchers tried to include all the children with ASD in this community. The limitations of this study were acknowledged by the authors and included the fact that:

The children were not randomly selected from the population. This may mean that the samples may not truly represent the groups of children that they were meant to represent (that is, children with ASD, children with special educational needs, or developmentally normal children).

The researchers could not obtain adequate blood samples from 100 children. If these children differed systematically from the children from whom blood samples were obtained, this could affect results.

The diagnosis of "possible colitis" was based mainly on current symptoms, because it was thought that it would not be possible for the parent or guardian or child to accurately recall whether the child had experienced these symptoms at the time of having the MMR vaccination (more than nine years previously).

This study adds to the pool of evidence that suggests that there is no causal link between the MMR vaccine and autism.

Sir Muir Gray adds...

Say no more.

Analysis by Bazian

Edited by NHS Choices

65

Argumentation and Moral Education

ANA MARÍAVICUÑA NAVARRO
Pontificia Universidad Católica de Chile, Chile
amvicuna@uc.cl

Moral education is indispensable for good quality education. Autonomy, the highest level of moral development, must be acquired through a socialization process that enables the person to take the decision to belong to a moral community. This is provided ideally by the building of a "community of inquiry". I explore the role of argumentation in the development of such a community.

KEYWORDS: argumentation, autonomy, community of inquiry, moral community, moral education, philosophy for children, pragma-dialectical rules for critical discussion

1. INTRODUCTION

The need for moral education is felt in our times from many different social environments. In the face of varied social problems, the general answer is to demand for more and better education. During the last years in Chile there has been a continued student movement of protest demanding for better education at elementary, high school and university levels and for a better access to it. These demands focus on universal access (public free education for everyone) and quality education. The students' demands have been supported by the majority of Chilean society and are now a major concern of the present administration. In fact, an important educational reform is under discussion in parliament.

My purpose in this paper is to reflect on some features which, in my opinion, would be required for an effective moral education, which is in turn an indispensable part of an education of good quality, just as are intellectual, aesthetic and responsible citizenship education. This reflection is based on twenty five years of experience (my own and my husband's) working as teacher trainers in the Philosophy for Children program in Chile and doing research on the subject.

D. Mohammed & M. Lewiński (eds.) (2016). *Argumentation and Reasoned Action: Proceedings of the 1st European Conference on Argumentation, Lisbon, 2015. Vol. II, 1021-1031.* London: College Publications.

To develop my argument, I center on Tugendhat's (1988) concept of a "moral community", relating it to the stages of moral development proposed by Kohlberg (1987) and the concept of a "community of inquiry" that is used in Philosophy for Children (e.g. Lipman, 1980). In Kohlberg's view, the highest level of moral development is the level of moral autonomy, where the person is ruled by principles that he understands and accepts autonomously. Now, since these principles are those of respect and cooperation with other human beings, it seems indisputable that they must be acquired through a socialization process that enables the person to make the decision to belong to a moral community, in the sense proposed by Tugendhat. I claim that the proper environment for this socialization process is provided ideally by the "community of inquiry" in the sense proposed by Lipman.

Next I explore the role that the teaching of argumentation has in the development of such a "community of inquiry" and defend an approach to the teaching of argumentative skills based on philosophical dialogue, the pragma-dialectical rules for critical discussion (van Eemeren, 2004) and the concept of "caring thinking" as it is used in Philosophy for Children.

2. THE NEED FOR MORAL EDUCATION

2.1 A double motivation: fear and hope

Our first concern with moral education originated in the special context of the Chilean return to democracy in 1990 after seventeen years of military dictatorship. Our fundamental motivation at that time was fear: we were afraid that the people in our country wouldn't be able to take care of their newly recovered democracy, that they wouldn't know how to treat each other with respect and, despite the high value they now placed on democracy and human rights –acquired through their many sufferings under the military coup of 1973 and the long subsequent military government–, they wouldn't know, in practice, how to deal with conflict in a reasonable way, without resourcing to violence. The best answer seemed to us to try to enhance reasonableness by working with children and training their teachers in the Philosophy for Children program reinforced with a program on informal logic. To this purpose we designed a research project and got funded by Fondecyt (the Chilean

Fund for the development of science and technology).[1] We worked with children of 4th to 7th grade at a school for socially deprived children in Santiago. Our intention was to show that doing philosophy with the children and training their teachers in the process would help develop in them democratic values, attitudes and behaviours (Fondecyt 0703-91). The results showed that we were right: both children and teachers not only developed reasoning skills indispensable to participate effectively in democracy but also attitudes and behaviours necessary for that, such as respect, tolerance, willingness to dialogue, and the like. (Vicuña & López, 1995)

From our work in this project stemmed an interest for the foundations of democracy and a new research project on the ethical foundations of human rights (Fondecyt 194-0687). We had realized that reasoning skills and democratic attitudes and behaviours were not enough to take care, develop and perfect our Chilean democracy in the long run. We came to the conclusion that moral education must be included as an indispensable ingredient in education for democracy, if we really want to prevent the repetition of the human rights' violations that our country -and many others in the world- have suffered. In this project we had the great fortune to be able to engage Professor Ernst Tugendhat and to benefit from his wisdom and his ethical theory. We worked with children of 7th and 8th grades at another low income school in Santiago and we also trained their teachers in the process. The results showed betterment in the children's ethical reasoning, as measured by an adapted test inspired in Kohlberg (1987).

Today, twenty five years after our first interest in moral education, the situation in Chile has changed. Our motivation is no longer fear, but hope, but -sadly- also frustration. There is hope, because most people in Chile agree now that human rights and democracy are to be respected and that it is possible and desirable to deal with conflict reasonably. Most people consider now, at least in their discourse, that negotiation and the search for agreement are more valuable than confrontation and violence. But, on the other hand, there is a sense of frustration, because our democracy is still incomplete; the gap between classes is enormous, both socially and economically. For instance, a successful professional can earn ten times more than a working class wage. And the gap could be even greater. People of different social classes don't mix; they live in different neighbourhoods,

---

[1] Fondecyt project 0703-91, "El diálogo filosófico como instrumento eficaz para desarrollar y fortalecer conductas y actitudes democráticas" [Philosophical dialogue as an effective tool for developing and strengthening democratic attitudes and behaviours] by Vicuña and López (1991).

apart from each other, go to different schools, and get an education of very different quality. The children of wealthy and upper middle class families go to expensive private schools while the children of the poor classes receive a poor quality education, which shows in the gap between the grades obtained by them in the national tests required to apply to continue studying at a university. Therefore, the benefits of democracy don't reach everyone, many people with the same capabilities are not offered the same opportunities and, even worse, the public school system is contributing to increase the gap instead of diminishing it. Justice, equality, wealth and respect for human rights are not distributed to all citizens equally.

2.2 Varied social problems demand for better education

Social problems associated with inequality, poverty and ignorance bring about social instability and an increase in the severity of the people's problems. For instance, drug abuse contributes to increase delinquency; teenage pregnancy contributes to enlarge the circle of poverty, and society at large suffers from insecurity and instability. But, as we know, the people who suffer the most are the poor.

The actual specific Chilean situation reflects these problems in a very patent way. Due to globalization and better access to information, through internet and social webs, the people are more aware of their own rights and the rights of minorities, such as the disabled, homosexuals, native population, and so on. There have been public demonstrations demanding better quality education and universal access to it for several years. The students' movement started to express publicly their demands by seizing their schools and organized marching in downtown Santiago almost ten years ago, in 2006, and this was replied in many of the country's major cities. And they have continued demonstrating over the years. Many other social demands have been conducted in a similar way, and they have been successful in installing their topic of interest and, sometimes, in getting solution to their problem.

The present Chilean government, led by President Michelle Bachelet, was elected on the campaign promise of bringing about major changes in education, political participation, the health care system and the retirement system that had been largely privatized during the military regime, with the result that access to these services has become more difficult for the poor people, who depend on the public system. It seemed that, having paid attention to the people's demands expressed in the social movements, the candidate and the "New Majority", the coalition of parties that support her, took these demands as their banner. Therefore, many people were full of hope and expected that long delayed changes would at last come to happen, after more than twenty years of democracy's recovery. In accordance with those programmatic goals, the government has presented proposals to congress to effect the desired changes, but they have met with the strong opposition not only of the right wing parties but also of the social movements. As it seems, the former think that the proposed reforms are too extreme and the latter that they are too slow or too narrow. Right now the situation is difficult; the president's popularity is at its lowest level, the economy is not growing as much as it was expected and there is a generalized feeling of discontent, aggravated by distrust in the political system caused by the scandal of politicians having received founding from enterprises which avoided paying taxes illegally.

Most of these problems that affect specifically the present Chilean situation are –and have always been– the problems caused by poverty and by ignorance everywhere in the world. But they are also caused by the selfishness and indifference of the powerful. Therefore, it seems that education is needed not only to free the poor classes of their ignorance but to teach all human beings to be more humane, that is to respect and to care for each other. For, how could anyone demand from the poor to keep clear from delinquency if the wealthy and powerful are plundering the state?

2.3 Good quality education must include moral education

Since every human being is entitled to reach his or her highest level of development, moral education is necessary for everyone. In Kohlberg's (1987) description of children's stages of moral reasoning, the highest stage is that of autonomy, where the individual is oriented by universal ethical principles selected by his/her own consciousness. These are universal principles of justice, reciprocity, equality and respect for human rights and for the dignity of every human being. The concept of *caring thinking* used in Philosophy for Children (Lipman, 1995) also seems to coincide with this description (I shall come back to this later).

In this connection, the distinction between an authoritarian and an autonomous morality, made by Ernst Tugendhat (1988), is relevant. According to him, the former is based on an appeal to "superior truths", such as the will of God or the tradition inherited from our ancestors, to determine what is right or wrong, whereas the latter is based on the person's own decision of willingly respecting certain rules.

As Tugendhat argues, only the second alternative is possible for us after the search for radical foundations characteristic of modernity, or as he puts it: "we cannot act as if we were back from modernity's pretension for foundation" (Tugendhat, 1988, p. 49).

An autonomous morality is even more necessary in our times with the globalization demands that cause the constant confrontation with different ways of life, different sets of values, and different beliefs. Therefore, a morally educated person must be prepared to justify his/her behaviour and also the demands he/she makes on other people. Since most of these demands have to do with the justice each one claims for himself and the right to be treated with respect, it is only natural that moral education should include understanding the obligation that each and everyone has towards the other, that is, the obligation to respect all human beings equally.

Therefore, a good quality education should include moral education, and this would contribute to produce better societies, since morally educated people would care for others and their well being.

3. HOW SHOULD MORAL EDUCATION BE ACHIEVED?

3.1 The problem of relativism and the foundation of morality

Since, as stated before, in modern times it is no longer possible to found morality in religious or traditional beliefs, we are left with the problem of finding a foundation that is both autonomous and able to overcome relativism.

I consider that Tugendhat's (1988) proposed foundation on a personal decision to belong to a moral community determined by universal equal respect is most appealing, because it complies with both requirements: it is autonomous and it is able to overcome moral relativism.

Since Tugendhat argues in several places (1988, 1992 & 1994) for this foundation, I shall limit myself to present his argument in just a summary way: moral reciprocal demands can no longer be founded on what Tugendhat calls "superior truths" (God, tradition). But, if we don't found ethical propositions, we would become ethical relativists in practice, for we would unwittingly accept violence as the way to resolve conflicts. Then, we must not renounce to found them. So, the only possible foundation left, according to him, is the individual's personal decision to belong to a moral community determined by universal equal respect. This means putting oneself under the rule of acting in accordance with Kant's categorical imperative, which is equivalent to the impartial application of the golden rule. Since it is in my own interest well understood to belong to such a moral community, I willingly put myself under the obligation of respecting others as I want to be respected myself.

Now, autonomous principles of respect and cooperation in a community ruled by universal equal respect must be acquired through a socialization process that enables the person to make the decision for the moral community. This decision should be totally free, without any appeal to threats or indoctrination, therefore the question for educators and leaders is how to motivate children, young people and adults to make this decision.

3.2 The "community of inquiry"

I think that the ideal setting for this moral education is provided by the "community of inquiry" advocated in Philosophy for Children, because it provides the living experience of partaking of such a community, role models of how to behave and an understanding of the principles that guide the community.

The concept of a "community of inquiry" is central to Philosophy for Children. It is the purpose of the teachers to build it, and it is the criterion by which to judge whether a class is working correctly in the program or not. Briefly stated, when we do philosophy with the children the classroom is converted into a "community of inquiry", that is, the group formed by the students and the teacher are "committed to the procedures of inquiry, to responsible search techniques that presuppose an openness to evidence and to reason."(Lipman, Sharp, & Oscanyan, 1980, p. 45)

The relationship between the members of the community is described by Lipman as follows:

> students listen to one another with respect, build on one another's ideas, challenge one another to supply reasons for otherwise unsupported opinions, assist each other in drawing inferences from what has been said, and seek to identify one another's assumptions. (Lipman, 1991, p. 15)

These descriptions show that the ideal model is characterized by intellectual pursuits as well as by affective ties. The members of the group care for each other and respect each other, but they also challenge each other to do the best in the pursuit of their goals. The clue to achieve this is "philosophical dialogue", another central concept in Philosophy for Children.

As Lipman explains (Lipman, Sharp, & Oscanyan, 1980), in philosophy itself we find the method by which this inquiry is conducted and this community is built: philosophical dialogue. This has been traditionally, since the beginning of philosophy, the preferred tool for stimulating thinking; through rigorous dialogue and questioning conducted Socrates the search for meaning and awakened the critical skills of his disciples, challenging them to think more clearly and to think for themselves. In a similar way, the Philosophy for Children teacher challenges the students to think for themselves and shows them how to do it, by modelling an inquisitive yet intellectually humble attitude and cooperative style of thinking that leads to construct meaning and understanding.

3.3 The pragma-dialectical rules for critical discussion

In the process of training a teacher in Philosophy for Children it is fundamental that he/she experiences directly what it is to partake of the building of such a community and to become skilled in conducting philosophical dialogues. We have seen in our practice that an important tool to help understand and give structure to such dialogues can be found in the pragma-dialectical approach to argumentation (van Eemeren & Grootendorst, 1992, 2004) and especially in the pragma-dialectical rules for critical discussion.

We have attempted to show how this works in practice in a forthcoming book in honor of Frans van Eemeren (López & Vicuña, 2015) in which we distinguish the main stages that can occur in a Philosophy for Children class and correlate them with the pragma-dialectical rules that could apply to each stage. It can be seen that moral rules and critical rules are interrelated and sometimes reinforce each other. For instance, Critical Rule 1: *Parties must not prevent each other from advancing standpoints or casting doubt on standpoints* (van Eemeren & Grootendorst, 1992, p. 208), also known as the "freedom rule", is a demand for egalitarian respect that is as indispensable to reach a resolution of the discussion as it is to have a civilized social behavior, so that in learning to respect these rules in dialogue, the students also learn how to treat each other with respect. A violation of this rule would result in a fallacy, for instance *ad hominem*, but this is not only an argumentatively wrong move; it is also a treatment that shows a lack of respect for the opponent.

In a previous paper (Vicuña, 2004), I have tried to show that the pragma-dialectical critical rules appeal to an ideal of reasonableness that refers not only to intellectual attitudes but also to moral ones. At the base of many of them, parallel to the intellectual demand for reasonableness, are moral ideals as mutual respect, tolerance, good will, intellectual honesty and truthfulness.

3.4 Caring Thinking

The previous remarks about the rules for critical discussion show a close relationship between them and the concept of "caring thinking", also a central concept in Philosophy for Children. This concept refers to empathy, respect, taking into consideration the results of our thinking, how our ways of thinking affect others, the planet, the universe, etc. It can better be expressed as caring for each other and caring for the world. This is one of the aspects of reasoning that is expected to be fostered in the community of inquiry.

Trough philosophical dialogue about things that matter to us we learn to reflect about our experiences in a cooperative way, and so we come to know each other better and to understand and care for each other's ideas, perspectives, and feelings. In Lipman's words this experience stimulates a "personal and an interpersonal growth" which is characterized by valuing and caring more for oneself and the others. In sharing one's thoughts and experiences with others in an atmosphere of respect, good will and empathy, one comes to know and appreciate oneself better and to value more others and their ideas and feelings. In this sense, caring thinking is something that can be learned, just as one can learn to appreciate music and other artistic activities.

According to Lipman (1995), "some types of feeling, such as caring, can be instances of thinking", so that it is possible to learn to think caringly:

> When we are thinking *caringly,* we attend to what we take to be important, to what we care about, to what demands, requires or needs us to think about it. Higher-order thinking, in other words, is not value-free. It has ethical and aesthetic aspects from which it is inseparable. To think about what can be done in the world is to have to take into account the environmental impact of so doing. To think about grandparents or children -people who may not be able to take care of themselves- is to have to take into account how they are to be cared for, and to give such thinking priority because of its importance. Caring thinking is not content merely to classify; it must rank and grade, assign priorities, distinguish between what is urgent and what is not. (Lipman, 1995, p. 7)

In my opinion, these examples clearly show how caring thinking works in practice and how it could and should be present in all disciplines and in everyday concerns, from family relations to the most serious problems of the world. Therefore, educating children –and young and old people as well– in such an environment of mutual respect and caring would certainly provide the ingredients both for quality education and for moral education.

4. CONCLUSION

I have argued that quality education must include moral or ethical education, that this is badly needed in our present times, both because people are not behaving in a way that shows respect and care for the suffering of many other humans beings, and because in our times we lack a strong foundation for morality and we need to found it to overcome relativism. I have also argued that a sound foundation can be found in Tugendhat's proposition of willingly submitting oneself to the rules of a moral community governed by universal equal respect. For the purpose of motivating the people to make this decision, philosophical dialogue and building a community of inquiry, as it is done in Philosophy for Children, provide the ideal setting. To achieve this goal, the pragma-dialectical rules for critical discussion and the cultivation of caring thinking are invaluable tools for training the teachers in the successful conduction of philosophical dialogues in their classrooms and to educate morally.

To this end we have contributed a philosophical novel (Tugendhat, López, & Vicuña, 1998) for the teaching of ethics, in the proposed lines, at high school level, which has been translated into German, Portuguese and Catalonian, and we are now preparing two more, one on argumentation and another one on citizenship.

REFERENCES

Eemeren, F. H., van, & Grootendorst, R. (1992). *Argumentation, communication and fallacies. A pragma-dialectical perspective.* Hillsdale, NJ: Lawrence Erlbaum Associates Publishers.

Eemeren, F. H., van, & Grootendorst, R. (2004). *A systematic theory of argumentation.* Cambridge: Cambridge University Press.

Kohlberg, L. (1987). *Child psychology and childhood education: A cognitive-developmental view.* New York: Longman.

Lipman, M. (1995). Caring as thinking. *Inquiry: Critical Thinking Across the Disciplines, 15*(1), 1-13.

Lipman, M., Sharp, A. M., & Oscanyan, F. (1980). *Philosophy in the Classroom* Philadelphia: Temple University Press.

López, C., & Vicuña, A. M. (2015). Las reglas de una discusión crítica y la formación de una Comunidad de Indagación. In F. Leal (Ed.), *Seamos razonables: Estudios en honor a Frans van Eemeren. (*Forthcoming*)*

Tugendhat, E. (1988). *Problemas de la Ética.* Barcelona: Crítica.

Tugendhat, E., López, C., & Vicuña, A. M. (1998). *Manuel y Camila se preguntan: ¿Cómo deberíamos vivir? Reflexiones sobre la moral.* Santiago: Planeta-UNAB.

Vicuña, A. M. (1999). Ethical education through philosophical discussion. *Thinking: The Journal of Philosophy for Children, 14*(2), 23–26.

Vicuña, A. M. (2004). Los criterios de aceptabilidad en el discurso argumentativo. In: A. Harvey (Ed.), *En torno al discurso. Contribuciones de América Latina* (pp. 46-55). Santiago: Ediciones Universidad Católica de Chile.

Vicuña, A. M., & López, C. (1995). Informe Final del Proyecto Fondecyt 0703-91 "El diálogo filosófico como instrumento eficaz para desarrollar y fortalecer conductas y actitudes democráticas" [Final Report of Fondecyt Project 0703-91 "Philosophical dialogue as an efficient tool for developing and strengthening democratic attitudes and behaviors"] (Unpublished final report)

Vicuña, A. M., López, C., & Tugendhat, E. (1997). Informe Final del Proyecto Fondecyt 194-0687 "La fundamentación ética de los derechos humanos y algunas proyecciones para una educación en el respeto a todos los hombres" [Final Report of Fondecyt Project 194-0687 "Ethical foudations of human rights and some projections for an education in respect for all human beings"] (Unpublished final report)

66

Mark my Words: Vindicating Moral and Legal Arguments

SHELDON WEIN
Saint Mary's University, Canada
sheldon.wein@gmail.com

The advice Jesus offers for interpreting social rules has implications for how we understand and vindicate moral and legal arguments. In particular Jesus's views are incompatible with even weak Dworkinian accounts of interpretation.

KEYWORDS: argumentation, circumstances of justice, Dworkin, interpretation, Jesus, law, *Mark*, morality

1. INTRODUCTION

Moral and legal arguments are often very similar. In this paper I consider some of the reasons standardly offered for this fact and (after briefly rejecting most) re-interpret some of them using insights recently developed by legal positivists. I also suggest that Jesus of Nazareth uses an account of interpretation at odds with that developed by Ronald Dworkin and widely accepted by non-Dworkinians.

The title of this paper comes from *Mark* 2:27 where Jesus (who had just been criticized for violating a social rule by working on a holy day) says, "The Sabbath was made for man, not man for the Sabbath."[1] Unfortunately for my purposes, what Jesus says next, at least as reported by Mark, ("So the Son of Man is Lord even of the Sabbath."), has been sufficiently distracting—because it is either so profound, so

---

[1] The background for the social rule about which Jesus offered interpretive advice is found at *Exodus* 20:8-11: "8 Remember the Sabbath day, to keep it holy. 9 Six days shalt thou labour, and do all thy work: 10 But the seventh day is the Sabbath of the Lord thy God: in it thou shalt not do any work, thou, nor thy son, nor thy daughter, thy manservant, nor thy maidservant, nor thy cattle, nor thy stranger that is within thy gates: 11 For in six days the Lord made heaven and earth, the sea, and all that in them is, and rested the seventh day: wherefore the Lord blessed the Sabbath day, and hallowed it." (King James Version)

D. Mohammed & M. Lewiński (eds.) (2016). *Argumentation and Reasoned Action: Proceedings of the 1st European Conference on Argumentation, Lisbon, 2015. Vol. II, 1033-1043.* London: College Publications.

confused, or so stupid—that philosophers have paid little attention to the original remark. Theologians have concentrated their attention on the second sentence and the apparent inference from the first sentence to the second, and consequently they too have failed to grasp the importance of the passage.[2]

I claim that the original statement, that the Sabbath was made for man rather than man for the Sabbath, at least as most naturally understood, says something important about legal and moral argumentation. It also offers support for a certain view of the nature of morality and of law and how we should best interpret the moral and legal rules governing our society.

2. MORAL AND LEGAL ARGUMENTATION

Many people have offered reasons for why legal and moral argumentation are similar. Most prominent among them are:

> 1. Law and morality are the same thing (in perhaps slightly different forms), so naturally argumentation within each is going to be the same.
> 2. All arguments are similar, so naturally moral and legal arguments are going to be similar.
> 3. Typically, moral and legal arguments appeal only to minimally rational agents so in this respect they are bound to be similar.
> 4. Typically, moral and legal arguments concern practical rather than theoretical reasoning.
> 5. Moral and legal arguments often cover the same subject matter—they often have the same content.
> 6. They both occur typically (only?) when humans are in the (Humean) circumstances of justice.
> 7. They often involve interpretation. (How should we interpret a general principle/rule? How does it apply to this particular case?) Interpretation in both cases adheres to the broad Dworkinian model of interpretation.
> 8. They serve the same social purposes in the same way.

I am going to ignore the first five points. The first is false (and usually only made by legal positivists who attribute the view to natural law theorists so they can knock down that straw man). The second through fifth points are all true but do not help differentiate legal and moral

[2] Theologians are not to blame for this. It is quite an interesting question why someone would think that anything about the Son of Man (or anyone else) follows from the first claim. But that matter is of no interest here.

arguments from those in a host of other fields, including even arguments about how to interpret rules in games and sports.[3]

3. THE CIRCUMSTANCES OF JUSTICE

The Humean circumstances of justice are those in which (to use Rawls's apt phrase) cooperation is both possible and necessary. This is standardly thought to include several factors, among them:

- shared vulnerability—the recognition that one's fellows can adversely affect one's well being
- moderate scarcity—the fact that resources are neither so limited that the only option is to fight for one's share or so plentiful that there is no need for rules governing distribution
- limited altruism and the presence of self bias
- the recognition that cooperative activities have externalities, many of which are positive but some of which are negative
- the recognition that these externalities can be variably distributed
- different conceptions of the good.

In these circumstances, moral rules can serve to reduce vulnerability, encourage cooperation, enhance (the effects of) altruism, minimize self bias, reduce the negative externalities that accompany human interaction, and allow us to better achieve that which each of us seeks in her life.

4. THE FORMAL CONDITIONS ON A MORAL CODE

A moral code is a system of more or less coherent rules which tell one what one ought to do (in Searle's terms, a set for rules distributing deontic powers). But not just any set of prescriptive rules counts as a moral code, as a morality. For a system of social rules to count as a morality, it must seek to meet the following formal conditions:[4]

---

[3] Though for similarities, see Chad M Oldfather.

[4] For an alternative account, see Rawls §23. The formal conditions I list should not be seen as strictly necessary conditions for something being a moral code. A moral theory might not meet one of the conditions. But in such a case the defenders of the theory owe us an explanation. Utilitarians, for example, do not have a strong commitment to the ought-implies-can condition because in principle they would accept a set of rules which required people to do what they literally could not do if they ever found themselves in the circumstances where *trying* to conform to a set of rules one could not literally conform to was more productive of utility than *trying* to conform to a set of rules one actually could follow. (Of course, utilitarians would include in the calculation the cost of

- the **rationality condition**: the rules of the code must be addressed to rational individuals, for only rational individuals are thought to be subject to the requirements of morality.[5]
- the **ought-implies-can condition**: the rules of morality must be confined to telling the individual to do things that she can do and to refrain from actions from which it is possible for her to abstain.
- the **constraint or sacrifice condition**: moral rules are necessarily rules which ask the individual to forgo the pursuit of her own interests or utility (usually for the sake of furthering the interests or utility of others).
- the **categorical condition**: moral imperatives must be categorical, rather than hypothetical. Moral imperatives apply to rational individuals just in virtue of the fact that they are rational individuals. Thus, one is not allowed to beg off the requirements of morality simply because one does not feel inclined to behave in the way morality requires.[6]
- the **simplicity condition**: the moral rules must be simple enough that all minimally rational members of the community, even young children, can understand.[7]
- the **benefit condition**: the rules of morality must be such that those who follow them will be better off than they would be if they did not follow them. Any set of social rules which did not make a community governed by those rules better off than it would be without those rules would not be a moral code.[8]

These conditions are defeasible but only with strong reasons and moral nihilism is simply the doctrine that there is no logically possible set of

---

having people adopt a rule they cannot follow and the subsequent disutility produced by the guilt and shame of not succeeding in living up to their moral requirements.) So these requirements must all be seen as achievement conditions rather than necessary conditions.

[5] This applies to moral practioners and not to moral patients (if the two groups are thought not to be co-extensive).

[6] I am indebted to Malcolm Murray for help in formulating this condition.

[7] This list may not be exhaustive. Both an **impartiality condition** and a **relative stability** condition may need to be added.

[8] Of course, not all actual moral codes have been to the benefit of the community at large. (Indeed, perhaps none of them have been!) This fact lends plausibility to the claims against morality made by Thrasymachus, Nietzsche, and Marx. But all moral codes must be such that it is reasonable to suppose that, at some point in time, using the code could reasonably have been seen to be in the interest of some significant segment of the community.

social rules which both should win the assent of rational individuals and meets all these conditions.[9]

5. THE CIRCUMSTANCES OF LEGALITY

James Madison famously (and falsely) said that "if men were angels, no government would be necessary."[10] It is often thought that because we humans are prone to weakness of the will and selfishness and anger and so forth, we need law and its coercive powers to supplement moral rules if we are to achieve the benefits that cooperative enterprises can bring us. Though I will not argue for it here, this is a mistake. Morality is actually more intimately connected with coercion than is the law. Moral rules are enforced diversely mostly by ostracizing those who violate the social norms. When humans first developed moral rules they lived in small bands, and in those circumstances being ostracized by one's fellows amounted to severe coercion—the result was likely death. Of course, every legal system we humans have ever developed has used coercion to enforce some of its rules. But one can easily see—simply by imagining a community of angels, ordinary humans without the character flaws mentioned above, who live in situations such as we find ourselves today—that one could have a legal system without coercion. Angels might not need criminal law but they would need patent law, contract law, tort law, and so on.

Thus according to Scott Shapiro

> to compensate of the deficiencies of nonlegal forms of planning by planning in the "right" way, namely, by adopting and applying morally sensible plans in a morally legitimate manner.

---

[9] Thrasymachus is usually thought to be the first moral nihilist. Nietzsche is the thinker who devoted the most care and energy to making this view a live option for philosophers. For an excellent contemporary defence, see Joyce 2001. J.L. Mackie calls this view "moral scepticism". A moral theory—one which meets the conditions set forth above—is a theory which succeeds in telling us how to construct a moral code worth having. Its output is a set of rules telling moral agents what they ought to do. It also serves to justify and defend the rules of the code. (Some moral theories do not meet all conditions outlined above. But when they do not, the defenders of the theory usually offer an explanation of why those conditions the theory does not meet cannot or should not be met. And it counts against a moral theory that it does not meet some of the conditions. Thus, a theory which met none of the conditions above would be a non-starter, not even worth our consideration.)

[10] James Madison, "The Federalist No. 51".

On this view

> *legal systems are institutions of social planning and their fundamental aim is to compensate for the deficiencies of alternative forms of planning in the circumstances of legality.* (p. 171)

He goes on to add that

> according to the Planning Theory, then, the fundamental problem to which law is a solution is not any *particular* moral quandary. Rather law is an answer to a higher-order problem, namely, the problem of how to solve moral quandaries *in general*. A community needs law whenever its moral problems—whatever they happen to be—are so numerous and serious, and their solutions so complex, contentious, or arbitrary, that nonlegal forms of ordering behaviour [such as moral codes] are inferior ways of guiding, coordinating, and monitoring conduct (p. 173).

Shapiro puts all this in terms of the law being a shared cooperative activity (in the sense worked out by Michael Bratman), but others, most notably Andrei Marmor, have tried to describe the law using (modified) versions of David Lewis's account of conventions. I think it is fair to say that philosophers have not yet figured out how best to characterize social practices.[11] But this failure need not detain us. We know that humans have created many social practices and work on the surface differences between these need not wait upon a full understanding of the underlying formal or logical structures of such practices.

6. DWORKINIAN INTERPRETATION

Ronald Dworkin argues that in hard cases—cases where, were all the facts, including all the facts of institutional history, known, reasonable people would still disagree about the proper outcome of the case—judges should decide the case by constructing theories which would explain how the existing law has been interpreted. Once they have collected all the theories which explain previous legal decisions in their jurisdictions (what Dworkin calls theories that *fit* the existing law), they should choose that theory which makes that law most attractive morally and apply that particular theory to the case before them.

---

[11] For doubts about how well either of these approaches might work, see Matthew Noah Smith, 2006.

What this directs judges to do is quite contentious. Dworkin himself favors those interpretations of existing law which come closest to showing individuals equal concern and respect, for, so he holds, only then will the law treat people with integrity. We need not go into the complex details that Dworkin uses in developing the core idea. We only need note that any account of how to interpret a social practice requires some guidelines regarding how the requirements (the deontic powers and constraints) of that practice are to be understood. Dworkin's theory is extremely holistic because how one should interpret a particular past decision or particular rule depends in part on what will put the entire legal system in the morally most attractive light.

It is worthwhile to distinguish here between what we might call a weakly Dworkinian account of interpretation and a strongly Dworkinian account. A weakly Dworkinian account holds that we should seek the interpretation which both best fits past practice and puts that practice in the (morally) best light. This is also true of strongly Dworkinian accounts of interpretation but the strongly Dworkinian accounts add that there has to be a general theory explaining and justifying the past practices. Thus, in his early works (particularly those essays collected in Dworkin, 1978, 1985), Dworkin was advocating a weakly Dworkinian account of interpretation. Dworkin 1986 offers three strongly Dworkinian accounts of interpretation: which he called conventionalism, pragmatism, and law as integrity. Dworkin favors law as integrity as the best of the three options—and he leans towards thinking these accounts are the only viable ones—but all three count as strongly Dworkinian in the sense I am using the term here.

Thus, there is a strongly holistic character to proper interpretation (both weak Dworkinian interpretation and especially strong Dworkinian interpretation). Shapiro points to this feature when he observes that

> The model for Dworkinian legal reasoning, therefore, is not geometry but morality. Just as moral reasoning is often thought to involve the careful balancing of competing moral principles in order to determine the morally correct course of action, legal reasoning employs the same highly discretionary weighing procedure to resolve difficult cases (p. 264).

Two points need to be noted here. The first is that legal positivists can easily accept—and most actually do accept—what I am calling the weak Dworkinian account of interpretation. (This is made clear by Hart in the Postscript to Hart and by Shapiro and Marmor.) That is to say, they hold that for many—though not necessary all—legal systems the interpretation of legal documents and cases meets and, indeed, is implicitly guided by the conditions that characterize weak Dworkinian interpretation. Second, I will be claiming that the best interpretation (be that "best" on either a strong or even a weak Dworkinian interpretive account) suggests the Jesus, in offering the interpretation he offers in *Mark* 2:27 rejects weak Dworkinian interpretation. It, of course, follows from this that Jesus also rejects strong Dworkinian interpretation. Thus, if I am right, on Dworkin's own view Jesus was an anti-Dworkinian.

7. JESUSIAN INTERPRETATION

Mark 2:27 is the first time that a major thinker endorsed the idea that a set of social rules must be understood as serving to further the interests of those to whom the rules apply. (Plato had Glaucon consider this in the *Republic*, but just for the purpose of mounting an attack on the position.) *Do not work on the Sabbath* is a social rule. Jesus's advice on how to understand/interpret the rule tells us something important about legal and moral argumentation versus argumentation in other fields. It is not clear whether folks understood this as a moral rule or a law or both. (I will not venture into the question of whether it was also a religious rule—as seems likely. All I need is the observation that it seems have been something people could understand as both a moral and a legal rule.) But it *is* clear that the people involved could distinguish between these two sorts of social rules and that a rule could be both part of the society's morality and its legal system.

Clearly this rule counts as a social construct in the Searlean sense. (This day and every subsequent 7th day count as a "Sabbath" and Sabbaths carry with them certain deontic constraints and powers, including the constraint not to work and the power to criticize those who do work.) The rule claims to give reasons for acting which are independent of one's beliefs and desires. (Even if you want to work on the Sabbath and you believe that the rule against doing so is stupid, you have a duty not to work on the Sabbath just because the rule tells you not to do so.)

If we see argumentation as a set of speech acts with various goals and purposes, we can illuminate the nature of the connections and differences between legal and moral argumentation. What is important about Jesus's advice is that it claims that we must see *each* rule—not just the system of rules—as meeting the benefit condition. If this is so then we should not—on Jesus's view—interpret the rule against working on the Sabbath in the context of all the other rules the society has governing (say) labour practices and ask, *given the totality of rules and principles accepted by the society*, which of the available interpretations best fits with the other rules and the way they have been understood, and then select among those interpretations the one that puts the morality or the legal system in the most favourable light. Rather, we should just ask which interpretation of this particular rule (apparently taken in isolation) puts it in the best light. There is no need to look at any other rules or practices. The only question Jesus seems to ask is (something like): Would interpreting this rule narrowly and strictly—so that no work can be done on the Sabbath no matter how much that work would benefit the members of society—or would interpreting the rule more generously—allowing for an exemption for health care workers and (perhaps) other essential service workers—be more likely to meet the benefit condition? Jesus takes it as obvious that the latter interpretation is the correct one and consequently there is no need for further argumentation on this point. (Which is presumably why he moves immediately to the further claim about his dominance over the Sabbath.)

This would, at first blush, seem to go against a Dworkinian model of interpretation for legal and moral rules. However, we must not be too quick here. It is open for a Dworkinian to agree with what I have argued is the most plausible interpretation of Jesus's remark (and of Jesusian interpretation) by claiming that any moral code or minimally decent legal system must meet the benefit condition on a rule by rule basis, rather than a more holistic basis. Only by doing so, the Dworkinian could argue, can we fully respect individuals. But, of course, that seems like a very difficult row to hoe—one Dworkin himself would never have chosen.

8. CONCLUSION

The key issue in interpreting *Mark* 2:27 is not that which has received most scholarly attention, *viz.*, the remarks that follow the claim that the Sabbath was made for man rather than man for the Sabbath. Whatever the importance of those remarks for understanding Christianity, they have nothing to do with the main philosophical insights concerning how to proceed when interpreting social rules. First, it is clear that Jesus is endorsing the benefit condition. On this view, any proper understanding of social rules must take into account how those rules might best benefit those who fall under such rules. This is both substantive but apparently weak. It is substantive because it requires any interpreter of social rules to (at least in principle) be able to offer an account of how those rules might be seen as meeting the benefit condition. It is also an apparently weak condition since, as we know, humans are rather adept at providing arguments that this or that set of rather obnoxious rules is actually beneficial to those who abide by them. (Think here of the justifications for slavery, oppression of women, bans on homosexuality, and so on.) Yet it is not as weak as it first appears. Arguments that something which is in fact of no benefit to people can be (and have been) countered with arguments showing that the first set of arguments are faulty in some way or other. In particular by showing that the benefits they supposedly provide are not really of benefit to the community which adopts those rules. That is one of the ways in which those who have advocated reform have always proceeded in helping us to construct better societies.

Second, there is a very real question regarding how to best understand "The Sabbath was made for man, not man for the Sabbath". On this, the text in *Mark* give us rather little to go on. However, on the most natural reading it suggests that Jesus's position is that each rule (taken individually) must meet the standard of being beneficial to those to whom it applies. And this adds some force to the benefit condition for it rules out any (even weak) Dworkinian interpretation that holds that it is the rules *taken collectively* that must be interpreted in a way that benefits the members of the community. And this question of interpretation can be divided into (at least) two separate issues: what was Jesus's position? and what is the most defensible position?[12] It is only the latter question that is of interest to argumentation theorists.

Here I have but one suggestion regarding where argumentation theorists and interpreters might turn for help. The question at issue—whether each rule taken severally or taken as part of a set of interconnected rules needs justification—is in some ways similar to the debate between rational choice contractarians and utilitarians. Their debate is over whether a set of social rules must benefit the individuals in the community taken severally or taken collectively. The many fine arguments provided in that debate might be redeployed to help us with this debate. Whether these arguments support a more or less holistic model of interpretation for moral and legal rules is an important question. For anyone who accepts the benefit condition—and surely we all should—it is a pressing issue for how best to understand and vindicate moral and legal arguments.

---

[12] Of course, the answer to these questions may be the same. (And for those who take Jesus to be God they perhaps necessarily are the same.) But logically they are distinct questions as can be seen by the fact that reasons one might give to support an answer to one of the questions could be inappropriate (*sans* controversial supplementation) as an answer to the other question.

ACKNOWLEDGEMENTS: I am grateful for the comments I received after presenting this paper to the first European Conference on Argumentation. I also benefited from advice from Thea E. Smith. Saint Mary's University provided generous funding supporting this research.

REFERENCES

Dworkin, R. M. (1978). *Taking rights seriously.* London: Duckworth.

Dworkin, R. M. (1985). *A matter of principle.* Cambridge, MA: Harvard University Press.

Dworkin, R. M. (1986). *Law's empire.* Cambridge, MA: Harvard University Press.

Hart, H.L.A. (2012). *The concept of law* (3rdEdition, with Postscript replying to R. M. Dworkin). New York: Oxford University Press.

Joyce, R. (2001). *The myth of morality.* New York: Cambridge University Press.

Madison, J. (2003). The Federalist No. 51. In Terence Ball (Ed.), *The Federalist,* 252. Cambridge: Cambridge University Press.

Marmor, A. (2009). *Social conventions: From language to law.* Princeton: Princeton University Press.

Oldfather, C.M. (no date). Of Umpires, Judges, and Metaphors: Adjudication in Aesthetic Sports and its Implications for Law, *Marquette University Law School Legal Studies Research Paper Series,* Research Paper No. 15-08.

Rawls, J. (1971). *A theory of justice.* Cambridge, MA: Harvard University Press.

Shapiro, S. (2011). *Legality.* Cambridge. MA: Harvard University Press.

Smith, M.N. (2006). The law as a social practice: Are shared activities at the foundations of law. *Legal Theory, 12,* 265-292.

67

Combinatorial Dialogue Games in Strategic Argumentation

SIMON WELLS
Edinburgh Napier University, UK
s.wells@napier.ac.uk

We introduce combinatorial dialogue games, an approach to strategizing within argumentative dialogue games where the moves played are interpreted as moves within an edge-addition and/or edge-removal combinatorial game. This enables an agent to reason about which move to make, regardless of the particular dialogue game that is being played. Our aim is to give agents the ability to play dialogue games better and to give researchers a clear framework within which to define new strategies.

KEYWORDS: argumentative dialogue, combinatorics, dialectical games, dialogue games, heuristics, strategy, tactics

1. INTRODUCTION

This paper introduces Combinatorial Dialogue Games (CDG), a formal approach to strategizing within argumentative dialogue games where the moves played within a dialogue game are interpreted as moves within an edge-addition and/or edge-removal combinatorial game played upon a graph.

Dialectical games have been used as interaction protocols to intelligent agents with the ability to engage in argumentative dialogue. As agents become more widespread, and move into the real world, existing in heterogeneous agent societies, it is important that they are able to act effectively to satisfy their, or their owners, goals. Thus effective play of dialectical games is increasingly important. Similarly there are increasing numbers of dialectical games, dialogue games and argument-based interaction protocols. Agents may not be confined to a playing a single game and thus must have the ability to choose an appropriate dialectical game, and subsequently to play it well. Thus a mechanism is required that enables an agent to play arbitrary dialectical

D. Mohammed & M. Lewiński (eds.) (2016). *Argumentation and Reasoned Action: Proceedings of the 1st European Conference on Argumentation, Lisbon, 2015. Vol. II, 1045-1053.* London: College Publications.

games but that avoids the approach of identifying specific guidelines for effective play of every possible individual game[1].

The goal of CDG is thus to enable an agent to reason about which move to make, regardless of the particular dialogue game that is being played. Thus the process of defining strategies related to dialogue outcome classes, selecting tactics for realising those strategies, and the definition and interpretation of heuristics for good, ideal, or merely societally responsible play are all abstracted away from the, potentially complex, underlying dialogue game. Such an approach increases the flexibility of the agents concerned, enabling them to better act within heterogeneous societies. However, this approach is not meant to replace existing methods for reasoning about arguments but to compliment them. CDG provides a mechanism for mapping from disparate dialogue game descriptions into a consistent framework for representing the state of the locutors arguments in terms of their commitments. Rather than providing strategies for each and every possible dialogue game, the CDG framework acts both as an interlingua, within which various strategies, tactics, and heuristics can be defined, and as a bridge to further, more specialist and established argument-oriented reasoning and evaluation mechanisms.

2. BACKGROUND

Dialectical games, due to (Hamblin, 1970), are multi-player (but usually two), turn-taking games in which the players take turns to make moves, that correspond to speech acts and locutional content, and the rules of the game regulate when a given move can be legally played.

Whilst many dialectical games have been developed to explore a range of problems in argumentative dialogue, there are a wide range of representational forms which made it difficult to produce computational implementations. This was mitigated by the introduction of the Dialogue Game Description Language (DGDL) (Wells & Reed, 2012). The DGDL is a domain specific language for producing syntactically correct descriptions of the rules and components of dialectical games. Underpinned by an extended Backus-Naur Form (EBNF) grammar, the DGDL supports the description of interpretations of a range of dialectical games from the literature including those from Hamblin (1970), Mackenzie (1979), Woods & Walton (1982), Walton & Krabbe (1995), McBurney & Parsons (2002), Bench-Capon (1998), Lorenzen (1978). Newer games such as those found in the MAgtALO system (Reed & Wells, 2007) and the Argument Blogging protocol

[1] This does not feel like an elegant solution.

(Wells *et al.*, 2009) are also supported. Additional features include support for dialectical shifts and embeddings (Wells, 2006) and Argumentation Schemes (Wells, 2014). A DGDL game description is comprised a composition, in which the game's component such as the participants, their roles, and commitments stores are defined, a set of rules, regulations that indirectly manipulate components, and a set of interactions, regulations that enable direct manipulations of game components by players. A simple example of a DGDL description is as follows:

```
simple{
        {turns,magnitude:single,ordering:strict}
        {players,min:2,max:2}
        {player,id:Player1}
        {player,id:Player2}
        {store,id:CStore,owner:Player1}
        {store,id:CStore,owner:Player2}
        {Assert,{p},"I assert that",
                {store(add, {p}, CStore, Speaker),
                store(add, {p}, CStore, Listener)
                }
        }
}
```

This description for the "simple" game defines a single move per turn, a strict ordering of turns, 2 players named Player1 and Player2 respectively, each with a single commitment store, and with a single "Assert" move that take a single piece of locutional content, is scaffolded with the phrase "I assert that" and which adds the content of the assertion to both the players commitment stores. The advantage of adopting the DGDL, from the perspective of this paper, is that it provides a single, consistent method for describing dialogue games and affords a game engine that enables agents to select the game to play at runtime by merely loading a new game description. Importantly, a DGDL description describes all of the relevant effects of playing a specific move, for example, adding "p" to the commitment store of both Speaker and Listener in the simple dialogue game above, in a way that is machine parseable. A set of DGDL descriptions can therefore be parsed by a, hypothetical, strategic engine to work out what the effects of playing a given move would be which is important in the context of determining which move is the best to make.

Given that reasonably arbitrary, with varying features and characteristics, dialectical games can be defined in the DGDL an issue arises of deciding what to say. Whilst the rules of a well formed game should always enable the players to determine which moves are permitted or legal it is not the role of either the game engine or rules to determine what members of the set of legal moves are relevant. Furthermore, there is a subset of relevant move that are strategically "good" and a subset of those that are optimal. Whilst determining the optimal set of moves to make at any given time is hard, the determination of good moves is necessary to enable agents to play dialectical games well.

3. COMBINATORIAL DIALOGUE GAMES

Moves within formal dialectical games usually take the form of a set of locutions that the players may utter in combination with content expressed in some knowledge representation language.

This combination of locution plus content constitutes a signature for the move and usually when considering a dialogue strategy we think in terms of sequences of moves that bring about some desirable state of the game's components. However, from the perspective of the arguments that are constructed, exposed, or otherwise elicited during a dialogue, the important aspect of making a move that must be considered is not the locution itself, but the effect upon the game's components of playing that move. If we consider the moves of a dialectical game as described using the DGDL, there are a limited number of identified effects that can occur. These include constraint over the legality of future moves, updates concerning the status of the game (or subgame), assignment of roles to the participants, and operations upon artifact stores, or, in other words *commitment state*[2].

If the entirety of the effect of playing a move is bound up in its effect on the game state then the locution itself can be considered merely to be a convenience, a label that references a set of effects. It follows therefore that locutions can be ignored temporarily and only the sets of effects need be examined and exploited. This approach enables an abstraction to be formulated, away from a game as a heterogeneous set of locutions that define a given "type" of dialogue, and towards a more homogeneous, condensed, intermediate structure that serves as an adjunct to the ongoing dialectical game.

[2] It should be noted however that the author does not consider this list of effects to be exhaustive but merely sufficient to enable the description of many of the games already introduced in the relevant literature, for a collection of which, (Wells, 2012) should be consulted.

At this point we can begin to sketch the framework of a Combinatorial Dialogue Game, a CDG, in terms of a private, directed-edge, addition/removal combinatorial game. If we ignore locutions and consider only commitment state then a CDG is a game in which incurring commitment adds node(s) and/or edge(s) to the CDG graph and retraction of commitment removes node(s) and/or edge(s) from the graph. Thus each time a player utters a locution the CDG associated with the dialogue is updated in line with the resultant changes to the player's joint public commitment state. Whilst the CDG is constructed based upon the public commitment state of the dialogue's participants, the CDG is not itself a public object but is private to a participant. It is conceived therefore that all players might build their own private representation of the commitment state of the dialogue that they use separately to determine a course of action in line with their own public and private dialogical goals.

Combinatorial games (Albert *et al.*, 2007; Conway, 1976) are traditionally studied as a single play in which there is a single, static graph that forms the board upon which the game is played until a win/loss/draw state is reached. However, because a change of commitment state alters the graph, the single-play assumption must be relaxed. This is because I don't know what my opponent knows, what their private goals are, although I can infer their public goals from the attitude that they have taken during the opening of the dialogue, and I don't know what they will day. Therefore I cannot know how the CDG graph will update after my opponent's turn. Therefore, depending upon whether the dialectical game is single or multiple move per turn and also whether the moves made have a commitment effect, the CDG graph may have to be updated and re-evaluated frequently, possibly after every turn.

One question that might be posed asks which features of extant dialectical games are necessary for a game to be amenable to the CDG approach? To address this we shall construct a characterisation of well formedness that places additional constraints on a dialectical game with the goal of making it more tractable within a computational context. Well-formed dialectical games must include a commitment model so that a CDG graph can be constructed and manipulated during a dialogue. However, a simplified model in which players may only incur commitment would be sufficient and necessite termination conditions based upon inconsistency in the players commitment store. Such an approach would not be amenable to the termination conditions based upon the "retraction of initial thesis" approach of, e.g. PPD_0 (Walton & Krabbe, 1995) but would allow a Socratic style of dialogue. Dialectical games that support an opening stage are useful as a way to force the establishment of the player's respective initial positions and theses that can define the motivations and respective goals of the participants. The addition of an adequate opening stage in which the player's positions are established also enables games to be played whose commitment models, after the opening stage, only allowed retraction of commitment. This would transform the CDG into a traditional edge removal combinatorial game and enable exploitation of game-playing strategies straight from the combinatorial game literature. In addition to a commitment model, a dialectical game should also incorporate termination conditions. These are useful because they give hard boundaries that help agents to determine when to stop talking and can be formulated in terms of commitment store contents that are, ideally, established during the opening phase. For example, the player who retracts commitment from their initial thesis loses and the other player wins. A game that has a commitment model in which at least one move enables the players to alter the commitment state of the dialogue and at least one termination condition that ascribes a win-loss state to one or another player should be sufficient to build a simple CDG graph on which subsequent moves, and their effects are recorded.

Having a well-formed dialectical game, and an ongoing associated dialogue, from both of which a CDG graph can be constructed, attention must turn to the identification of which commitments, or relations between them, that is which nodes and edges, to target. A simple, bottom-up offensive strategy suggests targeting those edges or nodes whose commitments underpin the opponent's position. Causing the opponent to retract those commitments will leave their thesis exposed. A top-down offensive strategy might target directly the opponent's position by providing a counter-argument position. Alternatively, a player might use their turns to execute a defensive strategy in which their own position is bolstered.

One aspect that has yet to be considered is how to return from a CDG to the locutional level of the associated dialectical game. This would appear to be tricky as locutions were the first element to be discarded in order to work with graphs constructed from pure commitment. Having identified the commitments to target, and whether they must be incurred or removed, there must be reification in terms of locutions, by identifying the things to say next. The simplest approach is to turn this into a search problem; dialogue profiles (Krabbe, 1999) must be generated until their pattern of commitment effects matches those of the CDG. This potentially still yields a large set of potential moves but rather than this set being the merely legal or relevant moves, they should constitute, at least part of the set of good moves. Heuristics may have a role to play in further reducing the size of the set of good moves.

Similarly, argumentation schemes may play a role in further restricting the set of available locutions by exploiting critical questions. As schemes, and the exploitation of critical questions, are supported by a minor DGDL extension (Wells, 2014) this may be a prudent approach

4. FURTHER WORK

The current approach assumes well-formed games, minimally, games that utilise artifact stores to record the various commitments of their players and that incorporate termination conditions.

One direction of future research is to relax these assumptions and to examine the utility of CDGs that are built from games that do not incorporate a commitment model within their rules. It would appear that the obvious route in this case would be to assume a commitment model and provide associate rules that map locutions in the game onto a commitment model that in turn yields a CDG. In (Wells, 2006) it was proposed that even without an explicit commitment model there exists an *essential cumulativeness* in dialogue because what is said cannot be unsaid, and even without commitment stores there is always a transcript of what has been said. This would, at least, extend the approach to those games that fall outside of the canon of games that can be traced back to Hamblin's influence and the genesis of the commitment store and related approaches. A similar solution may also enable adjunct termination conditions to be ascribed to dialectical games that do not define them, enabling support for goal-oriented behaviour.

Another interesting direction for research is to introduce additional graphs. In the CDGs described in this paper, we only consider a CDG that maps directly onto the current commitment state of the players. However, because a CDG is private to a given player, it need not merely model what has been said and how the players are currently positioned in relation to what has been said. It could, for example, be extended to handle my wider knowledge of the domain so as to include in the CDG nodes and edges that represent arguments I haven't yet made and arguments to which I might commit, or definitely not commit to in the future. Similarly, a CDG might be extended to cover what I know of my opponent's knowledge, things that they have not yet said in the current dialogue but that they have committed to in previous dialogues, and which I could therefore infer that they will still believe and will commit to if pressed. Such an approach enables a form of dialogical history to be utilised and recognises that what a rational agent says should remain consistent over time, unless of course, the agent has been persuaded to revise its beliefs at some intermediate juncture. Additionally, hypothetical or inferred knowledge might be included, to cover those things that I would expect my opponent to commit to based upon stereotype. Obviously, the further we step away from the actual utterances of the opponent, the more likelihood of a misstep, especially if too much reliance is placed upon increasingly tenuous strategizing. However, this would afford the CDG approach a richness of strategizing that accords with how many people choose and use their arguments in the real world.

5. CONCLUSION

In this paper we have introduced Combinatorial Dialogue Games as a means to effect game-agnostic purposive behaviour within dialectical game play. Elements of a mapping from dialectical game and ongoing dialogue to a form of directed-edge removal/addition combinatorial game have been presented together with an exploration of the reification required to transform a CDG into a set of desirable moves that the player should select from to play. It is argued that such an approach enables the specific locutions of a given dialectical game, and the associated game-specific playbook exploration of strategy to be avoided in favour of an approach that ignores locutions and thus enables strategic, tactical, and heuristic behaviours to be expressed within a single, consistent, extensible framework. The results are twofold; some progress has been made towards enabling agents to play dialogue games better without needing extensive prior analysis of a given dialectical game, and researchers have a clear framework within which to define, effect, and explore new strategies for better play.

REFERENCES

Albert, M. H. Nowakowski, R. J., & Wolfe, D. (2007). *Lessons in play: An introduction to combinatorial game theory*. A. K. Peters Ltd.

Bench-Capon, T. J. M. (1998). Specification and implementation of Toulmin dialogue game. *Proceedings of JURIX 98* (pp. 5-20).

Conway, J.H. (1976). *On numbers and games*. London: Academic Press Inc.

Hamblin, C. L. (1970). *Fallacies*. London: Methuen & Co. Ltd.

Krabbe, E. C. W. (2002). Profiles of dialogue as a dialectical tool. In F. H. van Eemeren, (Ed.), *Advances in Pragma-Dialectics*(pp. 153-167). Amsterdam/Newport News, VA: Sic Sat & Vale Press.

Lorenzen, P., & Lorenz, K. (1978) *Dialogische logik*. Dormstatdt: Wissenschftliche Buchgesellschaft.

Mackenzie, J. D. (1979). Question begging in non-cumulative systems. *Journal of Philosophical Logic, 8*, 117-133.

Reed, C., & Wells, S. (2007). Dialogical argument as an interface to complex debates. *IEEE Intelligent Systems Journal: Special Issue on Argumentation Technology, 22*(6), 60-65.

Walton, D., & Krabbe, E. C. W. (1995). *Commitment in dialogue.* Albany, NY: SUNY Press.

Wells, S. (2006). Cumulativeness in dialectical games. *Proceedings of the 6th International Conference on Argumentation* (ISSA 2006).

Wells, S. (2006). Knowing when to bargain. *Proceedings of the 1st Conference on Computational Models of Argument* (COMMA 2006).

Wells, S. (2014). Supporting argumentation schemes in argumentative dialogue games. *Studies in Logic, Grammar, & Rhetoric, 36*(1), 171-191.

Wells, S., Gourlay, C., & Reed, C. (2009). Argument blogging. *Proceedings of the 9th International Workshop on Computational Models of Natural Argument* (CMNA9).

Wells, S., & Reed, C. (2012). A domain specific language for describing diverse systems of dialogue. *Journal of Applied Logic,* 10(4), 309-329.

Woods, J., & Walton, D. N. (1982). Question-Begging and cumulativeness in dialectical games. *Nous, 16*, 585-605.

McBurney, P., & Parsons, S. (2002). Games that agents play: A formal framework for dialogues between autonomous agents. *Journal of Logic, Language and Information, 11*(3), 315-334.

68

Lost in Argumentation? China's Arguments in International Human Rights Treaty Bodies

JINGJING WU
Tilburg University, The Netherlands
j.wu_1@uvt.nl

By analysing Neil MacCormick's theory on legal argumentation and applying it on China's arguments in core international human rights treaty bodies, this paper argues that not only can legal argumentation theory be applied on the arguments in international human rights law, but more importantly, by giving the IHRL arguments a proper frame, it is possible to identify the trends and patterns of these arguments.

KEYWORDS: China, international human rights law, international human rights treaty bodies, legal argumentation, Neil MacCormick

1. INTRODUCTION

The temptation to bridge the legal argumentation theory with international human rights law (hereinafter as IHRL) comes from a curiosity about China's relationship with international human rights treaty bodies. China, according to some compliance theory study of IHRL, is one of the countries that "least likely to comply" (Kent, 2007, p. 2). However, little attention has been paid to what and how does China actually argue for its own position within the international human rights treaty bodies, especially during the State's sessions, in which the delegation of the State has an argumentative dialogue with the committee members on the issues within the mandate of the Convention. The rationale behind this scarce interest might be that IHRL is not considered "legal" enough to be taken as a proper example in legal argumentation (In Koskenniemi's famous book *From Apology to Utopia* (1989), he particularly excluded IHRL when he articulated the theory of "the structure of international legal argumentation"). However, this paper argues that not only can legal argumentation theory be applied on the IHRL, but more importantly, by giving the IHRL arguments a proper

D. Mohammed & M. Lewiński (eds.) (2016). *Argumentation and Reasoned Action: Proceedings of the 1st European Conference on Argumentation, Lisbon, 2015. Vol. II, 1055-1069.* London: College Publications.

frame, it is possible to identify the trends and patterns of these arguments. As for the countries like China, such observation about its IHRL arguments may reveal its relations in regards to the international human rights regimes in the way that other practical human rights studies (which mostly focus on compliance) could not grasp.

To first get a general impression of the role that legal argumentation plays in international human rights regimes, imagine the following scenario: a delegate from State X is presenting at her State's session in the Committee on the Rights of the Child (CRC). The committee members showed their concerns of the situation that State X had imbalanced gender ratio in the primary school wherein the girls only occupy 30 per cent of the attendance, which constituted a violation of article 2 and article 28 of the Convention on the principle of non-discrimination and children's right to education. The delegate, on the other hand, may argue that her State does not violate these treaty obligations by putting forward following possible arguments:

(1) The delegate denies the data that committee members held.
(2) The delegate points out that there are different forms of primary school in State X, and the committee only took certain type of primary school into consideration. When taking other kinds of primary school into account, the gender ratio is more balanced.
(3) The delegate argues that State X lacks the necessary resource (finance, human capital, infrastructure, etc.) to guarantee all children get into primary school. Moreover, the traditional biased culture in the country makes people think that girls do not need to be as educated as boys. All these elements put State X in a disadvantaged situation when it comes to fulfil girls' right to education. Nevertheless State X is making its best effort to improve this situation, which complies with the principle of "progressive realization" in article 2 of the International Covenant on Economic, Social and Cultural Rights (ICESCR).
(4) The delegate argues that State X does not think it is necessary to guarantee equal education opportunities for girls due to its religion or tradition. State X believes that different genders have their different roles for sustaining the society. In this sense, State X holds different views on the right to education that is deviated from CRC.

Above scenario to large extent reflects the real arguments a State makes in international human rights treaty bodies. To further study these arguments, the question follows: what are these arguments both in forms and essence; in other words, how to unpack these arguments in certain way rather than lumping them together. Hence, in the first part of this paper, I discuss the types of arguments and reasoning, which provides a guidance for unpacking arguments between States and international human rights committees. The discussion on the taxonomies of legal argumentation mainly refers to Neil MacCormick's book *Rhetoric and the Rule of Law* (2005), which gives a thorough and inspiring discussion on the types of legal argumentation and justification. In the second part, I apply the taxonomy of argumentation to reconstructing and grouping China's arguments in the core human rights treaty bodies[1] of which China is a member. This part not only exhibits the issues that are mostly discussed between China and core human rights treaty bodies, but also categorizes them under different argumentation types. Lastly, I conclude the observations of trends and patterns of China's argumentations, which could only be made explicit by applying to the types of legal argumentation and reasoning.

2. TYPES OF ARGUMENTATION AND REASONING: *RHETORIC AND THE RULE OF LAW* REVISIT

A starting point to unwrap MacCormick's legal argumentation theory is to identify the position for syllogism. A legal argumentation or judgement in a strict syllogistic fashion, as sprung in the legal systems in Enlightenment equipped by legal positivist language, is a deduction from law (as a major premise) and fact (as minor premise) to a verdict (as a conclusion) (la Torre, 1998, pp. 2-14). In this way, the syllogistic approach treats law as self-evident, cognitive operable knowledge. If give a legal argumentation (or decision) a syllogistic form, which could be represented as following:

> Whenever OF (as operative fact) then NC (as normative consequence), (legal rule)
> OF
> Therefore NC.[2]

---

[1] This paper focuses on core international human rights treaties bodies. According to United Nation Human Rights Office of the High Commissioner for Human Rights (OHCHR), there are in total 10 of them (9 human rights treaties and the Optional Protocol to the CAT); wherein China is a party of 6 of them, which are: International Convention on the Elimination of All Forms of Racial Discrimination (ICERD), International Covenant on Economic, Social and Cultural Rights (ICESCR), Convention on the Elimination of All Forms of Discrimination against Women (CEDAW), Convention against Torture and Other Cruel, Inhuman or Degrading Treatment or Punishment (CAT), Convention on the Rights of the Child (CRC), Convention on the Rights of Persons with Disabilities (CRPD). For the CRPD, due to lack of sessions, it won't be included in this paper.

[2] This representation is given in MacCormick, 2005, p. 32.

Comparing above form with the classic syllogism:

> All men are mortal
> Socrates is a man
> Socrates is mortal.

The question is: are these two logic structures the same? Legal scholars nowadays mostly would not think so for different considerations. One main reason is that a legal rule is in fact much different than the major premise in the second example ("all men are mortal") in the sense that the validity of the legal rule as major premise depends upon the existence of such norm (Kroatochwil, 1989, p. 212). Nonetheless, giving the fading position of syllogism in legal argumentation theory, MacCormick asked: "Is legal reasoning in any interesting sense syllogistic?" To this question, his answer is: "Yes—as least, yes with reservations and qualifications." (MacCormick, 2005, p. 32)

In MacCormick's argument, syllogism is still relevant, even "central" to legal reasoning in the sense that

> every invocation of any such legal text as a basis for a specific claim in a concrete situation can be expressed in the type of legal syllogism..., though much more informal representations of such a structure of reasoning are what is common in ordinary usage (MacCormick, 2005, p. 122).

However, although syllogism still holds its place in legal argumentation, MacCormick also recognized that syllogism alone is not a sufficient logic tool for framing legal argumentation, because: First of all, this type of argument has to start from some sort of assumption of a political or normative order, which means it has to presuppose a normative statement as the major premise. Second, since law is composed by human language, which is often vague and ambitious, interpretation is needed to give the law specific meanings. This process of interpretation cannot be realized by the form of syllogism, but the hermeneutic criteria that outside the deductive justification are required. Third, syllogism *per se* cannot decide if the norm as major premise is applicable (or in other words, relevant) to the fact considered (MacCormick, 2005, p. 69; La Torre, 1998, p. 14). In other words, the problem of syllogism when it comes to legal argumentation is that most of the defeasibilities are non-deductive and therefore cannot be revealed and solved within the syllogistic model.

Therefore, departing from the syllogism model, MacCormick concluded four ways that exhaust the possibilities to defeat the original syllogistic formula in a non-deductive (in other words, rhetorical or persuasive) way (MacCormick, 2005, p. 43). By generating these four non-deductive defeasibilities, MacCormick incorporated the persuasive (or rhetoric) reasoning into the syllogistic formula, which gave the legal argumentation a comprehensive theoretic framework. These four possibilities are: first, the "problem of proof", which argues that the existence of the alleged instance ("OF") has yet to be proven. It is to say that "no instance of 'OF' as alleged in indictment or pleadings has been proven (up to the required standard of proof) to have existed..." Second, the "problem of classification", which challenges that the alleged facts can be properly characterized as the instance ("OF") in the sense of proper to the law. Nonetheless, there is also another understanding of "problem of classification". That is to take classification as evaluation. In this case, a standard of judgment (such as "reasonable", "fair", "just", "proportionate" etc.) is introduced into the legal text, and the classification is to decide issues like what counts as being "reasonable" (or whatever) in the given context. Third, the "problem of interpretation", which raises the question that whether there is a misreading of the relevant provisions of the law. It is noted that there is a potential overlap between the "problem of interpretation" and "problem of classification", since both these types concern the interpretative reading of certain legal texts. However, the "problem of classification" particularly deals with the problem that whether "a given situation counts as belonging in a relevant category for the purposes of applying a legislative text" (MacCormick, 2005, p. 141). Therefore, it is still valuable to distinguish these two types of argument for a more accurate grab. Fourth, the "problem of relevancy", which deals with the problem that "whether there is a rule dealing with the alleged facts" (Sartor, 2006, p. 4). In other words,

> success in the claim or prosecution depends on reading authoritative legal materials as though they generated a rule "Whenever OF then NC" such that the allegations...are relevant given the facts alleged, or even the facts proven; but no such norm can properly be read out of the adduced materials as a reasonable concretization of them or determination from them (MacCormick, 2005, p. 43).

Although these four types of arguments exhaust the possibilities to challenge the original syllogism from persuasive perspective, (MacCormick, 2005, p. 43) these four types do not entail which justification are used. To further identify (and "weighing"[3]) the justification, MacCormick provided an impressively comprehensive answer. Among all the principles for justifying reasoning that have been discussed in the book, following are the ones that in my opinion relevant to argumentation in regards to IHRL:

(i) Universalized particular reasoning. A particular reasoning entails that as a justification, one may point out the particular characters of a certain case and argue that it is because of these (particular) characters, such and such rules do (not) apply. However, MacCormick emphasized that this particular reasoning, when used as a justification, should entail universalizability. This is to say that if a particular reasoning is used to justify one particular case, this reasoning should also be able to apply to all the cases that share the particular characteristics, otherwise the reasoning is not justifiable.

(ii) Judging by consequences. This kind of justification is made to claim that some value has to be to certain degree sacrificed for the sake of "a more thorough upholding of the other (value)" (MacCormick, 2005, p. 117). These different values are normative guidelines for achieving different consequences. Therefore choosing among different values are in fact deciding among the promises for different consequences. In this sense, utility reasoning is also included in this kind of justification.

(iii) For the interpretation argument, MacCormick summarized three kinds of justification that are mostly used for this type of argument: (a) Linguistic reasoning, which concerns the "plain meaning" of words used in "ordinary language" (MacCormick, 2005, p. 125). (b) Systemic reasoning, which takes the legal system under consideration as an integrated whole, therefore the interpretation of a legal text should be put into its context as part of a legal system.[4] Since this type of argument emphasizes the value of coherence in a legal system, it can also be understood under the principle of coherence. (c) Teleological-evaluative reasoning. It is about deployment of values embodied in the legislation to justify choosing one interpretation rather than another of a contested expression or phrase.

---

[3] It should be noted that MacCormick expressed his reservation on using the terminology "weighing" when it comes to legal reasoning and justification. Rf. MacCormick, 2005, p. 186.

[4] MacCormick also listed six types of systemic arguments: (1) Contextual harmonization; (2) Argument from precedent; (3) Argument from analogy; (4) Conceptual argument; (5) Argument from general principles; (6) Argument from history. Ibid. 127.

(iv) Being reasonable. This involves a particular value that holds much weight in legal discourse. A general insight is that when turning to "weighing" or "balancing" different reasons concerning different values, some standard like Adam Smith's "impartial spectator" comes into the picture (MacCormick, 2005, p. 186.). In this sense, the objectivity of reasonableness is achieved through the conditioned subjectivity.

In this way, MacCormick depicted a comprehensive contour for justifying a legal argument or decision that derives from the original syllogism formula while cannot be put into demonstrative reasoning within the syllogistic logic.

Moreover, within each principle of justification, a related type of reasoning (or justifying reasons) stands. This is to say that, under the particular reasoning principle, there is a particular reasoning, which is to use particulars as justifying reasons for certain argument. By the same token, there is also consequence reasoning (i.e. using consequence as justifying reasons), reasoning by reasonableness, and reasoning by coherence. It will be shown in the third part that these principles and reasoning types closely relate to China's arguments in the international human rights treaty bodies.

3. CHINA'S ARGUMENTS IN CORE HUMAN RIGHTS TREATY BODIES

In this part, I categorize China's arguments in the core human rights treaty bodies into the four types of argumentation discussed above. Although given the scope of this paper, this part cannot give each argument a comprehensive representation, it is suffice to say that almost all China's arguments can be put under these four types.

The fact argument (*i.e.* the problem of proof). The fact argument is used when China disagreed with the "fact" that committees held. In this case, China and the committees held inconsistent or incompatible facts as bases for their own arguments. This type of argument was used by Chinese delegation on the topic about the situation of the Panchen Lama (CAT/C/SR.846, para. 50); the alleged torture cases, especially in Tibet and Xinjiang autonomous provinces (CAT/C/SR.51, para. 7; CAT/C/SR.252/Add.1, para. 17); the claimed harassment of human rights defence lawyers, especially in cases involving ill-treatment of Tibetans and Uighurs (CAT/C/SR.846, para. 7). China also disagreed with the fact asserted by the committee that UNHCR did not have access to the border with North Korea (CAT/C/SR.844, para. 36; CEDAW/C/SR.744(B), para.40, 48), some information provided by the NGOs (CAT/C/SR.252/Add.1, para. 4, CAT/C/SR.846, para. 4), the claimed demographic change in ethnic minority areas (CERD/C/SR. 1469, para. 9, 15; CERD/C/SR. 1943, para. 9, 27), and the information about killing or abandoning female infants as a widespread problem (CRC/C/SR.299, para. 12). This type of arguments was also used when China considered the evidence was not sufficient. For instance, in one CAT session, responding to the case of Mr. Xu Zhiyong which was cited by the special rapporteur Ms. Sveaass, Chinese delegation argued that it was "not sufficiently substantiated for it to be concluded that they actually happened."(CAT/C/SR.846, para. 33)

The classification argument. One understanding of this type of argument concerns whether the "fact" can be classified properly as the terms used in the relevant IHRL. One typical case is about the situation of the immigrants from Democratic People's Republic of Korea (DPRK) in China. For all the five committees, they considered those people as asylum-seekers or refugees. On China's side, however, Chinese delegation took North Korean people entered China as illegal economic immigrants, who did not meet the criteria of refugees set in the Convention Relating to the Status of Refugees and its Protocol. Therefore, China claimed that their repatriation did not violate China's treaty obligation. (CERD/C/SR.1469, para. 3; CEDAW/C/SR.743(B), para. 42). The same type of argument was also used in China's position on Falun Gong (CRC/C/SR. 1064, para. 11; CRC/C/SR. 1834, para. 24), and the unemployment status of laid off workers from enterprises in the process of restructuring (E/C.12/2005/SR.7, para. 24). Another understanding of the classification argument, as aforementioned, regards evaluation. This type of argument was used by Chinese delegation in the case of religion freedom(CRC/C/SR.1064, para. 9), incomplete birth registration (CRC/C/SR.299, para. 4), discrimination against the disabled children within families (CRC/C/SR.1063, para. 36), forced abortion and sex-selected abortion (CEDAW/C/SR.1252, para. 54), especially when practicing family plan policy (CEDAW/C/SR.420, para. 11). In these cases, Chinese delegation argued that the relevant rights were protected in Chinese domestic laws. Violations of these rights, therefore, were illegal. Hence the instances of these kinds of human rights violations should not be classified (or evaluated) as violations of treaty obligation.

The interpretation argument. For this type of argument, although both Committees and China have a basic consent on the fact as well as its classification, as for whether such fact counts as complying with the treaty obligation, they hold different opinions. This type of arguments was used to defend China's position on the lack of definition of torture (CAT/C/SR.145/Add.2, para. 12; CAT/C/SR.251, para. 5; CAT/C/SR.252/Add.1, para. 8), the lack of definition of gender discrimination (CEDAW/C/SR.743 (B), para. 39), the lack of definition of racial discrimination (CERD/C/SR.1943, para. 3), involuntary disappearance failing out of the scope of article 1 of CAT (CAT/C/SR.844, para. 57, 63), no direct invoking of Conventions in domestic courts (CAT/C/SR.51, para. 2; CERD/C/SR.1164, para. 44; CERD/C/SR.1469, para. 4, 47; CERD/C/SR.1943, para. 4; CRC/C/SR.298, para. 15; CRC/C/SR.1062, para. 30; E/C.12/2014/SR.17, para. 37), policy on re-education through labour (CAT/C/SR.145/Add.2, para. 13; CAT/C/SR.146/Add.2, para. 16), Law on State Secrets (CAT/C/CHN/CO/4/Add.2, p. 10), religious freedom (CRC/C/SR.1064, para. 9), women's participation in government (CEDAW/SR.419, para. 53; CEDAW/C/SR.744.(B). para. 11), women's right to work (especially on equal pay for work of equal value) (CEDAW/C/SR.744 (B), para. 18), family policy and reservation on article 6 about child's right to life of CRC (CRC/C/SR.298, para. 14, 39; CRC/C/SR.1062, para. 27; CRC/C/SR.1833, para. 23), child's right to express and participation in decision making reservation on article 8 (1) of CESCR about right to organize trade union and right to strike (E/C.12/2005/SR.7, para. 27; E/C.12/2014/SR.17, para. 38; E/C.12/2014/SR.18, para. 18, 21, 28). On all above issues, China claimed to have different interpretations than committee members about whether China had complied with relevant Conventions.

The relevance argument. This type of argument was used in some early CAT sessions such as when Chinese delegation explained why China had not incorporated definition of torture into domestic courts (CAT/C/SR.51, para. 50), and the policy of re-education through labour (CAT/C/SR.51, para. 26). In those arguments, Chinese delegation held that China was entitled to its own views on these issues, which could be deviated from the norms of the relevant Conventions.

4. CHINA'S ARGUMENTATION IN CORE INTERNATIONAL HUMAN RIGHTS TREATY BODIES: CHANGING AND UNCHANGED ARGUMENTS

To unpack the findings of China's arguments in the core human rights committees, I distinguish two sorts of China's arguments: the one that has been replaced over time, and the one that stays during different sessions and even prevails across committees.

4.1 The changing arguments: a process of institutionalization

For the arguments that have been changed over time, there is a trend observed. That is most relevance arguments have changed into some other types of arguments—usually interpretation arguments. For instance, about whether should China incorporate core definitions into domestic laws. China's arguments in CAT committee changed from a relevant argument in the 1989 session saying that "definitions of torture varied, so that what in the Chinese view was positive might be considered negative elsewhere" (CAT/C/SR.51, para. 50), to interpretation arguments in later sessions (CAT sessions in 1993, 1996, 2008), in which Chinese delegation gave up their former arguments that China should be entitled for its own definition and view of torture, instead held that China was in compliance with article 1 of the Convention because the definition listed in article 1 was diffusely incorporated into different domestic laws of China (CAT/C/SR.145/Add.2; CAT/C/SR.252/Add.1, para. 8, 9). On the re-education through labour, China's arguments also started with a relevant argument, which stated that

> [r]rehabilitation through labour suited the Chinese situation and was in line with the needs of socialist construction. It was an important social-order measure and a major element in building a socialist legal system."(CAT/C/SR.51, para. 26).

This relevance argument turned into an interpretation (teleological) argument in later sessions stating that

> ([r]e-education through labour) constituted an effective means of educating and correcting repeat offenders against public security whose actions were not serious enough to warrant criminal sanctions. (CAT/C/SR.846, para. 27)

In fact, in recent China's sessions in those treaty bodies, relevance arguments almost disappeared. Hence, it is seen from these changes of argumentation types that there is a process of institutionalisation of China in regard to international human rights treaty bodies. To briefly explain this view, two points need to be illustrated: one is about what does relevance argument actually entail. The other is about the institutional perspective.

Relevance argument states that we cannot draw the legal rule (as "whenever OF then NC") from the given context. To put in the international human rights legal context, this type of argument is a way of rejecting both the norm entailed in the provision and the jurisdiction that assigns such norm. In this sense, the change of the argumentation that China had made from relevance arguments to other types of arguments (mostly interpretation arguments), is a change from rejecting the rules to resorting to the rules in argumentation and deliberation. Therefore this process should be taken as an evolvement of China's relationship in regards to IHRL and its treaty bodies. The meaning of this evolvement could be better grasped from the institutional perspective. In the institutional perspective of a legal system, legal reasoning is taken as one of the important features of a legal order, for that by conducting legal reasoning, one adopts the rules of a legal institute. The rules that constitute one legal order are usually exclusive from the ones in other legal orders, unless particular arrangement is done to make certain parts of the different legal orders compatible. Therefore, subjecting to the legal rules and conducting reasoning resorting to the rules represent that one is institutionalised in *this* legal system. In this sense, changing from relevance arguments, which mostly reject the norms and jurisdiction of IHRL and treaty bodies, to interpretation arguments, which defend the State's position by resorting to the rules of IHRL: this phenomenon shows the institutionalisation of China in regards to IHRL and its treaty bodies.

4.2. The unchanged arguments: the most used reasoning types

Nonetheless, there were large amount of arguments that stayed almost the same during different sessions and even were used in different committees. This sort of arguments represents China's continuous position on certain issues of IHRL, which were against Committees' opinions. It is observed that the arguments that have been stayed almost the same over time and across Committees mostly fall into following reasoning types: particular reasoning, consequence reasoning, and reasonableness.

Particular reasoning was used when China justified its position by resorting to China's legal culture, traditional culture, "developing country" status, large population, and limited resources. For instance, the particular legal culture was used as a justification in the argument of not to include core definition of several Conventions into Chinese domestic laws, which appeared in the discussions on article 1 of CAT, CEDAW, CRC, and CERD. For example, in one CEDAW session, Chinese delegation stated that "it was not the custom in Chinese legislation to include definition of terms."(CEDAW/C/SR.743 (B), para. 39). The traditional culture was used for arguing the unequal status between men and women especially on women's participation in government ("The appropriate number of women reflected the lower status of women in rural areas.") (CEDAW/C/SR.744 (B), para. 11), female infanticide and imbalanced sex-ratio ("(due to) deeply ingrained social traditions, especially in rural areas.")(CEDAW/C/SR.420, para. 65), different marriage consent age ("the two-year difference in the minimum legal marriageable age for men and women was not seen as discrimination in China")(CRC/C/SR.299, para. 40), and also on child's rights such as to justify the obstacles in practicing child's right to express and participation in decision making ("the principle of children's participation in decision-making ran counter to traditional Chinese culture, according to which children should above all be protected.")(CRC/C/SR.1063, para. 12). China's "developing country" status, its large population and its limited resource was used to justify the difficulty of collecting disaggregated data as mentioned in a CAT session ("China was a developing country with only limited resources and the task was made no easier by the size of its population.") (CAT/C/SR.846, para. 3) This type of reasoning was also used in China's defence for its reservation on article 6 of CRC about children's right to life ("(the reservation) had been dictated by the economic and social situation in China.")(CRC/C/SR.298, para. 14), and so forth. The key question concerning here is: does the factor that has been mentioned (given proved) "particular" enough to grant a deviation from the original rules in the given legal context? To "weigh" a particular reasoning, therefore, have to investigate the limit of the room for deviation under the particular rules and context, the "particularity" of the factor mentioned, and to what extent this factor could justify the deviation.

For consequence reasoning, there were two main values that were particularly concerned in China's arguments: one was the unity of sovereignty, the other was the economic and social development. For instance, for the sovereignty reasoning, on the policy of minority child, Chinese delegation said that "China's 56 nationalities and ethnic groups...had coexisted for 5000 years; they should continue to do so, for the sake of the unity of the nation." (CRC/C/SR.299, para. 6) On justifying the current Law on State Secret, China reasoned that

> [t]he ultimate goal of determining whether or not a matter constitutes a state secret is to safeguard national security and interests, regardless of whether the determination is made by executive or judicial means. (CAT/C/CHN/CO/4/Add.2, 5(c))

As for the economic and social development reasoning, it has shown up in the cases like defending the family plan policy and China's reservation on article 6 of CRC, wherein Chinese delegation argued that "China faced a problem of over-population which threatened to impede the economic and social development of the country." (CRC/C/SR.298, para. 39) Since a consequence principle is about choosing certain value among other values as a justification, the main question concerned here is that how much "weight" should be given to these values to justify a deviation from a given rule. Especially when it comes to sovereignty reasoning and economic development reasoning, these are two kinds of reasoning that are often used by developing countries to justify the constraints for realizing human rights.

Reasoning on reasonableness is spotted in the case that Chinese delegation resorted to relevant legislation and argued that either there was a reasonable gap between legislation and compliance or that the situation violating the Convention was also illegal according to Chinese domestic law therefore China did not violate its treaty obligations. The "reasonable gap" justification was used in the arguments such as incomplete birth registration ("Chinese law requires the registration of all citizens of the State...for the children that failed to be registered, it was mostly due to the ignorance of their parents.")(CRC/C/SR.299, para. 44), religion freedom ("religious freedom was guaranteed by the Constitution.")(CRC/C/SR.1064, para. 9), and so forth. The "illegal therefore non-violation" reasoning was used in the arguments about sex-selected abortion ("infanticide was an offence under the Criminal Code") (CEDAW/C/SR.1252, para. 54), discrimination against the disabled children within families ("allegations of discrimination against the disabled children within families concerned only isolated cases, since discrimination on the grounds of disability was unacceptable and was prohibited by law.")(CRC/C/SR.1063, para. 36), and so forth. It is understood that the gap between legislation and compliance is sometimes inevitable. However, this does not mean that having according legislation automatically marks a full compliance with the Convention. It is, therefore, a question about principle of reasonableness; in other words, given the mentioned circumstance and legal context, do the justification (such as resorting to legislation) be reasonable sufficient evidence to prove a State's compliance.

5. CONCLUSION

To conclude, by applying the perspective of legal argumentation theory to State's arguments in international human rights regimes, not only can we unpack State's arguments on different human rights issues by grouping them under the four types of arguments and identifying with different justifying reasons, but also reveal the trend and patterns of a State's positions on these issues. The case study of China illustrates that although China has been considered as one of the "non-cooperated" countries when it comes to IHRL, there is an optimistic change in China's arguments. This trend may as well motivates a further effort on articulating specific "weighing" of the reasoning that have been stalled between China and international human rights treaty bodies as discussed above. One way forward here may be that to put relevant principles of legal argumentation justification in its specific context.

REFERENCES

Committee on the Rights of the Child, Fortieth session, Summary record of the 1064th meeting (CRC/C/SR.1064) (3 October 2005).

Committee Against Torture, Forty-first session, Summary record of the 844th meeting (CAT/C/SR.844) (27 April 2009).

Committee Against Torture, Forty-first session, Summary record of the 846th meeting, (CAT/C/SR.846) (10 November 2008).

Committee Against Torture, Fourth session, Summary record of the 51st meeting (CAT/C/SR.51) (4 May 1990).

Committee Against Torture, Sixteenth session, Summary record of the 251st meeting (CAT/C/SR.251) (5 June 1996).

Committee Against Torture, Sixteenth session, Summary record of the public part of the 252nd meeting (CAT/C/SR.252/Add.1) (8 May 1996).

Committee Against Torture, Tenth session, Summary record of the third part (public) of the 145th meeting (CAT/C/SR.145/Add.2) (20 July 1993).

Committee Against Torture, Tenth session, Summary record of the third part (public) of the 146th meeting (CAT/C/SR.146/Add.2) (28 April 1992).

Committee on Economic, Social and Cultural Rights, Fifty-second session, Summary record of the 17th meeting (E/C.12/2014/SR.17) (16 May 2014).

Committee on Economic, Social and Cultural Rights, Fifty-second session, Summary record of the 18th meeting (E/C.12/2014/SR.18) (16 May 2014).

Committee on Economic, Social and Cultural Rights, Thirty-fourth session, Summary record of the 7th meeting (E/C.12/2005/SR.7) (3 May 2005).

Committee on the Elimination of Discrimination against Women, Fifty-ninth session, Summary record (partial) of the 1252nd meeting (CEDAW/C/SR.1252) (3 November 2014).

Committee on the Elimination of Discrimination against Women, Thirty-sixth session, Summary record of the 743rd meeting (Chamber B) (CEDAW/C/SR.743 (B)) (13 September 2006).

Committee on the Elimination of Discrimination against Women, Thirty-sixth session, Summary record of the 744th meeting (Chamber B) (CEDAW/C/SR.744 (B)) (24 August 2006).

Committee on the Elimination of Discrimination against Women, Twentieth session, Summary record of the 419th meeting (CEDAW/C/SR.419) (8 February 2002).

Committee on the Elimination of Discrimination against Women, Twentieth session, Summary record of the 420th meeting (CEDAW/C/SR.420) (8 February 2002).

Committee on the Elimination of Racial Discrimination, Fifty-ninth session, Summary record of the 1469th meeting (CERD/C/SR.1469) (17 August 2001).

Committee on the Elimination of Racial Discrimination, Forty-ninth session, Summary record of the 1164th meeting (CERD/C/SR.1164) (14 August 1996).

Committee on the Elimination of Racial Discrimination, Seventy-fifth session, Summary record of the 1943rd meeting (CERD/C/SR.1943) (7 December 2009).

Committee on the Rights of the Child, Fortieth session, Summary record of the 1062nd meeting (CRC/C/SR.1062) (27 September 2005).

Committee on the Rights of the Child, Fortieth session, Summary record of the 1063rd meeting (CRC/C/SR.1063) (20 December 2011).

Committee on the Rights of the Child, Sixty-fourth session, Summary record of the 1833rd meeting (CRC/C/SR.1833) (8 November 2013).

Committee on the Rights of the Child, Sixty-fourth session, Summary record of the 1834th meeting (CRC/C/SR.1834) (3 October 2013).

Committee on the Rights of the Child, Twelfth session, Summary record of the 298th meeting (CRC/C/SR.298) (19 June 1996).

Committee On the Rights of the Child, Twelfth session, Summary record of the 299th meeting (CRC/C/SR.299) (31 May 1996).

Consideration of reports submitted by States parties under article 19 of the Convention Comments by the Government of the People"s Republic of China concerning the concluding observations and recommendations of the Committee against Torture (CAT/C/CHN/CO/4) (18 December 2009).

Kent, A. (2007). *Beyond compliance: China, international organizations, and global security*. Stanford: Stanford University Press.

Kroatochwil, F. V. (1989). *Rules, norms, and decisions: On the conditions of practical and legal reasoning in international relations and domestic affairs*. Cambridge: Cambridge University Press.

La Torre, M. (1998). Theories of legal argumentation and concepts of law: An approximation. *EUI working paper, no.98/1*.

MacCormick, N. (2005). *Rhetoric and the rule of law*. Oxford: Oxford University Press.

Sartor, G. (2006). Syllogism and defeasibilty: A comment on Neil MacCormick's *Rhetoric and the Rule of Law. EUI working paper LAW, No. 2006/23.*

69

Are Inferences Concerning Action Formal or Material? An Inferentialist Perspective

TOMASZ ZARĘBSKI
University of Lower Silesia, Faculty of Education, Poland
tomasz.zarebski@dsw.edu.pl

The paper discusses, in the background of Robert Brandom's inferentialism, the problem of the status of practical inferences as to their correctness. In this context, the formal and the material account of such inferences will be juxtaposed and compared. The former assumes that the reasoning concerning successful action is based on a sort of deductive model, while the latter claims that this reasoning does not have, and actually cannot be based on, the logical form.

KEYWORDS: action, Brandom, Davidson, formal vs. material inferences, inferentialism, practical reasoning

1. INTRODUCTION

The below discussion is to juxtapose and confront two different, although related, views on reasoning concerning action: Donald Davidson's and Robert B. Brandom's. The former constitutes a more formal approach to practical inferences (i.e. those resulting in practice, or action), whereas the latter represent not a formal, but a material thinking about the relation between reasons and practice. In fact, the two perspectives in question stem from mostly different accounts of rationality in general, including the rationality of actions. As a result of the discussion, I will favour the Brandom's materially oriented stance.

2. DAVIDSON'S FORMAL APPROACH

The outline of the strategy of Davidson is sketched out in his essay *Actions, Reasons, and Causes* from 1963 (2001a), opening the book *Essays on Action and Events* (Davidson, 2001). In this text, the philosopher aims, first of all, at defending the notion of causality, both in explaining physical events and individual human actions, but at the

D. Mohammed & M. Lewiński (eds.) (2016). *Argumentation and Reasoned Action: Proceedings of the 1st European Conference on Argumentation, Lisbon, 2015. Vol. II, 1071-1080.* London: College Publications.

same time he also argues for the logical – albeit not nomological – pattern of our explanation of action. He opposes the strict division between the domain of *causes*, which we resort to when explaining physical phenomena, and the domain of *reasons*, to be referred to when justifying/explaining someone's action. In fact – Davidson insists – by explaining what one does we just indicate how the action was brought about, what has caused it. An explanation of one's action consists in showing logical connection between the cause and the effect.

Of course, it does not have to mean, and actually does not, that the sphere of human activity does not differ from the sphere of natural, or physical, world. In the physical world, the explanation of particular event is paradigmatically based on nomological-deductive model, where this event (the *explanandum*) is logically connected with a relevant general law (a part of the *explanans*). In addition, in nomological-deductive model, the general laws allow us also to predict some future events. In the scope of human activity the situation is different, since here there are no laws analogical to the physical ones, but at most some regularities to be hardly counted as laws. Accordingly, nomological-deductive model, as it stands, does not apply here. What Davidson aims at is to build a sort of explanation – or "rationalization" or "practical syllogism" as he calls it (Davidson, 2001, pp. 12-13) – of teleological actions, in which deductiveness will be salvaged and, instead of laws, some particular cause and effect relation will be employed. Thus, in the introduction to *Essays on Actions and Events* he writes:

> I accept the view that teleological explanation of action differs from explanation in the natural sciences in that laws are not essentially involved in the former but hold that both sorts of explanation can, and often must, invoke causal connections (Davidson, 2001, p. 8).

Still, it seems that in Davidson the main stress is put on the formal validity of practical syllogisms, that his decision to replace laws for cause and effect relations is to save, first and foremost, the formal character of it.

On the face of it, Davidson needs to reformulate the premises of his syllogism in the way that they ensure deductiveness of the whole inference. And so, what, in practical syllogism, constitutes the counterpart of *explanans* (from nomological-deductive model) is *intentions*. These, in turn, comprise two components: *beliefs* and *attitudes* – more precisely *pro-attitudes*, by which Davidson means:

> desires, wantings, urges, promptings, and a great variety of moral views, aesthetic principles, economic prejudices, social conventions, and public and private goals and values in so far as these can be interpreted as attitudes of an agent directed toward actions of a certain kind (Davidson, 2001a, p. 13).

As a result, both beliefs and attitudes in inferences concerning action are, respectively, the counterparts of premises in *explanans*, while the corresponding item of *explanandum* is, in Davidson, the action itself.

Davidson's argument in *Actions, Reasons, and Causes* has two distinct stages. The first is to explicate intentions as well as the way to explain someone's actions. The second is to show that our intentions are the causes of our actions. Below I will focus only on the first step.

As it was already said, our giving reasons that are to justify actions is called by Davidson "rationalization"[1] (Davidson, 2001a, p. 12) of them. Rationalizing consists in specifying actor's *intentions*, his or her *beliefs* and *desires* (pro-attitudes). Therefore, when rationalizing, we clarify that the agent did something because, first, he or she had *desired* doing it, and, second, that he or she had had the *belief* that undertaking such and such action would result in fulfilling their desire. In explaining someone's deed, we often limit ourselves only to indicating the *belief* that have been behind the action. Nevertheless, the full account of someone's intention requires knowing the both of its constituents. These two elements, beliefs and desires (pro-attitudes), Davidson sometimes refers to as to "primary reasons" (Davidson, 2001a, p. 13) for given action, which are also identical with intentions: "To know a primary reason why someone acted as he did is to know an intention with which the action was done" (Davidson, 2001a, p. 16). Primary reasons (intentions) are at the same time the ones that fully explain the action:

> In order to understand how a reason of any kind rationalizes an action it is necessary and sufficient that we see, at least in essential outline, how to construct a primary reason (Davidson, 2001a, p. 13).

Why did Davidson put so much stress on *primary reasons*? Because there are often also reasons which are not primary. Suppose – the author writes in *Actions, Reasons, and Causes* – that I enter the hose, flip the switch and turn on the light, thus illuminating the room.

---

[1] Obviously, it has nothing in common with rationalization in psychological sense (the excusing of what we had done), but only with rationalization as giving *rationale*, or reasons.

Unknowingly, by lighting the room, I also let a prowler to know that I am home – which, in consequence, puts him off robbing my house (Davidson, 2001a, p. 14). I performed the action (I flipped the switch), because I *wanted* to turn on the light and, at the same time, I *knew* that this was the right way to satisfy my wanting. The *wish* to turn on the light and the *belief* that flipping the switch will result in lighting the room – they altogether constitute the *primary reason* for my deed, i.e. lighting the room. However, it is not the primary reason for the act of alerting the prowler and scaring him off, despite the fact that it actually scared him off. In the letter case I act *with reasons*, not *for reasons*, whereas in the former – conversely – I act just *for these reasons*. According to Davidson, *intentional* action is such a one that is performed *for reasons*, the one which is based on a *primary reason*.

Obviously, in everyday life and quotidian conversations we justify what we do briefly and in a stylistically multiple way. Usually we do not present the both constituents of the primary reason, but restrict ourselves to mentioning only one, mostly the belief. When pouring a shot to a friend, it is enough to say "It will soothe your nerves" (*belief*) without supplementing it by the phrase "I want to do something to soothe your nerves" (*attitude, desire*) (Davidson, 2001a, p. 17); similarly, when I am going out and taking an umbrella, I can say "It's raining" without adding "I don't want to get wet". What is important, Davidson emphasizes, is that it would be possible to construct, or better *re*construct, a relevant practical syllogism:

> Corresponding to the belief and attitude of a primary reason for an action, we can always construct (with a little ingenuity) the premises of a syllogism from which it follows that the action has some 'desirability characteristic'. Thus there is a certain irreducible – though somewhat anaemic – sense in which every rationalization justifies: from the agent's point of view there was, when he acted, something to be said for the action (Davidson, 2001a, p. 17).

That any action comes about, one may say, it is only virtue of the logical, deductive connections between a relevant belief, pro-attitude and the action itself, irrespective of our capability of expressing it in a shortened, abridged way in the ordinary discourse.

3. BRANDOM'S MATERIAL APPROACH

Robert B. Brandom, despite drawing largely on Davidson's analysis as to the possibility of "rationalization" of an intentional action, opposes him as to the question of formal validity of practical inferences. Therefore, according to Brandom, the correctness of practical inferences does not depend, in the last resort, on formal properties of such inferences. What they are ultimately based on is the inferential materiality of reasoning concerning action.

2.1 The concept of material correctness of inferences

How to understand the mere concept of material correctness? The question is closely tied with two others issues: the normative account of discursive practice and the expressivist vision of logic. Both are described and rendered in Brandom's specific nomenclature. Hence, to explicate the notion of materiality will require from us to introduce both the Brandomian vocabulary and to sketch out his expressivist view on formality of logic.

The discourse and our dealing with any contentful utterances is possible thanks to the normative character of the linguistic practice. This normative aspect is manifested by two deontic statuses that come along any advanced assertion, i.e.: *commitment* and *entitlement*. In *Making It Explicit* Brandom sees it as follows. When I claim something, I thereby commit myself to what I am saying, and I do it paradigmatically not, or not only, to myself, but primarily before others. The other speakers, when talking to me, also take some normative stance towards my claim by ascribing me *commitment* to it, which means by treating me as committed to what I have said. Now, what does it specifically mean to be committed to a claim? This means my being obliged to justify my claim, to give some acceptable reason for the claim. If such a reason is sound and convincing in the eyes of my interlocutors, then they also treat me as entitled to commitment: they ascribe me entitlement to it. Such a process of attributing to each other deontic statuses is indispensably social in its core. For while committing myself to an assertion, I also commit myself, and am treated so, to all the inferential consequences of my saying – even those I am not aware of. Some of the consequences that are unknown to me I learn just from others. And the same refers to the entitlements: some of them I may learn from my interlocutors, and, conversely, some of what I find to be sound entitlements I will possibly have to relinquish after exchanging views with different speakers. Thus, the contentfulness of my assertion is dependent on, and constituted by, the claims that follow from it as well as the claims that justify it. And this is, in the last resort, the matter of social exchange of reasons (Brandom, 1998, pp. 157-163).

There is another sort of "hybrid", or heterogeneous, deontic status[2] – the one called *incompatibility.* In the above perspective it is possible that one undertake many different commitments. These do not have to be perfectly consistent and, given the fact that we are simply not always aware of every consequence of the view we declared to hold, actually the set of our beliefs is not consistent in every detail. This is where the incompatibility takes place, which is defined by Brandom as follows: "to say that two claims are incompatible is to say that *if* one is committed to the first, *then* one is not entitled to the second" (Brandom, 1998, p. 189). Here I treat this incompatibility relation as deontic status in this sense that when I am in the situation in which I maintain two incompatible beliefs, I am, in principle, obliged to withdraw one of them to make them conformed with each other.

What is typical of such an account is that the move from one assertion to another can proceed without any logical vocabulary, not least resorting to logical formal patterns. All what happens in inferring between two assertions is the practical move of attributing commitments to the speaker, say, from "This is a dog" to "This is a mammal". The move in question is correct, or incorrect, as it stands, in virtue of our ascribing commitments and vindicating entitlements. In a word, the source of the correctness of our inferences is normative and social "practices of giving and asking for reasons" (Brandom, 1998, p. 141) we play while speaking with each other. The proprieties of what we do when we play this game are, according to Brandom, embedded in our practice. Accordingly, these inferences, if correct, are *materially*, not formally, correct.

Then, what is the role of typical logical vocabulary and formal schemata? Their role comes down to expressing in a tidy and clear form the proprieties implicit in practice, to make them explicit. After introducing conditionals and other logical connectives, quantifiers etc., we are usually able to arrange our correct reasoning in a form of logical laws. But the primordial source both of the correctness of our inferences and, at a later stage, of the mere logical connectives is the practice of giving and asking for reasons[3] (Brandom, 2001, p. 85).

---

[2] It is an abridged way of speaking (that incompatibility is a sort of deontic *status*) I resort to here. To be precise, one would have to say that incompatibility is rather an inferential *relation*, not a status.

[3] It is another issue to explicate how formal logic can emerge from our discursive practice. Brandom's strategy aims at showing that the basic logical vocabulary can be defined in terms of the material inferences while the converse way is impossible (Brandom, 2001, p. 85-87).

Here, a formalistically oriented person would oppose Brandom and claim that, in the last resort, every good inference is a logical one, since material inferences can be treated as enthymematic: after supplementing the material reasoning with a relevant hidden premise, it simply turns into formally valid. One would say that the material move: from "It is a dog" to "It is a mammal" has an additional, enthymematic premise "If it is a dog, then it is a mammal", and, as a result, it assumes the logical form of *modus ponens*: $[p \wedge (p \rightarrow q)] \rightarrow q$. Brandom agrees that in thinking about inferential correctness one may choose one of the two options (i.e. to give priority to material or formal one), however he argues that the material correctness is more fundamental. He says explicitly:

> We need not treat all correct inferences as correct in virtue of their form, supplying implicit or suppressed premises involving logical vocabulary as needed. Instead, we can treat inferences such as that from ›Pittsburgh is to the west of Philadelphia‹ to ›Philadelphia is to the east of Pittsburgh‹, or from ›It is raining‹ to ›The streets will be wet‹, as *materially* good inferences – that is, inferences that are good because of the content of their *non*logical vocabulary (Brandom, 2001, p. 85)

This general approach is also extended on our practical reasoning.

2.2 The materiality of inferences concerning actions

While setting out his position in *Making It Explicit*, and later in *Articulating Reasons*, Brandom considers as paradigmatic three different examples of practical reasoning:

1. Only opening my umbrella will keep me dry, so I *shall* open my umbrella.
2. I am a bank employee going to work, so I *shall* wear a necktie.
3. Repeating the gossip would harm someone, to no purpose, so I *shall* not repeat the gossip (Brandom, 2001, pp. 84-85)[4].

The word "shall" is here important and especially relevant to practical inferences. "Shall", as Brandom stresses, "expresses the significance of

[4] Brandom here refers to Wilfrid Sellars' *Thought and Action* (see Brandom 1998, p. 679).

the conclusion as the acknowledging of a practical commitment" (Brandom, 1998, p. 245), i.e. expresses someone's undertaking an action, or being about to perform an action; while the weaker "will" may express the intention of our only conditionally undertaking a future action.

The above-mentioned examples, if considered in Davidsonian perspective, are valid only if they are appended with so far hidden, enthymematic premises that express a suitable pro-attitude. In view of that, one should have to present, respectively, the following, enthymematic premises:

1. Let me stay dry.
2. Bank employees are obliged (required) to wear neckties.
3. It is wrong to (one ought not) harm anyone to no purpose.

Yet, Brandom claims that there is no need to do this, at any case such an operation is only optional (Brandom, 2001, p. 85). The inferences in question are good/correct as they stand and may be treated as *materially valid*: valid not on account of their logical form, but rather on account of the content they confer. This content, in Brandom's account, is constituted in the practice of "giving and asking for reasons", of undertaking and ascribing commitments, of requiring from someone and their showing entitlements etc. Accordingly, in the practice of undertaking practical commitments (e.g. "I *shall* open my umbrella") and of displaying our entitlements to these commitments (e.g. "Only opening my umbrella will keep me dry") the shift from the entitlement to the commitment is legitimate and *materially* correct. Then, the supplementing it by so far unknown enthymematic premise, and making them explicit, does not contribute to its implicit correctness. What such a supplementation actually does is that it *expresses* its correctness in a "logically" clear way. Accordingly, the normative, volitional or evaluative vocabulary, embedded in Davidson's pro-attitudes, such as "want", "desire", "prefer", "be required", "be obliged", "ought to", "should" etc., are *expressive* in its character and function: they make *explicit* what is implicit in *practice*, they *express* the practical, material correctness in a tidy, deductive way of Davidson's practical syllogism.

An advocate of Davidson would undoubtedly say – which Brandom is aware of (Brandom, 2001, p. 87) – that the example 1. "Only opening my umbrella will keep me dry, so I *shall* open my umbrella" is, as it stands, incomplete, because the conclusion/action would not have flown from the premise for someone who wants to get wet; like Gene Kelly and Fred Astair in the movie *Singin' in the Rain*. It is true. Still –

Brandom would say – the fact that we are able to appendage this inference with the expression of desire of getting wet in the rain does not mean that this desire was not already, implicitly, embedded in this reasoning.

But the problem, for Brandom, is more basic. It comes down to the fact that material inferences, both theoretical (doxastic) and practical ones, are *nonmonotonic* – in opposition to the formal ones that are *monotonic*. Then, even if $p \rightarrow q$ is materially correct, it does not have to mean that $(p \wedge r) \rightarrow q$ or $(p \wedge r \wedge s) \rightarrow q$ are also materially correct. When some additional circumstances appear, then they may change the correct material inference into incorrect (in fact the latter would be another material inference), which would not deny the material correctness of the former. For example, the following one is surely correct (see Brandom, 2001, p. 88):

1. If I strike this dry, well-made match, then it will light; $p \rightarrow q$.

However, if some new circumstances appear, the conclusion q once will ensue, another time not:

2. If p and the match is in a strong electromagnetic field, then it will *not* light; $(p \wedge r) \rightarrow \sim q$.
3. If p and r and the match is in a Faraday cage, then it will light; $(p \wedge r \wedge s) \rightarrow q$.
4. If p and r and s and the room is evacuated of oxygen, then it will *not* light; $(p \wedge r \wedge s \wedge t) \rightarrow \sim q$.

Indeed, a strong proponent of formalism would further defend deductiveness by invoking *ceteris paribus* clauses, saying: "If I strike this dry, well-made match, then, all things being equal, it will light". But for Brandom that would not change anything. With *ceteris paribus* clause the sentence would simply mean: "q follows from p unless there is some *infirming* or *interfering* condition", which, in turn, means: "q follows from p except in the cases where for some reasons it doesn't" (Brandom, 2001, p. 88). In his view, *ceteris paribus* clause should be understood here as a factor that explicitly illustrates *nonmonotonicity* of material inferences instead of the factor that almost magically, *deus ex machina*, turns them into formal and monotonic ones (Brandom, 2001, pp. 88-89).

Thus, again, what Davidson calls pro-attitudes and what allows him to think about practical inferences in a deductive, formal way, this for Brandom plays purely the *expressive* role – it helps to state materially corrected, previously endorsed inferences in a clear logical way. As Brandom declared:

> Normative vocabulary (including expressions of preference) makes explicit endorsement (attributed or acknowledged) of material proprieties of practical reasoning (Brandom, 2001, p. 89).

Pro-attitudes allow articulating as assertions someone's endorsed desires, obligations, patterns social behaviour (the example with the "bank employee") or moral rules (the example with "repeating gossips").

4. CONCLUSION

The above discussion was aimed at showing some main controversies between Davidson's formal and Brandom's material (inferential) approach to inferences concerning action. In juxtaposition of the two stances, I argued that such inferences are better intelligible when thought of as material correct or incorrect, that the mere process of their formalization is, in an important sense, artificial; this artificial character of formalization is especially clear when we try to apply *ceteris paribus* clause to make them formally valid and thus to change nonmonotonic reasoning into monotonic one. As a result, the material account of correctness was favoured over the formal.

REFERENCES

Brandom, R. B. (1998). *Making It explicit.* Cambridge, MA: Harvard University Press.

Brandom R. B., (2001). *Articulating reasons.* Cambridge, MA: Harvard University Press.

Davidson, D. (2001). *Essays on actions and events.* New York: Oxford University Press Inc.

Davidson D. (2001a). Actions, reasons and causes. In D. Davidson (Ed.), *Essays on Actions and Events* (pp. 12-25). New York: Oxford University Press Inc.

70

Is Dialogue the Most Appropriate Model for Argumentation?

DAVID ZAREFSKY
Northwestern University (Emeritus), USA
d-zarefsky@northwestern.edu

Many contemporary approaches, including informal logic and pragma-dialectics, model argumentation as interpersonal dialogue. This presumes interactivity, permits commitment-based reasoning, and reduces ambiguity of positions. These conditions are lacking in more complex public argument, which ranges from speakers addressing large audiences, to mass media argumentation, to circulation of arguments through a culture. The paper will consider two alternative models, argumentation as public address and as debate, and will inquire whether interpersonal and public argumentation differ fundamentally.

KEYWORDS: argument models, debate, dialogue, discussion, interpersonal communication, normative argument, personal sphere, public sphere

1. INTRODUCTION

The inaugural edition of a new conference seems an appropriate place to reopen a fundamental question about the nature of argumentation.

The most important contribution of Stephen Toulmin's *The Uses of Argument* (1958), almost 60 years ago, was not the model that commonly bears his name. Rather, it was the argument that set up the model and established the need for it. As Toulmin noted, the then-reigning model came from formal logic. It was the analytic syllogism of the form "All A's are B's; all B's are C's; therefore, all A's are C's." Christened "Barbara," the syllogism featured a conclusion that followed from the premises with certainty. Such an ironclad inference was seen as the goal toward which all arguments should strive, and other kinds of argument should be criticized for the degree to which they fell short of the goal. Even everyday practical arguments were understood as

D. Mohammed & M. Lewiński (eds.) (2016). *Argumentation and Reasoned Action: Proceedings of the 1st European Conference on Argumentation, Lisbon, 2015. Vol. II, 1081-1091.* London: College Publications.

"applied formalism" (Cox & Willard, 1982, pp. xxii-xxv) and strove for the standards of the deductive syllogism even though that was an unattainable ideal.

Not so fast, Toulmin insisted. The analytic syllogism is not just a distant goal, he said, but a highly atypical form of argument. Other argument patterns differ from it not just in degree but in kind. They cannot meet its standards because they do not share its features. To make it our goal therefore assumes that we will have much fault-finding to do. We will not be able to establish standards of reasonableness for everyday argumentation or for argument in such specialized fields as ethics or politics. Dismissing these arguments from the category of the necessary, we will have no way to rescue them from the arbitrary. Yet such a result would offend common usage: both specialists and ordinary arguers do make claims all the time, do offer grounds for them, and do regard some grounds as better than others. We need a model that will allow us to describe and evaluate these activities rather than to consign them to irrelevance. And so the Toulmin model was born, eventually to be inscribed in textbooks in public speaking and introductory composition, although that was not its creator's purpose.

2. THE DIALOGUE MODEL AND ITS ASSUMPTIONS

My goal is not to defend or to criticize the Toulmin model; I will not assess whether it was necessary or whether it achieved its goal. Rather, I want to advance an argument analogous to the one Toulmin brought against the model of the analytic syllogism. My target is what we have often taken, implicitly or explicitly, as the normative model of argumentation: the interpersonal discussion. We have moved beyond our forebears of 60 years ago in that we do not regard argumentation only as a propositional structure but also as a mode of communication. Yet we have chosen as our model a particular instantiation of that communicative activity and made it a near-universal norm, just as Toulmin thought the formal logicians had done with the analytic syllogism. I am not going to suggest a different universal model, but will discuss what the discussion model leaves out and will propose an alternative for that.

Let me first establish that the model of an interpersonal discussion is prevalent. It is, of course, the normative model of pragma-dialectics, where the critical discussion – rule-governed interaction to resolve disagreements on the merits – is the ideal (van Eemeren & Grootendorst, 2004, pp. 42-68). It is the organizing principle of Douglas Walton's approach to informal logic (for example, Walton, 1995, pp. 18-26). He sees reasonableness related to context but, in identifying possible contexts, he spells out different types of dialogue – persuasion dialogue, information-seeking dialogue, eristic dialogue, etc. He also uses the term "critical discussion," although he does not mean exactly the same thing as van Eemeren and the pragma-dialecticians. But they do share the assumption that the paradigm of argumentation is face-to-face discussion between interlocutors. Democratic theorists (for example, Mansbridge, 1980; Leib, 2004) hold that democracy is about keeping the conversation going. Sometimes they refer to a conversation with many participants. Even so, however, this is a model of multiple simultaneous dyadic interactions. Likewise, some of the approaches to artificial intelligence presuppose a model of a protagonist and antagonist responding to each other's moves; this is the technological application of interpersonal encounters. Some of the early entries in the deliberative democracy literature hold that conversation is the essence of democracy (for example, Oakeshott, 1962). While they are not strictly limited to bilateral dyadic encounters, they build on that foundation. Even the influential American debate textbook by Ehninger and Brockriede (1963), which otherwise is grounded in different assumptions, begins its account of "the anatomy of a dispute" with the example of Mr. A and Mr. N each trying to convince the other in a dialogue that he should enjoy the small spot of shade as they both wait for a bus on a hot afternoon. And, for a final example, I recently reviewed a manuscript submitted to one of our major journals, in which the writer asserted matter-of-factly that *the* purpose of argumentation was "to convince your opponent," a statement that is true, clearly, only in an encounter between two people. These examples are hardly exhaustive, but they should make the point that argumentation often is modeled on an interpersonal dialogue in which each party seeks to convince the other.

Like any model, this one rests on certain assumptions or presuppositions, and the utility of the model depends on the widespread acceptance of its underpinnings. First, it assumes that the arguers and the evaluators are the same people. One's argument prevails if one's interlocutor yields his or her own position in the face of what he or she regards as a stronger argument. The adversary prevails if one yields one's own position for the same reason. Second, the process is interactive. Each party responds to claims made by the other. Interlocutors know what they are each committed to because, along the way, they indicate agreement, disagreement, or doubt. Parties can base their own reasoning on their opponent's implicit or explicit commitments. Reasoning proceeds as if it were deductive *for the particular person* to whom the argument is addressed. Third, argument is iterative. An advocate's entire case is not presented at once; each party has the opportunity to respond to the other at every point along the way, and an advocate can modify the presentation based on the interlocutor's responses, much as a computerized testing program will alter the difficulty level of a subsequent question based on whether the previous question was answered rightly or wrongly. And fourth, the disputants are assumed to being sincerely committed to resolving their difference of opinion. It is assumed that they mean what they say and are not taking positions playfully or out of a desire to frustrate the adversary. They are presumed to devote their time and energy to an uncertain interaction because they are genuinely committed to resolving a disagreement, whether because of its own merits, or for the sake of their relationship, or both.

These four assumptions – that arguers and evaluators are the same people, that the communication is interactive, that the presentation is iterative, and that resolution of disagreement is the goal – characterize interaction in what Goodnight (1982) has called the personal sphere. This is the sphere in which the argument engages and affects only those who engage in it, and in which they determine and apply standards for what counts as proper procedure and what counts as resolution of the disagreement. Others may be interested in the outcome, but no one else has standing to participate. Arguments between spouses, between intimate friends, between parents and children, and between team members are examples of argumentation in the personal sphere. Certainly it is far more common than is the analytic syllogism in logic. And certainly the model of argumentation as a kind of interpersonal dialogue fits it well. But is it prototypical of *all* argumentation, or sufficiently so that it can serve as a model for all argumentation?

3. QUESTIONING THE ASSUMPTIONS

Here there is reason for caution. One easily can imagine situations in which the assumptions underlying argumentation as interpersonal communication do not hold up. Often the evaluators of argumentation are different from those who participate in it. Examples come immediately to mind: mediation or arbitration in which a disagreement is settled by a third party; legal disputes in which a difference is settled by a judge or jury trying facts or applying law, election campaigns in which competing candidates appeal for votes from the general public, mass media campaigns in which the creators of argumentative messages seek affirmation or consumer response from a vast audience who did not participate in the campaign's design, visual argumentation in which an artist might be understood as making an argument and an amateur or professional critic, as evaluating it.

Second, much argumentation is not interactive in the sense that both parties express commitments and address the other's commitments. Each "party," in fact, could be a distillation of millions of people, and the cacophony of their discourse may prevent attention, much less response, to the expression of any particular commitments. This situation is even more likely in the case of what Lewiński (2014) has called "polylogues," discourses in which there are not two positions but many that are being tested, and in which they are engaging one another not in any sequential fashion but all at the same time.

Third, in many cases argumentation is not iterative. An arguer may need to present his or her entire case and only then find out how the interlocutor will respond. U.S. Supreme Court hearings are characterized by frequent interruptions and dialogues between individual justices and the attorneys appearing before them. But argumentation in newspaper editorials and letters to the editor, for example, are presented in their entirety. Only then might a response be forthcoming, and the response might or might not engage what the arguer regarded as the principal commitments. This is also true of lectures, presentations at academic conferences, many legislative debates, argumentative essays, and visual arguments. In these situations, it is necessary to anticipate an interlocutor's commitments and relevant concerns and to design one's own arguments with this image of the adversary's argumentative ground in mind. Whether one assesses the other correctly or not can have huge consequences yet can be known only after the fact. In these circumstances there is no opportunity to modify one's selection of arguments, or any of one's strategic maneuvers, based on the other's response.

Fourth, there are many instances of argumentation in which resolution of differences is not the principal goal and may not be the goal at all. When there is a third-party evaluator, in fact, arguers are very seldom trying to resolve the disagreement between themselves. They both think they are right and they seek the agreement of the third party. The party that prevails will consider itself vindicated. The "losing" party might re-examine its own position and conclude that the "winner" was correct after all. But it is just as likely that the losing party will express dissatisfaction with the decision, object after the fact to the decisor, resolve to strengthen his or her position so that it might prevail another day, or withdraw from the contest without admitting error.

Clearly, resolving disagreements is not the only reason people argue. Sometimes they seek to *clarify* differences of opinion by setting the competing positions side by side and identifying the *stasis* of the disagreement. Sometimes they seek to *intensify* differences of opinion by becoming more committed to their position in the act of arguing it before others. One defense of freedom of speech, for example, is that when one's views are challenged, one is called upon to defend them. This can result in hardening commitments which otherwise have been maintained casually or only as a result of prejudice. Sometimes people argue to further *polarize* differences of opinion, if they are worried about the prospect of a premature, incoherent, or politically damaging synthesis of competing positions that otherwise might emerge. In my country, for example, when the Obama administration sought a bipartisan synthesis on health care by grounding its proposal in ideas previously championed by Republicans, the Republicans changed their position and attacked ideas they previously had supported, without ever acknowledging that they had changed their mind, in order to maintain political polarization that they thought would be more to their advantage than bipartisan agreement. So even if argumentation seeks to *manage* differences of opinion, it does not always seek to *resolve* them. It may clarify, intensify, or polarize them.

And argumentation is not even always about managing disagreement. As Walton's classification of dialogue types (Walton, 1995, pp. 18-26) reminds us, there are yet other reasons for which people argue. Argumentation can be a means of seeking information, for example. Or, as I suggested many years ago, it can be an epistemic method: by testing hypotheses, we can determine the probable truth of contestable claims that by their nature are not susceptible to empirical verification (Zarefsky, 1979). Or we sometimes engage in argumentation for purely eristic reasons: for mental discipline, to exercise the rhetorical canon of invention, to play the devil's advocate "for the sake of the argument," or to value dissensus as an end in itself.

Each of the four assumptions underlying the model of argumentation as interpersonal dialogue, then, has been shown to have limitations when extended beyond the personal sphere. In addition, the very idea of conversation as the model for argument and deliberation in the public sphere is open to question. Although it is critical to some theories of deliberative democracy, Schudson persuasively maintains that this view is problematic (Schudson, 1997). Participating in public conversation is highly selective; many are driven away by fear of embarrassment, and there is no accounting for spectatorship in conversation. Furthermore, conversation frequently serves consummatory purposes for the participants, whereas the public sphere is about deciding and justifying practical action. And what makes the public sphere work, Schudson maintains, is not participating in a conversation but observing norms of civility and procedure (Schudson,

1997, p. 307). His article is a powerful warning against extending the normative model of interpersonal dialogue beyond the personal sphere and into the realm of public life.

4. CAVEATS

Let me be clear that I am not objecting to the interpersonal discussion normative model itself. It has helped to encourage and to organize substantial research in argumentation within the personal sphere. In particular, the model of the critical discussion in pragma-dialectics has nurtured a robust body of both descriptive research and normative evaluation, showing the heuristic value of a normative model in sustaining the most active research program in argumentation studies today. My objection is to extending this model beyond the personal sphere and making it the prototype for argumentation in general.

Nor am I prepared to offer an alternative universal model of argumentation. Although argumentation in any sphere is about the justification of claims, and although justification consists of providing reasons for the claims and making manifest that the reasons count as reasons, beyond that I am not convinced that we can say much about the process of arguing that is not bounded by context. Pursuing a universal model may be a worthy goal, but I for one do not think we have found it yet. Rather than supplanting the model of interpersonal dialogue, I believe that we should try to supplement it with a normative model that is more appropriate for arguments outside the personal sphere.

I also am aware that some have proposed that the dialogue model could be extended analogically beyond the context where it clearly fits, thereby making an additional model unnecessary. In the same way, Toulmin's early critics insisted that the syllogism could be reconfigured in order to encompass probabilistic and contingent arguments, making the Toulmin model unnecessary. So, for example, one could imagine regarding the argumentation in a televised presidential speech to the nation as just hundreds of millions of simultaneous critical discussions. One could do that, but it hardly characterizes the way in which the occasion is understood by its participants (with the possible exception of gifted communicators such as Franklin Roosevelt or Ronald Reagan, who had skills of empathy and could make viewers feel that they were being addressed individually). Moreover, few communications within the public sphere are dialogic, with the exception of events such as political interviews in which a large public audience essentially "overhears" a conversation between an interviewer and a subject that they both realize actually is conducted for the benefit of the viewing audience (for example, Andone, 2013). In fact, though, the public audience is addressed as a composite, and it is in its composite character that feedback is sought and obtained.

5. TOWARD A PARALLEL DEBATE MODEL

Rather than contort the model of interpersonal dialogue to make it fit the public sphere, it may be more sensible to try to develop a parallel model that is a more comfortable fit for public argument. While rhetorical considerations are present in interpersonal discussions, as is evident from the focus on strategic maneuvering, rhetoric finds its most traditional home in the public sphere, where collective decisions are required about exigent matters that are inherently uncertain. It would seem advisable, then, to develop a model of public argument that is rhetorically inflected. Because of rhetoric's focus, since Aristotle, on the available means of persuasion in the given case, rhetorical scholars often are reluctant to formulate generalizations that transcend the uniqueness of the particular case. Not to try, however, is to condemn rhetoric to the limited generalizability of the case study. Besides, rhetorical situations are not infinite; patterns of choice can be identified and theorized. And even individual case studies can profit from being evaluated by reference to a normative ideal that is adapted to context – analogously to the institution-specific studies of strategic maneuvering. This would help assure that rhetorical argumentation is evaluated with respect to rhetorical criteria, rather than the critic's personal preference or political ideology. That in turn should increase the replicability of rhetorical analysis and criticism.

With due trepidation and knowledge of the dangers involved, I want to hazard a beginning toward such a model. It takes as the unit of analysis the controversy and as the normative model of communication, the debate. A debate is an interaction in which opposed advocates present cases for and against a standpoint, and a third-party adjudicator (typically called a judge) determines which case is superior and hence what, on the basis of the arguments, he or she would believe or do. The debaters are not trying to resolve their difference of opinion; they remain committed advocates for their respective positions and the resolution takes place in the mind of the judge.

A debate involves a mix of competitive and cooperative motivations. Each debater wants the disagreement to be resolved in his or her favor, but each realizes that all the methods of argument that he or she uses will be available to the adversary as well. This combination gives each debater the incentive to find and present the strongest case possible, and this in turn enables the decision-maker to make a wise choice by exercising practical wisdom. Of course, one cannot be sure that the adversary will be sharp enough to spot one's weaknesses, or that the decision-maker will be free of bias. But the desire not to lose the debate, not to be proved wrong and to suffer loss of face, inclines the debater not to take the chance and instead to aim toward the highest standard. In this way, each debater's individual competitive motivation contributes to the common goal of making the best possible decision about matters that are uncertain.

Furthermore, a debate proceeds according to rules and norms, but they are themselves defeasible and subject to argument. Some rules are purely procedural, such as turn-taking and time limits. They can be modified, but not while a debate is in progress. Others are substantive, relating for example to the relative proof responsibilities of the competing parties. For example, in order to propose a solution, does one need to identify the cause of a problem? Or who gets the benefit of the doubt if the decision-maker is unable to decide a particular argument? Or when must a standpoint be proved in order to be carried, and when must it be disproved in order to be rejected? These are matters that the debaters will argue out, just as they argue the substance of their standpoints, and in fact normative and substantive disputes often will be intermixed. Strategic maneuvering will be employed to advocate the norms that will make it easier to prevail on the substantive matter in dispute. What preserves the rigor of the examination is the knowledge that an interlocutor is present and may be able to call one to account.

In a debate, multiple advocates represent each side of the standpoint, and the judge must select from an assortment of arguments the ones that ultimately are dispositive. This resembles argumentation in the public sphere, in which citizen-evaluators must assemble from fragments the case that they conclude is the more persuasive. If one wishes to simulate the cacophony of the public sphere, the image of a debate tournament, in which there are multiple simultaneous debates involving different issues and methods of refutation, comes close. This lacks the orderliness of an interpersonal discussion, but the public sphere is like that. The knowledge that one's adversary may deploy the same resources as oneself is what regulates the process and keeps it focused on the goal.

A debate can address only one standpoint at a time, and there usually are more than two sides to a question. But this is not a serious limitation. A debate is a yes-or-no choice with respect to X, not a choice between X and Y (unless, of course, the rejection of X entails commitment to Y). Affirming a standpoint implies commitment to it, at least for the time being, whereas rejecting a standpoint leaves other options open. Accordingly, one can imagine a series of successive debates which proceed until there is a standpoint to which the judge is prepared to commit. This represents in idealized form the process of considering multiple options until one settles on a policy concerning a given subject such as health care, economic growth, or foreign policy.

To be sure, I have described an idealized debate. To some degree, any actually existing debate will fall short of that ideal, just as any interpersonal argument will fall short of all the standards of critical discussion. But both models are aspirational, and both provide grounds for critique of argumentation that falls short. A debate model does for the public sphere what a discussion model does for the personal sphere but does not do so well universally.

6. CONCLUSION

This observation raises the obvious question of whether disagreements in the personal and public spheres are basically different, and what if anything unites them under the common rubric of argumentation. They are quite different, to be sure, since in one case the arguers and judges are the same people and the outcome affects no one other than the participants, while in the other case there is a third-party judge and the participants stand in for a larger public that is affected by the outcome. But the common thread is that both involve processes of justification of standpoints that are uncertain, and that this process involves relating standpoints to other claims that count as groundwork for them.

It is probably the case that argumentation in the public sphere is harder to systematize, because it is more unruly. This may be part of the reason, along with their commitment to the particulars of context, that rhetorical scholars of argumentation are reluctant to move beyond those individual case studies. But to resist systematization in principle is to deny the study of public argument the benefits that advances in computer technology and artificial intelligence could provide. The debate model, or some other more appropriate model, can be a point of departure, and nascent efforts at systematization should be encouraged, even though they represent only baby steps. There is plenty of work to do.

REFERENCES

Andone, C. (2013). *Argumentation in political interviews.* Amsterdam and Philadelphia: John Benjamins.

Cox, J.R., & Willard, C.A. (Ed.). (1982). *Advances in argumentation theory and research.* Carbondale: Southern Illinois University Press.

Eemeren, F.H. van, & Grootendorst, R. (2004). *A systematic theory of argumentation.* Cambridge: Cambridge University Press.

Ehninger, D., & Brockriede, W. (1963). *Decision by debate.* New York: Dodd, Mead.

Goodnight, G.T. (1982). The personal, technical, and public spheres of argument: A speculative inquiry into the art of public deliberation. *Argumentation and Advocacy, 18,* 214-227.

Leib, E.J. (2004). *Deliberative democracy in America.* University Park, PA: Pennsylvania State University Press.

Lewiński, M. (2014). Argumentative polylogues: Beyond dialectical understanding of fallacies. *Studies in Logic, Grammar and Rhetoric, 36*(1), 193-218.

Mansbridge, J.J. (1980). *Beyond adversary democracy.* Chicago: University of Chicago Press.

Oakeshott, M. (1962). *Rationalism in politics.* New York: Basic Books.

Schudson, M. (1997). Why conversation is not the soul of democracy. *Critical Studies in Media Communication, 14,* 297-309.

Toulmin, S. (1958). *The uses of argument.* Cambridge: Cambridge University Press.

Walton, D.N. (1995). *A pragmatic theory of fallacy.* Tuscaloosa: University of Alabama Press.

Zarefsky, D. (1979). Argument as hypothesis testing. In *Advanced debate,* D. Thomas (Ed.). Skokie, IL: National Textbook.

71

Meta-Reasoning in Making Moral Decisions under Normative Uncertainty

TOMASZ ŻURADZKI
Institute of Philosophy, Jagiellonian University, Poland
t.zuradzki@uj.edu.pl

In my paper I analyze recent discussions about making moral decisions under normative uncertainty. I discuss whether this kind of uncertainty should have practical consequences for decisions and whether there are reliable methods of reasoning that deal with the possibility that we are wrong about some moral issues. I defend a limited use of the decision theory model of reasoning in cases of normative uncertainty.

KEYWORDS: bioethics, decision theory, metaethics, moral uncertainty, the ethics of war

1. INTRODUCTION

The simple fact is that even when we feel quite certain about moral issues, we are susceptible to mistakes. For example, we have to act in the face of uncertainty about the facts, the consequences of our decisions, the identity of people involved, people's preferences, moral doctrines, specific moral duties, or the ontological status of some entities (belonging to some ontological class usually has serious implications for moral status). I want to analyze whether these kinds of uncertainties should have practical consequences for actions and whether there are reliable methods of reasoning that deal with the possibility that we understand some crucial moral issues wrong.

The most promising approach is to try to extend the decision theory model of reasoning to encompass normative uncertainty (Lockhart, 2000). But this model, when used to guide our action in moral terms, is highly controversial. It assumes that in the face of normative risk or uncertainty we are rational if and only if we maximize expected value (whatever it is). In this case two things would determine what we ought to do under normative uncertainty: 1) the probabilities

D. Mohammed & M. Lewiński (eds.) (2016). *Argumentation and Reasoned Action: Proceedings of the 1st European Conference on Argumentation, Lisbon, 2015. Vol. II, 1093-1104.* London: College Publications.

assigned to the various normative views; 2) the differences in values between the available actions, according to each of those views. This approach – if successful – would have some interesting applications both in metaethics (e.g. rejecting nihilism, see: Ross, 2006) and applied ethics in particular the ethics of war and bioethics (permissible killing people or animals, abortion, embryo research, see: Guerrero, 2007; Moller, 2011; Friberg-Fernros, 2014; Żuradzki, 2012, 2014). Moreover, the argument (from moral or normative uncertainty) is also used in other contexts (Henning, 2015).

The main problem that the supporters of this approach have to deal with is the question of intertheoretic comparisons of value. Normally, when we use expected utility calculus we use a common scale by which it is possible to measure the values attached to different outcomes (this is one reason why so many examples refer to money). But there doesn't seem to be any way of making this kind of intertheoretic comparison of moral values between different theories or doctrines. Moreover, recently a few philosophers presented other arguments against the attempts to extend expected value maximization style reasoning to encompass moral uncertainty (MacAskill, 2013; Weatherson, 2014; Harman, 2015; Nissan-Rozen, 2015, Hedden, forthcoming).

2. MANY FACES OF UNCERTAINTY

There are many ways in which you can be uncertain about the morally important aspects of your action. For example, you could be certain that some action can harm some people (or other beings with the moral significance), but uncertain about the identity and the number of people involved.

A combatant who fires indiscriminately from his gun at a place inhabited by noncombatants is in this position. He knows that there is a serious risk that some innocent people could be fatally shot, but he knows neither their identity nor number. Alternatively, you can know that there is a serious risk that some action will harm a person and you can know who is this person. Someone who plays Russian roulette with a prisoner of war is in this position.

Finally, you can be sure that some action will harm some identified beings, but not sure about its ontological or moral status. The real-life examples of beings with uncertain ontological or moral status (at least from the perspective of some people) include fetus, human embryos (in particular at the very early stages of development) or some products (real or only possible) of genetic engineering. Some philosophers have also argued that it is the case of at least some animals. Analogically to the two above examples, one could be tempted to say that someone who aborts fetus, destroys human embryos (during in vitro procedures or during scientific research) or kills animal is in a similar position as soldiers from the two above examples.

In all these cases there is a risk that something that has moral status will be harmed. So the popular argument says that in all these three cases there are strong reasons for agents not to act: not fire indiscriminately at area inhabited by noncombatants, not play Russian roulette with a prisoner of war, not harm human embryos or animals. And this popular argument adds that it is also a reason to condemn such actions no matter what their results are. It is wrong - says this argument - to impose a risk to someone who is not liable to be exposed, even if a potential victim is actually not harmed (review article about the problem of risk imposition: Hayenhjelm & Wolff, 2012). This argument - presented very often by Catholic preachers or scholars (but also, surprisingly, by defenders of animals rights) - treats all three above cases in a very similar way and is used to argue against the permissibility of abortion or destruction of early human embryos. It is usually presented in a form an analogy to hunting:

> *Example 1: Deer hunting*
> If I am hunting with a rifle, and I see something move in the trees but am unsure whether it is a deer or a person, I am obliged not to shoot until I establish that it is in fact a deer: better safe than sorry (Shaw, 2008, p. 219).

Catholic preachers or scholars (and vegetarians) argue that when someone is unsure whether some being has a full moral status or not, he should be obliged not to kill it. Since it is impossible to resolve empirically whether or not the target has full moral status, obligation not to kill is not time-limited. According to this view (see for example: Friberg-Fernros, 2014) the same argument can be applied to human embryos (in particular in early stages): since there are reasonable doubts about their personal status, morality requires that human embryos from conception be treated as persons.

3. FACTUAL UNCERTAINTY AND THE ETHICS OF WAR

In this chapter I will use some examples from the ethics of war to demonstrate the difference between cases of factual and normative uncertainty. In a war context the question of risk imposition is discussed usually in this kind of context:

> *Example 2: A security checkpoint*
> Imagine that you are a soldier, ordered to protect a military or diplomatic convoy as it passes through hostile territory, and you see a car stopped by the side of the road ahead. Or imagine that you are stationed at a security checkpoint and a car approaches despite signs and warnings directing it to stop. The occupants of the car may be civilians, but they also may be irregular forces waiting to attack (Haque, 2014, p. 65).

So the question in this case is how certain should be a soldier that people at the front of him are combatant, rather than noncombatants, before using deadly force? Surely, soldiers are in a different position than hunters, because the stake is much bigger: they risk their lives and they risk the case for which they fight. In a hunter case there is hardly anything valuable in killing a deer (except from hunter's pleasure, since I assume that it is not necessary hunting). In this sense a soldier's case is more similar to some cases of abortion or embryo research, where the stake also can be quite high (the well-being or health of woman; development of science during embryo research).

The obligation not to kill civilians is well-established in international law and international theory. For example Protocol Additional to the Geneva Conventions says:

> In order to ensure respect for and protection of the civilian population and civilian objects, the Parties to the conflict shall at all times distinguish between the civilian population and combatants and between civilian objects and military objectives and accordingly shall direct their operations only against military objectives (Protocol, 1977, art. 48).

So firstly, all participants in conflicts must determine whether an aim is legitimate or illegitimate target: they must distinguish between combatants and noncombatants (I will assume that combatants are people who are directly involved in hostilities, some of them can be civilians). And secondly, it is permissible to target only combatants (or other military objectives). Of course this description is highly idealized, and I we have seen in *A security checkpoint* case very often it unclear whether the target of operation are combatants or not. Surprisingly, international law proposes a very similar rule to this one that was proposed in *Deer hunting* case:

> In case of doubt whether a person is a civilian, that person shall be considered to be a civilian (Protocol, 1977, supra note 9).

> Those who plan or decide upon an attack shall (...) do everything feasible to verify that the objectives to be attacked are neither civilians nor civilian objects and are not subject to special protection but are military objectives (Protocol, 1977, supra note 5).

This point of the Protocol does not specify any the level of care with which a soldier must try to distinguish combatants from noncombatants. Neither this point specify how much effort should be put in verifying (soldiers should only do "everything feasible"). In its literal interpretations it means that if there is any doubt about the nature of a target proceeding is prohibited. It would mean for example that in cases like *A security checkpoint* when there is any doubt whether in an approaching car are noncombatants, soldiers are not allowed to use deadly force.

In many conflict situation this level of civilian protection would be too restrictive and some Western countries entered reservation about this provision (for example the UK states that it "applies only in cases of substantial doubt" - Declaration 2002). Haque noticed that despite literate meaning of this regulation in the literature or commentaries the most common approach to this problem is the balancing approach according to which "both the required level of certainty and the required level of risk vary with the balance of military and humanitarian considerations" (Haque, 2012, p. 63). This approach is visible in Walzer's *Just and Unjust Wars* when he writes about combatants' obligations:

> The degree of risk that is permissible is going to vary with nature of the target, the urgency of the moment, the available technology, and so on (Walzer, 2000, p. 156).

It is also proposed by commentators who underline that combatants should "balance" possible benefits and risk to civilians:

> The reasonable care rule is disquieting. It vests belligerents with considerable discretion in multifaceted balancing and legitimizes even large-scale injury to innocent civilians under certain circumstances (Waxman, 2012, p. 1393).

So the most common approach among scholars is quite contradictory to the literal meaning of the Protocol: now one expects that soldiers can act only if he is 100 percent certain in all cases. In some circumstances solders can attack even in the face of substantial doubts about the status of target (when the military stakes are very high), and in some circumstances soldiers must abstain from attack in the face of even very slight doubts.

It can be visible in the next hypothetical example, which is slightly modified version of the Example no. 1 *A security checkpoint*:

> *Example 3: A security checkpoint (modified)*
> Imagine that you are a soldier, ordered to protect a military or diplomatic convoy [consisting of: a) 5 people; b) 20 people] as it passes through hostile territory, and you see a car stopped by the side of the road ahead. Or imagine that you are stationed at a security checkpoint [consisting of: a) 5 soldiers; b) 20 soldiers] and a car approaches despite signs and warnings directing it to stop. The occupants of the car may be civilians, but they also may be irregular forces waiting to attack.

In this case we modified number of solders endangered: in the a) cases there are only 5 endangered soldiers, in the b) cases there 20 endangered soldiers. I assume that if the occupants of the car are irregular forces they will want to kill all soldiers. According to the balancing approach the way in which soldiers should precede depends on the number of solders endangered: they are permissible to act even if they have more serious doubts in the b-type cases than in the a-type cases.

There are serious objections to the balancing approach in case of the ethics of war (Haque, 2012), but I am not going to evaluate them. Instead, I want to demonstrate that this kind of balancing approach is even more difficult to use in the cases of moral uncertainty.

4. MORAL UNCERTAINTY AND BALANCING APPROACH

Let me start this part with a typical example discussed in the literature.

> *Example 4: Meat eating*
> Suppose, for example, that an agent is uncertain between two views about the morality of eating meat. In one view, eating meat is tantamount to murder; it is much, much worse, then, to eat meat than to abstain from it. In another view, it is ever so slightly better to eat meat than to abstain – better, perhaps,

> for reasons of health or pleasure. In the most plausible views of rationality under moral uncertainty, it is rational to avoid eating meat, even if one's belief in the second view is slightly higher (Sepielli, 2013, p. 581).

At first sight it seems that this example is very similar to the case of *Deer hunting*: on the one hand we can act in a risky way – we can have a small benefit (a deer is ours; we eat a nice meal), but there is a serious possibility that we do something very wrong: we kill a person; or we kill a being that have full moral status – the same or almost the same as an adult person. So it may seem that rationality requires to do the action with "the highest expected moral value". What does it mean? "An action's expected moral value is the probability-weighted sum of its moral values according to the various moral views or theories" (Sepielli, 2013, p. 581). Moreover, it seems that any plausible theory of rational decision making under uncertainty will care about moral stakes of decisions according to different moral theories. So, below I demonstrate that the expected moral value solution is only one of many possible solutions that were proposed recently.

It is worth noting that other examples related with decision-making under normative uncertainty (abortion, destroying early human embryos) are more complicated than meat eating because there are important values not on one side, but on both (so they are more similar to the security checkpoint cases). Abortions are usually defended because of some important values at stake (well-being or health of a woman, the low quality of life of possible child). The same is with destroying of early human embryos either because of reproductive purpose (in vitro fertilization) or because of scientific research. Below we will see the importance of this difference.

4.1 My Favorite Theory and its problems

Probably the most obvious proposition how to act under normative uncertainty is My Favorite Theory approach. It says that "a morally conscientious agent chooses an option that is permitted by the most credible moral theory" (Gustafsson & Torpman, 2014, p. 159). Even if you have doubts about meat eating, you are permitted to eat it if you believe that theory according to which animals have not any important moral status is more reliable than any other theory about animal status.

Although this approach looks very intuitive, there are interesting counter-examples. Consider the following case (adopted from (Gustafsson & Torpman, 2014) in which you can either choose action A or action B, and your credences are divided between two moral doctrines MD1 and MD2. You think that MD1 is slightly more reliable than MD2 (I introduce some numbers for convenience, for example that your credence in MD1 is 0,6 and your credence in MD2 is 0,4, but of course in real life this kind of precision is not necessary).

	A	B
MD1 (0,6)	Slightly wrong	Merely OK
MD2 (0,4)	Saintly	Morally terrible

Table 1

The descriptions of the results show the moral evaluation of possible outcomes (wrong, merely OK...). In this kind of cases it seems that despite the fact that an agent thinks that the moral theory MD1 is more reliable than MD2, she should act as if the theory MD2 were correct or right moral doctrine. It would mean that she should prefer action A over action B – against My Favorite Theory. Why? Because – analogically to our previous case of *Meat eating* – an agent may prefer not risking any serious moral wrongdoing. If she decided on B, which seems OK according to the MD1, the favorite theory for an agent, she risk a serious wrongdoing if she were not right about moral theories and in fact MD2 were the right doctrine.

As in our previous cases related to the ethics of war, this case also assumes the balancing account: the possibility of making comparisons and weighing different values on one scale. In recent literature there are two approaches to this problem: non-comparativism and comparativism.

This first position (Sepielli, 2013; Nissan-Rizan, 2014) assumes that there is no analogy between factual and normative uncertainties and it is impossible to make intertheoretic comparisons of values. It means that tables like Table 1 have no sense, because the moral evaluation of results (wrong, merely OK...) is different for MD1 and MD2. So it is meaningless to say that rationality requires that we avoid doing something morally terrible (according to MD1), since there is no common "currency" for both moral theories. This common currency is a necessary requirement

In the previous cases related to the ethics of war we compared the values of soldiers' lives, the importance of military target, and the value of noncombatants lives. Even if it is very hard to compare how much value "one civilian" versus some military target, the balancing account tries to do it. In cases of normative uncertainty this kind of comparisons are even harder: it seems that there is no way for making this kind of intertheoretic comparisons of moral values between different theories or doctrines.

4.2 Expected value approach and dominance principle

Comparativism in its strong form (Lockhart, 2000; Sepielli, 2009) is now an unpopular position. The main assumption of this view was to calculate the expected moral values of the available actions, relative to possible axiologies, and summing up those expected moral values, weighted by the degree of belief that the corresponding axiology is correct. There were two main propositions how to do it: the principle of equity among moral theories (Lockhart, 2000), the reactive-attitude approach (Sepielli, 2009). Both of them seem to be obviously mistaken (Sepielli, 2013).

Comparativism in its weak form seems to be much better proposition. It has weak form because it does not require any calculation of expected moral values of all available actions. It means that it can be applied only to very specific kinds of situations in which an agent's credences are not divided between two different moral doctrines, but between only one moral doctrine and some doctrine (or doctrines) that does not give any moral reasons. Its conclusion says that if some theories in which you have credence give you subjective reason to choose action A over action B, and no theories in which you have credence give you subjective reason to choose action B over action A, then you should (because of the requirements of rationality) choose A over B (Ross, 2006).

Let me introduce another example in which this type of reasoning should work perfectly well (this is a modified version of the example discussed by Ross, 2006). Suppose, for example, that John must decide whether to kill some being (that could have important moral status) or not. An agent strongly believes (it is doctrine no. 1 – D1) that it is highly probable that from the moral point of view it does not matter if he kills this type of being or not, but he is not absolutely certain of his normative views. Let me assume that his degree of credence is 0.99. This means that he thinks that there is a very small chance that another doctrine is the right one (D2). According to this second doctrine killing this kind of organism is in fact morally terrible (his degree of credence regarding this view is 0.01).

	A	B
D1 (0,99)	Does not matter	Does not matter
D2 (0,01)	Morally right	Morally terrible

Table 2

Accepting My Favorite Theory approach it does not matter what we choose because according to our favorite theory there are no reasons in favor or against both possible options. In this case we could – for example – decide by flipping a coin. But an agent accepts doctrine that says that there is very small chance (0,01) that it would be morally terrible to kill this type of organism and morally right not to kill it. If an agent wants to maximize the expected moral value of his decisions, he should choose A, even though he believes that the probability that killing this type of being is morally wrong is indeed extremely low.

Why this case is different than described in the Table 1? The application of expected moral value approach in this type of reasoning seems to be correct, because here there is no problem with intertheoretic comparisons of values. An agent does not have to compare in this situation any values or disvalues between different moral doctrines or views on the moral status of this living organism, since one of views says that everything he does in the situation is morally neutral. So it seems that in these types of cases, the ANU would indeed give a reason to prefer the safer option, only if the probability that this option is morally correct is greater than zero.

5. CONCLUSION

In this paper I sketched the problem of meta-reasoning in making moral decisions under normative uncertainty. I found that there is a promising type of meta-reasoning proposed by Ross (2006) that could be applied to some cases of normative uncertainty. However, it is worth noting that this kind of meta-reasoning has been recently extensively criticized. In this paper – because the lack of space – I have not considered some important arguments against meta-reasoning. In recent literature there are at least five such critiques. The first refers to the problem of "the infectiousness of nihilism" (MacAskill, 2013). This argument says that if we have nonzero credence in nihilism it is impossible to use expected value reasoning (also in this weak sense of comparativism) in the situations of normative uncertainty, because the expected value of all options is undefined. The second refers to the problem with accessibility. Even if descriptive facts may often be inaccessible to agents, we could assume that normative facts are a priori, then there is a sense in which any agent is in a position to know the moral truth (Hedden, forthcoming). In this case there would be no such situations as normative uncertainty. Thirdly, some authors say that although non-culpable factual ignorance is an excusing factor, it is not the case with normative ignorance which does not exculpate (Harman, 2014). The conclusion would be similar to the first critique: there is no such state as normative uncertainty. Fourthly, there is an argument from action-guiding and fetishism argument. It says that morally good people care non-derivatively about other people, their well-being and the like. Meta-reasoning would force agents to care not about people but about one thing: doing what they believe to be right, where this is read de dicto and not de re (Weatherson, 2014). If this critique is right probably the best theory under normative uncertainty would be My Favorite Theory account. And fifthly, as one author has just noticed there is an additional problem related with risk attitudes of agents, who are normatively uncertain. In many situations they could assign positive credence to several theories with different attitudes toward risk. In this kind of cases the intertheoretical comparisons of moral value would be meaningless in the same way that sentences like 'thunders are louder than honey is sweet' are meaningless. (Nissan, 2015, p. 358).

ACKNOWLEDGEMENTS: This work was supported by National Science Centre (NCN) grant SONATA number UMO-2011/03/D/HS5/01152.

REFERENCES

Declaration, (2002), United Kingdom Reservation / Declaration of July 2, 2002 to Protocol I, supra note 9, available at: http://www.icrc.org/ihl.nsf/NORM/0A9E03F0F2EE757CC125640200 3FB6D2?OpenDocument

Friberg-Fernros, H. (2014). Taking precautionary concerns seriously: A defense of a misused anti-abortion argument. *Journal of Medicine and Philosophy, 39*(3), 228-247.

Guerrero, A. A. (2007). Don't know, don't kill: Moral ignorance, culpability, and caution. *Philosophical Studies, 136*(1), 59-97.

Gustafsson, J. E., & Torpman O. (2014). In defence of My Favourite Theory. *Pacific Philosophical Quarterly, 95*(2), 159-174.

Harman, E. (2015). The irrelevance of moral uncertainty. In: R. Shaffer-Landau (Ed.), *Oxford Studies in Metaethics*, vol. 10 (pp. 53-78). Oxford: Oxford University Press.

Haque, A. A. (2012). Killing in the fog of war. *Southern California Law Review 86*(1), 63-116.

Hayenhjelm, M., & Wolff, J. (2012). The moral problem of risk impositions: A survey. *European Journal of Philosophy, 20*, 26-51.

Hedden, B. (forthcoming). Does MITE make right? On decision-making under normative uncertainty. In: R. Shaffer-Landau (Ed.), *Oxford Studies in Metaethics*, vol. 11, Oxford: Oxford University Press.

Henning, T. (2015). From choice to chance? Saving people, fairness, and lotteries. *Philosophical Review, 124*(2), 169-206.

Lockhart, T. (2000). *Moral uncertainty and its consequences*. Oxford: Oxford University Press.

MacAskill, W. (2013). The infectiousness of nihilism. *Ethics, 123*(3), 508-520.

Moller, D. (2011). Abortion and moral risk. *Philosophy, 86*(3), 425-443.

Nissan-Rozen, I. (2015). Against moral hedging. *Economics and Philosophy, 31*(3), 349-369.

Protocol (1977). Protocol Additional to the Geneva Conventions of 12 August 1949, and Relating to the Protection of Victims of International Armed Conflicts (Protocol I), June 8, 1977, available at: https://www.icrc.org/ihl/INTRO/470

Ross, J. (2006). Rejecting ethical deflationism. *Ethics, 116*(4), 742-768.

Sepielli, A. (2009). What to do when you don't know what to do. In: R. Shaffer-Landau (Ed.), *Oxford Studies in Metaethics*, vol. 4 (pp. 5-28). Oxford: Oxford University Press.

Sepielli, A. (2013), Moral uncertainty and the principle of equity among moral theories. *Philosophy and Phenomenological Research, 86*(3), 580-589.

Shaw, D. M. (2008). Moral qualms, future persons, and embryo research. *Bioethics, 22*(4), 218–223.

Walzer, M. (2000). *Just and unjust wars. A moral argument with historical illustrations* (3rd edition). London: Basic Books.

Waxman, M. C. (2008). Detention as targeting: Standards of certainty and detention of suspected terrorists, *Columbia Law Review, 108*(6), 1365-1430

Weatherson, B. (2014). Running risk morally. *Philosophical Studies, 167*(1), 141-163.

Żuradzki, T. (2012). Argument z niepewności normatywnej a etyczna ocena badań naukowych wykorzystujących ludzkie embriony. *Diametros, 32*, 131-159.

Żuradzki, T. (2014). Moral uncertainty in bioethical argumentation: A new understanding of the pro-life view on early human embryos. *Theoretical Medicine and Bioethics, 35*(6), 441-457.

www.ingramcontent.com/pod-product-compliance
Lightning Source LLC
Chambersburg PA
CBHW070851300426
44113CB00008B/796

* 9 7 8 1 8 4 8 9 0 2 1 2 1 *